Discrete-Time
Signal Processing

PRENTICE HALL SIGNAL PROCESSING SERIES

Alan V. Oppenheim, Editor

Discrete-Time Signal Processing

Alan V. Oppenheim
Ronald W. Schafer

PRENTICE HALL, Englewood Cliffs, New Jersey 07632

Library of Congress Cataloging-in-Publication Data

Oppenheim, Alan V.
  Discrete-time signal processing / Alan V. Oppenheim, Ronald W.
Schafer.
      p.   cm.—(Prentice Hall signal processing series)
  Bibliography: p.
  Includes index.
  ISBN 0-13-216292-X
  1. Signal processing—Mathematics.   2. Discrete-time systems.
I. Schafer, Ronald W.   II. Title.   III. Series.
TK5102.5.02452   1989
621.38′043—dc 19   88-25562
                                                           CIP

Editorial/production supervision: Barbara G. Flanagan
Interior design: Roger Brower
Cover design: Vivian Berman
Manufacturing buyer: Mary Noonan

© 1989 by Prentice-Hall, Inc.
A Division of Simon & Schuster
Englewood Cliffs, New Jersey 07632

Printed in the United States of America
10 9 8 7 6 5 4 3 2 1

ISBN   0-13-216292-X

Prentice-Hall International (UK) Limited, *London*
Prentice-Hall of Australia Pty. Limited, *Sydney*
Prentice-Hall Canada Inc., *Toronto*
Prentice-Hall Hispanoamericana, S.A., *Mexico*
Prentice-Hall of India Private Limited, *New Delhi*
Prentice-Hall of Japan, Inc., *Tokyo*
Simon & Schuster Asia Pte. Ltd., *Singapore*
Editora Prentice-Hall do Brasil, Ltda., *Rio de Janeiro*

To Phyllis, Jason, and Justine

To Dorothy, Bill, Kate, and Barbara
and in memory of John

Alan V. Oppenheim received the S.B. and S.M. degrees in 1961 and the Sc.D. degree in 1964, all in electrical engineering, from the Massachusetts Institute of Technology. In 1964 he joined the faculty at MIT, where he is currently Professor of Electrical Engineering and Computer Science. Since 1967 he has also been affiliated with MIT Lincoln Laboratory and since 1977 with Woods Hole Oceanographic Institution. His research interests are in the general area of signal processing and its applications to speech, image, and seismic data processing. He is coauthor of the widely used textbooks *Digital Signal Processing* and *Signals and Systems*. He is also editor of several advanced books on signal processing.

Dr. Oppenheim is a member of the National Academy of Engineering, a Fellow of the IEEE, and a member of Sigma Xi and Eta Kappa Nu. He has been a Guggenheim Fellow and a Sackler Fellow. He has also received a number of awards for outstanding research and teaching including the IEEE Education Medal, the IEEE Centennial Award, and the Society Award and Technical Achievement Award of the IEEE Society on Acoustics, Speech, and Signal Processing.

Ronald W. Schafer received the B.S.E.E. and M.S.E.E. degrees from the University of Nebraska in 1961 and 1962, respectively, and the Ph.D. degree from the Massachusetts Institute of Technology in 1968. From 1968 to 1974 he was a member of the Acoustics Research Department at Bell Laboratories, Murray Hill, New Jersey, and since 1974 he has been on the faculty of the Georgia Institute of Technology as Regents' Professor and holder of the John O. McCarty/Audichron Chair in the School of Electrical Engineering. His research interests are in discrete-time signal processing and its application to problems in speech communication and image analysis. He is coauthor of *Digital Signal Processing* and *Digital Processing of Speech Signals.*

Dr. Schafer is a Fellow of the IEEE and the Acoustical Society of America and a member of Sigma Xi, Eta Kappa Nu, and Phi Kappa Phi. He was awarded the Achievement Award and the Society Award of the IEEE Society on Acoustics, Speech, and Signal Processing, the IEEE Region III Outstanding Engineer Award, and the IEEE Centennial Award, and he shared the 1980 Emanuel R. Piore Award with L. R. Rabiner for their work in speech processing. He has received several awards for teaching, including the 1985 Class of 1934 Distinguished Professor Award at Georgia Tech.

Contents

Preface

This text has its origins in our initial thought several years ago of revising and updating our first text, *Digital Signal Processing*, which was published in 1975. The vitality of that book attests to the tremendous interest in and influence of signal processing, and it is clear that the field continues to grow in importance as the available technologies for implementing signal processing continue to develop. Shortly after beginning the revision, we realized that it would be more appropriate to develop a new textbook strongly based on our first one and at the same time continue to have the original text also available.

The title *Discrete-Time Signal Processing* was chosen for this new book for several reasons. In the mid-1960s digital signal processing emerged as a new branch of signal processing, driven by the potential and feasibility of implementing real-time signal processing using digital computers. The term *digital signal processing* refers specifically to processing based on digital technology, which inherently involves both time and amplitude quantization. However, the principal focus in essentially all texts on digital signal processing is on time quantization, i.e., the discrete-time nature of the signals. Furthermore, many signal processing technologies (e.g., charge transport devices and switched capacitor filters) are discrete time but not digital; i.e., signal values are clocked so that time is quantized but the signal amplitudes are represented in analog form.

In the mid-1970s, when the original text was published, courses on digital and discrete-time signal processing were available in only a few schools, and only at the graduate level. Now the basic principles are often taught at the undergraduate level, sometimes even as part of a first course on linear systems, or at a somewhat more advanced level in third-year, fourth-year, or beginning graduate subjects. Much of our thinking in planning this new text is in recognition of the importance of this material at the undergraduate level. In particular, we have considerably expanded the treatment of a number of topics, including linear systems, sampling, multirate signal processing, applications, and spectral analysis. In addition, a large number of examples are included to emphasize and illustrate important concepts. We have also removed and condensed some topics. This new text contains a rich set of more than 400 problems, and a solutions manual is available for course instructors.

It is assumed that the reader has a background of advanced calculus, including an introduction to complex variables, and an exposure to linear system theory for continuous-time signals, including Laplace and Fourier transforms, as taught in most undergraduate electrical and mechanical engineering curricula. With this background, the book is self-contained. In particular, no prior experience with discrete-time signals, z-transforms, discrete Fourier transforms, and the like is assumed. In later sections of some chapters, some topics such as quantization noise are included that assume a basic background in stochastic signals. A brief review of the background for these sections is included as Appendix A.

It has become common in many signal processing courses to include exercises to be done on a computer, and many of the homework problems in this book are easily turned into problems to be solved with the aid of a computer. With one or two exceptions, we have purposely avoided providing software to implement algorithms described in this book, for a variety of reasons. Foremost among them is that there are readily available a variety of inexpensive signal processing software packages for demonstrating and implementing signal processing on any of the popular personal computers and workstations. These packages are well documented and have excellent technical support, and many of them have excellent user interfaces that make them easily accessible to students. Furthermore, they are in a constant state of evolution, which strongly suggests that available software for classroom use should be constantly reviewed and updated. While we may have current favorites, these will no doubt change over time, and consequently our preference is for computer-based exercises to be independent of any specific software system or vendor. We have on occasion in this text illustrated points through the use of FORTRAN programs. We chose FORTRAN specifically for its general readability rather than with the implication that these specific programs are recommended for use in research or practical applications. Even though FORTRAN is often inefficient as an implementation of an algorithm, it can be a convenient language for communicating the structure of an algorithm.

The material in this book is organized in a way that provides considerable flexibility in its use at both the undergraduate and graduate level. A typical one-semester undergraduate elective might cover in depth Chapter 2, Sections 2.0–2.9; Chapter 3, Sections 3.0–3.6; Chapter 4; Chapter 5, Sections 5.0–5.3; Chapter 6, Sections 6.0–6.5; Chapter 7, Sections 7.0–7.2 and 7.4–7.5 and a brief overview of Sections 7.6–7.7. If students have studied discrete-time signals and systems in a general signals and systems course, it may be possible to move more quickly through the material of Chapters 2, 3, and 4, thus freeing time for covering Chapter 8. A first-year graduate course could augment the above topics with the remaining topics in Chapter 5, a brief exposure to the practical considerations in Section 3.7 and Sections 6.7–6.10, a discussion of optimal FIR filters as incorporated in Sections 7.6 and 7.7, and a thorough treatment of the discrete Fourier transform (Chapter 8) and its computation using the FFT (Chapter 9). The discussion of the DFT can be effectively augmented with many of the examples in Chapter 11. In a two-semester graduate course, the entire text together with a number of current advanced topics can be covered.

In Chapter 2, we introduce the basic class of discrete-time signals and systems and define basic system properties such as linearity, time invariance, stability, and causality. The primary focus of the book is on linear time-invariant systems because of the rich set of tools available for designing and analyzing this class of systems. In particular, in Chapter 2 we develop the time-domain representation of linear time-invariant systems through the convolution sum and introduce the class of linear time-invariant systems represented by linear constant-coefficient difference equations. In Chapter 6, we develop this class of systems in considerably more detail. Also in Chapter 2, we introduce the frequency-domain representation of signals and systems through the Fourier transform. The primary focus in Chapter 2 is on the representation of sequences in terms of the Fourier transform, i.e., as a linear combination of complex exponentials, and the development of the basic properties of the Fourier transform. We defer until Chapter 5 a detailed discussion of the analysis of linear time-invariant systems using the Fourier transform.

In Chapter 3, we carry out a detailed discussion of the relationship between continuous-time and discrete-time signals when the discrete-time signals are obtained through periodic sampling of continuous-time signals. This includes a development of the Nyquist sampling theorem. In addition, we discuss upsampling and downsampling of discrete-time signals, as used, for example, in multirate signal processing systems and for sampling rate conversion. The chapter concludes with a discussion of some of the practical issues encountered in conversion from continuous time to discrete time including prefiltering to avoid aliasing and modeling the effects of amplitude quantization when the discrete-time signals are represented digitally.

In Chapter 4, we develop the z-transform as a generalization of the Fourier transform. In Chapter 5, we carry out an extensive and detailed discussion of the use of the Fourier transform and the z-transform for the representation and analysis of linear time-invariant systems. In particular, in Chapter 5 we define the class of ideal, frequency-selective filters and develop the system function and pole-zero representation for systems described by linear constant-coefficient difference equations, a class of systems whose implementation is considered in detail in Chapter 6. Also in Chapter 5, we define and discuss group delay, phase response and phase distortion, and the relationships between the magnitude response and the phase response of systems, including a discussion of minimum-phase, allpass, and generalized linear phase systems.

In Chapter 6, we focus specifically on systems described by linear constant-coefficient difference equations and develop their representation in terms of block diagrams and linear signal flow graphs. Much of this chapter is concerned with developing a variety of the important system structures and comparing some of their properties. The importance of this discussion and the variety of filter structures relate to the fact that in a practical implementation of a discrete-time system, the effects of coefficient inaccuracies and arithmetic error can be very dependent on the specific structure used. While these basic issues are similar whether the technology used for implementation is digital or discrete-time analog, we illustrate them in this chapter in the context of a digital implementation through a discussion of the effects of coefficient quantization and arithmetic roundoff noise for digital filters.

While Chapter 6 is concerned with the representation and implementation of linear constant-coefficient difference equations, Chapter 7 is a discussion of the procedures for obtaining the coefficients of this class of difference equations to approximate a desired system response. The design techniques separate into those used for infinite impulse response (IIR) filters and those used for finite impulse response (FIR) filters.

In continuous-time linear system theory, the Fourier transform is primarily an analytical tool for representing signals and systems. In contrast, in the discrete-time case, many signal processing systems and algorithms involve the explicit computation of the Fourier transform. While the Fourier transform itself cannot be computed, a sampled version of it, the discrete Fourier transform (DFT), can be computed, and for finite-length signals the DFT is a complete Fourier representation of the signal. In Chapter 8, the discrete Fourier transform is introduced and its properties and relationship to the discrete-time Fourier transform are developed in detail. In Chapter 9, the rich and important variety of algorithms for computing or generating the discrete Fourier transform is introduced and discussed, including the Goertzel algorithm, the fast Fourier transform (FFT) algorithms, and the chirp transform.

In Chapter 10, we introduce the discrete Hilbert transform. This transform arises in a variety of practical applications, including inverse filtering, complex representations for real bandpass signals, single-sideband modulation techniques, and many others. It also has particular significance for the class of signal processing techniques referred to as cepstral analysis and homomorphic signal processing, as discussed in Chapter 12.

With the background developed in the earlier chapters and particularly Chapters 2, 3, 5, and 8, we focus in Chapter 11 on Fourier analysis of signals using the discrete Fourier transform. Without a careful understanding of the issues involved and the relationship between the DFT and the Fourier transform, using the DFT for practical signal analysis can often lead to confusions and misinterpretations. We address a number of these issues in Chapter 11. We also consider in some detail the Fourier analysis of signals with time-varying characteristics by means of the time-dependent Fourier transform.

In Chapter 12, we introduce a class of signal processing techniques referred to as cepstral analysis and homomorphic signal processing. This class of techniques, although nonlinear, is based on a generalization of the linear techniques that were the focus of the earlier chapters of the book.

In writing this book, we have been fortunate to receive valuable assistance, suggestions, and support from numerous colleagues, students, and friends. Over the years, a number of our colleagues at MIT and Georgia Institute of Technology have taught the material with us, and we have benefited greatly from their perspectives and input. These colleagues include Professors Jae Lim, Bruce Musicus, and Victor Zue at MIT and Professors Tom Barnwell, Mark Clements, Monty Hayes, Jim McClellan, Russ Mersereau, David Schwartz, and Mark Smith at Georgia Tech. Professors McClellan and Zue along with Jim Glass of MIT were also generous with their time in helping us to prepare several of the figures in the book.

In choosing and developing an effective and complete set of homework problems to include in this book, a number of students provided considerable help in sorting through, categorizing, and critiquing the large selection of potential homework problems that have accumulated over the years. We would particularly like to express our appreciation to Joseph Bondaryk, Dan Cobra, and Rosalind Wright for their indispensable help with this task as well as their further help with a variety of other aspects such as figure preparation and proofreading. The later stages of production of any text require the time-consuming and often tedious job of proofreading and scrutinizing the galley proofs and page proofs for errors, omissions, and last-minute improvements. We were extremely fortunate to have a long list of "volunteers" to help with this task. At MIT, Hiroshi Miyanaga and Patrick Velardo read a large portion of both the galley proofs and page proofs with exceptional care and dedication. Our sincere thanks also to MIT students Larry Candell and Avi Lele for meticulous reading of many chapters of the page proofs and to Michele Covell, Lee Hetherington, Paul Hillner, Tae Joo, Armando Rodriguez, Paul Shen, and Gregory Wornell for their help with galley proofs. Similarly, our thanks to Georgia Tech students Robert Bamberger, Jae Chung, Larry Heck, and David Pepper for careful reading of the page proofs. Cheung Au-Yeung, Beth Carlson, Kate Cummings, Brian George, Lois Hertz, David Mazel, Doug Reynolds, Craig Richardson, Janet Rutledge, and Kevin Tracy also gave valuable assistance with the galley proofs. We greatly appreciate the many valuable and perceptive suggestions made by all our students.

We would also like to express our thanks to Monica Dove and Deborah Gage at MIT and Cherri Dunn, Kayron Gilstrap, Pam Majors, and Stacy Schultz at Georgia Tech for their typing of various parts of the manuscript and for their help with the incredibly long list of details involved in the teaching and writing associated with the development of a textbook. We are particularly indebted to Barbara Flanagan, who served as production editor. Barbara's concern for perfection and meticulous attention to detail has contributed immeasurably to the quality of the final product. We would also like to express our appreciation to Vivian Berman, an artist and a friend, who helped us sort through many ideas for the cover design.

MIT and Georgia Tech have provided us with a stimulating environment for research and teaching throughout a major part of our technical careers and have provided significant encouragement and support for this project. In addition RWS particularly thanks the John and Mary Franklin Foundation for many years of support through the John O. McCarty/Audichron Chair.

Much of the structure and content of this book was shaped during the summer of 1985 when we were both guests in the Ocean Engineering Department at the Woods Hole Oceanographic Institution, and we wish to express our gratitude for this hospitality. In addition, AVO gives special thanks to the Woods Hole Oceanographic Institution and to our friends and summer neighbors the Wares and the Voses of Gansett Point for providing an exceptionally rejuvenating, productive, and enjoyable summer evironment since 1978.

We feel extremely fortunate to have worked with Prentice Hall on this project. Our relationship with Prentice Hall spans many years and many writing projects. The encouragement and support provided by Tim Bozik, Hank Kennedy, and many others at Prentice Hall enhance the enjoyment of writing and completing a project such as this one.

Alan V. Oppenheim
Ronald W. Schafer

Introduction

1

Signals are used to communicate between humans and between humans and machines; they are used to probe our environment to uncover details of structure and state not easily observable; and they are used to control and utilize energy and information. Signal processing is concerned with the representation, transformation, and manipulation of signals and the information they contain. For example, we may wish to separate two or more signals that have somehow been combined or to enhance some component or parameter of a signal model. For many decades signal processing has played a major role in such diverse fields as speech and data communication, biomedical engineering, acoustics, sonar, radar, seismology, oil exploration, instrumentation, robotics, consumer electronics, and many others.

Sophisticated signal processing algorithms and hardware are prevalent in a wide range of systems, from highly specialized military systems through industrial applications to low-cost, high-volume consumer electronics. Although we routinely take for granted the performance of home entertainment systems such as television and high-fidelity audio, these systems have always relied heavily on state-of-the-art signal processing. As another example, speech synthesis, which rapidly found its way into automatic voice response systems and consumer items such as learning aids and toys, moved at an astonishing pace from research literature to realization in military, industrial, and consumer systems.

The field of signal processing has always benefited from a close coupling between the theory, applications, and technologies for implementing signal processing systems. Prior to the 1960s, the technology for signal processing was almost exclusively continuous-time analog technology.† The rapid evolution of digital computers and microprocessors together with some important theoretical developments caused a major shift to digital technologies, giving rise to the field of digital signal processing. A fundamental aspect of digital signal processing is that it is based

---

† In a general context, we typically refer to the independent variable as "time" even though in specific contexts the independent variable may take on any of a broad range of possible dimensions. Consequently, continuous time and discrete time should be thought of as generic terms referring to a continuous independent variable and a discrete independent variable, respectively.

on processing of sequences of samples. This discrete-time nature of digital signal processing technology is also characteristic of other signal processing technologies such as surface acoustic wave (SAW) devices, charge-coupled devices (CCDs), charge transport devices (CTDs), and switched-capacitor technologies. In digital signal processing, signals are represented by sequences of finite-precision numbers, and processing is implemented using digital computation. The more general term *discrete-time signal processing* includes digital signal processing as a special case but also includes the possibility that sequences of samples (sampled data) are processed with other discrete-time technologies. Often the distinction between the terms discrete-time signal processing and digital signal processing is of minor importance, since both are concerned with discrete-time signals.

While there are many examples in which signals to be processed are inherently sequences, most applications involve the use of discrete-time technology for processing continuous-time signals. In this case, a continuous-time signal is converted into a sequence of samples, i.e., a discrete-time signal. After discrete-time processing, the output sequence is converted back to a continuous-time signal. Real-time operation is often desirable for such systems, meaning that the discrete-time system is implemented so that samples of the output are computed at the same rate at which the continuous-time signal is sampled. Discrete-time processing of continuous-time signals is commonplace in communication systems, radar and sonar, speech and video coding and enhancement, and biomedical engineering, to name just a few.

Much of traditional signal processing involves processing one signal to obtain another signal. Another important class of signal processing problems is *signal interpretation*. In such problems the objective of the processing is not to obtain an output signal but to obtain a characterization of the input signal. For example, in a speech recognition or understanding system, the objective is to interpret the input signal or extract information from it. Typically, such a system will apply preprocessing (filtering, parameter estimation, etc.) followed by a pattern recognition system to produce a symbolic representation such as a phonemic transcript of the speech. This symbolic output can in turn be the input to a symbolic processing system, such as a rule-based expert system, to provide the final signal interpretation.

Still another and relatively new category of signal processing involves the symbolic manipulation of signal processing expressions. This type of processing is particularly useful in signal processing workstations and for the computer-aided design of signal processing systems. In this class of processing, signals and systems are represented and manipulated as abstract data objects. Object-oriented programming languages such as LISP provide a convenient environment for manipulating signals and systems and signal processing expressions without explicitly evaluating the data sequences and provide the basis for this class of processing. The sophistication of systems designed to do signal expression processing is directly influenced by the incorporation of fundamental signal processing concepts, theorems, and properties, such as those that form the basis for this book. For example, a signal processing environment that incorporates the property that convolution in the time domain corresponds to multiplication in the frequency domain can explore a variety of rearrangements of filtering structures, including those involving the direct use of the discrete Fourier transform and the fast Fourier transform algorithm. Similarly, environments that incorporate the relationship between sampling rate and aliasing can make effective use of decimation and interpolation strategies for filter implementation. The development of object-oriented environments for computer-aided system design is still in its very early stages and any detailed discussion of it is beyond the scope of this text. However, it is important to recognize that the basic concepts that are the subject of this book should not be viewed as just theoretical in nature; they are likely to become an explicit integral part of computer-aided signal processing environments and workstations.

Signal processing problems are not confined, of course, to one-dimensional signals. Although there are some fundamental differences in the theories for one-dimensional and multidimensional signal processing, much of the material that we discuss in this text has a direct counterpart in multidimensional systems. The theory of multidimensional digital signal processing is presented in detail in Dudgeon and Mersereau (1983) and Lim (1989).† Many image processing applications require the use of two-dimensional signal processing techniques. This is the case in such areas as video coding, medical imaging, enhancement and analysis of aerial photographs, analysis of satellite weather photos, and enhancement of video transmissions from lunar and deep-space probes. Applications of multidimensional digital signal processing to image processing are discussed in Andrews and Hunt (1977), Castleman (1979), Pratt (1978), and Macovski (1983). Seismic data analysis as required in oil exploration, earthquake measurement, and nuclear test monitoring also utilizes multidimensional signal processing techniques. Seismic applications are discussed in Robinson and Treitel (1980) and Robinson and Durrani (1985).

Multidimensional signal processing is only one of many advanced and specialized topics that build on the fundamentals covered in this text. Spectral analysis based on the use of the discrete Fourier transform and the use of signal modeling is another particularly rich and important aspect of signal processing. We introduce many facets of this topic, focusing on the basic concepts and techniques relating to the use of the discrete Fourier transform. In addition to these techniques, a variety of spectral analysis methods rely in one way or another on specific signal models. For example, a class of high-resolution spectral analysis methods referred to as maximum entropy methods (MEM spectral analysis) is based on representing the signal to be analyzed as the response of a discrete-time linear time-invariant filter to either an impulse or to white noise. Spectral analysis is achieved by estimating the parameters (e.g., the difference equation coefficients) of the system and then evaluating the magnitude squared of the frequency response of the model filter. A thorough and detailed treatment of the issues and techniques of this approach to signal modeling and spectral analysis builds directly from the fundamentals in this text. Detailed discussions can be found in the texts by Kay (1988) and Marple (1987).

Signal modeling also plays an important role in data compression and coding, and again the fundamentals of difference equations provide the basis for understanding many of these techniques. For example, one class of signal coding techniques, referred to as linear predictive coding (LPC), exploits the notion that if a signal is the

† Authors' names and dates are used throughout the text to refer to books and papers listed in the Bibliography at the end of the book.

response of a certain class of discrete-time filters, the signal value at any time index is a linear function of (and thus linearly predictable from) previous values. Consequently, efficient signal representations can be obtained by estimating these prediction parameters and using them along with the prediction error to represent the signal. The signal can then be regenerated when needed from the model parameters. This class of signal coding techniques has been particularly effective in speech coding and is described in considerable detail in Jayant and Noll (1984), Markel and Gray (1976), and Rabiner and Schafer (1978).

Another advanced topic of considerable importance is adaptive signal processing. In this text the emphasis is almost entirely on linear time-invariant systems. Adaptive systems represent a particular class of time-varying and, in some sense, nonlinear systems with broad application and with established and effective techniques for their design and analysis. Again, many of these techniques build from the fundamentals of discrete-time signal processing covered in this text. Details of adaptive signal processing are given by Haykin (1986) and Widrow and Stearns (1985).

These represent only a few of the many advanced topics that extend from the topics covered in this text. Others include advanced and specialized filter design procedures, a variety of specialized algorithms for evaluation of the Fourier transform, specialized filter structures, and various advanced multirate signal processing techniques. An introduction to many of these advanced topics is contained in Lim and Oppenheim (1988).

It is often said that the purpose of a fundamental textbook should be to uncover rather than cover a subject, and in choosing the topics and depth of coverage in this book we have been guided by this philosophy. The preceding brief discussion of advanced topics and the Bibliography at the end of the book should be strongly suggestive of the rich variety of directions that these fundamentals begin to uncover.

Historical Perspective

Discrete-time signal processing has a rich history. It has advanced in uneven steps over a long period of time. Since the invention of calculus in the 17th century, scientists and engineers have developed models to represent physical phenomena in terms of functions of continuous variables and differential equations. Numerical techniques have been used to solve these equations when analytical solutions are not possible. Indeed, Newton used finite-difference methods that are special cases of some of the discrete-time systems that we present in this text. Mathematicians of the 18th century, such as Euler, Bernoulli, and Lagrange, developed methods for numerical integration and interpolation of functions of a continuous variable. Interesting historical research by Heideman, Johnson, and Burrus (1984) showed that Gauss discovered the fundamental principle of the fast Fourier transform (discussed in Chapter 9) as early as 1805 — even before the publication of Fourier's treatise on harmonic series representation of functions.

Until the early 1950s, signal processing as we have defined it was typically done with analog systems that were implemented with electronic circuits or even with mechanical devices. Even though digital computers were becoming available in business and scientific laboratories, they were expensive and had relatively limited capabilities. About that time, the need for more sophisticated signal processing in some application areas created considerable interest in discrete-time signal process- ing. One of the first uses of digital computers in digital signal processing was in oil prospecting, where seismic data could be recorded on magnetic tape for later processing. This type of signal processing could not generally be done in real time; minutes or even hours of computer time were often required to process only seconds of data. Even so, the flexibility of the digital computer made this alternative extremely inviting.

Also in the 1950s, the use of digital computers in signal processing arose in a different way. Because of the flexibility of digital computers, it was often useful to simulate a signal processing system on a digital computer before implementing it in analog hardware. In this way, a new signal processing algorithm or system could be studied in a flexible experimental environment before committing economic and engineering resources to constructing it. Typical examples of such simulations were the vocoder simulations carried out at Lincoln Laboratory and at Bell Laboratories. In the implementation of an analog channel vocoder, for example, the filter character- istics can affect the perceived quality of the coded speech signal in ways that are difficult to quantify objectively. Through computer simulations, these filter character- istics could be adjusted and the perceived quality of a speech coding system evaluated prior to construction of the analog equipment.

In all of these examples of signal processing using digital computers, the computer offered tremendous advantages in flexibility. However, the processing could not always be done in real time. Consequently, a prevalent attitude was that the digital computer was being used to *approximate*, or *simulate*, an analog signal processing system. In keeping with that style, early work on digital filtering was very much concerned with ways in which a filter could be programmed on a digital computer so that with analog-to-digital conversion of the signal, followed by the digital filtering, followed by digital-to-analog conversion, the overall system approxi- mated a good analog filter. The notion that digital systems might, in fact, be practical for the actual real-time implementation of signal processing in speech communication or radar processing or any of a variety of other applications seemed at the most optimistic times to be highly speculative. Speed, cost, and size were, of course, three of the important factors in favor of the use of analog components.

As signals were being processed on digital computers, researchers had a natural tendency to experiment with increasingly sophisticated signal processing algorithms. Some of these algorithms grew out of the flexibility of the digital computer and had no apparent practical implementation in analog equipment. Thus, many of these algor- ithms were treated as interesting, but somewhat impractical, ideas. An example of a class of algorithms of this type was the set of techniques referred to as cepstrum analysis and homomorphic filtering. It had been clearly demonstrated on digital computers that these techniques could be applied to advantage in speech bandwidth compression systems, deconvolution, and echo detection and removal. However, implementation of these techniques required the explicit evaluation of the inverse Fourier transform of the logarithm of the Fourier transform of the input, and the required accuracy and resolution of the Fourier transform were such that analog spectrum analyzers were not practical. The development of such signal processing algorithms made the notion of all-digital implementation of signal processing systems even more tempting. Active work began on the investigation of digital vocoders, digital spectrum analyzers, and other all-digital systems, with the hope that eventually such systems would become practical.

The evolution of a new point of view toward discrete-time signal processing was further accelerated by the disclosure by Cooley and Tukey (1965) of an efficient algorithm for computation of Fourier transforms. This class of algorithms has come to be known as the fast Fourier transform, or FFT. The FFT was significant for several reasons. Many signal processing algorithms that had been developed on digital computers required processing times several orders of magnitude greater than real time. Often this was because spectrum analysis was an important component of the signal processing and no efficient means were available for implementing it. The fast Fourier transform algorithm reduced the computation time of the Fourier transform by orders of magnitude, permitting the implementation of increasingly sophisticated signal processing algorithms with processing times that allowed interactive experimentation with the system. Furthermore, with the realization that the fast Fourier transform algorithms might, in fact, be implementable in special-purpose digital hardware, many signal processing algorithms that previously had appeared to be impractical began to appear to have practical implementations.

Another important implication of the fast Fourier transform algorithm was that it was an inherently discrete-time concept. It was directed toward the computation of the Fourier transform of a discrete-time signal or sequence and involved a set of properties and mathematics that was exact in the discrete-time domain—it was not simply an approximation to a continuous-time Fourier transform. This had the effect of stimulating a reformulation of many signal processing concepts and algorithms in terms of discrete-time mathematics, and these techniques then formed an exact set of relationships in the discrete-time domain. Following this shift away from the notion that signal processing on a digital computer was merely an approximation to analog signal processing techniques, there emerged a strong interest in discrete-time signal processing as an important field of investigation in its own right.

Another major development in the history of discrete-time signal processing occurred in the field of microelectronics. The invention and subsequent proliferation of the microprocessor paved the way for low-cost implementations of discrete-time signal processing systems. Although the first microprocessors were too slow to implement most discrete-time systems in real time, by the mid-1980s integrated circuit technology had advanced to a level that permitted the implementation of very fast fixed-point and floating-point microcomputers with architectures specially designed for implementing discrete-time signal processing algorithms. With this technology came the possibility for the first time of widespread application of discrete-time signal processing techniques.

Microelectronics engineers continue to strive for increased circuit densities and production yields, and as a result, the complexity and sophistication of microelectronic systems are continually increasing. As wafer-scale integration techniques become highly developed, very complex discrete-time signal processing systems will be implemented with low cost, miniature size, and low power consumption. Consequently, the importance of discrete-time signal processing will almost certainly continue to increase. Indeed, the future development of the field is likely to be even more dramatic than the course of development that we have just described. Discrete-time signal processing techniques are already promoting revolutionary advances in some fields of application. A notable example is in the area of telecommunications, where discrete-time signal processing techniques, microelectronic technology, and fiber optic transmission combine to change the nature of communication systems in truly revolutionary ways. A similar impact can be expected in many other areas of technology.

While discrete-time signal processing is a dynamic, rapidly growing field, its fundamentals are well formulated. Our goal in this book is to provide a coherent treatment of the theory of discrete-time linear systems, filtering, sampling, and discrete-time Fourier analysis. The topics presented should provide the reader with the knowledge necessary for an appreciation of the wide scope of applications for discrete-time signal processing and a foundation for contributing to future developments in this exciting field of technology.

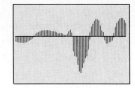

Discrete-Time Signals and Systems

2

2.0 INTRODUCTION

A signal can be defined as a function that conveys information, generally about the state or behavior of a physical system. Although signals can be represented in many ways, in all cases the information is contained in some pattern of variations. Signals are represented mathematically as functions of one or more independent variables. For example, a speech signal is represented mathematically as a function of time and a photographic image is represented as a brightness function of two spatial variables. It is a common convention, and one that will usually be followed in this book, to refer to the independent variable of the mathematical representation of a signal as time, although in specific examples the independent variable may in fact not represent time.

The independent variable in the mathematical representation of a signal may be either continuous or discrete. *Continuous-time signals* are defined along a continuum of times and thus are represented by a continuous independent variable. Continuous-time signals are often referred to as *analog signals*. *Discrete-time signals* are defined at discrete times and thus the independent variable has discrete values; i.e., discrete-time signals are represented as sequences of numbers. Signals such as speech or images may have either a continuous- or a discrete-variable representation, and if certain conditions hold, these representations are entirely equivalent. Besides the independent variables being either continuous or discrete, the signal amplitude may be either continuous or discrete. *Digital signals* are those for which both time and amplitude are discrete.

Signal processing systems may be classified along the same lines as signals. That is, continuous-time systems are systems for which both the input and the output are continuous-time signals, and discrete-time systems are those for which both the input and the output are discrete-time signals. Similarly, a digital system is one for which both the input and the output are digital signals. Digital signal processing, then, deals with the transformation of signals that are discrete in both amplitude and time. The principal focus in this book is on discrete-time (rather than digital) signals and systems. The effects of signal amplitude quantization are considered in Sections 3.7, 6.7–6.10, and 9.8.

Discrete-time signals may arise by sampling a continuous-time signal or they may be generated directly by some discrete-time process. Whatever the origin of the discrete-time signals, discrete-time signal processing systems have many attractive features. They can be realized with great flexibility with a variety of technologies such as charge transport devices, surface acoustic wave devices, general-purpose digital computers, or high-speed microprocessors. Complete signal processing systems can be implemented using VLSI techniques. Discrete-time systems can be used to simulate analog systems or, more important, to realize signal transformations that cannot be implemented with continuous-time hardware. Thus, discrete-time representations of signals are often desirable when sophisticated and flexible signal processing is required.

In this chapter we consider the fundamental concepts of discrete-time signals and signal processing systems for one-dimensional signals. We emphasize the class of linear time-invariant discrete-time systems. Many of the properties and results that we derive in this and subsequent chapters will be similar to properties and results for linear time-invariant continuous-time systems as presented in a variety of texts (see, for example, Papoulis, 1980; Oppenheim and Willsky, 1983). In fact, it is possible to approach the discussion of discrete-time systems by treating sequences as analog signals that are impulse trains. This approach, if implemented carefully, can lead to correct results and has formed the basis for much of the classical discussion of sampled data systems (see, for example, Raggazzini and Franklin, 1958; Freeman, 1965; Phillips and Nagle, 1984). However, not all sequences arise from sampling a continuous-time signal, and many discrete-time systems are not simply approximations to corresponding analog systems. Furthermore, there are important and fundamental differences between discrete- and continuous-time systems. Therefore, rather than attempt to force results from continuous-time system theory into a discrete-time framework, we will derive parallel results starting within a framework and with notation suitable to discrete-time systems. Discrete-time signals will be related to continuous-time signals only when it is necessary and useful to do so.

2.1 DISCRETE-TIME SIGNALS: SEQUENCES

Discrete-time signals are represented mathematically as sequences of numbers. A sequence of numbers x, in which the nth number in the sequence is denoted $x[n]$,† is

---

† A sequence is simply a function whose domain is the set of integers. Note that we use [ ] to enclose the independent variable of such functions, and we use ( ) to enclose the independent variable of continuous-variable functions.

formally written as

$$x = \{x[n]\}, \qquad -\infty < n < \infty, \tag{2.1}$$

where n is an integer. In a practical setting, such sequences can often arise from periodic sampling of an analog signal. In this case, the numeric value of the nth number in the sequence is equal to the value of the analog signal $x_a(t)$ at time nT; i.e.,

$$x[n] = x_a(nT), \qquad -\infty < n < \infty. \tag{2.2}$$

The quantity T is called the *sampling period* and its reciprocal is the *sampling frequency*. Although sequences do not always arise from sampling analog waveforms, it is convenient to refer to $x[n]$ as the "nth sample" of the sequence. Also, although strictly speaking $x[n]$ denotes the nth number in the sequence, the notation of Eq. (2.1) is often unnecessarily cumbersome, and it is convenient and unambiguous to refer to "the sequence $x[n]$" when we mean the entire sequence, just as we referred to "the analog signal $x_a(t)$." Discrete-time signals (i.e., sequences) are often depicted graphically as shown in Fig. 2.1. Although the abscissa is drawn as a continuous line, it is important to recognize that $x[n]$ is defined only for integer values of n. It is not correct to think of $x[n]$ as being zero for n not an integer; $x[n]$ is simply undefined for noninteger values of n.

As an example, Fig. 2.2(a) shows a segment of a speech signal corresponding to acoustic pressure variation as a function of time, and Figure 2.2(b) shows a sequence of samples of the speech signal. Although the original speech signal is defined at all values of time t, the sequence contains information about the signal only at discrete time instants. From the sampling theorem, discussed in Chapter 3, the original signal can be reconstructed as accurately as desired from a corresponding sequence of samples if the samples are taken frequently enough.

2.1.1 Basic Sequences and Sequence Operations

In the analysis of discrete-time signal processing systems, sequences are manipulated in several basic ways. The product and sum of two sequences $x[n]$ and $y[n]$ are defined as the sample-by-sample product and sum, respectively. Multiplication of a sequence $x[n]$ by a number α is defined as multiplication of each sample value by α. A

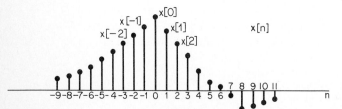

Figure 2.1 Graphical representation of a discrete-time signal.

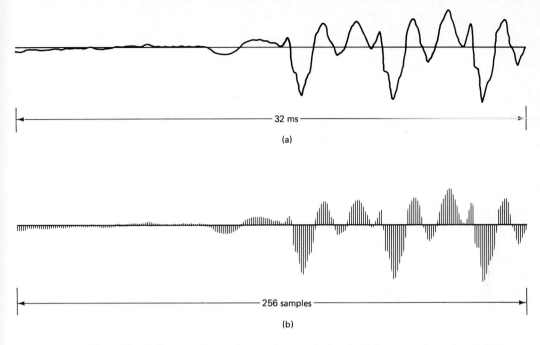

Figure 2.2   (a) Segment of a continuous-time speech signal. (b) Sequence of samples obtained from part (a) with $T = 125 \ \mu s$.

sequence $y[n]$ is said to be a delayed or shifted version of a sequence $x[n]$ if $y[n]$ has values

$$y[n] = x[n - n_0], \tag{2.3}$$

where n_0 is an integer.

In discussing the theory of discrete-time signals and systems, several basic sequences are of particular importance. These sequences are shown in Fig. 2.3 and discussed below.

The *unit sample sequence* (Fig. 2.3a), denoted by $\delta[n]$, is defined as the sequence with values

$$\delta[n] = \begin{cases} 0, & n \neq 0, \\ 1, & n = 0. \end{cases} \tag{2.4}$$

As we will see, the unit-sample sequence plays the same role for discrete-time signals and systems that the unit impulse function (Dirac delta function) does for continuous-time signals and systems. For convenience, the unit sample sequence is often referred to as a *discrete-time impulse*, or simply as an *impulse*. It is important to note that a discrete-time impulse does not suffer from the mathematical complications of the continuous-time impulse. Its definition is simple and precise.

As we will see in the discussion of linear systems, one of the important aspects of the impulse sequence is that an arbitrary sequence can be represented as a sum of

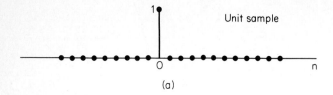

(a)

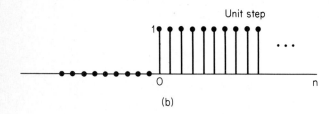

(b)

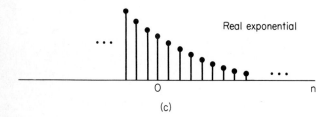

(c)

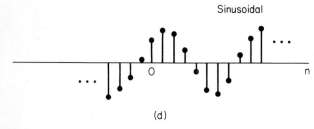

(d)

Figure 2.3 Some basic sequences. The sequences shown play important roles in the analysis and representation of discrete-time signals and systems.

scaled, delayed impulses. For example, the sequence $p[n]$ in Fig. 2.4 can be expressed as

$$p[n] = a_{-3}\delta[n + 3] + a_1\delta[n - 1] + a_2\delta[n - 2] + a_7\delta[n - 7]. \qquad (2.5)$$

More generally, any sequence can be expressed as

$$x[n] = \sum_{k = -\infty}^{\infty} x[k]\delta[n - k]. \qquad (2.6)$$

We will make specific use of Eq. (2.6) in discussing the representation of discrete-time linear systems.

The *unit step sequence* (Fig. 2.3b), denoted by $u[n]$, has values

$$u[n] = \begin{cases} 1, & n \geq 0, \\ 0, & n < 0. \end{cases} \qquad (2.7)$$

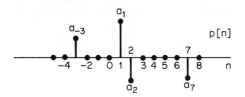

Figure 2.4   Example of a sequence to be represented as a sum of scaled and delayed impulses.

The unit step is related to the impulse by

$$u[n] = \sum_{k=-\infty}^{n} \delta[k], \tag{2.8}$$

i.e., the value of the unit step sequence at (time) index n is equal to the accumulated sum of the value at index n and all previous values of the impulse sequence. An alternative representation of the unit step in terms of the impulse is obtained by interpreting the unit step in Fig. 2.3(b) in terms of a sum of delayed impulses. This is expressed mathematically as

$$u[n] = \delta[n] + \delta[n-1] + \delta[n-2] + \cdots \tag{2.9a}$$

or

$$u[n] = \sum_{k=0}^{\infty} \delta[n-k]. \tag{2.9b}$$

Conversely, the impulse sequence can be expressed as the first backward difference of the unit step sequence, i.e.,

$$\delta[n] = u[n] - u[n-1]. \tag{2.10}$$

Exponential and *sinusoidal sequences* are extremely important in representing and analyzing linear time-invariant discrete-time systems. The general form for an exponential sequence is

$$x[n] = A\alpha^n. \tag{2.11}$$

If A and α are real numbers, then the sequence is real. If $0 < \alpha < 1$ and A is positive, then the sequence values are positive and decrease with increasing n, as in Fig. 2.3(c). For $-1 < \alpha < 0$, the sequence values alternate in sign but again decrease in magnitude with increasing n. If $|\alpha| > 1$, then the sequence grows in magnitude as n increases. A sinusoidal sequence has the general form

$$x[n] = A\cos(\omega_0 n + \phi) \qquad \text{for all } n, \tag{2.12}$$

with A real, and is illustrated in Fig. 2.3(d).

The exponential sequence $A\alpha^n$ with complex α has real and imaginary parts that are exponentially weighted sinusoids. Specifically, if $\alpha = |\alpha|e^{j\omega_0}$ and $A = |A|e^{j\phi}$, the sequence $A\alpha^n$ can be expressed in any of the following ways:

$$x[n] = A\alpha^n = |A|e^{j\phi}|\alpha|^n e^{j\omega_0 n}$$

$$= |A||\alpha|^n e^{j(\omega_0 n + \phi)} \qquad (2.13)$$

$$= |A||\alpha|^n \cos(\omega_0 n + \phi) + j|A||\alpha|^n \sin(\omega_0 n + \phi).$$

The sequence oscillates with an exponentially growing envelope if $|\alpha| > 1$ or with an exponentially decaying envelope if $|\alpha| < 1$. (As a simple example consider the case $\omega_0 = \pi$.)

When $|\alpha| = 1$, the sequence is referred to as a *complex exponential sequence* and has the form

$$x[n] = |A|e^{j(\omega_0 n + \phi)} = |A| \cos(\omega_0 n + \phi) + j|A| \sin(\omega_0 n + \phi), \qquad (2.14)$$

i.e., the real and imaginary parts of $e^{j\omega_0 n}$ vary sinusoidally with n. The fact that n is always an integer in Eq. (2.14) leads to some important differences between the properties of discrete-time and continuous-time complex exponential sequences and sinusoidal sequences. By analogy with the continuous-time case, the quantity ω_0 is called the *frequency* of the complex sinusoid or complex exponential and ϕ is called the *phase*. However, note that n is a dimensionless integer. Thus, the dimension of ω_0 must be radians. If we wish to maintain a closer analogy with the continuous-time case, we can specify the units of ω_0 to be radians/sample and the units of n to be samples.

A more interesting difference between continuous-time and discrete-time complex sinusoids is seen when we consider a frequency $(\omega_0 + 2\pi)$. In this case,

$$x[n] = Ae^{j(\omega_0 + 2\pi)n}$$

$$= Ae^{j\omega_0 n}e^{j2\pi n} = Ae^{j\omega_0 n}. \qquad (2.15)$$

Indeed, we can easily see that complex exponential sequences with frequencies $(\omega_0 + 2\pi r)$, where r is an integer, are indistinguishable from one another. An identical statement holds for sinusoidal sequences. Specifically, it is easily verified that

$$x[n] = A \cos[(\omega_0 + 2\pi r)n + \phi]$$

$$= A \cos(\omega_0 n + \phi). \qquad (2.16)$$

The implications of this property for sequences obtained by sampling sinusoids and other signals will be discussed in Chapter 3. For now, we simply conclude that when discussing complex exponential signals of the form $x[n] = Ae^{j\omega_0 n}$ or real sinusoidal signals of the form $x[n] = A \cos(\omega_0 n + \phi)$, we need only consider frequencies in a frequency interval of length 2π such as $-\pi \leq \omega_0 \leq \pi$ or $0 \leq \omega_0 < 2\pi$.

Another important difference between continuous-time and discrete-time complex exponentials and sinusoids concerns periodicity. In the continuous-time case, a sinusoidal signal and a complex exponential signal are both periodic with period equal to 2π divided by frequency. In the discrete-time case, a periodic sequence is one for which

$$x[n] = x[n + N], \qquad \text{for all } n, \tag{2.17}$$

where the period N is necessarily an integer. If this condition for periodicity is tested for the discrete-time sinusoid, then

$$A \cos(\omega_0 n + \phi) = A \cos(\omega_0 n + \omega_0 N + \phi), \tag{2.18}$$

which requires

$$\omega_0 N = 2\pi k, \tag{2.19}$$

where k is an integer. A similar statement holds for the complex exponential sequence $Ce^{j\omega_0 n}$, i.e., periodicity with period N requires

$$e^{j\omega_0(n+N)} = e^{j\omega_0 n}, \tag{2.20}$$

which is true only for $\omega_0 N = 2\pi k$, as in Eq. (2.19). Consequently, complex exponential and sinusoidal sequences are not necessarily periodic with period $(2\pi/\omega_0)$ and, depending on the value of ω_0, may not be periodic at all. For example, for $\omega_0 = 3\pi/4$, the smallest value of N for which Eq. (2.19) can be satisfied with k an integer is $N = 8$ (corresponding to $k = 3$). For $\omega_0 = 1$, there are no integer values of N or k for which Eq. (2.19) is satisfied.

When we combine the condition of Eq. (2.19) with our previous observation that ω_0 and $(\omega_0 + 2\pi r)$ are indistinguishable frequencies, then it is clear that there are N distinguishable frequencies for which the corresponding sequences are periodic with period N. One set of frequencies is $\omega_k = 2\pi k/N$, $k = 0, 1, \ldots, N - 1$. These properties of complex exponential and sinusoidal sequences are basic to both the theory and the design of computational algorithms for discrete-time Fourier analysis, and they will be discussed in more detail in Chapters 8 and 9.

Related to the preceding discussion is the fact that the interpretation of high and low frequencies is somewhat different for continuous-time and discrete-time sinusoidal and complex exponential signals. For a continuous-time sinusoidal signal $x(t) = A \cos(\Omega_0 t + \phi)$, as Ω_0 increases, $x(t)$ oscillates more and more rapidly. In contrast, for the discrete-time sinusoidal signal $x[n] = A \cos(\omega_0 n + \phi)$, as ω_0 increases from $\omega_0 = 0$ toward $\omega_0 = \pi$, $x[n]$ oscillates more and more rapidly. As ω_0 increases from $\omega_0 = \pi$ to $\omega_0 = 2\pi$, the oscillations become slower. This is illustrated in Fig. 2.5. In fact, because of the periodicity in ω_0 of sinusoidal and complex exponential sequences, $\omega_0 = 2\pi$ is indistinguishable from $\omega_0 = 0$ and, more generally, frequencies around $\omega_0 = 2\pi$ are indistinguishable from frequencies around $\omega_0 = 0$. As a consequence, for sinusoidal and complex exponential signals, values of ω_0 in the vicinity of $\omega_0 = 2\pi k$ for any integer value of k are typically referred to as low frequencies (relatively slow oscillations), while values of ω_0 in the vicinity of $\omega_0 = (\pi + 2\pi k)$ for any integer value of k are typically referred to as high frequencies (relatively rapid oscillations).

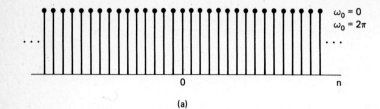

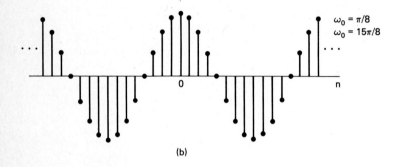

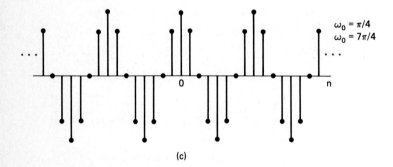

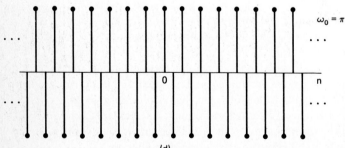

Figure 2.5 $\cos \omega_0 n$ for several different values of ω_0. As ω_0 increases from zero toward π (parts a–d), the sequence oscillates more rapidly. As ω_0 increases from π to 2π (parts d–a), the oscillations become slower.

2.2 DISCRETE-TIME SYSTEMS

A discrete-time system is defined mathematically as a transformation or operator that maps an input sequence with values $x[n]$ into an output sequence with values $y[n]$. This can be denoted as

$$y[n] = T\{x[n]\} \tag{2.21}$$

and is indicated pictorially in Fig. 2.6. Equation (2.21) represents a rule or formula for computing the output sequence values from the input sequence values. It should be emphasized that the value of the output sequence at each value of the index n may be a function of $x[n]$ at all values of n. The following examples illustrate some simple and useful systems.

Example 2.1   The Ideal Delay System

The ideal delay system is defined by the equation

$$y[n] = x[n - n_d], \qquad -\infty < n < \infty, \tag{2.22}$$

where n_d is a fixed positive integer called the delay of the system. In words, the ideal delay system simply shifts the input sequence to the right by n_d samples to form the output. If in Eq. (2.22) n_d is a fixed negative integer, then the system would shift the input to the left by n_d samples corresponding to a time advance.

In Example 2.1 only one sample of the input sequence is involved in determining a certain output sample. In the following example, this is not the case.

Example 2.2   Moving Average

The general moving average system is defined by the equation

$$y[n] = \frac{1}{M_1 + M_2 + 1} \sum_{k=-M_1}^{M_2} x[n - k]$$

$$= \frac{1}{M_1 + M_2 + 1} \{x[n + M_1] + x[n + M_1 - 1] + \cdots + x[n]$$

$$+ x[n - 1] + \cdots + x[n - M_2]\}. \tag{2.23}$$

This system computes the nth sample of the output sequence as the average of $(M_1 + M_2 + 1)$ samples of the input sequence around the nth sample. Figure 2.7 shows an input sequence plotted as a function of a dummy index k and the samples involved in the computation of the output sample $y[n]$ for $n = 7$, $M_1 = 0$, and $M_2 = 5$. The output sample $y[7]$ is equal to $\frac{1}{6}$ times the sum of all the samples between the vertical dotted lines. To compute $y[8]$, both dotted lines would move one sample to the right.

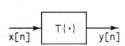

Figure 2.6   Representation of a discrete-time system, i.e., a transformation that maps an input sequence $x[n]$ into a unique output sequence $y[n]$.

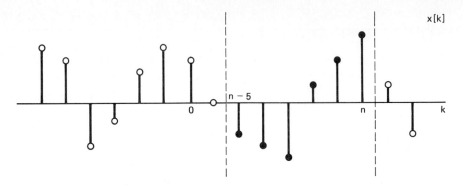

Figure 2.7   Sequence values involved in computing a causal moving average.

Classes of systems are defined by placing constraints on the properties of the transformation $T\{\cdot\}$. Doing so often leads to very general mathematical representations, as we will see. Of particular importance to consider are the system constraints and properties, discussed in Sections 2.2.1–2.2.5.

2.2.1 Memoryless Systems

A system is referred to as memoryless if the output $y[n]$ at every value of n depends only on the input $x[n]$ at the same value of n.

Example 2.3

An example of a memoryless system is one for which $x[n]$ and $y[n]$ are related by

$$y[n] = (x[n])^2 \quad \text{for each value of } n. \tag{2.24}$$

The system in Example 2.1 is not memoryless unless $n_d = 0$. In particular it is referred to as having "memory" whether n_d is positive (time delay) or negative (time advance). The system in Example 2.2 is not memoryless unless $M_1 = M_2 = 0$.

2.2.2 Linear Systems

The class of *linear systems* is defined by the principle of superposition. If $y_1[n]$ and $y_2[n]$ are the responses of a system when $x_1[n]$ and $x_2[n]$ are the respective inputs, then the system is linear if and only if

$$T\{x_1[n] + x_2[n]\} = T\{x_1[n]\} + T\{x_2[n]\} = y_1[n] + y_2[n] \tag{2.25a}$$

and

$$T\{ax[n]\} = aT\{x[n]\} = ay[n]. \tag{2.25b}$$

The first property is called the *additivity property* and the second is called the *homogeneity or scaling property*. These two properties can be combined into the principle of superposition, stated as

$$T\{ax_1[n] + bx_2[n]\} = aT\{x_1[n]\} + bT\{x_2[n]\} \tag{2.26}$$

for arbitrary constants a and b. This can be generalized to the superposition of many inputs. Specifically, if

$$x[n] = \sum_k a_k x_k[n], \tag{2.27a}$$

then

$$y[n] = \sum_k a_k y_k[n], \tag{2.27b}$$

where $y_k[n]$ is the system response to the input $x_k[n]$.

By using the definition of the principle of superposition, we can easily show that the systems of Examples 2.1 and 2.2 are linear systems. (See Problem 2.2.) An example of a nonlinear system is the system in Example 2.3.

Example 2.4

The system referred to as an accumulator and defined by the relation

$$y[n] = \sum_{k=-\infty}^{n} x[k] \tag{2.28}$$

is a linear system. The output at each sample instant n is equal to the accumulated sum of the value at n and all previous values of the input. Comparing Eq. (2.28) and Eq. (2.8), we note that if the input $x[n]$ is a unit impulse $\delta[n]$, then the output $y[n]$ is a unit step $u[n]$.

2.2.3 Time-Invariant Systems

A time-invariant system (equivalently often referred to as a shift-invariant system) is one for which a time shift or delay of the input sequence causes a corresponding shift in the output sequence. Specifically, suppose that a system transforms the input sequence with values $x[n]$ into the output sequence with values $y[n]$. The system is said to be time-invariant if for all n_0 the input sequence with values $x_1[n] = x[n - n_0]$ produces the output sequence with values $y_1[n] = y[n - n_0]$.

All of the systems in Examples 2.1–2.4 are time-invariant. The following example illustrates a system that is not time-invariant.

Example 2.5

The system defined by the relation

$$y[n] = x[Mn], \qquad -\infty < n < \infty, \tag{2.29}$$

with M a positive integer, is called a *compressor*. Specifically, it discards $(M - 1)$ samples out of M; i.e., it creates the output sequence by selecting every Mth sample. It should be clear simply from the description of its operation that this system is not time-invariant. We can see this mathematically by considering the response $y_1[n]$ to the input $x_1[n]$ defined as $x_1[n] = x[n - n_0]$. Then

$$y_1[n] = x_1[Mn] = x[Mn - n_0]. \tag{2.30}$$

But

$$y[n - n_0] = x[M(n - n_0)] \neq y_1[n]. \tag{2.31}$$

With $x_1[n]$ equal to $x[n]$ shifted by n_0, $y_1[n]$ is not $y[n]$ shifted by n_0, so the downsampler is not time-invariant, unless of course $M = 1$.

2.2.4 Causality

A system is causal if for every choice of n_0 the output sequence value at index $n = n_0$ depends only on the input sequence values for $n \leq n_0$. This implies that if $x_1[n] = x_2[n]$ for $n \leq n_0$, then $y_1[n] = y_2[n]$ for $n \leq n_0$. That is, the system is *nonanticipative*. The system of Example 2.1 is causal for $n_d \geq 0$ and is noncausal for $n_d < 0$. The system of Example 2.2 is causal if $M_1 \leq 0$ and $M_2 \geq -M_1$; otherwise it is noncausal. The system of Example 2.3 is causal, as is the accumulator of Example 2.4. However, the system of Example 2.5 is noncausal if $M > 1$ since $y[1] = x[M]$. Another noncausal system is given in the following example.

Example 2.6

The *forward difference system* is defined by the relation

$$y[n] = x[n + 1] - x[n]. \tag{2.32}$$

Obviously $y[n]$ depends on $x[n + 1]$; therefore the system is noncausal. However, the *backward difference system*, defined by

$$y[n] = x[n] - x[n - 1], \tag{2.33}$$

is causal.

2.2.5 Stability

A system is stable in the bounded-input bounded-output (BIBO) sense if and only if every bounded input sequence produces a bounded output sequence. The input $x[n]$ is bounded if there exists a fixed positive finite value B_x such that

$$|x[n]| \leq B_x < \infty \qquad \text{for all } n. \tag{2.34}$$

Stability requires that for every bounded input there exists a fixed positive finite value B_y such that

$$|y[n]| \leq B_y < \infty \qquad \text{for all } n. \tag{2.35}$$

Examples 2.1, 2.2, 2.3, and 2.5 are stable systems. The accumulator of Example 2.4 is unstable. To demonstrate this we need only specify one bounded input for which the output is unbounded. Thus, consider $x[n] = u[n]$, which is clearly bounded. For this input to the accumulator, the output is

$$y[n] = \sum_{k=-\infty}^{n} u[k]$$

$$= \begin{cases} 0, & n < 0, \\ (n + 1), & n \geq 0. \end{cases} \tag{2.36}$$

Although this response is finite if n is finite, it is nevertheless unbounded, i.e., there is no fixed positive finite value B_y such that $(n + 1) \leq B_y < \infty$ for all n.

It is important to emphasize that the properties we have defined in this section are properties of *systems*, not of the inputs to the system. That is, we may be able to find inputs for which the properties hold, but the existence of the property for some inputs does not mean that the system has the property. For the system to have the property, it must hold for all inputs. For example, an unstable system may have some bounded inputs for which the output is bounded, but for the system to have the property of stability, it must be true that for *all* bounded inputs the output is bounded. If, as we were able to do with the accumulator system of Example 2.4, we can find just one input for which the system property does not hold, then we have shown that the system does *not* have that property.

2.3 LINEAR TIME-INVARIANT SYSTEMS

A particularly important class of systems consists of those that are linear and time-invariant. These two properties in combination lead to especially convenient representations for these systems. Most important, this class of systems has significant signal processing applications. The class of linear systems is defined by the principle of superposition in Eq. (2.26). If the linearity property is combined with the representation of a general sequence as a linear combination of delayed impulses as in Eq. (2.6), it follows that a linear system can be completely characterized by its impulse response. Specifically, let $h_k[n]$ be the response of the system to $\delta[n - k]$, an impulse occurring at $n = k$. Then from Eq. (2.6),

$$y[n] = T\left\{ \sum_{k=-\infty}^{\infty} x[k]\delta[n-k] \right\}. \tag{2.37}$$

From the principle of superposition in Eq. (2.26), we can write

$$y[n] = \sum_{k=-\infty}^{\infty} x[k]T\{\delta[n-k]\} = \sum_{k=-\infty}^{\infty} x[k]h_k[n]. \tag{2.38}$$

According to Eq. (2.38), the system response to any input can be expressed in terms of the response of the system to $\delta[n - k]$. If only linearity is imposed, $h_k[n]$ will depend on both n and k, in which case the computational usefulness of Eq. (2.38) is limited. We obtain a more useful result if we impose the additional constraint of time invariance.

The property of time invariance implies that if $h[n]$ is the response to $\delta[n]$, then the response to $\delta[n - k]$ is $h[n - k]$. With this additional constraint, Eq. (2.38) becomes

$$y[n] = \sum_{k=-\infty}^{\infty} x[k]h[n-k]. \tag{2.39}$$

As a consequence of Eq. (2.39), a linear time-invariant system (which we will sometimes abbreviate as LTI) is completely characterized by its impulse response, $h[n]$, in the sense that, given $h[n]$, it is possible to use Eq. (2.39) to compute the output $y[n]$ due to *any* input $x[n]$.

Equation (2.39) is commonly called the *convolution sum.* If $y[n]$ is a sequence whose values are related to the values of two sequences $h[n]$ and $x[n]$ as in Eq. (2.39), we say that $y[n]$ is the convolution of $x[n]$ with $h[n]$ and represent this by the notation

$$y[n] = x[n] * h[n]. \tag{2.40}$$

The operation of discrete-time convolution takes two sequences $x[n]$ and $h[n]$ and produces a third sequence $y[n]$. Equation (2.39) expresses each sample of the output sequence in terms all of the samples of the input and the impulse response sequences.

The derivation of Eq. (2.39) suggests the interpretation that the input sample at $n = k$, represented as $x[k]\delta[n - k]$, is transformed by the system into an output sequence $x[k]h[n - k]$, $-\infty < n < \infty$ and that these sequences for each k are superimposed to form the overall output sequence. This interpretation is illustrated in Fig. 2.8, which shows an impulse response, a simple input sequence having three nonzero samples, the individual outputs due to each sample, and the composite output due to all the samples in the input sequences. Specifically, $x[n]$ can be decomposed as the sum of the three sequences $x[-2]\delta[n + 2]$, $x[0]\delta[n]$, and $x[3]\delta[n - 3]$ representing the three nonzero values in the sequence $x[n]$. The sequences $x[-2]h[n + 2]$, $x[0]h[n]$, and $x[3]h[n - 3]$ are the system responses to $x[-2]\delta[n + 2]$, $x[0]\delta[n]$, and $x[3]\delta[n - 3]$, respectively. The response to $x[n]$ is then the sum of these three individual responses.

Although the convolution-sum expression is analogous to the convolution integral of continuous-time linear system theory, the convolution sum should not be thought of as an approximation to the convolution integral. The convolution integral plays mainly a theoretical role in continuous-time linear system theory; we will see that the convolution sum, in addition to its theoretical importance, often serves as an explicit realization of a discrete-time linear system. Thus it is important to gain some insight into the properties of the convolution sum in actual calculations.

The preceding interpretation of Eq. (2.39) emphasizes that the convolution sum is a direct result of linearity and time invariance. However, a slightly different way of looking at Eq. (2.39) leads to a particularly useful computational interpretation. When viewed as a formula for computing a single value of the output sequence, Eq. (2.39) dictates that $y[n]$ (i.e., the nth value of the output) is obtained by multiplying the input sequence (expressed as a function of k) by the sequence whose values are $h[n - k]$, $-\infty < k < \infty$, and then, for any fixed value of n, summing all the values of the products $x[k]h[n - k]$, with k a counting index in the summation process. The operation of convolving two sequences thus involves doing the computation for all values of n, thus generating the complete output sequence $y[n]$, $-\infty < n < \infty$. The key to carrying out the computations of Eq. (2.39) to obtain $y[n]$ is understanding how to form the sequence $h[n - k]$, $-\infty < k < \infty$, for all values of n that are of interest. To this end, it is useful to note that

$$h[n - k] = h[- (k - n)]. \tag{2.41}$$

The interpretation of Eq. (2.41) is best done with an example.

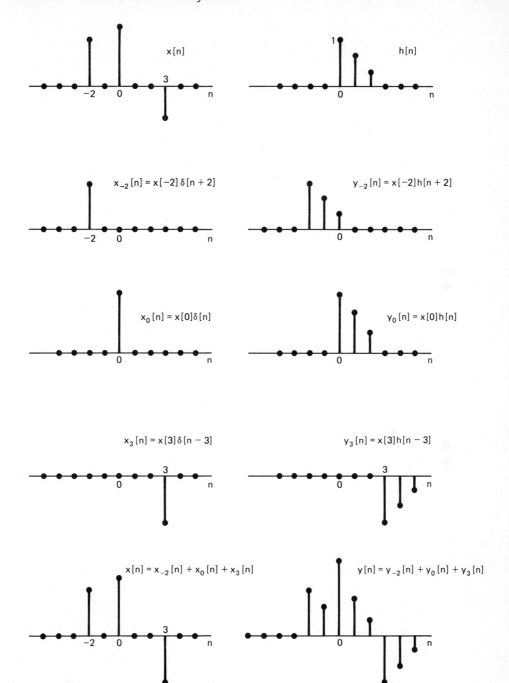

Figure 2.8   Representation of the output of a linear time-invariant system as the superposition of responses to individual samples of the input.

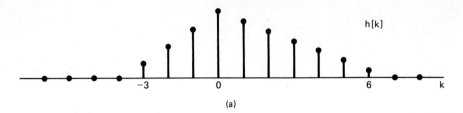

(a)

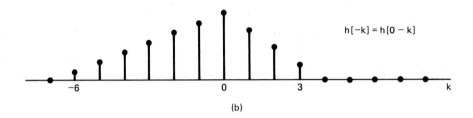

(b)

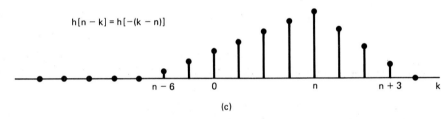

(c)

Figure 2.9 Forming the sequence $h[n - k]$. (a) The sequence $h[k]$ as a function of k. (b) The sequence $h[-k]$ as a function of k. (c) The sequence $h[n - k]$ as a function of k for $n = 4$.

Example 2.7

   Suppose $h[k]$ is the sequence shown in Fig. 2.9(a). First consider $h[-k]$ plotted against k; $h[-k]$ is simply $h[k]$ reflected or "flipped" about $k = 0$, as shown in Fig. 2.9(b). Replacing k by $k - n$, where n is a fixed integer, leads to a shift of the origin of the sequence $h[-k]$ to $k = n$, as shown in Fig. 2.9(c) for the case $n = 4$.

Generalizing Example 2.7, it should be clear that in general the sequence $h[n - k]$, $-\infty < k < \infty$, is obtained by

   1. reflecting $h[k]$ about the origin to obtain $h[-k]$;
   2. shifting the origin of the reflected sequence to $k = n$.

To implement discrete-time convolution, the two sequences $x[k]$ and $h[n - k]$ are multiplied together for $-\infty < k < \infty$ and the products are summed to compute the output sample $y[n]$. To obtain another output sample, the origin of the sequence

$h[-k]$ is shifted to the new sample position and the process is repeated. This computational procedure applies whether the computations are carried out numerically on sampled data or analytically with sequences for which the sample values have simple formulas. The following example illustrates discrete-time convolution for the latter case.

Example 2.8

Consider a system with impulse response

$$h[n] = u[n] - u[n - N]$$

$$= \begin{cases} 1, & 0 \le n \le N - 1, \\ 0, & \text{otherwise.} \end{cases}$$

The input is

$$x[n] = a^n u[n].$$

To find the output at a particular index n, we must form the sums over all k of the product $x[k]h[n - k]$. In this case, we can find formulas for $y[n]$ for different sets of values of n. For example, Fig. 2.10(a) shows the sequences $x[k]$ and $h[n - k]$ plotted for n a negative integer. Clearly, all negative values of n give a similar picture; i.e., the nonzero portions of the sequences $x[k]$ and $h[n - k]$ do not overlap, so

$$y[n] = 0, \qquad n < 0.$$

Figure 2.10(b) illustrates the two sequences when $0 \le n$ and $n - N + 1 \le 0$. These two conditions can be combined into the single condition $0 \le n \le N - 1$. By considering Fig. 2.10(b), we see that since

$$x[k]h[n - k] = a^k,$$

then

$$y[n] = \sum_{k=0}^{n} a^k \qquad \text{for } 0 \le n \le N - 1. \tag{2.42}$$

The limits on the sum are determined directly from Fig. 2.10(b). Equation (2.42) shows that $y[n]$ is the sum of $n + 1$ terms of a geometric series where the ratio of terms is a. This sum can be expressed in closed form using the general formula

$$\sum_{k=N_1}^{N_2} \alpha^k = \frac{\alpha^{N_1} - \alpha^{N_2 + 1}}{1 - \alpha}, \qquad N_2 \ge N_1. \tag{2.43}$$

Applying this to Eq. (2.42), we obtain

$$y[n] = \frac{1 - a^{n+1}}{1 - a}, \qquad 0 \le n \le N - 1. \tag{2.44}$$

Finally, Fig. 2.10(c) shows the two sequences when $0 < n - N + 1$ or $N - 1 < n$. As before,

$$x[k]h[n - k] = a^k, \qquad n - N + 1 < k \le n,$$

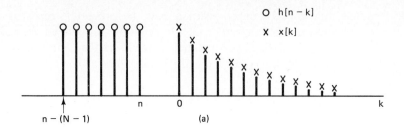

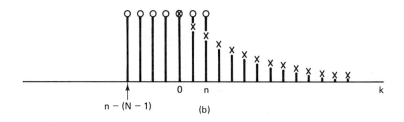

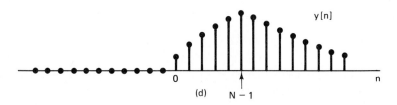

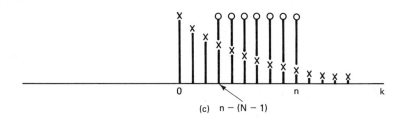

Figure 2.10 Sequences involved in computing a discrete convolution. (a)–(c) The sequences $x[k]$ and $h[n-k]$ as a function of k for different values of n. (Only nonzero samples are shown.) (d) Corresponding output sequence as a function of n.

but now the lower limit on the sum is $n - N + 1$, as seen in Fig. 2.10(c). Thus

$$y[n] = \sum_{k=n-N+1}^{n} a^k \qquad \text{for } N - 1 < n. \tag{2.45}$$

Using Eq. (2.43) we obtain

$$y[n] = \frac{a^{n-N+1} - a^{n+1}}{1-a}$$

or

$$y[n] = a^{n-N+1}\left(\frac{1-a^N}{1-a}\right).$$ (2.46)

Thus, because of the piecewise-exponential nature of both the input and the unit sample response, we have been able to obtain the following closed-form expression for $y[n]$ as a function of the index n:

$$y[n] = \begin{cases} 0, & n < 0, \\ \dfrac{1-a^{n+1}}{1-a}, & 0 \le n \le N-1, \\ a^{n-N+1}\left(\dfrac{1-a^N}{1-a}\right), & N-1 < n. \end{cases}$$ (2.47)

This sequence is shown in Fig. 2.10(d).

Example 2.8 illustrates how the convolution sum can be computed analytically when the input and the impulse response are given by simple formulas. In such cases, the sums may have a compact form that may be derived using the formula for the sum of a geometric series or other "closed-form" formulas.† When no simple form is available, the convolution sum can still be evaluated numerically using the technique illustrated in Example 2.8 whenever the sums are finite, which will be the case if either the input sequence or the impulse response is of finite length, i.e., has a finite number of nonzero samples. We cannot overemphasize that neatly drawn and labeled sketches such as those in Fig. 2.10 are invaluable to ensure that the limits on the sum are correctly determined.

2.4 PROPERTIES OF LINEAR TIME-INVARIANT SYSTEMS

Since all linear time-invariant systems are described by the convolution sum of Eq. (2.39), the properties of this class of systems are defined by the properties of discrete-time convolution. Therefore the impulse response is a complete characterization of the properties of a specific linear time-invariant system.

Some general properties of the class of linear time-invariant systems can be found by considering properties of the convolution operation. For example, the convolution operation is commutative:

$$x[n] * h[n] = h[n] * x[n].$$ (2.48)

This can be shown by applying a substitution of variables to Eq. (2.39). Specifically, with $m = n - k$,

$$y[n] = \sum_{m=-\infty}^{\infty} x[n-m]h[m],$$ (2.49)

† Such results are discussed, for example, in Grossman (1982). A very useful book of sums is Jolley (1961).

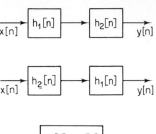

Figure 2.11 Three linear time-invariant systems with identical impulse responses.

so the roles of $x[n]$ and $h[n]$ in the summation are interchanged. That is, the order of the sequences in a convolution is unimportant, and hence the system output is the same if the roles of the input and impulse response are reversed. A linear time-invariant system with input $x[n]$ and impulse response $h[n]$ will have the same output as a linear time-invariant system with input $h[n]$ and impulse response $x[n]$. The convolution operation also distributes over addition, i.e.,

$$x[n] * (h_1[n] + h_2[n]) = x[n] * h_1[n] + x[n] * h_2[n].$$

This follows in a straightforward way from Eq. (2.39).

In a *cascade connection* of systems, the output of the first system is the input to the second, the output of the second is the input to the third, etc. The output of the last system is the overall output. Two linear time-invariant systems in cascade correspond to a linear time-invariant system with an impulse response that is the convolution of the impulse responses of the two systems. This is illustrated in Fig. 2.11. In the upper block diagram, the output of the first system will be $h_1[n]$ if $x[n] = \delta[n]$. Thus the output of the second system (and the overall impulse response of the system) will be

$$h[n] = h_1[n] * h_2[n]. \tag{2.50}$$

As a consequence of the commutative property of convolution, it follows that the impulse response of a cascade combination of linear time-invariant systems is independent of the order in which they are cascaded. This result is summarized in Fig. 2.11, where the three systems all have the same impulse response.

In a *parallel connection*, the systems have the same input and their outputs are summed to produce an overall output. It follows from the distributive property of convolution that two linear time-invariant systems in parallel are equivalent to a single system whose impulse response is the sum of the individual impulse responses, i.e.,

$$h[n] = h_1[n] + h_2[n]. \tag{2.51}$$

This is depicted in Fig. 2.12.

The constraints of linearity and time invariance define a class of systems with very special properties. Stability and causality represent additional properties, and it is often important to know whether a linear time-invariant system is stable and whether it is causal. Recall from Section 2.2.5 that a stable system is one for which

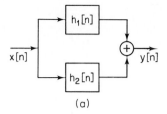

(a)

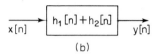

Figure 2.12   (a) Parallel combination of linear time-invariant systems. (b) An equivalent system.

(b)

every bounded input produces a bounded ouput. Linear time-invariant systems are stable if and only if the impulse response is absolutely summable, i.e., if

$$S = \sum_{k=-\infty}^{\infty} |h[k]| < \infty. \tag{2.52}$$

This can be shown as follows. From Eq. (2.49),

$$|y[n]| = \left| \sum_{k=-\infty}^{\infty} h[k]x[n-k] \right| \le \sum_{k=-\infty}^{\infty} |h[k]||x[n-k]|. \tag{2.53}$$

If $x[n]$ is bounded so that

$$|x[n]| \le B_x,$$

then

$$|y[n]| \le B_x \sum_{k=-\infty}^{\infty} |h[k]|. \tag{2.54}$$

Thus $y[n]$ is bounded if Eq. (2.52) holds; i.e., Eq. (2.52) is a sufficient condition for stability. To show that it is also a necessary condition, we must show that if $S = \infty$, then a bounded input can be found that will cause an unbounded output. Such an input is the sequence with values

$$x[n] = \begin{cases} \dfrac{h^*[-n]}{|h[-n]|}, & h[n] \ne 0, \\[2mm] 0, & h[n] = 0, \end{cases} \tag{2.55}$$

where $h^*[n]$ is the complex conjugate of $h[n]$. The sequence $x[n]$ is clearly bounded by unity. However, the value of the output at $n = 0$ is

$$y[0] = \sum_{k=-\infty}^{\infty} x[-k]h[k] = \sum_{k=-\infty}^{\infty} \frac{|h[k]|^2}{|h[k]|} = S. \tag{2.56}$$

Thus if $S = \infty$, it is possible for a bounded input sequence to produce an unbounded output sequence.

The class of causal systems was defined in Section 2.2.4 as those systems for which the output $y[n_0]$ depends only on the input samples $x[n]$, for $n \le n_0$. It follows from Eq. (2.39) or Eq. (2.49) that this definition implies the condition

$$h[n] = 0, \qquad n < 0, \tag{2.57}$$

for causality of linear time-invariant systems. (See Problem 2.12.) For this reason it is sometimes convenient to refer to a sequence that is zero for $n < 0$ as a *causal sequence*, meaning that it could be the impulse response of a causal system.

To illustrate how the properties of linear time-invariant systems are reflected in the impulse response, let us reconsider some of the systems defined in Examples 2.1–2.6. First note that only the systems of Examples 2.1, 2.2, 2.4, and 2.6 are linear and time-invariant. Although the impulse response of nonlinear or time-varying systems can be found, it is generally of limited interest since the convolution-sum formula and Eqs. (2.52) and (2.57), expressing stability and causality, do not apply to such systems.

First, let us find the impulse responses of the systems in Examples 2.1, 2.2, 2.4, and 2.6. We can do this by simply computing the response of each system to $\delta[n]$ using the defining relationship for the system. The resulting impulse responses are as follows:

Ideal Delay (Example 2.1)

$$h[n] = \delta[n - n_d], \qquad n_d \text{ a positive fixed integer} \tag{2.58}$$

Moving Average (Example 2.2)

$$h[n] = \frac{1}{M_1 + M_2 + 1} \sum_{k=-M_1}^{M_2} \delta[n - k]$$

$$= \begin{cases} \dfrac{1}{M_1 + M_2 + 1}, & -M_1 \le n \le M_2 \\ 0, & \text{otherwise} \end{cases} \tag{2.59}$$

Accumulator (Example 2.4)

$$h[n] = \sum_{k=-\infty}^{n} \delta[k]$$

$$= \begin{cases} 1, & n \ge 0 \\ 0, & n < 0 \end{cases} \tag{2.60}$$

$$= u[n]$$

Forward Difference (Example 2.6)

$$h[n] = \delta[n + 1] - \delta[n] \tag{2.61}$$

Backward Difference (Example 2.6)

$$h[n] = \delta[n] - \delta[n-1] \tag{2.62}$$

Next let us test the stability of each system by computing the sum

$$S = \sum_{n=-\infty}^{\infty} |h[n]|.$$

For the ideal delay, moving average, forward difference, and backward difference examples, it is clear that $S < \infty$ since the impulse response has only a finite number of nonzero samples. Such systems are called *finite-duration impulse response* (FIR) systems. Clearly FIR systems will always be stable as long as each of the impulse response values is finite in magnitude. The accumulator, however, is unstable because

$$S = \sum_{n=0}^{\infty} u[n] = \infty.$$

In Section 2.2.5 we also demonstrated the instability of the accumulator by giving an example of a bounded input (the unit step) for which the output is unbounded.

The impulse response of the accumulator is infinite in duration. This is an example of the class of systems referred to as *infinite-duration impulse response* (IIR) systems. An example of an IIR system that is stable is one having the impulse response $h[n] = a^n u[n]$ with $|a| < 1$. In this case

$$S = \sum_{n=0}^{\infty} |a|^n. \tag{2.63}$$

If $|a| < 1$, the formula for the sum of the terms of an infinite geometric series gives

$$S = \frac{1}{1 - |a|} < \infty. \tag{2.64}$$

If, on the other hand, $|a| \geq 1$, the sum is infinite and the system is unstable.

To test causality of the linear time-invariant systems in Examples 2.1, 2.2, 2.4, and 2.6, we can check to see if $h[n] = 0$ for $n < 0$. As discussed in Section 2.2.4, the ideal delay ($n_d \geq 0$ in Eq. 2.22) is causal. If $n_d < 0$, the system is noncausal. For the moving average, causality requires that $-M_1 \geq 0$ and $M_2 \geq 0$. The accumulator and backward difference systems are causal, and the forward difference system is noncausal.

The property of convolution can be used to describe and analyze interconnections of linear time-invariant systems. Consider the system of Fig. 2.13(a), which consists of a forward difference system cascaded with an ideal delay of one sample. According to the commutative property of convolution, the order in which systems are cascaded does not matter as long as they are linear and time-invariant. Therefore we obtain the same result when we compute the forward difference of a sequence and delay the result (Fig. 2.13a) as when we delay the sequence first and then compute the forward difference (Fig. 2.13b). Also, it follows from Eq. (2.50) that the

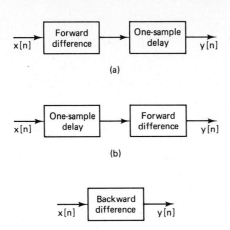

(a)

(b)

(c)

Figure 2.13 Equivalent systems found by using the commutative property of convolution.

overall impulse response of each cascade system is the convolution of the individual impulse responses. Therefore

$$h[n] = (\delta[n+1] - \delta[n]) * \delta[n-1]$$

$$= \delta[n-1] * (\delta[n+1] - \delta[n])$$

$$= \delta[n] - \delta[n-1]. \tag{2.65}$$

Thus $h[n]$ is identical to the impulse response of the backward difference system, i.e., the cascaded systems of Figs. 2.13(a) and 2.13(b) can be replaced by a backward difference system, as shown in Fig. 2.13(c).

Note that the noncausal forward difference systems in Figs. 2.13(a) and (b) have been converted to causal systems by cascading them with a delay. In general, any noncausal FIR system can be made causal by cascading it with a sufficiently long delay.

Another example of cascaded systems introduces the concept of an *inverse system*. Consider the cascade of systems in Fig. 2.14. The impulse response of the cascade system is

$$h[n] = u[n] * (\delta[n] - \delta[n-1])$$

$$= u[n] - u[n-1]$$

$$= \delta[n]. \tag{2.66}$$

That is, the cascade combination of an accumulator followed by a backward difference (or vice versa) yields a system whose overall impulse response is the impulse. Thus the output of the cascade system will always be equal to the input since

Figure 2.14 An accumulator in cascade with a backward difference. Since the backward difference is the inverse system for the accumulator, the cascade combination is equivalent to the identity system.

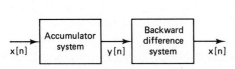

$x[n] * \delta[n] = x[n]$. In this case, the backward difference system compensates exactly for (or inverts) the effect of the accumulator, i.e., the backward difference system is the *inverse system* for the accumulator. From the commutative property of convolution, the accumulator is likewise the inverse system for the backward difference system. Also note that this example provides a system interpretation of Eqs. (2.8) and (2.10). In general, if a linear time-invariant system has impulse response $h[n]$, then its inverse system, if it exists, has impulse response $h_i[n]$ defined by the relation

$$h[n] * h_i[n] = h_i[n] * h[n] = \delta[n]. \tag{2.67}$$

Inverse systems are useful in many situations where it is necessary to compensate for the effects of a linear system. In general, it is difficult to solve Eq. (2.67) directly for $h_i[n]$ given $h[n]$. However, in Chapter 5 we will see that the Fourier transform and more generally the z-transform provide a straightforward method of finding an inverse system.

2.5 LINEAR CONSTANT-COEFFICIENT DIFFERENCE EQUATIONS

An important subclass of linear time-invariant systems consists of those systems for which the input $x[n]$ and the output $y[n]$ satisfy an Nth-order linear constant-coefficient difference equation of the form

$$\sum_{k=0}^{N} a_k y[n-k] = \sum_{n=0}^{M} b_n x[n-n]. \tag{2.68}$$

The properties discussed in Section 2.4 and some of the analysis techniques introduced there can be used to find difference equation representations for some of the linear time-invariant systems that we have defined.

Example 2.9

An example of the class of linear constant-coefficient difference equations is the accumulator (Example 2.4). To show that its input and output satisfy such a difference equation, consider Fig. 2.14 again. If $y[n]$ represents the output of the accumulator, then we have seen that the backward difference of $y[n]$ is equal to $x[n]$ since the backward difference system is the inverse of the accumulator; i.e.,

$$y[n] - y[n-1] = x[n]. \tag{2.69}$$

Thus we have shown that in addition to the defining relationship of Eq. (2.28), which is in the form of a convolution, the input and output also satisfy a linear constant-coefficient difference equation of the form of Eq. (2.68), where $N = 1$, $a_0 = 1$, $a_1 = -1$, $M = 0$, and $b_0 = 1$.

The difference equation in Eq. (2.69) as a representation of the accumulator can be better understood by rewriting it as

$$y[n] = y[n-1] + x[n]. \tag{2.70}$$

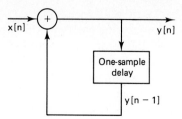

Figure 2.15 Block diagram of a recursive difference equation representing an accumulator.

According to Eq. (2.70), for each value of n we add the current input value $x[n]$ to the previously accumulated sum $y[n-1]$. This interpretation of the accumulator is represented in block diagram form in Fig. 2.15. Equation (2.70) and the block diagram in Fig. 2.15 are referred to as a *recursion representation* of the system since each value is computed using previously computed values. This general notion will be explored in more detail later in this section.

Example 2.10

Let us consider the moving average system of Example 2.2, with $M_1 = 0$ so that the system is causal. In this case, from Eq. (2.59) the impulse response is

$$h[n] = \frac{1}{(M_2 + 1)} (u[n] - u[n - M_2 - 1]), \tag{2.71}$$

from which it follows that

$$y[n] = \frac{1}{(M_2 + 1)} \sum_{k=0}^{M_2} x[n - k], \tag{2.72}$$

which is a special case of Eq. (2.68), with $N = 0$, $a_0 = 1$, and $b_k = 1/(M_2 + 1)$ for $0 \le k \le M_2$. The impulse response can also be expressed as

$$h[n] = \frac{1}{(M_2 + 1)} (\delta[n] - \delta[n - M_2 - 1]) * u[n], \tag{2.73}$$

which suggests that the causal moving average system can be represented as the cascade system of Fig. 2.16. We can obtain a difference equation for this block diagram by noting first that

$$x_1[n] = \frac{1}{(M_2 + 1)} (x[n] - x[n - M_2 - 1]). \tag{2.74}$$

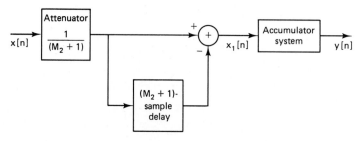

Figure 2.16 Block diagram of the recursive form of a moving average system.

From Eq. (2.69) of Example 2.9, the output of the accumulator satisfies the difference equation

$$y[n] - y[n - 1] = x_1[n]$$

so that

$$y[n] - y[n - 1] = \frac{1}{(M_2 + 1)} (x[n] - x[n - M_2 - 1]). \tag{2.75}$$

Again, we have a difference equation in the form of Eq. (2.68), but this time $N = 1$, $a_0 = 1$, $a_1 = -1$, and $b_0 = -b_{M_2+1} = 1/(M_2 + 1)$, and $b_k = 0$ otherwise.

Interestingly, there are several distinct forms for the representation of the moving average system in terms of difference equations. In Chapter 6 we will see that an unlimited number of distinct difference equations can be used to represent a given linear time-invariant input/output relation.

Just as in the case of linear constant-coefficient differential equations for continuous-time systems, without additional constraints or information a linear constant-coefficient difference equation for discrete-time systems does not provide a unique specification of the output for a given input. Specifically, suppose that for a given input $x_p[n]$ we have determined by some means one output sequence $y_p[n]$ so that an equation of the form of Eq. (2.68) is satisfied. Then the same equation with the same input is satisfied by any output of the form

$$y[n] = y_p[n] + y_h[n], \tag{2.76}$$

where $y_h[n]$ is any solution to Eq. (2.68) with $x[n] = 0$, i.e., to the equation

$$\sum_{k=0}^{N} a_k y_h[n - k] = 0. \tag{2.77}$$

Equation (2.77) is referred to as the *homogeneous equation* and $y_h[n]$ the *homogeneous solution*. The sequence $y_h[n]$ is in fact a member of a family of solutions of the form

$$y_h[n] = \sum_{m=1}^{N} A_m z_m^n. \tag{2.78}$$

Substituting Eq. (2.78) into Eq. (2.77) shows that the complex numbers z_m must be roots of the polynomial

$$\sum_{k=0}^{N} a_k z^{-k} = 0. \tag{2.79}$$

Equation (2.78) assumes that all N roots of the polynomial in Eq. (2.79) are distinct. The form of terms associated with multiple roots is slightly different, but there are always N undetermined coefficients. An example of the homogeneous solution with multiple roots is considered in Problem 2.13.

Since $y_h[n]$ has N undetermined coefficients, a set of N auxiliary conditions is required for unique specification of $y[n]$ for a given $x[n]$. These auxiliary conditions might consist of specifying fixed values of $y[n]$ at specific values of n, such as $y[-1]$,

$y[-2]$, ..., $y[-N]$, and then solving a set of N linear equations for the N undetermined coefficients.

Alternatively, if the auxiliary conditions are a set of auxiliary values of $y[n]$, the other values of $y[n]$ can be generated by rewriting Eq. (2.68) as a recurrence formula, i.e., in the form

$$y[n] = -\sum_{k=1}^{N} \frac{a_k}{a_0} y[n-k] + \sum_{r=0}^{M} \frac{b_r}{a_0} x[n-r]. \tag{2.80}$$

If the input $x[n]$ together with a set of auxiliary values, say $y[-1]$, $y[-2]$, ..., $y[-N]$, is specified, then $y[0]$ can be determined from Eq. (2.80). With $y[0]$, $y[-1]$, ..., $y[-N+1]$ available, $y[1]$ can then be calculated, and so on. Using this procedure, $y[n]$ is said to be computed *recursively*; i.e., the output computation involves not only the input sequence but also previous values of the output sequence.

To generate values of $y[n]$ for $n < -N$ (again assuming that the values $y[-1]$, $y[-2]$, ..., $y[-N]$ are given as auxiliary conditions), we can rearrange Eq. (2.68) in the form

$$y[n-N] = -\sum_{k=0}^{N-1} \frac{a_k}{a_N} y[n-k] + \sum_{r=0}^{M} \frac{b_r}{a_N} x[n-r], \tag{2.81}$$

from which $y[-N-1]$, $y[-N-2]$, ..., can be computed recursively. The following example illustrates this recursive procedure.

Example 2.11

The difference equation satisfied by the input and output of a system is

$$y[n] = ay[n-1] + x[n]. \tag{2.82}$$

Consider the input $x[n] = K\delta[n]$, where K is an arbitrary number, and the auxiliary condition $y[-1] = c$. Beginning with this value, the output for $n > -1$ can be computed recursively as follows:

$$y[0] = ac + K,$$

$$y[1] = ay[0] + 0$$

$$= a(ac + K) = a^2 c + aK,$$

$$y[2] = a(a^2 c + aK) = a^3 c + a^2 K,$$

$$y[3] = a(a^3 c + a^2 K) = a^4 c + a^3 K,$$

$$\vdots \qquad\qquad \vdots$$

For this simple case, we can see that for $n \geq 0$, $y[n]$ is given by

$$y[n] = a^{n+1} c + a^n K \qquad \text{for } n \geq 0. \tag{2.83}$$

To determine the output for $n < 0$, we express the difference equation in the form

$$y[n-1] = a^{-1}(y[n] - x[n]), \tag{2.84a}$$

or

$$y[n] = a^{-1}(y[n+1] - x[n+1]). \tag{2.84b}$$

Using the auxiliary condition $y[-1] = c$, we can compute $y[n]$ for $n < -1$ as follows:

$$y[-2] = a^{-1}\{(y[-1] - x[-1])\}$$
$$= a^{-1}c,$$
$$y[-3] = a^{-1}a^{-1}c = a^{-2}c,$$
$$y[-4] = a^{-1}a^{-2}c = a^{-3}c.$$

Clearly,

$$y[n] = a^{n+1}c \qquad \text{for } n \leq -1. \tag{2.85}$$

Thus, in summary, combining Eqs. (2.83) and (2.85) we obtain as the result of the recursive computation

$$y[n] = a^{n+1}c + Ka^{n}u[n] \qquad \text{for all } n. \tag{2.86}$$

Several important points are illustrated by the solution of Example 2.11. First note that we implemented the system by recursively computing the output in both positive and negative directions beginning with $n = -1$. Clearly this procedure is noncausal. Also note that when $K = 0$, the input is zero, but $y[n] = a^{n+1}c$. A linear system requires that the output be zero for all time when the input is zero for all time (see Problem 2.1). Consequently, this system is not linear. Furthermore, if the input were shifted by n_0 samples, i.e., $x_1[n] = K\delta[n - n_0]$, the output would be

$$y_1[n] = a^{n+1}c + Ka^{n-n_0}u[n - n_0], \tag{2.87}$$

and the system therefore is not time-invariant.

Our principal interest in this text is in systems that are linear and time-invariant, in which case the auxiliary conditions must be consistent with these additional requirements. In Chapter 4, when we discuss the solution of difference equations using the z-transform, we implicitly incorporate conditions of linearity and time invariance. As we will see in that discussion, even with the additional constraints of linearity and time invariance, the solution to the difference equation, and therefore the system, is not uniquely specified. In particular, there are in general both causal and noncausal linear time-invariant systems consistent with a given difference equation.

If a system is characterized by a linear constant-coefficient difference equation and is further specified to be linear, time-invariant, and causal, the solution is unique. In this case the auxiliary conditions are often stated as *initial-rest conditions*. In other words, the auxiliary information is that if the input $x[n]$ is zero for n less than some time n_0, then the output $y[n]$ is constrained to be zero for n less than n_0. This then provides sufficient initial conditions to obtain $y[n]$ for $n \geq n_0$ recursively using Eq. (2.80).

To summarize, for a system for which the input and output satisfy a linear constant-coefficient difference equation:

- The output for a given input is not uniquely specified. Auxiliary information or conditions are required.

- If the auxiliary information is in the form of N sequential values of the output, later values can be obtained by rearranging the difference equation as a recursive relation running forward in n, and prior values can be obtained by rearranging the difference equation as a recursive relation running backward in n.

- Linearity, time invariance, and causality of the system will depend on the auxiliary conditions. If an additional condition is that the system is initially at rest, then the system will be linear, time-invariant, and causal.

With the preceding discussion in mind, let us now reconsider Example 2.11, but with initial-rest conditions. With $x[n] = K\delta[n]$, $y[-1] = 0$ since $x[n] = 0$, $n < 0$. Consequently from Eq. (2.86),

$$y[n] = Ka^n u[n]. \tag{2.88}$$

If the input is instead $K\delta[n - n_0]$, again with initial-rest conditions, then the recursive solution is carried out using the initial condition $y[n] = 0$, $n < n_0$. Note that for $n_0 < 0$, initial rest implies that $y[-1] \neq 0$. That is, initial rest does not always mean $y[-1] = \cdots = y[-N] = 0$. It does mean that $y[n_0 - 1] = \cdots = y[n_0 - N] = 0$ if $x[n] = 0$ for $n < n_0$. Note also that the impulse response for the example is $h[n] = a^n u[n]$, i.e., it is zero for $n < 0$, consistent with the causality imposed by the assumption of initial rest.

The preceding discussion assumed that $N \geq 1$ in Eq. (2.68). If $N = 0$ in Eq. (2.68), no recursion is required to use the difference equation to compute the output and therefore no auxiliary conditions are required. That is,

$$y[n] = \sum_{k=0}^{M} \left(\frac{b_k}{a_0}\right) x[n - k]. \tag{2.89}$$

Equation (2.89) is in the form of a convolution, and by setting $x[n] = \delta[n]$ we see that the impulse response is

$$h[n] = \sum_{k=0}^{M} \left(\frac{b_k}{a_0}\right) \delta[n - k],$$

or

$$h[n] = \begin{cases} \left(\dfrac{b_k}{a_0}\right), & 0 \leq n \leq M, \\ 0, & \text{otherwise.} \end{cases} \tag{2.90}$$

The impulse response is obviously finite in duration. Indeed, the output of any FIR system can be computed nonrecursively using the difference equation of Eq. (2.89), where the coefficients are the values of the impulse sequence. The moving average system of Example 2.10 with $M_1 = 0$ is an example of a causal FIR system. An interesting feature of that system was that we also found a recursive equation for the output. In Chapter 6 we will show that there are many possible ways of implementing a desired signal transformation using difference equations. Advantages of one method over another depend on practical considerations such as numerical accuracy, data storage, and the number of multiplications and additions required to compute each sample of the output.

2.6 FREQUENCY-DOMAIN REPRESENTATION OF DISCRETE-TIME SIGNALS AND SYSTEMS

In the previous sections we have introduced some of the fundamental concepts of the theory of discrete-time signals and systems. For linear time-invariant systems we saw that a representation of the input sequence as a weighted sum of delayed impulses leads to a representation of the output as a weighted sum of delayed responses. As with continuous-time signals, discrete-time signals may be represented in a number of different ways. For example, sinusoidal and complex exponential sequences play a particularly important role in representing discrete-time signals. This is because complex exponential sequences are eigenfunctions of linear time-invariant systems, and the response to a sinusoidal input is sinusoidal with the same frequency as the input and with amplitude and phase determined by the system. This fundamental property of linear time-invariant systems makes representations of signals in terms of sinusoids or complex exponentials (i.e., Fourier representations) very useful in linear system theory.

To demonstrate the eigenfunction property of complex exponentials for discrete-time systems, consider an input sequence $x[n] = e^{j\omega n}$ for $-\infty < n < \infty$, i.e., a complex exponential of radian frequency ω. From Eq. (2.49), the corresponding output of a linear time-invariant system with impulse response $h[n]$ is

$$y[n] = \sum_{k=-\infty}^{\infty} h[k]e^{j\omega(n-k)}$$

$$= e^{j\omega n}\left(\sum_{k=-\infty}^{\infty} h[k]e^{-j\omega k}\right). \tag{2.91}$$

If we define

$$H(e^{j\omega}) = \sum_{k=-\infty}^{\infty} h[k]e^{-j\omega k}, \tag{2.92}$$

Eq. (2.91) becomes

$$y[n] = H(e^{j\omega})e^{j\omega n}. \tag{2.93}$$

Consequently, $e^{j\omega n}$ is an eigenfunction of the system, and the associated eigenvalue is $H(e^{j\omega})$. From Eq. (2.93) we see that $H(e^{j\omega})$ describes the change in complex amplitude of a complex exponential as a function of the frequency ω. The eigenvalue $H(e^{j\omega})$

is called the *frequency response* of the system. In general, $H(e^{j\omega})$ is complex and can be expressed in terms of its real and imaginary parts as

$$H(e^{j\omega}) = H_R(e^{j\omega}) + jH_I(e^{j\omega}) \tag{2.94}$$

or in terms of magnitude and phase as

$$H(e^{j\omega}) = |H(e^{j\omega})|e^{j \angle H(e^{j\omega})}. \tag{2.95}$$

Example 2.12

As a simple example of how we can find the frequency response of a linear time-invariant system, consider the ideal delay system defined by

$$y[n] = x[n - n_d], \tag{2.96}$$

where n_d is a fixed integer. If we consider $x[n] = e^{j\omega n}$ as input to this system, from Eq. (2.96) we have

$$y[n] = e^{j\omega(n - n_d)} = e^{-j\omega n_d}e^{j\omega n}.$$

Thus for any given value of ω, we obtain an output that is the input multiplied by a complex constant, the value of which depends on ω. The frequency response of the ideal delay is therefore

$$H(e^{j\omega}) = e^{-j\omega n_d}. \tag{2.97}$$

As an alternative method of obtaining the frequency response, recall that $h[n] = \delta[n - n_d]$ for the ideal delay system. Using Eq. (2.92) we obtain

$$H(e^{j\omega}) = \sum_{n=-\infty}^{\infty} \delta[n - n_d]e^{-j\omega n} = e^{-j\omega n_d}.$$

Using the Euler relation, the real and imaginary parts of the frequency response are

$$H_R(e^{j\omega}) = \cos \omega n_d, \tag{2.98a}$$

$$H_I(e^{j\omega}) = -\sin \omega n_d. \tag{2.98b}$$

The magnitude and phase are

$$|H(e^{j\omega})| = 1, \tag{2.99a}$$

$$\angle H(e^{j\omega}) = -\omega n_d. \tag{2.99b}$$

In Section 2.7 we will show that a broad class of signals can be represented as a linear combination of complex exponentials in the form

$$x[n] = \sum_k \alpha_k e^{j\omega_k n}. \tag{2.100}$$

From the principle of superposition, the corresponding output of a linear time-invariant system is

$$y[n] = \sum_k \alpha_k H(e^{j\omega_k})e^{j\omega_k n}. \tag{2.101}$$

Thus if we can find a representation of $x[n]$ as a superposition of complex exponential sequences, as in Eq. (2.100), then we can find the output using Eq. (2.101) if we know the frequency response of the system. The following simple example illustrates this fundamental property of linear time-invariant systems.

Example 2.13

Since it is simple to express a sinusoid as a linear combination of complex exponentials, let us consider a sinusoidal input

$$x[n] = A\cos(\omega_0 n + \phi) = \frac{A}{2}e^{j\phi}e^{j\omega_0 n} + \frac{A}{2}e^{-j\phi}e^{-j\omega_0 n}. \qquad (2.102)$$

From Eq. (2.93), the response to $x_1[n] = (A/2)e^{j\phi}e^{j\omega_0 n}$ is

$$y_1[n] = H(e^{j\omega_0})\frac{A}{2}e^{j\phi}e^{j\omega_0 n}. \qquad (2.103a)$$

The response to $x_2[n] = (A/2)e^{-j\phi}e^{-j\omega_0 n}$ is

$$y_2[n] = H(e^{-j\omega_0})\frac{A}{2}e^{-j\phi}e^{-j\omega_0 n}. \qquad (2.103b)$$

Thus the total response is

$$y[n] = \frac{A}{2}[H(e^{j\omega_0})e^{j\phi}e^{j\omega_0 n} + H(e^{-j\omega_0})e^{-j\phi}e^{-j\omega_0 n}]. \qquad (2.104)$$

If $h[n]$ is real, it can be shown (see Problem 2.26) that $H(e^{-j\omega_0}) = H^*(e^{j\omega_0})$. Consequently

$$y[n] = A|H(e^{j\omega_0})|\cos(\omega_0 n + \phi + \theta), \qquad (2.105)$$

where $\theta = \measuredangle H(e^{j\omega_0})$ is the phase of the system at frequency ω_0.

For the simple example of the ideal delay, $|H(e^{j\omega_0})| = 1$ and $\theta = -\omega_0 n_d$, as we determined in Example 2.12. Therefore

$$y[n] = A\cos(\omega_0 n + \phi - \omega_0 n_d)$$
$$= A\cos[\omega_0(n - n_d) + \phi], \qquad (2.106)$$

which is consistent with what we would obtain directly using the definition of the ideal delay system.

The concept of the frequency response of linear time-invariant systems is essentially the same for continuous-time and discrete-time systems. However, an important distinction arises because the frequency response of discrete-time linear time-invariant systems is *always* a periodic function of the frequency variable ω with period 2π. To show this, we substitute $\omega + 2\pi$ into Eq. (2.92) to obtain

$$H[e^{j(\omega + 2\pi)}] = \sum_{n=-\infty}^{\infty} h[n]e^{-j(\omega + 2\pi)n}. \qquad (2.107)$$

Using the fact that $e^{\pm j2\pi n} = 1$ for n an integer, we have

$$e^{-j(\omega + 2\pi)n} = e^{-j\omega n}e^{-j2\pi n} = e^{-j\omega n}.$$

Therefore we see that

$$H(e^{j(\omega + 2\pi)}) = H(e^{j\omega}) \tag{2.108}$$

and, more generally,

$$H(e^{j(\omega + 2\pi r)}) = H(e^{j\omega}) \qquad \text{for } r \text{ an integer,} \tag{2.109}$$

i.e., $H(e^{j\omega})$ is periodic with period 2π. Note that this is obviously true for the ideal delay system since $e^{-j(\omega + 2\pi)n_d} = e^{-j\omega n_d}$, when n_d is an integer.

The reason for this periodicity is related directly to our earlier observation that the sequence

$$\{e^{j\omega n}\}, \qquad -\infty < n < \infty,$$

is indistinguishable from the sequence

$$\{e^{j(\omega + 2\pi)n}\}, \qquad -\infty < n < \infty.$$

Since these two sequences have identical values for all n, the system must respond identically to both input sequences. This requires that Eq. (2.108) hold.

Since $H(e^{j\omega})$ is periodic with period 2π, and since frequencies ω and $\omega + 2\pi$ are indistinguishable, it follows that we need only specify $H(e^{j\omega})$ over an interval of length 2π, e.g., $0 \le \omega < 2\pi$ or $-\pi < \omega \le \pi$. The inherent periodicity defines the frequency response everywhere outside the chosen interval. For simplicity and for consistency with the continuous-time case, it is generally convenient to specify $H(e^{j\omega})$ over the interval $-\pi < \omega \le \pi$. With respect to this interval, the low frequencies are frequencies close to zero, while the "high frequencies" are frequencies close to $\pm\pi$. Recalling that frequencies differing by an integer multiple of 2π are indistinguishable, we might generalize the preceding statement as follows: The "low frequencies" are those that are close to an even multiple of π, while the "high frequencies" are those that are close to an odd multiple of π, consistent with our earlier discussion in Section 2.1.

Example 2.14

An important class of linear time-invariant systems includes those systems for which the frequency response is unity over a certain range of frequencies and is zero at the remaining frequencies. These correspond to *ideal frequency-selective filters*. In keeping with the preceding discussion, the frequency response of an ideal lowpass filter is shown in Figure 2.17(a). Because of the inherent periodicity of the discrete-time frequency response, it has the appearance of a multiband filter since frequencies around $\omega = 2\pi$ are indistinguishable from frequencies around $\omega = 0$. In effect, however, the frequency response passes only low frequencies and rejects high frequencies. Since the frequency response is completely specified by its behavior over the interval $-\pi < \omega < \pi$, the ideal lowpass filter frequency response is more typically shown only in the interval $-\pi < \omega < \pi$ as in Fig. 2.17(b). The frequency response for ideal highpass, bandstop, and bandpass filters are shown in Figures 2.18(a), (b), and (c), respectively.

Example 2.15

The impulse response of the moving average system of Example 2.2 is

$$h[n] = \begin{cases} \dfrac{1}{M_1 + M_2 + 1}, & -M_1 \le n \le M_2, \\ 0, & \text{otherwise.} \end{cases}$$

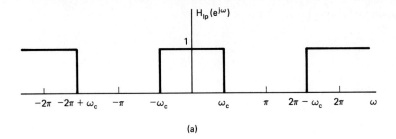

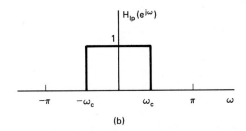

(b)

Figure 2.17   Ideal lowpass filter showng (a) periodicity of the frequency response, (b) one period of the periodic frequency response.

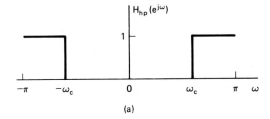

(a)

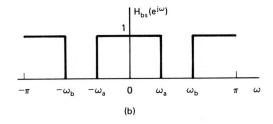

(b)

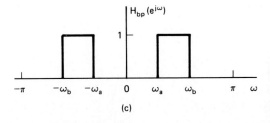

(c)

Figure 2.18   Ideal frequency-selective filters. (a) Highpass filter. (b) Bandstop filter. (c) Bandpass filter. In each case, the frequency response is periodic with period 2π. Only one period is shown.

Therefore the frequency response is

$$H(e^{j\omega}) = \frac{1}{M_1 + M_2 + 1} \sum_{n=-M_1}^{M_2} e^{-j\omega n}. \tag{2.110}$$

Equation (2.110) can be expressed in closed form by using Eq. (2.43) so that

$$H(e^{j\omega}) = \frac{1}{M_1 + M_2 + 1} \frac{e^{j\omega M_1} - e^{-j\omega(M_2+1)}}{1 - e^{-j\omega}}$$

$$= \frac{1}{M_1 + M_2 + 1} e^{-j\omega(M_2 - M_1 + 1)/2} \frac{e^{j\omega(M_1 + M_2 + 1)/2} - e^{-j\omega(M_1 + M_2 + 1)/2}}{1 - e^{-j\omega}}$$

$$= \frac{1}{M_1 + M_2 + 1} e^{-j\omega(M_2 - M_1)/2} \frac{e^{j\omega(M_1 + M_2 + 1)/2} - e^{-j\omega(M_1 + M_2 + 1)/2}}{e^{j\omega/2} - e^{-j\omega/2}}$$

$$= \frac{1}{M_1 + M_2 + 1} e^{-j\omega(M_2 - M_1)/2} \frac{\sin[\omega(M_1 + M_2 + 1)/2]}{\sin(\omega/2)}. \tag{2.111}$$

The magnitude and phase of $H(e^{j\omega})$ are plotted in Fig. 2.19 for $M_1 = 0$, $M_2 = 4$. Note that $H(e^{j\omega})$ is periodic as required of the frequency response of a discrete-time system. Also note that $|H(e^{j\omega})|$ falls off at "high frequencies" and $\angle H(e^{j\omega})$, i.e., the phase of $H(e^{j\omega})$, varies linearly with ω. This attenuation of the high frequencies suggests that the system will smooth out rapid variations in the input sequence; i.e., it is a rough approximation to a lowpass filter. This is consistent with what we would intuitively expect about the behavior of the moving average system.

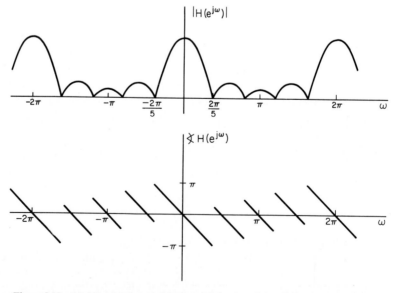

Figure 2.19 (a) Magnitude and (b) phase of the frequency response of the moving average system for the case $M_1 = 0$ and $M_2 = 4$.

One of the advantages of the frequency-response representation of a linear time-invariant system is that interpretations of system behavior such as the one we made in Example 2.15 often follow easily. We will elaborate on this point in considerably more detail in Chapter 5. At this point, however, let us return to the question of how we may find representations of the form of Eq. (2.100) for an arbitrary input sequence.

2.7 REPRESENTATION OF SEQUENCES BY FOURIER TRANSFORMS

Many sequences can be represented by a Fourier integral of the form

$$x[n] = \frac{1}{2\pi} \int_{-\pi}^{\pi} X(e^{j\omega})e^{j\omega n} \, d\omega, \tag{2.112}$$

where $X(e^{j\omega})$ is given by

$$X(e^{j\omega}) = \sum_{n=-\infty}^{\infty} x[n]e^{-j\omega n}. \qquad \textbf{DFT} \tag{2.113}$$

Equations (2.112) and (2.113) together form a *Fourier representation* for the sequence. Equation (2.112), the *inverse Fourier transform*, is a *synthesis* formula. That is, it represents $x[n]$ as a superposition of infinitesimally small complex sinusoids of the form

$$\frac{1}{2\pi} X(e^{j\omega})e^{j\omega n} \, d\omega,$$

with ω ranging over an interval of length 2π and with $X(e^{j\omega})$ determining the relative amount of each complex sinusoidal component. Although in writing Eq. (2.112) we have chosen the range of values for ω between $-\pi$ and $+\pi$, any interval of length 2π can be used. Equation (2.113), the *Fourier transform*,† is an expression for computing $X(e^{j\omega})$ from the sequence $x[n]$, i.e., for *analyzing* the sequence $x[n]$ to determine how much of each frequency component is required to synthesize $x[n]$ using Eq. (2.112).

In general, the Fourier transform is a complex-valued function of ω. As with the frequency response, we sometimes express $X(e^{j\omega})$ in rectangular form as

$$X(e^{j\omega}) = X_R(e^{j\omega}) + jX_I(e^{j\omega}) \tag{2.114a}$$

or in polar form as

$$X(e^{j\omega}) = |X(e^{j\omega})|e^{j\angle X(e^{j\omega})}. \tag{2.114b}$$

The quantities $|X(e^{j\omega})|$ and $\angle X(e^{j\omega})$ are called the *magnitude* and *phase*, respectively, of the Fourier transform. The Fourier transform is also sometimes referred to as the

---

† Sometimes we will refer to Eq. (2.113) more explicitly as the discrete-time Fourier transform, particularly when it is important to distinguish it from the continuous-time Fourier transform.

Fourier spectrum or simply the *spectrum.* Also, the terminology *magnitude spectrum* or *amplitude spectrum* is sometimes used to refer to $|X(e^{j\omega})|$ and the angle or phase $\not{\!\angle} X(e^{j\omega})$ is sometimes called the *phase spectrum.*

The phase $\not{\!\angle} X(e^{j\omega})$ is not uniquely specified by Eq. (2.114b) since any integer multiple of 2π may be added to $\not{\!\angle} X(e^{j\omega})$ at any value of ω without affecting the result of the complex exponentiation. When we specifically want to refer to the principal value, i.e., $\not{\!\angle} X(e^{j\omega})$ restricted to the range of values between $-\pi$ and $+\pi$, we will denote this as $\mathrm{ARG}[X(e^{j\omega})]$. In Chapter 5 we will also find it useful to refer to a phase function that is a continuous function of ω for $0 < \omega < \pi$, and this will be denoted $\arg[X(e^{j\omega})]$.

By comparing Eqs. (2.92) and (2.113) we can see that the frequency response of a linear time-invariant system is simply the Fourier transform of the impulse response and that therefore the impulse response can be obtained from the frequency response by applying the inverse Fourier transform integral; i.e.,

$$h[n] = \frac{1}{2\pi} \int_{-\pi}^{\pi} H(e^{j\omega})e^{j\omega n}\, d\omega. \tag{2.115}$$

As discussed previously, the frequency response is a periodic function. Likewise, the Fourier transform is periodic with period 2π. Indeed, Eq. (2.113) is of the form of a Fourier series for the continuous-variable periodic function $X(e^{j\omega})$, and Eq. (2.112), which expresses the sequence values $x[n]$ in terms of the periodic function $X(e^{j\omega})$, is of the form of the integral that would be used to obtain the coefficients in the Fourier series. Our use of Eqs. (2.112) and (2.113) focuses on the representation of the sequence $x[n]$. Nevertheless it is useful to be aware of the equivalence between the Fourier series representation of continuous-variable periodic functions and the Fourier transform representation of discrete-time signals, since all the familiar properties of Fourier series can be applied, with appropriate interpretation of variables, to the Fourier transform representation of a sequence.

We have not yet shown explicitly that Eqs. (2.112) and (2.113) are inverses of each other or considered the question of how broad a class of signals can be represented in the form of Eq. (2.112). To demonstrate that Eq. (2.112) is the inverse of Eq. (2.113), we can find $X(e^{j\omega})$ using Eq. (2.113) and then substitute the result into Eq. (2.112). Specifically, consider

$$\frac{1}{2\pi} \int_{-\pi}^{\pi} \left(\sum_{m=-\infty}^{\infty} x[m]e^{-j\omega m} \right) e^{j\omega n}\, d\omega = \hat{x}[n], \tag{2.116}$$

where tentatively we have used $\hat{x}[n]$ to denote the result of the Fourier synthesis. We wish to show that $\hat{x}[n] = x[n]$ if $X(e^{j\omega})$ can be found using Eq. (2.113). Note that the "dummy index" of summation has been changed to m to distinguish it from n, the variable index in Eq. (2.112). If the infinite sum converges uniformly† for all ω, then

---

† Uniform convergence means that for every value of ω,

$$\lim_{M \to \infty} |X(e^{j\omega}) - X_M(e^{j\omega})| = 0.$$

we can interchange the order of integration and summation to obtain

$$\hat{x}[n] = \sum_{m=-\infty}^{\infty} x[m]\left(\frac{1}{2\pi}\int_{-\pi}^{\pi} e^{j\omega(n-m)}\,d\omega\right). \qquad (2.117)$$

Evaluating the integral within the parentheses gives

$$\frac{1}{2\pi}\int_{-\pi}^{\pi} e^{j\omega(n-m)}\,d\omega = \frac{\sin\pi(n-m)}{\pi(n-m)}$$

$$= \begin{cases} 1, & m = n, \\ 0, & m \neq n, \end{cases}$$

$$= \delta[n-m].$$

Thus

$$\hat{x}[n] = \sum_{m=-\infty}^{\infty} x[m]\delta[n-m] = x[n],$$

which is what we set out to show.

The question of determining the class of signals that can be represented by Eq. (2.112) is equivalent to considering the convergence of the infinite sum in Eq. (2.113). That is, we are concerned with the conditions that must be satisfied by the terms in the sum in Eq. (2.113) such that

$$|X(e^{j\omega})| < \infty \qquad \text{for all } \omega,$$

where $X(e^{j\omega})$ is the limit as $M \to \infty$ of the finite sum

$$X_M(e^{j\omega}) = \sum_{n=-M}^{M} x[n]e^{-j\omega n}. \qquad (2.118)$$

A sufficient condition for convergence can be found as follows:

$$|X(e^{j\omega})| = \left|\sum_{n=-\infty}^{\infty} x[n]e^{-j\omega n}\right| < \infty$$

$$\leq \sum_{n=-\infty}^{\infty} |x[n]||e^{-j\omega n}| < \infty$$

$$\leq \sum_{n=-\infty}^{\infty} |x[n]| < \infty.$$

Thus, if $x[n]$ is *absolutely summable*, then $X(e^{j\omega})$ exists. Furthermore, in this case the series can be shown to converge uniformly to a continuous function of ω.

Since a stable sequence is, by definition, absolutely summable, all stable sequences have Fourier transforms. It also follows, then, that any stable *system* will have a finite and continuous frequency response.

Absolute summability is a sufficient condition for existence of a Fourier transform representation. In Examples 2.12 and 2.15, we computed the Fourier transforms of the sequences $\delta[n - n_d]$ and $[1/(M_1 + M_2 + 1)](u[n + M_1] - u[n - M_2 - 1])$. These sequences are absolutely summable since they are finite in length. Clearly, any finite-length sequence is absolutely summable and thus will have a Fourier transform representation. In the context of linear time-invariant systems, any FIR system will be stable and therefore will have a finite, continuous frequency response. When a sequence has infinite length, we must be concerned about convergence of the infinite sum. The following example illustrates this case.

Example 2.16

Let $x[n] = a^n u[n]$. The Fourier transform of this sequence is

$$X(e^{j\omega}) = \sum_{n=0}^{\infty} a^n e^{-j\omega n} = \sum_{n=0}^{\infty} (ae^{-j\omega})^n$$

$$= \frac{1}{1 - ae^{-j\omega}} \qquad \text{if } |ae^{-j\omega}| < 1 \quad \text{or} \quad |a| < 1.$$

Clearly the condition $|a| < 1$ is the condition for absolute summability of $x[n]$; i.e.,

$$\sum_{n=0}^{\infty} |a|^n = \frac{1}{1 - |a|} < \infty \qquad \text{if } |a| < 1. \tag{2.119}$$

Absolute summability is a *sufficient* condition for existence of a Fourier transform representation, and it also guarantees uniform convergence. Some sequences are not absolutely summable but are square summable, i.e.,

$$\sum_{n=-\infty}^{\infty} |x[n]|^2 < \infty. \tag{2.120}$$

Such sequences can be represented by a Fourier transform if we are willing to relax the condition of uniform convergence of the infinite sum defining $X(e^{j\omega})$. Specifically, in this case we have mean-square convergence, i.e., with

$$X(e^{j\omega}) = \sum_{n=-\infty}^{\infty} x[n]e^{-j\omega n} \tag{2.121a}$$

and

$$X_M(e^{j\omega}) = \sum_{n=-M}^{M} x[n]e^{-j\omega n}, \tag{2.121b}$$

then

$$\lim_{M \to \infty} \int_{-\pi}^{\pi} |X(e^{j\omega}) - X_M(e^{j\omega})|^2 \, d\omega \to 0. \tag{2.122}$$

In other words, the error $|X(e^{j\omega}) - X_M(e^{j\omega})|$ may not approach zero at each value of ω as $M \to \infty$, but the total "energy" in the error does. Example 2.17 illustrates this case.

Example 2.17

Let us determine the impulse response of the ideal lowpass filter discussed in Example 2.14. The frequency response is

$$H_{lp}(e^{j\omega}) = \begin{cases} 1, & |\omega| < \omega_c, \\ 0, & \omega_c < |\omega| \le \pi, \end{cases} \qquad (2.123)$$

with periodicity 2π also understood. The impulse response $h_{lp}[n]$ can be found using the Fourier transform synthesis equation (2.112):

$$h_{lp}[n] = \frac{1}{2\pi} \int_{-\omega_c}^{\omega_c} e^{j\omega n}\, d\omega$$

$$= \frac{\sin \omega_c n}{\pi n}, \qquad -\infty < n < \infty. \qquad (2.124)$$

We note that since $h_{lp}[n]$ is nonzero for $n < 0$, the ideal lowpass filter is noncausal. Also, $h_{lp}[n]$ is *not* absolutely summable. The sequence values approach zero as $n \to \infty$, but only as $1/n$. This is because $X(e^{j\omega})$ is discontinuous at $\omega = \omega_c$. Since $x[n]$ is not absolutely summable,

$$\sum_{n=-\infty}^{\infty} \frac{\sin \omega_c n}{\pi n} e^{-j\omega n}$$

does not converge uniformly for all values of ω. To obtain an intuitive feeling for this, let us consider $H_M(e^{j\omega})$ as the sum of a finite number of terms:

$$H_M(e^{j\omega}) = \sum_{n=-M}^{M} \frac{\sin \omega_c n}{\pi n} e^{-j\omega n}. \qquad (2.125)$$

We can show that $H_M(e^{j\omega})$ can be expressed as

$$H_M(e^{j\omega}) = \frac{1}{2\pi} \int_{-\omega_c}^{\omega_c} \frac{\sin[(2M+1)(\omega-\theta)/2]}{\sin[(\omega-\theta)/2]}\, d\theta.$$

The function $H_M(e^{j\omega})$ is evaluated in Fig. 2.20 for several values of M. Note that as M increases, the oscillatory behavior at $\omega = \omega_c$ (often referred to as the Gibbs phenomenon) is more rapid, but the size of the ripples does not decrease. In fact, it can be shown that as $M \to \infty$ the maximum amplitude of the oscillations does not go to zero but the oscillations converge in location toward the point $\omega = \omega_c$. Thus, the infinite sum does not converge uniformly to the discontinuous function $H_{lp}(e^{j\omega})$ of Eq. (2.123). However, $h_{lp}[n]$ as given in Eq. (2.124) is square summable, and correspondingly $H_M(e^{j\omega})$ converges in the mean-square sense to $X(e^{j\omega})$, i.e.,

$$\lim_{M\to\infty} \int_{-\pi}^{\pi} |H(e^{j\omega}) - H_M(e^{j\omega})|^2\, d\omega \to 0.$$

Although the error between $\lim_{M\to\infty} H_M(e^{j\omega})$ and $H_{lp}(e^{j\omega})$ might seem unimportant because the two functions differ only at $\omega = \omega_c$, we will see in Chapter 7 that the behavior of finite sums has important implications in the design of discrete-time systems for filtering.

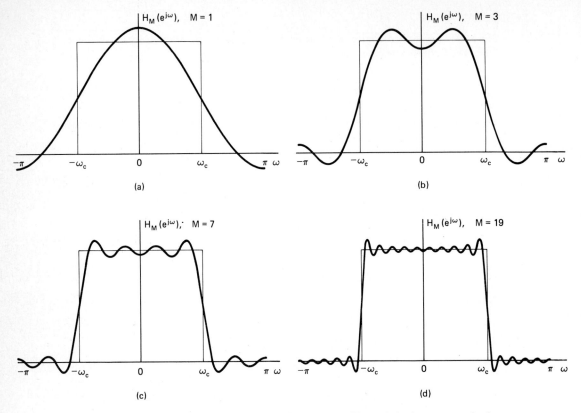

Figure 2.20   Convergence of the Fourier transform. The oscillatory behavior at $\omega = \omega_c$ is often called the Gibbs phenomenon.

It is sometimes useful to have a Fourier transform representation for certain sequences that are neither absolutely summable nor square summable. We illustrate several of these in the following examples.

Example 2.18

Consider the sequence $x[n] = 1$ for all n. This sequence is neither absolutely nor square summable, and Eq. (2.113) does not converge in either the uniform or mean-square sense for this case. However, it is possible and useful to define the Fourier transform of this sequence to be the periodic impulse train

$$X(e^{j\omega}) = \sum_{r=-\infty}^{\infty} 2\pi\delta(\omega + 2\pi r). \tag{2.126}$$

The impulses in this case are functions of a continuous variable and therefore are of "infinite height, zero width, and unit area," consistent with the fact that Eq. (2.113) does not converge. The use of Eq. (2.126) as a Fourier representation of the sequence $x[n] = 1$ is justified principally because formal substitution of Eq. (2.126) into Eq. (2.112) leads to the correct result. Example 2.19 represents a generalization of this example.

Example 2.19

Consider a sequence $x[n]$ whose Fourier transform is the periodic impulse train

$$X(e^{j\omega}) = \sum_{r=-\infty}^{\infty} 2\pi\delta(\omega - \omega_0 + 2\pi r). \tag{2.127}$$

Substituting $X(e^{j\omega})$ into the inverse Fourier transform integral of Eq. (2.112) and noting that the integration involves only one period of $X(e^{j\omega})$, we have

$$x[n] = \frac{1}{2\pi} \int_{-\pi}^{\pi} 2\pi\delta(\omega - \omega_0)e^{j\omega n}\, d\omega. \tag{2.128}$$

Using the definition of the impulse function, it follows that

$$x[n] = e^{j\omega_0 n} \qquad \text{for any } n.$$

For $\omega_0 = 0$ this reduces to the sequence considered in Example 2.18.

Clearly, $x[n]$ in Example 2.19 is not absolutely summable, nor is it square summable, and $|X(e^{j\omega})|$ is not finite for all ω. Thus the mathematical statement

$$\sum_{n=-\infty}^{\infty} e^{j\omega_0 n} e^{-j\omega n} = \sum_{r=-\infty}^{\infty} 2\pi\delta(\omega - \omega_0 + 2\pi r) \tag{2.129}$$

must be interpreted in a special way. Such an interpretation is provided by the theory of generalized functions (Lighthill, 1958). Using that theory, the concept of a Fourier transform representation can be extended rigorously to the class of sequences that can be expressed as a sum of discrete frequency components such as

$$x[n] = \sum_{k} a_k e^{j\omega_k n}, \qquad -\infty < n < \infty. \tag{2.130}$$

From the result of Example 2.19, it follows that

$$X(e^{j\omega}) = \sum_{r=-\infty}^{\infty} \sum_{k} 2\pi a_k \delta(\omega - \omega_k + 2\pi r) \tag{2.131}$$

is a consistent Fourier transform representation of $x[n]$ in Eq. (2.130).

Another sequence that is neither absolutely nor square summable is the unit step sequence $u[n]$. Although it is not completely straightforward to show, this sequence can be represented by the following Fourier transform:

$$U(e^{j\omega}) = \frac{1}{1 - e^{-j\omega}} + \sum_{r=-\infty}^{\infty} \pi\delta(\omega + 2\pi r). \tag{2.132}$$

2.8 SYMMETRY PROPERTIES OF THE FOURIER TRANSFORM

In using Fourier transforms, it is useful to have a detailed knowledge of the way that properties of the sequence manifest themselves in the Fourier transform and vice versa. In this section and Section 2.9 we discuss and summarize a number of such properties.

Symmetry properties of the Fourier transform are often very useful. The following discussion presents these properties, and the proofs are considered in Problems 2.29 and 2.30. Before presenting the properties, we begin with some definitions.

A *conjugate-symmetric sequence* $x_e[n]$ is defined as a sequence for which $x_e[n] = x_e^*[-n]$, and a *conjugate-antisymmetric sequence* $x_o[n]$ is defined as a sequence for which $x_o[n] = -x_o^*[-n]$, where * denotes complex conjugation. Any sequence $x[n]$ can be expressed as a sum of a conjugate-symmetric and conjugate-antisymmetric sequence. Specifically,

$$x[n] = x_e[n] + x_o[n], \tag{2.133a}$$

where

$$x_e[n] = \tfrac{1}{2}(x[n] + x^*[-n]) \tag{2.133b}$$

and

$$x_o[n] = \tfrac{1}{2}(x[n] - x^*[-n]). \tag{2.133c}$$

A real sequence that is conjugate-symmetric such that $x_e[n] = x_e[-n]$ is called an *even sequence*, and a real sequence that is conjugate-antisymmetric such that $x_o[n] = -x_o[-n]$ is called an *odd sequence*.

A Fourier transform $X(e^{j\omega})$ can be decomposed into a sum of conjugate-symmetric and conjugate-antisymmetric functions as

$$X(e^{j\omega}) = X_e(e^{j\omega}) + X_o(e^{j\omega}), \tag{2.134a}$$

where

$$X_e(e^{j\omega}) = \tfrac{1}{2}[X(e^{j\omega}) + X^*(e^{-j\omega})] \tag{2.134b}$$

and

$$X_o(e^{j\omega}) = \tfrac{1}{2}[X(e^{j\omega}) - X^*(e^{-j\omega})]. \tag{2.134c}$$

Clearly, $X_e(e^{j\omega})$ is conjugate-symmetric and $X_o(e^{j\omega})$ is conjugate-antisymmetric; i.e.,

$$X_e(e^{j\omega}) = X_e^*(e^{-j\omega}) \tag{2.135a}$$

and

$$X_o(e^{j\omega}) = -X_o^*(e^{-j\omega}). \tag{2.135b}$$

TABLE 2.1   SYMMETRY PROPERTIES OF THE FOURIER TRANSFORM

Sequence $x[n]$	Fourier Transform $X(e^{j\omega})$
1. $x^*[n]$	$X^*(e^{-j\omega})$
2. $x^*[-n]$	$X^*(e^{j\omega})$
3. $\mathcal{R}e\{x[n]\}$	$X_e(e^{j\omega})$ (conjugate-symmetric part of $X(e^{j\omega})$)
4. $j\,\mathcal{I}m\{x[n]\}$	$X_o(e^{j\omega})$ (conjugate-antisymmetric part of $X(e^{j\omega})$)
5. $x_e[n]$ (conjugate-symmetric part of $x[n]$)	$X_R(e^{j\omega})$
6. $x_o[n]$ (conjugate-antisymmetric part of $x[n]$)	$jX_I(e^{j\omega})$

The following properties apply only when $x[n]$ is real.

7. Any real $x[n]$	$X(e^{j\omega}) = X^*(e^{-j\omega})$ (Fourier transform is conjugate-symmetric)
8. Any real $x[n]$	$X_R(e^{j\omega}) = X_R(e^{-j\omega})$ (real part is even)
9. Any real $x[n]$	$X_I(e^{j\omega}) = -X_I(e^{-j\omega})$ (imaginary part is odd)
10. Any real $x[n]$	$|X(e^{j\omega})| = |X(e^{-j\omega})|$ (magnitude is even)
11. Any real $x[n]$	$\measuredangle X(e^{j\omega}) = -\measuredangle X(e^{-j\omega})$ (phase is odd)
12. $x_e[n]$ (even part of $x[n]$)	$X_R(e^{j\omega})$
13. $x_0[n]$ (odd part of $x[n]$)	$jX_I(e^{j\omega})$

As with sequences, if a real function of a continuous variable is conjugate-symmetric, it is referred to as an *even function*, and a real conjugate-antisymmetric function of a continuous variable is referred to as an *odd function*.

The symmetry properties of the Fourier transform are summarized in Table 2.1. The first six properties apply for a general complex sequence $x[n]$ with Fourier transform $X(e^{j\omega})$. Properties 1 and 2 are considered in Problem 2.29. Property 3 follows from properties 1 and 2 together with the fact that the Fourier transform of the sum of two sequences is the sum of their Fourier transforms. Specifically, the Fourier transform of $\mathcal{R}e\{x[n]\} = \frac{1}{2}(x[n] + x^*[n])$ is the conjugate-symmetric part of $X(e^{j\omega})$, or $X_e(e^{j\omega})$. Similarly, $j\mathcal{I}m\{x[n]\} = \frac{1}{2}(x[n] - x^*[n])$ or, equivalently, $j\mathcal{I}m\{x[n]\}$ has a Fourier transform that is the conjugate-antisymmetric component $X_o(e^{j\omega})$ corresponding to property 4. By considering the Fourier transform of $x_e[n]$ and $x_o[n]$, the conjugate-symmetric and conjugate-antisymmetric components of $x[n]$, properties 5 and 6 follow.

If $x[n]$ is a real sequence, these symmetry properties become particularly straightforward and useful. Specifically, for a real sequence the Fourier transform is conjugate-symmetric; i.e., $X(e^{j\omega}) = X^*(e^{-j\omega})$ (property 7). Expressing $X(e^{j\omega})$ in terms of its real and imaginary parts as

$$X(e^{j\omega}) = X_R(e^{j\omega}) + jX_I(e^{j\omega}), \qquad (2.136)$$

properties 8 and 9 then follow, specifically,

$$X_R(e^{j\omega}) = X_R(e^{-j\omega}) \tag{2.137a}$$

and

$$X_I(e^{j\omega}) = -X_I(e^{-j\omega}). \tag{2.137b}$$

In other words, the real part of the Fourier transform is an even function, and the imaginary part is an odd function if the sequence is real. In a similar manner, by expressing $X(e^{j\omega})$ in polar form,

$$X(e^{j\omega}) = |X(e^{j\omega})|e^{j \angle X(e^{j\omega})}, \tag{2.138}$$

it follows that for a real sequence $x[n]$, the magnitude of the Fourier transform $|X(e^{j\omega})|$ is an even function of ω and the phase, $\angle X(e^{j\omega})$, can be chosen to be an odd function of ω (properties 10 and 11). Also for a real sequence, the even part of $x[n]$ transforms to $X_R(e^{j\omega})$ and the odd part of $x[n]$ transforms to $jX_I(e^{j\omega})$ (properties 12 and 13).

Example 2.20

To illustrate a few of these properties, let us return to the sequence of Example 2.16, where we showed that the Fourier transform of the real sequence $x[n] = a^n u[n]$ is

$$X(e^{j\omega}) = \frac{1}{1 - ae^{-j\omega}} \qquad \text{if } |a| < 1. \tag{2.139}$$

Then

$$X^*(e^{-j\omega}) = \frac{1}{1 - ae^{-j\omega}} = X(e^{j\omega}), \qquad \text{(property 7)}$$

$$X_R(e^{j\omega}) = \frac{1 - a\cos\omega}{1 + a^2 - 2a\cos\omega} = X_R(e^{-j\omega}), \qquad \text{(property 8)}$$

$$X_I(e^{j\omega}) = \frac{a\sin\omega}{1 + a^2 - 2a\cos\omega} = -X_I(e^{-j\omega}), \qquad \text{(property 9)}$$

$$|X(e^{j\omega})| = \frac{1}{(1 + a^2 - 2a\cos\omega)^{1/2}} = |X(e^{-j\omega})|, \qquad \text{(property 10)}$$

$$\angle X(e^{j\omega}) = \tan^{-1}\left(\frac{a\sin\omega}{1 - a\cos\omega}\right) = -\angle X(e^{-j\omega}). \qquad \text{(property 11)}$$

These functions are plotted in Fig. 2.21 for $a > 0$, specifically $a = 0.9$ (solid curve) and $a = 0.5$ (dashed curve). In Problem 2.27 we consider the corresponding plots for $a < 0$.

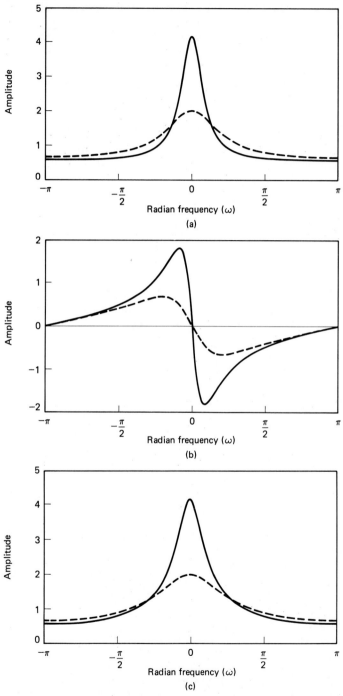

Figure 2.21   Frequency response for a system with impulse response $h[n] = a^n u[n]$. (a) Real part. (b) Imaginary part. (c) Magnitude. $a > 0$; $a = 0.9$ (solid curve) and $a = 0.5$ (dashed curve).

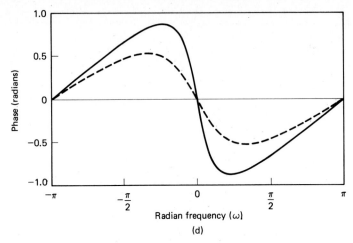

Figure 22.1 (*continued*) (d) Phase.

2.9 FOURIER TRANSFORM THEOREMS

In addition to the symmetry properties, a variety of theorems (presented in Sections 2.9.1–2.9.7) relate operations on the sequence to operations on the Fourier transform. We will see that these theorems are quite similar in most cases to corresponding theorems for continuous-time signals and their Fourier transforms. To facilitate the statement of these theorems, we introduce the following operator notation:

$$X(e^{j\omega}) = \mathscr{F}\{x[n]\},$$

$$x[n] = \mathscr{F}^{-1}\{X(e^{j\omega})\},$$

$$x[n] \overset{\mathscr{F}}{\longleftrightarrow} X(e^{j\omega}).$$

That is, $\mathscr{F}$ denotes the operation of "taking the Fourier transform of $x[n]$," and $\mathscr{F}^{-1}$ is the inverse of that operation. Most of the theorems will be stated without proof. The proofs, which are left as exercises (Problem 2.31), generally involve only simple manipulations of variables of summation or integration. The theorems in this section are summarized in Table 2.2.

2.9.1 Linearity of the Fourier Transform

If

$$x_1[n] \overset{\mathscr{F}}{\longleftrightarrow} X_1(e^{j\omega})$$

and

$$x_2[n] \overset{\mathscr{F}}{\longleftrightarrow} X_2(e^{j\omega}),$$

TABLE 2.2 FOURIER TRANSFORM THEOREMS

Sequence $x[n]$ $y[n]$	Fourier Transform $X(e^{j\omega})$ $Y(e^{j\omega})$
1. $ax[n] + by[n]$	$aX(e^{j\omega}) + bY(e^{j\omega})$
2. $x[n - n_d]$ (n_d an integer)	$e^{-j\omega n_d} X(e^{j\omega})$
3. $e^{j\omega_0 n} x[n]$	$X(e^{j(\omega - \omega_0)})$
4. $x[-n]$	$X(e^{-j\omega})$ if $x[n]$ real $X^*(e^{j\omega})$
5. $nx[n]$	$j\dfrac{dX(e^{j\omega})}{d\omega}$
6. $x[n] * y[n]$	$X(e^{j\omega})Y(e^{j\omega})$
7. $x[n]y[n]$	$\dfrac{1}{2\pi}\displaystyle\int_{-\pi}^{\pi} X(e^{j\theta})Y(e^{j(\omega - \theta)})d\theta$

Parseval's Theorem

$$8. \quad \sum_{n=-\infty}^{\infty} |x[n]|^2 = \frac{1}{2\pi}\int_{-\pi}^{\pi} |X(e^{j\omega})|^2 \, d\omega$$

$$9. \quad \sum_{n=-\infty}^{\infty} x[n]y^*[n] = \frac{1}{2\pi}\int_{-\pi}^{\pi} X(e^{j\omega})Y^*(e^{j\omega}) \, d\omega$$

then

$$ax_1[n] + bx_2[n] \overset{\mathscr{F}}{\longleftrightarrow} aX_1(e^{j\omega}) + bX_2(e^{j\omega}). \tag{2.140}$$

2.9.2 Time Shifting and Frequency Shifting

If

$$x[n] \overset{\mathscr{F}}{\longleftrightarrow} X(e^{j\omega}),$$

then for the time-shifted sequence,

$$x[n - n_d] \overset{\mathscr{F}}{\longleftrightarrow} e^{-j\omega n_d} X(e^{j\omega}) \tag{2.141}$$

and for the frequency-shifted Fourier transform,

$$e^{j\omega_0 n} x[n] \overset{\mathscr{F}}{\longleftrightarrow} X(e^{j(\omega - \omega_0)}). \tag{2.142}$$

2.9.3. Time Reversal

If

$$x[n] \overset{\mathscr{F}}{\longleftrightarrow} X(e^{j\omega}),$$

then if the sequence is time reversed,

$$x[-n] \xleftrightarrow{\mathcal{F}} X(e^{-j\omega}). \tag{2.143}$$

If $x[n]$ is real, this theorem reduces to

$$x[-n] \xleftrightarrow{\mathcal{F}} X^*(e^{j\omega}). \tag{2.144}$$

2.9.4 Differentiation in Frequency

If

$$x[n] \xleftrightarrow{\mathcal{F}} X(e^{j\omega}),$$

then

$$nx[n] \xleftrightarrow{\mathcal{F}} j\frac{dX(e^{j\omega})}{d\omega}. \tag{2.145}$$

2.9.5 Parseval's Theorem

If

$$x[n] \xleftrightarrow{\mathcal{F}} X(e^{j\omega}),$$

then

$$E = \sum_{n=-\infty}^{\infty} |x[n]|^2 = \frac{1}{2\pi}\int_{-\pi}^{\pi} |X(e^{j\omega})|^2 \, d\omega. \tag{2.146}$$

The function $|X(e^{j\omega})|^2$ is called the *energy density spectrum* since it determines how the energy is distributed in frequency. Necessarily, the energy density spectrum is defined only for finite-energy signals. A more general form of Parseval's theorem is shown in Problem 2.34.

2.9.6 The Convolution Theorem

If

$$x[n] \xleftrightarrow{\mathcal{F}} X(e^{j\omega})$$

and

$$h[n] \xleftrightarrow{\mathcal{F}} H(e^{j\omega})$$

and if

$$y[n] = \sum_{k=-\infty}^{\infty} x[k]h[n-k] = x[n] * h[n], \tag{2.147}$$

then

$$Y(e^{j\omega}) = X(e^{j\omega})H(e^{j\omega}). \tag{2.148}$$

Thus convolution of sequences implies multiplication of the corresponding Fourier transforms. Note that the time shifting property is a special case of the convolution property since

$$\delta[n - n_d] \overset{\mathscr{F}}{\longleftrightarrow} e^{-j\omega n_d} \tag{2.149}$$

and if $h[n] = \delta[n - n_d]$, then $y[n] = x[n] * \delta[n - n_d]$. Therefore,

$$H(e^{j\omega}) = e^{-j\omega n_d} \quad \text{and} \quad Y(e^{j\omega}) = e^{-j\omega n_d}X(e^{j\omega}).$$

A formal derivation of the convolution theorem is easily achieved by applying the definition of the Fourier transform to $y[n]$ as expressed in Eq. (2.147). This theorem can also be interpreted as a direct consequence of the eigenfunction property of complex exponentials for linear time-invariant systems. Recall that $H(e^{j\omega})$ is the frequency response of the linear time-invariant system whose impulse response is $h[n]$. Also recall that if

$$x[n] = e^{j\omega n},$$

then

$$y[n] = H(e^{j\omega})e^{j\omega n}.$$

That is, complex exponentials are *eigenfunctions* of linear time-invariant systems where $H(e^{j\omega})$, the Fourier transform of $h[n]$, is the eigenvalue. From the definition of integration, the Fourier transform synthesis equation corresponds to the representation of a sequence $x[n]$ as a superposition of complex exponentials of infinitesimal size, i.e.,

$$x[n] = \frac{1}{2\pi} \int_{-\pi}^{\pi} X(e^{j\omega})e^{j\omega n}\, d\omega = \lim_{\Delta\omega \to 0} \frac{1}{2\pi} \sum_k X(e^{jk\Delta\omega})e^{jk\Delta\omega n}\, \Delta\omega.$$

By the eigenfunction property of linear systems and by the principle of superposition, the corresponding output will be

$$y[n] = \lim_{\Delta\omega \to 0} \frac{1}{2\pi} \sum_k H(e^{jk\Delta\omega})X(e^{jk\Delta\omega})e^{jk\Delta\omega n}\, \Delta\omega = \frac{1}{2\pi} \int_{-\pi}^{\pi} H(e^{j\omega})X(e^{j\omega})e^{j\omega n}\, d\omega.$$

Thus we conclude that

$$Y(e^{j\omega}) = H(e^{j\omega})X(e^{j\omega}),$$

as in Eq. (2.148).

2.9.7 The Modulation or Windowing Theorem

If

$$x[n] \overset{\mathscr{F}}{\longleftrightarrow} X(e^{j\omega})$$

and

$$w[n] \overset{\mathscr{F}}{\longleftrightarrow} W(e^{j\omega})$$

and if

$$y[n] = x[n]w[n], \tag{2.150}$$

then

$$Y(e^{j\omega}) = \frac{1}{2\pi} \int_{-\pi}^{\pi} X(e^{j\theta})W(e^{j(\omega-\theta)}) \, d\theta. \tag{2.151}$$

Equation (2.151) is a periodic convolution, i.e., it is a convolution of two periodic functions with the limits of integration extending over only one period. The duality inherent in most Fourier transform theorems is evident when we compare the convolution and modulation theorems. However, in contrast to the continuous-time case, where this duality is complete, in the discrete-time case fundamental differences arise because the Fourier transform is a sum while the inverse transform is an integral with a periodic integrand. While for continuous time we can state that convolution in the time domain is represented by multiplication in the frequency domain and vice versa, in discrete-time this statement must be modified somewhat. Specifically, discrete-time convolution of sequences (the convolution sum) is equivalent to multiplication of corresponding periodic Fourier transforms, and multiplication of sequences is equivalent to *periodic* convolution of corresponding Fourier transforms.

The theorems of this section and a number of fundamental Fourier transform pairs are summarized in Tables 2.2 and 2.3, respectively. One of the ways that knowledge of Fourier transform theorems and properties is useful is in determining Fourier transforms or inverse transforms. Often by using the theorems and known transform pairs it is possible to represent a sequence in terms of operations on other sequences for which the transform is known, thereby simplifying an otherwise difficult or tedious problem. Examples 2.21–2.24 illustrate this.

Example 2.21

Let $x[n] = a^n u[n-5]$. This sequence can be expressed as

$$x[n] = a^5 a^{n-5} u[n-5]. \tag{2.152}$$

Therefore, using Theorems 1 and 2 of Table 2.2 and the transform pair 4 of Table 2.3, we obtain

$$X(e^{j\omega}) = \frac{a^5 e^{-j\omega 5}}{1 - ae^{-j\omega}}. \tag{2.153}$$

Example 2.22

Suppose that

$$X(e^{j\omega}) = \frac{1}{(1 - ae^{-j\omega})(1 - be^{-j\omega})}. \tag{2.154}$$

Direct substitution of $X(e^{j\omega})$ into Eq. (2.112) leads to an integral that is difficult to evaluate by ordinary real integration techniques. However, using the technique of partial

TABLE 2.3 FOURIER TRANSFORM PAIRS

Sequence	Fourier Transform
1. $\delta[n]$	1
2. $\delta[n - n_0]$	$e^{-j\omega n_0}$
3. $1 \quad (-\infty < n < \infty)$	$\displaystyle\sum_{k=-\infty}^{\infty} 2\pi\delta(\omega + 2\pi k)$
4. $a^n u[n] \quad (\lvert a \rvert < 1)$	$\dfrac{1}{1 - ae^{-j\omega}}$
5. $u[n]$	$\dfrac{1}{1 - e^{-j\omega}} + \displaystyle\sum_{k=-\infty}^{\infty} \pi\delta(\omega + 2\pi k)$
6. $(n + 1)a^n u[n] \quad (\lvert a \rvert < 1)$	$\dfrac{1}{(1 - ae^{-j\omega})^2}$
7. $\dfrac{r^n \sin \omega_p(n + 1)}{\sin \omega_p} u[n] \quad (\lvert r \rvert < 1)$	$\dfrac{1}{1 - 2r \cos \omega_p e^{-j\omega} + r^2 e^{-j2\omega}}$
8. $\dfrac{\sin \omega_c n}{\pi n}$	$X(e^{j\omega}) = \begin{cases} 1, & \lvert \omega \rvert < \omega_c, \\ 0, & \omega_c < \lvert \omega \rvert \le \pi \end{cases}$
9. $x[n] = \begin{cases} 1, & 0 \le n \le M \\ 0, & \text{otherwise} \end{cases}$	$\dfrac{\sin[\omega(M + 1)/2]}{\sin(\omega/2)} e^{-j\omega M/2}$
10. $e^{j\omega_0 n}$	$\displaystyle\sum_{k=-\infty}^{\infty} 2\pi\delta(\omega - \omega_0 + 2\pi k)$
11. $\cos(\omega_0 n + \phi)$	$\pi \displaystyle\sum_{k=-\infty}^{\infty} [e^{j\phi}\delta(\omega - \omega_0 + 2\pi k) + e^{-j\phi}\delta(\omega + \omega_0 + 2\pi k)]$

fraction expansion, which we discuss in detail in Chapter 4, $X(e^{j\omega})$ can be expanded into the form

$$X(e^{j\omega}) = \frac{a/(a - b)}{1 - ae^{-j\omega}} - \frac{b/(a - b)}{1 - be^{-j\omega}}. \qquad (2.155)$$

Using Theorem 1 of Table 2.2 and the transform pair 4 of Table 2.3, it follows that

$$x[n] = \left(\frac{a}{a - b}\right)a^n u[n] - \left(\frac{b}{a - b}\right)b^n u[n]. \qquad (2.156)$$

Example 2.23

Consider the Fourier transform

$$X(e^{j\omega}) = \begin{cases} e^{-j\omega n_d}, & \omega_c < \lvert \omega \rvert < \pi, \\ 0, & \lvert \omega \rvert < \omega_c, \end{cases} \qquad (2.157)$$

with period 2π understood. This Fourier transform can be expressed as

$$X(e^{j\omega}) = e^{-j\omega n_d}(1 - R(e^{j\omega})),$$

where $R(e^{j\omega})$ is periodic with period 2π and

$$R(e^{j\omega}) = \begin{cases} 1, & |\omega| < \omega_c, \\ 0, & \omega_c < |\omega| < \pi. \end{cases}$$

Using the result of Example 2.17 to obtain the inverse transform of $R(e^{j\omega})$, together with properties 1 and 2 of Table 2.2, we have for $x[n]$

$$x[n] = \delta[n - n_d] - r[n - n_d]$$

$$= \delta[n - n_d] - \frac{\sin \omega_c(n - n_d)}{\pi(n - n_d)}.$$

Example 2.24

Let us determine the impulse response for a stable linear time-invariant system for which the input $x[n]$ and output $y[n]$ satisfy the linear constant-coefficient difference equation

$$y[n] - \tfrac{1}{2}y[n - 1] = x[n] - \tfrac{1}{4}x[n - 1]. \tag{2.158}$$

In Chapter 5 we will see that the z-transform is more useful than the Fourier transform for dealing with difference equations. However, this example offers a hint of the utility of transform methods for analysis of linear systems. To find the impulse response, we set $x[n] = \delta[n]$; with $h[n]$ denoting the impulse response, Eq. (2.158) becomes

$$h[n] - \tfrac{1}{2}h[n - 1] = \delta[n] - \tfrac{1}{4}\delta[n - 1]. \tag{2.159}$$

Applying the Fourier transform to both sides of Eq. (2.159) and using properties 1 and 2 of Table 2.2, we obtain

$$H(e^{j\omega}) - \tfrac{1}{2}e^{-j\omega}H(e^{j\omega}) = 1 - \tfrac{1}{4}e^{-j\omega} \tag{2.160}$$

or

$$H(e^{j\omega}) = \frac{1 - \tfrac{1}{4}e^{-j\omega}}{1 - \tfrac{1}{2}e^{-j\omega}}. \tag{2.161}$$

To obtain $h[n]$, we want to determine the inverse Fourier transform of $H(e^{j\omega})$. Toward this end, we rewrite Eq. (2.161) as

$$H(e^{j\omega}) = \frac{1}{1 - \tfrac{1}{2}e^{-j\omega}} - \frac{\tfrac{1}{4}e^{-j\omega}}{1 - \tfrac{1}{2}e^{-j\omega}}. \tag{2.162}$$

From transform 4 of Table 2.3,

$$(\tfrac{1}{2})^n u[n] \overset{\mathcal{F}}{\longleftrightarrow} \frac{1}{1 - \tfrac{1}{2}e^{-j\omega}}.$$

Combining this transform with property 3 of Table 2.2,

$$-(\tfrac{1}{4})(\tfrac{1}{2})^{n-1} u[n - 1] \overset{\mathcal{F}}{\longleftrightarrow} -\frac{\tfrac{1}{4}e^{-j\omega}}{1 - \tfrac{1}{2}e^{-j\omega}}. \tag{2.163}$$

Based on property 1 of Table 2.2, then,

$$h[n] = (\tfrac{1}{2})^n u[n] - (\tfrac{1}{4})(\tfrac{1}{2})^{n-1} u[n - 1]. \tag{2.164}$$

2.10 DISCRETE-TIME RANDOM SIGNALS   (skip)

The preceding sections have focused on mathematical representations of discrete-time signals and systems and the insights that derive from such mathematical representations. We have seen that discrete-time signals and systems have both a time-domain and a frequency-domain representation, each with an important place in the theory and design of discrete-time signal processing systems. Until now we have assumed that the signals are deterministic, i.e., that each value of a sequence is uniquely determined by a mathematical expression, a table of data, or a rule of some type.

In many situations, the processes that generate signals are so complex as to make precise description of a signal extremely difficult or undesirable, if not impossible. In such cases, modeling the signal as a stochastic process is analytically useful. As an example we will see in Chapter 6 that many of the effects encountered in implementing digital signal processing algorithms with finite register length can be represented by additive noise, i.e., a stochastic sequence. Many mechanical systems generate acoustic or vibratory signals that can be processed to diagnose potential failure; again, signals of this type are often best modeled in terms of stochastic signals. Speech signals to be processed for automatic recognition or bandwidth compression and music to be processed for quality enhancement are two more of many examples.

A stochastic signal is considered to be a member of an ensemble of discrete-time signals that is characterized by a set of probability density functions. In other words, for a specific signal at a particular time, the amplitude of the signal sample at that time is assumed to have been determined by an underlying scheme of probabilities.

The key to the mathematical representation of such signals lies in their description in terms of averages. While stochastic signals are not absolutely summable or square summable and consequently do not directly have Fourier transforms, many (but not all) of the properties of such signals can be summarized in terms of the *autocorrelation* or *autocovariance sequence*, for which the Fourier transform often exists. As we will discuss in this section, the Fourier transform of the autocovariance sequence has a useful interpretation in terms of the frequency distribution of the power in the signal. The use of the autocovariance sequence and its transform has another important advantage: the effect of processing stochastic signals with a discrete-time linear system can be conveniently described in terms of the effect of the system on the autocovariance sequence.

In the following discussion, we assume that the reader is familiar with the basic concepts of stochastic processes such as averages, correlation and covariance functions, and power spectrum. A brief review and summary of notation and concepts is provided in Appendix A. A more detailed presentation of the theory of random signals can be found in a variety of excellent texts, such as Davenport (1970) and Papoulis (1984). Our primary objective in this section is to present a specific set of results pertaining to representation of stochastic signals that will be useful in subsequent chapters. Therefore, we focus on wide-sense stationary random signals and their representation in the context of processing with linear time-invariant systems. Although for simplicity we assume that $x[n]$ and $h[n]$ are real-valued, the results are easily generalized to the complex case.

Consider a stable linear time-invariant system with impulse response $h[n]$. Let $x[n]$ be a real-valued sequence that is a sample sequence of a wide-sense stationary discrete-time random process. Then the output of the linear system is a sample function of an output random process related to the input process by the linear transformation

$$y[n] = \sum_{k=-\infty}^{\infty} h[n-k]x[k] = \sum_{k=-\infty}^{\infty} h[k]x[n-k].$$

As we have shown, since the system is stable, $y[n]$ will be bounded if $x[n]$ is bounded. We will see below that if the input is stationary,† then so is the output. The input signal may be characterized by its mean m_x and its autocorrelation function $\phi_{xx}[m]$, or we may also have additional information about first- or even second-order probability distributions. In characterizing the output random process $y[n]$ we desire similar information. For many applications, it is sufficient to characterize both the input and output in terms of simple averages, such as the mean, variance, and autocorrelation. Therefore, we will derive input-output relationships for these quantities.

The means of the input and output processes are

$$m_{x_n} = \mathscr{E}\{\mathbf{x}_n\}, \qquad m_{y_n} = \mathscr{E}\{\mathbf{y}_n\}, \tag{2.165}$$

where $\mathscr{E}\{\cdot\}$ denotes the expected value. For notational convenience we will be less careful about distinguishing between the random variables $\mathbf{x}_n$ and $\mathbf{y}_n$ and their values $x[n]$ and $y[n]$, so that Eqs. (2.165) will alternatively be written

$$m_x[n] = \mathscr{E}\{x[n]\}, \qquad m_y[n] = \mathscr{E}\{y[n]\}. \tag{2.166}$$

If $x[n]$ is stationary, then $m_x[n]$ is independent of n and will be written as m_x, with similar notation for $m_y[n]$ if $y[n]$ is stationary.

The mean of the output process is

$$m_y[n] = \mathscr{E}\{y[n]\} = \sum_{k=-\infty}^{\infty} h[k]\mathscr{E}\{x[n-k]\},$$

where we have used the fact that the expected value of a sum is the sum of the expected values. Since the input is stationary, $m_x[n-k] = m_x$ and consequently

$$m_y[n] = m_x \sum_{k=-\infty}^{\infty} h[k]. \tag{2.167}$$

From Eq. (2.167) we see that the mean of the output is also constant. An equivalent expression to Eq. (2.167) in terms of the frequency response is

$$m_y = H(e^{j0})m_x. \tag{2.168}$$

---

† In the remainder of the text, we will use the term *stationary* to mean "wide-sense stationary."

Assuming temporarily that the output is nonstationary, the autocorrelation function of the output process for a real input is

$$\phi_{yy}[n, n + m] = \mathscr{E}\{y[n]y[n + m]\}$$

$$= \mathscr{E}\left\{\sum_{k=-\infty}^{\infty} \sum_{r=-\infty}^{\infty} h[k]h[r]x[n - k]x[n + m - r]\right\}$$

$$= \sum_{k=-\infty}^{\infty} h[k] \sum_{r=-\infty}^{\infty} h[r]\mathscr{E}\{x[n - k]x[n + m - r]\}.$$

Since $x[n]$ is assumed to be stationary, $\mathscr{E}\{x[n - k]x[n + m - r]\}$ depends only on the time difference $m + k - r$. Therefore,

$$\phi_{yy}[n, n + m] = \sum_{k=-\infty}^{\infty} h[k] \sum_{r=-\infty}^{\infty} h[r]\phi_{xx}[m + k - r] = \phi_{yy}[m]. \qquad (2.169)$$

That is, the output autocorrelation sequence also depends only on the time difference m. Thus for a linear time-invariant system that is excited by a wide-sense stationary input, the output is also wide-sense stationary.

By making the substitution $\ell = r - k$, Eq. (2.169) can be expressed as

$$\phi_{yy}[m] = \sum_{\ell=-\infty}^{\infty} \phi_{xx}[m - \ell] \sum_{k=-\infty}^{\infty} h[k]h[\ell + k]$$

$$= \sum_{\ell=-\infty}^{\infty} \phi_{xx}[m - \ell]c[\ell], \qquad (2.170)$$

where we have defined

$$c[\ell] = \sum_{k=-\infty}^{\infty} h[k]h[\ell + k]. \qquad (2.171)$$

A sequence of the form of $c[\ell]$ is often called an *aperiodic autocorrelation sequence* or simply the *autocorrelation sequence of* $h[n]$. It should be emphasized that $c[\ell]$ is the autocorrelation of an aperiodic, i.e., finite-energy, sequence and should not be confused with the autocorrelation of an infinite-energy sequence. Indeed, it can be seen that $c[\ell]$ is simply the discrete convolution of $h[n]$ with $h[-n]$. Equation (2.170), then, can be interpreted to mean that the autocorrelation of the output of a linear system is the convolution of the input autocorrelation with the aperiodic autocorrelation of the system impulse response.

Equation (2.170) suggests that Fourier transforms may be useful in characterizing the response of a linear time-invariant system to a stochastic input. Assume for convenience that $m_x = 0$; i.e., the autocorrelation and autocovariance sequences are identical. Then, with $\Phi_{xx}(e^{j\omega})$, $\Phi_{yy}(e^{j\omega})$, and $C(e^{j\omega})$ denoting the Fourier transforms of $\phi_{xx}[m]$, $\phi_{yy}[m]$, and $c[\ell]$, respectively, from Eq. (2.170)

$$\Phi_{yy}(e^{j\omega}) = C(e^{j\omega})\Phi_{xx}(e^{j\omega}). \qquad (2.172)$$

Also, from Eq. (2.171)

$$C(e^{j\omega}) = H(e^{j\omega})H^*(e^{j\omega})$$

$$= |H(e^{j\omega})|^2,$$

so

$$\Phi_{yy}(e^{j\omega}) = |H(e^{j\omega})|^2\Phi_{xx}(e^{j\omega}). \tag{2.173}$$

Equation (2.173) provides the motivation for the term *power density spectrum*. Specifically,

$$\mathscr{E}\{y^2[n]\} = \phi_{yy}[0] = \frac{1}{2\pi}\int_{-\pi}^{\pi}\Phi_{yy}(e^{j\omega})\,d\omega \tag{2.174}$$

$$= \text{total average power in output.}$$

Substituting Eq. (2.173) into Eq. (2.174), we have

$$\mathscr{E}\{y^2[n]\} = \phi_{yy}[0] = \frac{1}{2\pi}\int_{-\pi}^{\pi}|H(e^{j\omega})|^2\Phi_{xx}(e^{j\omega})\,d\omega. \tag{2.175}$$

Suppose that $H(e^{j\omega})$ is an ideal bandpass filter, as shown in Fig. 2.18(c). We recall that $\phi_{xx}[m]$ is an even sequence, so

$$\Phi_{xx}(e^{j\omega}) = \Phi_{xx}(e^{-j\omega}).$$

Likewise, $|H(e^{j\omega})|^2$ is an even function of ω. Therefore, we can write

$$\phi_{yy}[0] = \text{average power in output}$$

$$= \frac{1}{\pi}\int_{\omega_a}^{\omega_b}\Phi_{xx}(e^{j\omega})\,d\omega + \frac{1}{\pi}\int_{-\omega_b}^{-\omega_a}\Phi_{xx}(e^{j\omega})\,d\omega. \tag{2.176}$$

Thus the area under $\Phi_{xx}(e^{j\omega})$ for $\omega_a \leq |\omega| \leq \omega_b$ can be taken to represent the mean-square value of the input in that frequency band. We observe that the output power must remain nonnegative, so

$$\lim_{(\omega_b - \omega_a) \to 0} \phi_{yy}[0] \geq 0.$$

This result, together with Eq. (2.176), implies that

$$\Phi_{xx}(\omega) \geq 0. \tag{2.177}$$

Thus we note that the power density function of a real signal is real, even, and nonnegative.

Another important result concerns the cross-correlation between the input and output of a linear time-invariant system:

$$\phi_{xy}[m] = \mathscr{E}\{x[n]y[n + m]\}$$

$$= \mathscr{E}\left\{x[n]\sum_{k=-\infty}^{\infty}h[k]x[n + m - k]\right\}$$

$$= \sum_{k=-\infty}^{\infty}h[k]\phi_{xx}[m - k]. \tag{2.178}$$

In this case we note that the cross-correlation between input and output is the convolution of the impulse response with the input autocorrelation sequence.

The Fourier transform of Eq. (2.178) is

$$\Phi_{xy}(e^{j\omega}) = H(e^{j\omega})\Phi_{xx}(e^{j\omega}). \tag{2.179}$$

This result has a useful application when the input is white noise; i.e., $\phi_{xx}[m] = \sigma_x^2 \delta[m]$. Substituting into Eq. (2.176), we note that

$$\phi_{xy}[m] = \sigma_x^2 h[m], \tag{2.180}$$

i.e., for a zero-mean white-noise input, the cross-correlation between input and output of a linear system is proportional to the impulse response of the system. Similarly, the power spectrum of a white-noise input is

$$\Phi_{xx}(e^{j\omega}) = \sigma_x^2, \qquad -\pi \le \omega \le \pi. \tag{2.181}$$

Thus from Eq. (2.179)

$$\Phi_{xy}(e^{j\omega}) = \sigma_x^2 H(e^{j\omega}), \tag{2.182}$$

i.e., the cross power spectrum is in this case proportional to the frequency response of the system. Equations (2.180) and (2.182) may serve as the basis for estimating the impulse response or frequency response of a linear time-invariant system if it is possible to observe the output of the system in response to a white-noise input.

2.11 SUMMARY

In this chapter we have considered a number of basic definitions relating to discrete-time signals and systems. We considered the definition of a set of basic sequences, the definition and representation of linear time-invariant systems in terms of the convolution sum, and some implications of stability and causality. The class of systems for which the input and output satisfy a linear constant-coefficient difference equation with initial rest conditions was shown to be an important subclass of linear time-invariant systems. The recursive solution of such difference equations was discussed and the classes of FIR and IIR systems defined.

An important means for the analysis and representation of linear time-invariant systems lies in their frequency-domain representation. The response of a system to a complex exponential input was considered, leading to the definition of the frequency response. The relation between impulse response and frequency response was then interpreted as a Fourier transform pair.

We called attention to many properties of Fourier transform representations and discussed a variety of useful Fourier transform pairs. Tables 2.1 and 2.2 summarize the properties and theorems and Table 2.3 contains some useful Fourier transform pairs.

The chapter concludes with an introduction to discrete-time random signals. These basic ideas and results will be developed further and used in later chapters.

Although the material in this chapter was presented without direct reference to continuous-time signals, an important class of discrete-time signal processing problems arises from sampling such signals. Consequently, in Chapter 3 we consider the relationship between continuous-time signals and sequences obtained by periodic sampling.

PROBLEMS

2.1. Consider an arbitrary linear system with input $x[n]$ and output $y[n]$. Show that if $x[n] = 0$ for all n, then $y[n]$ must also be zero for all n.

2.2. Using the definition of linearity (Eqs. 2.25), show that the ideal delay system (Example 2.1) and the moving average system (Example 2.2) are both linear systems.

2.3. For each of the following systems, determine whether the system is (1) stable, (2) causal, (3) linear, (4) time-invariant, and (5) memoryless.

    (a) $T(x[n]) = g[n]x[n]$

    (b) $T(x[n]) = \sum_{k=n_0}^{n} x[k]$

    (c) $T(x[n]) = \sum_{k=n-n_0}^{n+n_0} x[k]$

    (d) $T(x[n]) = x[x - n_0]$

    (e) $T(x[n]) = e^{x[n]}$

    (f) $T(x[n]) = ax[n] + b$

    (g) $T(x[n]) = x[-n]$

    (h) $T(x[n]) = x[n] + 3u[n + 1]$

2.4. The system T in Fig. P2.4 is known to be *time-invariant*. When the inputs to the system are $x_1[n]$, $x_2[n]$, and $x_3[n]$, the responses of the system are $y_1[n]$, $y_2[n]$, and $y_3[n]$ as shown.

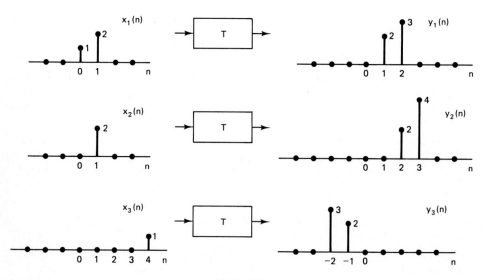

Figure P2.4

(a) Determine whether the system T could be linear.

(b) If the input $x[n]$ to the system T is $\delta[n]$, what is the system response $y[n]$?

(c) Determine all possible inputs $x[n]$ for which the response of the system T can be determined from the given information alone.

2.5. The system L in Fig. P2.5 is known to be *linear*. Shown are three output signals, $y_1[n]$, $y_2[n]$, and $y_3[n]$, in response to the input signals $x_1[n]$, $x_2[n]$, and $x_3[n]$, respectively.

(a) Determine whether the system L could be time-invariant.

(b) If the input $x[n]$ to the system L is $\delta[n]$, what is the system response $y[n]$?

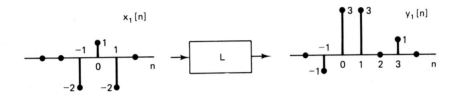

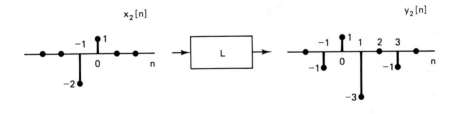

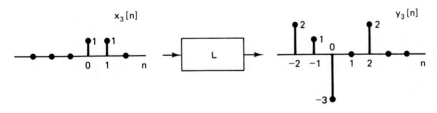

Figure P2.5

2.6. For each of the pairs of sequences in Fig. P2.6, use discrete convolution to find the response to the input $x[n]$ of the linear time-invariant system with impulse response $h[n]$.

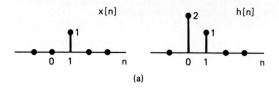

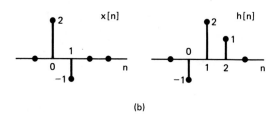

(a)

(b)

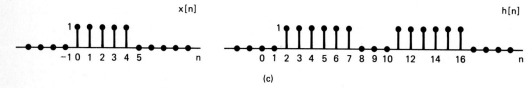

(c)

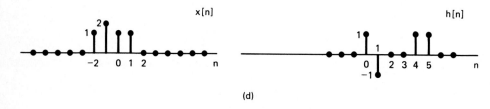

(d)

Figure P2.6

2.7. The impulse response of a linear time-invariant system is shown in Fig. P2.7. Determine and carefully sketch the response of this system to the input $x[n] = u[n - 4]$.

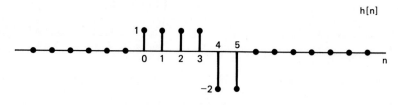

Figure P2.7

2.8. Consider a discrete-time linear time-invariant system with impulse response $h[n]$. If the input $x[n]$ is a periodic sequence with period N, i.e., $x[n] = x[n + N]$, show that the output $y[n]$ is also a periodic sequence with period N.

2.9. A linear time-invariant system has impulse response

$$h[n] = \begin{cases} 1, & n \geq 0, \\ 0, & n < 0, \end{cases}$$

$$= u[n].$$

Determine the response of this system to the input $x[n]$ shown in Fig. P2.9 and described as follows:

$$x[n] = \begin{cases} 0, & n < 0, \\ a^n, & 0 \leq n \leq N_1, \\ 0, & N_1 < n < N_2, \\ a^{n - N_2}, & N_2 \leq n \leq N_2 + N_1, \\ 0, & N_2 + N_1 < n. \end{cases} \qquad 0 < a < 1$$

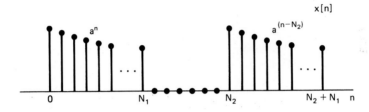

Figure P2.9

2.10. **(a)** The impulse response of a linear time-invariant system is known to be zero except in the interval $N_0 \leq n \leq N_1$. The input $x[n]$ is known to be zero except in the interval $N_2 \leq n \leq N_3$. As a result, the output is constrained to be zero except in some interval $N_4 \leq n \leq N_5$. Determine N_4 and N_5 in terms of N_0, N_1, N_2, and N_3.

(b) If $x[n]$ is zero except for N consecutive points and $h[n]$ is zero except for M consecutive points, what is the maximum number of consecutive points for which $y[n]$ can be nonzero?

2.11. By direct evaluation of the convolution sum, determine the step response of a linear time-invariant system whose impulse response $h[n]$ is given by

$$h[n] = a^{-n}u[-n], \qquad 0 < a < 1.$$

2.12. Causality of a system was defined in Section 2.2.4. From this definition show that for a linear time-invariant system, causality implies that the impulse response $h[n]$ is zero for $n < 0$. One approach is to show that if $h[n]$ is *not* zero for $n < 0$, then the system *cannot* be causal. Show also that if the impulse response is zero for $n < 0$, then the system will necessarily be causal.

2.13. In Section 2.5, we stated that the solution to the homogeneous difference equation

$$\sum_{k=0}^{N} a_k y_h[n - k] = 0 \tag{P2.13-1}$$

is of the form

$$y_h[n] = \sum_{m=1}^{N} A_m z_m^n, \tag{P2.13-2}$$

with the A_m's arbitrary and the z_m's the N roots of the polynomial

$$\sum_{k=0}^{N} a_k z^{-k} = 0, \tag{P2.13-3}$$

i.e.,

$$\sum_{k=0}^{N} a_k z^{-k} = \prod_{m=1}^{N} (1 - z_m z^{-1}). \tag{P2.13-4}$$

(a) Determine the general form of the homogeneous solution to the difference equation

$$y[n] - \tfrac{3}{4}y[n-1] + \tfrac{1}{8}y[n-2] = 2x[n-1] \tag{P2.13-5}$$

(b) Determine the coefficients A_m in the homogeneous solution if $y[-1] = 1$ and $y[0] = 0$.

(c) Now consider the difference equation

$$y[n] - y[n-1] + \tfrac{1}{4}y[n-2] = 2x[n-1]. \tag{P2.13-6}$$

If the homogeneous solution contains only terms of the form of Eq. (P2.13-2), show that the initial conditions $y[-1] = 1$ and $y[0] = 0$ cannot be satisfied.

(d) If Eq. (P2.13-3) has two roots that are identical, then in place of Eq. (P2.13-2), $y_h[n]$ will take the form

$$y_h[n] = \sum_{m=1}^{N-1} A_m z_m^n + nB_1 z_1^n, \tag{P2.13-7}$$

where we have assumed that the double root is z_1. Using Eq. (P2.13-7), determine the general form of $y_h[n]$ for Eq. (P2.13-6). Explicitly verify that your answer satisfies Eq. (P2.13-6) with $x[n] = 0$.

(e) Determine the coefficients A_1 and B_1 in the homogeneous solution obtained in part (d) if $y[-1] = 1$ and $y[0] = 0$.

2.14. Consider the following linear constant-coefficient difference equation:

$$y[n] - \tfrac{3}{4}y[n-1] + \tfrac{1}{8}y[n-2] = 2x[n-1].$$

Determine $y[n]$ when $x[n] = \delta[n]$ and $y[n] = 0$, $n < 0$.

2.15. Consider a system with input $x[n]$ and output $y[n]$ that satisfy the difference equation

$$y[n] = ny[n-1] + x[n].$$

The system is causal and satisfies initial-rest conditions; i.e., if $x[n] = 0$ for $n < n_0$, then $y[n] = 0$ for $n < n_0$.

(a) If $x[n] = \delta[n]$, determine $y[n]$ for all n.

(b) Is the system linear? Justify your answer.

(c) Is the system time-invariant? Justify your answer.

2.16. Consider a system with input $x[n]$ and output $y[n]$. The input/output relation for the system is defined by the following two properties:

1. $y[n] - ay[n-1] = x[n]$

2. $y[0] = 1$

(a) Determine whether the system is time-invariant.

(b) Determine whether the system is linear.

(c) Assume that the difference equation (property 1) remains the same but the value $y[0]$ is specified to be zero. Does this change your answer to either part (a) or part (b)?

2.17. A causal linear time-invariant system is described by the following difference equation:

$$y[n] - 5y[n-1] + 6y[n-2] = 2x[n-1].$$

(a) Determine the homogeneous response of the system, i.e., the possible outputs if $x[n] = 0$ for all n.

(b) Determine the impulse response of the system.

(c) Determine the step response of the system.

2.18. (a) Find the frequency response $H(e^{j\omega})$ of the linear time-invariant system whose input and output satisfy the difference equation

$$y[n] - \tfrac{1}{2}y[n-1] = x[n] + 2x[n-1] + x[n-2].$$

(b) Write a difference equation that characterizes a system whose frequency response is

$$H(e^{j\omega}) = \frac{1 - \tfrac{1}{2}e^{-j\omega} + e^{-j3\omega}}{1 + \tfrac{1}{2}e^{-j\omega} + \tfrac{3}{4}e^{-j2\omega}}.$$

2.19. Which of the following discrete-time signals could be eigenfunctions of an LTI system?

(a) $5^n u[n]$

(b) $e^{j2\omega n}$

(c) $e^{j\omega n} + e^{j2\omega n}$

(d) 5^n

(e) $5^n \cdot e^{j2\omega n}$

2.20. Consider the linear time-invariant system with impulse response

$$h[n] = \left(\frac{j}{2}\right)^n u[n], \qquad \text{where } j = \sqrt{-1}$$

Determine the steady-state response, i.e., the response for large n, to the excitation

$$x[n] = \cos(\pi n)u[n].$$

2.21. Three systems A, B, and C have input and output indicated in Fig. P2.21. Determine if each system could be LTI. If your answer is yes, specify if there could be more than one LTI system with the given input/output pair. Clearly explain your answer.

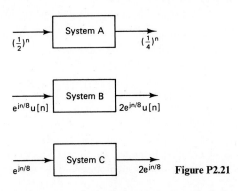

Figure P2.21

2.22. In the nonlinear system in Fig. P2.22, the output M is

$$M = \max_{n} \left| \sum_{k=0}^{10} \left(\frac{1}{2}\right)^{k} x^2[n-k] \right|,$$

where $|\cdot|$ denotes magnitude. Since M is equal to the maximum value over all time (i.e., all n), it is a constant.

Suppose that $x[n]$ is a complex exponential of the form $x[n] = e^{j\omega_0 n}$. For different choices of ω_0, M will be different; i.e., for this class of inputs, M is a function of ω_0 that we denote as $M(\omega_0)$.

Determine whether $M(\omega_0)$ is periodic in ω_0 and, if it is, determine the period.

$x[n] = e^{j\omega_0 n}$ Nonlinear system M **Figure P2.22**

2.23. A linear time-invariant system has frequency response

$$H(e^{j\omega}) = \begin{cases} e^{-j\omega 3}, & |\omega| < \dfrac{2\pi}{16}\left(\dfrac{3}{2}\right), \\[2mm] 0, & \dfrac{2\pi}{16}\left(\dfrac{3}{2}\right) \le |\omega| \le \pi. \end{cases}$$

The input to the system is a periodic unit-impulse train with period $N = 16$; i.e.,

$$x[n] = \sum_{k=-\infty}^{\infty} \delta[n + 16k].$$

Find the output of the system.

2.24. A commonly used numerical operation called the *first difference* is defined as

$$y[n] = \nabla(x[n]) = x[n] - x[n-1],$$

where $x[n]$ is the input and $y[n]$ is the output of the first-difference system.
(a) Show that this system is linear and time-invariant.
(b) Find the impulse response of this system.
(c) Find and sketch the frequency response (magnitude and phase).
(d) Show that if

$$x[n] = f[n] * g[n],$$

then

$$\nabla(x[n]) = \nabla(f[n]) * g[n] = f[n] * \nabla(g[n]),$$

where $*$ denotes discrete convolution.
(e) Find the impulse response of a system that could be cascaded with the first-difference system to recover the input; i.e. find $h_i[n]$ where

$$h_i[n] * \nabla(x[n]) = x[n].$$

2.25. Consider the system in Fig. P2.25.
(a) Find the impulse response $h[n]$ of the overall system.
(b) Find the frequency response of the overall system.
(c) Specify a difference equation that relates the output $y[n]$ to the input $x[n]$.
(d) Is this system causal? Under what condition would the system be stable?

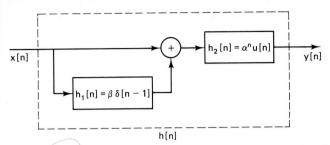

Figure P2.25

2.26. Let $H(e^{j\omega})$ denote the frequency response of an LTI system with impulse response $h[n]$, where $h[n]$ is in general complex.

(a) Using Eq. (2.92), show that $H^*(e^{-j\omega})$ is the frequency response of a system with impulse response $h^*[n]$, where * denotes complex conjugation.

(b) Show that if $h[n]$ is real, the frequency response is conjugate symmetric, i.e., $H(e^{-j\omega}) = H^*(e^{j\omega})$.

2.27. For $X(e^{j\omega}) = 1/(1 - ae^{-j\omega})$, with $-1 < a < 0$, determine and sketch the following as a function of ω.

(a) $\mathcal{R}e\{X(e^{j\omega})\}$

(b) $\mathcal{I}m\{X(e^{j\omega})\}$

(c) $|X(e^{j\omega})|$

(d) $\sphericalangle X(e^{j\omega})$

2.28. (a) Determine the Fourier transform of the sequence

$$r[n] = \begin{cases} 1, & 0 \le n \le M, \\ 0, & \text{otherwise.} \end{cases}$$

(b) Consider the sequence

$$w[n] = \begin{cases} \dfrac{1}{2}\left(1 + \cos\dfrac{2\pi n}{M}\right), & 0 \le n \le M, \\ 0, & \text{otherwise.} \end{cases}$$

Sketch $w[n]$ and express the Fourier transform of $w[n]$ in terms of the Fourier transform of $r[n]$. *Hint*: First express $w[n]$ in terms of $r[n]$ and the complex exponentials

$$e^{j(2\pi n/M)} \qquad \text{and} \qquad e^{-j(2\pi n/M)}.$$

(c) Sketch the magnitude of $R(e^{j\omega})$ and $W(e^{j\omega})$.

2.29. Let $X(e^{j\omega})$ denote the Fourier transform of $x[n]$. Using the Fourier transform synthesis or analysis equations (Eqs. 2.112 and 2.113), show that

(a) the Fourier transform of $x^*[n]$ is $X^*(e^{-j\omega})$;

(b) the Fourier transform of $x^*[-n]$ is $X^*(e^{j\omega})$.

2.30. Show that for $x[n]$ real, property 7 in Table 2.1 follows from property 1 and that properties 8–11 follow from property 7.

2.31. In Section 2.9 we stated a number of Fourier transform theorems without proof. Using the Fourier analysis or synthesis equations (Eqs. 2.112 and 2.113), demonstrate the validity of Theorems 1–5 in Table 2.2.

2.32. In Section 2.9.6 it was argued intuitively that

$$Y(e^{j\omega}) = H(e^{j\omega})X(e^{j\omega}), \tag{P2.32-1}$$

when $Y(e^{j\omega})$, $H(e^{j\omega})$, and $X(e^{j\omega})$ are the Fourier transforms of the output $y[n]$, impulse response $h[n]$, and input $x[n]$ of a linear time-invariant system; i.e.,

$$y[n] = \sum_{k=-\infty}^{\infty} x[k]h[n-k]. \tag{P2.32-2}$$

Verify Eq. (P2.32-1) by applying the Fourier transform to the convolution sum given in Eq. (P2.32-2).

2.33. By applying the Fourier synthesis Eq. (2.112) to Eq. (2.151) and using Theorem 3 in Table 2.2, demonstrate the validity of the modulation theorem (Theorem 7, Table 2.2).

2.34. Let $x[n]$ and $y[n]$ denote complex sequences and $X(e^{j\omega})$ and $Y(e^{j\omega})$ their Fourier transforms.

(a) By using the convolution theorem (Theorem 6 in Table 2.2) and appropriate properties from Table 2.1, determine in terms of $x[n]$ and $y[n]$ the sequence whose Fourier transform is $X(e^{j\omega})Y^*(e^{j\omega})$.

(b) Using the result in part (a), show that

$$\sum_{n=-\infty}^{\infty} x[n]y^*[n] = \frac{1}{2\pi} \int_{-\pi}^{\pi} X(e^{j\omega})Y^*(e^{j\omega})\, d\omega. \tag{P2.34}$$

Equation (P2.34) is a more general form of Parseval's theorem as given in Section 2.9.5.

(c) Using Eq. (P2.34), determine the numerical value of the sum

$$\sum_{n=-\infty}^{\infty} \frac{\sin(\pi n/4)}{2\pi n} \frac{\sin(\pi n/6)}{5\pi n}.$$

2.35. Let $X(e^{j\omega})$ denote the Fourier transform of the signal $x[n]$ shown in Fig. P2.35. Perform the following calculations without explicitly evaluating $X(e^{j\omega})$.

(a) Evaluate $X(e^{j\omega})|_{\omega=0}$.

(b) Find $\sphericalangle X(e^{j\omega})$.

(c) Evaluate $\displaystyle\int_{-\pi}^{\pi} X(e^{j\omega})\, d\omega$.

(d) Determine and sketch the signal whose Fourier transform is $\mathcal{R}e\{X(e^{j\omega})\}$.

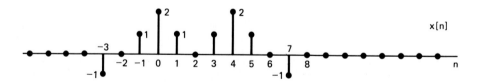

Figure P2.35

2.36. Let $x[n]$ and $X(e^{j\omega})$ represent a sequence and its Fourier transform, respectively. Determine, in terms of $X(e^{j\omega})$, the transforms of $y_s[n]$, $y_d[n]$, and $y_e[n]$. In each case, sketch $Y(e^{j\omega})$ for $X(e^{j\omega})$ as shown in Fig. P2.36.

(a) Sampler:

$$y_s[n] = \begin{cases} x[n], & n \text{ even,} \\ 0, & n \text{ odd} \end{cases}$$

(b) Compressor:

$$y_d[n] = x[2n]$$

(c) Expander:

$$y_e[n] = \begin{cases} x[n/2], & n \text{ even,} \\ 0, & n \text{ odd} \end{cases}$$

Note that $y_s[n] = \frac{1}{2}\{x[n] + (-1)^n x[n]\}$ and $-1 = e^{j\pi}$.

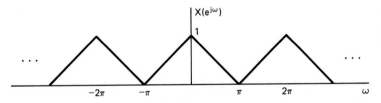

Figure P2.36

2.37. For the system in Fig. P2.37, determine the output $y[n]$ when the input $x[n]$ is $\delta[n]$ and $H(e^{j\omega})$ is an ideal lowpass filter as indicated, i.e.,

$$H(e^{j\omega}) = \begin{cases} 1, & |\omega| < \pi/2, \\ 0, & \pi/2 < |\omega| \le \pi. \end{cases}$$

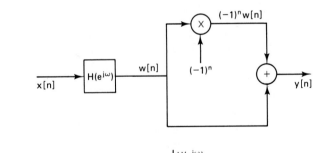

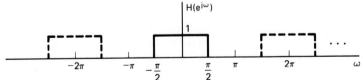

Figure P2.37

2.38. Let $x[n]$ and $y[n]$ be stationary, uncorrelated random signals. Show that if

$$w[n] = x[n] + y[n],$$

then

$$m_w = m_x + m_y \quad \text{and} \quad \sigma_w^2 = \sigma_x^2 + \sigma_y^2.$$

2.39. Let $e[n]$ denote a white-noise sequence and let $s[n]$ denote a sequence that is uncorrelated with $e[n]$. Show that the sequence

$$y[n] = s[n]e[n]$$

is white, i.e.,

$$E\{y[n]y[n+m]\} = A\delta[m],$$

where A is a constant.

2.40. Let $x[n]$ be a real, stationary, white-noise process, with zero mean and variance σ_x^2. Let $y[n]$ be the corresponding output when $x[n]$ is the input to a linear time-invariant system with impulse response $h[n]$. Show that
  (a) $E\{x[n]y[n]\} = h[0]\sigma_x^2$
  (b) $\sigma_y^2 = \sigma_x^2 \sum_{n=-\infty}^{\infty} h^2[n]$

2.41. Let $x[n]$ be a real stationary white-noise sequence, with zero mean and variance σ_x^2. Let $x[n]$ be the input to the cascade of two causal linear discrete-time time-invariant systems as shown in Fig. P2.41.
  (a) Is $\sigma_y^2 = \sigma_x^2 \sum_{k=0}^{\infty} h_1^2[k]$?
  (b) Is $\sigma_w^2 = \sigma_y^2 \sum_{k=0}^{\infty} h_2^2[k]$?

  (c) Let $h_1[n] = a^n u[n]$ and $h_2[n] = b^n u[n]$. Determine the impulse response of the overall system in Fig. P2.41, and from this determine σ_w^2. If your answer to part (b) was yes, is that consistent with your answer to part (c)?

$$x[n] \quad\quad h_1[n] \quad\quad y[n] \quad\quad h_2[n] \quad\quad w[n] \quad\quad \textbf{Figure P2.41}$$

2.42. Sometimes we are interested in the statistical behavior of a linear time-invariant system when the input is a suddenly applied random signal. Such a situation is depicted in Fig. P2.42.

   Let $x[n]$ be a stationary white-noise process. Then the input to the system, $w[n]$, given by

$$w[n] = \begin{cases} x[n], & n \geq 0, \\ 0, & n < 0, \end{cases}$$

is a nonstationary process, as is the output $y[n]$.
  (a) Derive an expression for the mean of the output in terms of the mean of the input.
  (b) Derive an expression for the autocorrelation sequence $\phi_{yy}[n_1, n_2]$ of the output.
  (c) Show that for large n the formulas derived in parts (a) and (b) approach the results for stationary inputs.
  (d) Assume that $h[n] = a^n u[n]$. Find the mean and mean-square values of the output in terms of the mean and mean-square values of the input. Sketch the dependence of these parameters as a function of n.

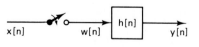

$$x[n] \quad\quad w[n] \quad\quad h[n] \quad\quad y[n]$$

(Switch closed at n = 0)   **Figure P2.42**

2.43. Consider a random process $x[n]$ that is the response of the linear time-invariant system shown in Fig. P2.43. In the figure, $w[n]$ represents a real zero-mean stationary white-noise process with $E\{w^2[n]\} = \sigma_w^2$.
  (a) Express $\mathcal{E}\{x^2[n]\}$ in terms of $\phi_{xx}[n]$ or $\Phi_{zz}(e^{j\omega})$.
  (b) Determine $I_{xx}(e^{j\omega})$, the power density spectrum of $x[n]$.
  (c) Determine $\phi_{xx}[n]$, the correlation function of $x[n]$.

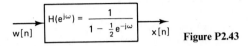

$$H(e^{j\omega}) = \frac{1}{1 - \frac{1}{2}e^{-j\omega}}$$

$w[n]$ $x[n]$ **Figure P2.43**

2.44. Consider a linear time-invariant system whose impulse response is real and is given by $h[n]$. Suppose the responses of the system to the two inputs $x[n]$ and $v[n]$ are $y[n]$ and $z[n]$ as shown in Fig. P2.44.

The inputs $x[n]$ and $v[n]$ in the figure represent real zero-mean stationary random processes with autocorrelation functions $\phi_{xx}[n]$ and $\phi_{vv}[n]$, cross-correlation function $\phi_{xv}[n]$, power spectra $\Phi_{xx}(e^{j\omega})$ and $\Phi_{vv}(e^{j\omega})$, and cross power spectrum $\Phi_{xv}(e^{j\omega})$.
  (a) Given $\phi_{xx}[n]$, $\phi_{vv}[n]$, $\phi_{xv}[n]$, $\Phi_{xx}(e^{j\omega})$, $\Phi_{vv}(e^{j\omega})$, and $\Phi_{xv}(e^{j\omega})$ determine $\Phi_{yz}(e^{j\omega})$, the cross power spectrum of $y[n]$ and $z[n]$, where $\Phi_{yz}(e^{j\omega})$ is defined by

$$\phi_{yz}[n] \xleftrightarrow{\mathcal{F}} \Phi_{yz}(e^{j\omega}),$$

with $\phi_{yz}[n] = E\{y[k]z[k-n]\}$.

  (b) Is the cross power spectrum $\Phi_{xv}(e^{j\omega})$ always nonnegative, i.e., is $\Phi_{xv}(e^{j\omega}) \geq 0$ for all ω? Justify your answer.

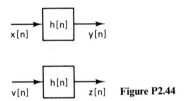

$x[n]$ $h[n]$ $y[n]$

$v[n]$ $h[n]$ $z[n]$ **Figure P2.44**

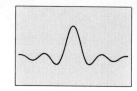

Sampling of Continuous-Time Signals

3

3.0 INTRODUCTION

Discrete-time signals can arise in many ways, but they most commonly occur as representations of continuous-time signals. This is partly due to the fact that processing of continuous-time signals is often carried out by discrete-time processing of sequences obtained by sampling. It is somewhat remarkable that under reasonable constraints, a continuous-time signal can be adequately represented by samples. In this chapter we discuss the process of periodic sampling in some detail, including the issue of aliasing when the signal is not bandlimited or the sampling rate is not high enough. Of particular importance is the fact that continuous-time signal processing can be implemented through a process of sampling, discrete-time processing, and subsequent reconstruction of a continuous-time signal.

3.1 PERIODIC SAMPLING

Although other possibilities exist (see Steiglitz, 1965; Oppenheim and Johnson, 1972), the typical method of obtaining a discrete-time representation of a continuous-time signal is through periodic sampling, wherein a sequence of samples $x[n]$ is obtained from a continuous-time signal $x_c(t)$ according to the relation

$$x[n] = x_c(nT), \qquad -\infty < n < \infty. \qquad (3.1)$$

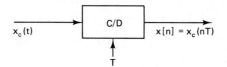

Figure 3.1   Block diagram representation of an ideal continuous-to-discrete (C/D) converter.

In Eq. (3.1), T is the *sampling period,* and its reciprocal, $f_s = 1/T$, is the *sampling frequency,* in samples per second.

We refer to a system that implements the operation of Eq. (3.1) as an *ideal continuous-to-discrete-time (C/D) converter,* and we depict it in block diagram form as indicated in Fig. 3.1. As an example of the relationship between $x_c(t)$ and $x[n]$, in Fig. 2.2 we illustrated a continuous-time speech waveform and the corresponding sequence of samples.

In a practical setting, the operation of sampling is often implemented by an analog-to-digital (A/D) converter. Such systems can be viewed as approximations to the ideal C/D converter. Important considerations in the implementation or choice of an A/D converter include quantization of the output samples, linearity, the need for sample-and-hold circuits, and limitations on the sampling rate. The effects of quantization are briefly discussed in Sections 3.7.2 and 3.7.3. Other practical issues of A/D conversion are outside the scope of this text.

The sampling operation is generally not invertible; i.e., given the output $x[n]$, it is not possible in general to reconstruct $x_c(t)$, the input to the sampler, since many continuous-time signals can produce the same output sequence of samples. The inherent ambiguity in sampling is of primary concern in signal processing. Fortunately it is possible to remove the ambiguity by restricting the class of input signals to the sampler.

It is convenient to mathematically represent the sampling process in the two stages depicted in Figure 3.2(a). This consists of an impulse train modulator followed by conversion of the impulse train to a sequence. Figure 3.2(b) illustrates two continuous-time signals and the results of impulse train sampling. Figure 3.2(c) depicts the corresponding output sequences. The essential difference between $x_s(t)$ and $x[n]$ is that $x_s(t)$ is, in a sense, a continuous-time signal (specifically an impulse train) that is zero except at integer multiples of T. The sequence $x[n]$ on the other hand is indexed on the integer variable n, which in effect introduces a time normalization; i.e, $x[n]$ contains no explicit information about the sampling rate. Furthermore, the samples of $x_c(t)$ are represented by finite numbers in $x[n]$ rather than as the areas of impulses as in $x_s(t)$.

It is important to emphasize that Fig. 3.2(a) is a mathematical representation of sampling, not a representation of any physical circuits or systems designed to implement the sampling operation. Whether a piece of hardware can be construed to be an approximation to the block diagram of Fig. 3.2(a) is a secondary issue at this point. We have introduced this representation of the sampling operation because it leads to a simple derivation of a key result and because this approach leads to a number of important insights that are difficult to obtain from a more formal derivation based on manipulation of Fourier transform formulas.

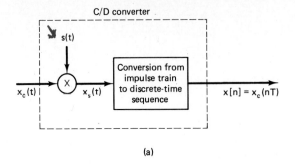

(a)

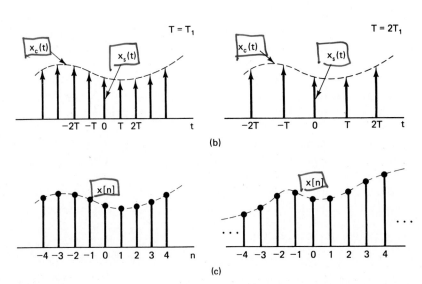

(b)

(c)

Figure 3.2 Sampling with a periodic impulse train followed by conversion to a discrete-time sequence. (a) Overall system. (b) $x_s(t)$ for two sampling rates. The dashed envelope represents $x_c(t)$. (c) The output sequence for the two different sampling rates.

3.2 FREQUENCY-DOMAIN REPRESENTATION OF SAMPLING

To derive the frequency-domain relation between the input and output of an ideal C/D converter, let us first consider the conversion of $x_c(t)$ to $x_s(t)$ through impulse train modulation. The modulating signal $s(t)$ is a periodic impulse train

$$s(t) = \sum_{n=-\infty}^{\infty} \delta(t - nT), \tag{3.2}$$

where $\delta(t)$ is the unit impulse function or Dirac delta function. Consequently,

$$x_s(t) = x_c(t)s(t)$$

$$= x_c(t) \sum_{n=-\infty}^{\infty} \delta(t - nT). \tag{3.3}$$

Through the "sifting property" of the impulse function, $x_s(t)$ can be expressed as

$$x_s(t) = \sum_{n=-\infty}^{\infty} x_c(nT)\delta(t - nT). \tag{3.4}$$

Let us now consider the Fourier transform of $x_s(t)$. Since from Eq. (3.3), $x_s(t)$ is the product of $x_c(t)$ and $s(t)$, the Fourier transform of $x_s(t)$ is the convolution of the Fourier transforms $X_c(j\Omega)$ and $S(j\Omega)$. The Fourier transform of a periodic impulse train is a periodic impulse train (Oppenheim and Willsky, 1983). Specifically, $S(j\Omega)$ is

$$S(j\Omega) = \frac{2\pi}{T} \sum_{k=-\infty}^{\infty} \delta(\Omega - k\Omega_s), \tag{3.5}$$

where $\Omega_s = 2\pi/T$ is the sampling frequency in radians/s. Since

$$X_s(j\Omega) = \frac{1}{2\pi} X_c(j\Omega) * S(j\Omega),$$

where $*$ denotes the operation of convolution, it follows that

$$X_s(j\Omega) = \frac{1}{T} \sum_{k=-\infty}^{\infty} X_c(j\Omega - kj\Omega_s). \tag{3.6}$$

Equation (3.6) provides the relationship between the Fourier transforms of the input and the output of the impulse train modulator in Fig. 3.2(a). We see from Eq. (3.6) that the Fourier transform of $x_s(t)$ consists of periodically repeated copies of the Fourier transform of $x_c(t)$. The copies of $X_c(j\Omega)$ are shifted by integer multiples of the sampling frequency and then superimposed to produce the periodic Fourier transform of the impulse train of samples. Figure 3.3 depicts the frequency-domain representation of impulse train sampling. Figure 3.3(a) represents a bandlimited Fourier transform where the highest nonzero frequency component in $X_c(j\Omega)$ is at Ω_N. Figure 3.3(b) shows the periodic impulse train $S(j\Omega)$, and Fig. 3.3(c) shows $X_s(j\Omega)$, the result of convolving $X_c(j\Omega)$ with $S(j\Omega)$. From Fig. 3.3(c) it is evident that when

$$\Omega_s - \Omega_N > \Omega_N, \quad \text{or} \quad \Omega_s > 2\Omega_N, \tag{3.7}$$

the replicas of $X_c(j\Omega)$ do not overlap and therefore, when they are added together in Eq. (3.6), there remains (to within a scale factor of $1/T$) a replica of $X_c(j\Omega)$ at each integer multiple of Ω_s. Consequently $x_c(t)$ can be recovered from $x_s(t)$ with an ideal lowpass filter. This is depicted in Fig. 3.4(a), which shows the impulse train modulator followed by a linear time-invariant system with frequency response $H_r(j\Omega)$. For $X_c(j\Omega)$ as in Fig. 3.4(b), $X_s(j\Omega)$ would be as shown in Fig. 3.4(c), where it is assumed that $\Omega_s > 2\Omega_N$. Since

$$X_r(j\Omega) = H_r(j\Omega)X_s(j\Omega), \tag{3.8}$$

it follows that if $H_r(j\Omega)$ is an ideal lowpass filter with gain T and cutoff frequency Ω_c such that

$$\Omega_N < \Omega_c < (\Omega_s - \Omega_N), \tag{3.9}$$

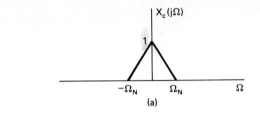

(a)

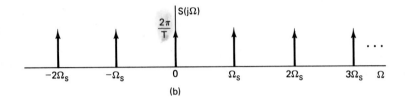

(b)

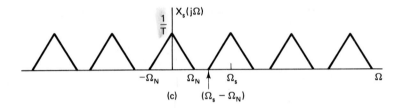

(c) $(\Omega_s - \Omega_N)$

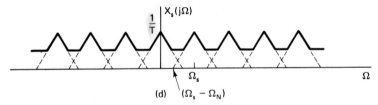

(d) $(\Omega_s - \Omega_N)$

Figure 3.3 Effect in the frequency domain of sampling in the time domain. (a) Spectrum of the original signal. (b) Spectrum of sampling function. (c) Spectrum of sampled signal with $\Omega_S > 2\Omega_N$. (d) Spectrum of sampled signal with $\Omega_S < 2\Omega_N$.

then

$$X_r(j\Omega) = X_c(j\Omega), \tag{3.10}$$

as depicted in Fig. 3.4(e).

If the inequality (3.7) does not hold, i.e., if $\Omega_s \leq 2\Omega_N$, the copies of $X_c(j\Omega)$ overlap so that when they are added together $X_c(j\Omega)$ is no longer recoverable by lowpass filtering. This is illustrated in Fig. 3.3(d). In this case, the reconstructed output $x_r(t)$ in Fig. 3.4(a) is related to the original continuous-time input through a distortion referred to as *aliasing*. Figure 3.5 illustrates aliasing in the frequency domain for the simple case of a cosine signal. Figure 3.5(a) shows the Fourier transform of the signal

$$x_c(t) = \cos \Omega_0 t. \tag{3.11}$$

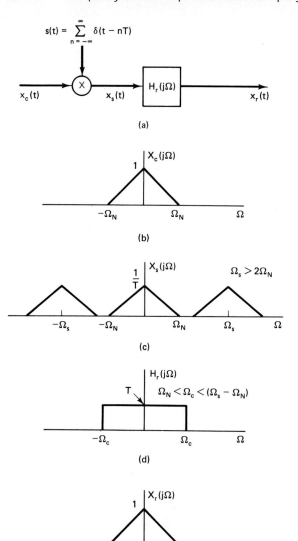

(a)

(b)

(c)

(d)

(e)

Figure 3.4 Exact recovery of a continuous-time signal from its samples using an ideal lowpass filter.

Part (b) shows the Fourier transform of $x_s(t)$ with $\Omega_0 < \Omega_s/2$, and part (c) shows the Fourier transform of $x_s(t)$ with $\Omega_0 > \Omega_s/2$. Parts (d) and (e) correspond to the Fourier transform of the lowpass filter output for $\Omega_0 < \Omega_s/2 = \pi/T$ and $\Omega_0 > \pi/T$, respectively, with $\Omega_c = \Omega_s/2$. Figures 3.5(c) and (e) correspond to the case of aliasing. With no aliasing (b and d), the reconstructed output $x_r(t)$ is

$$x_r(t) = \cos \Omega_0 t. \tag{3.12}$$

With aliasing, the reconstructed output is

$$x_r(t) = \cos(\Omega_s - \Omega_0)t, \tag{3.13}$$

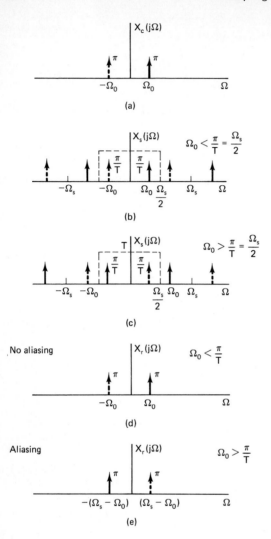

Figure 3.5　The effect of aliasing in the sampling of a cosine signal.

i.e., the higher-frequency signal $\cos \Omega_0 t$ has taken on the identity (alias) of the lower-frequency signal $\cos(\Omega_s - \Omega_0)t$ as a consequence of the sampling and reconstruction. This discussion is the basis for the Nyquist sampling theorem (Nyquist, 1928; Shannon, 1949), stated as follows.

Nyquist Sampling Theorem.　Let $x_c(t)$ be a bandlimited signal with

$$X_c(j\Omega) = 0 \qquad \text{for } |\Omega| > \Omega_N. \tag{3.14a}$$

Then $x_c(t)$ is uniquely determined by its samples $x[n] = x_c(nT)$, $n = 0, \pm 1, \pm 2, \ldots$, if

$$\Omega_s = \frac{2\pi}{T} > 2\Omega_N. \tag{3.14b}$$

The frequency Ω_N is commonly referred as the *Nyquist frequency*, and the frequency $2\Omega_N$ that must be exceeded by the sampling frequency is called the *Nyquist rate*.

Thus far we have considered only the impulse train modulator in Fig. 3.2(a). Our eventual objective is to express $X(e^{j\omega})$, the discrete-time Fourier transform of the sequence $x[n]$, in terms of $X_s(j\Omega)$ and $X_c(j\Omega)$. To this end, let us consider an alternative expression for $X_s(j\Omega)$. Applying the Fourier transform to Eq. (3.4), we obtain

$$X_s(j\Omega) = \sum_{n=-\infty}^{\infty} x_c(nT)e^{-j\Omega Tn}. \tag{3.15}$$

Since

$$x[n] = x_c(nT) \tag{3.16}$$

and

$$X(e^{j\omega}) = \sum_{n=-\infty}^{\infty} x[n]e^{-j\omega n}, \tag{3.17}$$

it follows that

$$X_s(j\Omega) = X(e^{j\omega})|_{\omega=\Omega T} = X(e^{j\Omega T}). \tag{3.18}$$

Consequently from Eqs. (3.6) and (3.18),

$$X(e^{j\Omega T}) = \frac{1}{T}\sum_{k=-\infty}^{\infty} X_c(j\Omega - jk\Omega_s), \tag{3.19}$$

or, equivalently,

$$X(e^{j\omega}) = \frac{1}{T}\sum_{k=-\infty}^{\infty} X_c\left(j\frac{\omega}{T} - j\frac{2\pi k}{T}\right). \tag{3.20}$$

From Eqs. (3.18)–(3.20) we see that $X(e^{j\omega})$ is simply a frequency-scaled version of $X_s(j\Omega)$ with the frequency scaling specified by $\omega = \Omega T$. This scaling can alternatively be thought of as a normalization of the frequency axis so that the frequency $\Omega = \Omega_s$ in $X_s(j\Omega)$ is normalized to $\omega = 2\pi$ for $X(e^{j\omega})$. The fact that there is a frequency scaling or normalization in the transformation from $X_s(j\Omega)$ to $X(e^{j\omega})$ is directly associated with the fact that there is a time normalization in the transformation from $x_s(t)$ to $x[n]$. Specifically, as we see in Fig. 3.2, $x_s(t)$ retains a spacing between samples equal to the sampling period T. In contrast, the "spacing" of sequence values $x[n]$ is always unity, i.e., the time axis is normalized by a factor of T. Correspondingly in the frequency domain, the frequency axis is normalized by a factor of $f_s = 1/T$.

3.3 RECONSTRUCTION OF A BANDLIMITED SIGNAL FROM ITS SAMPLES

According to the sampling theorem, samples of a continuous-time bandlimited signal taken frequently enough are sufficient to represent the signal exactly in the sense that the signal can be recovered from the samples and from knowledge of the sampling period. Impulse train modulation provides a convenient means for understanding the process of reconstructing the continuous-time bandlimited signal from its samples.

In Section 3.2 we saw that if the conditions of the sampling theorem are met and if the modulated impulse train is filtered by an appropriate lowpass filter, then the Fourier transform of the filter output will be identical to the Fourier transform of the original continuous-time signal $x_c(t)$, and thus the output of the filter will be $x_c(t)$. If we are given a sequence of samples $x[n]$, we can form an impulse train $x_s(t)$ in which successive impulses are assigned an area equal to successive sequence values, i.e.,

$$x_s(t) = \sum_{n=-\infty}^{\infty} x[n]\delta(t - nT). \tag{3.21}$$

The nth sample is associated with the impulse at $t = nT$, where T is the sampling period associated with the sequence $x[n]$. If this impulse train is the input to an ideal lowpass continuous-time filter with frequency response $H_r(j\Omega)$ and impulse response $h_r(t)$, then the output of the filter will be

$$x_r(t) = \sum_{n=-\infty}^{\infty} x[n]h_r(t - nT). \tag{3.22}$$

A block diagram representation of this signal reconstruction process is shown in Fig. 3.6(a). Recall that the ideal reconstruction filter has a gain of T (to compensate for the factor of $1/T$ in Eq. 3.19 or 3.20) and a cutoff frequency Ω_c between Ω_N and $\Omega_s - \Omega_N$. A convenient and commonly used choice of the cutoff frequency is $\Omega_c = \Omega_s/2 = \pi/T$. This choice is appropriate for any relationship between Ω_s and Ω_N that avoids aliasing (i.e., so long as $\Omega_s > 2\Omega_N$). Figure 3.6(b) shows the frequency response of the ideal reconstruction filter. The corresponding impulse response, $h_r(t)$, is the inverse Fourier transform of $H_r(j\Omega)$, and for cutoff frequency π/T is given by

$$h_r(t) = \frac{\sin \pi t/T}{\pi t/T}. \tag{3.23}$$

This impulse response is shown in Figure 3.6(c). From substituting Eq. (3.23) into Eq. (3.22) it follows that

$$x_r(t) = \sum_{n=-\infty}^{\infty} x[n] \frac{\sin[\pi(t - nT)/T]}{\pi(t - nT)/T}. \tag{3.24}$$

From the frequency-domain argument of Section 3.2 we saw that if $x[n] = x_c(nT)$, where $X_c(j\Omega) = 0$ for $|\Omega| \geq \pi/T$, then $x_r(t)$ is equal to $x_c(t)$. It is not immediately obvious that this is true by considering Eq. (3.24) alone. However, useful insight is gained by looking at Eq. (3.24) more closely. First let us consider the function $h_r(t)$ given by Eq. (3.23). We note that

$$h_r(0) = 1. \tag{3.25a}$$

This follows from l'Hôpital's rule. In addition,

$$h_r(nT) = 0 \qquad \text{for } n = \pm 1, \pm 2, \dots. \tag{3.25b}$$

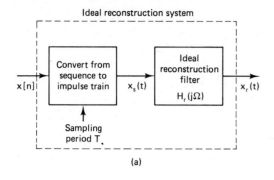

(a)

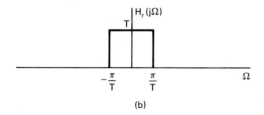

(b)

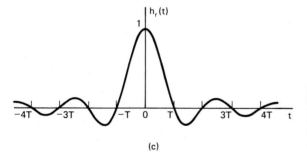

(c)

Figure 3.6 (a) Block diagram of an ideal bandlimited signal reconstruction system. (b) Frequency response of an ideal reconstruction filter. (c) Impulse response of an ideal reconstruction filter.

It follows from Eqs. (3.25) and Eq. (3.22) that if $x[n] = x_c(nT)$, then

$$x_r(mT) = x_c(mT) \tag{3.26}$$

for all integer values of m. That is, the signal that is reconstructed by Eq. (3.24) has the same values at the sampling times as the original continuous-time signal, independent of the sampling period T.

In Fig. 3.7 we show a continuous-time signal $x_c(t)$ and the corresponding modulated impulse train. Figure 3.7(c) shows several of the terms

$$x[n] \frac{\sin[\pi(t - nT)/T]}{\pi(t - nT)/T}$$

and the resulting reconstructed signal $x_r(t)$. As suggested by this figure, the ideal lowpass filter *interpolates* between the impulses of $x_s(t)$ to construct a continuous-time signal $x_r(t)$. From Eq. (3.26), the resulting signal is an exact reconstruction of $x_c(t)$ at

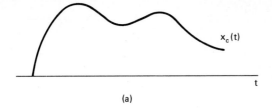

(a)

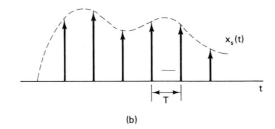

(b)

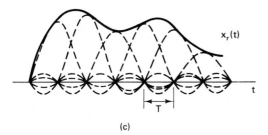

(c)

Figure 3.7 Ideal bandlimited interpolation.

the sampling times. The fact that the lowpass filter interpolates the correct recon-struction between the samples, if there is no aliasing, follows from our frequency-domain analysis of the sampling and reconstruction process.

It is useful to formalize the preceding discussion by defining an ideal system for reconstructing a bandlimited signal from a sequence of samples. We will call this system the *ideal discrete-to-continuous-time (D/C) converter*. The desired system is depicted in Fig. 3.8. As we have seen, the ideal reconstruction process can be represented as the conversion of the sequence to an impulse train as in Eq. (3.21) followed by filtering with an ideal lowpass filter, resulting in the output given by Eq. (3.24). The intermediate step of conversion to an impulse train is a mathematical convenience in deriving Eq. (3.24) and in understanding the signal reconstruction process. However, once we are familiar with this process, it is useful to define a more compact representation, as depicted in Fig. 3.8(b), where the input is the sequence $x[n]$ and the output is the continuous-time signal $x_r(t)$ given by Eq. (3.24).

The properties of the ideal D/C converter are most easily seen in the frequency domain. To derive an input/output relation in the frequency domain, consider the Fourier transform of Eq. (3.24), which is

$$X_r(j\Omega) = \sum_{n=-\infty}^{\infty} x[n]H_r(j\Omega)e^{-j\Omega Tn}.$$

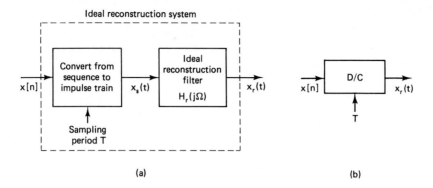

Figure 3.8   (a) Ideal bandlimited signal reconstruction. (b) Equivalent representation as an ideal D/C converter.

By factoring $H_r(j\Omega)$ out of the sum, we can write

$$X_r(j\Omega) = H_r(j\Omega)X(e^{j\Omega T}). \tag{3.27}$$

Equation (3.27) provides a frequency-domain description of the ideal D/C converter. According to Eq. (3.27), $X(e^{j\omega})$ is frequency scaled (i.e., ω is replaced by ΩT). The ideal lowpass filter $H_r(j\Omega)$ selects the base period of the resulting periodic Fourier transform $X(e^{j\Omega T})$ and compensates for the $1/T$ scaling inherent in sampling. Thus, if the sequence $x[n]$ has been obtained by sampling a bandlimited signal at the Nyquist rate or higher, then the reconstructed signal $x_r(t)$ will be equal to the original bandlimited signal. In any case, it is also clear from Eq. (3.27) that the output of the ideal D/C converter is always bandlimited to at most the cutoff frequency of the lowpass filter, which is typically taken to be one-half the sampling frequency.

3.4 DISCRETE-TIME PROCESSING OF CONTINUOUS-TIME SIGNALS

A major application of discrete-time systems is in the processing of continuous-time signals. This is accomplished by a system of the general form depicted in Fig. 3.9. The system is a cascade of a C/D converter followed by a discrete-time system followed by

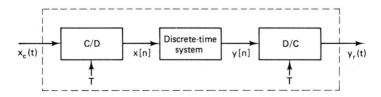

Figure 3.9   Discrete-time processing of continuous-time signals.

a D/C converter. The block diagram of Fig. 3.9 represents a large class of systems since the sampling rate and the discrete-time system can be chosen as we wish. Note that the overall system is equivalent to a continuous-time system since it transforms the continuous-time input signal $x_c(t)$ into the continuous-time output signal $y_r(t)$. However, the properties of the system are dependent on the choice of the discrete-time system and the sampling rate. We assume in Fig. 3.9 that the C/D and D/C converters have the same sampling rate. This is not essential, and later sections of this chapter and some of the problems at the end of the chapter consider systems where the input and output sampling rates are not the same.

The previous sections of this chapter have been devoted to understanding the C/D and D/C conversion operations in Fig. 3.9. For convenience and as a first step in understanding the overall system of Fig. 3.9, we summarize the mathematical representations of these operations.

The C/D converter produces a discrete-time signal

$$x[n] = x_c(nT), \tag{3.28}$$

i.e., a sequence of samples of the continuous-time input signal $x_c(t)$. The Fourier transform of this sequence is related to the Fourier transform of the continuous-time input signal as

$$X(e^{j\omega}) = \frac{1}{T} \sum_{k=-\infty}^{\infty} X_c\left(j\frac{\omega}{T} - j\frac{2\pi k}{T}\right). \tag{3.29}$$

The D/C converter creates a continuous-time output signal of the form

$$y_r(t) = \sum_{n=-\infty}^{\infty} y[n] \frac{\sin[\pi(t-nT)/T]}{\pi(t-nT)/T}, \tag{3.30}$$

where the sequence $y[n]$ is the output of the discrete-time system when the input to the system is $x[n]$. From Eq. (3.27), $Y_r(j\Omega)$, the continuous-time Fourier transform of $y_r(t)$, and $Y(e^{j\omega})$, the discrete-time Fourier transform of $y[n]$, are related by

$$Y_r(j\Omega) = H_r(j\Omega)Y(e^{j\Omega T})$$

$$= \begin{cases} TY(e^{j\Omega T}), & |\Omega| < \pi/T, \\ 0, & \text{otherwise.} \end{cases} \tag{3.31}$$

Next let us relate the output sequence $y[n]$ to the input sequence $x[n]$ or equivalently $Y(e^{j\omega})$ to $X(e^{j\omega})$. A simple example is the identity system, i.e., $y[n] = x[n]$. This is the case that we have studied in detail already. We know that if $x_c(t)$ has a bandlimited Fourier transform such that $X_c(j\Omega) = 0$ for $|\Omega| \geq \pi/T$ and if the discrete-time system in Fig. 3.9 is the identity system such that $y[n] = x[n] = x_c(nT)$, then the output will be $y_r(t) = x_c(t)$. Recall that in proving this result, we utilized the frequency-domain representations of the continuous-time and discrete-time signals, since the key concept of aliasing is most easily understood in the frequency domain. Likewise, when we deal with systems more complicated than the identity system, we generally carry out the analysis in the frequency domain. If the discrete-time system is nonlinear or time-varying, it is usually difficult to obtain a general relationship between the Fourier transforms of the input and the output of the system. (In Problem 3.16 we consider an example of the system of Fig. 3.9 in which the discrete-time system is nonlinear.) However, the linear time-invariant case leads to a rather simple and very useful result.

3.4.1 Linear Time-Invariant Discrete-Time Systems

If the discrete-time system in Fig. 3.9 is linear and time-invariant, we then have

$$Y(e^{j\omega}) = H(e^{j\omega})X(e^{j\omega}), \tag{3.32}$$

where $H(e^{j\omega})$ is the frequency response of the system or, equivalently, the Fourier transform of the unit sample response, and $X(e^{j\omega})$ and $Y(e^{j\omega})$ are the Fourier transforms of the input and output, respectively. Combining Eqs. (3.31) and (3.32), we obtain

$$Y_r(j\Omega) = H_r(j\Omega)H(e^{j\Omega T})X(e^{j\Omega T}). \tag{3.33}$$

Next, using Eq. (3.29) with $\omega = \Omega T$, we have

$$Y_r(j\Omega) = H_r(j\Omega)H(e^{j\Omega T})\frac{1}{T}\sum_{k=-\infty}^{\infty} X_c\left(j\Omega - j\frac{2\pi k}{T}\right). \tag{3.34}$$

If $X_c(j\Omega) = 0$ for $|\Omega| \geq \pi/T$, then the ideal lowpass reconstruction filter $H_r(j\Omega)$ cancels the factor $1/T$ and selects only the term in Eq. (3.34) for $k = 0$, i.e.,

$$Y_r(j\Omega) = \begin{cases} H(e^{j\Omega T})X_c(j\Omega), & |\Omega| < \pi/T, \\ 0, & |\Omega| \geq \pi/T. \end{cases} \tag{3.35}$$

Thus if $X_c(j\Omega)$ is bandlimited and the sampling rate is above the Nyquist rate, the output is related to the input through an equation of the form

$$Y_r(j\Omega) = H_{eff}(j\Omega)X_c(j\Omega), \tag{3.36}$$

where

$$H_{eff}(j\Omega) = \begin{cases} H(e^{j\Omega T}), & |\Omega| < \pi/T, \\ 0, & |\Omega| \geq \pi/T. \end{cases} \tag{3.37}$$

That is, the overall continuous-time system is equivalent to a linear time-invariant system whose *effective* frequency response is given by Eq. (3.37).

It is important to emphasize that the linear and time-invariant behavior of the system of Fig. 3.9 depends on two factors. First, the discrete-time system must be linear and time-invariant. Second, the input signal must be bandlimited and the sampling rate must be high enough so that any aliased components are removed by the discrete-time system. As a simple illustration of this second condition being violated, consider the case when $x_c(t)$ is a single unit-amplitude pulse whose duration is less than the sampling period. If the pulse is unity at $t = 0$, then $x[n] = \delta[n]$. However, it is clearly possible to shift the pulse so that it is not aligned with any of the sampling times, i.e., $x[n] = 0$ for all n. Obviously such a pulse, being timelimited, is not bandlimited. Even if the discrete-time system is the identity system, such that $y[n] = x[n]$, the overall system will not be time-invariant. In general, if the discrete-time system in Fig. 3.9 is linear and time-invariant and the sampling frequency is above the Nyquist rate associated with the bandwidth of the input $x_c(t)$, then the overall system is guaranteed to be equivalent to a linear time-invariant continuous-time system with an effective frequency response given by Eq. (3.37). Furthermore, Eq. (3.37) is valid even if some aliasing occurs in the C/D converter as long as $H(e^{j\omega})$ does not pass the aliased components. Example 3.1 is a simple illustration of this.

Example 3.1   The Ideal Lowpass Filter

Consider Fig. 3.9, with the linear time-invariant discrete-time system having frequency response

$$H(e^{j\omega}) = \begin{cases} 1, & |\omega| < \omega_c, \\ 0, & \omega_c < |\omega| \leq \pi. \end{cases} \tag{3.38}$$

This frequency response is, of course, periodic with period 2π, as shown in Fig. 3.10(a). For bandlimited inputs sampled above the Nyquist rate, it follows from Eq. (3.37) that the overall system of Fig. 3.9 will behave as a linear time-invariant continuous-time system with frequency response

$$H_{eff}(j\Omega) = \begin{cases} 1, & |\Omega T| < \omega_c \text{ or } |\Omega| < \omega_c/T, \\ 0, & |\Omega T| > \omega_c \text{ or } |\Omega| > \omega_c/T. \end{cases} \tag{3.39}$$

As shown in Fig. 3.10(b), this effective frequency response is that of an ideal lowpass filter with cutoff frequency $\Omega_c = \omega_c/T$.

As an interpretation of this result, consider the graphical illustration given in Fig. 3.11. Figure 3.11(a) indicates the Fourier transform of a bandlimited signal. Figure 3.11(b) shows the Fourier transform of the intermediate modulated impulse train, which is identical to $X(e^{j\Omega T})$, the discrete-time Fourier transform of the sequence of samples evaluated for $\omega = \Omega T$. In Fig. 3.11(c), the discrete-time Fourier transform of the sequence of samples and the frequency response of the discrete-time system are both plotted as a function of the normalized discrete-time frequency variable ω. Figure 3.11(d) shows $Y(e^{j\omega}) = H(e^{j\omega})X(e^{j\omega})$, the Fourier transform of the output of the discrete-time

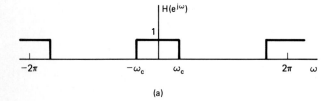

(a)

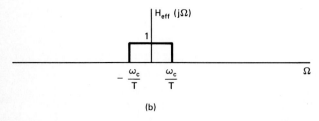

(b)

Figure 3.10 (a) Frequency response of the discrete-time system in Fig. 3.9. (b) Corresponding frequency response for bandlimited inputs.

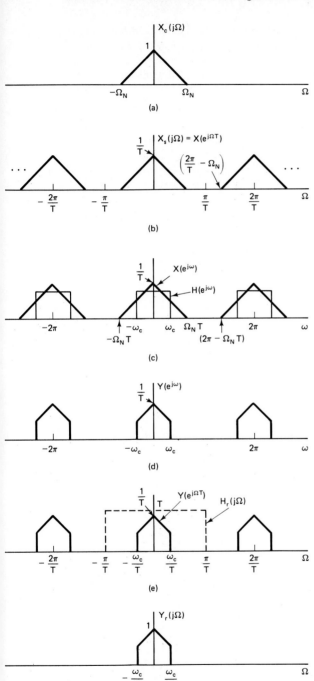

Figure 3.11 (a) Fourier transform of a bandlimited input signal. (b) Fourier transform of sampled input plotted as a function of continuous-time frequency Ω. (c) Fourier transform $X(e^{j\omega})$ of sequence of samples and frequency response $H(e^{j\omega})$ of discrete-time system plotted vs. ω. (d) Fourier transform of output of discrete-time system. (e) Fourier transform of output of discrete-time system and frequency response of ideal reconstruction filter plotted vs. Ω. (f) Fourier transform of output.

system. Figure 3.11(e) illustrates the Fourier transform of the output of the discrete-time system as a function of continuous-time frequency Ω together with the frequency response of the ideal reconstruction filter $H_r(j\Omega)$ of the D/C converter. Finally, Fig. 3.11(f) shows the resulting Fourier transform of the output of the D/C converter. By comparing Figs. 3.11(a) and 3.11(f), we see that the system behaves as a linear time-invariant system with frequency response given by Eq. (3.39) and plotted in Fig. 3.10(b).

Several important points are illustrated in this example. First note that the ideal lowpass discrete-time filter with discrete-time cutoff frequency ω_c has the effect of an ideal lowpass filter with cutoff frequency $\Omega_c = \omega_c/T$ when used in the configuration of Fig. 3.9. This cutoff frequency depends on both ω_c and T. In particular, by using a fixed discrete-time lowpass filter but varying the sampling period T, an equivalent continuous-time lowpass filter with a variable cutoff frequency can be implemented. For example, if T were chosen so that $\Omega_N T < \omega_c$, then the output of the system of Fig. 3.9 would be $y_r(t) = x_c(t)$. Also, as illustrated in Problem 3.11, Eq. (3.39) will be valid even if some aliasing is present in Figs. 3.11(b) and (c), as long as these distorted (aliased) components are eliminated by the filter $H(e^{j\omega})$. In particular, from Fig. 3.11(c), we see that for no aliasing to be present in the output, we require that

$$(2\pi - \Omega_N T) > \omega_c \tag{3.40}$$

as compared with the Nyquist requirement that

$$(2\pi - \Omega_N T) > \Omega_N T. \tag{3.41}$$

As another example of continuous-time processing using a discrete-time system, let us consider the implementation of an ideal differentiator for bandlimited signals.

Example 3.2   Ideal Bandlimited Differentiator

The ideal continuous-time differentiator system is defined by

$$y_c(t) = \frac{d}{dt}[x_c(t)], \tag{3.42}$$

with corresponding frequency response

$$H_c(j\Omega) = j\Omega. \tag{3.43}$$

Since we are considering a realization in the form of Fig. 3.9, the inputs are restricted to be bandlimited. For processing bandlimited signals, it is sufficient that

$$H_{eff}(j\Omega) = \begin{cases} j\Omega, & |\Omega| < \pi/T, \\ 0, & |\Omega| \geq \pi/T, \end{cases} \tag{3.44}$$

as depicted in Fig. 3.12(a). The corresponding discrete-time system has frequency response

$$H(e^{j\omega}) = j\omega/T, \qquad |\omega| < \pi, \tag{3.45}$$

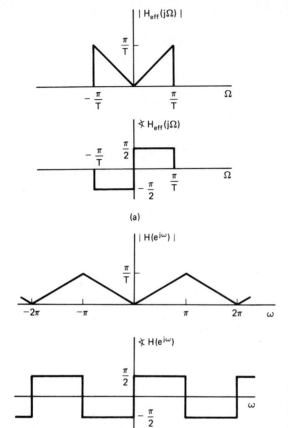

Figure 3.12 (a) Frequency response of a continuous-time ideal bandlimited differentiator $H_c(j\Omega) = j\Omega$, $|\Omega| < \Omega_c$. (b) Frequency response of discrete-time filter to implement a continuous-time bandlimited differentiator.

and is periodic with period 2π. This frequency response is plotted in Fig. 3.12(b). The corresponding impulse response can be shown to be

$$h[n] = \frac{\pi n \cos \pi n - \sin \pi n}{\pi n^2 T}, \qquad -\infty < n < \infty,$$

$$h[n] = \begin{cases} 0, & n = 0, \\ \dfrac{\cos \pi n}{nT}, & n \neq 0. \end{cases} \qquad (3.46)$$

Thus, if a discrete-time system with this impulse response were used in the configuration of Fig. 3.9, the output for every bandlimited input would be the derivative of the input.

3.4.2 Impulse Invariance

We have shown that the cascade system of Fig. 3.9 can be equivalent to a linear time-invariant system for bandlimited input signals. Let us now assume that, as depicted in

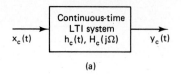

(a)

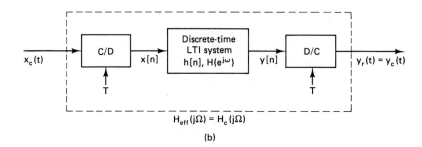

(b)

Figure 3.13   (a) Continuous-time LTI system. (b) Equivalent system for bandlimited inputs.

Figure 3.13, we are given a desired continuous-time system that we wish to implement in the form of Fig. 3.9. With $H_c(j\Omega)$ bandlimited, Eq. (3.37) specifies how to choose $H(e^{j\omega})$ so that $H_{eff}(j\Omega) = H_c(j\Omega)$. Specifically,

$$H(e^{j\omega}) = H_c(j\omega/T), \qquad |\omega| < \pi, \tag{3.47}$$

with the further requirement that T be chosen such that

$$H_c(j\Omega) = 0, \qquad |\Omega| \geq \pi/T. \tag{3.48}$$

Under the constraints of Eqs. (3.47) and (3.48), there is also a straightforward and useful relationship between the continuous-time impulse response $h_c(t)$ and the discrete-time impulse response $h[n]$. In particular, as we verify shortly,

$$h[n] = Th_c(nT), \tag{3.49}$$

i.e., the impulse response of the discrete-time system is a scaled, sampled version of $h_c(t)$. When $h[n]$ and $h_c(t)$ are related through Eq. (3.49), the discrete-time system is said to be an *impulse-invariant* version of the continuous-time system.

Equation (3.49) is a direct consequence of the discussion in Section 3.2. Specifically, with $x[n]$ and $x_c(t)$ replaced by $h[n]$ and $h_c(t)$ in Eq. (3.16), i.e.,

$$h[n] = h_c(nT), \tag{3.50}$$

Eq. (3.20) becomes

$$H(e^{j\omega}) = \frac{1}{T} \sum_{k=-\infty}^{\infty} H_c\left(j\frac{\omega}{T} - j\frac{2\pi k}{T}\right) \tag{3.51}$$

or, equivalently,

$$H(e^{j\omega}) = \frac{1}{T} H_c\left(j\frac{\omega}{T}\right), \qquad |\omega| < \pi. \tag{3.52}$$

Modifying Eqs. (3.50) and (3.52) to account for the scale factor of T in Eq. (3.49), we have

$$h[n] = Th_c(nT), \tag{3.53}$$

$$H(e^{j\omega}) = H_c\left(j\frac{\omega}{T}\right), \qquad |\omega| < \pi. \tag{3.54}$$

3.5 CONTINUOUS-TIME PROCESSING OF ~~skip~~ DISCRETE-TIME SIGNALS

In Section 3.4 we discussed and analyzed the use of discrete-time systems for processing continuous-time signals in the configuration of Fig. 3.9. In this section we consider the complementary situation depicted in Fig. 3.14, which is appropriately referred to as continuous-time processing of discrete-time signals. While the system of Fig. 3.14 is not typically used to implement discrete-time systems, it provides a useful interpretation of certain discrete-time systems.

From the definition of the ideal D/C converter, $X_c(j\Omega)$ and therefore also $Y_c(j\Omega)$ will necessarily be zero for $|\Omega| \geq \pi/T$. Thus the C/D converter samples $y_c(t)$ without aliasing, and we can express $x_c(t)$ and $y_c(t)$ as

$$x_c(t) = \sum_{n=-\infty}^{\infty} x[n] \frac{\sin[\pi(t - nT)/T]}{\pi(t - nT)/T} \tag{3.55}$$

and

$$y_c(t) = \sum_{n=-\infty}^{\infty} y[n] \frac{\sin[\pi(t - nT)/T]}{\pi(t - nT)/T}, \tag{3.56}$$

where $x[n] = x_c(nT)$ and $y[n] = y_c(nT)$. The frequency-domain relationships for Fig. 3.14 are

$$X_c(j\Omega) = TX(e^{j\Omega T}), \qquad |\Omega| < \pi/T, \tag{3.57a}$$

$$Y_c(j\Omega) = H_c(j\Omega)X_c(j\Omega), \qquad |\Omega| < \pi/T, \tag{3.57b}$$

$$Y(e^{j\omega}) = \frac{1}{T} Y_c(j\omega/T), \qquad |\omega| < \pi. \tag{3.57c}$$

Therefore, by substituting Eqs. (3.57a) and (3.57b) into Eq. (3.57c) it follows that the overall system behaves as a discrete-time system whose frequency response is

$$H(e^{j\omega}) = H_c(j\omega/T), \qquad |\omega| < \pi, \tag{3.58}$$

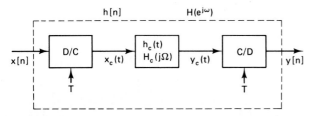

Figure 3.14   Continuous-time processing of discrete-time signals.

or, equivalently, the overall frequency response of the system in Fig. 3.14 will be equal to a given $H(e^{j\omega})$ if the frequency response of the continuous-time system is

$$H_c(j\Omega) = H(e^{j\Omega T}), \qquad |\Omega| < \pi/T. \tag{3.59}$$

Since $X_c(j\Omega) = 0$ for $|\Omega| \geq \pi/T$, $H_c(j\Omega)$ may be chosen arbitrarily above π/T. A convenient, but arbitrary, choice is $H_c(j\Omega) = 0$ for $|\Omega| \geq \pi/T$.

    With this representation of a discrete-time system we can focus on the equivalent effect of the continuous-time system on the bandlimited continuous-time signal $x_c(t)$. This is illustrated in Examples 3.3 and 3.4.

Example 3.3   Noninteger Delay

    Let us consider a discrete-time system with frequency response $H(e^{j\omega})$ given by

$$H(e^{j\omega}) = e^{-j\omega\Delta}, \qquad |\omega| < \pi. \tag{3.60}$$

When Δ is an integer, this system has a straightforward interpretation as a delay of Δ, i.e.,

$$y[n] = x[n - \Delta]. \tag{3.61}$$

When Δ is not an integer, Eq. (3.61) has no formal meaning. However, using the system of Fig. 3.14, a useful time-domain interpretation can be applied to the system specified by Eq. (3.60). Consider $H_c(j\Omega)$ in Fig. 3.14 chosen as

$$H_c(j\Omega) = H(e^{j\Omega T}) = e^{-j\Omega\Delta T}. \tag{3.62}$$

Then from Eq. (3.59), the overall discrete-time system in Fig. 3.14 will have the frequency response given by Eq. (3.60), whether or not Δ is an integer. To interpret the system of Eq. (3.60), we note that from Eq. (3.62),

$$y_c(t) = x_c(t - \Delta T). \tag{3.63}$$

Furthermore, $x_c(t)$ is the bandlimited interpolation of $x[n]$, and $y[n]$ is $y_c(t)$ sampled. Thus, for example, if $\Delta = \frac{1}{2}$, $y[n]$ would be the values of the bandlimited interpolation halfway between the input sequence values. This is illustrated in Fig. 3.15. Formally, from Eqs. (3.56), (3.63), and (3.55a),

$$y[n] = x_c(nT - \Delta T) = \sum_{k=-\infty}^{\infty} x[k] \left. \frac{\sin[\pi(t - \Delta T - kT)/T]}{\pi(t - \Delta T - kT)/T} \right|_{t=nT}. \tag{3.64}$$

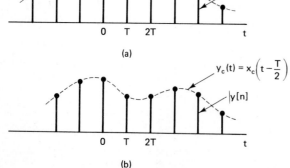

(a)

(b)

Figure 3.15 Continuous-time processing of the discrete-time sequence in part (a) can produce a new sequence with a "half-sample" delay, as in part (b).

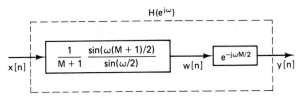

Figure 3.16 The moving average system represented as a cascade of two systems.

The noninteger delay represented by Eq. (3.64) has considerable practical significance since such a factor often arises in the frequency-domain representation of systems. When such a term is found in the frequency response of a causal discrete-time system, it can be interpreted in the light of this example. This interpretation is illustrated in Example 3.4.

Example 3.4   Moving Average

In Example 2.15 we considered the moving average system and obtained its frequency response. For the case $M_1 = 0$ and $M_2 = M$, the frequency response is

$$H(e^{j\omega}) = \frac{1}{(M+1)} \frac{\sin[\omega(M+1)/2]}{\sin(\omega/2)} e^{-j\omega M/2}, \qquad |\omega| < \pi. \tag{3.65}$$

Let us interpret this as the cascade of two systems, as indicated in Fig. 3.16. The first system imposes a frequency-domain amplitude weighting. The second system represents the linear phase term in Eq. (3.65). If M is an even integer (meaning the moving average of an odd number of samples), then the linear phase term corresponds to an integer delay, i.e.,

$$y[n] = w[n - M/2]. \tag{3.66}$$

However, if M is odd, the linear phase term corresponds to a noninteger delay, specifically, an integer plus one-half sample interval. This noninteger delay can be interpreted in terms of the discussion in Example 3.3; i.e., $y[n]$ is equivalent to bandlimited interpolation of $w[n]$ followed by a continuous-time delay of $MT/2$ (where T is the assumed but arbitrary sampling period associated with the D/C interpolation of $w[n]$) followed by C/D conversion again with sampling period T.

3.6   CHANGING THE SAMPLING RATE USING DISCRETE-TIME PROCESSING

We have seen that a continuous-time signal $x_c(t)$ can be represented by a discrete-time signal consisting of a sequence of samples

$$x[n] = x_c(nT). \tag{3.67}$$

Alternatively, our previous discussion has shown that even if $x[n]$ was not obtained originally by sampling, we can always use Eq. (3.24) to find a continuous-time bandlimited signal $x_r(t)$ whose samples are $x[n] = x_c(nT)$.

It is often necessary to change the sampling rate of a discrete-time signal, i.e., to obtain a new discrete-time representation of the underlying continuous-time signal of the form

$$x'[n] = x_c(nT'), \tag{3.68}$$

where $T' \neq T$. One approach to obtaining the sequences $x'[n]$ from $x[n]$ is to reconstruct $x_c(t)$ from $x[n]$ using Eq. (3.24) and then resample $x_c(t)$ with period T' to obtain $x'[n]$. Often this is not a desirable approach because of the nonideal analog reconstruction filter, D/A converter, and A/D converter that would be used in a practical implementation. Thus, it is of interest to consider methods of changing the sampling rate that involve only discrete-time operations.

3.6.1 Sampling Rate Reduction by an Integer Factor

The sampling rate of a sequence can be reduced by "sampling" it, i.e., by defining a new sequence

$$x_d[n] = x[nM] = x_c(nMT). \tag{3.69}$$

Equation (3.69) defines the system depicted in Fig. 3.17, which is called a *sampling rate compressor* (Crochiere and Rabiner, 1983) or simply a *compressor*. From Eq. (3.69) it is clear that $x_d[n]$ could be obtained directly from $x_c(t)$ by sampling with period $T' = MT$. Furthermore, if $X_c(j\Omega) = 0$ for $|\Omega| > \Omega_N$, then $x_d[n]$ is an exact representation of $x_c(t)$ if $\pi/T' = \pi/(MT) > \Omega_N$. That is, the sampling rate can be reduced by a factor of M without aliasing if the original sampling rate was at least M times the Nyquist rate or if the bandwidth of the sequence is first reduced by a factor of M by discrete-time filtering. In general, the operation of reducing the sampling rate (including any prefiltering) will be called *downsampling*.

As in the case of sampling a continuous-time signal, it is useful to obtain a frequency-domain relation between the input and output of the compressor. This time, however, it will be a relationship between discrete-time Fourier transforms. Although several methods can be used to derive the desired result, we will base our derivation on the results already obtained for sampling continuous-time signals. First recall that the discrete-time Fourier transform of $x[n] = x_c(nT)$ is

$$X(e^{j\omega}) = \frac{1}{T} \sum_{k=-\infty}^{\infty} X_c\left(j\frac{\omega}{T} - j\frac{2\pi k}{T}\right). \tag{3.70}$$

Similarly, the discrete-time Fourier transform of $x_d[n] = x[nM] = x_c(nT')$ with $T' = MT$ is

$$X_d(e^{j\omega}) = \frac{1}{T'} \sum_{r=-\infty}^{\infty} X_c\left(j\frac{\omega}{T'} - j\frac{2\pi r}{T'}\right). \tag{3.71}$$

Now, since $T' = MT$, we can write Eq. (3.71) as

$$X_d(e^{j\omega}) = \frac{1}{MT} \sum_{r=-\infty}^{\infty} X_c\left(j\frac{\omega}{MT} - j\frac{2\pi r}{MT}\right). \tag{3.72}$$

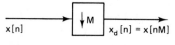

$x[n]$

Sampling
period T

$x_d[n] = x[nM]$

Sampling
period $T' = MT$

Figure 3.17   Representation of downsampler or discrete-time sampler.

To see the relationship between Eqs. (3.72) and (3.70), note that the summation index r in Eq. (3.72) can be expressed as

$$r = i + kM, \tag{3.73}$$

where k and i are integers such that $-\infty < k < \infty$ and $0 \le i \le M - 1$. Clearly r is still an integer ranging from $-\infty$ to ∞, but now Eq. (3.72) can be expressed as

$$X_d(e^{j\omega}) = \frac{1}{M} \sum_{i=0}^{M-1} \left[\frac{1}{T} \sum_{k=-\infty}^{\infty} X_c \left(j\frac{\omega}{MT} - j\frac{2\pi k}{T} - j\frac{2\pi i}{MT} \right) \right]. \tag{3.74}$$

The term inside the square brackets in Eq. (3.74) is recognized from Eq. (3.70) as

$$X(e^{j(\omega - 2\pi i)/M}) = \frac{1}{T} \sum_{k=-\infty}^{\infty} X_c \left(j\frac{\omega - 2\pi i}{MT} - j\frac{2\pi k}{T} \right). \tag{3.75}$$

Thus we can express Eq. (3.74) as

$$X_d(e^{j\omega}) = \frac{1}{M} \sum_{i=0}^{M-1} X(e^{j(\omega/M - 2\pi i/M)}). \tag{3.76}$$

The analogy between Eqs. (3.70) and (3.76) is clear. Equation (3.70) expresses the Fourier transform of the sequence of samples $x[n]$ (period T) in terms of the Fourier transform of the continuous-time signal $x_c(t)$. Equation (3.76) expresses the Fourier transform of the discrete-time sampled sequence $x_d[n]$ (sampling period M) in terms of the Fourier transform of the sequence $x[n]$. If we compare Eqs. (3.71) and (3.76) we see that $X_d(e^{j\omega})$ can be thought of as being composed of either an infinite set of copies of $X_c(j\Omega)$, frequency scaled through $\omega = \Omega T'$ and shifted by integer multiples of $2\pi/T'$ (Eq. 3.71), or M copies of the periodic Fourier transform $X(e^{j\omega})$, frequency scaled by M and shifted by integer multiples of $2\pi/M$ (Eq. 3.76). Either interpretation makes it clear that $X_d(e^{j\omega})$ is periodic with period 2π (as are all discrete-time Fourier transforms) and that aliasing can be avoided by ensuring that $X(e^{j\omega})$ is bandlimited, i.e.,

$$X(e^{j\omega}) = 0, \qquad \omega_N \le |\omega| \le \pi, \tag{3.77}$$

and $2\pi/M \ge 2\omega_N$.

Downsampling is illustrated in Fig. 3.18. Figure 3.18(a) shows the Fourier transform of a bandlimited continuous-time signal, and Fig. 3.18(b) shows the Fourier transform of the impulse train of samples obtained with sampling period T. Figure 3.18(c) shows $X(e^{j\omega})$ and is related to Fig. 3.18(b) through Eq. (3.18). As we have already seen, Figs. 3.18(b) and (c) differ only in a relabeling of the frequency axis. Figure 3.18(d) shows the discrete-time Fourier transform of the downsampled sequence when $M = 2$. We have plotted this Fourier transform as a function of normalized frequency $\omega = \Omega T'$. Finally, Fig. 3.18(e) shows the discrete-time Fourier transform of the downsampled sequence plotted as a function of the continuous-time

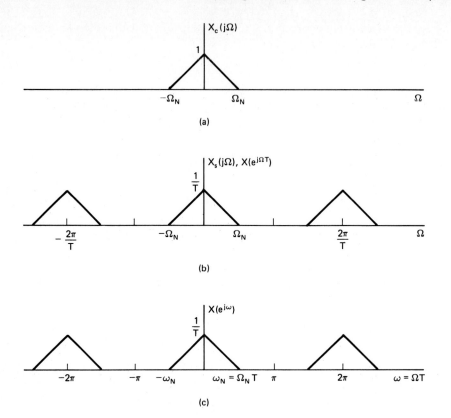

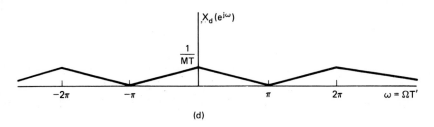

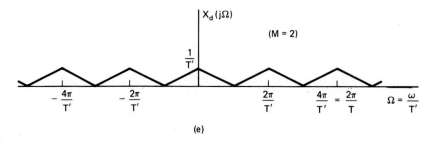

Figure 3.18   Frequency-domain illustration of downsampling.

frequency variable Ω. Figure 3.18(e) is identical to Fig. 3.18(d) except for the relabeling of the frequency axis through the relation $\Omega = \omega/T'$.

In this example, $2\pi/T = 4\Omega_N$, i.e., the original sampling rate is exactly twice the minimum rate to avoid aliasing. Thus, when the original sampled sequence is downsampled by a factor of $M = 2$, no aliasing results. If the downsampling factor is more than 2 in this case, aliasing will result, as illustrated in Fig. 3.19.

Figure 3.19(a) shows the continuous-time Fourier transform of $x_c(t)$, and Fig. 3.19(b) shows the discrete-time Fourier transform of the sequence $x[n] = x_c(nT)$, when $2\pi/T = 4\Omega_N$. Thus, $\omega_N = \Omega_N T = \pi/2$. Now if we downsample by a factor of $M = 3$, we obtain the sequence $x_d[n] = x[3n] = x_c(n3T)$ whose discrete-time Fourier transform is plotted in Fig. 3.19(c) with normalized frequency $\omega = \Omega T'$. Note that because $M\omega_N = 3\pi/2$, which is greater than π, aliasing occurs. In general, to avoid aliasing in downsampling by a factor of M requires that

$$\omega_N M < \pi \quad \text{or} \quad \omega_N < \pi/M. \tag{3.78}$$

If this condition does not hold, aliasing occurs, but it may be tolerable for some applications. In other cases, downsampling can be done without aliasing if we are willing to reduce the bandwidth of the signal $x[n]$ before downsampling. Thus, if $x[n]$ is filtered by an ideal lowpass filter with cutoff frequency π/M, then the output $\tilde{x}[n]$ can be decimated without aliasing, as illustrated in Figs. 3.19(d), (e), and (f). Note that the sequence $\tilde{x}_d[n] = \tilde{x}[nM]$ no longer represents the original underlying continuous-time signal $x_c(t)$. Rather $\tilde{x}_d[n] = \tilde{x}_c(nT')$, where $T' = MT$, and $\tilde{x}_c(t)$ is obtained from $x_c(t)$ by lowpass filtering with cutoff frequency $\Omega_c = \pi/T' = \pi/(MT)$.

From the preceding discussion we see that a general system for downsampling by a factor of M is the one shown in Fig. 3.20. Such a system is called a *decimator*, and downsampling by lowpass filtering followed by compression has been termed *decimation* (Crochiere and Rabiner, 1983).

3.6.2 Increasing the Sampling Rate by an Integer Factor

We have seen that reduction of the sampling rate of a discrete-time signal by an integer factor involves sampling the sequence in a manner analogous to sampling a continuous-time signal. Not surprisingly, increasing the sampling rate involves operations analogous to D/C conversion. To see this, consider a signal $x[n]$ whose sampling rate we wish to increase by a factor of L. If we consider the underlying continuous-time signal $x_c(t)$, the objective is to obtain samples

$$x_i[n] = x_c(nT'), \tag{3.79}$$

where $T' = T/L$, from the sequence of samples

$$x[n] = x_c(nT). \tag{3.80}$$

We will refer to the operation of increasing the sampling rate as *upsampling*.

It is clear from Eqs. (3.79) and (3.80) that

$$x_i[n] = x[n/L] = x_c(nT/L), \quad n = 0, \pm L, \pm 2L, \ldots. \tag{3.81}$$

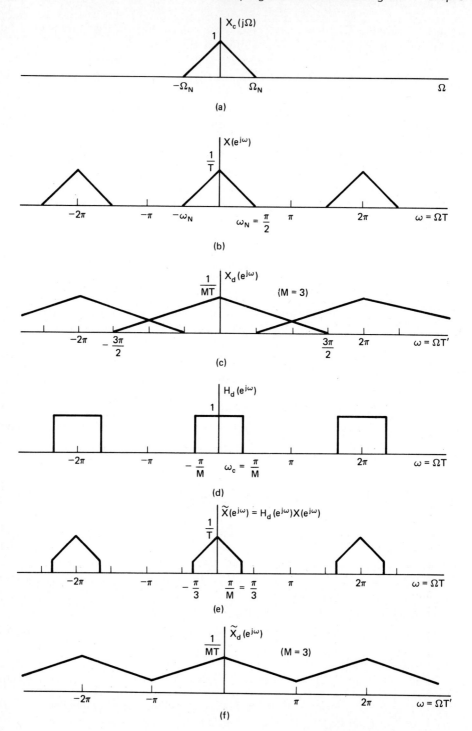

Figure 3.19   (a)–(c) Downsampling with aliasing. (d)–(f) Downsampling with prefiltering to avoid aliasing.

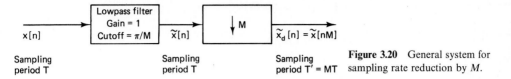

Figure 3.20 General system for sampling rate reduction by M.

Figure 3.21 shows a system for obtaining $x_i[n]$ from $x[n]$ using only discrete-time processing. The system on the left is called a *sampling rate expander* (Crochiere and Rabiner, 1983) or simply an *expander*. Its output is

$$x_e[n] = \begin{cases} x[n/L], & n = 0, \pm L, \pm 2L, \ldots, \\ 0, & \text{otherwise}, \end{cases} \tag{3.82}$$

or, equivalently,

$$x_e[n] = \sum_{k=-\infty}^{\infty} x[k]\delta[n - kL]. \tag{3.83}$$

The system on the right is a lowpass discrete-time filter with cutoff frequency π/L and gain L. This system plays a role similar to the ideal D/C converter in Fig. 3.8(b). First we create a discrete-time impulse train $x_e[n]$ and then we lowpass filter to reconstruct the sequence.

The operation of the system in Fig. 3.21 is most easily understood in the frequency domain. The Fourier transform of $x_e[n]$ can be expressed as

$$X_e(e^{j\omega}) = \sum_{n=-\infty}^{\infty} \left(\sum_{k=-\infty}^{\infty} x[k]\delta[n - kL] \right) e^{-j\omega n}$$

$$\tag{3.84}$$

$$= \sum_{k=-\infty}^{\infty} x[k]e^{-j\omega Lk} = X(e^{j\omega L}).$$

Thus the Fourier transform of the output of the expander is a frequency-scaled version of the Fourier transform of the input, i.e., ω is replaced by ωL so that ω is now normalized by

$$\omega = \Omega T'. \tag{3.85}$$

This effect is illustrated in Fig. 3.22. Figure 3.22(a) shows a bandlimited continuous-time Fourier transform, and Fig. 3.22(b) shows the discrete-time Fourier transform of the sequence $x[n] = x_c(nT)$, where $\pi/T = \Omega_N$. Figure 3.22(c) shows $X_e(e^{j\omega})$ according to Eq. (3.84), with $L = 2$, and Fig. 3.22(e) shows the Fourier transform of the desired signal $x_i[n]$. We see that $X_i(e^{j\omega})$ can be obtained from $X_e(e^{j\omega})$ by correcting the amplitude scale from $1/T$ to $1/T'$ and by removing all the frequency-scaled images of $X_c(j\Omega)$ except at integer multiples of 2π. For the case depicted in Fig. 3.22 this requires a lowpass filter with a gain of 2 and cutoff frequency $\pi/2$, as shown in Fig.

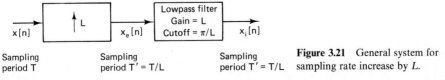

Figure 3.21 General system for sampling rate increase by L.

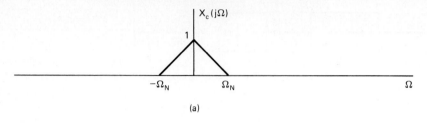

(a)

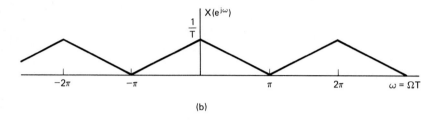

(b)

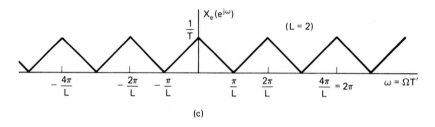

(c)

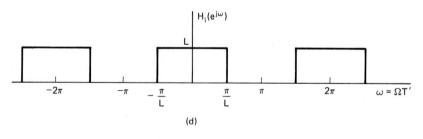

(d)

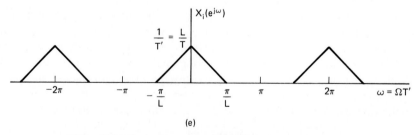

(e)

Figure 3.22   Frequency-domain illustration of interpolation.

3.22(d). In general, the required gain would be L, since $L(1/T) = [1/(T/L)] = 1/T'$, and the cutoff frequency would be π/L.

This example shows that the system of Fig. 3.21 does indeed give an output satisfying Eq. (3.79) if the input sequence $x[n] = x_c(nT)$ was obtained by sampling without aliasing. That system is therefore called an *interpolator* since it fills in the missing samples, and the operation of upsampling is therefore considered to be synonymous with *interpolation*.

As in the case of the D/C converter, it is possible to obtain an interpolation formula for $x_i[n]$ in terms of $x[n]$. First note that the impulse response of the lowpass filter in Fig. 3.21 is

$$h_i[n] = \frac{\sin(\pi n/L)}{\pi n/L}.$$ (3.86)

Using Eq. (3.83), we obtain

$$x_i[n] = \sum_{k=-\infty}^{\infty} x[k] \frac{\sin[\pi(n-kL)/L]}{\pi(n-kL)/L}.$$ (3.87)

The impulse response $h_i[n]$ has the properties

$$h_i[0] = 1,$$
$$h_i[n] = 0, \qquad n = \pm L, \pm 2L, \dots.$$ (3.88)

Thus for the ideal lowpass interpolation filter we have

$$x_i[n] = x[n/L] = x_c(nT/L) = x_c(nT'), \qquad n = 0, \pm L, \pm 2L, \dots,$$ (3.89)

as desired. The fact that $x_i[n] = x_c(nT')$ for all n follows from our frequency-domain argument.

In practice, ideal lowpass filters cannot be implemented exactly, but we will see in Chapter 7 that very good approximations can be designed. (Also see Schafer and Rabiner, 1973, and Oetken et al., 1975.) In some cases very simple interpolation procedures are adequate. Since linear interpolation is often used, although generally it is not very accurate, it is worthwhile to examine linear interpolation in the present context.

Linear interpolation can be accomplished by the system of Fig. 3.21 if the filter has impulse response

$$h_{lin}[n] = \begin{cases} 1 - |n|/L, & |n| \le L, \\ 0, & \text{otherwise}, \end{cases}$$ (3.90)

as shown in Fig. 3.23 for $L = 5$. With this filter, the interpolated output will be

$$x_{lin}[n] = \sum_{k=-\infty}^{\infty} x_e[k]h_{lin}[n-k] = \sum_{k=-\infty}^{\infty} x[k]h_{lin}[n-kL].$$ (3.91)

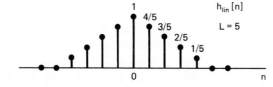

Figure 3.23 Impulse response for linear interpolation.

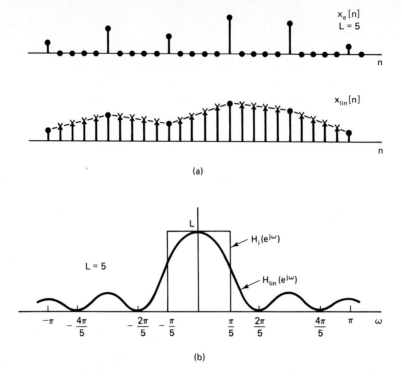

Figure 3.24 (a) Illustration of linear interpolation by filtering. (b) Frequency response of linear interpolator compared with ideal lowpass interpolation filter.

Figure 3.24(a) depicts $x_e[n]$ and $x_{lin}[n]$ for the case $L = 5$. From this figure we see that $x_{lin}[n]$ is identical to the sequence obtained by linear interpolation between the samples. Note that

$$h_{lin}[0] = 1,$$
$$h_{lin}[n] = 0, \qquad n = \pm L, \pm 2L, \ldots, \tag{3.92}$$

so that

$$x_{lin}[n] = x[n/L] \quad \text{at} \quad n = 0, \pm L, \pm 2L, \ldots. \tag{3.93}$$

The amount of distortion in the intervening samples can be gauged by comparing the frequency response of the linear interpolator to the ideal lowpass interpolator for a factor-of-L interpolation. It can be shown (see Problem 3.22) that

$$H_{lin}(e^{j\omega}) = \frac{1}{L}\left[\frac{\sin(\omega L/2)}{\sin(\omega/2)}\right]^2. \tag{3.94}$$

This function is plotted in Fig. 3.24(b) for $L = 5$ together with the ideal lowpass interpolation filter. It is clear from this figure that if the original signal is sampled at the Nyquist rate, linear interpolation will not be very good since the output of the filter will contain considerable energy in the band $\pi/L < |\omega| \le \pi$. However, if the original sampling rate is much higher than the Nyquist rate, then the linear interpolator will be more successful in removing the frequency-scaled images of $X_c(j\Omega)$ at multiples of $2\pi/L$. This is intuitively reasonable since if the original sampling rate greatly exceeds the Nyquist rate, the signal will not vary significantly between samples, and thus linear interpolation should be reasonably accurate.

3.6.3 Changing the Sampling Rate by a Noninteger Factor

We have shown how to increase or decrease the sampling rate of a sequence by an integer factor. By combining decimation and interpolation it is possible to change the sampling rate by a noninteger factor. Specifically, consider Fig. 3.25(a), which shows an interpolator that decreases the sampling period from T to T/L, followed by a decimator, which increases the sampling period by M, producing an output sequence $\tilde{x}_d[n]$ that has an effective sampling period of $T' = TM/L$. By choosing L and M appropriately, we can approach arbitrarily close to any desired ratio of sampling periods. For example, if $L = 100$ and $M = 101$, then $T' = 1.01T$.

If $M > L$ there is a net increase in the sampling period (decrease in sampling rate) and if $M < L$, the opposite is true. From Fig. 3.25(a) we see that the interpolation and decimation filters are in cascade, so that they can be combined as shown in Fig. 3.25(b) into one lowpass filter with gain L and cutoff equal to the minimum of π/L and π/M. If $M > L$, then π/M is the dominant cutoff frequency and there is a net reduction in sampling rate. As pointed out in Section 3.6.1, if $x[n]$ was obtained by sampling at the Nyquist rate, the sequence $\tilde{x}_d[n]$ will represent a lowpass-filtered version of the original underlying bandlimited signal if we are to

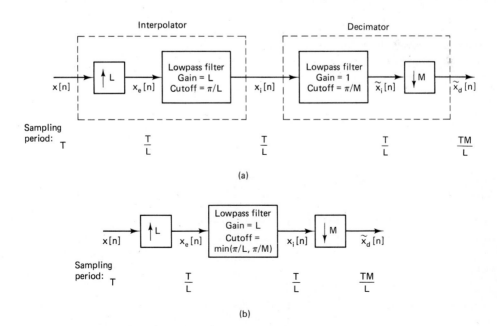

Figure 3.25   (a) System for changing the sampling rate by a noninteger factor. (b) Simplified system in which the decimation and interpolation filters are combined.

avoid aliasing. On the other hand, if $M < L$, then π/L is the dominant cutoff frequency and there will be no need to further limit the bandwidth of the signal below the original Nyquist frequency.

3.7 PRACTICAL CONSIDERATIONS

So far in this chapter our discussions of the representation of continuous-time signals by discrete-time signals have focused on idealized models of periodic sampling and bandlimited interpolation. We have formalized our discussions in terms of an idealized sampling system that we have called the *ideal continuous-to-discrete (C/D) converter* and an idealized bandlimited interpolator system called the *ideal discrete-to-continuous (D/C) converter*. These idealized conversion systems allow us to concentrate on the essential mathematical details of the relationship between a bandlimited signal and its samples. For example, in Section 3.4 we used the idealized C/D and D/C conversion systems to show that linear time-invariant discrete-time systems can be used in the configuration of Fig. 3.26(a) to implement linear time-invariant continuous-time systems if the input is bandlimited and the sampling rate exceeds the Nyquist rate. In a practical setting, continuous-time signals are not precisely band-limited, ideal filters cannot be realized, and the ideal C/D and D/C converters can only be approximated. The block diagram of Fig. 3.26(b) shows a more realistic model for digital processing of continuous-time (analog) signals. In this section we will examine some of the imperfections introduced by each of the components of the system in Fig. 3.26(b). While it is difficult to carry out an exact analysis, our approximate analysis will indicate the effects of each imperfection and suggest that the system of Fig. 3.26(b) can very closely approximate the system of Fig. 3.26(a).

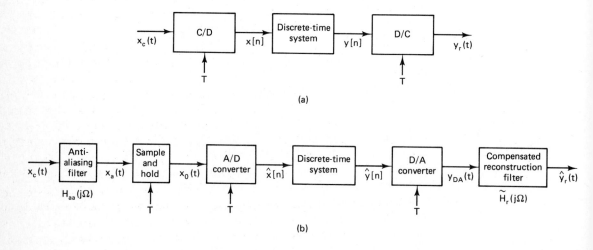

Figure 3.26 (a) Discrete-time filtering of continuous-time signals. (b) A more realistic model.

3.7.1 Prefiltering to Avoid Aliasing

In most cases, it is desirable to minimize the sampling rate of a discrete-time system for processing analog systems. This is because the amount of arithmetic processing required to implement the system is proportional to the number of samples to be processed. If the input is not bandlimited or if the Nyquist frequency of the input is too high, prefiltering is often used. An example of such a situation occurs in processing speech signals, where often only the low-frequency band up to about 3–4 kHz is required for intelligibility even though the speech signal may have significant frequency content in the 4–20 kHz range. Also, even if the signal is naturally bandlimited, wideband additive noise may fill in the higher-frequency range, and as a result of sampling, these noise components would be aliased into the low-frequency band. If we wish to avoid aliasing, the input signal must be forced to be bandlimited to frequencies below one-half the desired sampling rate. This can be accomplished by lowpass filtering the continuous-time signal prior to C/D conversion, as shown in Fig. 3.27. In this context the lowpass filter that precedes the C/D converter is called an *anti-aliasing filter*. Ideally, the frequency response of the anti-aliasing filter would be

$$H_{aa}(j\Omega) = \begin{cases} 1, & |\Omega| < \Omega_c < \pi/T, \\ 0, & |\Omega| > \Omega_c. \end{cases} \tag{3.95}$$

From the discussion of Section 3.4.1, it follows that the overall system from the output of the anti-aliasing filter $x_a(t)$ to the output $y_r(t)$ will always behave as a linear time-invariant system, since the input to the C/D converter $x_a(t)$ is forced by the anti-aliasing filter to be bandlimited to frequencies below π/T radians/s. Thus the overall effective frequency response of Fig. 3.27 will be the product of $H_{aa}(j\Omega)$ and the effective frequency response from $x_a(t)$ to $y_r(t)$. Combining Eqs. (3.95) and (3.37) gives

$$H_{eff}(j\Omega) = \begin{cases} H(e^{j\Omega T}), & |\Omega| < \pi/T, \\ 0, & |\Omega| > \pi/T. \end{cases} \tag{3.96}$$

Thus for an ideal lowpass anti-aliasing filter, the system of Fig. 3.27 behaves as a linear time-invariant system with frequency response given by Eq. (3.96) even when $X_c(j\Omega)$ is not bandlimited. In practice, the frequency response $H_{aa}(j\Omega)$ cannot be ideally bandlimited, but $H_{aa}(j\Omega)$ can be made small for $|\Omega| > \pi/T$ so that aliasing is minimized. In this case, the overall frequency response of the system in Fig. 3.27 would be approximately

$$H_{eff}(j\Omega) \approx H_{aa}(j\Omega)H(e^{j\Omega T}). \tag{3.97}$$

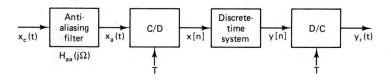

Figure 3.27   Use of prefiltering to avoid aliasing.

To achieve negligibly small frequency response above π/T it would be necessary for $H_{aa}(j\Omega)$ to begin to "roll off," i.e., begin to introduce attenuation, at frequencies below π/T. Equation (3.97) suggests that the rolloff of the anti-aliasing filter (and other linear time-invariant distortions to be discussed later) could be at least partially compensated for by taking them into account in the design of the discrete-time system. This is illustrated in Problem 3.32.

3.7.2 Analog-to-Digital (A/D) Conversion

An ideal C/D converter converts a continuous-time signal into a discrete-time signal, where each sample is known with infinite precision. As an approximation to this for digital signal processing, the system of Fig. 3.28 converts a continuous-time (analog) signal into a digital signal, i.e., a sequence of finite-precision or quantized samples. The two systems in Fig. 3.28 correspond to systems that are available as physical devices. The A/D converter is a physical device that converts a voltage or current amplitude at its input into a binary code representing a quantized amplitude value closest to the amplitude of the input. Under the control of an external clock, the A/D converter can be caused to start and complete an A/D conversion every T seconds. However, the conversion is not instantaneous, and for this reason a high-performance A/D system typically includes a sample and hold as in Fig. 3.28. The ideal sample-and-hold system is the system whose output is

$$x_0(t) = \sum_{n=-\infty}^{\infty} x[n]h_0(t - nT), \tag{3.98}$$

where $x[n] = x_a(nT)$ are the ideal samples of $x_a(t)$, and $h_0(t)$ is the impulse response of the zero-order-hold system, i.e.,

$$h_0(t) = \begin{cases} 1, & 0 < t < T, \\ 0, & \text{otherwise.} \end{cases} \tag{3.99}$$

If we note that Eq. (3.98) has the equivalent form

$$x_0(t) = h_0(t) * \sum_{n=-\infty}^{\infty} x_a(nT)\delta(t - nT), \tag{3.100}$$

we see that the ideal sample and hold is equivalent to impulse train modulation followed by linear filtering with the zero-order-hold system as depicted in Fig. 3.29(a). The relationship between the Fourier transform of $x_0(t)$ and the Fourier transform of $x_a(t)$ can be worked out following the style of analysis of Section 3.2, and

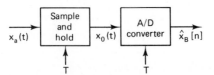

Figure 3.28 Physical configuration for analog-to-digital conversion.

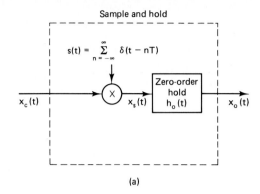

Sample and hold

$$s(t) = \sum_{n=-\infty}^{\infty} \delta(t - nT)$$

Zero-order hold $h_o(t)$

$x_c(t)$ × $x_s(t)$ $x_o(t)$

(a)

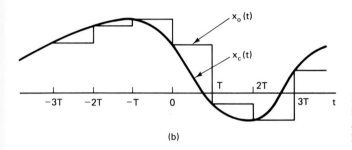

$x_o(t)$

$x_c(t)$

−3T   −2T   −T   0   T   2T   3T   t

(b)

Figure 3.29 (a) Representation of an ideal sample and hold. (b) Typical input and output signals for the sample and hold.

we will do a similar analysis when we discuss the D/A converter. However, the analysis is unnecessary at this point since everything we need to know about the behavior of the system can be seen from the time-domain expression. Specifically, the output of the zero-order hold is a staircase waveform where the sample values are held constant during the sampling period of T seconds. This is illustrated in Fig. 3.29(b). Physical sample-and-hold circuits are designed to sample $x_a(t)$ as nearly instantaneously as possible and to hold the sample value as nearly constant as possible until the next sample is taken. The purpose of this is to provide the constant input voltage (or current) required by the A/D converter. The details of the wide variety of A/D conversion processes and the details of sample-and-hold and A/D circuit implementations are outside the scope of this book. Many practical issues arise in obtaining a sample and hold that samples quickly and holds the sample value constant with no decay or "glitches." Likewise, many practical concerns dictate the speed and accuracy of conversion of A/D converter circuits. Such questions are considered in Hnatek (1976) and Schmid (1976), and details of performance of a specific product are available in manufacturers' specification and data sheets. Our concern in this section is the analysis of the quantization effects in A/D conversion.

Since the purpose of the sample and hold in Fig. 3.28 is to implement ideal sampling and to hold the sample value for quantization by the A/D converter, we can represent the system of Fig. 3.28 by the system of Fig. 3.30, where the ideal C/D converter represents the sampling performed by the sample and hold, and, as we will describe later, the quantizer and coder together represent the operation of the A/D converter.

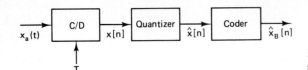

Figure 3.30 Mathematical representation of the system in Fig. 3.28.

   The quantizer is a nonlinear system whose purpose is to transform the input sample $x[n]$ into one of a finite set of prescribed values. We represent this operation as

$$\hat{x}[n] = Q(x[n]) \tag{3.101}$$

and refer to $\hat{x}[n]$ as the *quantized sample*. Quantizers can be defined with either uniformly or nonuniformly spaced quantization levels; however, in the context of signal processing the quantization steps usually are uniform. Figure 3.31 shows a typical uniform quantizer characteristic† in which the sample values are *rounded* to the nearest quantization level.

   Several features of Fig. 3.31 should be emphasized. First note that this quantizer would be appropriate for a signal whose samples are both positive and negative (bipolar). If it is known that the input samples are always positive (or negative), then a different distribution of the quantization levels would be appropriate. Next note that the quantizer of Fig. 3.31 has an even number of quantization levels. With an even number of levels, it is not possible to have a quantization level at zero amplitude and also have an equal number of positive and negative quantization levels. Generally the number of quantization levels will be a power of 2, but the number will be much greater than eight so this difference is usually inconsequential.

   Figure 3.31 also depicts coding of the quantization levels. Since there are eight quantization levels, we can label them by a binary code of 3 bits. (In general 2^{B+1} levels can be coded with a $(B + 1)$-bit binary code.) In principle, any assignment of symbols can be used, and many binary coding schemes exist, each with its own advantages and disadvantages depending on the application. For example, the right-hand column of binary numbers in Fig. 3.31 illustrates the *offset binary* coding scheme, in which the binary symbols are assigned in numeric order starting with the most negative quantization level. However, in digital signal processing, we generally wish to use a binary code that permits us to do arithmetic directly with the code words as scaled representations of the quantized samples.

   The left-hand column in Fig. 3.31 shows an assignment according to the two's complement binary number system. This very convenient system for representing signed numbers is used in most computers and microprocessors and thus it is perhaps the most convenient labeling of the quantization levels. Note incidentally that the offset binary code can be converted to two's complement code by simply complementing the most significant bit.

   In the two's-complement system, the leftmost or most significant bit is considered as the sign bit, and we can consider the remaining bits as representing either binary integers or fractions. We will assume the latter; i.e., we assume a binary

---

   † Such quantizers are also called *linear* quantizers because of the linear progression of quantization steps.

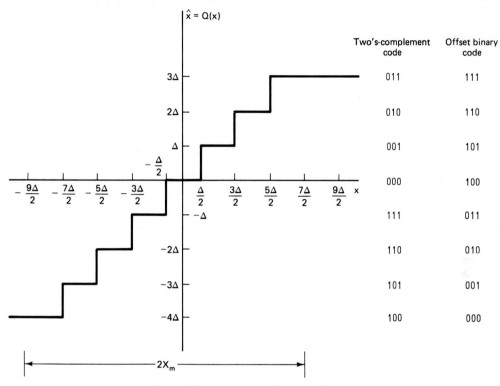

Figure 3.31   Typical quantizer for A/D conversion.

fraction point between the two most significant bits. Then for the two's-complement interpretation, the binary symbols have the following meaning for $B = 2$:

Binary Symbol	Numeric Value, $\hat{x}_B$
$0_\wedge 1\ 1$	3/4
$0_\wedge 1\ 0$	1/2
$0_\wedge 0\ 1$	1/4
$0_\wedge 0\ 0$	0
$1_\wedge 1\ 1$	- 1/4
$1_\wedge 1\ 0$	$-1/2$
$1_\wedge 0\ 1$	$-3/4$
$1_\wedge 0\ 0$	-1

In general if we have a $(B + 1)$-bit binary two's-complement fraction of the form

$$a_0 a_1 a_2 \ldots a_B,$$

then its value is

$$-a_0 2^0 + a_1 2^{-1} + a_2 2^{-2} + \cdots + a_B 2^{-B}.$$

The relationship between the code words and the quantized levels depends on the parameter X_m in Fig. 3.31. In general this parameter is called the full-scale level of the A/D converter. Typical values are 10, 5, or 1 volt. From Fig. 3.31 we see that the step size Δ of the quantizer would in general be

$$\Delta = \frac{2X_m}{2^{B+1}} = \frac{X_m}{2^B}.$$ (3.102)

The smallest quantization levels ($\pm\Delta$) correspond to the least significant bit of the binary code word. Furthermore, the numeric relationship between the code words and the quantized samples is

$$\hat{x}[n] = X_m \hat{x}_B[n]$$ (3.103)

since we have assumed that $\hat{x}_B[n]$ is a binary number such that $-1 \leq \hat{x}_B[n] < 1$ (for two's complement). In this scheme the binary coded samples $\hat{x}_B[n]$ are directly proportional to the quantized samples (in two's complement binary), and therefore they can be used as a numeric representation of the amplitude of the samples. Indeed, it is generally appropriate to assume that the input signal is normalized to X_m, so that the numeric values of $\hat{x}[n]$ and $\hat{x}_B[n]$ are identical and there is no need to distinguish between the quantized samples and the binary coded samples.

Figure 3.32 shows a simple example of quantization and coding of the samples of a sine wave using a 3-bit quantizer. The unquantized samples $x[n]$ are shown with solid dots and the quantized samples $x[n]$ are shown with open circles. Also shown is

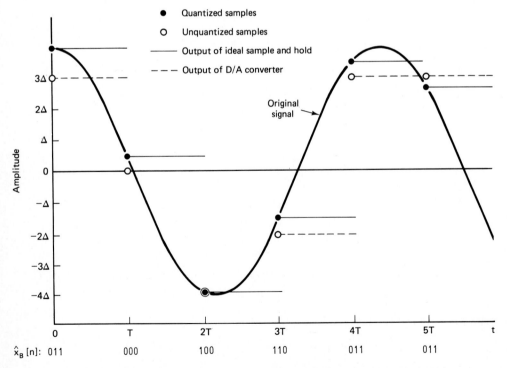

Figure 3.32 Sampling, quantization, coding, and D/A conversion with a 3-bit quantizer.

the output of an ideal sample and hold. The dotted lines labeled "output of D/A converter" will be discussed later. Figure 3.32 also shows the 3-bit code words that represent each sample. Note that since the analog input $x_a(t)$ exceeds the full-scale value of the quantizer, some of the positive samples are "clipped."

Although much of the preceding discussion pertains to two's-complement coding of the quantization levels, the basic principles of quantization and coding in A/D conversion are the same regardless of the binary code used to represent the samples. A more detailed discussion of the binary arithmetic systems used in digital computing can be found in texts on computer arithmetic (see, for example, Knuth, 1981; Cavanagh, 1984). We now turn to an analysis of the effect of quantization. Since this analysis does not depend on the assignment of binary code words, it will lead to rather general conclusions.

3.7.3 Analysis of Quantization Errors

From Figs. 3.31 and 3.32 we see that the quantized sample $\hat{x}[n]$ will in general be different from the true sample value $x[n]$. The difference between them is the *quantization error*, defined as

$$e[n] = \hat{x}[n] - x[n]. \tag{3.104}$$

For example, for the 3-bit quantizer of Fig. 3.31, if $\Delta/2 < x[n] \le 3\Delta/2$, then $\hat{x}[n] = \Delta$, and it follows that

$$-\Delta/2 < e[n] \le \Delta/2. \tag{3.105}$$

In the case of Fig. 3.31, Eq. (3.105) holds whenever

$$-9\Delta/2 < x[n] \le 7\Delta/2. \tag{3.106}$$

In the general case of a $(B + 1)$-bit quantizer with Δ given by Eq. (3.102), the quantization error satisfies Eq. (3.105) whenever

$$(-X_m - \Delta/2) < x[n] \le (X_m - \Delta/2). \tag{3.107}$$

If $x[n]$ is outside this range, as it is for the sample at $t = 0$ in Fig. 3.32, then the quantization error is larger in magnitude than $\Delta/2$ and such samples are said to be clipped.

A simplified but useful model of the quantizer is depicted in Fig. 3.33. In this model the quantization error samples are thought of as an additive noise signal. This model is exactly equivalent to the quantizer if we know $e[n]$. In most cases $e[n]$ is not known, and a statistical model based on Fig. 3.33 may then be useful in representing

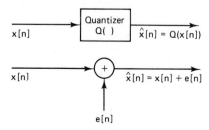

Figure 3.33   Additive noise model for quantizer.

the effects of quantization. We will also use such a model in Chapters 6 and 9 to describe the effects of quantization in signal processing algorithms. The statistical representation of quantization errors is based on the following assumptions:

1. The error sequence $e[n]$ is a sample sequence of a stationary random process.
2. The error sequence is uncorrelated with the sequence $x[n]$.
3. The random variables of the error process are uncorrelated; i.e., the error is a white-noise process.
4. The probability distribution of the error process is uniform over the range of quantization error.

As we will see, these assumptions lead to a rather simple analysis of quantization effects. It is easy to find situations where these assumptions are clearly not valid. For example, if $x_a(t)$ is a step function, the assumptions would not be justified. However, when the signal is a complicated signal, such as speech or music where the signal fluctuates rapidly in a somewhat unpredictable manner, these assumptions are more realistic. Experiments have shown that as the signal becomes more complicated, the measured correlation between the signal and the quantization error decreases, and the error also becomes uncorrelated (see Bennett, 1948; Widrow, 1956, 1961). In a heuristic sense, the assumptions of the statistical model appear to be valid if the signal is sufficiently complex and the quantization steps sufficiently small so that the amplitude of the signal is likely to traverse many quantization steps from sample to sample.

As an illustration, Fig. 3.34(a) shows a sequence of unquantized samples of a speech waveform. Figures 3.34(b) and 3.34(c) show the quantization-noise samples after quantization of the samples of Fig. 3.34(a) with an 8-bit and a 3-bit quantizer, respectively. In the 8-bit case, the quantization noise has been magnified by a factor of 32. Comparison of the noise sequences with the signal supports the preceding assertions about the quantization-noise properties in the finely quantized (8-bit) case. In Chapter 11 this will be demonstrated further when we calculate the power density spectrum and autocorrelation of a quantization noise sequence.

For quantizers that round the sample value to the nearest quantization level, as shown in Fig. 3.31, the amplitude of the quantization noise is in the range

$$-\Delta/2 < e[n] \le \Delta/2. \tag{3.108}$$

For small Δ, it is reasonable to assume that $e[n]$ is a random variable uniformly distributed from $-\Delta/2$ to $\Delta/2$. Therefore the first-order probability density for the quantization noise is as shown in Fig. 3.35. (If truncation rather than rounding is used in implementing quantization, then the error would always be negative and we would assume a uniform probability density from $-\Delta$ to 0.) To complete the statistical model for the quantization noise, we assume that successive noise samples are uncorrelated with each other and that $e[n]$ is uncorrelated with $x[n]$. Thus $e[n]$ is assumed to be a uniformly distributed white-noise sequence. The mean value of $e[n]$ is zero and its variance is

$$\sigma_e^2 = \frac{\Delta^2}{12}. \tag{3.109}$$

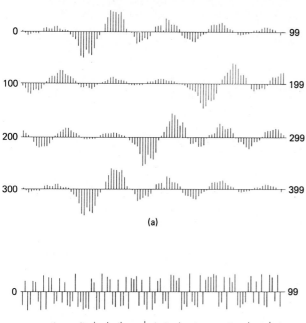

(a)

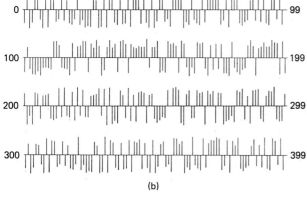

(b)

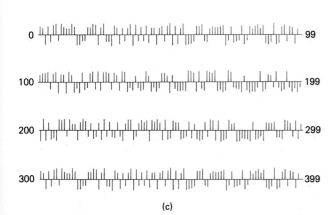

(c)

Figure 3.34 Example of quantization noise. Quantization of the speech waveform in part (a) with an 8-bit and 3-bit quantizer results in the quantization noise waveforms shown in parts (b) and (c), respectively. The quantization noise in (b) is magnified by a factor of 5 relative to (a). The quantization noise in (c) is magnified by a factor of 100 relative to (a).

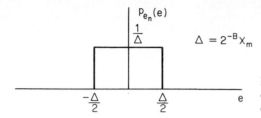

Figure 3.35 Probability density function of quantization error for a rounding quantizer such as in Fig. 3.31.

For a $(B + 1)$-bit quantizer with full-scale value X_m, the noise variance or power is

$$\sigma_e^2 = \frac{2^{-2B}X_m^2}{12}. \tag{3.110}$$

A common measure of the amount of degradation of a signal by additive noise is the signal-to-noise ratio, defined as the ratio of signal variance (power) to noise variance. Expressed in decibels, the signal-to-noise ratio of a $(B + 1)$-bit quantizer is

$$\text{SNR} = 10 \log_{10}\left(\frac{\sigma_x^2}{\sigma_e^2}\right) = 10 \log_{10}\left(\frac{12 \cdot 2^{2B}\sigma_x^2}{X_m^2}\right)$$

$$= 6.02B + 10.8 - 20 \log_{10}\left(\frac{X_m}{\sigma_x}\right). \tag{3.111}$$

From Eq. (3.111) we see that the signal-to-noise ratio increases approximately 6 dB for each bit added to the wordlength of the quantized samples, i.e., for each doubling of the number of quantization levels. It is particularly instructive to consider the term

$$- 20 \log_{10}\left(\frac{X_m}{\sigma_x}\right) \tag{3.112}$$

in Eq. (3.111). First recall that X_m is a parameter of the quantizer and it would usually be fixed in a practical system. The quantity σ_x is the rms value of the signal amplitude, and it would necessarily be less than the peak amplitude of the signal. For example, if $x_a(t)$ is a sine wave of peak amplitude X_p, then $\sigma_x = X_p/\sqrt{2}$. If σ_x is too large, the peak signal amplitude will exceed the full-scale amplitude X_m of the A/D converter. In this case Eq. (3.111) is no longer valid and severe distortion results. If, on the other hand, σ_x is too small, then the term in Eq. (3.112) will become large and negative, thereby decreasing the signal-to-noise ratio in Eq. (3.111). In fact, it is easily seen that when σ_x is halved, SNR decreases by 6 dB. Thus it is very important that the signal amplitude be carefully matched to the full-scale amplitude of the A/D converter.

For analog signals such as speech or music, the distribution of amplitudes tends to be concentrated about zero and falls off rapidly with increasing amplitude. In such cases, the probability that the magnitude of a sample will exceed three or four times the rms value is very low. For example, if the signal amplitude has a Gaussian distribution, only 0.064 percent of the samples would have an amplitude greater than $4\sigma_x$. Thus, to avoid peak clipping (as assumed in our statistical model), we might set the gain of filters and amplifiers preceding the A/D converter so that $\sigma_x = X_m/4$. Using this value of σ_x in Eq. (3.111) gives

$$\text{SNR} \approx 6B - 1.25 \text{ dB}. \tag{3.113}$$

For example, obtaining a signal-to-noise ratio of about 90–96 dB for use in high-quality music recording and playback requires 16-bit quantization, but it should be remembered that such performance is obtained only if the input signal is carefully matched to the full-scale range of the A/D converter.

This tradeoff between peak signal amplitude and absolute size of the quantization noise is fundamental to any quantization process. We will see its importance again in Chapter 6 when we discuss roundoff noise in implementing discrete-time linear systems.

3.7.4 D/A Conversion

In Section 3.3 we discussed how a bandlimited signal can be reconstructed from a sequence of samples using ideal lowpass filtering. In terms of Fourier transforms, the reconstruction is represented as

$$X_r(j\Omega) = X(e^{j\Omega T})H_r(j\Omega), \tag{3.114}$$

where $X(e^{j\omega})$ is the discrete-time Fourier transform of the sequence of samples and $X_r(j\Omega)$ is the Fourier transform of the reconstructed continuous-time signal. The ideal reconstruction filter is

$$H_r(j\Omega) = \begin{cases} T, & |\Omega| < \pi/T, \\ 0, & |\Omega| > \pi/T. \end{cases} \tag{3.115}$$

In the time domain the corresponding relation between $x_r(t)$ and $x[n]$, where $H_r(j\Omega)$ is the ideal reconstruction filter, is

$$x_r(t) = \sum_{n=-\infty}^{\infty} x[n] \frac{\sin[\pi(t - nT)/T]}{\pi(t - nT)/T}. \tag{3.116}$$

The system that takes the sequence $x[n]$ as input and produces $x_r(t)$ as output is called the *ideal D/C converter*. A physically realizable counterpart to the ideal D/C converter is a *digital-to-analog converter* (D/A converter) followed by an approximate lowpass filter. As depicted in Fig. 3.36(a), a D/A converter takes a sequence of binary code words as its input and produces a continuous-time output of the form

$$x_{DA}(t) = \sum_{n=-\infty}^{\infty} X_m \hat{x}_B[n] h_0(t - nT)$$

$$= \sum_{n=-\infty}^{\infty} \hat{x}[n] h_0(t - nT), \tag{3.117}$$

where $h_0(t)$ is the impulse response of the zero-order hold given by Eq. (3.99). The dotted lines in Fig. 3.32 show the output of a D/A converter for the quantized examples of the sine wave. Note that the D/A converter holds the quantized sample

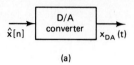

(a)

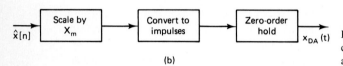

(b)

Figure 3.36 (a) Block diagram of D/A converter. (b) Representation in terms of a zero-order hold.

for one sample period in the same way that the sample and hold holds the unquantized input sample. If we use the additive-noise model to represent the effects of quantization, Eq. (3.117) becomes

$$x_{DA}(t) = \sum_{n=-\infty}^{\infty} x[n]h_0(t - nT) + \sum_{n=-\infty}^{\infty} e[n]h_0(t - nT). \qquad (3.118)$$

To simplify our discussion, we define

$$x_0(t) = \sum_{n=-\infty}^{\infty} x[n]h_0(t - nT), \qquad (3.119a)$$

$$e_0(t) = \sum_{n=-\infty}^{\infty} e[n]h_0(t - nT), \qquad (3.119b)$$

so that Eq. (3.118) can be written as

$$x_{DA}(t) = x_0(t) + e_0(t). \qquad (3.120)$$

The signal component $x_0(t)$ is related to the input signal $x_a(t)$ since $x[n] = x_a(nT)$. The noise signal $e_0(t)$ depends on the quantization-noise samples $e[n]$ in the same way that $x_0(t)$ depends on the unquantized signal samples. The Fourier transform of Eq. (3.119a) is

$$X_0(j\Omega) = \sum_{n=-\infty}^{\infty} x[n]H_0(j\Omega)e^{-j\Omega nT}$$

$$= \left(\sum_{n=-\infty}^{\infty} x[n]e^{-j\Omega Tn} \right) H_0(j\Omega) \qquad (3.121)$$

$$= X(e^{j\Omega T})H_0(j\Omega),$$

Now since

$$X(e^{j\Omega T}) = \frac{1}{T} \sum_{k=-\infty}^{\infty} X_a(j\Omega + j2\pi k/T), \qquad (3.122)$$

it follows that

$$X_0(j\Omega) = \left[\frac{1}{T} \sum_{k=-\infty}^{\infty} X_a(j\Omega + j2\pi k/T) \right] H_0(j\Omega). \qquad (3.123)$$

If $X_a(j\Omega)$ is bandlimited to frequencies below π/T, the shifted copies of $X_a(j\Omega)$ do not overlap in Eq. (3.123), and if we define a compensated reconstruction filter as

$$\tilde{H}_r(j\Omega) = \frac{H_r(j\Omega)}{H_0(j\Omega)}, \tag{3.124}$$

then the output of the filter will be $x_a(t)$ if the input is $x_0(t)$. The frequency response of the zero-order-hold filter is easily shown to be

$$H_0(j\Omega) = \frac{2\sin(\Omega T/2)}{\Omega} e^{-j\Omega T/2}. \tag{3.125}$$

Therefore the compensated reconstruction filter is

$$\tilde{H}_r(j\Omega) = \begin{cases} \dfrac{\Omega T/2}{\sin(\Omega T/2)} e^{j\Omega T/2}, & |\Omega| < \pi/T, \\ 0, & |\Omega| > \pi/T. \end{cases} \tag{3.126}$$

Figure 3.37(a) shows $|H_0(j\Omega)|$ as given by Eq. (3.125) compared with the ideal interpolation filter $H_r(j\Omega)$ as given by Eq. (3.115). Both filters have a gain of T at $\Omega = 0$, but the zero-order hold, although lowpass in nature, does not cut off sharply at $\Omega = \pi/T$. Figure 3.37(b) shows the magnitude of the frequency response of the ideal compensated reconstruction filter to be used following a zero-order-hold reconstruction system such as a D/A converter. The phase response would ideally correspond to an advance time shift of $T/2$ s to compensate for the delay of that amount introduced

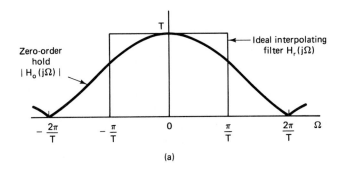

(a)

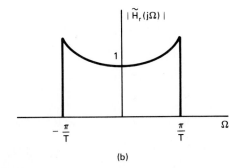

(b)

Figure 3.37 (a) Frequency response of zero-order hold compared with ideal interpolating filter. (b) Ideal compensated reconstruction filter for use with a zero-order hold output.

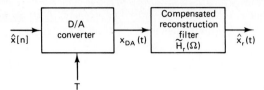

Figure 3.38 Physical configuration for digital-to-analog conversion.

by the zero-order hold. Since such time advance cannot be realized in practical real-time approximations to the ideal compensated reconstruction filter, only the magnitude response would normally be compensated, and often even this compensation is neglected since the gain of the zero-order hold drops to only $2/\pi$ (or -4 dB) at $\Omega = \pi/T$.

Figure 3.38 shows a D/A converter followed by an ideal compensated reconstruction filter. As can be seen from the preceding discussion, with the ideal compensated reconstruction filter following the D/A converter, the reconstructed output signal would be

$$\hat{x}_r(t) = \sum_{n=-\infty}^{\infty} \hat{x}[n] \frac{\sin[\pi(t-nT)/T]}{\pi(t-nT)/T}$$

$$= \sum_{n=-\infty}^{\infty} x[n] \frac{\sin[\pi(t-nT)/T]}{\pi(t-nT)/T} + \sum_{n=-\infty}^{\infty} e[n] \frac{\sin[\pi(t-nT)/T]}{\pi(t-nT)/T}. \qquad (3.127)$$

In other words, the output would be

$$\hat{x}_r(t) = x_a(t) + e_a(t), \qquad (3.128)$$

where $e_a(t)$ would be a bandlimited white-noise signal.

Returning to a consideration of Fig. 3.26(b), we are now in a position to understand the behavior of systems for digital processing of analog signals. If we assume that the output of the anti-aliasing filter is bandlimited to frequencies below π/T, that $\tilde{H}_r(j\Omega)$ is similarly bandlimited, and that the discrete-time system is linear and time-invariant, then the output of the overall system will be of the form

$$\hat{y}_r(t) = y_a(t) + e_a(t), \qquad (3.129)$$

where

$$Y_a(j\Omega) = \tilde{H}_r(j\Omega)H_0(j\Omega)H(e^{j\Omega T})H_{aa}(j\Omega)X_c(j\Omega), \qquad (3.130)$$

where $H_{aa}(j\Omega)$, $H_0(j\Omega)$ and $\tilde{H}_r(j\Omega)$ are the frequency responses of the anti-aliasing filter, the zero-order hold, and the reconstruction lowpass filter, respectively. $H(e^{j\Omega T})$ is the frequency response of the discrete-time system. Similarly, assuming that the quantization noise introduced by the A/D converter is white noise with variance $\sigma_e^2 = \Delta^2/12$, it can be shown that the power spectrum of the output noise is

$$P_{e_a}(j\Omega) = |\tilde{H}_r(j\Omega)H_0(j\Omega)H(e^{j\Omega T})|^2\sigma_e^2, \qquad (3.131)$$

i.e., the input quantization noise is changed by the successive stages of discrete- and continuous-time filtering. From Eq. (3.130) it follows that under the assumption of negligible aliasing, the overall effective frequency response from $x_c(t)$ to $\hat{y}_r(t)$ is

$$H_{eff}(j\Omega) = \tilde{H}_r(j\Omega)H_0(j\Omega)H(e^{j\Omega T})H_{aa}(j\Omega). \qquad (3.132)$$

If the anti-aliasing filter is ideal, as in Eq. (3.95), and if the compensation of the reconstruction filter is ideal, as in Eq. (3.126), then the effective frequency response is as given in Eq. (3.96). Otherwise Eq. (3.132) provides a reasonable model for the effective response. Note that Eq. (3.132) suggests that compensation for imperfections in any of the four terms in Eq. (3.130) can in principle be included in any of the other terms; e.g., the discrete-time system can include appropriate compensation for the anti-aliasing filter or the zero-order hold or the reconstruction filter or all of these.

   In addition to the filtering supplied by Eq. (3.132), Eq. (3.129) reminds us that the output will also be contaminated by the filtered quantization noise. In Chapter 6 we will see that noise can also be introduced in the implementation of the discrete-time linear system.

3.7.5 Application of Decimation and Interpolation to A/D and D/A Conversion

Throughout this section we have discussed idealized filters for preventing aliasing and for reconstructing a smoothly interpolated analog signal from the output of a D/A converter. Our underlying assumption has been that these filters could be adequately approximated, and indeed it is true that such sharp cutoff analog filters can be realized using active networks and integrated circuits. However, in applications involving powerful but inexpensive digital processors, the continuous-time filters may account for a major part of the cost of a system for discrete-time processing of analog signals. Sharp cutoff filters are difficult and expensive to implement, and if the system is to operate with a variable sampling rate, adjustable filters would be required. Thus, it is desirable to eliminate the continuous-time filters or simplify the requirements on them.

   One approach is depicted in Fig. 3.39. First, the input continuous-time signal is sampled at a very high sampling rate—far above the Nyquist rate of the signal. Often such high sampling rates permit very inexpensive A/D conversion for low-bandwidth signals, and a very simple lowpass filter is sufficient to avoid aliasing. Then, as indicated in Fig. 3.39(a), the sampling rate can be reduced to a sampling rate as low as the Nyquist rate by the decimation techniques discussed in Section 3.6. As specified in Fig. 3.20, the sampling rate reduction must include a sharp cutoff digital filter. Any desired digital processing can then be done at the low sampling rate to minimize computation, as depicted in Fig. 3.39(b). If D/A conversion is required, the output of the discrete-time processor can be interpolated to a much higher sampling rate and then converted to a continuous-time output by using a D/A converter followed by a very simple reconstruction filter, since the discrete-time interpolation filter ensures that the output will be bandlimited to frequencies much lower than the output sampling frequency. This is depicted in Fig. 3.39(c). Note also that the compensation

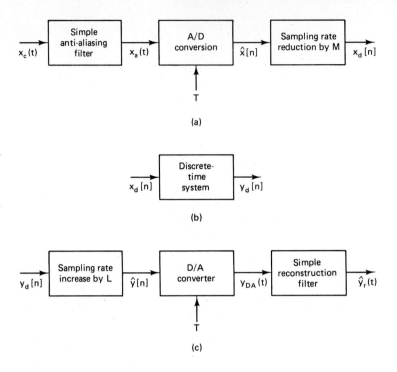

Figure 3.39 The requirements on the analog lowpass filters in sampling and reconstruction can be relaxed by use of a sampling rate much higher than the Nyquist rate. The sampling rate reduction in part (a) is done as indicated in Fig. 3.20. The sampling rate increase in part (c) is done as indicated in Fig. 3.21.

for the zero-order hold and the rolloff of the simple reconstruction filter can be in the filter required for the sampling rate increase.

The use of decimation in A/D conversion is illustrated in Fig. 3.40. Figure 3.40(a) shows the Fourier transform of a signal that occupies the band $|\Omega| < \Omega_N$ plus the Fourier transform of what might correspond to high-frequency "noise." Also shown (dotted line) is the frequency response of a simple anti-aliasing filter, which does not cut off sharply but eventually falls to zero at frequencies above a frequency Ω_c. Figure 3.40(b) shows the Fourier transform of the output of this filter. If the signal $x_a(t)$ is sampled with period T such that $(2\pi/T - \Omega_c) > \Omega_N$, then the discrete-time Fourier transform of the sequence $\hat{x}[n]$ will be as shown in Fig. 3.40(c). Note that the "noise" will be aliased, but aliasing will not affect the signal band $|\omega| < \omega_N = \Omega_N T$. Now if T and T' are chosen so that $T' = MT$ and $\pi/T' = \Omega_N$, then $\hat{x}[n]$ can be filtered by a sharp cutoff discrete-time filter (shown idealized in Fig. 3.40c) with unity gain and cutoff frequency π/M. The output of the discrete-time filter can be downsampled by M to obtain the sampled sequence $x_d[n]$ whose Fourier transform is shown in Fig. 3.40(d). Thus all the sharp cutoff filtering has been done by a discrete-time system, and only nominal continuous-time filtering is required.

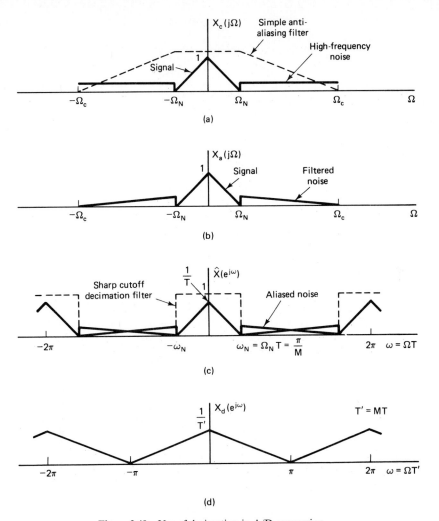

Figure 3.40   Use of decimation in A/D conversion.

To simplify the continuous-time filtering in D/A conversion, we can use the system of Fig. 3.39(c). Suppose that the input $y_d[n]$ has discrete-time Fourier transform as in Fig. 3.41(a), where Nyquist rate sampling (with period $T' = MT$) is assumed. Figure 3.41(b) shows the Fourier transform of the interpolated sequence (sampling period $T'' = T'/L$) together with the frequency response of the discrete-time interpolation filter, perhaps including compensation for a zero-order hold or reconstruction filtering. Figure 3.41(c) shows a simple reconstruction filter that will suffice to select the desired central image, producing an output with Fourier transform shown in Fig. 3.41(d). Note that we require only $\Omega_c < (2\pi/T'' - \Omega_N)$. Thus, we have used a sharp cutoff discrete-time system to provide the bandlimiting necessary for accurate reconstruction of the continuous-time output signal.

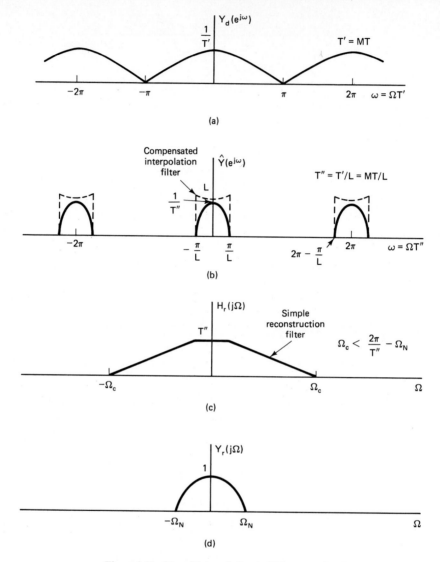

Figure 3.41   Use of interpolation in D/A conversion.

3.8 SUMMARY

In this chapter, we developed and explored the relationship between continuous-time signals and the discrete-time sequence obtained by periodic sampling. The fundamental theorem that allows the continuous-time signal to be represented by a sequence of samples is the Nyquist theorem, which states that for a bandlimited signal, periodic samples are a sufficient representation as long as the sampling rate is sufficiently high relative to the highest frequency in the continuous-time signal. Under this condition, the continuous-time signal can be reconstructed from the samples by lowpass filtering, corresponding to bandlimited interpolation. If the sampling rate is too low relative to the bandwidth of the signal, then aliasing results.

The ability to represent signals by sampling permits the discrete-time processing of continuous-time signals. This is accomplished by first sampling, then applying discrete-time processing, and reconstructing a continuous-time signal from the result. Examples given were lowpass filtering and differentiation.

A particularly important class of processing is the class corresponding to sampling rate changes. Downsampling a discrete-time signal corresponds in the frequency domain to a replication of the discrete-time spectrum and rescaling of the frequency axis, which may require additional bandlimiting to avoid aliasing. Upsampling corresponds to effectively increasing the sampling rate and is also represented in the frequency domain by a rescaling of the frequency axis. By combining upsampling and downsampling by integer amounts, noninteger sampling rate conversion can be achieved.

In the final sections of the chapter, we explored a number of practical considerations associated with the discrete-time processing of continuous-time signals, including the use of prefiltering to avoid aliasing, quantization error in analog-to-digital conversion, and some issues associated with the filtering used in sampling and reconstruction of the continuous-time signals.

PROBLEMS

3.1. The sequence

$$x[n] = \cos\left(\frac{\pi}{4}n\right), \qquad -\infty < n < \infty,$$

was obtained by sampling an analog signal

$$x_c(t) = \cos(\Omega_0 t), \qquad -\infty < t < \infty,$$

at a sampling rate of 1000 samples/s. What are two possible values of Ω_0 that could have resulted in the sequence $x[n]$?

3.2. Let $h_c(t)$ denote the impulse response of a linear time-invariant continuous-time filter and $h_d[n]$ the impulse response of a linear time-invariant discrete-time filter.
   (a) If

$$h_c(t) = \begin{cases} e^{-at}, & t \geq 0, \\ 0, & t < 0, \end{cases}$$

   determine the continuous-time filter frequency response and sketch its magnitude.
   (b) If $h_d[n] = Th_c(nT)$ with $h_c(t)$ as in part (a), determine the discrete-time filter frequency response and sketch its magnitude.
   (c) For a given value of a, determine as a function of T the minimum magnitude of the discrete-time filter frequency response.

3.3. A simple model of a multipath communication channel is indicated in Fig. P3.3-1. Assume that $s_c(t)$ is bandlimited such that $S_c(j\Omega) = 0$ for $|\Omega| \geq \pi/T$ and that $x_c(t)$ is

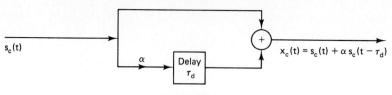

Figure P3.3-1

sampled with a sampling period T to obtain the sequence

$$x[n] = x_c(nT).$$

(a) Determine the Fourier transform of $x_c(t)$ and the Fourier transform of $x[n]$ in terms of $S_c(j\Omega)$.

(b) We want to simulate the multipath system with a discrete-time system by choosing $H(e^{j\omega})$ in Fig. P3.3-2 so that the output $r[n] = x_c(nT)$ when the input is $s[n] = s_c(nT)$. Determine $H(e^{j\omega})$ in terms of T and τ_d.

(c) Determine the impulse response $h[n]$ in Fig. P3.3-2 when (i) $\tau_d = T$ and (ii) $\tau_d = T/2$.

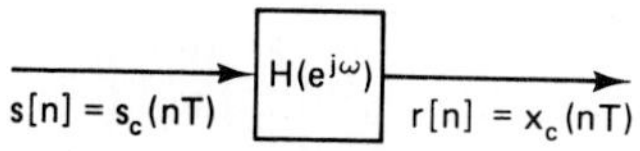

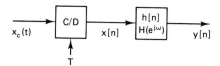

Figure P3.3-2

3.4. Consider the system in Fig. P3.4 with the following relations:

$$X_c(j\Omega) = 0, \qquad |\Omega| \ge 2\pi \times 10^4, \tag{P3.4-1}$$

$$x[n] = x_c(nT), \tag{P3.4-2}$$

$$y[n] = T \sum_{k=-\infty}^{n} x[k]. \tag{P3.4-3}$$

Figure P3.4

(a) For this system, what is the maximum allowable value of T if aliasing is to be avoided, i.e., so that $x_c(t)$ can be recovered from $x[n]$.

(b) Determine $h[n]$.

(c) In terms of $X(e^{j\omega})$, what is the value of $y[n]$ for $n = \infty$?

(d) Determine if there is any value of T for which

$$y[n]\Big|_{n=\infty} = \int_{-\infty}^{\infty} x_c(t)\, dt. \tag{P3.4-4}$$

If there is such a value for T, determine the *maximum* value. If there is not, explain and specify how T would be chosen so that the equality in Eq. (P3.4-4) is best approximated.

3.5. A complex bandpass analog signal $x_c(t)$ has the Fourier transform shown in Fig. P3.5, where $(\Omega_2 - \Omega_1) = \Delta\Omega$. This signal is sampled to produce the sequence $x[n] = x_c(nT)$.

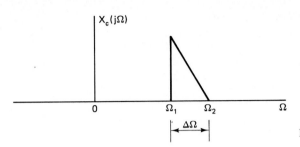

Figure P3.5

(a) Sketch the Fourier transform $X(e^{j\omega})$ of the sequence $x[n]$ for $T = \pi/\Omega_2$.
(b) What is the *lowest* sampling frequency that can be used without incurring any aliasing distortion, i.e., so that $x_c(t)$ can be recovered from $x[n]$?
(c) Draw the block diagram of a system that can be used to recover $x_c(t)$ from $x[n]$ if the sampling rate is greater than or equal to the rate determined in part (b). Assume that (complex) ideal filters are available.

3.6. An analog signal $x_c(t)$, with Fourier transform $X_c(j\Omega)$ shown in Fig. P3.6, is sampled with sampling period $T = 2\pi/\Omega_0$ to form the sequence $x[n] = x_c(nT)$.

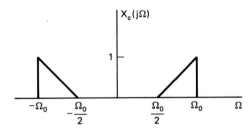

Figure P3.6

(a) Sketch the Fourier transform $X(e^{j\omega})$ for $|\omega| < \pi$.
(b) The signal $x[n]$ is to be transmitted across a digital channel. At the receiver, the original signal $x_c(t)$ must be recovered. Draw a block diagram of the recovery system and specify its characteristics. Assume that ideal filters are available.
(c) In terms of Ω_0, for what range of values for T can $x_c(t)$ be recovered from $x[n]$?

3.7. In many applications discrete-time random signals arise through periodic sampling of continuous-time random signals. We are concerned in this problem with a derivation of the sampling theorem for random signals. Consider a continuous-time, stationary, random process defined by the random variables $\{x_a(t)\}$, where t is a continuous variable. The autocorrelation function is defined as

$$\phi_{x_c x_c}(\tau) = \mathcal{E}\{x(t)x^*(t + \tau)\}$$

and the power density spectrum is

$$P_{x_c x_c}(\Omega) = \int_{-\infty}^{\infty} \phi_{x_c x_c}(\tau)e^{-j\Omega\tau}\, d\tau.$$

A discrete-time random process obtained by periodic sampling is defined by the set of random variables $\{x[n]\}$, where $x[n] = x_a(nT)$ and T is the sampling period.

(a) What is the relationship between $\phi_{xx}[n]$ and $\phi_{x_c x_c}(\tau)$?

(b) Express the power density spectrum of the discrete-time process in terms of the power density spectrum of the continuous-time process.

(c) Under what condition is the discrete-time power density spectrum a faithful representation of the continuous-time power density spectrum?

3.8. Consider a continuous-time random process $x_c(t)$ with a bandlimited power density spectrum $P_{x_c x_c}(\Omega)$ as depicted in Fig. P3.8-1. Suppose that we sample $x_c(t)$ to obtain the discrete-time random process $x[n] = x_c(nT)$.

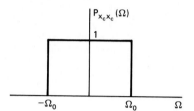

Figure P3.8-1

(a) What is the autocorrelation sequence of the discrete-time random process?

(b) For the continuous-time power density spectrum in Fig. P3.8-1, how should T be chosen so that the discrete-time process is white, i.e., so that the power spectrum is constant for all ω?

(c) If the continuous-time power density spectrum is as shown in Fig. P3.8-2, how should T be chosen so that the discrete-time process is white?

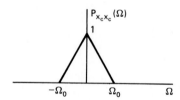

Figure P3.8-2

(d) What is the general requirement on the continuous-time process and the sampling period such that the discrete-time process is white?

3.9. In Fig. P3.9, assume that $X_c(j\Omega) = 0$, $|\Omega| \geq \pi/T_1$. For the general case in which $T_1 \neq T_2$ in the system, express $y_c(t)$ in terms of $x_c(t)$. Is the basic relationship different for $T_1 > T_2$ and $T_1 < T_2$?

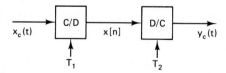

Figure P3.9

3.10. In the system of Fig. P3.10, $X_c(j\Omega)$ and $H(e^{j\omega})$ are as shown. Sketch and label the Fourier transform of $y_c(t)$ for each of the following cases:

(a) $1/T_1 = 1/T_2 = 10^4$

(b) $1/T_1 = 1/T_2 = 2 \times 10^4$

(c) $1/T_1 = 2 \times 10^4$, $1/T_2 = 10^4$

(d) $1/T_1 = 10^4$, $1/T_2 = 2 \times 10^4$

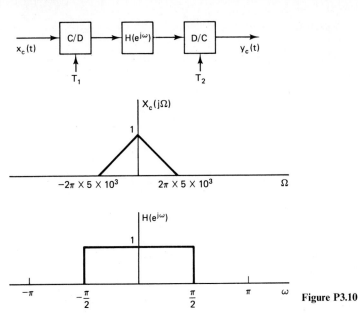

Figure P3.10

3.11. Figure P3.11-1 shows the overall system for filtering a continuous-time signal using a discrete-time filter. The frequency responses of the reconstruction filter $H_r(j\Omega)$ and the discrete-time filter $H(e^{j\omega})$ are shown in Fig. P3.11-2.

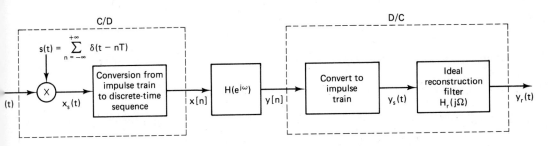

Figure P3.11-1

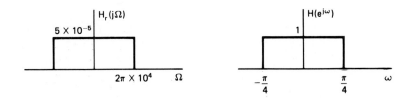

Figure P3.11-2

(a) For $X_c(j\Omega)$ as shown in Fig. P3.11-3 and $1/T = 20$ kHz, sketch $X_s(j\Omega)$ and $X(e^{j\omega})$.

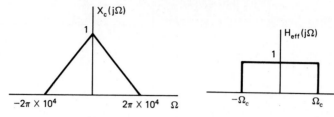

Figure P3.11-3   **Figure P3.11-4**

For a certain range of values for T the overall system, with input $x_c(t)$ and output $y_c(t)$, is equivalent to a continuous-time lowpass filter with frequency response $H_{eff}(j\Omega)$ sketched in Fig. P3.11-4.

(b) Determine the range of values of T for which this is true.

(c) For the range of values determined in (b), sketch Ω_c as a function of $1/T$.

Note: This is one way of implementing a variable-cutoff continuous-time filter using fixed continuous-time and discrete-time filters and a variable sampling rate.

3.12. In the system of Fig. 3.9, assume that the discrete-time system is linear and time-invariant and

$$X_c(j\Omega) = 0 \quad \text{for} \quad |\Omega| \geq 4000\pi.$$

Determine the largest possible value for T and the corresponding frequency response $H(e^{j\omega})$ for the discrete-time system so that

$$Y_c(j\Omega) = \begin{cases} |\Omega| X_c(j\Omega), & 1000\pi < |\Omega| < 2000\pi, \\ 0, & \text{otherwise.} \end{cases}$$

3.13. In the system of Fig. 3.9, assume that $X_c(j\Omega) = 0$ for $|\Omega| > \pi/T$. Determine and plot the magnitude and phase of the discrete-time LTI system such that the output $y_r(t)$ is the running integral of the input, i.e.,

$$y_r(t) = \int_{-\infty}^{t} x_c(\tau)\, d\tau.$$

3.14. Consider the system of Fig. 3.9, with the discrete-time system an ideal lowpass filter with cutoff frequency $\pi/8$ radians/s.

(a) If $x_c(t)$ is bandlimited to 5 kHz, what is the maximum value of T to avoid aliasing in the C/D converter?

(b) If $1/T = 10$ kHz, what will the cutoff frequency of the effective continuous-time filter be?

(c) Repeat part (b) for $1/T = 20$ kHz.

3.15. A bandlimited continuous-time signal is known to contain a 60-Hz component, which we want to remove by processing with the system of Fig. 3.9 where $T = 10^{-4}$.

(a) What is the highest frequency that the analog signal can contain if aliasing is to be avoided?

(b) The discrete-time system to be used has frequency response

$$H(e^{j\omega}) = \frac{[1 - e^{-j(\omega - \omega_0)}][1 - e^{-j(\omega + \omega_0)}]}{[1 - 0.9e^{-j(\omega - \omega_0)}][1 - 0.9e^{-j(\omega + \omega_0)}]}.$$

Sketch the magnitude and phase of $H(e^{j\omega})$.

(c) What value should be chosen for ω_0 to eliminate the 60-Hz component?

3.16. Consider the system in Fig. 3.9 with

$$X_c(j\Omega) = 0 \qquad \text{for} \qquad |\Omega| \geq 2\pi(1000)$$

and the discrete-time system a squarer, i.e., $y[n] = x^2[n]$. What is the largest value of T such that $y_c(t) = x_c^2(t)$?

3.17. This problem explores the effect of interchanging the order of two operations on a signal, namely, sampling and performing a zero-memory nonlinear operation.

(a) Consider the two signal processing systems in Fig. P3.17-1, where the C/D and D/C converters are ideal. The mapping $g[x] = x^2$ represents a zero-memory nonlinear device. For the two systems in Fig. P3.17-1, sketch the signal spectra at points 1, 2, and 3 when the sampling rate is selected to be $1/T = 2f_m$ samples/s and $x_c(t)$ has the spectral characteristic (bandlimited to f_m hertz) shown in Fig. P3.17-2. Is $y_1(t) = y_2(t)$? If not, why not? Is $y_1(t) = x_c^2(t)$? Explain your answer.

System 1:

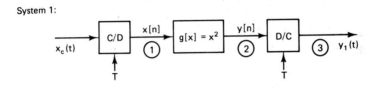

System 2:

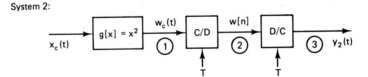

Figure P3.17-1

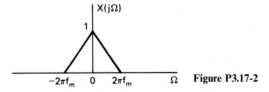

Figure P3.17-2

(b) Consider system 1, and let $x(t) = A \cos 30\pi t$. Let the sampling rate be $1/T = 40$ samples/s. Is $y_1(t) = x_c^2(t)$? Explain why or why not.

(c) Consider the signal processing system shown in Fig. P3.17-3, where $g[x] = x^3$ and

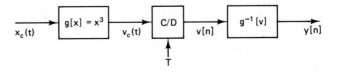

Figure P3.17-3

$g^{-1}[v]$ is the (unique) inverse, i.e., $g^{-1}[g(x)] = x$. Let $x(t) = A \cos 30\pi t$ and $1/T = 40$ samples/s. Express $v[n]$ in terms of $x[n]$. Is there spectral aliasing? Express $y[n]$ in terms of $x[n]$. What conclusion can you reach from this example? (You may want to use the identity

$$\cos^3 \Omega_0 t = \tfrac{3}{4} \cos \Omega_0 t + \tfrac{1}{4} \cos 3\Omega_0 t.)$$

(d) One practical problem is that of digitizing a signal having a large dynamic range. Suppose we compress the dynamic range by passing the signal through a zero-memory nonlinear device prior to A/D conversion and then expand it back after A/D conversion. What is the impact of the nonlinear operation prior to the A/D converter in our choice of the sampling rate?

3.18. For the LTI system in Fig. P3.18-1, $H(e^{j\omega})$ is

$$H(e^{j\omega}) = e^{-j\omega/2}, \qquad |\omega| \le \pi \quad \text{(half-sample delay)}.$$

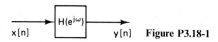

$x[n]$ $y[n]$ **Figure P3.18-1**

(a) Determine a choice for T and $h_c(t)$ in the system of Fig. 3.14 so that the system in Fig. 3.14 is equivalent to the system in Fig. P3.18-1 with $H(e^{j\omega})$ as specified.

(b) Determine and sketch $y[n]$ when the input sequence is

$$x[n] = \cos\left(\frac{5\pi}{2} n - \frac{\pi}{4}\right)$$

as sketched in Fig. P3.18-2.

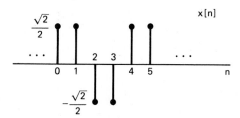

Figure P3.18-2

3.19. Consider the system of Fig. 3.14 with the continuous-time LTI system causal and characterized by the linear constant-coefficient differential equation

$$\frac{d^2 y_c(t)}{dt^2} + 4 \frac{dy_c(t)}{dt} + 3y_c(t) = x_c(t).$$

The overall system is equivalent to a causal discrete-time linear time-invariant system. Determine the impulse response $h[n]$ of the equivalent discrete-time system.

3.20. Consider the sequence $x[n]$ whose Fourier transform $X(e^{j\omega})$ is shown in Fig. P3.20. Sketch $X_s(e^{j\omega})$ and $X_d(e^{j\omega})$ for $M = 3$ and for $\omega_H = \pi/2$ and $\omega_H = \pi/4$, where

$$x_s[n] = \begin{cases} x[n], & n = Mk, \quad k = 0, \pm 1, \pm 2, \dots, \\ 0, & \text{otherwise,} \end{cases}$$

and

$$x_d[n] = x_s[Mn] = x[Mn].$$

What is the maximum value of ω_H to avoid aliasing when $M = 3$?

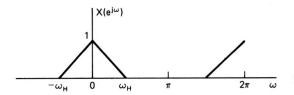

Figure P3.20

3.21. In Fig. P3.21, $x[n] = x_c(nT)$ and $y[n] = x[2n]$.

(a) In the system shown in Fig. P3.21, if $x_c(t)$ has a Fourier transform such that $X_c(j\Omega) = 0, |\Omega| > 2\pi \times 100$, what value of T is required so that

$$X(e^{j\omega}) = 0, \qquad \frac{\pi}{2} < |\omega| \le \pi?$$

(b) How should T' be chosen so that $y_c(t) = x_c(t)$?

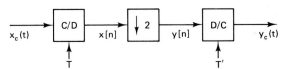

Figure P3.21

3.22. Figure 3.21 depicts a system for interpolating a signal by a factor of L, where

$$x_e[n] = \begin{cases} x[n/L], & n = 0, \ \pm L, \ \pm 2L, \ \text{etc.}, \\ 0, & \text{otherwise,} \end{cases}$$

and the lowpass filter interpolates between the nonzero values of $x_e[n]$ to generate the upsampled or interpolated signal $x_i[n]$. When the lowpass filter is ideal, the interpolation is referred to as bandlimited interpolation. As indicated in Section 3.6.2, simple interpolation procedures are often adequate. Two simple procedures often used are zero-order-hold and linear interpolation. For zero-order-hold interpolation, each value of $x[n]$ is simply repeated L times, i.e.,

$$x_i[n] = \begin{cases} x_e[0], & n = 0, \ 1, \ \ldots, \ L-1 \\ x_e[L], & n = L, \ L+1, \ \ldots, \ 2L-1 \\ x_e[2L], & n = 2L, \ 2L+1, \ \ldots, \\ \ \ \vdots & \end{cases}$$

Linear interpolation is described in Section 3.6.2.

(a) Determine an appropriate choice for the impulse response of the lowpass filter in Fig. 3.21 to implement zero-order-hold interpolation. Also determine the corresponding frequency response.

(b) Equation (3.90) specifies the impulse response for linear interpolation. Determine the corresponding frequency response. (You may find it helpful to use the fact that $h_{\text{lin}}[n]$ is triangular and consequently corresponds to the convolution of two rectangular sequences.)

(c) Sketch the magnitude of the filter frequency response for zero-order-hold and linear interpolation. Which is a better approximation to ideal bandlimited interpolation?

3.23. Suppose you obtained a sequence $s[n]$ by filtering a speech signal $s_c(t)$ with a continuous-time lowpass filter with a cutoff frequency of 5 kHz and then sampling it at a 10-kHz rate as shown in Fig. P3.23-1. Unfortunately, the speech signal $s_c(t)$ was destroyed once the

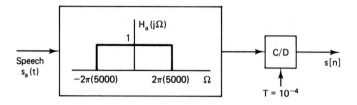

Figure P3.23-1

sequence $s[n]$ was stored on magnetic tape. Later, you find that what you should have done is followed the process shown in Fig. P3.23-2. Develop a method to obtain $s_1[n]$ from $s[n]$ using discrete-time processing. Your method may require a very large amount of computation but should *not* require a C/D or D/C converter. If your method uses a discrete-time filter, you should specify the frequency response of the filter.

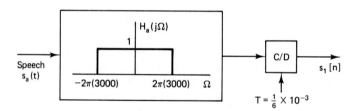

Figure P3.23-2

3.24. Consider the system shown in Fig. P3.24-1, where

$$H(e^{j\omega}) = \begin{cases} 1, & |\omega| < \pi/L, \\ 0, & \pi/L < |\omega| \le \pi. \end{cases}$$

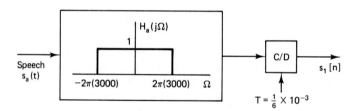

Figure P3.24-1

Sketch $Y_c(j\Omega)$ if $X_c(j\Omega)$ is as shown in Fig. P3.24-2.

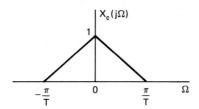

Figure P3.24-2

3.25. We wish to compute the autocorrelation function of an upsampled signal, as indicated in Fig. P3.25-1. It is suggested that this can equivalently be accomplished with the system of Fig. P3.25-2. Can $H_2(e^{j\omega})$ be chosen so that $\phi_3[n] = \phi_1[n]$? If not, why not? If so, specify $H_2(e^{j\omega})$.

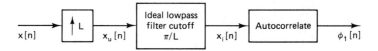

Figure P3.25-1

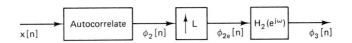

Figure P3.25-2

3.26. The system shown in Fig. P3.26 approximately interpolates the sequence $x[n]$ by a factor L. Suppose that the linear filter has impulse response $h[n]$ such that $h[n] = h[-n]$ and $h[n] = 0$ for $|n| > (RL - 1)$, where R and L are integers; i.e., the impulse response is symmetric and of length $(2RL - 1)$ samples.

Figure P3.26

(a) In answering the following, do not be concerned about the causality of the system. It can be made causal by including some delay. Specifically, how much delay must be inserted to make the system causal?

(b) What conditions must be satisfied by $h[n]$ in order that $y[n] = x[n/L]$ for $n = 0, \pm L, \pm 2L, \pm 3L, \ldots$?

(c) By exploiting the symmetry of the impulse response, show that each sample of $y[n]$ can be computed with no more than RL multiplications.

(d) By taking advantage of the fact that multiplications by zero need not be done, show that only $2R$ multiplications per output sample are required.

3.27. We are interested in upsampling a sequence by a factor of 2, using a system of the form of Fig. 3.21. However, the lowpass filter in Fig. 3.21 is to be approximated by a 5-point filter with impulse response $h[n]$ indicated in Fig. P3.27-1. In this system the output $y_1[n]$ is obtained by direct convolution of $h[n]$ with $w[n]$.

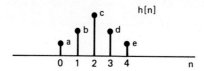

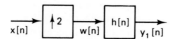

Figure P3.27-1

(a) A proposed implementation of the system with this choice of $h[n]$ is shown in Fig. P3.27-2. The three impulse responses $h_1[n]$, $h_2[n]$, and $h_3[n]$ are all restricted to be zero outside the range $0 \leq n \leq 2$. Determine and clearly justify a choice for $h_1[n]$, $h_2[n]$, and $h_3[n]$ so that $y_1[n] = y_2[n]$ for any $x[n]$, i.e., so that the two systems are identical.

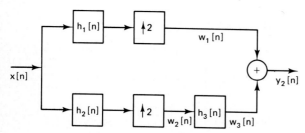

Figure P3.27-2

(b) Determine the number of multiplications per output point required in the system of Fig. P3.27-1 and in the system of Fig. P3.27-2. You should find that the system of Fig. P3.27-2 is more efficient.

3.28. In the system in Fig. P3.28,

$$X_c(j\Omega) = 0, \qquad |\Omega| \geq \pi/T,$$

and

$$H(e^{j\omega}) = \begin{cases} e^{-j\omega}, & |\omega| < \pi/L, \\ 0, & \pi/L < |\omega| \leq \pi. \end{cases}$$

How is $y[n]$ related to the input signal $x_c(t)$?

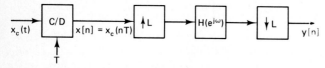

Figure P3.28

3.29. Consider the analysis/synthesis system shown in Fig. P3.29-1. The lowpass filter $h_0[n]$ is identical in the analyzer and synthesizer, and the highpass filter $h_1[n]$ is identical in the analyzer and synthesizer. The Fourier transforms of $h_0[n]$ and $h_1[n]$ are related by

$$H_1(e^{j\omega}) = H_0(e^{j(\omega+\pi)}).$$

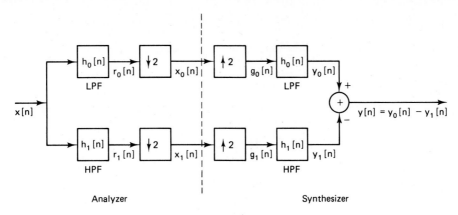

Figure P3.29-1

(a) If $X(e^{j\omega})$ and $H_0(e^{j\omega})$ are as shown in Fig. P3.29-2, sketch (to within a scale factor) $X_0(e^{j\omega})$, $G_0(e^{j\omega})$, and $Y_0(e^{j\omega})$.

(b) Write a general expression for $G_0(e^{j\omega})$ in terms of $X(e^{j\omega})$ and $H_0(e^{j\omega})$. Do *not* assume that $X(e^{j\omega})$ and $H_0(e^{j\omega})$ are as shown in Fig. P3.29-2.

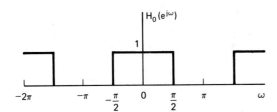

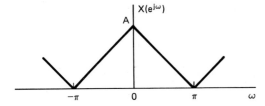

Figure P3.29-2

(c) Determine a set of conditions on $H_0(e^{j\omega})$ that is as general as possible and that will guarantee that $y[n]$ is proportional to $x[n - n_d]$ for any stable input $x[n]$.

Note: Analyzer/synthesizer filter banks of the form developed in this problem are called quadrature mirror filter banks. For further reading, see Crochiere and Rabiner (1983), pp. 378–392.

3.30. Consider a real-valued sequence $x[n]$ for which

$$X(e^{j\omega}) = 0, \qquad \frac{\pi}{3} \leq |\omega| \leq \pi.$$

One sequence value of $x[n]$ may have been corrupted and we would like to approximately or exactly recover it. With $\hat{x}[n]$ denoting the corrupted signal,

$$\hat{x}[n] = x[n] \qquad \text{for} \qquad n \neq n_0$$

and $\hat{x}[n_0]$ is real but not related to $x[n_0]$. In each of the following three cases, specify a practical algorithm for exactly or approximately recovering $x[n]$ from $\hat{x}[n]$.

(a) The value of n_0 is known.

(b) The exact value of n_0 is *not* known, but we know that n_0 is an even number.

(c) Nothing about n_0 is known.

3.31. Communication systems often require conversion from time-division multiplexing (TDM) to frequency-division multiplexing (FDM). In this problem we examine a simple example of such a system. The block diagram of the system to be studied is shown in Fig. P3.31-1. The TDM input is assumed to be the sequence of interleaved samples

$$w[n] = \begin{cases} x_1[n/2] & \text{for } n \text{ an even integer,} \\ x_2[(n-1)/2] & \text{for } n \text{ an odd integer.} \end{cases}$$

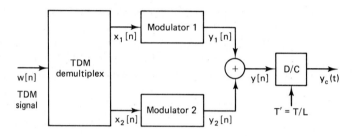

Figure P3.31-1

Assume that the sequences $x_1[n] = x_{c1}(nT)$ and $x_2[n] = x_{c2}(nT)$ have been obtained by sampling without aliasing the continuous-time signals $x_{c1}(t)$ and $x_{c2}(t)$, respectively. Also assume that these two signals have the same highest frequency, Ω_N, and that the sampling period is $T = \pi/\Omega_N$.

(a) Draw a block diagram of a system that produces $x_1[n]$ and $x_2[n]$ as outputs, i.e., obtain a system for TDM demultiplexing using simple operations. State whether or not your system is linear, time-invariant, causal, and stable.

The kth modulator system ($k = 1$ or 2) is defined by the block diagram in Fig. P3.31-2. The lowpass filter $H_i(e^{j\omega})$, which is the same for both channels, has gain L and cutoff frequency π/L, and the highpass filters $H_k(e^{j\omega})$ have unity gain and cutoff frequency ω_k. The modulator frequencies are such that

$$\omega_2 = \omega_1 + \pi/L \quad \text{and} \quad \omega_2 + \pi/L \le \pi \quad (\text{assume } \omega_1 > \pi/2).$$

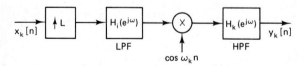

Figure P3.31-2

(b) Find ω_1 and L so that after ideal D/C conversion with sampling period T/L, the Fourier transform of $y_c(t)$ is zero except in the band of frequencies

$$2\pi \times 10^5 \le |\Omega| \le 2\pi \times 10^5 + 2\Omega_N.$$

Assume that $\Omega_N = 2\pi \times 5 \times 10^3$.

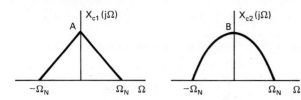

Figure P3.31-3

(c) Assume that the continuous-time Fourier transforms of the two original input signals are as sketched in Fig. P3.31-3. Sketch the Fourier transforms at each point in the system. This may help in determining the desired parameters in part (b).

(d) Based on your solution to parts (a)–(c), discuss how the system could be generalized to handle M equal-bandwidth channels.

3.32. In Section 3.7.1 we considered the use of prefiltering to avoid aliasing. In practice, the anti-aliasing filter cannot be ideal. However, the nonideal characteristics can be at least partially compensated for with a discrete-time system applied to the sequence $x[n]$ which is the output of the C/D converter.

Consider the two systems in Fig. P3.32-1. The anti-aliasing filters $H_{ideal}(j\Omega)$ and $H_{aa}(j\Omega)$ are shown in Fig. P3.32-2. $H(e^{j\omega})$ in Fig. P3.32-1 is to be specified to compensate for the nonideal characteristics of $H_{aa}(j\Omega)$.

Sketch $H(e^{j\omega})$ so that the two sequences $x[n]$ and $w[n]$ are identical.

System 1:

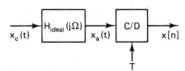

System 2:

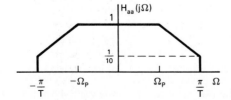

Figure P3.32-1

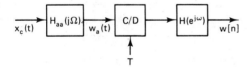

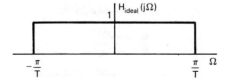

Figure P3.32-2

3.33. In this problem we consider some of the properties of the various number representations discussed in the text. Consider a number x such that $|x| < 1$ and further assume that $|x|$ can be represented as a binary fraction with b bits to the right of the binary point. We will find it convenient to introduce the notation $\simeq$ to mean "is represented by."

For sign and magnitude,

$$x \simeq \begin{cases} |x|, & x \geq 0, \\ 1 + |x|, & x \leq 0 \end{cases}$$

For one's complement,

$$x \simeq \begin{cases} |x|, & x \geq 0 \\ 2 - 2^{-b} - |x|, & x \leq 0 \end{cases}$$

For two's complement,

$$x \simeq \begin{cases} |x|, & x \geq 0 \\ 2 - |x|, & x < 0 \end{cases}$$

(a) Show that in all three representations, if x is represented by a $(B + 1)$-bit binary number, the bit immediately to the left of the binary point (the sign bit) is 0 if $x > 0$ and 1 if $x < 0$.

(b) Show that the following algorithm is sufficient to obtain the one's-complement representation of a negative number x from its magnitude $|x|$: Obtain the negative of $|x|$ by changing every 1 to a 0 and every 0 to a 1, including the sign bit.

(c) Show that the following algorithms serve to generate the two's-complement representation of a negative number x from its magnitude, $|x|$.

    (i) Find the one's-complement representation and add 2^{-B}.

    (ii) Starting at the right, examine the bits of $|x|$ in turn. For each 0 in $|x|$, place a 0 in x. When the first 1 is encountered in $|x|$, place a 1 in x. Thereafter, change each 0 to a 1 and each 1 to a zero, for all bits including the sign bit.

3.34. The two's-complement representation of binary numbers leads to very attractive simplifications of arithmetic processes. To illustrate this, consider some details of two's-complement addition. We use the notation $\simeq$ to mean "is represented by." Thus a number x has the two's-complement representation

$$x \simeq \begin{cases} |x|, & x \geq 0, \\ 2 - |x|, & x < 0, \end{cases}$$

where $|x| < 1$ and $(B + 1)$ bits are used in the representation of x. Two's-complement addition is performed as follows:

1. All numbers are treated as $(B + 1)$-bit *unsigned* binary numbers.

2. Addition is simple binary addition.

3. Carries past the sign bit are ignored; i.e., if the sum is greater than 2, the carry is ignored. Thus addition is implemented modulo 2.

(a) Using the above notation and definitions, write complete expressions for the two's-complement addition of two numbers x_1 and x_2, where $|x_1|$ and $|x_2| < 1$. Consider all possible cases; i.e., note that x_1 and x_2 can each be either $+$ or $-$ and also that $|x_1|$ can be either greater than or less than $|x_2|$.

(b) Note that when two numbers of like sign are added, the resultant magnitude may be greater than 1. This condition is called *overflow*. Show that overflow is indicated whenever the result of adding two numbers of like sign has the opposite sign.

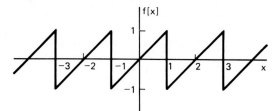

Figure P3.34

(c) Show that two's-complement addition of x_1 and x_2 is equivalent to $f[x_1 + x_2]$, where $f[x]$ is as depicted in Fig. P3.34.

(d) Suppose that $x_1 = 5/8$, $x_2 = 3/4$, and $x_3 = -(1/2)$. Find the two's-complement representation of each number, and add the two's-complement representations together in the sequence $(x_1 + x_2) + x_3$. Note that overflow occurs in the addition $(x_1 + x_2)$, but the final result is correct. Show that in general any number of overflows can occur in the process of accumulating the sum of three or more two's-complement numbers, and the result will be correct if the correct sum has magnitude less than unity.

3.35. In a floating-point representation of numbers, the number is represented in the form

$$F = 2^c M, \qquad \tfrac{1}{2} \le |M| < 1,$$

where c is commonly called the characteristic and M the mantissa. Since $1/2 \le |M| < 1$, M is a fixed-point fraction. Negative floating-point numbers can be represented by representing the mantissa as a floating-point fraction in either sign-and-magnitude, one's-complement, or two's-complement notation.

Consider a floating-point number F that is to be quantized by quantizing the mantissa to B bits, excluding sign, so that the value of the least significant bit in the mantissa is 2^{-B}.

Let $Q[F]$ represent the quantized value. It is convenient to represent $Q[F]$ as $Q[F] = F(1 + \varepsilon)$, so that the error $E = Q[F] - F$ is

$$E = \varepsilon F.$$

(a) Assuming that F is a positive number, show that for rounding, $-2^{-B} < \varepsilon \le 2^{-B}$ and for truncation of the mantissa, $-2 \cdot 2^{-B} < \varepsilon \le 0$.

(b) Assuming that F is a negative number, determine an upper and lower bound on ε for (i) rounding, (ii) sign-and-magnitude truncation, (iii) one's-complement truncation, and (iv) two's-complement truncation.

3.36. As discussed in Section 3.7.2, to process sequences on a digital computer, we must quantize the amplitude of the sequence to a set of discrete levels. This quantization can be expressed in terms of passing the input sequence $x[n]$ through a quantizer $Q(x)$ that has an input/output relation as shown in Fig. 3.31.

As discussed in Section 3.7.3, if the quantization interval Δ is small compared with changes in the level of the input sequence, we can assume that the output of the quantizer $y[n]$ is of the form

$$y[n] = x[n] + e[n],$$

where $e[n] = Q(x[n]) - x[n]$ and $e[n]$ is a stationary random process with a first-order probability density uniformly distributed between $-\Delta/2$ and $\Delta/2$, uncorrelated from sample to sample and uncorrelated with $x[n]$, so that $\mathscr{E}\{e[n]x[m]\} = 0$ for all m and n. Let $x[n]$ be a stationary white-noise process, with zero mean and variance σ_x^2.

(a) Find the mean, variance, and autocorrelation sequence of $e[n]$.

(b) What is the signal-to-quantizing-noise ratio σ_x^2/σ_e^2?

(c) The quantized signal $y[n]$ is to be filtered by a digital filter with impulse response $h[n] = \frac{1}{2}[a^n + (-a)^n]u[n]$. Determine the variance of the noise produced at the output due to the input quantization noise and determine the signal-to-noise ratio at the output.

In some cases we may want to use nonlinear quantization steps, for example, logarithmically spaced quantization steps. This can be accomplished by applying uniform quantization to the logarithm of the input as depicted in Fig. P3.36, where $Q[\cdot]$ is a uniform quantizer as specified in Fig. 3.31. In this case, if we assume that Δ is small compared with changes in the sequence $\ln(x[n])$, then we can assume that the output of the quantizer is

$$\ln(y[n]) = \ln(x[n]) + e[n].$$

Thus

$$y[n] = x[n] \cdot \exp(e[n]).$$

For small e we can approximate $\exp(e[n])$ by $(1 + e[n])$, so that

$$y[n] \simeq x[n](1 + e[n]) = x[n] + f[n]. \tag{P3.36}$$

This equation will be used to describe the effect of logarithmic quantization. We assume $e[n]$ to be a stationary random process, uncorrelated from sample to sample, independent of the signal $x[n]$, and with first-order probability density uniformly distributed between $\pm \Delta/2$.

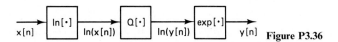

Figure P3.36

(d) Determine the mean, variance, and autocorrelation sequence of the *additive* noise $f[n]$ defined in Eq. (P3.36).

(e) What is the signal-to-quantizing-noise ratio σ_x^2/σ_f^2? Note that in this case σ_x^2/σ_f^2 is independent of σ_x^2. Within the limits of our assumption, therefore, the signal-to-quantizing-noise ratio is independent of the input signal level, whereas for linear quantization the ratio σ_x^2/σ_e^2 depends directly on σ_x^2.

(f) The quantized signal $y[n]$ is to be filtered by means of a digital filter with impulse response $h[n] = \frac{1}{2}[a^n + (-a)^n]u[n]$. Determine the variance of the noise produced at the output due to the input quantization noise and determine the signal-to-noise ratio at the output.

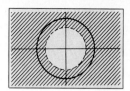

The *z*-Transform

4

4.0 INTRODUCTION

We have seen that the Fourier transform plays a key role in representing and analyzing discrete-time signals and systems. In this chapter, we consider a generalization of the Fourier transform referred to as the *z-transform*. The *z*-transform for discrete-time signals is the counterpart of the Laplace transform for continuous-time signals, and they each have a similar relationship to the corresponding Fourier transform. A principal motivation for introducing this generalization is that the Fourier transform does not converge for all sequences and it is useful to have a generalization of the Fourier transform that encompasses a broader class of signals. A second advantage is that in analytical problems the *z*-transform notation is often more convenient than the Fourier transform notation. Still another advantage is that the *z*-transform allows us to bring the power of complex variable theory to bear on problems of discrete-time signals and systems.

In this chapter, we define the *z*-transform representation of a sequence and study how the properties of a sequence are related to the properties of its *z*-transform. In discussing the *z*-transform we use a number of results from the theory of complex variables. In applying these results we will be precise, but we will not attempt to maintain a high degree of mathematical rigor.

4.1 *z*-TRANSFORM

The Fourier transform $X(e^{j\omega})$ of a sequence $x[n]$ was defined in Chapter 2 as

$$X(e^{j\omega}) = \sum_{n=-\infty}^{\infty} x[n]e^{-j\omega n}. \qquad (4.1)$$

The *z*-transform $X(z)$ of a sequence $x[n]$ is defined as

$$X(z) = \sum_{n=-\infty}^{\infty} x[n]z^{-n}. \qquad (4.2)$$

Equation (4.2) is in general an infinite sum or infinite power series, with z being a complex variable. Sometimes it is useful to consider Eq. (4.2) as an operator that transforms a sequence into a function, and we will refer to the *z-transform operator* $\mathcal{Z}\{\cdot\}$ defined as

$$\mathcal{Z}\{x[n]\} = \sum_{n=-\infty}^{\infty} x[n]z^{-n} = X(z). \qquad (4.3)$$

With this interpretation, the z-transform operator is seen to transform the sequence $x[n]$ into the function $X(z)$, where z is a continuous complex variable. The correspondence between a sequence and its z-transform is indicated by the notation

$$x[n] \overset{z}{\longleftrightarrow} X(z). \qquad (4.4)$$

The z-transform as we have defined it in Eq. (4.2) is often referred to as the *two-sided* or *bilateral z-transform*, in contrast to the *one-sided* or *unilateral z-transform*, which is defined as

$$\mathcal{X}(z) = \sum_{n=0}^{\infty} x[n]z^{-n}. \qquad (4.5)$$

Clearly the bilateral and unilateral transforms are equivalent only if $x[n] = 0$ for $n < 0$. In this book, we will use the bilateral transform almost exclusively. However, at the end of this chapter we provide a brief introduction to the unilateral z-transform and show how it is particularly useful for solving linear constant-coefficient difference equations with nonzero auxiliary conditions.

It is evident from a comparison of Eqs. (4.1) and (4.2) that there is a close relationship between the Fourier transform and the z-transform. In particular, if we replace the complex variable z in Eq. (4.2) with the complex variable $e^{j\omega}$, then the z-transform reduces to the Fourier transform. This is one motivation for the notation $X(e^{j\omega})$ for the Fourier transform; when it exists, the Fourier transform is simply $X(z)$ with $z = e^{j\omega}$. This corresponds to restricting z to have unity magnitude, i.e., for $|z| = 1$, the z-transform corresponds to the Fourier transform. More generally, we can express the complex variable z in polar form as

$$z = re^{j\omega}.$$

With z expressed in this form, Eq. (4.2) becomes

$$X(re^{j\omega}) = \sum_{n=-\infty}^{\infty} x[n](re^{j\omega})^{-n}, \qquad \text{or}$$

$$X(re^{j\omega}) = \sum_{n=-\infty}^{\infty} (x[n]r^{-n})e^{-j\omega n}. \qquad (4.6)$$

Equation (4.6) can be interpreted as the Fourier transform of the product of the original sequence $x[n]$ and the exponential sequence r^{-n}. Obviously, for $r = 1$, Eq. (4.6) reduces to the Fourier transform of $x[n]$.

Since the z-transform is a function of a complex variable, it is convenient to describe and interpret it using the complex z-plane. In the z-plane, the contour corresponding to $|z| = 1$ is a circle of unit radius, as illustrated in Fig. 4.1. This

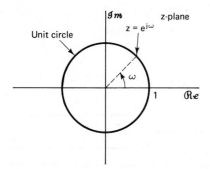

Figure 4.1 The unit circle in the complex z-plane.

contour is referred to as the *unit circle*. The z-transform evaluated on the unit circle corresponds to the Fourier transform. Note that ω is the angle between the vector to a point z on the unit circle and the real axis of the complex z-plane. If we evaluate $X(z)$ at points on the unit circle in the z-plane beginning at $z = 1$ (i.e., $\omega = 0$) through $z = j$ ($\omega = \pi/2$) to $z = -1$ ($\omega = \pi$), we obtain the Fourier transform for $0 \leq \omega \leq \pi$. Continuing around the unit circle would correspond to examining the Fourier transform from $\omega = \pi$ to $\omega = 2\pi$ or, equivalently, from $\omega = -\pi$ to $\omega = 0$. In Chapter 2, the Fourier transform was displayed on a linear frequency axis. Interpreting the Fourier transform as the z-transform on the unit circle in the z-plane corresponds conceptually to wrapping the linear frequency axis around the unit circle with $\omega = 0$ at $z = 1$ and $\omega = \pi$ at $z = -1$. With this interpretation, the inherent periodicity in frequency of the Fourier transform is captured naturally since a change of angle of 2π radians in the z-plane corresponds to traversing the unit circle once and returning to exactly the same point.

As we discussed in Chapter 2, the power series representing the Fourier transform does not converge for all sequences, i.e., the infinite sum may not always be finite. Similarly, the z-transform does not converge for all sequences or for all values of z. For any given sequence, the set of values of z for which the z-transform converges is called the *region of convergence*, which we abbreviate ROC. As we stated in Sec. 2.7, uniform convergence of the Fourier transform requires that the sequence be absolutely summable. Applying this to Eq. (4.6) leads to the condition

$$\sum_{n=-\infty}^{\infty} |x[n]r^{-n}| < \infty \tag{4.7}$$

for absolute convergence of the z-transform. It should be clear from Eq. (4.7) that because of the multiplication of the sequence by the real exponential r^{-n}, it is possible for the z-transform to converge even if the Fourier transform does not. For example, the sequence $x[n] = u[n]$ is not absolutely summable, and therefore the Fourier transform does not converge. However, $r^{-n}u[n]$ is absolutely summable if $r > 1$. This means that the z-transform for the unit step exists with a region of convergence $|z| > 1$.

Convergence of the power series of Eq. (4.2) is dependent only on $|z|$; i.e., since $|X(z)| < \infty$ if

$$\sum_{n=-\infty}^{\infty} |x[n]||z|^{-n} < \infty, \tag{4.8}$$

the region of convergence of the power series in Eq. (4.2) consists of all values of z such that the inequality in Eq. (4.8) holds. Thus, if some value of z, say $z = z_1$, is in the ROC, then all values of z on the circle defined by $|z| = |z_1|$ will also be in the ROC. As one consequence of this, the region of convergence will consist of a ring in the z-plane centered about the origin. Its outer boundary will be a circle (or the ROC may extend outward to infinity) and its inner boundary will be a circle (or it may extend inward to include the origin). This is illustrated in Fig. 4.2. If the ROC includes the unit circle, this of course implies convergence of the z-transform for $|z| = 1$ or, equivalently, the Fourier transform of the sequence converges. Conversely, if the ROC does not include the unit circle, the Fourier transform does not converge absolutely.

A power series of the form of Eq. (4.2) is a Laurent series. Therefore, a number of elegant and powerful theorems from the theory of functions of a complex variable can be employed in the study of the z-transform (see, for example, Churchill and Brown, 1984). A Laurent series, and therefore the z-transform, represents an analytic function at every point inside the region of convergence; hence the z-transform and all its derivatives must be continuous functions of z within the region of convergence. This implies that if the region of uniform convergence includes the unit circle, then the Fourier transform and all its derivatives with respect to ω must be continuous functions of ω. Also, from the discussion in Section 2.7 the sequence must be absolutely summable, i.e., a stable sequence.

Uniform convergence of the z-transform requires absolute summability of the exponentially weighted sequence, as stated in Eq. (4.7). Neither of the sequences

$$x_1[n] = \frac{\sin \omega_c n}{\pi n}, \qquad -\infty < n < \infty,$$

and

$$x_2[n] = \cos \omega_0 n, \qquad -\infty < n < \infty,$$

is absolutely summable. Furthermore, neither of these sequences multiplied by r^{-n} would be absolutely summable for any value of r. Thus, these sequences do not have a z-transform that satisfies absolute convergence. However, we showed in Section 2.7 that even though sequences such as $x_1[n]$ are not absolutely summable, they do have finite energy, and the Fourier transform converges in the mean-square sense to a discontinuous periodic function. Similarly, the sequence $x_2[n]$ is neither absolutely

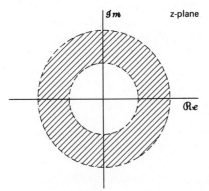

Figure 4.2 The region of convergence (ROC) as a ring in the z-plane. For specific cases, the inner boundary can extend inward to the origin, and the ROC becomes a disc. For other cases, the outer boundary can extend outward to infinity.

nor square summable, but a useful Fourier transform for $x_2[n]$ can be defined using impulses. In both cases the Fourier transforms are not continuous, infinitely differentiable functions, so they cannot result from evaluating a z-transform on the unit circle. Thus in such cases it is not strictly correct to think of the Fourier transform as being the z-transform evaluated on the unit circle, although we nevertheless use a notation that implies this.

The z-transform is most useful when the infinite sum can be expressed in closed form, i.e., when it can be "summed" and expressed as a simple mathematical formula. Among the most important and useful z-transforms are those for which $X(z)$ is a rational function inside the region of convergence, i.e.,

$$X(z) = \frac{P(z)}{Q(z)}, \tag{4.9}$$

where $P(z)$ and $Q(z)$ are polynomials in z. The values of z for which $X(z) = 0$ are called the *zeros* of $X(z)$, and the values of z for which $X(z)$ is infinite are referred to as the *poles* of $X(z)$. The poles of $X(z)$ for finite values of z are the roots of the denominator polynomial. In addition, poles may occur at $z = 0$ or $z = \infty$. For rational z-transforms a number of important relationships exist between the location of poles of $X(z)$ and the region of convergence of the z-transform. We discuss these more specifically in Section 4.2. First, however, we illustrate the z-transform with several examples.

Example 4.1

Consider the signal $x[n] = a^n u[n]$. From Eq. (4.2),

$$X(z) = \sum_{n=-\infty}^{\infty} a^n u[n] z^{-n} = \sum_{n=0}^{\infty} (az^{-1})^n.$$

For convergence of $X(z)$ we require that

$$\sum_{n=0}^{\infty} |az^{-1}|^n < \infty.$$

Thus, the region of convergence is the range of values of z for which $|az^{-1}| < 1$ or, equivalently, $|z| > |a|$. Inside the region of convergence the infinite series converges to

$$X(z) = \sum_{n=0}^{\infty} (az^{-1})^n = \frac{1}{1 - az^{-1}} = \frac{z}{z - a}, \qquad |z| > |a|. \tag{4.10}$$

Here we have used the familiar formula for the sum of terms of a geometric series. The z-transform converges for any finite value of $|a|$. The Fourier transform of $x[n]$, on the other hand, converges only if $|a| < 1$. For $a = 1$, $x[n]$ is the unit step sequence with z-transform

$$X(z) = \frac{1}{1 - z^{-1}}, \qquad |z| > 1. \tag{4.11}$$

In Example 4.1, the infinite sum is equal to a rational function of z inside the region of convergence; for most purposes, this rational function is a much more convenient representation than the infinite sum. We will see that any sequence that

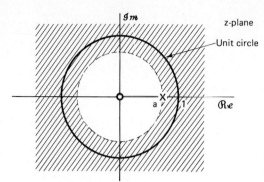

Figure 4.3 Pole-zero plot and region of convergence for Example 4.1.

can be represented as a sum of exponentials can be represented by a rational z-transform. Such a z-transform is determined to within a constant multiplier by its zeros and its poles. For this example, there is one zero, at $z = 0$, and one pole, at $z = a$. The pole-zero plot and the region of convergence for Example 4.1 are shown in Fig. 4.3. For $|a| > 1$, the ROC does not include the unit circle, consistent with the fact that for these values of a, the Fourier transform of the exponentially growing sequence $a^n u[n]$ does not converge.

Example 4.2

Now let $x[n] = -a^n u[-n-1]$. Then

$$X(z) = - \sum_{n=-\infty}^{\infty} a^n u[-n-1] z^{-n} = - \sum_{n=-\infty}^{-1} a^n z^{-n}$$

$$= - \sum_{n=1}^{\infty} a^{-n} z^n = 1 - \sum_{n=0}^{\infty} (a^{-1} z)^n. \tag{4.12}$$

If $|a^{-1} z| < 1$ or, equivalently, $|z| < |a|$, the sum in Eq. (4.12) converges and

$$X(z) = 1 - \frac{1}{1 - a^{-1} z} = \frac{1}{1 - a z^{-1}} = \frac{z}{z - a}, \qquad |z| < |a|. \tag{4.13}$$

The pole-zero plot and region of convergence for this example are shown in Fig. 4.4. Note that for $|a| < 1$ the sequence $- a^n u[-n-1]$ would grow exponentially as $n \to -\infty$ and, thus, the Fourier transform would not exist.

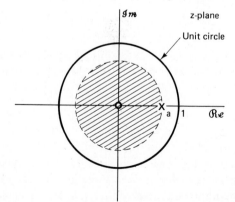

Figure 4.4 Pole-zero plot and region of convergence for Example 4.2.

Comparing Eqs. (4.10) and (4.13) and Figs. 4.3 and 4.4, we see that the sequences and therefore the infinite sums are different; however, the algebraic expressions for $X(z)$ and the corresponding pole-zero plots are identical in Examples 4.1 and 4.2. The z-transforms differ only in the region of convergence. This emphasizes the need for specifying both the algebraic expression and the region of convergence for the z-transform of a given sequence. Also, in both examples the sequences were exponentials and the resulting z-transforms were rational. In fact, as further suggested by the next example, $X(z)$ will be rational whenever $x[n]$ is a linear combination of real or complex exponentials.

Example 4.3

Consider a signal that is the sum of two real exponentials:

$$x[n] = (\tfrac{1}{2})^n u[n] + (-\tfrac{1}{3})^n u[n]. \tag{4.14}$$

The z-transform is then

$$X(z) = \sum_{n=-\infty}^{\infty} \{(\tfrac{1}{2})^n u[n] + (-\tfrac{1}{3})^n u[n]\} z^{-n}$$

$$= \sum_{n=-\infty}^{\infty} (\tfrac{1}{2})^n u[n] z^{-n} + \sum_{n=-\infty}^{\infty} (-\tfrac{1}{3})^n u[n] \, z^{-n} \tag{4.15}$$

$$= \sum_{n=0}^{\infty} (\tfrac{1}{2} z^{-1})^n + \sum_{n=0}^{\infty} (-\tfrac{1}{3} z^{-1})^n$$

$$= \frac{1}{1 - \tfrac{1}{2} z^{-1}} + \frac{1}{1 + \tfrac{1}{3} z^{-1}} = \frac{2(1 - \tfrac{1}{12} z^{-1})}{(1 - \tfrac{1}{2} z^{-1})(1 + \tfrac{1}{3} z^{-1})}$$

$$= \frac{2z(z - \tfrac{1}{12})}{(z - \tfrac{1}{2})(z + \tfrac{1}{3})}. \tag{4.16}$$

For convergence of $X(z)$, both sums in Eq. (4.15) must converge, which requires that both $|\tfrac{1}{2} z^{-1}| < 1$ and $|(-\tfrac{1}{3}) z^{-1}| < 1$ or, equivalently, $|z| > \tfrac{1}{2}$ and $|z| > \tfrac{1}{3}$. Thus, the region of convergence is $|z| > \tfrac{1}{2}$. The pole-zero plot and ROC for the z-transform of each of the individual terms and for the combined signal are shown in Fig. 4.5.

In each of the preceding examples, we started with the definition of the sequence and manipulated each of the infinite sums into a form whose sum could be recognized. When the sequence is recognized as a sum of exponential sequences of the form of Examples 4.1 and 4.2, the z-transform can be obtained much more simply using the fact that the z-transform operator is linear. Specifically, from the definition of the z-transform, Eq. (4.2), if $x[n]$ is the sum of two terms, then $X(z)$ will be the sum of the corresponding z-transforms of the individual terms. The ROC will be the intersection of the individual regions of convergence, i.e., the values of z for which both individual sums converge. We have already used this fact in obtaining Eq. (4.15) in Example 4.3. Example 4.4 shows how the z-transform in Example 4.3 can be obtained in a much more straightforward manner.

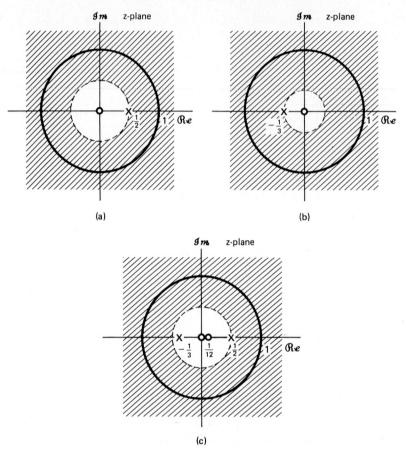

Figure 4.5 Pole-zero plot and region of convergence for the individual terms and the sum of terms in Examples 4.3 and 4.4. (a) $1/(1 - \frac{1}{2}z^{-1})$, $|z| > \frac{1}{2}$. (b) $1/(1 + \frac{1}{3}z^{-1})$, $|z| > \frac{1}{3}$. (c) $1/(1 - \frac{1}{2}z^{-1}) + 1/(1 + \frac{1}{3}z^{-1})$, $|z| > \frac{1}{2}$.

Example 4.4

Again let $x[n]$ be given by Eq. (4.14). Then the z-transforms of the two individual terms are

$$(\tfrac{1}{2})^n u[n] \xrightarrow{\;z\;} \frac{1}{1 - \tfrac{1}{2}z^{-1}}, \qquad |z| > \tfrac{1}{2}, \tag{4.17}$$

$$(-\tfrac{1}{3})^n u[n] \xrightarrow{\;z\;} \frac{1}{1 + \tfrac{1}{3}z^{-1}}, \qquad |z| > \tfrac{1}{3}, \tag{4.18}$$

and, consequently,

$$(\tfrac{1}{2})^n u[n] + (-\tfrac{1}{3})^n u[n] \xrightarrow{\;z\;} \frac{1}{1 - \tfrac{1}{2}z^{-1}} + \frac{1}{1 + \tfrac{1}{3}z^{-1}}, \qquad |z| > \tfrac{1}{2}, \tag{4.19}$$

as we had determined in Example 4.3. The pole-zero plot and ROC for the z-transform of each of the individual terms and for the combined signal are shown in Fig. 4.5.

All the major points of the Examples 4.1–4.5 are summarized by Example 4.5.

Example 4.5

Consider the sequence $x[n]$ given by

$$x[n] = (-\tfrac{1}{3})^n u[n] - (\tfrac{1}{2})^n u[-n-1].\qquad(4.20)$$

Note that this sequence grows exponentially as $n \to -\infty$. Using the general result of Example 4.1,

$$(-\tfrac{1}{3})^n u[n] \;\overset{z}{\longleftrightarrow}\; \frac{1}{1+\frac{1}{3}z^{-1}},\qquad |z| > \tfrac{1}{3},$$

and using the result of Example 4.2,

$$-(\tfrac{1}{2})^n u[-n-1] \;\overset{z}{\longleftrightarrow}\; \frac{1}{1-\frac{1}{2}z^{-1}},\qquad |z| < \tfrac{1}{2}.$$

Thus, by the linearity of the z-transform,

$$X(z) = \frac{1}{1+\frac{1}{3}z^{-1}} + \frac{1}{1-\frac{1}{2}z^{-1}},\qquad \tfrac{1}{3} < |z|,\;\; |z| < \tfrac{1}{2},$$

$$(4.21)$$

$$= \frac{2(1-\frac{1}{12}z^{-1})}{(1+\frac{1}{3}z^{-1})(1-\frac{1}{2}z^{-1})} = \frac{2z(z-\frac{1}{12})}{(z+\frac{1}{3})(z-\frac{1}{2})}.$$

In this case the ROC is the annular region $\tfrac{1}{3} < |z| < \tfrac{1}{2}$. Note that the rational function in this example is identical to the rational function in Examples 4.3 and 4.4, but the ROC is different in the two cases. The pole-zero plot and the ROC for this example are shown in Fig. 4.6. Note that in this case the ROC does not contain the unit circle, so the sequence in Eq. (4.20) does not have a Fourier transform.

In each of the preceding examples, we expressed the z-transform both as a ratio of polynomials in z and also as a ratio of polynomials in z^{-1}. From the form of the definition of the z-transform as given in Eq. (4.2), we see that for sequences that are zero for $n < 0$, $X(z)$ involves only negative powers of z. Thus for this class of signals it is particularly convenient for $X(z)$ to be expressed in terms of polynomials in z^{-1} rather than z; however, even when $x[n]$ is nonzero for $n < 0$, $X(z)$ can still be expressed in terms of factors of the form $(1 - az^{-1})$. It should be remembered that

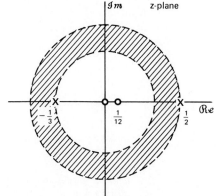

Figure 4.6 Pole-zero plot and region of convergence for Example 4.5.

such a factor introduces both a pole and a zero, as illustrated by the algebraic expressions for the preceding examples.

From these examples it is easily seen that infinitely long exponential sequences have z-transforms that can be expressed as rational functions of either z or z^{-1}. The case where the sequence has finite length also has a rather simple form. If the sequence is nonzero only in the interval $N_1 \le n \le N_2$, the z-transform

$$X(z) = \sum_{n=N_1}^{N_2} x[n]z^{-n} \tag{4.22}$$

has no problems of convergence as long as each of the terms $|x[n]z^{-n}|$ is finite. In general it may not be possible to express the sum of a finite set of terms in a closed form but in such cases it may be unnecessary. For example, if $x[n] = \delta[n] + \delta[n-5]$, then it is easily seen that $X(z) = 1 + z^{-5}$, which is finite for $|z| > 0$. An example of a case where a finite number of terms can be summed to produce a more compact representation of the z-transform is given in Example 4.6.

Example 4.6

Consider the signal

$$x[n] = \begin{cases} a^n, & 0 \le n \le N-1, \\ 0, & \text{otherwise.} \end{cases}$$

Then

$$X(z) = \sum_{n=0}^{N-1} a^n z^{-n} = \sum_{n=0}^{N-1} (az^{-1})^n$$

$$= \frac{1 - (az^{-1})^N}{1 - az^{-1}} = \frac{1}{z^{N-1}} \frac{z^N - a^N}{z - a}. \tag{4.23}$$

The ROC is determined by the set of values of z for which

$$\sum_{n=0}^{N-1} |az^{-1}|^n < \infty.$$

Since there are only a finite number of nonzero terms, the sum will be finite as long as $|az^{-1}|$ is finite, which in turn requires only that $|a| < \infty$ and $z \ne 0$. Thus, assuming that $|a|$ is finite, the ROC includes the entire z-plane with the exception of the origin

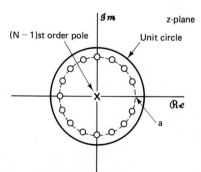

Figure 4.7 Pole-zero plot for Example 4.6 with $N = 16$ and a real such that $0 < a < 1$. The region of convergence for this example consists of all values of z except $z = 0$.

($z = 0$). The pole-zero plot for this example, with $N = 16$ and a real and between zero and unity, is shown in Fig. 4.7. Specifically, the N roots of the numerator polynomial are at

$$z_k = ae^{j(2\pi k/N)}, \qquad k = 0, 1, \ldots, N - 1. \tag{4.24}$$

(When $a = 1$, these complex values are the Nth roots of unity.) The zero at $k = 0$ cancels the pole at $z = a$. Consequently, there are no poles other than at the origin. The remaining zeros are at

$$z_k = ae^{j(2\pi k/N)}, \qquad k = 1, \ldots, N - 1. \tag{4.25}$$

The transform pairs corresponding to some of the preceding examples, as well as a number of other commonly encountered z-transform pairs, are summarized in Table 4.1. We will see that these basic transform pairs are very useful in finding z-transforms given a sequence or, conversely, in finding the sequence corresponding to a given z-transform.

TABLE 4.1 SOME COMMON z-TRANSFORM PAIRS

Sequence	Transform	ROC
1. $\delta[n]$	1	All z
2. $u[n]$	$\dfrac{1}{1 - z^{-1}}$	$|z| > 1$
3. $-u[-n-1]$	$\dfrac{1}{1 - z^{-1}}$	$|z| < 1$
4. $\delta[n-m]$	z^{-m}	All z except 0 (if $m > 0$) or ∞ (if $m < 0$)
5. $a^n u[n]$	$\dfrac{1}{1 - az^{-1}}$	$|z| > |a|$
6. $-a^n u[-n-1]$	$\dfrac{1}{1 - az^{-1}}$	$|z| < |a|$
7. $na^n u[n]$	$\dfrac{az^{-1}}{(1 - az^{-1})^2}$	$|z| > |a|$
8. $-na^n u[-n-1]$	$\dfrac{az^{-1}}{(1 - az^{-1})^2}$	$|z| < |a|$
9. $[\cos \omega_0 n]u[n]$	$\dfrac{1 - [\cos \omega_0]z^{-1}}{1 - [2\cos \omega_0]z^{-1} + z^{-2}}$	$|z| > 1$
10. $[\sin \omega_0 n]u[n]$	$\dfrac{[\sin \omega_0]z^{-1}}{1 - [2\cos \omega_0]z^{-1} + z^{-2}}$	$|z| > 1$
11. $[r^n \cos \omega_0 n]u[n]$	$\dfrac{1 - [r\cos \omega_0]z^{-1}}{1 - [2r\cos \omega_0]z^{-1} + r^2 z^{-2}}$	$|z| > r$
12. $[r^n \sin \omega_0 n]u[n]$	$\dfrac{[r\sin \omega_0]z^{-1}}{1 - [2r\cos \omega_0]z^{-1} + r^2 z^{-2}}$	$|z| > r$
13. $\begin{cases} a^n, & 0 \le n \le N-1, \\ 0, & \text{otherwise} \end{cases}$	$\dfrac{1 - a^N z^{-N}}{1 - az^{-1}}$	$|z| > 0$

4.2 PROPERTIES OF THE REGION OF CONVERGENCE FOR THE z-TRANSFORM

The examples of the previous section suggest that the properties of the region of convergence depend on the nature of the signal. These properties are summarized below, followed by some discussion and intuitive justification. We assume specifically that the algebraic expression for the z-transform is a rational function and that $x[n]$ has finite amplitude except possibly at $n = \infty$ or $n = -\infty$.

> **PROPERTY 1:**   The ROC is a ring or disk in the z-plane centered at the origin, i.e., $0 \le r_R < |z| < r_L \le \infty$.

> **PROPERTY 2:**   The Fourier transform of $x[n]$ converges absolutely if and only if the ROC of the z-transform of $x[n]$ includes the unit circle.

> **PROPERTY 3:**   The ROC cannot contain any poles.

> **PROPERTY 4:**   If $x[n]$ is a *finite-duration sequence*, i.e., a sequence that is zero except in a finite interval $-\infty < N_1 \le n \le N_2 < \infty$, then the ROC is the entire z-plane except possibly $z = 0$ or $z = \infty$.

> **PROPERTY 5:**   If $x[n]$ is a *right-sided sequence*, i.e., a sequence that is zero for $n < N_1 < \infty$, the ROC extends outward from the *outermost* (i.e., largest magnitude) finite pole in $X(z)$ to (and possibly including) $z = \infty$.

> **PROPERTY 6:**   If $x[n]$ is a *left-sided sequence*, i.e., a sequence that is zero for $n > N_2 > -\infty$, the ROC extends inward from the *innermost* (smallest magnitude) nonzero pole in $X(z)$ to (and possibly including) $z = 0$.

> **PROPERTY 7:**   A *two-sided sequence* is an infinite-duration sequence that is neither right-sided nor left-sided. If $x[n]$ is a two-sided sequence, the ROC will consist of a ring in the z-plane, bounded on the interior and exterior by a pole, and, consistent with property 3, not containing any poles.

> **PROPERTY 8:**   The ROC must be a connected region.

As discussed in Section 4.1, property 1 results from the fact that convergence of Eq. (4.2) for a given $x[n]$ is dependent only on $|z|$, and property 2 is a consequence of the fact that Eq. (4.2) reduces to the Fourier transform when $|z| = 1$. Property 3 follows from the recognition that $X(z)$ is infinite at a pole and therefore by definition does not converge.

Properties 4 through 7 can all be developed more or less directly from the interpretation of the z-transform as the Fourier transform of the original sequence modified by an exponential weighting. Let us first consider property 4. Figure 4.8 shows a finite-duration sequence and the exponential sequence r^{-n} for $1 < r$ (a decaying exponential) and for $0 < r < 1$ (a growing exponential). Convergence of the z-transform requires absolute summability of the sequence $x[n]|z|^{-n}$ or, equivalently, $x[n]r^{-n}$. It should be evident from Fig. 4.8 that since $x[n]$ has only a finite number of nonzero values, as long as each of these values is finite $x[n]$ will be absolutely

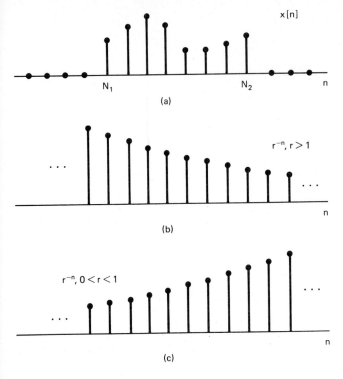

$x[n]$

N_1 N_2 n

(a)

$r^{-n}, r > 1$

n

(b)

$r^{-n}, 0 < r < 1$

n

(c)

Figure 4.8 Finite-length sequence and weighting sequences implicit in convergence of the z-transform. (a) The finite-length sequence $x[n]$. (b) Weighting sequence r^{-n} for $1 < r$. (c) Weighting sequence r^{-n} for $0 < r < 1$.

summable. Furthermore, this will not be affected by the exponential weighting if the weighting sequence has finite amplitude in the interval where $x[n]$ is nonzero, i.e., $N_1 \leq n \leq N_2$. Therefore, for a finite-duration sequence, $x[n]r^{-n}$ will be absolutely summable for $0 < r < \infty$. The only possible complication arises for $r = 0$ or for $r = \infty$. If $x[n]$ is nonzero for any positive values of n (i.e., if $N_2 > 0$) and if r, or equivalently $|z|$, is zero, then $x[n]r^{-n}$ will be infinite for $0 < n \leq N_2$. Correspondingly, if $x[n]$ is nonzero for any negative values of n (i.e., if $N_1 < 0$), then $x[n]r^{-n}$ will be infinite for $N_1 \leq n < 0$ if r, or equivalently $|z|$, is infinite.

   Property 5 can be interpreted in a somewhat similar manner. Figure 4.9 illustrates a right-sided sequence and the exponential sequence r^{-n} for two different values of r. A right-sided sequence is zero prior to some value of n, say N_1. If the circle $|z| = r_0$ is in the ROC, then $x[n]r_0^{-n}$ is absolutely summable or, equivalently, the Fourier transform of $x[n]r_0^{-n}$ converges. Since $x[n]$ is right-sided, the term $x[n]$ multiplied by any real exponential sequence which, with increasing n, decays faster than r_0^{-n} will also be absolutely summable. Specifically, as illustrated in Fig. 4.9, this more rapid exponential decay will further attenuate sequence values for positive values of n and cannot cause sequence values for negative values of n to become unbounded since $x[n]z^{-n} = 0$ for $n < N_1$. Based on this, we can conclude that for a right-sided sequence, the ROC extends outward from some circle in the z-plane, concentric with the origin. This circle in fact is at the outermost pole in $X(z)$. To see this, assume that the poles occur at $z = d_1, \ldots, d_N$, with d_1 having the smallest magnitude, i.e., corresponding to the innermost pole, and d_N having the largest

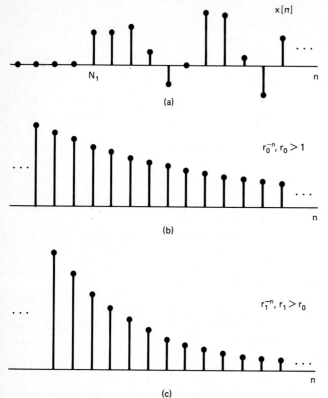

Figure 4.9 Right-sided sequence and weighting sequences implicit in convergence of the z-transform. (a) The right-sided sequence $x[n]$. (b) Weighting sequence r_0^{-n} for $1 < r_0$. (c) Weighting sequence r_1^{-n} for $r_1 > r_0$.

magnitude, i.e., corresponding to the outermost pole. To simplify the argument, we will assume that all the poles are simple poles although the argument can be easily generalized for multiple-order poles. As we will see in Section 4.3, for $N_1 \leq n$, $x[n]$ will consist of a sum of exponentials of the form

$$x[n] = \sum_{k=1}^{N} A_k(d_k)^n, \qquad n \geq N_1. \tag{4.26}$$

The least rapidly increasing of these exponentials, as n increases, is the one corresponding to the innermost pole, i.e., d_1, and the most slowly decaying (or most rapidly growing) is the one corresponding to the outermost pole, i.e., d_N. Now let us consider $x[n]$ with the exponential weighting r^{-n} applied, i.e.,

$$x[n]r^{-n} = r^{-n} \sum_{k=1}^{N} A_k(d_k)^n, \qquad n \geq N_1, \tag{4.27}$$

$$= \sum_{k=1}^{N} A_k(d_k r^{-1})^n, \qquad n \geq N_1. \tag{4.28}$$

Absolute summability of $x[n]r^{-n}$ requires that each exponential in Eq. (4.28) be absolutely summable, i.e.,

$$\sum_{n=N_1}^{\infty} |d_k r^{-1}|^n| < \infty, \qquad k = 1, \ldots, N, \tag{4.29}$$

or, equivalently,

$$|r| > |d_k|, \qquad k = 1, \ldots, N. \tag{4.30}$$

Since the outermost pole, d_N, is the one with the largest absolute value,

$$|r| > |d_N|, \tag{4.31}$$

i.e., the ROC is outside the outermost pole, extending to infinity. If $N_1 < 0$, the ROC will not include $|z| = \infty$ since r^{-n} is infinite for r infinite and n negative.

As suggested by the preceding discussion, it is possible to be very precise about property 5 (as well as the associated properties 6 and 7). The essence of the argument, however, is that for a sum of right-sided exponential sequences with an exponential weighting applied, the exponential weighting must be restricted so that all of the exponentially weighted terms decay with increasing n.

For property 6, which is concerned with left-sided sequences, an exactly parallel argument can be carried out. Here, however, $x[n]$ will consist of a sum of exponentials of the same form as Eq. (4.28) but for $n \leq N_2$, i.e.,

$$x[n] = \sum_{k=1}^{N} A_k(d_k)^n, \qquad n \leq N_2, \tag{4.32}$$

or, with exponential weighting,

$$x[n]r^{-n} = \sum_{k=1}^{N} A_k(d_k r^{-1})^n, \qquad n \leq N_2. \tag{4.33}$$

Since $x[n]$ now extends to $-\infty$ along the negative n-axis, r must be restricted so that for each d_k, the exponential sequence $(d_k r^{-1})^n$ decays to zero as n *decreases* toward $-\infty$. Equivalently,

$$|r| < |d_k|, \qquad k = 1, \ldots, N,$$

or, since d_1 has the smallest magnitude,

$$|r| < |d_1|, \tag{4.34}$$

i.e., the ROC is inside the innermost pole. If the left-sided sequence has nonzero values for positive values of n, then the ROC will not include the origin, $z = 0$.

For right-sided sequences, the ROC is dictated by the exponential weighting required to have all exponential terms decay to zero for increasing n; for left-sided sequences, the exponential weighting must be such that all exponential terms decay to zero for decreasing n. For two-sided sequences, the exponential weighting needs to be balanced since if it decays too fast for increasing n it may grow too quickly for decreasing n and vice versa. More specifically, for two-sided sequences some of the poles contribute only for $n > 0$ and the rest only for $n < 0$. The region of convergence

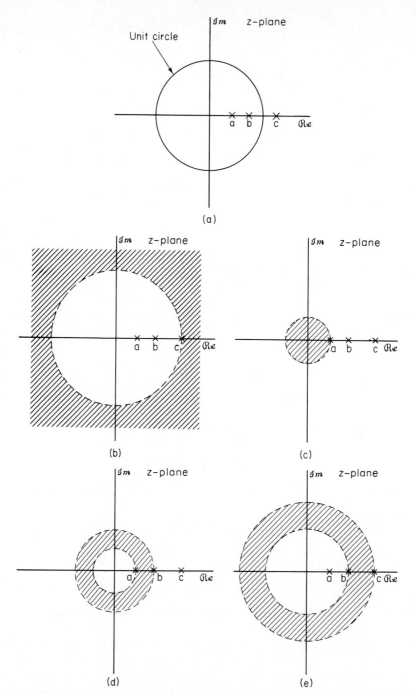

Figure 4.10 Examples of four z-transforms with the same pole-zero locations, illustrating the different possibilities for the region of convergence. Each corresponds to a different sequence: (b) to a right-sided sequence, (c) to a left-sided sequence, (d) to a two-sided sequence, and (e) to a two-sided sequence.

is bounded on the inside by the pole with the largest magnitude that contributes for $n > 0$ and on the outside by the pole with the smallest magnitude that contributes for $n < 0$.

Property 8 is somewhat more difficult to develop formally but, at least intuitively, it is strongly suggested by our discussion of properties 4 through 7. Any infinite two-sided sequence can be represented as a sum of a right-sided part (say for $n \geq 0$) and a left-sided part that includes everything not included in the right-sided part. The right-sided part will have an ROC given by Eq. (4.31) while the ROC of the left-sided part will be given by Eq. (4.34). The ROC of the entire two-sided sequence must be the intersection of these two regions. Thus, if such an intersection exists, it will always be a simply connected annular region of the form

$$r_R < |z| < r_L.$$

There is a possibility of no overlap between the regions of convergence of the right- and left-sided parts; i.e., $r_L < r_R$. An example is the sequence $x[n] = (\frac{1}{2})^n u[n] - (-\frac{1}{3})^n u[-n-1]$. In this case, the z-transform of the sequence simply does not exist. If such cases arise, however, it may still be possible to use the z-transform by considering the sequence to be the sum of two sequences each of which has a z-transform, but the two transforms cannot be combined in algebraic expressions since they have no common ROC.

To illustrate some of the properties of the ROC, consider the pole-zero pattern shown in Fig. 4.10(a). From properties 1, 3, and 8, there are only four possible choices for the ROC. These are indicated in Figs. 4.10(b), (c), (d), and (e), each being associated with a different sequence. Specifically, Fig. 4.10(b) corresponds to a right-sided sequence, Fig. 4.10(c) to a left-sided sequence, and Figs. 4.10(d) and 4.10(e) to two different two-sided sequences. If we assume, as indicated in Fig. 4.10(a), that the unit circle falls between the pole at $z = b$ and the pole at $z = c$, then the only one of the four cases for which the Fourier transform would converge is that in Fig. 4.10(e).

4.3 THE INVERSE z-TRANSFORM

One of the important roles of the z-transform is in the analysis of discrete-time linear systems. Often this analysis involves finding the z-transform of sequences and, after some manipulation of the algebraic expressions, finding the inverse z-transform. There are a number of formal and informal ways of determining the inverse z-transform from a given algebraic expression and associated region of convergence. In Section 4.5 we will develop the formal inverse z-transform expression based on the Cauchy integral theorem and illustrate how it can be evaluated using contour integration and residue theory. For the typical kinds of sequences and z-transforms that we will encounter in the analysis of discrete linear time-invariant systems, less formal procedures are sufficient and preferable. In Sections 4.3.1–4.3.3 we consider some of these procedures, specifically the inspection method, partial fraction expansion, and power series expansion.

4.3.1 Inspection Method

This method consists simply of becoming familiar with, or recognizing "by inspection," certain transform pairs. For example, in Section 4.1 we evaluated the z-transform for sequences of the form $x[n] = a^n u[n]$, where a can be either real or complex. Sequences of this form arise quite frequently, and consequently it is particularly useful to make direct use of the transform pair

$$a^n u[n] \overset{z}{\longleftrightarrow} \frac{1}{1 - az^{-1}}, \qquad |z| > |a|. \tag{4.35}$$

If we need to find the inverse z-transform of

$$X(z) = \left(\frac{1}{1 - \frac{1}{2}z^{-1}} \right), \qquad |z| > \tfrac{1}{2}, \tag{4.36}$$

and we recall the z-transform pair of Eq. (4.35), we would recognize "by inspection" the associated sequence as $x[n] = (\tfrac{1}{2})^n u[n]$. If the ROC associated with $X(z)$ in Eq. (4.36) had been $|z| < \tfrac{1}{2}$, we can recall transform pair 6 in Table 4.1 to find by inspection that $x[n] = -(\tfrac{1}{2})^n u[-n-1]$.

Tables of z-transforms, such as Table 4.1, are invaluable in applying the inspection method. If the table is extensive, it may be possible to express a given z-transform as a sum of terms each of whose inverse is given in the table. If so, the inverse transform (i.e., the corresponding sequence) can be written from the table. More extensive tables of z-transforms are given in Jury (1964).

4.3.2 Partial Fraction Expansion

As already described, inverse z-transforms can be found by inspection if the z-transform expression is recognized or tabulated. Sometimes $X(z)$ may not be given explicitly in an available table, but it may be possible to obtain an alternative expression for $X(z)$ as a sum of simpler terms, each of which is tabulated. This is the case for any rational function, since we can obtain a partial fraction expansion and easily identify the sequences corresponding to the individual terms.

To see how to obtain a partial fraction expansion, let us assume that $X(z)$ is expressed as a ratio of polynomials in z^{-1}; i.e.,

$$X(z) = \frac{\displaystyle\sum_{k=0}^{M} b_k z^{-k}}{\displaystyle\sum_{k=0}^{N} a_k z^{-k}}. \tag{4.37}$$

Such z-transforms arise frequently in the study of linear time-invariant systems. An equivalent expression is

$$X(z) = \frac{z^N \displaystyle\sum_{k=0}^{M} b_k z^{M-k}}{z^M \displaystyle\sum_{k=0}^{N} a_k z^{N-k}}. \tag{4.38}$$

Equation (4.38) points out that for such functions, there will be M zeros and N poles at nonzero locations in the z-plane. In addition, there will be either $M - N$ poles at $z = 0$ if $M > N$ or $N - M$ zeros at $z = 0$ if $N > M$. In other words, z-transforms of the form of Eq. (4.37) always have the same number of poles and zeros in the finite z-plane, and there are no poles or zeros at $z = \infty$. To obtain the partial fraction expansion of $X(z)$ in Eq. (4.37), it is most convenient to note that $X(z)$ could be expressed in the form

$$X(z) = \frac{b_0}{a_0} \frac{\prod\limits_{k=1}^{M}(1 - c_k z^{-1})}{\prod\limits_{k=1}^{N}(1 - d_k z^{-1})}, \tag{4.39}$$

where the c_k's are the nonzero zeros of $X(z)$ and the d_k's are the nonzero poles of $X(z)$. If $M < N$ and the poles are all first order, then $X(z)$ can be expressed as

$$X(z) = \sum_{k=1}^{N} \frac{A_k}{1 - d_k z^{-1}}. \tag{4.40}$$

Obviously the common denominator of the fractions in Eq. (4.40) is the same as the denominator in Eq. (4.39). Multiplying both sides of Eq. (4.40) by $(1 - d_k z^{-1})$ and evaluating for $z = d_k$ shows that the coefficients, A_k, can be found from

$$A_k = (1 - d_k z^{-1}) X(z) \Big|_{z = d_k}. \tag{4.41}$$

(In making the substitution $z = d_k$, it is often more convenient to substitute $z^{-1} = d_k^{-1}$.) Clearly, the numerator that would result from adding the terms in Eq. (4.40) would be at most of degree $(N - 1)$ in the variable z^{-1}. If $M \geq N$, then a polynomial must be added to the right-hand side of Eq. (4.40), the order of which is $(M - N)$. Thus for $M \geq N$, the complete partial fraction expansion would have the form

$$X(z) = \sum_{r=0}^{M-N} B_r z^{-r} + \sum_{k=1}^{N} \frac{A_k}{1 - d_k z^{-1}}. \tag{4.42}$$

If we are given a rational function of the form of Eq. (4.37), with $M \geq N$, the B_r's can be obtained by long division of the numerator by the denominator, with the division process terminating when the remainder is of lower degree than the denominator. The A_k's can still be obtained with Eq. (4.41).

If $X(z)$ has multiple-order poles and $M \geq N$, Eq. (4.42) must be further modified. In particular, if $X(z)$ has a pole of order s at $z = d_i$, then Eq. (4.42) becomes

$$X(z) = \sum_{r=0}^{M-N} B_r z^{-r} + \sum_{k=1,k\neq i}^{N} \frac{A_k}{1 - d_k z^{-1}} + \sum_{m=1}^{s} \frac{C_m}{(1 - d_i z^{-1})^m}. \tag{4.43}$$

The coefficients A_k and B_r are obtained as above. The coefficients C_m are obtained from the equation

$$C_m = \frac{1}{(s-m)!\,(-d_i)^{s-m}} \left\{ \frac{d^{s-m}}{dw^{s-m}} \left[(1 - d_i w)^s X(w^{-1}) \right] \right\}_{w = d_i^{-1}}. \tag{4.44}$$

Equation (4.43) gives the most general form for the partial fraction expansion of a rational z-transform expressed as a function of z^{-1} for the case $M \geq N$ and for d_i a pole of order s. If there are several multiple-order poles, there will be a term like the third sum in Eq. (4.43) for each multiple-order pole. If there are no multiple-order poles, Eq. (4.43) reduces to Eq. (4.42). If the order of the numerator is less than the order of the denominator $(M < N)$, then the polynomial term disappears from Eqs. (4.42) and (4.43).

It should be emphasized that we could have achieved the same results by assuming that the rational z-transform was expressed as a function of z instead of z^{-1}. That is, instead of factors of the form $(1 - az^{-1})$ we could have considered factors of the form $z - a$. This would lead to a set of equations similar in form to Eqs. (4.39)–(4.44) that would be convenient for use with a table of z-transforms expressed in terms of z. Since Table 4.1 is expressed in terms of z^{-1}, the development we pursued is more useful.

To see how to find the sequence corresponding to a given rational z-transform, let us suppose that $X(z)$ has only first-order poles, so that Eq. (4.42) is the most general form of the partial fraction expansion. To find $x[n]$ we first note that the z-transform operation is linear, so that the inverse transform of individual terms can be found and then added together to form $x[n]$.

The terms $B_r z^{-r}$ correspond to shifted and scaled impulse sequences, i.e., terms of the form $B_r \delta[n - r]$. The fractional terms correspond to exponential sequences. To decide whether a term

$$\frac{A_k}{1 - d_k z^{-1}}$$

corresponds to $(d_k)^n u[n]$ or $-(d_k)^n u[-n-1]$, we must use the properties of the region of convergence that were discussed in Section 4.2. From that discussion it follows that if $X(z)$ has only simple poles and the ROC is of the form $r_R < |z| < r_L$, then a given pole d_k will correspond to a right-sided exponential $(d_k)^n u[n]$ if $|d_k| < r_R$ and it will correspond to a left-sided exponential if $|d_k| > r_L$. Thus, the region of convergence can be used to sort the poles. Multiple-order poles also are divided into left-sided and right-sided contributions in the same way. The use of the region of convergence in finding inverse z-transforms from the partial fraction expansion is illustrated by the following examples.

Example 4.7

Suppose that a sequence $x[n]$ has z-transform

$$X(z) = \frac{1 + 2z^{-1} + z^{-2}}{1 - \frac{3}{2}z^{-1} + \frac{1}{2}z^{-2}} = \frac{1 + 2z^{-1} + z^{-2}}{(1 - \frac{1}{2}z^{-1})(1 - z^{-1})} \tag{4.45}$$

with region of convergence $|z| > 1$. The pole-zero plot for $X(z)$ is shown in Fig. 4.11. From the region of convergence, it is clear that $x[n]$ is a right-sided sequence. Since

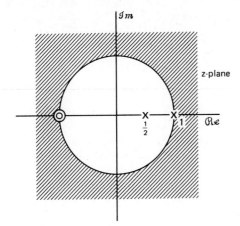

Figure 4.11   Pole-zero plot for the z-transform in Example 4.7.

$M = N = 2$ and the poles are all first order, we know that $X(z)$ can be represented as

$$X(z) = B_0 + \frac{A_1}{1 - \frac{1}{2}z^{-1}} + \frac{A_2}{1 - z^{-1}}.$$

The constant B_0 can be found by long division:

$$
\begin{array}{r}
2 \\
\tfrac{1}{2}z^{-2} - \tfrac{3}{2}z^{-1} + 1 \overline{\smash{\big)}\, z^{-2} + 2z^{-1} + 1} \\
\underline{z^{-2} - 3z^{-1} + 2} \\
5z^{-1} - 1
\end{array}
$$

Since the remainder after one step of long division is of degree 1 in the variable z^{-1}, it is not necessary to continue to divide. Thus, $X(z)$ can be expressed as

$$X(z) = 2 + \frac{-1 + 5z^{-1}}{(1 - \frac{1}{2}z^{-1})(1 - z^{-1})}.$$

Now the coefficients A_1 and A_2 can be found using Eq. (4.41):

$$A_1 = \left. \frac{1 + 2z^{-1} + z^{-2}}{1 - z^{-1}} \right|_{z^{-1}=2} = \frac{1 + 4 + 4}{1 - 2} = -9,$$

$$A_2 = \left. \frac{1 + 2z^{-1} + z^{-2}}{1 - \frac{1}{2}z^{-1}} \right|_{z^{-1}=1} = \frac{1 + 2 + 1}{\frac{1}{2}} = 8.$$

Therefore $X(z)$ is

$$X(z) = 2 - \frac{9}{1 - \frac{1}{2}z^{-1}} + \frac{8}{1 - z^{-1}}. \qquad (4.46)$$

From Table 4.1 we see that

$$2 \quad \overset{z}{\longleftrightarrow} \quad 2\delta[n],$$

$$\frac{1}{1 - \frac{1}{2}z^{-1}} \quad \overset{z}{\longleftrightarrow} \quad (\tfrac{1}{2})^n u[n],$$

$$\frac{1}{1 - z^{-1}} \quad \overset{z}{\longleftrightarrow} \quad u[n].$$

Thus, from the linearity of the z-transform,

$$x[n] = 2\delta[n] - 9(\tfrac{1}{2})^n u[n] + 8u[n].$$

Example 4.8

In this example we consider all possible sequences for which $X(z)$ is given by Eq. (4.45). From the pole-zero diagram of Fig. 4.11 and the properties of the region of convergence, it follows that $X(z)$ can have three different regions of convergence: (1) $|z| > 1$ as in Example 4.7; (2) $|z| < \tfrac{1}{2}$; and (3) $\tfrac{1}{2} < |z| < 1$. Case 1 was shown to be a right-sided sequence; case 2 corresponds to a left-sided sequence; and case 3 corresponds to a two-sided sequence. Since the partial fraction expansion of $X(z)$ depends only on the algebraic form of $X(z)$, it will be the same for all three cases. By inspection it follows that the following three sequences all have the z-transform given in Eqs. (4.45) and (4.46) but with different regions of convergence:

CASE 1: $x[n] = 2\delta[n] - 9(\tfrac{1}{2})^n u[n] + 8u[n]$

CASE 2: $x[n] = 2\delta[n] + 9(\tfrac{1}{2})^n u[-n-1] - 8u[-n-1]$

CASE 3: $x[n] = 2\delta[n] - 9(\tfrac{1}{2})^n u[n] - 8u[-n-1]$

In Section 4.4 we will discuss and illustrate a number of properties of the z-transform that, in combination with the partial fraction expansion, provide a means for determining the inverse z-transform from a given rational algebraic expression and associated ROC even when $X(z)$ is not exactly in the form of Eq. (4.39).

4.3.3 Power Series Expansion

The defining expression for the z-transform is a power series where the sequence values $x[n]$ are the coefficients of z^{-n}. Thus, if the z-transform is given as a power series in the form

$$X(z) = \sum_{n=-\infty}^{\infty} x[n]z^{-n}$$

$$= \cdots + x[-2]z^2 + x[-1]z + x[0] + x[1]z^{-1} + x[2]z^{-2} + \cdots,$$

(4.47)

we can determine any particular value of the sequence by finding the coefficient of the appropriate power of z^{-1}. We have already used this approach in finding the inverse transform of the polynomial part of the partial fraction expansion when $M \geq N$. This approach is also very useful for finite-length sequences where $X(z)$ may have no simpler form than a polynomial in z^{-1}.

Example 4.9

Suppose $X(z)$ is given in the form

$$X(z) = z^2(1 - \tfrac{1}{2}z^{-1})(1 + z^{-1})(1 - z^{-1}).$$

(4.48)

Although $X(z)$ is obviously a rational function, its only poles are at $z = 0$, so a partial fraction expansion according to the technique of Section 4.3.2 is not appropriate. However, by multiplying the factors of Eq. (4.48), we can express $X(z)$ as

$$X(z) = z^2 - \tfrac{1}{2}z - 1 + \tfrac{1}{2}z^{-1}.$$

Therefore, by inspection, $x[n]$ is seen to be

$$x[n] = \begin{cases} 1, & n = -2, \\ -\tfrac{1}{2}, & n = -1, \\ -1, & n = 0, \\ \tfrac{1}{2}, & n = 1, \\ 0, & \text{otherwise.} \end{cases}$$

Equivalently,

$$x[n] = \delta[n+2] - \tfrac{1}{2}\delta[n+1] - \delta[n] + \tfrac{1}{2}\delta[n-1].$$

In finding z-transforms of a sequence we generally seek to sum the power series of Eq. (4.47) to obtain a simpler mathematical expression, e.g., a rational function. If we wish to use the power series to find the sequence corresponding to a given $X(z)$ expressed in closed form we must expand $X(z)$ back into a power series. Many power series have been tabulated for transcendental functions such as log, sin, sinh, etc. (see, for example, Jolley, 1961). In some cases such power series can have a useful interpretation as z-transforms, as we illustrate in Example 4.10. For rational z-transforms, a power series expansion can be obtained by long division, as illustrated in Examples 4.11 and 4.12.

Example 4.10

Consider the z-transform

$$X(z) = \log(1 + az^{-1}), \qquad |z| > |a|. \tag{4.49}$$

Using the power series expansion for $\log(1 + x)$, with $|x| < 1$, we obtain

$$X(z) = \sum_{n=1}^{\infty} \frac{(-1)^{n+1}a^n z^{-n}}{n}.$$

Therefore, $x[n]$ is seen to be

$$x[n] = \begin{cases} (-1)^{n+1}\dfrac{a^n}{n}, & n \geq 1, \\ 0, & n \leq 0. \end{cases} \tag{4.50}$$

Example 4.11

Consider the z-transform

$$X(z) = \frac{1}{1 - az^{-1}}, \qquad |z| > |a|. \tag{4.51}$$

Since the region of convergence is the exterior of a circle, the sequence is a right-sided sequence. Furthermore, since $X(z)$ approaches a finite constant as z approaches infinity, it is a causal sequence. Thus we divide so as to obtain a series in powers of z^{-1}. Carrying out the long division, we obtain

$$
1 - az^{-1} \overline{\left) \begin{array}{l} 1 + az^{-1} + a^2z^{-2} + \cdots \\ 1 \\ \\ \end{array} \right.}
$$

$$
\begin{array}{l}
\underline{1 - az^{-1}} \\
az^{-1} \\
\underline{az^{-1} - a^2z^{-2}} \\
a^2z^{-2} \cdots
\end{array}
$$

or

$$
\frac{1}{1 - az^{-1}} = 1 + az^{-1} + a^2z^{-2} + \cdots ,
$$

so $x[n] = a^n u[n]$.

Example 4.12

As another example we can consider the same ratio of polynomials as in Eq. (4.51) but with a different region of convergence:

$$
X(z) = \frac{1}{1 - az^{-1}}, \qquad |z| < |a|. \tag{4.52}
$$

Because of the region of convergence, the sequence is a left-sided sequence and since $X(z)$ at $z = 0$ is finite, the sequence is zero for $n > 0$. Thus we divide so as to obtain a series in powers of z as follows:

$$
-a + z \overline{\left) \begin{array}{l} -a^{-1}z - a^{-2}z^2 - \cdots \\ z \end{array} \right.}
$$

$$
\begin{array}{l}
\underline{z - a^{-1}z^2} \\
a^{-1}z^2
\end{array}
$$

Therefore, $x[n] = -a^n u[-n - 1]$.

4.4 *z*-TRANSFORM PROPERTIES

Many of the properties of the z-transform are particularly useful in studying discrete-time signals and systems. For example, these properties are often used in conjunction with the inverse z-transform techniques discussed in Sec. 4.3, to obtain the inverse z-transform of more complicated expressions. In Chapter 5 we will see that these properties also form the basis for transforming linear constant-coefficient difference equations to algebraic equations in terms of the transform variable z, the solution to which can then be obtained using the inverse z-transform. In this section we consider some of the most frequently used properties. Additional theorems and properties can be found in a variety of references (see, for example, Ragazzini and Franklin, 1958;

Jury, 1964). In the following discussion, $X(z)$ denotes the z-transform of $x[n]$, and the ROC of $X(z)$ is indicated by R_x, i.e.,

$$x[n] \xleftrightarrow{z} X(z), \qquad \text{ROC} = R_x.$$

As we have seen, R_x represents a set of values of z such that $r_R < |z| < r_L$. For properties that involve two sequences and associated z-transforms, the transform pairs will be denoted as

$$x_1[n] \xleftrightarrow{z} X_1(z), \qquad \text{ROC} = R_{x_1},$$
$$x_2[n] \xleftrightarrow{z} X_2(z), \qquad \text{ROC} = R_{x_2}.$$

4.4.1 Linearity

The linearity property states that

$$ax_1[n] + bx_2[n] \xleftrightarrow{z} aX_1(z) + bX_2(z), \qquad \text{ROC contains } R_{x_1} \cap R_{x_2},$$

and follows directly from the z-transform definition, Eq. (4.2). As indicated, the region of convergence is at least the intersection of the individual regions of convergence. For sequences with rational z-transforms, if the poles of $aX_1(z) + bX_2(z)$ consist of all the poles of $X_1(z)$ and $X_2(z)$, i.e., there is no pole-zero cancellation, then the region of convergence will be exactly equal to the overlap of the individual regions of convergence. If the linear combination is such that some zeros are introduced that cancel poles, then the region of convergence may be larger. A simple example of this occurs when $x_1[n]$ and $x_2[n]$ are of infinite duration but the linear combination is of finite duration. In this case the region of convergence of the linear combination is the entire z-plane, with the possible exception of $z = 0$ or $z = \infty$. An example of this was given in Example 4.6, where $x[n]$ can be expressed as

$$x[n] = a^n u[n] - a^n u[n - N].$$

Both $a^n u[n]$ and $a^n u[n - N]$ are infinite-extent right-sided sequences, and their z-transforms have a pole at $z = a$. Therefore, their individual regions of convergence would both be $|z| > |a|$. However, as shown in Example 4.6, the pole at $z = a$ is canceled by a zero at $z = a$, and therefore the ROC extends to the entire z-plane with the exception of $z = 0$.

We have already exploited the linearity property in our previous discussion of the use of the partial fraction expansion for evaluating the inverse z-transform. With that procedure, $X(z)$ is expanded into a sum of simpler terms, and through linearity the inverse z-transform is the sum of the inverse transforms of each of these terms.

4.4.2 Time Shifting

$$x[n - n_0] \xleftrightarrow{z} z^{-n_0} X(z), \qquad \begin{array}{l} \text{ROC} = R_x \text{ (except for the} \\ \text{possible addition or} \\ \text{deletion of } z = 0 \text{ or } z = \infty) \end{array}$$

The quantity n_0 is an integer. If it is positive, the original sequence $x[n]$ is shifted right, and if n_0 is negative, $x[n]$ is shifted left. As in the case of linearity, the ROC can be changed since the factor z^{-n_0} can alter the number of poles at $z = 0$ or $z = \infty$.

The derivation of this property follows directly from the z-transform expression in Eq. (4.2). Specifically, if $y[n] = x[n - n_0]$, the corresponding z-transform is

$$Y(z) = \sum_{n=-\infty}^{\infty} x[n - n_0]z^{-n}.$$

With the substitution of variables $m = n - n_0$,

$$Y(z) = \sum_{m=-\infty}^{\infty} x[m]z^{-(m+n_0)}$$

$$= z^{-n_0} \sum_{m=-\infty}^{\infty} x[m]z^{-m}$$

or

$$Y(z) = z^{-n_0} X(z).$$

The time-shifting property is often useful, in conjunction with other properties and procedures, for obtaining the inverse z-transform. We illustrate with an example.

Example 4.13

Consider the z-transform

$$X(z) = \frac{1}{z - \frac{1}{4}}, \qquad |z| > \tfrac{1}{4}.$$

From the ROC we identify this as corresponding to a right-sided sequence. We can first rewrite $X(z)$ in the form

$$X(z) = \frac{z^{-1}}{1 - \frac{1}{4}z^{-1}}, \qquad |z| > \tfrac{1}{4}. \tag{4.53}$$

This z-transform is of the form of Eq. (4.39) with $M = N = 1$, and its expansion in the form of Eq. (4.42) is

$$X(z) = -4 + \frac{4}{1 - \frac{1}{4}z^{-1}}. \tag{4.54}$$

From Eq. (4.54) it follows that $x[n]$ can be expressed as

$$x[n] = -4\delta[n] + 4(\tfrac{1}{4})^n u[n]. \tag{4.55}$$

An expression for $x[n]$ can be obtained more directly by applying the time-shifting property. First, $X(z)$ can be written as

$$X(z) = z^{-1}\left(\frac{1}{1 - \frac{1}{4}z^{-1}}\right), \qquad |z| > \tfrac{1}{4}. \tag{4.56}$$

From the time-shifting property, we recognize the factor z^{-1} in Eq. (4.56) as being associated with a time shift of one sample to the right of the sequence $(\frac{1}{4})^n u[n]$; i.e.,

$$x[n] = (\tfrac{1}{4})^{n-1} u[n-1] \tag{4.57}$$

Although Eqs. (4.55) and (4.57) appear to be different sequences, it is easily verified that they are the same for all values of n.

4.4.3 Multiplication by an Exponential Sequence

$$z_0^n x[n] \xleftrightarrow{z} X(z/z_0), \qquad \text{ROC} = |z_0| R_x$$

The notation $\text{ROC} = |z_0| R_x$ denotes that the ROC is R_x scaled by $|z_0|$; i.e., if R_x is the set of values of z such that $r_R < |z| < r_L$, then $|z_0| R_x$ is the set of values of z such that $|z_0| r_R < |z| < |z_0| r_L$.

This property is easily shown by simply substituting $z_0^n x[n]$ into Eq. (4.2). As a consequence of this property, all the pole-zero locations are scaled by a factor z_0 since if $X(z)$ has a pole at $z = z_1$, then $X(z_0^{-1} z)$ will have a pole at $z = z_0 z_1$. If z_0 is a positive real number, this can be interpreted as a shrinking or expanding of the z-plane; i.e., the pole and zero locations change along radial lines in the z-plane. If z_0 is complex with unity magnitude, so that $z_0 = e^{j\omega_0}$, the scaling corresponds to a rotation in the z-plane by an angle of ω_0; i.e., the pole and zero locations change in position along circles centered at the origin. This in turn can be interpreted as a frequency shift or translation, associated with the modulation in the time domain by the complex exponential sequence $e^{j\omega_0 n}$. That is, if the Fourier transform exists, this property has the form

$$e^{j\omega_0 n} x[n] \xleftrightarrow{\mathcal{F}} X(e^{j(\omega - \omega_0)})$$

Example 4.14

Suppose that we know that

$$u[n] \xleftrightarrow{z} \frac{1}{1 - z^{-1}}, \qquad |z| > 1, \tag{4.58}$$

and we wish to determine the z-transform of

$$x[n] = r^n \cos(\omega_0 n)\, u[n] \tag{4.59}$$

using Eq. (4.58) and the exponential multiplication property. First $x[n]$ is expressed as

$$x[n] = \tfrac{1}{2}(re^{j\omega_0})^n u[n] + \tfrac{1}{2}(re^{-j\omega_0})^n u[n].$$

Then using Eq. (4.58) and the exponential multiplication property, we see that

$$\tfrac{1}{2}(re^{j\omega_0})^n u[n] \xleftrightarrow{z} \frac{\tfrac{1}{2}}{1 - re^{j\omega_0} z^{-1}}, \qquad |z| > r,$$

$$\tfrac{1}{2}(re^{-j\omega_0})^n u[n] \xleftrightarrow{z} \frac{\tfrac{1}{2}}{1 - re^{-j\omega_0} z^{-1}}, \qquad |z| > r.$$

From the linearity property it follows that

$$X(z) = \frac{\frac{1}{2}}{1 - re^{j\omega_0}z^{-1}} + \frac{\frac{1}{2}}{1 - re^{-j\omega_0}z^{-1}}, \qquad |z| > r$$

$$= \frac{(1 - r\cos\omega_0 z^{-1})}{1 - 2r\cos\omega_0 z^{-1} + r^2 z^{-2}}, \qquad |z| > r. \qquad (4.60)$$

4.4.4 Differentiation of X(z)

$$nx[n] \xleftrightarrow{\ z\ } -z\frac{dX(z)}{dz}, \qquad \begin{array}{l} \text{ROC} = R_x(\text{except for the} \\ \text{possible addition} \\ \text{or deletion of } z = 0 \\ \text{or } z = \infty) \end{array}$$

This property is verified by differentiating the z-transform expression of Eq. (4.2); i.e., since

$$X(z) = \sum_{n=-\infty}^{\infty} x[n]z^{-n},$$

$$-z\frac{dX(z)}{dz} = -z\sum_{n=-\infty}^{\infty} (-n)x[n]z^{-n-1}$$

$$= \sum_{n=-\infty}^{\infty} nx[n]z^{-n} = \mathcal{Z}\{nx[n]\}$$

We illustrate the use of the differentiation property with two examples.

Example 4.15

In this example we use the differentiation property together with the time-shifting property to determine the inverse z-transform considered in Example 4.10. With $X(z)$ given by

$$X(z) = \log(1 + az^{-1}), \qquad |z| > |a|,$$

we first differentiate to obtain a rational expression:

$$\frac{dX(z)}{dz} = \frac{-az^{-2}}{1 + az^{-1}}.$$

From the differentiation property,

$$nx[n] \xleftrightarrow{\ z\ } -z\frac{dX(z)}{dz} = \frac{az^{-1}}{1 + az^{-1}}, \qquad |z| > |a|. \qquad (4.61)$$

The inverse transform of Eq. (4.61) can be obtained by the combined use of the z-transform pair of Example 4.1, the differentiation property, the linearity property, and the time-shifting property. Specifically, we can express $nx[n]$ as

$$nx[n] = a(-a)^{n-1}u[n-1].$$

Therefore

$$x[n] = (-1)^{n+1} \frac{a^n}{n} u[n-1] \xleftrightarrow{z} \log(1 + az^{-1}), \qquad |z| > |a|.$$

Example 4.16

As another example of the use of the differentiation property, let us determine the z-transform of the sequence

$$x[n] = na^n u[n] = n(a^n u[n]).$$

From the z-transform pair of Example 4.1 and the differentiation property, it follows that

$$X(z) = -z \frac{d}{dz}\left(\frac{1}{1 - az^{-1}}\right), \qquad |z| > |a|$$

$$= \frac{az^{-1}}{(1 - az^{-1})^2}, \qquad |z| > |a|.$$

Therefore

$$na^n u[n] \xleftrightarrow{z} \frac{az^{-1}}{(1 - az^{-1})^2}, \qquad |z| > |a|.$$

4.4.5 Conjugation of a Complex Sequence

$$x^*[n] \xleftrightarrow{z} X^*(z^*), \qquad \text{ROC} = R_x$$

This property follows in a straightforward manner from the definition of the z-transform, the details of which are left as an exercise (Problem 4.29).

4.4.6 Time Reversal

$$x[-n] \xleftrightarrow{z} X(1/z), \qquad \text{ROC} = \frac{1}{R_x}$$

The notation ROC $= 1/R_x$ implies that R_x is inverted; i.e., if R_x is the set of values of z such that $r_R < |z| < r_L$, then the ROC is the set of values of z such that $1/r_L < |z| < 1/r_R$. Thus, if z_0 is in the ROC for $x[n]$, then $1/z_0$ is in the ROC for the z-transform of $x[-n]$.

As with Property 4.4.5, this property follows easily from the definition of the z-transform and the details are left as an exercise (Problem 4.29).

Example 4.17

As a simple example of the use of the property of time reversal, consider the sequence

$$x[n] = a^{-n} u[-n],$$

which is a time-reversed version of $a^n u[n]$. From the time-reversal property it follows that

$$X(z) = \frac{1}{1 - az} = \frac{-a^{-1}z^{-1}}{1 - a^{-1}z^{-1}}, \qquad |z| < |a^{-1}|.$$

4.4.7 Convolution of Sequences

$$x_1[n] * x_2[n] \xleftrightarrow{\ z\ } X_1(z)X_2(z), \qquad \text{ROC contains } R_{x_1} \cap R_{x_2}$$

To derive this property formally we consider

$$y[n] = \sum_{k=-\infty}^{\infty} x_1[k]x_2[n-k]$$

so that

$$Y(z) = \sum_{n=-\infty}^{\infty} y[n]z^{-n}$$

$$= \sum_{n=-\infty}^{\infty} \left\{ \sum_{k=-\infty}^{\infty} x_1[k]x_2[n-k] \right\} z^{-n}.$$

If we interchange the order of summation,

$$Y(z) = \sum_{k=-\infty}^{\infty} x_1[k] \sum_{n=-\infty}^{\infty} x_2[n-k]z^{-n}.$$

Changing the index of summation in the second sum from n to $m = n - k$, we obtain

$$Y(z) = \sum_{k=-\infty}^{\infty} x_1[k] \left\{ \sum_{m=-\infty}^{\infty} x_2[m]z^{-m} \right\} z^{-k}.$$

Thus, for values of z inside the regions of convergence of both $X_1(z)$ and $X_2(z)$ we can write

$$Y(z) = X_1(z)X_2(z),$$

where the region of convergence includes the intersection of the regions of convergence of $X_1(z)$ and $X_2(z)$. If a pole that borders on the region of convergence of one of the z-transforms is canceled by a zero of the other, then the region of convergence of $Y(z)$ may be larger. As we develop and exploit in Chapter 5, the convolution property plays a particularly important role in the analysis of LTI systems. Specifically, as a consequence of this property, the z-transform of the output of an LTI system is the product of the z-transform of the input and the z-transform of the system impulse response.

Example 4.18

Let $x_1[n] = a^n u[n]$ and $x_2[n] = u[n]$. The corresponding z-transforms are

$$X_1(z) = \sum_{n=0}^{\infty} a^n z^{-n} = \frac{1}{1 - az^{-1}}, \qquad |z| > |a|$$

and

$$X_2(z) = \sum_{n=0}^{\infty} z^{-n} = \frac{1}{1 - z^{-1}}, \qquad |z| > 1.$$

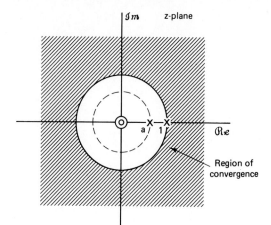

Figure 4.12 Pole-zero plot for the z-transform of the convolution of the sequences $u[n]$ and $a^n u[n]$.

If $|a| < 1$, the z-transform of the convolution of $x_1[n]$ with $x_2[n]$ is then

$$Y(z) = \frac{1}{(1 - az^{-1})(1 - z^{-1})} = \frac{z^2}{(z - a)(z - 1)}, \qquad |z| > 1. \tag{4.62}$$

The poles and zeros of $Y(z)$ are plotted in Fig. 4.12 and the region of convergence is seen to be the overlap region. The sequence $y[n]$ can be obtained by determining the inverse z-transform. By expanding $Y(z)$ in Eq. (4.62) in a partial fraction expansion,

$$Y(z) = \frac{1}{1 - a}\left(\frac{1}{1 - z^{-1}} - \frac{a}{1 - az^{-1}}\right), \qquad |z| > 1.$$

Therefore

$$y[n] = \frac{1}{1 - a}(u[n] - a^{n+1}u[n]).$$

4.4.8 Initial Value Theorem

If $x[n]$ is zero for $n < 0$, i.e., $x[n]$ is right-sided, then

$$x[0] = \lim_{z \to \infty} X(z).$$

This theorem is shown by considering the limit of each term in the series of Eq. (4.2) (see Problem 4.32).

4.4.9 Summary of Some z-Transform Properties

We have presented and discussed a number of the theorems and properties of z-transforms, many of which are useful in manipulating z-transforms. These properties and a number of others are summarized for convenient reference in Table 4.2. Also included in Table 4.2 are the complex convolution property and Parseval's relation, which are discussed in Sections 4.6 and 4.7. These properties are generalizations of corresponding Fourier transform properties discussed in Chapter 2.

TABLE 4.2 SOME z-TRANSFORM PROPERTIES

Property Number	Section Reference	Sequence	Transform	ROC
1		$x[n]$ $x_1[n]$ $x_2[n]$	$X(z)$ $X_1(z)$ $X_2(z)$	R_x R_{x_1} R_{x_2}
2	4.4.1	$ax_1[n] + bx_2[n]$	$aX_1(z) + bX_2(z)$	Contains $R_{x_1} \cap R_{x_2}$
3	4.4.2	$x[n - n_0]$	$z^{-n_0}X(z)$	R_x except for the possible addition or deletion of the origin or ∞
4	4.4.3	$z_0^n x[n]$	$X(z/z_0)$	$|z_0|R_x$
5	4.4.4	$nx[n]$	$-z\dfrac{dX(z)}{dz}$	R_x except for the possible addition or deletion of the origin or ∞
6	4.4.5	$x^*[n]$	$X^*(z^*)$	R_x
7		$\mathcal{R}e\{x[n]\}$	$\dfrac{1}{2}[X(z) + X^*(z^*)]$	Contains R_x
		$\mathcal{I}m\{x[n]\}$	$\dfrac{1}{2j}[X(z) - X^*(z^*)]$	Contains R_x
8	4.4.6	$x[-n]$	$X(1/z)$	$1/R_x$
9	4.4.7	$x_1[n] * x_2[n]$	$X_1(z)X_2(z)$	Contains $R_{x_1} \cap R_{x_2}$
10	4.4.8	Initial value theorem: $x[n] = 0, \quad n < 0$	$\displaystyle\lim_{z \to \infty} X(z) = x[0]$	
11	4.6	$x_1[n]x_2[n]$	$\dfrac{1}{2\pi j}\oint_c X_1(v)X_2(z/v)v^{-1}\,dv$	Contains $R_{x_1}R_{x_2}$
12	4.7	Parseval's relation:		

$$\sum_{n=-\infty}^{\infty} x_1[n]x_2^*[n] = \frac{1}{2\pi j}\oint_c X_1(v)X_2^*(1/v^*)v^{-1}\,dv$$

4.5 THE INVERSE z-TRANSFORM USING CONTOUR INTEGRATION

In Section 4.3 we discussed a number of procedures for obtaining the sequence associated with a given z-transform expression. In this section we develop a formal expression for the inverse z-transform. This expression is used in Sections 4.6 and 4.7 to develop two additional properties of the z-transform, specifically the complex convolution theorem and Parseval's relation for the z-transform.

A formal inverse z-transform relation can be derived using the Cauchy integral theorem (see Churchill and Brown, 1984). This theorem states that

$$\frac{1}{2\pi j} \oint_C z^{-k} \, dz = \begin{cases} 1, & k = 1, \\ 0, & k \neq 1, \end{cases} \tag{4.63}$$

where C is a counterclockwise contour that encircles the origin.

The z-transform relation is

$$X(z) = \sum_{n=-\infty}^{\infty} x[n] z^{-n}. \tag{4.64}$$

Multiplying both sides of Eq. (4.64) by z^{k-1} and integrating with a contour integral for which the contour of integration encloses the origin and lies entirely in the region of convergence of $X(z)$, we obtain

$$\frac{1}{2\pi j} \oint_C X(z) z^{k-1} \, dz = \frac{1}{2\pi j} \oint_C \sum_{n=-\infty}^{\infty} x[n] z^{-n+k-1} \, dz. \tag{4.65}$$

Interchanging the order of integration and summation on the right-hand side of Eq. (4.65) (valid if the series is convergent), we obtain

$$\frac{1}{2\pi j} \oint_C X(z) z^{k-1} \, dz = \sum_{n=-\infty}^{\infty} x[n] \frac{1}{2\pi j} \oint_C z^{-n+k-1} \, dz, \tag{4.66}$$

which from Eq. (4.63) becomes

$$\frac{1}{2\pi j} \oint_C X(z) z^{k-1} \, dz = x[k].$$

Therefore, the inverse z-transform relation is given by the contour integral

$$x[n] = \frac{1}{2\pi j} \oint_C X(z) z^{n-1} \, dz, \tag{4.67}$$

where C is a counterclockwise closed contour in the region of convergence of $X(z)$ and encircling the origin of the z-plane. It should be stressed that in deriving Eq. (4.67) we made no assumption about whether k in Eq. (4.65) or n in Eq. (4.67) was positive or negative, and consequently Eq. (4.67) is valid for both positive and negative values of n.

Equation (4.67) is the formal inverse z-transform expression. If the region of convergence includes the unit circle and if the contour of integration is taken to be the unit circle, then on this contour, $X(z)$ reduces to the Fourier transform and Eq. (4.67) reduces to the inverse Fourier transform expression, Eq. (2.112). Specifically, with $z = e^{j\omega}$, Eq. (4.67) becomes

$$x[n] = \frac{1}{2\pi} \int_{-\pi}^{\pi} X(e^{j\omega}) e^{j\omega n} \, d\omega, \tag{4.68}$$

where we have used the fact that integrating in z counterclockwise around the unit circle is equivalent to integrating in ω from $-\pi$ to $+\pi$ and that $dz = je^{j\omega} \, d\omega$.

Contour integrals of the form of Eq. (4.67) are often conveniently evaluated using Cauchy's residue theorem which, when applied to Eq. (4.67), gives

$$x[n] = \frac{1}{2\pi j} \oint_C X(z) z^{n-1} \, dz$$

$$= \sum [\text{residues of } X(z) z^{n-1} \text{ at the poles inside } C]. \tag{4.69}$$

Equation (4.69) is valid for any proper z-transform $X(z)$, but finding residues of nonrational functions is often difficult. However, if $X(z)z^{n-1}$ is a rational function of z, it may be expressed as

$$X(z) z^{n-1} = \frac{\psi(z)}{(z - d_0)^s}, \tag{4.70}$$

where $X(z)z^{n-1}$ has s poles at $z = d_0$ and $\psi(z)$ has no poles at $z = d_0$. The residue of $X(z)z^{n-1}$ at $z = d_0$ is given by

$$\text{Res}[X(z) z^{n-1} \text{ at } z = d_0] = \frac{1}{(s-1)!} \left[\frac{d^{s-1}\psi(z)}{dz^{s-1}} \right]_{z=d_0}. \tag{4.71}$$

In particular, if there is only a first-order pole at $z = d_0$, i.e., if $s = 1$, then

$$\text{Res}[X(z) z^{n-1} \text{ at } z = d_0] = \psi(d_0). \tag{4.72}$$

It is clear from a comparison of Eq. (4.44) and Eq. (4.71) that finding residues of $X(z)z^{n-1}$ is much like finding coefficients for a partial fraction expansion of $X(z)$. Generally, for rational fuctions it will be easier to use the method developed in Section 4.3.2; however, we will illustrate the use of Eq. (4.69) with the simple rational function of Example 4.19. We will also find the general expression of Eq. (4.67) to be useful in later chapters.

Example 4.19

As an example of the use of contour integration to evaluate the inverse transform relation, let us consider the inverse transform of

$$X(z) = \frac{1}{1 - az^{-1}}, \qquad |z| > |a|.$$

Using Eq. (4.67), we obtain

$$x[n] = \frac{1}{2\pi j} \oint_C \frac{z^{n-1}}{1 - az^{-1}} \, dz = \frac{1}{2\pi j} \oint_C \frac{z^n \, dz}{z - a},$$

where the contour of integration, C, is a circle of radius greater than a. For $n \geq 0$, the contour of integration encloses only one pole at $z = a$. Consequently, for $n \geq 0$, $x[n]$ is given by

$$x[n] = a^n, \qquad n \geq 0.$$

For $n < 0$, there is a multiple-order pole at $z = 0$ whose order depends on n. For $n = -1$, the pole is first order with a residue of $-a^{-1}$. The residue of the pole at $z = a$ is a^{-1}. Consequently, the sum of the residues is zero, and hence $x[-1] = 0$. For $n = -2$,

$$\mathrm{Res}\left[\frac{1}{z^2(z-a)} \text{ at } z = a\right] = a^{-2}$$

and

$$\mathrm{Res}\left[\frac{1}{z^2(z-a)} \text{ at } z = 0\right] = -a^{-2},$$

and thus $x[-2] = 0$. By continuing this procedure it can be verified that for this example $x[n] = 0$, $n < 0$. As n becomes more negative, the evaluation of the residue of the multiple-order pole at $z = 0$ becomes increasingly tedious.

While Eq. (4.67) is valid for all n, its use for $n < 0$ is often cumbersome because of the multiple-order poles at $z = 0$.

This can be avoided by modifying Eq. (4.67) by means of a substitution of variables, making the residue theorem easier to apply for $n < 0$. Specifically, consider the substitution of variables $z = p^{-1}$, so that Eq. (4.67) becomes

$$x[n] = \frac{-1}{2\pi j} \oint_{C'} X(1/p) p^{-n+1} p^{-2} \, dp. \qquad (4.73)$$

Observe that since the contour in Eq. (4.67) is a counterclockwise contour, the contour in Eq. (4.73) is a clockwise contour. Multiplying by -1 to reverse the direction of the contour, the above substitution of variables then leads to the expression

$$x[n] = \frac{1}{2\pi j} \oint_{C'} X(1/p) p^{-n-1} \, dp$$

$$= \sum [\text{residues of } X(1/p) p^{-n-1} \text{ at the poles inside } C']. \qquad (4.74)$$

If the contour C in Eq. (4.67) is a circle of radius r in the z-plane, then the contour C' in Eq. (4.74) is a circle of radius $1/r$ in the p-plane. The poles of $X(z)$ that were outside the contour C correspond now to poles of $X(1/p)$ that are inside the contour C', and vice versa. Additional poles or zeros may or may not appear at the origin or at infinity, but that is not crucial to the argument. For this specific example, $x[n]$ can also be expressed as

$$x[n] = \frac{1}{2\pi j} \oint_{C'} \frac{p^{-n-1}}{1 - ap} \, dp. \qquad (4.75)$$

The contour of integration C' is now a circle of radius less than $1/a$. For $n < 0$ there are no singularities inside the contour of integration, so $x[n] = 0$ for $n < 0$. Just as Eq. (4.69) was cumbersome (although certainly valid) for evaluating $x[n]$ for $n < 0$, the expression in Eq. (4.74) is likewise cumbersome (but still valid) for evaluating $x[n]$ for $n \geq 0$ because of the multiple-order poles that appear at the origin.

4.6 THE COMPLEX CONVOLUTION THEOREM

In Section 2.9.7, we discussed the periodic convolution property for Fourier transforms. Specifically, the Fourier transform of a product of sequences is the periodic convolution of their Fourier transforms. A generalization of that property is the complex convolution theorem relating the z-transform of a product of sequences to the z-transforms of the individual sequences.

To derive the complex convolution theorem, let

$$w[n] = x_1[n]x_2[n]$$

so that

$$W(z) = \sum_{n=-\infty}^{\infty} x_1[n]x_2[n]z^{-n}.$$

With $X_2(z)$ denoting the z-transform of $x_2[n]$,

$$x_2[n] = \frac{1}{2\pi j} \oint_{C_2} X_2(v)v^{n-1}\, dv,$$

where C_2 is a counterclockwise contour within the region of convergence of $X_2(v)$. Then

$$W(z) = \frac{1}{2\pi j} \sum_{n=-\infty}^{\infty} x_1[n] \oint_{C_2} X_2(v)(z/v)^{-n}\, v^{-1}\, dv$$

$$= \frac{1}{2\pi j} \oint_{C_2} \left[\sum_{n=-\infty}^{\infty} x_1[n](z/v)^{-n} \right] X_2(v)v^{-1}\, dv$$

or

$$W(z) = \frac{1}{2\pi j} \oint_{C_2} X_1(z/v)X_2(v)v^{-1}\, dv, \qquad (4.76)$$

where C_2 is a closed contour in the overlap of the regions of convergence of $X_1(z/v)$ and $X_2(v)$. Alternatively, $W(z)$ can be expressed as

$$W(z) = \frac{1}{2\pi j} \oint_{C_1} X_1(v)X_2(z/v)v^{-1}\, dv, \qquad (4.77)$$

where C_1 is a closed contour in the overlap of the regions of convergence of $X_1(v)$ and $X_2(z/v)$. To determine the region of convergence to associate with $W(z)$, let us assume the regions of convergence for $X_1(z)$ and $X_2(z)$, respectively, to be

$$R_{x_1}: \quad r_{R1} < |z| < r_{L1},$$

$$R_{x_2}: \quad r_{R2} < |z| < r_{L2}.$$

Then in Eq. (4.76), C_2 is a contour such that

$$r_{R2} < |v| < r_{L2}$$

and

$$r_{R1} < \left| \frac{z}{v} \right| < r_{L1}.$$

Combining these two expressions, we require that the ROC of $W(z)$ be

$$r_{R1} r_{R2} < |z| < r_{L1} r_{L2}.$$

This ROC is denoted $R_{x_1} R_{x_2}$. Sometimes the region of convergence may be larger than this, but it will always include the region defined above and then extend inward or outward to the nearest poles, i.e., the ROC of $W(z)$ contains $R_{x_1} R_{x_2}$.

As we would expect, Eqs. (4.76) and (4.77) reduce to periodic convolutions of the Fourier transforms when the contour C_2 or C_1 is taken to be the unit circle and $W(z)$ is evaluated on the unit circle. To see this, let

$$v = e^{j\theta} \qquad \text{and} \qquad z = e^{j\omega}.$$

Equation (4.77) becomes

$$W(e^{j\omega}) = \frac{1}{2\pi} \int_{-\pi}^{\pi} X_1(e^{j\theta}) X_2(e^{j(\omega - \theta)}) \, d\theta, \tag{4.78}$$

which is identical to the periodic convolution relationship in Section 2.9.7.

In using the complex convolution theorem of Eq. (4.76) or (4.77), one of the primary sources of difficulty is in determining which poles of the integrand are inside the contour of integration and which are outside. A simple example in the use of the complex convolution theorem should serve to illustrate the procedure.

Example 4.20

Let $x_1[n] = a^n u[n]$ and $x_2[n] = b^n u[n]$. Then the z-transforms $X_1(z)$ and $X_2(z)$ are, respectively,

$$X_1(z) = \frac{1}{1 - az^{-1}}, \qquad |z| > |a|,$$

$$X_2(z) = \frac{1}{1 - bz^{-1}}, \qquad |z| > |b|.$$

Substituting into Eq. (4.76), we obtain

$$W(z) = \frac{1}{2\pi j} \oint_{C_2} \frac{-(z/a)}{(v - z/a)} \frac{1}{v - b} \, dv.$$

The integrand has two poles, one located at $v = b$ and the second at $v = z/a$. Since the contour of integration of this expression must be within the region of convergence of $X_2(v)$, it must enclose the pole at $v = b$. To determine whether it encloses the pole at $v = z/a$, we note that the ROC of $X_1(z)$ is $|z| > |a|$. Therefore, the ROC for $X_1(z/v)$ is $|z/v| > |a|$. Thus if

$$\left| \frac{z}{v} \right| > |a|,$$

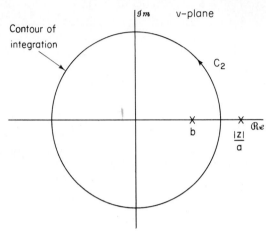

Contour of integration

Figure 4.13 **Figure 4.13** Poles of the integrand and the contour of integration in applying the complex convolution theorem in Example 4.20.

then

$$\left| \frac{z}{a} \right| > |v|.$$

Consequently, the pole must always lie outside the contour of integration C_2. The location of the poles and the contour of integration are indicated in Fig. 4.13, with a and b assumed to be real. Using the Cauchy residue theorem to evaluate $W(z)$, we obtain

$$W(z) = \frac{-z/a}{b - z/a}$$

$$= \frac{1}{1 - abz^{-1}}, \qquad |z| \geq |ab|.$$

Observe that this expression results only from evaluating the residue at the pole inside the contour of integration. It is easily verified that if we had mistakenly considered the pole at z/a to be inside the contour of integration, the result of evaluating the integral would have been identically zero.

4.7 PARSEVAL'S RELATION

In Section 2.9.5, we considered Parseval's relation for the Fourier transform. The generalization of this relation to the z-transform follows from the complex convolution theorem. In particular we consider two complex sequences, $x_1[n]$ and $x_2[n]$. Parseval's relation states that

$$\sum_{n=-\infty}^{\infty} x_1[n] x_2^*[n] = \frac{1}{2\pi j} \oint_C X_1(v) X_2^*(1/v^*) v^{-1} \, dv, \qquad (4.79)$$

where the contour of integration must be in the overlap of the regions of convergence of $X_1(v)$ and $X_2^*(1/v^*)$. The relation of Eq. (4.79) can be derived by defining a sequence $y[n]$ as

$$y[n] = x_1[n] x_2^*[n] \qquad (4.80a)$$

and noting that

$$\sum_{n=-\infty}^{\infty} y[n] = Y(z)\Big|_{z=1}. \tag{4.80b}$$

From property 5 in Table 4.2 and the complex convolution theorem,

$$Y(z) = \sum_{n=-\infty}^{\infty} x_1[n]x_2^*[n]z^{-n} = \frac{1}{2\pi j}\oint_C X_1(v)X_2^*(z^*/v^*)v^{-1}\,dv. \tag{4.81}$$

Now evaluating Eq. (4.81) for $z = 1$ gives Eq. (4.79). Note that Eq. (4.80b) assumes that the ROC of $Y(z)$ contains the unit circle.

If $X_1(z)$ and $X_2(z)$ both converge on the unit circle, we can choose $v = e^{j\omega}$, and Eq. (4.79) becomes

$$\sum_{n=-\infty}^{\infty} x_1[n]x_2^*[n] = \frac{1}{2\pi}\int_{-\pi}^{\pi} X_1(e^{j\omega})X_2^*(e^{j\omega})\,d\omega, \tag{4.82}$$

which is Parseval's relation in terms of the Fourier transform, as discussed in Section 2.9.5.

Parseval's relation in the form of Eq. (4.82) is often difficult to evaluate. For example, when $X_1(z)$ and $X_2(z)$ are rational functions, the integrand in Eq. (4.82) will be a complicated function of $e^{j\omega}$ even in the simplest case. However, Eq. (4.79) can be evaluated straightforwardly using the Cauchy residue theorem.

A special case of Eq. (4.79) is when $x_1[n] = x_2[n] = x[n]$, where $x[n]$ is a real sequence. In this case Eq. (4.79) becomes

$$\sum_{n=-\infty}^{\infty} x^2[n] = \frac{1}{2\pi j}\oint_C X(v)X(v^{-1})v^{-1}\,dv, \tag{4.83}$$

where C is a closed contour in the ROC of $X(z)$. As mentioned above, if the sum on the left is to be finite, the ROC of $X(z)$ must include the unit circle.

Equation (4.83) gives an expression for the energy in the sequence $x[n]$ in terms of the z-transform. This equation is applied in Chapter 6 to find the noise power in the output of a digital filter due to roundoff noise. The following example illustrates the use of the Cauchy residue theorem in evaluating Eq. (4.83).

Example 4.21

Consider a right-sided real sequence $x[n]$ with z-transform

$$X(z) = \frac{1}{(1 - az^{-1})(1 - bz^{-1})},$$

where $|a|$ and $|b|$ are less than 1. Then

$$\sum_{n=-\infty}^{\infty} x^2[n] = \frac{1}{2\pi j}\oint_C \frac{dv}{(1 - av^{-1})(1 - bv^{-1})(1 - av)(1 - bv)v}$$

$$= \frac{1}{2\pi j}\oint_C \frac{v\,dv}{(v - a)(v - b)(1 - av)(1 - bv)}, \tag{4.84}$$

where C is taken to be the unit circle. The poles of the integrand are at $v = a, b, a^{-1}$, and b^{-1}. Of these, only those at $v = a$ and $v = b$ are inside the contour of integration. Therefore the Cauchy residue theorem gives

$$\sum_{n=-\infty}^{\infty} x^2[n] = \frac{a}{(a-b)(1-a^2)(1-ba)} + \frac{b}{(b-a)(1-ab)(1-b^2)}$$

$$= \frac{a(1-b^2) - b(1-a^2)}{(a-b)(1-a^2)(1-b^2)(1-ab)}.$$

Note that the energy could have been computed by obtaining $x[n]$ using a partial fraction expansion of $X(z)$, squaring the sum of exponentials, and then summing the resulting exponential series. However, this would be quite tedious even for the second-order example. On the other hand, if $X(z)$ has zeros or if there are many more poles, calculation of the energy using Eq. (4.83) is only slightly more difficult than for this two-pole example.

4.8 THE UNILATERAL z-TRANSFORM

The z-transform, as defined in Eq. (4.2) and as considered so far in this chapter, is often referred to as the *bilateral z-transform*. In contrast, the *unilateral z-transform* is defined as

$$\mathscr{X}(z) = \sum_{n=0}^{\infty} x[n]z^{-n}. \tag{4.85}$$

The unilateral z-transform differs from the bilateral z-transform in that it incorporates only values of $x[n]$ for $n \geq 0$. That is, the lower limit of the sum in Eq. (4.85) is zero regardless of the values of $x[n]$ for $n < 0$.

If $x[n]$ is zero for $n < 0$, the unilateral and bilateral z-transforms will be identical, whereas if $x[n]$ is not zero for $n < 0$, they will be different. In either case, the properties of the region of convergence of the unilateral z-transform will be the same as the properties of the region of convergence of the bilateral z-transform of right-sided sequences. That is, the ROC for all unilateral z-transforms will be of the form $|z| > r_R$, and for rational unilateral z-transforms the boundary of the ROC will be defined by the pole that is farthest from the origin of the z-plane.

Example 4.22

Consider the unit impulse sequence

$$x[n] = \delta[n].$$

The bilateral z-transform is

$$X(z) = \sum_{n=-\infty}^{\infty} x[n]z^{-n} = \sum_{n=-\infty}^{\infty} \delta[n]z^{-n}$$

$$= 1.$$

The unilateral z-transform is

$$\mathscr{X}(z) = \sum_{n=0}^{\infty} x[n]z^{-n} = \sum_{n=0}^{\infty} \delta[n]z^{-n}$$

$$= 1.$$

Since in this case $x[n] = 0$, $n < 0$, the unilateral and bilateral z-transforms are the same.

Example 4.23

Consider the shifted unit impulse sequence

$$x[n] = \delta[n + 1].$$

In this case, the unit sample occurs at $n = -1$ and consequently $x[n]$ is zero for $n \geq 0$. The bilateral z-transform is

$$X(z) = \sum_{n=-\infty}^{\infty} \delta[n + 1]z^{-n} = z.$$

This is, of course, consistent with Example 4.22 and the time-shifting property 4.4.2. In contrast,

$$\mathscr{X}(z) = \sum_{n=0}^{\infty} \delta[n + 1]z^{-n}$$

$$= 0.$$

The principal use of the unilateral z-transform is in analyzing systems described by linear constant-coefficient difference equations with non-initial rest conditions. Such systems are generally implemented recursively with causality assumed. Thus, the output is computed for $n \geq 0$ with prescribed initial conditions. In applying the unilateral z-transform to problems of this type, the linearity property and the time-shifting property are particularly important. The linearity property for the unilateral z-transform is identical to the linearity property (property 1 in Table 4.2) for the bilateral z-transform. The time-shifting property is different. To develop this property, consider a sequence $x[n]$ with unilateral z-transform $\mathscr{X}(z)$ and let

$$y[n] = x[n - 1]. \tag{4.86}$$

Then

$$\mathscr{Y}(z) = \sum_{n=0}^{\infty} x[n - 1]z^{-n}$$

$$= x[-1] + \sum_{n=1}^{\infty} x[n - 1]z^{-n}.$$

With the substitution of variables $m = n - 1$ we can rewrite $\mathscr{Y}(z)$ as

$$\mathscr{Y}(z) = x[-1] + \sum_{m=0}^{\infty} x[m]z^{-(m+1)}$$

so that

$$\mathscr{Y}(z) = x[-1] + z^{-1}\mathscr{X}(z). \tag{4.87}$$

Similarly, as considered in Problem 4.45, the signal

$$y[n] = x[n-2]$$

has unilateral z-transform

$$\mathscr{Y}(z) = x[-2] + x[-1]z^{-1} + z^{-2}\mathscr{X}(z). \tag{4.88}$$

More generally, if

$$y[n] = x[n-m]$$

for $m > 0$, then

$$\mathscr{Y}(z) = x[-m] + x[-m+1]z^{-1} + \cdots + x[-1]z^{-m+1} + z^{-m}\mathscr{X}(z)$$

$$= \sum_{k=1}^{m} x[k-1-m]z^{-k+1} + z^{-m}\mathscr{X}(z). \tag{4.89}$$

The procedure for using the unilateral z-transform to obtain the solution to difference equations with specified initial conditions and with inputs specified for $n \geq 0$ is best illustrated with an example.

Example 4.24

Consider a system for which the input and output satisfy the linear constant-coefficient difference equation

$$y[n] - \tfrac{1}{2}y[n-1] = x[n]. \tag{4.90}$$

The input for $n \geq 0$ is

$$x[n] = 1, \qquad n \geq 0.$$

and the initial condition at $n = -1$ is

$$y[-1] = 1.$$

Applying the unilateral z-transform to Eq. (4.90) and using the linearity property as well as the time-shifting property of Eq. (4.89), we have

$$\mathscr{Y}(z) - \tfrac{1}{2}\{y[-1] + z^{-1}\mathscr{Y}(z)\} = \mathscr{X}(z) = \frac{1}{1 - z^{-1}},$$

from which it follows that

$$\mathscr{Y}(z) = \frac{1}{1 - \tfrac{1}{2}z^{-1}}\left\{\tfrac{1}{2}y[-1] + \frac{1}{1 - z^{-1}}\right\},$$

or

$$\mathscr{Y}(z) = \frac{1/2}{1 - \tfrac{1}{2}z^{-1}} + \frac{1}{(1 - \tfrac{1}{2}z^{-1})(1 - z^{-1})}.$$

Performing a partial fraction expansion of the second term gives

$$\mathscr{Y}(z) = \frac{1/2}{1 - \frac{1}{2}z^{-1}} - \frac{1}{1 - \frac{1}{2}z^{-1}} + \frac{2}{1 - z^{-1}}$$

$$= \frac{2}{1 - z^{-1}} - \frac{1/2}{1 - \frac{1}{2}z^{-1}}.$$

(4.91)

Since the unilateral z-transform of the sequence

$$x[n] = a^n, \qquad n \geq 0,$$

is

$$\mathscr{X}(z) = \frac{1}{1 - az^{-1}},$$

$y[n]$ is seen to be

$$y[n] = 2 - \frac{1}{2}\left(\frac{1}{2}\right)^n, \qquad n \geq 0.$$

4.9 SUMMARY

In this chapter we have defined the z-transform of a sequence and shown that it is a generalization of the Fourier transform. The discussion of this chapter focused on the properties of the z-transform and techniques for obtaining the z-transform of a sequence and vice versa. Specifically, we showed that the defining power series of the z-transform may converge when the Fourier transform does not. We explored in detail the dependence of the shape of the region of convergence on the properties of the sequence. A full understanding of the properties of the region of convergence is essential for successful use of the z-transform. This is particularly true in developing techniques for finding the sequence that corresponds to a given z-transform, i.e., finding inverse z-transforms. Much of the discussion focused on z-transforms that are rational functions in their region of convergence. For such functions, we described a technique of inverse transformation based on the partial fraction expansion of $X(z)$. We also discussed other techniques for inverse transformation such as the use of tabulated power series expansions and long division.

An important part of the chapter was a discussion of some of the many properties of the z-transform that make it useful in analyzing discrete-time signals and systems. A variety of examples demonstrated how these properties can be used to find direct and inverse z-transforms.

The latter part of the chapter showed how the theory of functions of a complex variable can be applied to the z-transform. In particular, the Cauchy residue theorem was applied to evaluate inverse transforms and contour integrals arising in complex convolution and Parseval's theorem. The chapter concluded with a brief discussion of the unilateral z-transform.

PROBLEMS

4.1. Determine the z-transform, including the region of convergence, for each of the following sequences:
   (a) $(\frac{1}{2})^n u[n]$
   (b) $-(\frac{1}{2})^n u[-n-1]$
   (c) $(\frac{1}{2})^n u[-n]$
   (d) $\delta[n]$
   (e) $\delta[n-1]$
   (f) $\delta[n+1]$
   (g) $(\frac{1}{2})^n (u[n] - u[n-10])$

4.2. Determine the z-transform of the sequence

$$x[n] = \begin{cases} n, & 0 \le n \le N-1, \\ N, & N \le n. \end{cases}$$

4.3. Determine the z-transform of each of the following sequences. Include with your answer the region of convergence in the z-plane and a sketch of the pole-zero plot. Express all sums in closed form; α can be complex.
   (a) $x_a[n] = \alpha^{|n|}, \quad 0 < |\alpha| < 1$

   (b) $x_b[n] = \begin{cases} 1 & 0 \le n \le N-1, \\ 0, & N \le n, \\ 0. & n < 0 \end{cases}$

   (c) $x_c[n] = \begin{cases} n, & 0 \le n \le N, \\ 2N - n, & N+1 \le n \le 2N, \\ 0, & 2N \le n, \\ 0, & n < 0 \end{cases}$

   [*Hint:* Note that $x_b[n]$ is a rectangular sequence and $x_c[n]$ is a triangular sequence. First express $x_c[n]$ in terms of $x_b[n]$.]

4.4. (a) Without explicitly solving for $X(z)$, find the region of convergence of the z-transform for each of the following sequences:
   (i) $x[n] = [(\frac{1}{2})^n + (\frac{3}{4})^n]u[n-10]$
   (ii) $x[n] = \begin{cases} 1, & -10 \le n \le 10 \\ 0, & \text{otherwise} \end{cases}$
   (iii) $x[n] = 2^n u[-n]$

   (b) For which of the above sequences do the Fourier transforms converge?

4.5. Let $x[n]$ denote a causal sequence; i.e., $x[n] = 0$, $n < 0$. Furthermore, assume that $x[0] \ne 0$.
   (a) Show that there are no poles or zeros of $X(z)$ at $z = \infty$, i.e., that $\lim_{z \to \infty} X(z)$ is nonzero and finite.
   (b) Show that the number of poles in the finite z-plane equals the number of zeros in the finite z-plane. (The finite z-plane excludes $z = \infty$.)

4.6. Consider the z-transform $X(z)$ whose pole-zero plot is as shown in Fig. P4.6.
   (a) Determine the region of the convergence of $X(z)$ if it is known that the Fourier

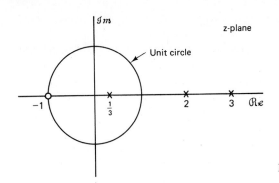

Figure P4.6

transform exists. For this case, determine whether the corresponding sequence is right-sided, left-sided, or two-sided.

(b) How many possible two-sided sequences have the pole-zero plot shown in Fig. P4.6?

(c) Is it possible for the pole-zero plot in Fig. P4.6 to be associated with a sequence that is both stable and causal? If so, give the appropriate region of convergence.

4.7. Determine the sequence $x[n]$ with z-transform

$$X(z) = (1 + 2z)(1 + 3z^{-1})(1 - z^{-1})$$

4.8. Listed below are several z-transforms. For each, determine the inverse z-transform using both methods—partial fraction expansion and power series expansion—discussed in Section 4.3.

(a) $X(z) = \dfrac{1}{1 + \frac{1}{2}z^{-1}}, \qquad |z| > \frac{1}{2}$

(b) $X(z) = \dfrac{1}{1 + \frac{1}{2}z^{-1}}, \qquad |z| < \frac{1}{2}$

(c) $X(z) = \dfrac{1 - \frac{1}{2}z^{-1}}{1 + \frac{3}{4}z^{-1} + \frac{1}{8}z^{-2}}, \qquad |z| > \frac{1}{2}$

(d) $X(z) = \dfrac{1 - \frac{1}{2}z^{-1}}{1 - \frac{1}{4}z^{-2}}, \qquad |z| > \frac{1}{2}$

(e) $X(z) = \dfrac{1 - az^{-1}}{z^{-1} - a}, \qquad |z| > |1/a|$

4.9. Given below are four z-transforms. Determine which ones *could* be the z-transform of a *causal* sequence. Do not evaluate the inverse transform. You should be able to give the answer by inspection. Clearly state your reasons in each case.

(a) $\dfrac{(1 - z^{-1})^2}{(1 - \frac{1}{2}z^{-1})}$

(b) $\dfrac{(z - 1)^2}{(z - \frac{1}{2})}$

(c) $\dfrac{(z - \frac{1}{4})^5}{(z - \frac{1}{2})^6}$

(d) $\dfrac{(z - \frac{1}{4})^6}{(z - \frac{1}{2})^5}$

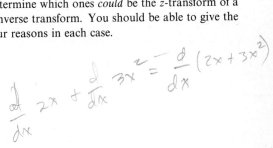

4.10. Consider a sequence with z-transform $X(z) = P(z)/Q(z)$; where $P(z)$ and $Q(z)$ are polynomials in z. If the sequence is absolutely summable and if all the roots of $Q(z)$ are inside the unit circle, is the sequence necessarily causal? If your answer is yes, clearly explain. If your answer is no, give a counterexample.

4.11. Consider a right-sided sequence $x[n]$ with z-transform

$$X(z) = \frac{1}{(1 - az^{-1})(1 - bz^{-1})} = \frac{z^2}{(z - a)(z - b)}.$$

In Section 4.3 we considered the determination of $x[n]$ by carrying out a partial fraction expansion, with $X(z)$ considered as a ratio of polynomials in z^{-1}. Carry out a partial fraction expansion of $X(z)$ considered as a ratio of polynomials in z, and from this expansion determine $x[n]$.

√4.12. Determine the inverse z-transform of each of the following. In parts (a)–(c), use the methods specified. In part (d), use any method you prefer.

    (a) Long division:

$$X(z) = \frac{1 - \frac{1}{3}z^{-1}}{1 + \frac{1}{3}z^{-1}}, \qquad x[n] \text{ a right-sided sequence}$$

    (b) Partial fraction:

$$X(z) = \frac{3}{z - \frac{1}{4} - \frac{1}{8}z^{-1}}, \qquad x[n] \text{ stable}$$

    (c) Power series:

$$X(z) = \ln(1 - 4z), \qquad |z| < \frac{1}{4}$$

    (d) $X(z) = \dfrac{1}{1 - \frac{1}{3}z^{-3}}, \qquad |z| > (3)^{-1/3}$

4.13. Determine the inverse z-transform for each of the following using any method.

    (a) $X(z) = \dfrac{1}{(1 + \frac{1}{2}z^{-1})^2(1 - 2z^{-1})(1 - 3z^{-1})}, \qquad$ stable sequence

    (b) $X(z) = e^{z^{-1}}$

    (c) $X(z) = \dfrac{z^3 - 2z}{z - 2}, \qquad$ left-sided sequence

4.14. Determine the inverse z-transform of each of the following. You should find the z-transform properties in Section 4.4 helpful.

    (a) $X(z) = \dfrac{3z^{-3}}{(1 - \frac{1}{4}z^{-1})^2}, \qquad x[n]$ left-sided

    (b) $X(z) = \sin(z), \qquad$ ROC includes $|z| = 1$

    (c) $X(z) = \dfrac{z^7 - 2}{1 - z^{-7}}, \qquad |z| > 1$

4.15. Determine a sequence $x[n]$ whose z-transform is $X(z) = e^z + e^{1/z}, z \neq 0$. (Consider using a power series expansion.)

4.16. Determine the inverse z-transform of
$$X(z) = \log 2(\tfrac{1}{2} - z), \qquad |z| < \tfrac{1}{2}$$
by

(a) using the power series
$$\log(1 - x) = -\sum_{i=1}^{\infty} \frac{x^i}{i}, \qquad |x| < 1;$$

(b) first differentiating $X(z)$ and then using this to recover $x[n]$.

4.17. Let $x[n]$ be a causal stable sequence with a z-transform $X(z)$. In Chapter 12 we will define the *complex cepstrum* $\hat{x}[n]$ as the inverse transform of the logarithm of $X(z)$, i.e.,
$$\hat{X}(z) = \log X(z) \overset{z}{\longleftrightarrow} \hat{x}[n],$$
where the ROC of $\hat{X}(z)$ includes the unit circle. (Strictly speaking, taking the logarithm of a complex number requires some careful considerations. Furthermore, the logarithm of a valid z-transform may not be a valid z-transform. We will postpone the justifications of this operation until Chapter 12. For now, we assume it is valid.)

Determine the complex cepstrum for the following sequence.
$$x[n] = \delta[n] + a\delta[n - N], \qquad \text{where } |a| < 1.$$

4.18. For each of the following sequences, determine the z-transform and region of convergence, and sketch the pole-zero diagram.

(a) $x[n] = a^n u[n] + b^n u[n] + c^n u[-n - 1], \qquad |a| < |b| < |c|$

(b) $x[n] = n^2 a^n u[n]$

(c) $x[n] = e^{n^4}\left(\cos\frac{\pi}{12}n\right)u[n] - e^{n^4}\left(\cos\frac{\pi}{12}n\right)u[n - 1]$

4.19. The pole-zero diagram in Fig. P4.19 corresponds to the z-transform $X(z)$ of a causal sequence $x[n]$. Sketch the pole-zero diagram of $Y(z)$, where $y[n] = x[-n + 3]$. Also specify the region of convergence for $Y(z)$.

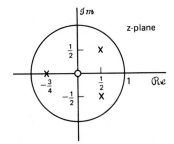

Poles: $z = \tfrac{1}{2} + j\tfrac{1}{2}, z = \tfrac{1}{2} - j\tfrac{1}{2}, z = -\tfrac{3}{4}$

Zero: $z = 0$

Figure P4.19

4.20. Let $x[n]$ be the sequence with the pole-zero plot shown in Fig. P4.20. Sketch the pole-zero plot for:

(a) $y[n] = (\tfrac{1}{2})^n x[n]$

(b) $w[n] = \cos\left(\frac{\pi n}{2}\right)x[n]$

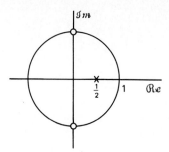

Figure P4.20

√**4.21.** Consider a linear time-invariant system with impulse response $h[n]$ and input $x[n]$ given by

$$h[n] = \begin{cases} a^n, & n \geq 0, \\ 0, & n < 0, \end{cases} \quad a \leq 1 \text{ or } h(n) \to \infty$$

$$x[n] = \begin{cases} 1, & 0 \leq n \leq (N-1), \\ 0, & \text{otherwise.} \end{cases}$$

(a) Determine the output $y[n]$ by explicitly evaluating the discrete convolution of $x[n]$ and $h[n]$.

(b) Determine the output $y[n]$ by computing the inverse z-transform of the product of the z-transforms of the input and the impulse response.

4.22. Consider an LTI system which is stable and for which $H(z)$, the z-transform of the impulse response, is given by

$$H(z) = \frac{3 - 7z^{-1} + 5z^{-2}}{1 - \frac{5}{2}z^{-1} + z^{-2}}.$$

Suppose $x[n]$, the input to the system, is a unit step sequence.

(a) Find the output $y[n]$ by evaluating the discrete convolution of $x[n]$ and $h[n]$.

(b) Find the output $y[n]$ by computing the inverse z-transform of $Y(z)$.

4.23. Determine the unit step response of the causal system for which the z-transform of the impulse response is

$$H(z) = \frac{1 - z^3}{1 - z^4}$$

4.24. If the input $x[n]$ to an LTI system is $x[n] = u[n]$, the output $y[n]$ is

$$y[n] = (\tfrac{1}{2})^{n-1} u[n + 1]$$

(a) Find $H(z)$, the z-transform of the system impulse response, and plot its pole-zero diagram.

(b) Find the impulse response $h[n]$.

(c) Is the system stable?

(d) Is the system causal?

4.25. The input to a causal linear time-invariant system is

$$x[n] = u[-n - 1] + (\tfrac{1}{2})^n u[n].$$

The z-transform of the output of this system is

$$Y(z) = \frac{-\frac{1}{2}z^{-1}}{(1 - \frac{1}{2}z^{-1})(1 + z^{-1})}.$$

(a) Determine $H(z)$, the z-transform of the system impulse response. Be sure to specify the region of convergence.
(b) What is the region of convergence for $Y(z)$?
(c) Determine $y[n]$.

4.26. The system function of a causal linear time-invariant system is

$$H(z) = \frac{1 - z^{-1}}{1 + \frac{3}{4}z^{-1}}.$$

The input to this system is

$$x[n] = (\tfrac{1}{2})^n u[n] + u[-n - 1].$$

(a) Find the impulse response of the system for all values of n.
(b) Find the output $y[n]$ for all values of n.
(c) Is the system stable, i.e., is $h[n]$ absolutely summable?

4.27. A causal LTI system has impulse response $h[n]$, for which the z-transform is

$$H(z) = \frac{1 + z^{-1}}{(1 - \frac{1}{2}z^{-1})(1 + \frac{1}{4}z^{-1})}.$$

(a) What is the region of convergence of $H(z)$?
(b) Is the system stable? Explain.
(c) Find the z-transform $X(z)$ of an input $x[n]$ that will produce the output

$$y[n] = -\tfrac{1}{3}(-\tfrac{1}{4})^n u[n] - \tfrac{4}{3}(2)^n u[-n - 1].$$

(d) Find the impulse response $h[n]$ of the system.

4.28. Assume that $x[n]$ is real and even, i.e., $x[n] = x[-n]$. Further assume that z_0 is a zero of $X(z)$; i.e., $X(z_0) = 0$.
(a) Show that $1/z_0$ is also a zero of $X(z)$.
(b) Are there other zeros of $X(z)$ implied by the information given?

4.29. Using the definition of the z-transform (Eq. 4.2), show that, if $X(z)$ is the z-transform of $x[n]$, then
(a) $x^*[n] \xleftrightarrow{z} X^*(z^*)$
(b) $x[-n] \xleftrightarrow{z} X(1/z)$
(c) $\mathcal{R}e\{x[n]\} \xleftrightarrow{z} \tfrac{1}{2}[X(z) + X^*(z^*)]$
(d) $\mathcal{I}m\{x[n]\} \xleftrightarrow{z} \dfrac{1}{2j}[X(z) - X^*(z^*)]$

4.30. Consider a *real* sequence $x[n]$ that has all the poles and zeros of its z-transform inside the unit circle. Determine in terms of $x[n]$ a *real* sequence $x_1[n]$ not equal to $x[n]$ but for which $x_1[0] = x[0]$, $|x_1[n]| = |x[n]|$, and the z-transform of $x_1[n]$ has all its poles and zeros inside the unit circle.

4.31. A real finite-duration sequence whose z-transform has no zeros at conjugate reciprocal pair locations and no zeros on the unit circle is uniquely specified to within a positive scale factor by its Fourier transform phase (Hayes et al., 1980).

An example of zeros at conjugate reciprocal pair locations is $z = a$ and $(a^*)^{-1}$. Even though we can generate sequences that do not satisfy the above set of conditions, almost any sequence of practical interest satisfies the conditions and therefore is uniquely specified to within a positive scale factor by its Fourier phase.

Consider a sequence $x[n]$ which is real, which is zero outside $0 \leq n \leq N - 1$, and whose z-transform has no zeros at conjugate reciprocal pair locations and no zeros on the unit circle. We wish to develop an algorithm that reconstructs $cx[n]$ from $\not\prec X(e^{j\omega})$, the Fourier transform phase of $x[n]$, where c is a positive scale factor.

(a) Specify a set of $(N - 1)$ linear equations, the solution to which will provide the recovery of $x[n]$ to within a positive or negative scale factor from $\tan\{\not\prec X(e^{j\omega})\}$. You do not have to prove that the set of $(N - 1)$ linear equations has a unique solution. Also show that if we know $\not\prec X(e^{j\omega})$ rather than just $\tan\{\not\prec X(e^{j\omega})\}$, the sign of the scale factor can also be determined.

(b) Suppose $x[n]$ is given by

$$
x[n] = \begin{cases}
0, & n < 0 \\
1, & n = 0 \\
2, & n = 1 \\
3, & n = 2 \\
0, & n \geq 3
\end{cases}
$$

Using the approach developed in part (a), demonstrate that $cx[n]$ can be determined from $\not\prec X(e^{j\omega})$, where c is a positive scale factor.

4.32. For a sequence $x[n]$ that is zero for $n < 0$, show that $\lim_{z \to \infty} X(z) = x[0]$. (Consider the limit of each term in the z-transform definition, Eq. 4.2.) What is the corresponding theorem if the sequence is zero for $n > 0$?

4.33. Consider a sequence $x[n]$ for which the z-transform is

$$
X(z) = \frac{\frac{1}{3}}{1 - \frac{1}{2}z^{-1}} + \frac{\frac{1}{4}}{1 - 2z^{-1}}
$$

and for which the region of convergence includes the unit circle. Use the theorems derived in Problem 4.32 to determine $x[0]$.

4.34. The aperiodic autocorrelation function for a real-valued stable sequence $x[n]$ is defined as

$$
c_{xx}[n] = \sum_{k=-\infty}^{\infty} x[k]x[n + k].
$$

(a) Show that the z-transform of $c_{xx}[n]$ is

$$
C_{xx}(z) = X(z)X(z^{-1}).
$$

Determine the region of convergence for $C_{xx}(z)$.

(b) Suppose that $x[n] = a^n u[n]$. Show the pole-zero plot for $C_{xx}(z)$, including the region of convergence. Also find $c_{xx}[n]$ by evaluating the inverse z-transform of $C_{xx}(z)$.

(c) Specify another sequence, $x_1[n]$, that is not equal to $x[n]$ in part (b) but that has the same autocorrelation function, $c_{xx}[n]$, as $x[n]$ in part (b).

(d) Specify a third sequence, $x_2[n]$, that is not equal to $x[n]$ or $x_1[n]$ but that has the same autocorrelation function as $x[n]$ in part (b).

4.35. Determine whether or not the function $X(z) = z^*$ can correspond to the z-transform of a sequence. Clearly explain your reasoning.

4.36. For each z-transform listed below, determine the inverse z-transform using contour integration.

(a) $X(z) = \dfrac{1}{1 + \frac{1}{4}z^{-1}}, \quad |z| > \frac{1}{4}$

(b) $X(z) = \dfrac{1}{1 + \frac{1}{4}z^{-1}}, \quad |z| < \frac{1}{4}$

(c) $X(z) = \dfrac{1 - \frac{1}{2}z^{-1}}{1 + \frac{3}{4}z^{-1} + \frac{1}{8}z^{-2}}, \quad |z| > \frac{1}{2}$

(d) $X(z) = \dfrac{1 - \frac{1}{4}z^{-1}}{1 - \frac{1}{9}z^{-2}}, \quad |z| > \frac{1}{3}$

(e) $X(z) = \dfrac{1 - az^{-1}}{z^{-1} - a}, \quad |z| > |1/a|$

4.37. Consider a stable linear time-invariant system. The z-transform of the impulse response is

$$H(z) = \frac{z^{-1} + z^{-2}}{(1 - \frac{1}{2}z^{-1})(1 + \frac{1}{3}z^{-1})}.$$

Suppose $x[n]$, the input to the system, is $2u[n]$. Determine $y[n]$ at $n = 1$.

4.38. Suppose $X(z)$, the z-transform of $x[n]$, is given by

$$X(z) = \frac{z^{10}}{(z - \frac{1}{2})(z - \frac{3}{2})^{10}(z + \frac{3}{2})^2(z + \frac{5}{2})(z + \frac{7}{2})}.$$

It is also known that $x[n]$ is a stable sequence.
(a) Determine the region of convergence of $X(z)$.
(b) Determine $x[n]$ at $n = -8$. (Consider using contour integration as discussed in Section 4.5.)

4.39. Suppose $X(z)$, the z-transform of $x[n]$, is given by

$$X(z) = \frac{z^{20}}{(z - \frac{1}{2})(z + \frac{5}{2})^2(z + \frac{7}{2})(z - \frac{9}{2})}.$$

(a) Suppose $x[n]$ is a stable sequence. Determine $x[n]$ at $n = -18$.
(b) Suppose $x[n]$ is a left-sided sequence. Determine $x[n]$ at $n = -18$.
(Consider using contour integration as discussed in Section 4.5.)

4.40. In Section 4.5 we discussed the fact that for $n < 0$ it is often more convenient to evaluate the inverse transform relation of Eq. (4.67) by using a substitution of variables $z = p^{-1}$ to obtain Eq. (4.74). If the contour of integration is the unit circle, for example, this has the effect of mapping the outside of the unit circle to the inside and vice versa. It is worth convincing yourself, however, that without this substitution, we can obtain $x[n]$ for $n < 0$ by evaluating the residues at the multiple-order poles at $z = 0$. Let

$$X(z) = \frac{1}{1 - \frac{1}{2}z}$$

where the region of convergence includes the unit circle.
(a) Determine $x[0]$, $x[-1]$, and $x[-2]$ by evaluating Eq. (4.67) explicitly, obtaining the residue for the poles at $z = 0$.
(b) Determine $x[n]$ for $n > 0$ by evaluating Eq. (4.67) and for $n \le 0$ by evaluating Eq. (4.74).

4.41. Let $X(z)$ denote a ratio of polynomials in z, i.e.,

$$X(z) = \frac{B(z)}{A(z)}.$$

Show that if $X(z)$ has a first-order pole at $z = z_0$, then the residue of $X(z)$ at $z = z_0$ is equal to

$$\frac{B(z_0)}{A'(z_0)},$$

where $A'(z_0)$ denotes the derivative of $A(z)$ evaluated at $z = z_0$.

4.42. Suppose that $F(z)$ is a rational function, i.e.,

$$F(z) = \frac{B(z)}{A(z)},$$

where $B(z)$ and $A(z)$ are polynomials. Moreover, assume that $F(z)$ has no poles or zeros of multiplicity greater than unity. Let C be a simple counterclockwise closed contour, and let Z and P be, respectively, the number of zeros and poles of $F(z)$ enclosed by the contour. (Assume that there are no poles or zeros on C.)

(a) Show that

$$\frac{1}{2\pi j} \oint_C \frac{F'(z)}{F(z)} \, dz = Z - P,$$

where $F'(z)$ is the derivative of $F(z)$.

(b) With $F(z)$ expressed in polar form as $F(z) = |F(z)|e^{j \, \measuredangle F(z)}$, show that the change in $\measuredangle F(z)$ as the contour C is traversed exactly once is $2\pi(Z - P)$. (It can be shown that these results generalize in the case of multiple-order poles and zeros by counting poles and zeros according to their multiplicity; i.e., a second-order pole is counted twice.)

4.43. Let $X(z)$ denote the z-transform of a sequence that is both causal and stable:

$$X(z) = \frac{1}{1 - \sum_{k=1}^{N} a_k z^{-k}}.$$

Determine a numerical value for the following expression for the system. (This expression can be interpreted as the mean value of the magnitude squared on a logarithmic scale.)

$$\frac{1}{2\pi} \int_{-\pi}^{\pi} \log[|X(e^{j\omega})|^2] \, d\omega$$

[*Hint:* Think of the expression as evaluating the *contour* integral of $\log|X(z)|^2$ on a particular contour and apply the Cauchy integral theorem.]

4.44. Determine the unilateral z-transform $\mathcal{X}(z)$ for each of the following sequences.
(a) $\delta[n]$
(b) $\delta[n-1]$
(c) $\delta[n+1]$
(d) $(\frac{1}{2})^n u[n]$
(e) $-(\frac{1}{2})^n u[-n-1]$
(f) $(\frac{1}{2})^n u[-n]$
(g) $\{(\frac{1}{2})^n + (\frac{1}{4})^n\} u[n]$
(h) $(\frac{1}{2})^{n-1} u[n-1]$

4.45. If $\mathscr{X}(z)$ denotes the unilateral z-transform of $x[n]$, determine, in terms of $\mathscr{X}(z)$, the unilateral z-transform of the following.

(a) $x[n + 1]$

(b) $x[n - 2]$

(c) $\displaystyle\sum_{k=-\infty}^{n} x[k]$

4.46. For each of the following difference equations and associated input and initial conditions, determine the response $y[n]$ by using the unilateral z-transform.

(a) $y[n] + 3y[n - 1] = x[n]$
$$x[n] = (\tfrac{1}{2})^n u[n]$$
$$y[-1] = 1$$

(b) $y[n] - \tfrac{1}{2}y[n - 1] = x[n] - \tfrac{1}{2}x[n - 1]$
$$x[n] = u[n]$$
$$y[-1] = 0$$

(c) $y[n] - \tfrac{1}{2}y[n - 1] = x[n] - \tfrac{1}{2}x[n - 1]$
$$x[n] = u[n]$$
$$y[-1] = 1$$

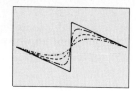

Transform Analysis of Linear Time-Invariant Systems

5

5.0 INTRODUCTION

In Chapter 2 we developed the Fourier transform representation of discrete-time signals and systems, and in Chapter 4 we extended that representation to the z-transform. In both chapters, the emphasis was on the transforms and their properties with only a brief preview of the details of their use in the analysis of linear time-invariant (LTI) systems. In this chapter we develop in more detail the representation and analysis of LTI systems using the Fourier and z-transforms. The material in this chapter is essential background for our discussion in Chapter 6 of the implementation of LTI systems and in Chapter 7 of the design of LTI systems.

As developed in Chapter 2, an LTI system can be completely characterized in the time domain by its impulse response $h[n]$, with the output $y[n]$ due to a given input $x[n]$ specified through the convolution sum

$$y[n] = x[n] * h[n] = \sum_{k=-\infty}^{\infty} x[k]h[n-k]. \tag{5.1}$$

Alternatively, as discussed in Section 2.7, since the frequency response and impulse response are directly related through the Fourier transform, the frequency response, assuming it exists (i.e. converges), provides an equally complete characterization of LTI systems. In Chapter 4, we developed the z-transform as a generalization of the

202

$h[n]$ = Impulse response
$H(e^{j\omega})$ = frequency response
$H(z)$ = System function

Sec. 5.1   The Frequency Response of LTI Systems   **203**

Fourier transform, and we showed that $Y(z)$, the z-transform of the output of an LTI system, is related to $X(z)$, the z-transform of the input, and $H(z)$, the z-transform of the system impulse response, by

$$Y(z) = H(z)X(z) \tag{5.2}$$

with an appropriate region of convergence. $H(z)$ is referred to as the *system function*. Since the z-transform and a sequence form a unique pair, it follows that any LTI system is completely characterized by its system function, again assuming convergence.

As we will see in this chapter, both the frequency response and the system function are extremely useful in the analysis and representation of LTI systems because we can readily infer many properties of the system response from them.

5.1 THE FREQUENCY RESPONSE OF LTI SYSTEMS

The frequency response $H(e^{j\omega})$ of an LTI system was defined in Section 2.6 as the complex gain (eigenvalue) that the system applies to the complex exponential input (eigenfunction) $e^{j\omega n}$. Furthermore, in Section 2.9.6 we developed the fact that since the Fourier transform of a sequence represents a decomposition as a linear combination of complex exponentials, the Fourier transforms of the system input and output are related by

$$Y(e^{j\omega}) = H(e^{j\omega})X(e^{j\omega}), \tag{5.3}$$

where $X(e^{j\omega})$ and $Y(e^{j\omega})$ are the Fourier transforms of the system input and output, respectively. With the frequency response expressed in polar form, the magnitude and phase of the Fourier transforms of the system input and output are related by

$$|Y(e^{j\omega})| = |H(e^{j\omega})| \cdot |X(e^{j\omega})|, \tag{5.4a}$$

$$\sphericalangle Y(e^{j\omega}) = \sphericalangle H(e^{j\omega}) + \sphericalangle X(e^{j\omega}). \tag{5.4b}$$

$|H(e^{j\omega})|$ is referred to as the *magnitude* response or the *gain* of the system, and $\sphericalangle H(e^{j\omega})$ is referred to as the *phase response* or *phase shift* of the system.

The magnitude and phase effects represented by Eqs. (5.4) can be either desirable, if the input signal is modified in a useful way, or undesirable, if the input signal is changed in a deleterious manner. In the latter case, we often refer to the effects of an LTI system on a signal, as represented by Eqs. (5.4a) and (5.4b), as *magnitude* and *phase distortions*, respectively.

5.1.1 Ideal Frequency-Selective Filters

An important implication of Eq. (5.4a) is that frequency components of the input are suppressed in the output if $|H(e^{j\omega})|$ is small at those frequencies. Whether this suppression of Fourier components is viewed as desirable or undesirable depends on the specific problem. Example 2.14 formalized the general notion of frequency-selective filters through the definition of certain ideal frequency responses. For example, the ideal lowpass filter was defined as the discrete-time linear time-invariant system whose frequency response is

$$H_{lp}(e^{j\omega}) = \begin{cases} 1, & |\omega| < \omega_c, \\ 0, & \omega_c < |\omega| \le \pi, \end{cases} \tag{5.5}$$

and, of course, is also periodic with period 2π. The ideal lowpass filter selects the low-frequency components of the signal and rejects the high-frequency components. The corresponding impulse response was shown in Example 2.17 to be

$$h_{lp}[n] = \frac{\sin \omega_c n}{\pi n}, \qquad -\infty < n < \infty. \tag{5.6}$$

Correspondingly the *ideal highpass filter* is defined as

$$H_{hp}(e^{j\omega}) = \begin{cases} 0, & |\omega| < \omega_c, \\ 1, & \omega_c < |\omega| \le \pi, \end{cases} \tag{5.7}$$

and its impulse response is

$$h_{hp}[n] = \delta[n] - h_{lp}[n] = \delta[n] - \frac{\sin \omega_c n}{\pi n}. \tag{5.8}$$

The ideal highpass filter passes the frequency band $\omega_c < \omega \le \pi$ undistorted and rejects frequencies below ω_c. Other ideal frequency-selective filters were defined in Example 2.14.

The ideal lowpass filter is noncausal and its impulse response extends from $-\infty$ to $+\infty$. Therefore it is not possible to compute the output of the ideal lowpass filter either recursively or nonrecursively; i.e., the system is not *computationally realizable*.

Another important property of the ideal lowpass filter as defined in Eq. (5.5) is that the phase response is specified to be zero. If it were not zero, the low-frequency band selected by the filter would also have phase distortion. It will become clear later in this chapter that causal approximations to ideal frequency-selective filters must have nonzero phase response.

5.1.2 Phase Distortion and Delay

To understand the effect of the phase of a linear system, let us first consider the ideal delay system. The impulse response is

$$h_{id}[n] = \delta[n - n_d] \tag{5.9}$$

and the frequency response is

$$H_{id}(e^{j\omega}) = e^{-j\omega n_d} \tag{5.10}$$

or

$$|H_{id}(e^{j\omega})| = 1, \tag{5.11a}$$

$$\measuredangle H_{id}(e^{j\omega}) = -\omega n_d, \qquad |\omega| < \pi, \tag{5.11b}$$

with periodicity 2π in ω assumed. For now we will assume that n_d is an integer.

In many applications, delay distortion would be considered a rather mild form of phase distortion since its effect is only to shift the sequence in time. Often this would be inconsequential or it could easily be compensated for by introducing delay in other parts of a larger system. Thus, in designing approximations to ideal filters and other linear time-invariant systems, we often are willing to accept linear phase response rather than zero phase response as our ideal. For example, an ideal lowpass filter with linear phase would be defined as

$$H_{lp}(e^{j\omega}) = \begin{cases} e^{-j\omega n_d}, & |\omega| < \omega_c, \\ 0, & \omega_c < |\omega| \le \pi. \end{cases} \tag{5.12}$$

Its impulse response is

$$h_{lp}[n] = \frac{\sin \omega_c(n - n_d)}{\pi(n - n_d)}, \qquad -\infty < n < \infty. \tag{5.13}$$

In a similar manner we could define other ideal frequency-selective filters with linear phase. These filters would have the desired effect of isolating a band of frequencies in the input signal as well as the additional effect of delaying the output by n_d. Note, however, that no matter how large we make n_d, the ideal lowpass filter is always noncausal.

A convenient measure of the linearity of the phase is the *group delay*. The basic concept of group delay relates to the effect of the phase on a narrowband signal. Specifically, consider the output of a system with frequency response $H(e^{j\omega})$ for a narrowband input $x[n] = s[n]\cos(\omega_0 n)$. Since it is assumed that $X(e^{j\omega})$ is nonzero only around $\omega = \omega_0$, the effect of the phase of the system can be approximated around $\omega = \omega_0$ as the linear approximation

$$\measuredangle H(e^{j\omega}) \simeq -\phi_0 - \omega n_d. \tag{5.14}$$

With this approximation it can be shown (see Problem 5.3) that the response $y[n]$ to $x[n] = s[n]\cos(\omega_0 n)$ is approximately $y[n] = s[n - n_d]\cos(\omega_0 n - \phi_0 - \omega_0 n_d)$. Consequently, the time delay of the envelope $s[n]$ of the narrowband signal $x[n]$ with Fourier transform centered at ω_0 is given by the negative of the slope of the phase at x_0. In considering the linear approximation to $\measuredangle H(e^{j\omega})$ around $\omega = \omega_0$, as given in Eq. (5.14), we must consider the phase response as a continuous function of ω. The phase response specified in this way will be denoted as $\arg[H(e^{j\omega})]$ and is referred to as the *continuous phase of* $H(e^{j\omega})$. This representation of Fourier transform phase as a continuous function of ω for $0 < |\omega| < \pi$ will also play a particularly important role in Chapter 12.

With phase specified as a continuous function of ω, the group delay of a system is defined as

$$\tau(\omega) = \text{grd}[H(e^{j\omega})] = -\frac{d}{d\omega}\{\arg[H(e^{j\omega})]\}. \tag{5.15}$$

The deviation of the group delay away from a constant indicates the degree of nonlinearity of the phase.

5.2 SYSTEM FUNCTIONS FOR SYSTEMS CHARACTERIZED BY LINEAR CONSTANT-COEFFICIENT DIFFERENCE EQUATIONS

While ideal frequency-selective filters are useful conceptually, they cannot be implemented with finite computation. Therefore it is of interest to consider a class of systems that can be implemented as approximations to ideal frequency-selective filters.

In Section 2.5 we considered the class of systems whose input and output satisfy a linear constant-coefficient difference equation of the form

$$\sum_{k=0}^{N} a_k y[n-k] = \sum_{k=0}^{M} b_k x[n-k]. \tag{5.16}$$

We showed that if we further assume that the system is causal, the difference equation can be used to compute the output recursively. If the auxiliary conditions correspond to initial rest, the system will be causal, linear, and time-invariant.

The properties and characteristics of LTI systems for which the input and output satisfy a linear constant-coefficient difference equation are best developed through the z-transform. Applying the z-transform to both sides of Eq. (5.16) and using the linearity property (Section 4.4.1) and the time-shifting property (Section 4.4.2), we obtain

$$\sum_{k=0}^{N} a_k z^{-k} Y(z) = \sum_{k=0}^{M} b_k z^{-k} X(z)$$

or, equivalently,

$$\left(\sum_{k=0}^{N} a_k z^{-k}\right) Y(z) = \left(\sum_{k=0}^{M} b_k z^{-k}\right) X(z). \tag{5.17}$$

From Eq. (5.2) and Eq. (5.17) it follows that for a system whose input and output satisfy a difference equation of the form of Eq. (5.16), the system function has the algebraic form

$$H(z) = \frac{Y(z)}{X(z)} = \frac{\displaystyle\sum_{k=0}^{M} b_k z^{-k}}{\displaystyle\sum_{k=0}^{N} a_k z^{-k}}. \tag{5.18}$$

$H(z)$ in Eq. (5.18) is a ratio of polynomials in z^{-1} because Eq. (5.16) consists of a linear combination of delay terms. Although Eq. (5.18) can, of course, be rewritten so that the polynomials are expressed as powers of z rather than of z^{-1}, common practice is not to do so. Also, it is often convenient to express Eq. (5.18) in factored form as

$$H(z) = \left(\frac{b_0}{a_0}\right) \frac{\displaystyle\prod_{k=1}^{M} (1 - c_k z^{-1})}{\displaystyle\prod_{k=1}^{N} (1 - d_k z^{-1})}. \tag{5.19}$$

Each of the factors $(1 - c_k z^{-1})$ in the numerator contributes a zero at $z = c_k$ and a pole at $z = 0$. Similarly, each of the factors $(1 - d_k z^{-1})$ in the denominator contributes a zero at $z = 0$ and a pole at $z = d_k$.

There is a straightforward relationship between the difference equation and the corresponding algebraic expression for the system function. Specifically, the numerator polynomial in Eq. (5.18) has the same coefficients as the right-hand side of Eq. (5.16) (the terms of the form $x[n - k]$) while the denominator polynomial in Eq. (5.18) has the same coefficients as the left-hand side of Eq. (5.16) (the terms of the form $y[n - k]$). Thus, given either the system function in the form of Eq. (5.18) or the difference equation in the form of Eq. (5.16), it is straightforward to obtain the other.

Example 5.1

Suppose that the system function of a linear time-invariant system is

$$H(z) = \frac{(1 + z^{-1})^2}{(1 - \frac{1}{2}z^{-1})(1 + \frac{3}{4}z^{-1})}. \tag{5.20}$$

To find the difference equation that is satisfied by the input and output of this system, we express $H(z)$ in the form of Eq. (5.18) by multiplying the numerator and denominator factors to obtain

$$H(z) = \frac{1 + 2z^{-1} + z^{-2}}{1 + \frac{1}{4}z^{-1} - \frac{3}{8}z^{-2}} = \frac{Y(z)}{X(z)}. \tag{5.21}$$

Thus,

$$(1 + \tfrac{1}{4}z^{-1} - \tfrac{3}{8}z^{-2})Y(z) = (1 + 2z^{-1} + z^{-2})X(z),$$

and the difference equation is

$$y[n] + \tfrac{1}{4}y[n - 1] - \tfrac{3}{8}y[n - 2] = x[n] + 2x[n - 1] + x[n - 2]. \tag{5.22}$$

5.2.1 Stability and Causality

To obtain Eq. (5.18) from Eq. (5.16) we assumed that the system was linear and time-invariant so that Eq. (5.2) applied, but we made no further assumption about stability or causality. Correspondingly, from the difference equation we can obtain the algebraic expression for the system function, but not the region of convergence. Specifically, the region of convergence of $H(z)$ is not determined from the derivation leading to Eq. (5.18) since all that is required for Eq. (5.17) to hold is that $X(z)$ and $Y(z)$ have overlapping regions of convergence. This is consistent with the fact that, as we saw in Chapter 2, the difference equation does not uniquely specify the impulse response of a linear time-invariant system. For the system function of Eq. (5.18) or (5.19) a number of choices for the region of convergence satisfy the requirement that it be a ring bounded by, but not containing, poles. For a given ratio of polynomials, each possible choice for the region of convergence will lead to a different impulse response, but they will all correspond to the same difference equation. However, if we assume that the system is causal, it follows that $h[n]$ must be a right-sided sequence and therefore the region of convergence of $H(z)$ must be outside the outermost pole. Alternatively, if we assume the system is stable, then from the discussion in Section 2.4, the impulse response must be absolutely summable, i.e.,

$$\sum_{n=-\infty}^{\infty} |h[n]| < \infty. \tag{5.23}$$

Since Eq. (5.23) is identical to the condition that

$$\sum_{n=-\infty}^{\infty} |h[n]z^{-n}| < \infty \tag{5.24}$$

for $|z| = 1$, the condition for stability is equivalent to the condition that the ROC of $H(z)$ include the unit circle.

Example 5.2

Consider the LTI system with input and output related through the difference equation

$$y[n] - \tfrac{5}{2}y[n-1] + y[n-2] = x[n]. \tag{5.25}$$

From the previous discussions, $H(z)$ is given by

$$H(z) = \frac{1}{1 - \tfrac{5}{2}z^{-1} + z^{-2}} = \frac{1}{(1 - \tfrac{1}{2}z^{-1})(1 - 2z^{-1})}. \tag{5.26}$$

The pole-zero plot for $H(z)$ is indicated in Fig. 5.1. There are three possible choices for the ROC. If the system is assumed to be causal, then the ROC is outside the outermost pole, i.e., $|z| > 2$. In this case the system will not be stable since the ROC does not include the unit circle. If we assume the system is stable, then the ROC will be $\tfrac{1}{2} < |z| < 2$. For the third possible choice of ROC, $|z| < \tfrac{1}{2}$, the system will be neither stable nor causal.

As Example 5.2 suggests, causality and stability are not necessarily compatible requirements. In order for a linear time-invariant system whose input and output satisfy a difference equation in the form of Eq. (5.16) to be both causal and stable, the ROC of the corresponding system function must be outside the outermost pole *and* include the unit circle. Clearly this requires that all the poles of the system function be inside the unit circle.

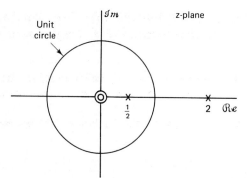

Figure 5.1 Pole-zero plot for Example 5.2.

5.2.2 Inverse Systems

For a given linear time-invariant system with system function $H(z)$, the corresponding inverse system is defined to be the system with system function $H_i(z)$ such that if it is cascaded with $H(z)$, the overall effective system function is unity; i.e.,

$$G(z) = H(z)H_i(z) = 1. \tag{5.27}$$

This implies that

$$H_i(z) = \frac{1}{H(z)}. \tag{5.28}$$

The time-domain condition equivalent to Eq. (5.27) is

$$g[n] = h[n] * h_i[n] = \delta[n]. \tag{5.29}$$

From Eq. (5.28), the frequency response of the inverse system, if it exists, is

$$H_i(e^{j\omega}) = \frac{1}{H(e^{j\omega})}, \tag{5.30}$$

i.e., $H_i(e^{j\omega})$ is the reciprocal of $H(e^{j\omega})$. Equivalently, the log magnitude, phase, and group delay of the inverse system are negatives of the corresponding functions for the original system. Not all systems have an inverse. For example, the ideal lowpass filter does not have an inverse. There is no way to recover the frequency components above the cutoff frequency that are set to zero by the ideal lowpass filter.

Many systems do have inverses, and the class of systems with rational system functions provides a very useful and interesting example. Specifically, consider

$$H(z) = \left(\frac{b_0}{a_0}\right) \frac{\displaystyle\prod_{k=1}^{M}(1 - c_k z^{-1})}{\displaystyle\prod_{k=1}^{N}(1 - d_k z^{-1})}, \tag{5.31}$$

with zeros at $z = c_k$ and poles at $z = d_k$ in addition to possible zeros and/or poles at $z = 0$ and $z = \infty$. Then

$$H_i(z) = \left(\frac{a_0}{b_0}\right) \frac{\displaystyle\prod_{k=1}^{N}(1 - d_k z^{-1})}{\displaystyle\prod_{k=1}^{M}(1 - c_k z^{-1})}, \tag{5.32}$$

i.e., the poles of $H_i(z)$ are the zeros of $H(z)$ and vice versa. The question arises as to what region of convergence to associate with $H_i(z)$. The answer is provided by the convolution theorem, expressed in this case by Eq. (5.29). For Eq. (5.29) to hold, the regions of convergence of $H(z)$ and $H_i(z)$ must overlap. If $H(z)$ is causal, its region of convergence is

$$|z| > \max_{k}|d_k|. \tag{5.33}$$

Thus any appropriate region of convergence for $H_i(z)$ that overlaps with the region specified by Eq. (5.33) is a valid region of convergence for $H_i(z)$. Some simple examples will illustrate some of the possibilities.

Example 5.3

Let $H(z)$ be

$$H(z) = \frac{1 - 0.5z^{-1}}{1 - 0.9z^{-1}},$$

with ROC $|z| > 0.9$. $H_i(z)$ is

$$H_i(z) = \frac{1 - 0.9z^{-1}}{1 - 0.5z^{-1}}.$$

Since $H_i(z)$ has only one pole, there are only two possibilities for its ROC; specifically, $|z| > 0.5$ and $|z| < 0.5$. Obviously only the choice $|z| > 0.5$ overlaps with $|z| > 0.9$. Therefore, the impulse response of the inverse system is

$$h_i[n] = (0.5)^n u[n] - 0.9(0.5)^{n-1} u[n-1].$$

In this case the inverse system is both causal and stable.

Example 5.4

Suppose that $H(z)$ is

$$H(z) = \frac{z^{-1} - 0.5}{1 - 0.9z^{-1}}, \qquad |z| > 0.9.$$

The inverse system function is

$$H_i(z) = \frac{1 - 0.9z^{-1}}{z^{-1} - 0.5} = \frac{-2 + 1.8z^{-1}}{1 - 2z^{-1}}.$$

As before, there are two possible regions of convergence: $|z| < 2$ and $|z| > 2$. In this case, however, *both* regions overlap with $|z| > 0.9$, so both are valid inverse systems. The corresponding impulse response for an ROC $|z| < 2$ is

$$h_{i1}[n] = 2(2)^n u[-n-1] - 1.8(2)^{n-1} u[-n]$$

and for an ROC $|z| > 2$ is

$$h_{i2}[n] = -2(2)^n u[n] + 1.8(2)^{n-1} u[n-1].$$

We see that $h_{i1}[n]$ is stable and noncausal while $h_{i2}[n]$ is unstable and causal.

An obvious generalization from Examples 5.3 and 5.4 is that if $H(z)$ is a causal system with zeros at $c_k, k = 1, \ldots, M$, then its inverse system will be causal if and only if we associate the region of convergence

$$|z| > \max_k |c_k|$$

with $H_i(z)$. If we also require that the inverse system be stable, then the region of convergence of $H_i(z)$ must include the unit circle. Therefore it must be true that

$$\max_k |c_k| < 1,$$

i.e., all the zeros of $H(z)$ must be inside the unit circle. Thus a linear time-invariant system is stable and causal and also has a stable and causal inverse if and only if both the poles and the zeros of $H(z)$ are inside the unit circle. Such systems are referred to as *minimum-phase* systems and will be discussed in more detail in Section 5.6.

5.2.3 Impulse Response for Rational System Functions

The discussion of the partial fraction expansion technique for finding inverse z-transforms (Section 4.3.2) can be applied to the system function $H(z)$ to obtain a general expression for the impulse response of a system that has a rational system function as in Eq. (5.19). Recall that any rational function of z^{-1} with only first-order poles can be expressed in the form

$$H(z) = \sum_{r=0}^{M-N} B_r z^{-r} + \sum_{k=1}^{N} \frac{A_k}{1 - d_k z^{-1}}, \tag{5.34}$$

where the terms in the first summation would be obtained by long division of the denominator into the numerator and would be present only if $M \geq N$. The coefficients A_k in the second set of terms are obtained using Eq. (4.41). If $H(z)$ has multiple-order poles, its partial fraction expansion would have the form of Eq. (4.43). If the system is assumed to be causal, then the ROC is outside all of the poles in Eq. (5.34) and it follows that

$$h[n] = \sum_{r=0}^{M-N} B_r \delta[n - r] + \sum_{k=1}^{N} A_k d_k^n u[n], \tag{5.35}$$

where the first summation is included only if $M \geq N$.

In discussing LTI systems, it is useful to identify two classes. In the first class, at least one nonzero pole of $H(z)$ is not canceled by a zero. In this case there will be at least one term of the form $A_k(d_k)^n u[n]$, and $h[n]$ will not be of finite length, i.e., will not be zero outside a finite interval. Systems of this class are therefore called *infinite impulse response* (IIR) systems. A simple IIR system is discussed in the following example.

Example 5.5

Consider a causal system whose input and output satisfy the difference equation

$$y[n] - ay[n - 1] = x[n]. \tag{5.36}$$

The system function is (by inspection)

$$H(z) = \frac{1}{1 - az^{-1}}. \tag{5.37}$$

Figure 5.2 shows the pole-zero plot of $H(z)$. The region of convergence is $|z| > |a|$, and the condition for stability is $|a| < 1$. The inverse z-transform of $H(z)$ is

$$h[n] = a^n u[n]. \tag{5.38}$$

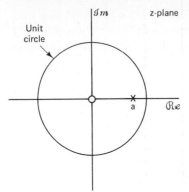

Figure 5.2   Pole-zero plot for Example 5.5.

For the second class of systems, $H(z)$ has no poles except at $z = 0$, i.e., $N = 0$ in Eqs. (5.16) and (5.18). Thus a partial fraction expansion is not possible; $H(z)$ is simply a polynomial in z^{-1} of the form

$$H(z) = \sum_{k=0}^{M} b_k z^{-k}.$$ (5.39)

(We assume without loss of generality that $a_0 = 1$.) In this case, $H(z)$ is determined to within a constant multiplier by its zeros. From Eq. (5.39), $h[n]$ is seen by inspection to be

$$h[n] = \sum_{k=0}^{M} b_k \delta[n - k] = \begin{cases} b_n, & 0 \le n \le M, \\ 0, & \text{otherwise.} \end{cases}$$ (5.40)

In this case the impulse response is finite in length, i.e., it is zero outside a finite interval. Consequently these systems are called *finite impulse response* (FIR) systems. Note that for FIR systems the difference equation of Eq. (5.16) is identical to the convolution sum, i.e.,

$$y[n] = \sum_{k=0}^{M} b_k x[n - k].$$ (5.41)

Example 5.6 gives a simple example of an FIR system.

Example 5.6

Consider an impulse response that is a truncation of the impulse response of Example 5.5:

$$h[n] = \begin{cases} a^n, & 0 \le n \le M, \\ 0 & \text{otherwise.} \end{cases}$$

Then the system function is

$$H(z) = \sum_{n=0}^{M} a^n z^{-n} = \frac{1 - a^{M+1} z^{-M-1}}{1 - az^{-1}}.$$ (5.42)

Since the zeros of the numerator are at

$$z_k = a e^{j2\pi k/(M+1)}, \qquad k = 0, 1, \ldots, M,$$ (5.43)

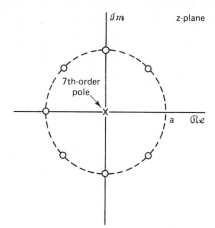

Figure 5.3   Pole-zero plot for Example 5.6.

where a is assumed real and positive, the pole at $z = a$ is canceled by a zero. The pole-zero plot for the case $M = 7$ is shown in Fig. 5.3.

The difference equation satisfied by the input and output of the linear time-invariant system is the discrete convolution

$$y[n] = \sum_{k=0}^{M} a^k x[n - k]. \tag{5.44}$$

However, Eq. (5.42) suggests that the input and output also satisfy the difference equation

$$y[n] - ay[n - 1] = x[n] - a^{M+1} x[n - M - 1]. \tag{5.45}$$

These two equivalent difference equations result from the two equivalent forms of $H(z)$ in Eq. (5.42).

5.3 FREQUENCY RESPONSE FOR RATIONAL SYSTEM FUNCTIONS

If a stable linear time-invariant system has a rational system function (i.e., if its input and output satisfy a difference equation as in Eq. 5.18), then its frequency response (the system function evaluated on the unit circle) has the form

$$H(e^{j\omega}) = \frac{\displaystyle\sum_{k=0}^{M} b_k e^{-j\omega k}}{\displaystyle\sum_{k=0}^{N} a_k e^{-j\omega k}}. \tag{5.46}$$

That is, $H(e^{j\omega})$ is a ratio of polynomials in the variable $e^{-j\omega}$.

To determine the magnitude, phase, and group delay associated with the frequency response of such systems, it is useful to express $H(e^{j\omega})$ in terms of the poles and zeros of $H(z)$. Such an expression results from substituting $z = e^{j\omega}$ into Eq. (5.19):

$$H(e^{j\omega}) = \left(\frac{b_0}{a_0}\right) \frac{\prod_{k=1}^{M}(1 - c_k e^{-j\omega})}{\prod_{k=1}^{N}(1 - d_k e^{-j\omega})}. \tag{5.47}$$

From Eq. (5.47) it follows that $|H(e^{j\omega})|$ is

$$|H(e^{j\omega})| = \left|\frac{b_0}{a_0}\right| \frac{\prod_{k=1}^{M}|1 - c_k e^{-j\omega}|}{\prod_{k=1}^{N}|1 - d_k e^{-j\omega}|}. \tag{5.48}$$

Sometimes it is convenient to consider the magnitude squared rather than the magnitude of the system function. The magnitude-squared function is

$$|H(e^{j\omega})|^2 = H(e^{j\omega})H^*(e^{j\omega}),$$

where * denotes complex conjugation, and for $H(e^{j\omega})$ as in Eq. (5.47),

$$|H(e^{j\omega})|^2 = \left(\frac{b_0}{a_0}\right)^2 \frac{\prod_{k=1}^{M}(1 - c_k e^{-j\omega})(1 - c_k^* e^{j\omega})}{\prod_{k=1}^{N}(1 - d_k e^{-j\omega})(1 - d_k^* e^{j\omega})}. \tag{5.49}$$

From Eq. (5.48) we see that $|H(e^{j\omega})|$ is the product of the magnitudes of all the zero factors of $H(z)$ evaluated on the unit circle, divided by the product of the magnitudes of all the pole factors evaluated on the unit circle. It is common practice to transform these products into a corresponding sum of terms by considering $20 \log_{10}|H(e^{j\omega})|$ instead of $|H(e^{j\omega})|$. The logarithm of Eq. (5.48) is

$$20 \log_{10}|H(e^{j\omega})| = 20 \log_{10}\left|\frac{b_0}{a_0}\right| + \sum_{k=1}^{M} 20 \log_{10}|1 - c_k e^{-j\omega}|$$
$$- \sum_{k=1}^{N} 20 \log_{10}|1 - d_k e^{-j\omega}|. \tag{5.50}$$

The function $20 \log_{10}|H(e^{j\omega})|$ is referred to as the *log magnitude* of $H(e^{j\omega})$ expressed in *decibels* (dB); sometimes this quantity is called the *gain in dB*; i.e.,

$$\text{Gain in dB} = 20 \log_{10}|H(e^{j\omega})|. \tag{5.51}$$

Note that zero dB corresponds to a value of $|H(e^{j\omega})| = 1$, while $|H(e^{j\omega})| = 10^m$ is $20m$ dB. Also, $|H(e^{j\omega})| = 2^m$ is approximately $6m$ dB. When $|H(e^{j\omega})| < 1$, the quantity $20 \log_{10}|H(e^{j\omega})|$ is negative. This would be the case, for example, in the stopband of a frequency-selective filter. It is common practice to define

$$\text{Attenuation in dB} = -20 \log_{10}|H(e^{j\omega})|$$
$$= -\text{Gain in dB}. \tag{5.52}$$

The attenuation is therefore a positive number when the magnitude response is less than 1. For example, 60 dB attenuation at a given frequency means that at that frequency $|H(e^{j\omega})| = 0.001$.

Another advantage to expressing the magnitude in decibels stems from Eq. (5.4a), which, after taking logarithms of both sides, becomes

$$20 \log_{10}|Y(e^{j\omega})| = 20 \log_{10}|H(e^{j\omega})| + 20 \log_{10}|X(e^{j\omega})|, \qquad (5.53)$$

so the frequency response in dB is added to the log magnitude of the input Fourier transform to find the log magnitude of the output Fourier transform. If Eq. (5.53) replaces Eq. (5.4a), then the effects of both magnitude and phase are additive.

The phase response for a rational system function has the form

$$\measuredangle H(e^{j\omega}) = \measuredangle \left[\frac{b_0}{a_0} \right] + \sum_{k=1}^{M} \measuredangle [1 - c_k e^{-j\omega}] - \sum_{k=1}^{N} \measuredangle [1 - d_k e^{-j\omega}]. \qquad (5.54)$$

As in Eq. (5.50), the zero factors contribute with a plus sign and the pole factors contribute with a minus sign.

The corresponding group delay for a rational system function is

$$\mathrm{grd}[H(e^{j\omega})] = \sum_{k=1}^{N} \frac{d}{d\omega} (\arg[1 - d_k e^{-j\omega}]) - \sum_{k=1}^{M} \frac{d}{d\omega} (\arg[1 - c_k e^{-j\omega}]), \qquad (5.55)$$

or

$$\mathrm{grd}[H(e^{j\omega})] = \sum_{k=1}^{N} \frac{|d_k|^2 - \mathscr{R}e\{d_k e^{-j\omega}\}}{1 + |d_k|^2 - 2\mathscr{R}e\{d_k e^{-j\omega}\}} - \sum_{k=1}^{M} \frac{|c_k|^2 - \mathscr{R}e\{c_k e^{-j\omega}\}}{1 + |c_k|^2 - 2\mathscr{R}e\{c_k e^{-j\omega}\}}. \qquad (5.56)$$

In Eq. (5.54) as written, the phase of each of the terms is ambiguous, i.e., any integer multiple of 2π can be added to each term at each value of ω without changing the value of the complex number. The expression for group delay, on the other hand, involves differentiating the continuous phase.

When the angle of a complex number is computed, using an arctangent subroutine on a calculator or with a computer system subroutine, the principal value is obtained. The principal value of the phase of $H(e^{j\omega})$ is denoted as $\mathrm{ARG}[H(e^{j\omega})]$, where

$$-\pi < \mathrm{ARG}[H(e^{j\omega})] \le \pi. \qquad (5.57)$$

Any other angle that gives the correct complex value of the function $H(e^{j\omega})$ can be represented in terms of the principal value as

$$\measuredangle H(e^{j\omega}) = \mathrm{ARG}[H(e^{j\omega})] + 2\pi r(\omega), \qquad (5.58)$$

where $r(\omega)$ is a positive or negative integer that can be different at each value of ω. Similarly, in calculating any of the terms in Eq. (5.54) we would typically obtain the principal value.

If the principal value is used to compute the phase response as a function of ω, then $\mathrm{ARG}[H(e^{j\omega})]$ may be a discontinuous function. The discontinuities introduced by taking the principal value will be jumps of 2π radians. This is illustrated in Fig. 5.4(a), which shows a continuous-phase function $\arg[H(e^{j\omega})]$ and its principal value

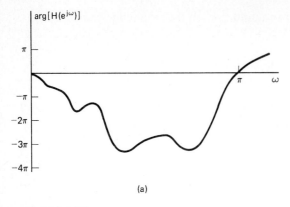

(a)

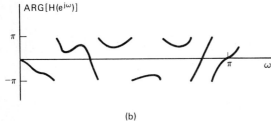

(b)

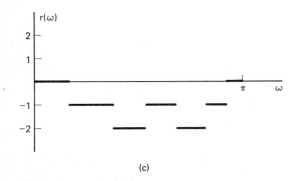

(c)

Figure 5.4 (a) Continuous-phase curve for a system function evaluated on the unit circle. (b) Principal value of the phase curve in part (a). (c) Integer multiples of 2π to be added to $\text{ARG}[H(e^{j\omega})]$ to obtain $\arg[H(e^{j\omega})]$.

$\text{ARG}[H(e^{j\omega})]$ plotted over the range $0 \leq \omega \leq \pi$. The phase function plotted in Fig. 5.4(a) exceeds the range $-\pi$ to $+\pi$. The principal value, shown in Fig. 5.4(b), has jumps of 2π due to the integer multiples of 2π that must be subtracted in certain regions to bring the phase curve within the principal value range. Figure 5.4(c) shows the corresponding value of $r(\omega)$ in Eq. (5.58).

Let us consider Eq. (5.54) when the principal value is used to compute the individual contributions to the phase. It is not difficult to see that

$$\text{ARG}[H(e^{j\omega})] = \text{ARG}\left[\frac{b_0}{a_0}\right] + \sum_{k=1}^{M} \text{ARG}[1 - c_k e^{-j\omega}]$$

$$- \sum_{k=1}^{N} \text{ARG}[1 - d_k e^{-j\omega}] + 2\pi r(\omega),$$

(5.59)

where r is an integer that can be different at each value of ω. The last term, $+2\pi r$, is required because, in general, the principal value of a sum of angles is *not* equal to the sum of the principal values of the individual angles. This is of considerable importance in the theory of cepstral analysis and homomorphic systems, discussed in Chapter 12. However, it presents no problem in plotting phase functions, since the principal value can be used to compute the phase function for each pole and zero and an appropriate multiple of 2π can then be added or subtracted as in Eq. (5.59) to obtain the principal value of the total phase function.

The principal value of the phase function can be computed using Eq. (5.59). Alternatively, we can use the relation

$$\text{ARG}[H(e^{j\omega})] = \arctan\left[\frac{H_R(e^{j\omega})}{H_I(e^{j\omega})}\right], \tag{5.60}$$

where $H_R(e^{j\omega})$ and $H_I(e^{j\omega})$ are the real and imaginary parts, respectively, of $H(e^{j\omega})$. However, when computing the group delay function of Eq. (5.15) it is the derivative of the continuous phase function, i.e., $\arg[H(e^{j\omega})]$, in which we are interested:

$$\text{grd}[H(e^{j\omega})] = -\frac{d}{d\omega}\{\arg[H(e^{j\omega})]\}. \tag{5.61}$$

Except at the discontinuities of $\text{ARG}[H(e^{j\omega})]$ corresponding to jumps between $+\pi$ and $-\pi$,

$$\frac{d}{d\omega}\{\arg[H(e^{j\omega})]\} = \frac{d}{d\omega}\{\text{ARG}[H(e^{j\omega})]\}. \tag{5.62}$$

Consequently, the group delay can be obtained from the principal value by differentiating except at the discontinuities. Similarly, we can express the group delay in terms of the ambiguous phase $\angle H(e^{j\omega})$ as

$$\text{grd}[H(e^{j\omega})] = -\frac{d}{d\omega}[\angle H(e^{j\omega})], \tag{5.63}$$

with the interpretation of Eq. (5.63) that impulses caused by discontinuities of 2π in $\angle H(e^{j\omega})$ are ignored.

5.3.1 Frequency Response of a Single Zero or Pole

Equations (5.50), (5.54), and (5.56) represent the magnitude in dB, the phase, and the group delay, respectively, as a sum of contributions from each of the poles and zeros of the system function. To obtain further insight into the properties of frequency responses of stable linear time-invariant systems with rational system functions, it is worthwhile to first examine the properties of a single factor of the form $(1 - re^{j\theta}e^{-j\omega})$, where r is the radius and θ is the angle of the pole or zero in the z-plane. This factor is typical of either a pole or a zero at a radius r and angle θ in the z-plane.

The magnitude squared of such a factor is

$$|1 - re^{j\theta}e^{-j\omega}|^2 = (1 - re^{j\theta}e^{-j\omega})(1 - re^{-j\theta}e^{j\omega}) = 1 + r^2 - 2r\cos(\omega - \theta). \tag{5.64}$$

Since for any complex quantity C,

$$10 \log_{10}|C|^2 = 20 \log_{10}|C|,$$

the log magnitude in dB is

$$20 \log_{10}|1 - re^{j\theta}e^{-j\omega}| = 10 \log_{10}[1 + r^2 - 2r\cos(\omega - \theta)]. \qquad (5.65)$$

The phase for such a factor is

$$\mathrm{ARG}[1 - re^{j\theta}e^{-j\omega}] = \arctan\left[\frac{r\sin(\omega - \theta)}{1 - r\cos(\omega - \theta)}\right]. \qquad (5.66)$$

Differentiating the right-hand side of Eq. (5.66) (except at discontinuities) gives the group delay of such a factor as

$$\mathrm{grd}[1 - re^{j\theta}e^{-j\omega}] = \frac{r^2 - r\cos(\omega - \theta)}{1 + r^2 - 2r\cos(\omega - \theta)} = \frac{r^2 - r\cos(\omega - \theta)}{|1 - re^{j\theta}e^{-j\omega}|^2}. \qquad (5.67)$$

The functions in Eqs. (5.64)–(5.67) are, of course, periodic in ω with period 2π. Figure 5.5(a) shows a plot of Eq. (5.65) as a function of ω over one period ($0 \le \omega < 2\pi$) for several values of θ with $r = 0.9$. Note that the function dips sharply in the vicinity of $\omega = 0$. Also, we note from Eq. (5.65) that when r is fixed, the log magnitude is a function of $(\omega - \theta)$, so as θ changes it is shifted in frequency. In general, the maximum value of Eq. (5.65) occurs at $(\omega - \theta) = \pi$ and is

$$10 \log_{10}(1 + r^2 + 2r) = 20 \log_{10}(1 + r),$$

which for $r = 0.9$ is equal to 5.57 dB. Similarly, the minimum value of Eq. (5.65), occurring at $\omega = \theta$, is

$$10 \log_{10}(1 + r^2 - 2r) = 20 \log_{10}|1 - r|,$$

which is equal to -20 dB for $r = 0.9$. Note that the plot of the magnitude-squared function in Eq. (5.64) would look similar to Fig. 5.5(a) except that it would have a much wider relative range of values. Hence its plot would appear much sharper for the same value of r.

Figure 5.5(b) shows the phase function in Eq. (5.66) as a function of ω for $r = 0.9$ and several values of θ. Note that the phase is zero at $\omega = \theta$ and that for fixed r, the function simply shifts with θ. Figure 5.5(c) shows the group delay function in Eq. (5.67) for the same conditions on r and θ. Note that the high positive slope of the phase around $\omega = \theta$ corresponds to a large negative peak in the group delay function at $\omega = \theta$.

A simple geometric construction is often very useful in approximate sketching of frequency-response functions directly from the pole-zero plot. The procedure is based on the facts that the frequency response corresponds to the system function evaluated on the unit circle in the z-plane and that the complex value of each pole and zero factor can be represented by a vector in the z-plane from the pole or zero to a point on

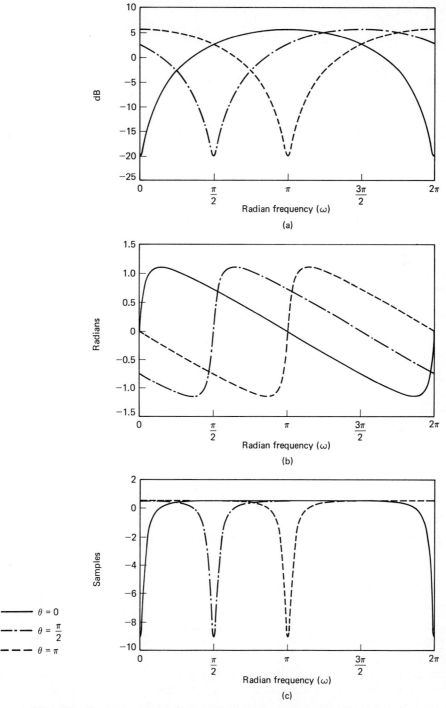

Figure 5.5   Frequency response for a single zero, with $r = 0.9$ and the three values of θ shown on the figure. (a) Log magnitude. (b) Phase. (c) Group delay.

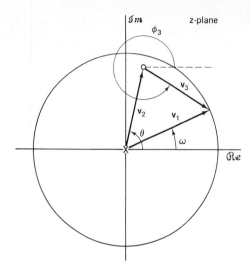

Figure 5.6 z-plane vectors for a first-order system function evaluated on the unit circle, with $r < 1$.

the unit circle. Let us first illustrate the procedure for a first-order system function of the form

$$H(z) = (1 - re^{j\theta}z^{-1}) = \frac{(z - re^{j\theta})}{z}, \qquad r < 1. \tag{5.68}$$

In Section 5.3.2 we will consider higher-order examples. Such a factor has a pole at $z = 0$ and a zero at $z = re^{j\theta}$ as illustrated in Fig. 5.6. Also indicated in this figure are the vectors $\mathbf{v}_1$, $\mathbf{v}_2$, and $\mathbf{v}_3 = \mathbf{v}_1 - \mathbf{v}_2$, representing the complex numbers $e^{j\omega}$, $re^{j\theta}$, and $(e^{j\omega} - re^{j\theta})$, respectively. In terms of these vectors, the magnitude of the complex number

$$\frac{e^{j\omega} - re^{j\theta}}{e^{j\omega}}$$

is the ratio of the magnitudes of the vectors $\mathbf{v}_3$ and $\mathbf{v}_1$, i.e.,

$$|1 - re^{j\theta}e^{-j\omega}| = \left| \frac{e^{j\omega} - re^{j\theta}}{e^{j\omega}} \right| = \frac{|\mathbf{v}_3|}{|\mathbf{v}_1|}, \tag{5.69}$$

or, since $|\mathbf{v}_1| = 1$, Eq. (5.69) is just equal to $|\mathbf{v}_3|$. The corresponding phase is

$$\measuredangle(1 - re^{j\theta}e^{-j\omega}) = \measuredangle(e^{j\omega} - re^{j\theta}) - \measuredangle(e^{j\omega}) = \measuredangle(\mathbf{v}_3) - \measuredangle(\mathbf{v}_1)$$

$$= \phi_3 - \phi_1 = \phi_3 - \omega. \tag{5.70}$$

Typically a vector such as $\mathbf{v}_3$ from a zero to the unit circle is referred to as a zero vector, and a vector from a pole to the unit circle is called a pole vector. Thus the contribution of a single zero factor $(1 - re^{j\theta}z^{-1})$ to the magnitude function at frequency ω is the length of the zero vector $\mathbf{v}_3$ from the zero to the point $z = e^{j\omega}$ on the unit circle. The vector has minimum length when $\omega = \theta$. This accounts for the sharp dip in the magnitude function at $\omega = \theta$ in Fig. 5.5(a). Note that the pole vector $\mathbf{v}_1$ from the pole at $z = 0$ to $z = e^{j\omega}$ always has unit length. Thus it does not have any effect on the magnitude response. Equation (5.70) states that the phase function is

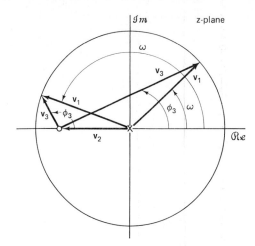

Figure 5.7 z-plane vectors for a first-order system function evaluated on the unit circle, with $\theta = \pi$, $r < 1$. The pole vector $\mathbf{v}_1$ and the zero vector $\mathbf{v}_3$ are shown for two different values of ω.

equal to the difference between the angle of the zero vector from the zero at $re^{j\theta}$ to the point $z = e^{j\omega}$ and the angle of the pole vector from the pole at $z = 0$ to the point $z = e^{j\omega}$.

The pole-zero plot for the case $\theta = \pi$ is shown in Fig. 5.7, and the pole and zero vectors are shown for two different values of ω. Clearly, as ω increases from zero the magnitude of the vector $\mathbf{v}_3$ decreases until it reaches a minimum at $\omega = \pi$, thereby accounting for the shape of the curve corresponding to $\theta = \pi$ in Fig. 5.5(a). The angle of vector $\mathbf{v}_3$ in Fig. 5.7 increases more slowly than ω at first so that the phase curve starts out negative; then when ω is close to π, the angle of vector $\mathbf{v}_3$ increases more rapidly than ω, thereby accounting for the steep positive slope of the phase function around $\omega = \pi$. Note that when $\omega = \pi$ the angles of vectors $\mathbf{v}_3$ and $\mathbf{v}_1$ are equal, so the net phase is zero.

The dependence of the frequency-response contributions of a single factor $(1 - re^{j\theta}e^{-j\omega})$ on the radius r is shown in Fig. 5.8 for $\theta = \pi$ and several values of r. Note that the log magnitude function plotted in Fig. 5.8(a) dips more sharply as r becomes closer to 1; indeed, the magnitude in dB approaches $-\infty$ at $\omega = \theta$ as $r \to 1$. The phase function plotted in Fig. 5.8(b) has positive slope around $\omega = \theta$, which becomes infinite as $r \to 1$. Thus for $r = 1$, the phase function is discontinuous with a jump of π radians at $\omega = \theta$. Away from $\omega = \theta$ the slope of the phase function is negative. Since the group delay is the negative of the slope of the phase curve, the group delay is negative around $\omega = \theta$ and it dips sharply as $r \to 1$. Figure 5.8(c) shows that as we move away from $\omega = \theta$, the group delay becomes positive and relatively flat. When $r = 1$, the group delay is equal to $\frac{1}{2}$ everywhere except at $\omega = \theta$, where it is undefined.

The geometric construction for a zero on the unit circle at $z = -1$ is shown in Fig. 5.9. Indicated are vectors for two different frequencies, $\omega = (\pi - \varepsilon)$ and $\omega = (\pi + \varepsilon)$, where ε is small. Two observations can be made. First, the length of the vector $\mathbf{v}_3$ approaches zero as ω approaches the angle of the zero vector ($\varepsilon \to 0$). Therefore the multiplicative contribution to the frequency response is zero ($-\infty$ dB). Second, the vector $\mathbf{v}_3$ changes its angle discontinuously by π radians as ω goes from $(\pi - \varepsilon)$ to $(\pi + \varepsilon)$.

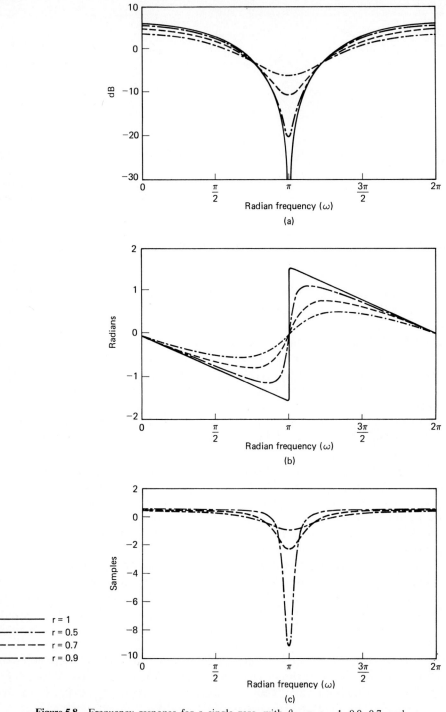

Figure 5.8 Frequency response for a single zero, with $\theta = \pi$, $r = 1$, 0.9, 0.7, and 0.5. (a) Log magnitude. (b) Phase. (c) Group delay for $r = 0.9$, 0.7, and 0.5.

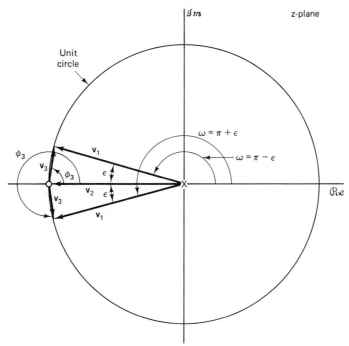

Figure 5.9 z-plane vectors for a zero at $z = -1$ for two different frequencies close to π ($\omega = \pi - \varepsilon$ and $\omega = \pi + \varepsilon$).

Figures 5.5 and 5.8 were restricted to $r \leq 1$. If $r > 1$, the log magnitude function behaves similarly to the case $r < 1$, i.e., it dips more sharply as $r \to 1$, as shown in Fig. 5.10(a). The phase function in Fig. 5.10(b) shows a discontinuity of 2π radians at $\omega = \theta$ for all values of $r > 1$. The source of this discontinuity can be seen from Fig. 5.11, which shows vectors for $\omega = (\pi - \varepsilon)$ and $\omega = (\pi + \varepsilon)$. Note that the pole vector $\mathbf{v}_1$ has an angle of ω, which varies continuously from $\omega = 0$ to $\omega = 2\pi$. The angle of the zero vector $\mathbf{v}_3$ is labeled ϕ_3 in Fig. 5.11. If this angle is measured positively in the counterclockwise direction, Fig. 5.11 shows that ϕ_3 jumps from zero to 2π radians as ω goes from $(\pi - \varepsilon)$ to $(\pi + \varepsilon)$. This jump of 2π radians is evident in Fig. 5.10(b). The discontinuity of 2π radians can also be interpreted as being due to computing the principal-value phase function. The angle ϕ_3 can also be seen to be positive for $\omega = (\pi - \varepsilon)$ and negative for $\omega = (\pi + \varepsilon)$. With this interpretation the angle is continuous at $\omega = \theta$. However, since the total angle of the factor $(1 - re^{j\theta}e^{-j\omega})$ is less than $-\pi$ radians at $\omega = (\pi + \varepsilon)$, the principal value would appear as in Fig. 5.10(b).

The phase curves in Fig. 5.10(b) all have negative slope. Therefore the group delay function for $r > 1$ is positive for all ω. This is also easily seen by considering Eq. (5.67) for $r > 1$.

The preceding discussion and Figs. 5.5, 5.8, and 5.10 all pertain to a single factor of the form $(1 - re^{j\theta}e^{-j\omega})$. If the factor represents a zero of $H(z)$, then the curves of Figs. 5.5, 5.8, and 5.10 will contribute to the frequency-response functions with positive algebraic sign. If the factor represents a pole of $H(z)$, then all the contributions will enter with opposite sign. Thus the contribution of a pole at $z = re^{j\theta}$ would

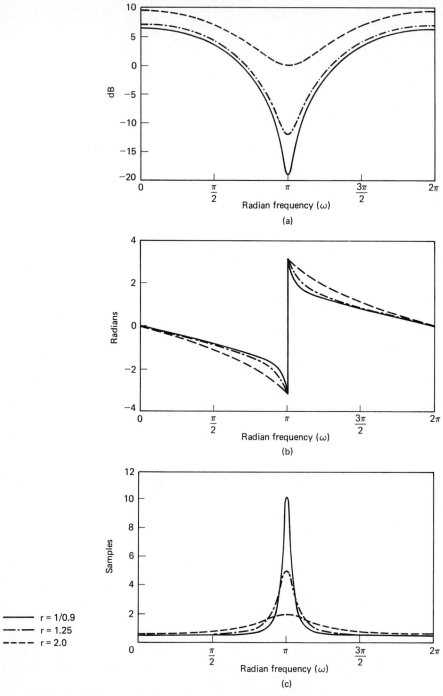

Figure 5.10 Frequency response for a single real zero outside the unit circle with $\theta = \pi$, $r = 1/0.9$, 1.25, 2. (a) Log magnitude. (b) Phase (principal value). (c) Group delay.

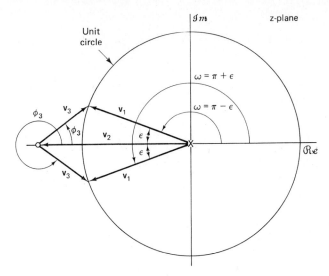

Figure 5.11 z-plane vectors for a single zero evaluated on the unit circle, with $\theta = \pi, r > 1$.

be the negative of the curves in Figs. 5.5 and 5.8. Instead of dipping toward zero ($-\infty$ dB), the magnitude function would peak around $\omega = \theta$. The dependence on r would be the same as for a zero; i.e., the closer r is to 1, the more peaked the contribution to the magnitude function. For stable and causal systems there will, of course, be no poles outside the unit circle, i.e., r will always be less than 1.

5.3.2 Examples with Multiple Poles and Zeros

In this section we illustrate the use of the results of Section 5.3.1 to determine the frequency response of systems with rational system functions.

Example 5.7

Consider the second-order system

$$H(z) = \frac{1}{(1 - re^{j\theta}z^{-1})(1 - re^{-j\theta}z^{-1})} = \frac{1}{1 - 2r\cos\theta z^{-1} + r^2 z^{-2}}. \qquad (5.71)$$

The difference equation satisfied by the input and output of the system is

$$y[n] - 2r\cos\theta y[n-1] + r^2 y[n-2] = x[n].$$

Using the partial fraction expansion technique, the impulse response of a causal system with this system function can be shown to be

$$h[n] = \frac{r^n \sin[\theta(n+1)]}{\sin\theta} u[n]. \qquad (5.72)$$

The system function in Eq. (5.71) has a pole at $z = re^{j\theta}$ and at the conjugate location, $z = re^{-j\theta}$, and two zeros at $z = 0$. The pole-zero plot is shown in Fig. 5.12. From our discussion in Section 5.3.1,

$$20\log_{10}|H(e^{j\omega})| = -10\log_{10}[1 + r^2 - 2r\cos(\omega - \theta)]$$
$$\qquad\qquad - 10\log_{10}[1 + r^2 - 2r\cos(\omega + \theta)] \qquad (5.73a)$$

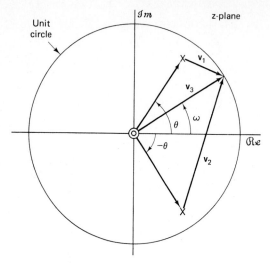

Figure 5.12   Pole-zero plot for Example 5.7.

and

$$\sphericalangle H(e^{j\omega}) = -\arctan\left[\frac{r\sin(\omega - \theta)}{1 - r\cos(\omega - \theta)}\right] - \arctan\left[\frac{r\sin(\omega + \theta)}{1 - r\cos(\omega + \theta)}\right] \quad (5.73b)$$

and

$$\mathrm{grd}[H(e^{j\omega})] = -\frac{r^2 - r\cos(\omega - \theta)}{1 + r^2 - 2r\cos(\omega - \theta)} - \frac{r^2 - r\cos(\omega + \theta)}{1 + r^2 - 2r\cos(\omega + \theta)}. \quad (5.73c)$$

These functions are plotted in Fig. 5.13 for $r = 0.9$ and $\theta = \pi/4$.

Figure 5.12 shows the pole and zero vectors $\mathbf{v}_1$, $\mathbf{v}_2$, and $\mathbf{v}_3$. The magnitude response is the product of the lengths of the zero vectors (which in this case are always unity) divided by the product of the lengths of the pole vectors. That is,

$$|H(e^{j\omega})| = \frac{|\mathbf{v}_3|^2}{|\mathbf{v}_1|\cdot|\mathbf{v}_2|} = \frac{1}{|\mathbf{v}_1|\cdot|\mathbf{v}_2|}. \quad (5.74)$$

When $\omega \approx \theta$, the length of the vector $\mathbf{v}_1 = e^{j\omega} - re^{j\theta}$ becomes small and changes significantly as ω varies about θ, while the length of the vector $\mathbf{v}_2 = e^{j\omega} - re^{-j\theta}$ changes only slightly as ω varies around $\omega = 0$. Thus the pole at angle θ dominates the frequency response around $\omega = \theta$, as is evident from Fig. 5.13. By symmetry, the pole at angle $-\theta$ dominates the frequency response around $\omega = -\theta$.

Example 5.8

Consider an FIR system whose impulse response is

$$h[n] = \delta[n] - 2r\cos\theta\,\delta[n - 1] + r^2\delta[n - 2]. \quad (5.75)$$

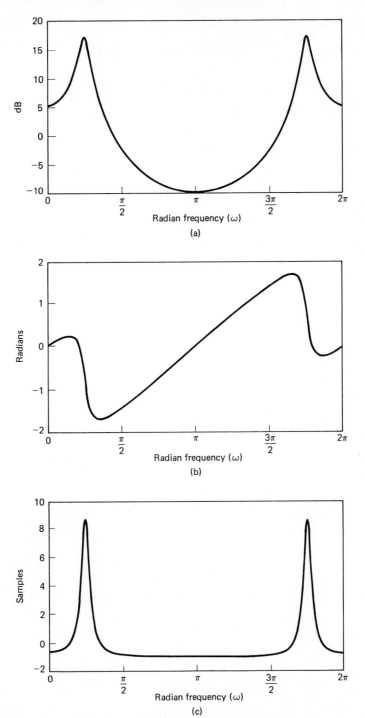

Figure 5.13   Frequency response for a complex-conjugate pair of poles as in Example 5.7, with $r = 0.9$, $\theta = \pi/4$. (a) Log magnitude. (b) Phase. (c) Group delay.

The corresponding system function is

$$H(z) = 1 - 2r \cos \theta \, z^{-1} + r^2 z^{-2}, \tag{5.76}$$

which is the reciprocal of the system function in Example 5.7. Therefore the frequency-response plots for this FIR system are simply the negative of the plots in Fig. 5.13. Note that the pole and zero locations are interchanged in the reciprocal.

Example 5.9

As an example of a system function with both poles and zeros, consider

$$H(z) = \frac{0.05634(1 + z^{-1})(1 - 1.0166z^{-1} + z^{-2})}{(1 - 0.683z^{-1})(1 - 1.4461z^{-1} + 0.7957z^{-2})}. \tag{5.77}$$

The zeros of this system function are at

Radius	Angle
1	π rad
1	± 1.0376 rad (59.45°)

The poles are at

Radius	Angle
0.683	0
0.892	± 0.6257 rad (35.85°)

The pole-zero plot for this system is shown in Fig. 5.14. Figure 5.15 shows the log magnitude, phase, and group delay of the system. The effect of the zeros that are on the unit circle at $\omega = \pm 1.0376$ and π is clearly evident. However, the poles are placed so that rather than peaking for frequencies close to their angles, the total log magnitude remains close to 0 dB over a band from $\omega = 0$ to $\omega = 0.2\pi$ (and, by symmetry, from $\omega = 1.8\pi$ to $\omega = 2\pi$), and then it drops abruptly and remains below -25 dB from about $\omega = 0.3\pi$ to

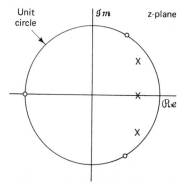

Figure 5.14   Pole-zero plot for the lowpass filter of Example 5.9.

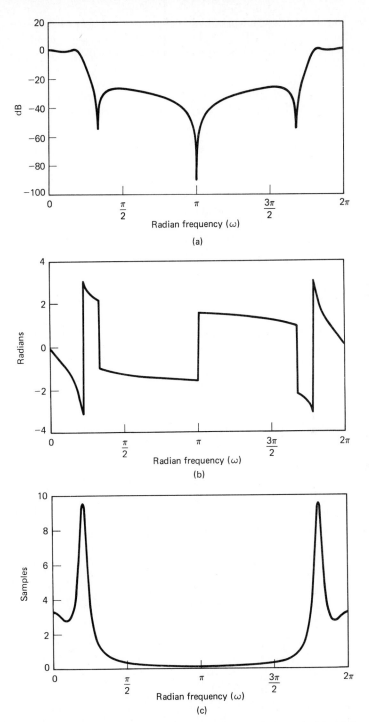

Figure 5.15 Frequency response for the lowpass filter of Example 5.9. (a) Log magnitude. (b) Phase. (c) Group delay.

1.7π. This system is a lowpass filter designed by an approximation method to be described in Chapter 7. The example suggests that useful approximations to frequency-selective filter responses can be achieved using poles to build up the magnitude response and zeros to suppress it.

In this example we see both types of discontinuities in the plotted phase curve. At $\omega \approx 0.22\pi$ there is a discontinuity of 2π due to the use of the principal value in plotting. At $\omega = \pm 1.0376$ and $\omega = \pi$, the discontinuities of π are due to the zeros on the unit circle.

5.4 RELATIONSHIP BETWEEN MAGNITUDE AND PHASE

The frequency response of a linear time-invariant system is the Fourier transform of the impulse response. In general, knowledge about the magnitude provides no information about the phase, and vice versa. However, for systems described by linear constant-coefficient difference equations, i.e., rational system functions, there is some constraint between magnitude and phase. In particular, as we discuss in this section, if the magnitude of the frequency response and the number of poles and zeros are known, then there are only a finite number of choices for the associated phase. Similarly, if the number of poles and zeros and the phase are known, then to within a scale factor there are only a finite number of choices for the magnitude. Furthermore, under a constraint referred to as minimum phase, the frequency-response magnitude specifies the phase uniquely, and the frequency-response phase specifies the magnitude to within a scale factor.

To explore the possible choices of system function given the magnitude squared of the system frequency response, we consider $|H(e^{j\omega})|^2$ expressed as

$$
\begin{aligned}
|H(e^{j\omega})|^2 &= H(e^{j\omega})H^*(e^{j\omega}) \\
&= H(z)H^*(1/z^*)|_{z=e^{j\omega}}.
\end{aligned}
\tag{5.78}
$$

Restricting the system function $H(z)$ to be rational in the form of Eq. (5.19), i.e.

$$
H(z) = \frac{b_0}{a_0} \frac{\displaystyle\prod_{k=1}^{M} (1 - c_k z^{-1})}{\displaystyle\prod_{k=1}^{N} (1 - d_k z^{-1})},
\tag{5.79}
$$

then $H^*(1/z^*)$ in Eq. (5.78) is

$$
H^*\left(\frac{1}{z^*}\right) = \frac{b_0}{a_0} \frac{\displaystyle\prod_{k=1}^{M} (1 - c_k^* z)}{\displaystyle\prod_{k=1}^{N} (1 - d_k^* z)},
\tag{5.80}
$$

where we have assumed that a_0 and b_0 are real. Therefore, Eq. (5.78) states that the magnitude squared of the frequency response is the evaluation on the unit circle of the z-transform $C(z)$ given by

$$C(z) = H(z)H^*(1/z^*) \tag{5.81}$$

$$= \left(\frac{b_0}{a_0}\right)^2 \frac{\displaystyle\prod_{k=1}^{M}(1 - c_k z^{-1})(1 - c_k^* z)}{\displaystyle\prod_{k=1}^{N}(1 - d_k z^{-1})(1 - d_k^* z)}. \tag{5.82}$$

If we are given $|H(e^{j\omega})|^2$, then by replacing $e^{j\omega}$ by z we can construct $C(z)$. From $C(z)$ we would like to infer as much as possible about $H(z)$. We first note that for each pole d_k of $H(z)$, there is a pole of $C(z)$ at d_k and $(d_k^*)^{-1}$. Similarly, for each zero c_k of $H(z)$ there is a zero of $C(z)$ at c_k and $(c_k^*)^{-1}$. Consequently, the poles and zeros of $C(z)$ occur in conjugate reciprocal pairs, with one element of each pair associated with $H(z)$ and one element of each pair associated with $H^*(1/z^*)$. Furthermore, if one element of each pair is inside the unit circle, then the other (i.e., the conjugate reciprocal) will be outside the unit circle. The only other alternative is for both to be on the unit circle in the same location.

If $H(z)$ is assumed to correspond to a causal, stable system, then all its poles must lie inside the unit circle. With this constraint, the poles of $H(z)$ can be identified from the poles of $C(z)$. However, with this constraint alone, the zeros of $H(z)$ cannot be uniquely identified from the zeros of $C(z)$. This can be seen from the following example.

Example 5.10

Consider two stable systems with system functions

$$H_1(z) = \frac{2(1 - z^{-1})(1 + 0.5z^{-1})}{(1 - 0.8e^{j\pi/4}z^{-1})(1 - 0.8e^{-j\pi/4}z^{-1})} \tag{5.83}$$

and

$$H_2(z) = \frac{(1 - z^{-1})(1 + 2z^{-1})}{(1 - 0.8e^{j\pi/4}z^{-1})(1 - 0.8e^{-j\pi/4}z^{-1})}. \tag{5.84}$$

The pole-zero plots for these systems are shown in Figs. 5.16(a) and 5.16(b), respectively.
Now

$$C_1(z) = H_1(z)H_1^*(1/z^*)$$

$$= \frac{2(1 - z^{-1})(1 + 0.5z^{-1})2(1 - z)(1 + 0.5z)}{(1 - 0.8e^{j\pi/4}z^{-1})(1 - 0.8e^{-j\pi/4}z^{-1})(1 - 0.8e^{-j\pi/4}z)(1 - 0.8e^{j\pi/4}z)} \tag{5.85}$$

and

$$C_2(z) = H_2(z)H_2^*(1/z^*)$$

$$= \frac{(1 - z^{-1})(1 + 2z^{-1})(1 - z)(1 + 2z)}{(1 - 0.8e^{j\pi/4}z^{-1})(1 - 0.8e^{-j\pi/4}z^{-1})(1 - 0.8e^{-j\pi/4}z)(1 - 0.8e^{j\pi/4}z)}. \tag{5.86}$$

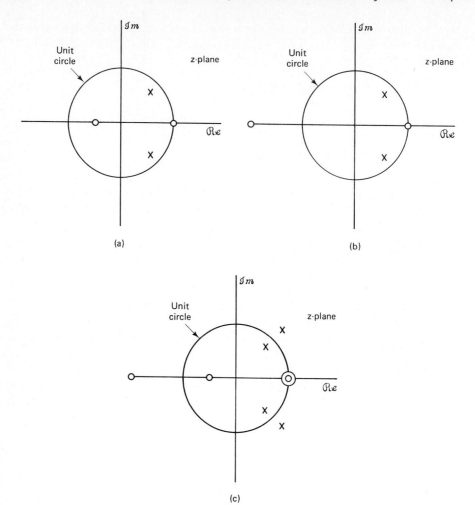

Figure 5.16   Pole-zero plots for two system functions and their common magnitude-squared function. (a) $H_1(z)$. (b) $H_2(z)$. (c) $C_1(z)$, $C_2(z)$.

Using the fact that

$$4(1 + \tfrac{1}{2}z^{-1})(1 + \tfrac{1}{2}z) = (1 + 2z^{-1})(1 + 2z), \tag{5.87}$$

we see that $C_1(z) = C_2(z)$. The pole-zero plot for $C_1(z)$ and $C_2(z)$ is shown in Fig. 5.16(c).

The system functions $H_1(z)$ and $H_2(z)$ in Example 5.10 differ only in the location of the zeros. In the example, the factor $2(1 + 0.5z^{-1}) = (z^{-1} + 2)$ contributes the same to the magnitude squared of the frequency response as the factor $(1 + 2z^{-1})$, and consequently $|H_1(e^{j\omega})|$ and $|H_2(e^{j\omega})|$ are equal. However, the phase functions for these two frequency responses are different.

Example 5.11

Suppose we are given the pole-zero plot for $C(z)$ in Fig. 5.17 and want to determine the poles and zeros to associate with $H(z)$. The conjugate reciprocal pairs of poles and zeros for which one element of each is associated with $H(z)$ and one with $H^*(1/z^*)$ are

$$
\begin{array}{ll}
\text{Pole pair 1:} & (P_1, P_4) \\
\text{Pole pair 2:} & (P_2, P_5) \\
\text{Pole pair 3:} & (P_3, P_6) \\
\text{Zero pair 1:} & (Z_1, Z_4) \\
\text{Zero pair 2:} & (Z_2, Z_5) \\
\text{Zero pair 3:} & (Z_3, Z_6)
\end{array}
$$

Knowing that $H(z)$ corresponds to a stable, causal system, we must choose the poles from each pair that are inside the unit circle, i.e. P_1, P_2, and P_3. No such constraint is imposed on the zeros. However, if we assume that the coefficients a_k and b_k are real in Eqs. (5.16) and (5.18), the zeros (and poles) either are real or occur in complex conjugate pairs. Consequently the zeros to associate with $H(z)$ are

$$
Z_3 \quad \text{or} \quad Z_6
$$
$$
\text{and} \quad (Z_1, Z_2) \quad \text{or} \quad (Z_4, Z_5).
$$

Therefore, there are a total of four different stable, causal systems with three poles and three zeros for which the pole-zero plot of $C(z)$ is that shown in Fig. 5.17 and, equivalently, for which the frequency-response magnitude is the same. If we had not assumed that the coefficients a_k and b_k were real, the number of choices would be greater. Furthermore, if the number of poles and zeros of $H(z)$ were not restricted, the number of choices for $H(z)$ would be unlimited. To see this, assume that $H(z)$ has a factor of the form

$$
\frac{z^{-1} - a^*}{1 - az^{-1}},
$$

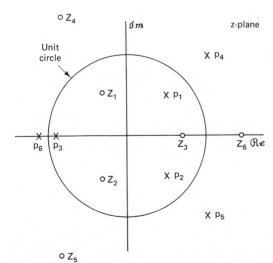

Figure 5.17   Pole-zero plot for the magnitude-squared function in Example 5.11.

i.e.,

$$H(z) = H_1(z) \frac{z^{-1} - a^*}{1 - az^{-1}}. \tag{5.88}$$

Factors of this form are referred to as *allpass factors* since they have unity magnitude on the unit circle; they are discussed in more detail in Section 5.5. It is easily verified that

$$C(z) = H(z)H^*(1/z^*) = H_1(z)H_1^*(1/z^*), \tag{5.89}$$

i.e., allpass factors cancel in $C(z)$ and therefore would not be identifiable from the pole-zero plot of $C(z)$. Consequently, if the number of poles and zeros of $H(z)$ is unspecified, then given $C(z)$, any choice for $H(z)$ can be cascaded with an arbitrary number of allpass factors with poles inside the unit circle (i.e., $|a| < 1$).

5.5 ALLPASS SYSTEMS

As indicated in the discussion of Example 5.11, a stable system function of the form

$$H_{ap}(z) = \frac{z^{-1} - a^*}{1 - az^{-1}} \tag{5.90}$$

has a frequency-response magnitude that is independent of ω. This can be seen by writing $H_{ap}(e^{j\omega})$ in the form

$$\begin{aligned} H_{ap}(e^{j\omega}) &= \frac{e^{-j\omega} - a^*}{1 - ae^{-j\omega}} \\ &= e^{-j\omega} \frac{1 - a^* e^{j\omega}}{1 - ae^{-j\omega}}. \end{aligned} \tag{5.91}$$

In the expression in Eq. (5.91), the term $e^{-j\omega}$ has unity magnitude and the remaining numerator and denominator factors are complex conjugates of each other and therefore have the same magnitude. Consequently $|H_{ap}(e^{j\omega})| = 1$. Such a system is called an *allpass system* since the system passes all of the frequency components of its input with magnitude gain of unity. The most general form for the system function of an allpass system with a real-valued impulse response is a product of factors like Eq. (5.90) with complex poles being paired with their conjugates; i.e.,

$$H_{ap}(z) = \prod_{k=1}^{M_r} \frac{z^{-1} - d_k}{1 - d_k z^{-1}} \prod_{k=1}^{M_c} \frac{(z^{-1} - e_k^*)(z^{-1} - e_k)}{(1 - e_k z^{-1})(1 - e_k^* z^{-1})}, \tag{5.92}$$

where the d_k's are the real poles and the e_k's are the complex poles of $H_{ap}(z)$. For causal and stable allpass systems, $|d_k| < 1$ and $|e_k| < 1$. In terms of our general

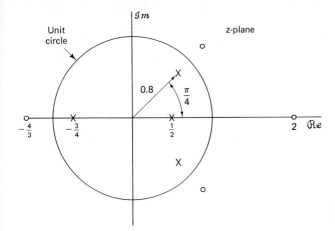

Figure 5.18 Typical pole-zero plot for an allpass system.

notation for system functions, allpass systems have $M = N = 2M_c + M_r$ poles and zeros. Figure 5.18 shows a typical pole-zero plot for an allpass system. In this case $M_r = 2$ and $M_c = 1$. Note that each pole of $H_{ap}(z)$ is paired with a conjugate reciprocal zero.

The frequency response for a general allpass system can be expressed in terms of the frequency responses of first-order allpass systems like that specified in Eq. (5.90), each of which consists of a single pole inside the unit circle and a zero at the conjugate reciprocal location. The magnitude response for such a term is, as we have shown, unity. Thus the log magnitude in dB is zero. Similarly the log magnitude for the general allpass system is zero with $a = re^{j\theta}$. The phase function for Eq. (5.90) is

$$\measuredangle\left[\frac{e^{-j\omega} - re^{-j\theta}}{1 - re^{j\theta}e^{-j\omega}}\right] = -\omega - 2\arctan\left[\frac{r\sin(\omega - \theta)}{1 - r\cos(\omega - \theta)}\right]. \qquad (5.93)$$

Likewise, the phase of a second-order allpass system with poles at $z = re^{j\theta}$ and $z = re^{-j\theta}$ is

$$\measuredangle\left[\frac{(e^{-j\omega} - re^{-j\theta})(e^{-j\omega} - re^{j\theta})}{(1 - re^{j\theta}e^{-j\omega})(1 - re^{-j\theta}e^{-j\omega})}\right] = -2\omega - 2\arctan\left[\frac{r\sin(\omega - \theta)}{1 - r\cos(\omega - \theta)}\right]$$

$$- 2\arctan\left[\frac{r\sin(\omega + \theta)}{1 - r\cos(\omega + \theta)}\right]. \qquad (5.94)$$

Figure 5.19 shows plots of log magnitude, phase, and group delay for two first-order allpass systems, one with a real pole at $z = 0.9$ ($\theta = 0, r = 0.9$) and another with a pole at $z = -0.9$ ($\theta = \pi, r = 0.9$). For both systems, the radii of the poles are $r = 0.9$. Likewise, Fig. 5.20 shows the same functions for a second-order allpass system with poles at $z = 0.9e^{j\pi/4}$ and $z = 0.9e^{-j\pi/4}$. Figures 5.19(b) and 5.20(b) illustrate a general property of allpass systems. In Fig. 5.19(b) we see that the phase is

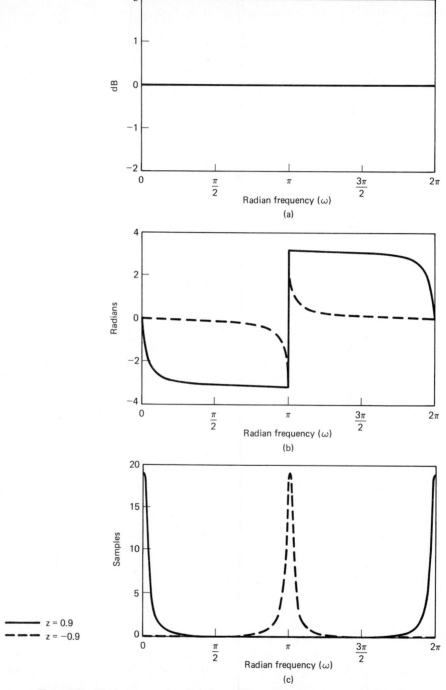

Figure 5.19 Frequency response for allpass filters with real poles at $z = 0.9$ (solid line) and $z = -0.9$ (dashed line). (a) Log magnitude. (b) Phase (principal value). (c) Group delay.

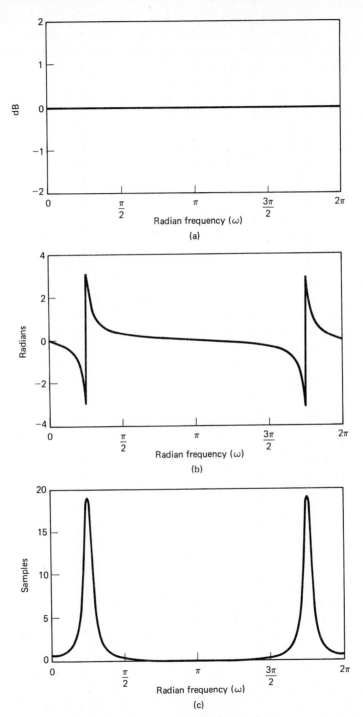

Figure 5.20 Frequency response of second-order allpass system with poles at $z = 0.9e^{\pm j\pi/4}$. (a) Log magnitude. (b) Phase (principal value). (c) Group delay.

nonpositive for $0 < \omega < \pi$. Similarly, in Fig. 5.20(b) if the discontinuity of 2π resulting from the computation of principal value is removed, the resulting contin- uous-phase curve is nonpositive for $0 < \omega < \pi$. Since the more general allpass system given by Eq. (5.92) is just a product of such first- and second-order factors, we can conclude that the (continuous) phase, $\arg[H_{ap}(e^{j\omega})]$, of an allpass system is always nonpositive for $0 < \omega < \pi$. This may not appear to be true if the principal value is plotted, as is illustrated in Fig. 5.21, which shows the log magnitude, phase, and group delay for an allpass system with poles and zeros as in Fig. 5.18. However, we can establish this result by first considering the group delay.

The group delay of the simple one-pole allpass system of Eq. (5.90) is the negative derivative of the phase given by Eq. (5.93). With a small amount of algebra it can be shown that

$$\text{grd}\left[\frac{e^{-j\omega} - re^{-j\theta}}{1 - re^{j\theta}e^{-j\omega}}\right] = \frac{1 - r^2}{1 + r^2 - 2r\cos(\omega - \theta)} = \frac{1 - r^2}{|1 - re^{j\theta}e^{-j\omega}|^2}. \tag{5.95}$$

Since $r < 1$ for a stable and causal allpass system, from Eq. (5.95) the group delay contributed by a single allpass factor is always positive. Since the group delay of a higher-order allpass system will be a sum of positive terms as in Eq. (5.95), it is true in general that the group delay of a rational allpass system is always positive. This is confirmed by Figs. 5.19(c), 5.20(c), and 5.21(c), which show the group delay for first- order, second-order, and third-order allpass systems, respectively.

The positivity of the group delay of an allpass system is the basis for a simple proof of the negativity of the phase of an allpass system. First note that

$$\arg[H_{ap}(e^{j\omega})] = -\int_0^\omega \text{grd}[H_{ap}(e^{j\phi})]d\phi + \arg[H_{ap}(e^{j0})] \tag{5.96}$$

for $0 \le \omega \le \pi$. From Eq. (5.92) it follows that

$$H_{ap}(e^{j0}) = \prod_{k=1}^{M_r} \frac{1 - d_k}{1 - d_k} \prod_{k=1}^{M_c} \frac{|1 - e_k|^2}{|1 - e_k|^2} = 1. \tag{5.97}$$

Therefore $\arg[H_{ap}(e^{j0})] = 0$ and since

$$\text{grd}[H_{ap}(e^{j\omega})] \ge 0, \tag{5.98}$$

it follows from Eq. (5.96) that

$$\arg[H_{ap}(e^{j\omega})] \le 0 \qquad \text{for } 0 \le \omega < \pi. \tag{5.99}$$

The positivity of the group delay and the nonpositivity of the continuous phase are important properties of allpass systems.

Allpass systems have many uses. They can be used as compensators for phase (or group delay) distortion, as we will see in Chapter 7, and they are useful in the theory of minimum-phase systems, as we will see in Section 5.6. They are also useful

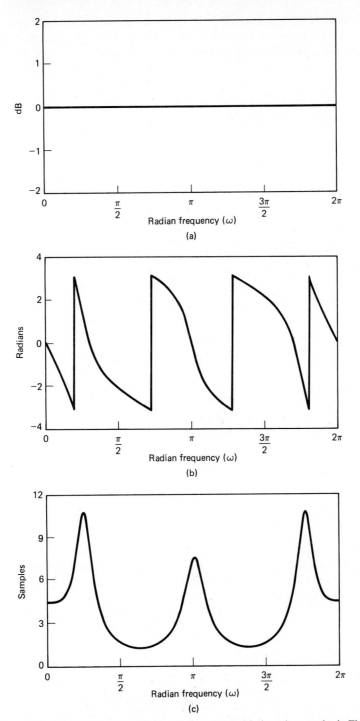

Figure 5.21 Frequency response for an allpass system with the pole-zero plot in Fig. 5.18. (a) Log magnitude. (b) Phase (principal value). (c) Group delay.

in transforming frequency-selective lowpass filters into other frequency-selective forms and in obtaining variable-cutoff frequency-selective filters. These applications are discussed in Chapter 7 and applied in the problems.

SKIP 5.6 MINIMUM-PHASE SYSTEMS

In Section 5.4 we showed that the frequency-response magnitude for an LTI system with rational system function does not uniquely characterize the system. If the system is stable and causal, the poles must be inside the unit circle, but stability and causality place no such restriction on the zeros. For certain classes of problems it is useful to impose the additional restriction that the inverse system (one with system function $1/H(z)$) also be stable and causal. As discussed in Section 5.2.2, this then restricts the zeros as well as the poles to be inside the unit circle since the poles of $1/H(z)$ are the zeros of $H(z)$. Such systems are commonly referred to as *minimum-phase* systems. The name *minimum-phase* comes from a property of the phase response, which is not obvious from the above definition. This and other fundamental properties that we discuss are unique to this class of systems and therefore any one of them could be taken as the definition of the class. These properties are developed in Section 5.6.2.

If we are given a magnitude-squared function in the form of Eq. (5.82) and we know that the system is a minimum-phase system, then $H(z)$ is uniquely determined; it will consist of all the poles and zeros of $C(z) = H(z)H^*(1/z^*)$ that lie inside the unit circle.† This approach is often followed in filter design when only the magnitude response is determined by the design method used (see Chapter 7).

In Section 5.4 we showed that from the magnitude squared of the frequency response alone, we could not uniquely determine the system function $H(z)$ since any choice that had the given frequency-response magnitude could be cascaded with arbitrary allpass factors without affecting the magnitude. A related observation is that any rational system function‡ can be expressed as

$$H(z) = H_{min}(z)H_{ap}(z), \qquad (5.100)$$

where $H_{min}(z)$ is a minimum-phase system and $H_{ap}(z)$ is an allpass system.

To show this, suppose that $H(z)$ has one zero outside the unit circle at $z = 1/c^*$, where $|c| < 1$, and the remaining poles and zeros are inside the unit circle. Then $H(z)$ can be expressed as

$$H(z) = H_1(z)(z^{-1} - c^*), \qquad (5.101)$$

---

† We have assumed that $C(z)$ has no poles or zeros on the unit circle. Strictly speaking, systems with poles on the unit circle are unstable and are generally to be avoided in practice. Zeros on the unit circle, however, often occur in practical filter designs. By our definition, such systems are nonminimum phase, but many of the properties of minimum-phase systems hold even in this case.

‡ Somewhat for convenience, we will restrict the discussion to stable, causal systems, although the observation applies more generally.

where, by definition, $H_1(z)$ is minimum phase. An equivalent expression for $H(z)$ is

$$H(z) = H_1(z)(1 - cz^{-1})\frac{z^{-1} - c^*}{1 - cz^{-1}}. \qquad (5.102)$$

Since $|c| < 1$, the factor $H_1(z)(1 - cz^{-1})$ also is minimum phase, and it differs from $H(z)$ only in that the zero of $H(z)$ that was outside the unit circle at $z = 1/c^*$ is reflected inside the unit circle to the conjugate reciprocal location $z = c$. The term $(z^{-1} - c^*)/(1 - cz^{-1})$ is allpass. This example can be generalized in a straightforward way to include more zeros outside the unit circle, thereby showing that in general any system function can be expressed as

$$H(z) = H_{min}(z)H_{ap}(z), \qquad (5.103)$$

where $H_{min}(z)$ contains the poles and zeros of $H(z)$ that lie inside the unit circle plus zeros that are the conjugate reciprocals of the zeros of $H(z)$ that lie outside the unit circle. $H_{ap}(z)$ comprises all the zeros of $H(z)$ that lie outside the unit circle together with poles to cancel the reflected conjugate reciprocal zeros in $H_{min}(z)$.

Using Eq. (5.103), we can form a nonminimum-phase system from a minimum-phase system by reflecting one or more zeros lying inside the unit circle to their conjugate reciprocal locations outside the unit circle, or, conversely, we can form a minimum-phase system from a nonminimum-phase system by reflecting all the zeros lying outside the unit circle to their conjugate reciprocal locations inside. In either case, both the minimum-phase and the nonminimum-phase systems will have the same frequency-response magnitudes.

5.6.1 Frequency-Response Compensation

In many signal processing contexts, we would like to compensate a signal that has been processed by an LTI system with an undesirable frequency response, as indicated in Fig. 5.22. This situation may arise, for example, in transmitting signals over a communications channel. If perfect compensation is achieved, then $s_c[n] = s[n]$, i.e., $H_c(z)$ is the inverse of $H_d(z)$. However, if we assume that the distorting system is stable and causal and require the compensating system to be stable and causal, then perfect compensation is possible only if $H_d(z)$ is a minimum-phase system so that it has a stable, causal inverse.

Based on the previous discussions, assuming $H_d(z)$ is known or approximated as a rational system function, we can form a minimum-phase system $H_{dmin}(z)$ by reflecting all the zeros of $H_d(z)$ that are outside the unit circle to their conjugate

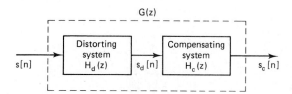

Figure 5.22   Illustration of distortion compensation by linear filtering.

reciprocal locations inside the unit circle. $H_d(z)$ and $H_{dmin}(z)$ have the same frequency-response magnitude and are related through an allpass system $H_{ap}(z)$, i.e.,

$$H_d(z) = H_{dmin}(z)H_{ap}(z). \qquad (5.104)$$

Choosing the compensating filter to be

$$H_c(z) = \frac{1}{H_{dmin}(z)}, \qquad (5.105)$$

the overall system function $G(z)$ relating $s[n]$ and $s_c[n]$ is

$$G(z) = H_d(z)H_c(z) = H_{ap}(z), \qquad (5.106)$$

i.e., it corresponds to an allpass system. Consequently, the frequency-response
magnitude is exactly compensated, while the phase response is modified to
$\measuredangle H_{ap}(e^{j\omega})$.

The following example illustrates the discussion for the case of an FIR system.

Example 5.12

Consider the system function

$$H(z) = (1 - 0.9e^{j0.6\pi}z^{-1})(1 - 0.9e^{-j0.6\pi}z^{-1})$$
$$\times (1 - 1.25e^{j0.8\pi}z^{-1})(1 - 1.25e^{-j0.8\pi}z^{-1}). \qquad (5.107)$$

The pole-zero plot is shown in Fig. 5.23. Since $H(z)$ has only zeros (all poles are at $z = 0$),
it follows that the system has a finite-duration impulse response. Therefore the
system is stable; and since $H(z)$ is a polynomial with only negative powers of z, the system
is causal. However, since two of the zeros are outside the unit circle, the system is
nonminimum phase. Figure 5.24 shows the log magnitude, phase, and group delay for
$H(e^{j\omega})$.

The corresponding minimum-phase system is obtained by reflecting the zeros that
occur at $z = 1.25e^{\pm j0.8\pi}$ to their conjugate reciprocal locations inside the unit circle. If
we express $H(z)$ as

$$H(z) = (1 - 0.9e^{j0.6\pi}z^{-1})(1 - 0.9e^{-j0.6\pi}z^{-1})(1.25)^2$$
$$\times (z^{-1} - 0.8e^{-j0.8\pi})(z^{-1} - 0.8e^{j0.8\pi}), \qquad (5.108)$$

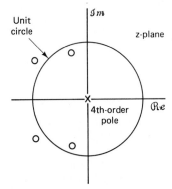

Figure 5.23 Pole-zero plot of FIR
system in Example 5.12.

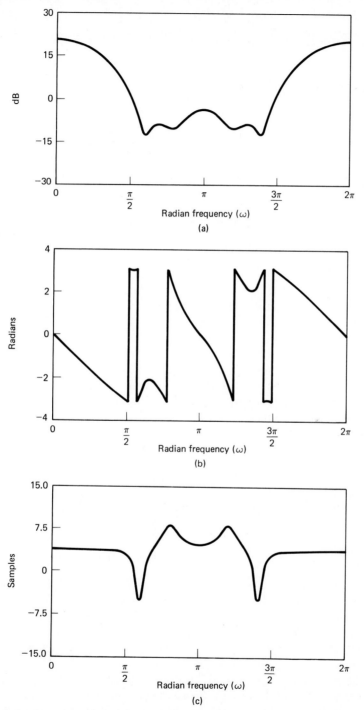

Figure 5.24 Frequency response for FIR system with pole-zero plot in Fig. 5.23. (a) Log magnitude. (b) Phase (principal value). (c) Group delay.

then

$$H_{min}(z) = (1.25)^2(1 - 0.9e^{j0.6\pi}z^{-1})(1 - 0.9e^{-j0.6\pi}z^{-1})$$

$$\times (1 - 0.8e^{-j0.8\pi}z^{-1})(1 - 0.8e^{j0.8\pi}z^{-1}), \quad (5.109)$$

and the allpass system that relates $H_{min}(z)$ and $H(z)$ is

$$H_{ap}(z) = \frac{(z^{-1} - 0.8e^{-j0.8\pi})(z^{-1} - 0.8e^{j0.8\pi})}{(1 - 0.8e^{j0.8\pi}z^{-1})(1 - 0.8e^{-j0.8\pi}z^{-1})}. \quad (5.110)$$

The log magnitude, phase, and group delay of $H_{min}(e^{j\omega})$ are shown in Fig. 5.25. Figs. 5.24(a) and 5.25(a) are, of course, identical. The log magnitude, phase, and group delay for $H_{ap}(e^{j\omega})$ are plotted in Fig. 5.26.

Note that the inverse system for $H(z)$ would have poles at $z = 1.25e^{\pm j0.8\pi}$ and at $z = 0.9e^{\pm j0.6\pi}$, and thus the causal inverse would be unstable. The minimum-phase inverse would be the reciprocal of $H_{min}(z)$ as given by Eq. (5.109), and if this inverse were used in the cascade system of Fig. 5.22, the overall effective system function would be $H_{ap}(z)$ as given in Eq. (5.110).

5.6.2 Properties of Minimum-Phase Systems

We have been using the term "minimum phase" to refer to systems that are causal and stable and that have a causal and stable inverse. This choice of name is motivated by a property of the phase function which, while not obvious, follows from the above definition. In this section we develop a number of interesting and important properties of minimum-phase systems relative to all other systems that have the same frequency-response magnitude.

The Minimum Phase-Lag Property.   The use of the terminology "minimum phase" as a descriptive name for a system having all its poles and zeros inside the unit circle is suggested by Example 5.12. Recall that as a consequence of Eq. (5.100), the continuous phase, i.e., $\arg[H(e^{j\omega})]$, of any nonminimum-phase system can be expressed as

$$\arg[H(e^{j\omega})] = \arg[H_{min}(e^{j\omega})] + \arg[H_{ap}(e^{j\omega})]. \quad (5.111)$$

Therefore the continuous phase that would correspond to the principal-value phase of Fig. 5.24(b) is the sum of the continuous phase associated with the minimum-phase function of Fig. 5.25(b) and the continuous phase of the allpass system associated with the principal-value phase shown in Fig. 5.26(b). As was shown in Section 5.5 and as indicated by the principal-value phase curves of Figs. 5.19(b), 5.20(b), 5.21(b), and 5.26(b), the continuous-phase curve of an allpass system is negative for $0 \le \omega \le \pi$. Thus the reflection of zeros of $H_{min}(z)$ from inside the unit circle to conjugate reciprocal locations outside always *decreases* the (continuous) phase or *increases* the negative of the phase, which is called the *phase-lag* function. Thus, the causal, stable system that has $|H_{min}(e^{j\omega})|$ as its magnitude response and also has all its zeros (and, of

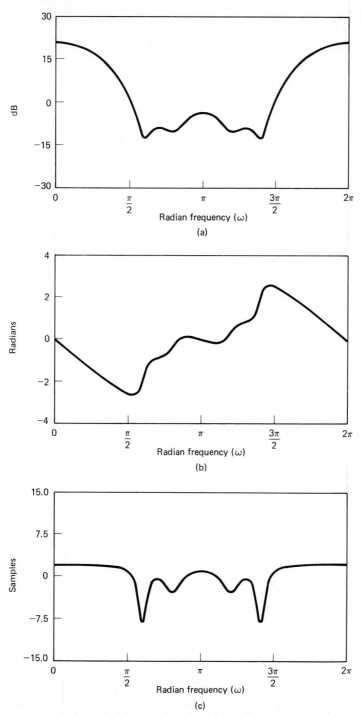

Figure 5.25 Frequency response for minimum-phase system in Example 5.12.
(a) Log magnitude. (b) Phase. (c) Group delay.

course, poles) inside the unit circle has the minimum phase-lag function (for $0 \leq \omega < \pi$) of all the systems having that same magnitude response. Therefore a more precise terminology is *minimum phase-lag* system, but minimum phase is historically the established terminology.

To make the interpretation of minimum phase-lag systems more precise, it is necessary to impose the additional constraint that $H(e^{j\omega})$ be positive at $\omega = 0$, i.e.,

$$H(e^{j0}) = \sum_{n=-\infty}^{\infty} h[n] > 0. \tag{5.112}$$

Note that $H(e^{j0})$ will be real if we restrict $h[n]$ to be real. The condition of Eq. (5.112) is necessary because a system with impulse response $-h[n]$ has the same poles and zeros for its system function as a system with impulse response $h[n]$. However, multiplying by -1 would alter the phase by π radians. Thus to remove this ambiguity, we must impose the condition of Eq. (5.112) to ensure that a system with all its poles and zeros inside the unit circle also has the minimum phase-lag property. However, this constraint is often of little significance and our definition at the beginning of Section 5.6, which does not include it, is the generally accepted definition of the class of minimum-phase systems.

The Minimum Group Delay Property.   Example 5.12 illustrates another property of systems whose poles and zeros are all inside the unit circle. First note that the group delay for the systems that have the same magnitude response is

$$\text{grd}[H(e^{j\omega})] = \text{grd}[H_{min}(e^{j\omega})] + \text{grd}[H_{ap}(e^{j\omega})]. \tag{5.113}$$

The group delay for the minimum-phase system shown in Fig. 5.25(c) is always less than the group delay for the nonminimum-phase system shown in Fig. 5.24(c). This is because, as Fig. 5.26(c) shows, the allpass system that converts the minimum-phase system into the nonminimum-phase system has positive group delay. In Section 5.5 we showed this to be a general property of allpass systems; they always have positive group delay for all ω. Thus, if we again consider all the systems that have a given magnitude response $|H_{min}(e^{j\omega})|$, the one that has all its poles and zeros inside the unit circle has the minimum group delay. An equally appropriate name for such systems would therefore be *minimum group delay* systems, but this terminology is not generally used.

The Minimum Energy Delay Property.   In Example 5.12 there are a total of four causal FIR systems with real impulse responses that have the same frequency response as the system in Eq. (5.107). The associated pole-zero plots are shown in Fig. 5.27, where Fig. 5.27(d) corresponds to Eq. (5.107) and Fig. 5.27(a) to the minimum-phase system of Eq. (5.109). The impulse responses for these four cases are plotted in Fig. 5.28. If we compare the four sequences in Fig. 5.28, we observe that the

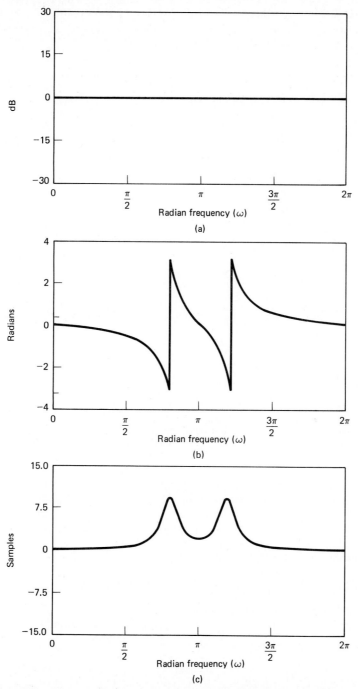

Figure 5.26 Frequency response of allpass system of Example 5.12. (The sum of corresponding curves in Figs. 5.25 and 5.26 equals the corresponding curve in Fig. 5.24 with the sum of the phase curves taken modulo 2π.) (a) Log magnitude. (b) Phase (principal value). (c) Group delay.

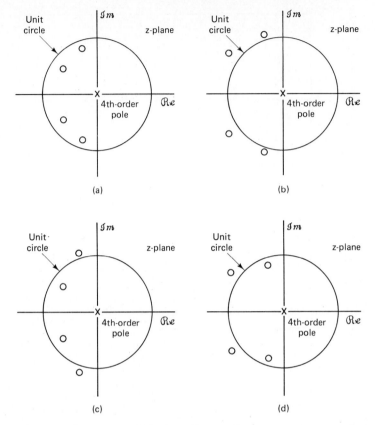

Figure 5.27   Four systems, all having the same frequency-response magnitude. Zeros are at all combinations of $0.9e^{\pm j0.6\pi}$ and $0.8e^{\pm j0.8\pi}$ and their reciprocals.

minimum-phase sequence appears to have larger samples at its left-hand end than all the other sequences. Indeed, it is true for this example and in general that

$$|h[0]| < |h_{min}[0]| \qquad (5.114)$$

for any causal, stable sequence $h[n]$ for which

$$|H(e^{j\omega})| = |H_{min}(e^{j\omega})|. \qquad (5.115)$$

A proof of this property is suggested in Problem 5.35.

All the impulse responses whose magnitude response is equal to $|H_{min}(e^{j\omega})|$ have the same total energy as $h_{min}[n]$, since by Parseval's theorem

$$\sum_{n=0}^{\infty} |h[n]|^2 = \frac{1}{2\pi} \int_{-\pi}^{\pi} |H(e^{j\omega})|^2 \, d\omega = \frac{1}{2\pi} \int_{-\pi}^{\pi} |H_{min}(e^{j\omega})|^2 \, d\omega$$

$$\qquad (5.116)$$

$$= \sum_{n=0}^{\infty} |h_{min}[n]|^2.$$

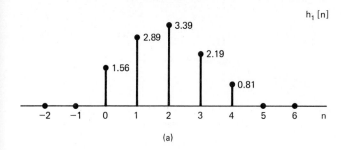

(a)

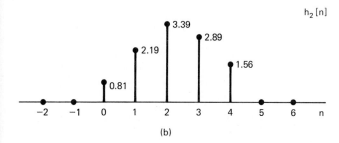

(b)

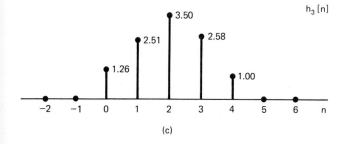

(c)

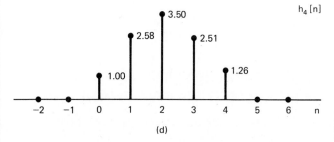

(d)

Figure 5.28 Sequences corresponding to the pole-zero plots of Fig. 5.27.

If we define the partial energy of the impulse response as

$$E[n] = \sum_{m=0}^{n} |h[m]|^2,$$ (5.117)

then it can be shown that (see Problem 5.36)

$$\sum_{m=0}^{n} |h[m]|^2 \le \sum_{m=0}^{n} |h_{min}[m]|^2$$ (5.118)

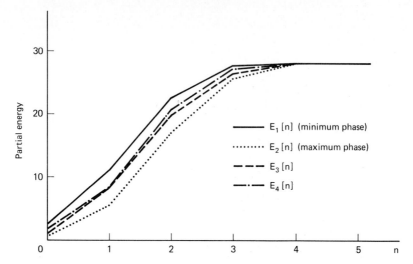

Figure 5.29   Partial energies for the four sequences of Fig. 5.27. (Note that $E_1[n]$ is for the minimum-phase sequence $h_1[n]$, and $E_2[n]$ is for the maximum-phase sequence $h_2[n]$.)

for all impulse responses $h[n]$ belonging to the family of systems that have magnitude response given by Eq. (5.115). According to Eq. (5.118), the partial energy of the minimum-phase system is most concentrated around $n = 0$, i.e., the energy of the minimum-phase system is delayed the least of all systems having the same magnitude response function. For this reason, minimum-phase (lag) systems are also called *minimum energy delay*, or simply *minimum-delay systems*. This delay property is illustrated by Fig. 5.29, which shows plots of the partial energy for the four sequences in Fig. 5.28. We note for this example, and it is true in general, that the minimum energy delay occurs for the system that has all its zeros *inside* the unit circle (i.e., the minimum-phase system) and the maximum energy delay occurs for the system that has all its zeros *outside* the unit circle. Maximum energy delay systems are also often called maximum-phase systems.

5.7 LINEAR SYSTEMS WITH GENERALIZED LINEAR PHASE

In designing filters and other signal processing systems that pass some portion of the frequency band undistorted, it is desirable to have approximately constant frequency-response magnitude and zero phase in that band. For causal systems, zero phase is not attainable and consequently some phase distortion must be allowed. As we saw in Section 5.1.2, the effect of linear phase with integer (negative) slope is a simple delay. A nonlinear phase, on the other hand, can have a major effect on the shape of a signal, even when the frequency-response magnitude is constant. Thus in many situations it is particularly desirable to design systems to have exactly or approximately linear phase. In this section we consider a formalization and generalization of the notions of linear phase and ideal time delay by considering the class of systems that have constant group delay. We begin by reconsidering the concept of delay in a discrete-time system.

5.7.1 Systems with Linear Phase

Consider an LTI system whose frequency response over one period is

$$H_{id}(e^{j\omega}) = e^{-j\omega\alpha}, \qquad |\omega| < \pi, \tag{5.119}$$

where α is a real number, and not necessarily an integer. Note that this system has constant magnitude response, linear phase, and constant group delay, i.e.,

$$|H_{id}(e^{j\omega})| = 1, \tag{5.120a}$$

$$\measuredangle H_{id}(e^{j\omega}) = -\omega\alpha, \tag{5.120b}$$

$$\mathrm{grd}[H_{id}(e^{j\omega})] = \alpha. \tag{5.120c}$$

The inverse Fourier transform of $H_{id}(e^{j\omega})$ is the impulse response

$$h_{id}[n] = \frac{\sin\pi(n-\alpha)}{\pi(n-\alpha)}, \qquad -\infty < n < \infty. \tag{5.121}$$

The output of this system for an input $x[n]$ is

$$y[n] = x[n] * \frac{\sin\pi(n-\alpha)}{\pi(n-\alpha)} = \sum_{k=-\infty}^{\infty} x[k]\frac{\sin\pi(n-k-\alpha)}{\pi(n-k-\alpha)}. \tag{5.122}$$

If $\alpha = n_d$, where n_d is an integer, then as considered in Section 5.1.2,

$$h_{id}[n] = \delta[n - n_d] \tag{5.123}$$

and

$$y[n] = x[n] * \delta[n - n_d] = x[n - n_d], \tag{5.124}$$

i.e., if $\alpha = n_d$ is an integer, the system with linear phase and unity gain in Eq. (5.119) simply shifts the input sequence by n_d samples. If α is not an integer, the most straightforward interpretation is the one developed in Example 3.3 in Chapter 3. Specifically, a representation of the system of Eq. (5.119) is that shown in Fig. 5.30, with $h_c(t) = \delta(t - \alpha T)$ and $H_c(j\Omega) = e^{-j\Omega\alpha T}$ so that

$$H(e^{j\omega}) = e^{-j\omega\alpha}, \qquad |\omega| < \pi. \tag{5.125}$$

In this representation the choice of T is irrelevant and could simply be normalized to unity. It is important to stress again that this representation is valid whether or not $x[n]$ was originally obtained by sampling a continuous-time signal. According to the representation in Fig. 5.30, $y[n]$ is the sequence of samples of the time-shifted bandlimited interpolation of the input sequence $x[n]$, i.e., $x[n] = x_c(nT - \alpha T)$. The system of Eq. (5.119) is called an *ideal delay system* and is said to have α samples of delay, even if α is not an integer.

   This discussion of the ideal delay system also provides a useful interpretation of linear phase when it is associated with a nonconstant magnitude response. For

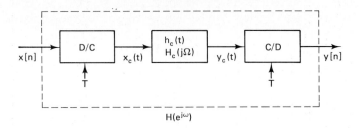

Figure 5.30   Interpretation of noninteger delay in discrete-time systems.

example, consider a more general frequency response with linear phase, i.e., $H(e^{j\omega})$ of the form

$$H(e^{j\omega}) = |H(e^{j\omega})|e^{-j\omega\alpha}, \qquad |\omega| < \pi. \tag{5.126}$$

Equation (5.126) suggests the interpretation of Fig. 5.31. The signal $x[n]$ is filtered by the zero-phase frequency response $|H(e^{j\omega})|$ and the filtered output is then "delayed" by the (integer or noninteger) amount α. Suppose, for example, that $H(e^{j\omega})$ is the linear phase ideal lowpass filter $H_{lp}(e^{j\omega})$ given by

$$H_{lp}(e^{j\omega}) = \begin{cases} e^{-j\omega\alpha}, & |\omega| < \omega_c, \\ 0, & \omega_c < |\omega| \leq \pi. \end{cases} \tag{5.127}$$

The corresponding impulse response is

$$h_{lp}[n] = \frac{\sin \omega_c(n - \alpha)}{\pi(n - \alpha)}. \tag{5.128}$$

Note that Eq. (5.121) is obtained if $\omega_c = \pi$.

The impulse response of the ideal lowpass filter illustrates some interesting properties of linear phase systems. Figure 5.32(a) shows $h_{lp}[n]$ for $\omega_c = 0.4\pi$ and $\alpha = n_d = 5$. Note that when α is an integer, the impulse response is symmetric about $n = n_d$, i.e.,

$$h_{lp}[2n_d - n] = \frac{\sin \omega_c(2n_d - n - n_d)}{\pi(2n_d - n - n_d)}$$

$$= \frac{\sin \omega_c(n_d - n)}{\pi(n_d - n)} \tag{5.129}$$

$$= h_{lp}[n].$$

In this case we could define a *zero-phase system*

$$\hat{H}_{lp}(e^{j\omega}) = H_{lp}(e^{j\omega})e^{j\omega n_d} = |H_{lp}(e^{j\omega})|, \tag{5.130}$$

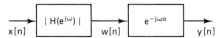

Figure 5.31   Representation of a linear phase LTI system as a cascade of a magnitude filter and a delay.

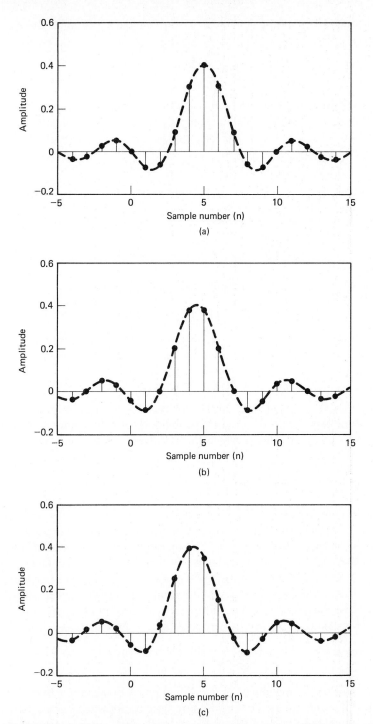

Figure 5.32   Ideal lowpass filter impulse responses, with $\omega_c = 0.4\pi$. (a) Delay $= \alpha = 5$.
(b) Delay $= \alpha = 4.5$. (c) Delay $= \alpha = 4.3$.

where the impulse response is shifted to the left by n_d samples yielding an even sequence

$$\hat{h}_{lp}[n] = \frac{\sin \omega_c n}{\pi n} = \hat{h}_{lp}[-n]. \tag{5.131}$$

Figure 5.32(b) shows $h_{lp}[n]$ for $\omega_c = 0.4\pi$ and $\alpha = 4.5$. This is typical of the case when the linear phase corresponds to an integer plus one-half. As in the integer delay case, it is easily shown that if α is an integer plus one-half (or 2α is an integer), then

$$h_{lp}[2\alpha - n] = h_{lp}[n]. \tag{5.132}$$

In this case, the point of symmetry is α, which is not an integer. Therefore, since the symmetry is not about a point of the sequence, it is not possible to shift the sequence to obtain an even sequence that has zero phase. This is similar to the case of Example 3.4 with M odd.

    Figure 5.32(c) represents a third case, in which there is no symmetry at all. In this case $\omega_c = 0.4\pi$ and $\alpha = 4.3$.

    In general a linear phase system has frequency response

$$H(e^{j\omega}) = |H(e^{j\omega})|e^{-j\omega\alpha}. \tag{5.133}$$

From the preceding discussion, if 2α is an integer, i.e., if α is an integer or an integer plus one-half, the corresponding impulse response has even symmetry about α, i.e.,

$$h[2\alpha - n] = h[n]. \tag{5.134}$$

Equation (5.134) is a sufficient condition for the system to have linear phase. However, this symmetry condition by itself is obviously not necessary since Fig. 5.32(c) shows a linear phase system for which the impulse response is *not* symmetric.

5.7.2 Generalized Linear Phase

In the discussion in Section 5.7.1, we considered a class of systems whose frequency response is of the form of Eq. (5.126), i.e., a real-valued nonnegative function of ω multiplied by a linear phase term $e^{-j\omega\alpha}$. For a frequency response of this form, the phase of $H(e^{j\omega})$ is entirely associated with the linear phase factor $e^{-j\omega\alpha}$, i.e., $\angle H(e^{j\omega}) = -\omega\alpha$, and consequently systems in this class are referred to as linear phase systems. In the moving average of Example 3.4 the frequency response in Eq. (3.65) is a real-valued function of ω multiplied by a linear phase term, but the system is not, strictly speaking, a linear phase system since at frequencies for which the factor

$$\frac{1}{M+1} \frac{\sin[\omega(M+1)/2]}{\sin(\omega/2)}$$

is negative, this term contributes an additional phase of π radians to the total phase.

    Many of the advantages of linear phase systems apply to systems with frequency response having the form of Eq. (3.65) as well, and consequently it is useful to generalize somewhat the definition and concept of linear phase. Specifically, a system will be said to be a *generalized linear phase system* if its frequency response can be expressed in the form

$$H(e^{j\omega}) = A(e^{j\omega})e^{-j\alpha\omega + j\beta}, \tag{5.135}$$

where α and β are constants and $A(e^{j\omega})$ is a real (possibly bipolar) function of ω. For the linear phase system of Eq. (5.127) and the moving average filter of Example 3.4, $\beta = 0$. We see, however, that the bandlimited differentiator of Example 3.2 has the form of Eq. (5.135) with $\alpha = 0$, $\beta = \pi/2$, and $A(e^{j\omega}) = \omega/T$.

A system whose frequency response has the form of Eq. (5.135) is called a generalized linear phase system, since the phase of such a system consists of constant terms added to the linear function $-\omega\alpha$. However, if we ignore any discontinuities that result from addition of constant phase over all or part of the band $|\omega| < \pi$, then such a system can be characterized by constant group delay. That is, the class of systems such that

$$\tau(\omega) = \text{grd}[H(e^{j\omega})] = -\frac{d}{d\omega}\{\arg[H(e^{j\omega})]\} = \alpha \tag{5.136}$$

have linear phase of the more general form

$$\arg[H(e^{j\omega})] = \beta - \omega\alpha, \qquad 0 < \omega < \pi, \tag{5.137}$$

where β and α are both real constants.

Recall that we showed in Section 5.7.1 that the impulse responses of linear phase systems may have symmetry about α if 2α is an integer. To see the implication of this for generalized linear phase systems, it is useful to derive an equation that must be satisfied by $h[n]$, α, and β for constant group delay systems. This equation is derived by noting that for constant group delay systems the frequency response can be expressed as

$$H(e^{j\omega}) = A(e^{j\omega})e^{j(\beta - \alpha\omega)}$$

$$= A(e^{j\omega})\cos(\beta - \omega\alpha) + jA(e^{j\omega})\sin(\beta - \omega\alpha) \tag{5.138}$$

or, equivalently, as

$$H(e^{j\omega}) = \sum_{n=-\infty}^{\infty} h[n]e^{-j\omega n}$$

$$= \sum_{n=-\infty}^{\infty} h[n]\cos \omega n - j \sum_{n=-\infty}^{\infty} h[n]\sin \omega n, \tag{5.139}$$

where we have assumed that $h[n]$ is real. The tangent of the phase angle of $H(e^{j\omega})$ can be expressed as

$$\tan(\beta - \omega\alpha) = \frac{\sin(\beta - \omega\alpha)}{\cos(\beta - \omega\alpha)} = \frac{-\sum\limits_{n=-\infty}^{\infty} h[n]\sin \omega n}{\sum\limits_{n=-\infty}^{\infty} h[n]\cos \omega n}.$$

Cross-multiplying and combining terms with a trigonometric identity leads to the following equation:

$$\sum_{n=-\infty}^{\infty} h[n]\sin(\omega(n-\alpha)+\beta) = 0. \tag{5.140}$$

This equation, which must hold for all ω, is a necessary condition on $h[n]$, α, and β for the system to have constant group delay. It is not a sufficient condition and due to its implicit nature it does not tell us how to find a linear phase system. For example, it can be shown that one set of conditions that satisfy Eq. (5.140) is

$$\beta = 0 \text{ or } \pi, \tag{5.141a}$$

$$2\alpha = M = \text{an integer}, \tag{5.141b}$$

$$h[2\alpha - n] = h[n]. \tag{5.141c}$$

With $\beta = 0$ or π, Eq. (5.140) becomes

$$\sum_{n=-\infty}^{\infty} h[n]\sin[\omega(n-\alpha)] = 0, \tag{5.142}$$

from which it can be shown that if 2α is an integer, terms in Eq. (5.142) can be paired so that each pair of terms is identically zero for all ω. These conditions in turn imply that the corresponding frequency response has the form of Eq. (5.135) with $\beta = 0$ or π and $A(e^{j\omega})$ an even (and of course real) function of ω.

Alternatively, if $\beta = \pi/2$ or $3\pi/2$, then Eq. (5.140) becomes

$$\sum_{n=-\infty}^{\infty} h[n]\cos[\omega(n-\alpha)] = 0, \tag{5.143}$$

and it can be shown that

$$\beta = \pi/2 \text{ or } 3\pi/2, \tag{5.144a}$$

$$2\alpha = M = \text{an integer}, \tag{5.144b}$$

$$h[2\alpha - n] = -h[n] \tag{5.144c}$$

satisfy Eq. (5.143) for all ω. Equations (5.144) imply that the frequency response has the form of Eq. (5.135) with $\beta = \pi/2$ and $A(e^{j\omega})$ an odd function of ω.

Note that Eqs. (5.141) and (5.144) give two sets of conditions that guarantee generalized linear phase or constant group delay, but as we have already seen in Fig. 5.32(c), there are other systems that satisfy Eq. (5.135) without these symmetry conditions.

5.7.3 Causal Generalized Linear Phase Systems

If the system is causal, then Eq. (5.140) becomes

$$\sum_{n=0}^{\infty} h[n]\sin[\omega(n-\alpha)+\beta] = 0 \qquad \text{for all } \omega. \tag{5.145}$$

Causality and the conditions in Eq. (5.141) and (5.144) imply that

$$h[n] = 0, \qquad n < 0 \text{ and } n > M,$$

i.e., causal FIR systems have generalized linear phase if they have impulse response length $(M + 1)$ and satisfy either Eq. (5.141c) or (5.144c). Specifically, if

$$h[n] = \begin{cases} h[M - n], & 0 \le n \le M, \\ 0, & \text{otherwise,} \end{cases} \tag{5.146a}$$

then it can be shown that

$$H(e^{j\omega}) = A_e(e^{j\omega})e^{-j\omega M/2}, \tag{5.146b}$$

where $A_e(e^{j\omega})$ is a real, even, periodic function of ω. Similarly, if

$$h[n] = \begin{cases} -h[M - n], & 0 \le n \le M, \\ 0, & \text{otherwise,} \end{cases} \tag{5.147a}$$

then it follows that

$$H(e^{j\omega}) = jA_o(e^{j\omega})e^{-j\omega M/2} = A_o(e^{j\omega})e^{-j\omega M/2 + j\pi/2}, \tag{5.147b}$$

where $A_o(e^{j\omega})$ is a real, odd, periodic function of ω. Note that in both cases, the length of the impulse response is $(M + 1)$ samples.

The conditions in Eqs. (5.146a) and (5.147a) are sufficient to guarantee a causal system with generalized linear phase. However, they are not necessary conditions. Clements and Pease (1988) have shown that causal infinite-duration impulse responses can also have Fourier transforms with generalized linear phase. The corresponding system functions, however, are not rational and thus the systems cannot be implemented with difference equations.

Expressions for the frequency response of FIR linear phase systems are useful in filter design and in understanding some of the properties of such systems. In deriving these expressions it turns out that significantly different expressions result, depending on the type of symmetry and whether M is an even or odd integer. For this reason it is generally useful to define four types of FIR generalized linear phase systems.

Type I FIR linear phase systems.   A type I system is defined as one that has a symmetric impulse response

$$h[n] = h[M - n], \qquad 0 \le n \le M, \tag{5.148}$$

with M an even integer. The delay $M/2$ is an integer. The frequency response is

$$H(e^{j\omega}) = \sum_{n=0}^{M} h[n]e^{-j\omega n}. \tag{5.149}$$

By applying the symmetry condition, Eq. (5.148), it is possible to manipulate the sum in Eq. (5.149) into the form

$$H(e^{j\omega}) = e^{-j\omega M/2}\left(\sum_{k=0}^{M/2} a[k]\cos \omega k\right), \tag{5.150a}$$

where

$$a[0] = h[M/2], \tag{5.150b}$$

$$a[k] = 2h[(M/2) - k], \qquad k = 1, 2, \ldots, M/2. \tag{5.150c}$$

Thus from Eq. (5.150a) we see that $H(e^{j\omega})$ has the form of Eq. (5.146b), and in particular, β in Eq. (5.135) is either 0 or π.

Type II FIR linear phase systems. A type II system has a symmetric impulse response as in Eq. (5.148), with M an odd integer. $H(e^{j\omega})$ for this case can be expressed as

$$H(e^{j\omega}) = e^{-j\omega M/2} \left\{ \sum_{k=1}^{(M+1)/2} b[k]\cos[\omega(k - \tfrac{1}{2})] \right\}, \tag{5.151a}$$

where

$$b[k] = 2h[(M+1)/2 - k], \qquad k = 1, 2, \ldots, (M+1)/2. \tag{5.151b}$$

Again, $H(e^{j\omega})$ has the form of Eq. (5.146b) with a time delay of $M/2$, which in this case is an integer plus one-half, and β in Eq. (5.135) is either 0 or π.

Type III FIR linear phase systems. If the system has an antisymmetric impulse response

$$h[n] = -h[M - n], \qquad 0 \le n \le M, \tag{5.152}$$

with M an even integer, then $H(e^{j\omega})$ has the form

$$H(e^{j\omega}) = je^{-j\omega M/2} \left[\sum_{k=1}^{M/2} c[k]\sin \omega k \right], \tag{5.153a}$$

where

$$c[k] = 2h[(M/2) - k], \qquad k = 1, 2, \ldots, M/2. \tag{5.153b}$$

In this case $H(e^{j\omega})$ has the form of Eq. (5.147b) with a delay of $M/2$, which is an integer, and β in Eq. (5.135) is $\pi/2$ or $3\pi/2$.

Type IV FIR linear phase systems. If the impulse response is antisymmetric as in Eq. (5.152) and M is odd, then

$$H(e^{j\omega}) = je^{-j\omega M/2} \left[\sum_{k=1}^{(M+1)/2} d[k]\sin[\omega(k - \tfrac{1}{2})] \right], \tag{5.154a}$$

where

$$d[k] = 2h[(M+1)/2 - k], \qquad k = 1, 2, \ldots, (M+1)/2 \tag{5.154b}$$

As in the case of type III systems, $H(e^{j\omega})$ has the form of Eq. (5.147b) with delay $M/2$, which is an integer plus one-half and β in Eq. (5.135) is $\pi/2$ or $3\pi/2$.

Examples of FIR linear phase systems.   Figure 5.33 shows an example of each of the four types of FIR linear phase systems. The associated frequency responses are given in Examples 5.13–5.16.

Example 5.13   Type I

If the impulse response is

$$h[n] = \begin{cases} 1, & 0 \leq n \leq 4, \\ 0, & \text{otherwise,} \end{cases} \qquad (5.155)$$

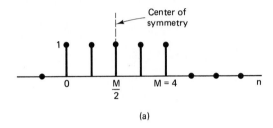

(a)

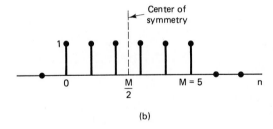

(b)

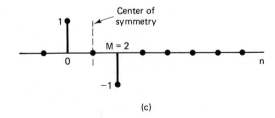

(c)

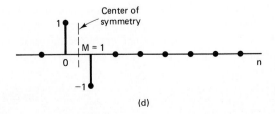

(d)

Figure 5.33   Examples of FIR linear phase systems. (a) Type I, M even, $h[n] = h[M - n]$. (b) Type II, M odd, $h[n] = h[M - n]$. (c) Type III, M even, $h[n] = -h[M - n]$. (d) Type IV, M odd, $h[n] = -h[M - n]$.

as shown in Fig. 5.33(a), the system satisfies the condition of Eq. (5.148). The frequency response is

$$H(e^{j\omega}) = \sum_{n=0}^{4} e^{-j\omega n} = \frac{1 - e^{-j\omega 5}}{1 - e^{-j\omega}}$$

$$= e^{-j\omega 2} \frac{\sin(5\omega/2)}{\sin(\omega/2)}. \tag{5.156}$$

The magnitude, phase, and group delay of the system are shown in Fig. 5.34. Since $M = 4$ is even, the group delay is an integer, i.e., $\alpha = 2$.

Example 5.14   Type II

If the length of the impulse response of the previous example is extended by one sample, we obtain the impulse response of Fig. 5.33(b), which has frequency response

$$H(e^{j\omega}) = e^{-j\omega 5/2} \frac{\sin(3\omega)}{\sin(\omega/2)}. \tag{5.157}$$

The frequency-response functions for this system are shown in Fig. 5.35. Note that the group delay in this case is constant with $\alpha = 5/2$.

Example 5.15   Type III

If the impulse response is

$$h[n] = \delta[n] - \delta[n - 2] \tag{5.158}$$

as in Fig. 5.33(c), then

$$H(e^{j\omega}) = 1 - e^{-j2\omega}$$

$$= j[2 \sin(\omega)]e^{-j\omega}. \tag{5.159}$$

The frequency-response plots for this example are given in Fig. 5.36. Note that the group delay in this case is constant with $\alpha = 1$.

Example 5.16   Type IV

In this case (Fig. 5.33d) the impulse response is

$$h[n] = \delta[n] - \delta[n - 1], \tag{5.160}$$

for which the frequency response is

$$H(e^{j\omega}) = 1 - e^{-j\omega}$$

$$= j[2 \sin(\omega/2)]e^{-j\omega/2}. \tag{5.161}$$

The frequency response for this system is shown in Fig. 5.37. Note that the group delay is equal to $\frac{1}{2}$ for all ω.

Zero locations for FIR linear phase systems.   The preceding examples illustrate the properties of the impulse response and the frequency response for all four types of FIR linear phase systems. It is also instructive to consider the locations of the zeros of the system function for FIR linear phase systems. The system function is

$$H(z) = \sum_{n=0}^{M} h[n]z^{-n}. \tag{5.162}$$

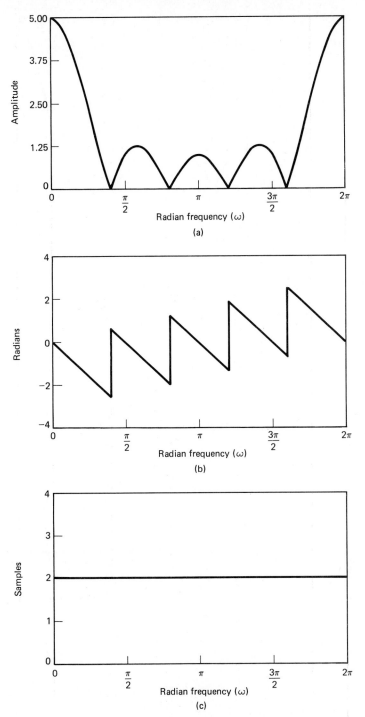

Figure 5.34 Frequency response of type I system of Example 5.13. (a) Magnitude.
(b) Phase. (c) Group delay.

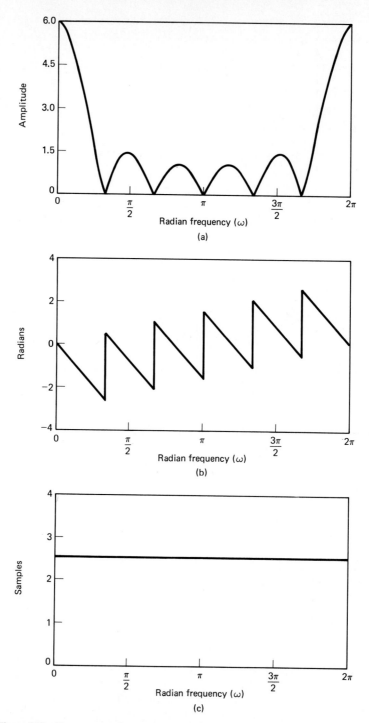

Figure 5.35 Frequency response of type II system of Example 5.14. (a) Magnitude. (b) Phase. (c) Group delay.

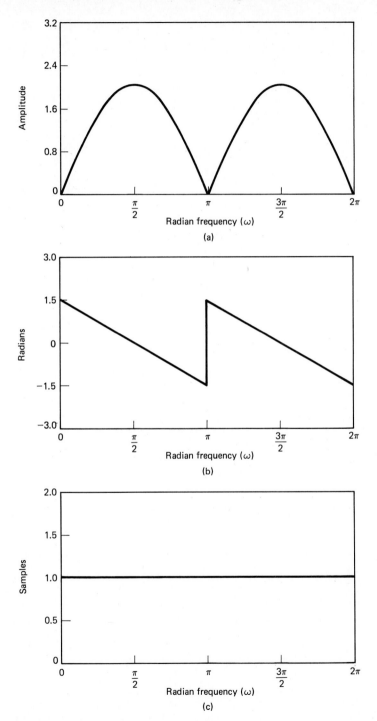

Figure 5.36 Frequency response of type III system of Example 5.15. (a) Magnitude.
(b) Phase. (c) Group delay.

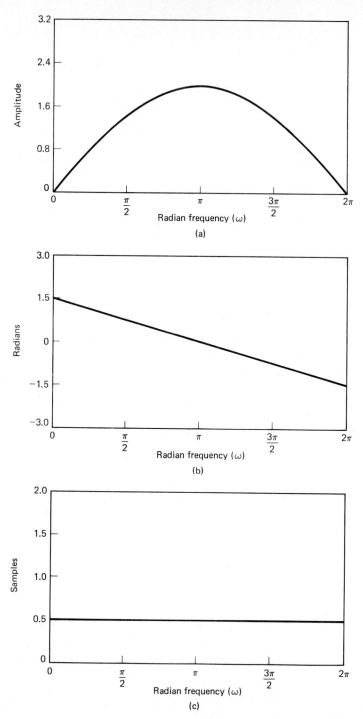

Figure 5.37   Frequency response of type IV system of Example 5.16. (a) Magnitude.
(b) Phase. (c) Group delay.

In the symmetric cases (types I and II), we can use Eq. (5.148) to express $H(z)$ as

$$H(z) = \sum_{n=0}^{M} h[M-n]z^{-n} = \sum_{k=M}^{0} h[k]z^k z^{-M}$$

$$= z^{-M} H(z^{-1}).$$

(5.163)

From Eq. (5.163) we conclude that if z_0 is a zero of $H(z)$, then

$$H(z_0) = z_0^{-M} H(z_0^{-1}) = 0.$$

(5.164)

This implies that if $z_0 = re^{j\theta}$ is a zero of $H(z)$, then $z_0^{-1} = r^{-1}e^{-j\theta}$ is also a zero of $H(z)$. When $h[n]$ is real and z_0 is a zero of $H(z)$, then $z_0^* = re^{-j\theta}$ will also be a zero of $H(z)$ and, by the preceding argument, so will $(z_0^*)^{-1} = r^{-1}e^{j\theta}$. Therefore, when $h[n]$ is real, each complex zero not on the unit circle will be part of a set of four conjugate reciprocal zeros of the form

$$(1 - re^{j\theta}z^{-1})(1 - re^{-j\theta}z^{-1})(1 - r^{-1}e^{j\theta}z^{-1})(1 - r^{-1}e^{-j\theta}z^{-1}).$$

If a zero of $H(z)$ is on the unit circle, i.e., $z_0 = e^{j\theta}$, then $z_0^{-1} = e^{-j\theta} = z_0^*$, so zeros on the unit circle come in pairs of the form

$$(1 - e^{j\theta}z^{-1})(1 - e^{-j\theta}z^{-1}).$$

If a zero of $H(z)$ is real and not on the unit circle, the reciprocal will also be a zero of $H(z)$, and $H(z)$ will have factors of the form

$$(1 \pm rz^{-1})(1 \pm r^{-1}z^{-1}).$$

Finally, a zero of $H(z)$ at $z = \pm 1$ can appear by itself since ± 1 is its own reciprocal and its own conjugate. Thus we may also have factors of $H(z)$ of the form

$$(1 \pm z^{-1}).$$

The case of a zero at $z = -1$ is particularly important. From Eq. (5.163),

$$H(-1) = (-1)^M H(-1).$$

If M is even, we have a simple identity, but if M is odd, $H(-1) = -H(-1)$, so $H(-1)$ must be zero. Thus for symmetric impulse responses with M odd, the system function *must* have a zero at $z = -1$. Figures 5.38(a) and 5.38(b) show typical zero locations for type I (M even) and type II (M odd) systems, respectively.

If the impulse response is antisymmetric (types III and IV), then following the approach used to obtain Eq. (5.163) we can show that

$$H(z) = -z^{-M} H(z^{-1}).$$

(5.165)

This equation can be used to show that the zeros of $H(z)$ for the antisymmetric case are constrained in the same way as the zeros for the symmetric case. In the antisymmetric case, however, both $z = 1$ and $z = -1$ are of special interest. If $z = 1$, Eq. (5.165) becomes

$$H(1) = -H(1).$$

(5.166)

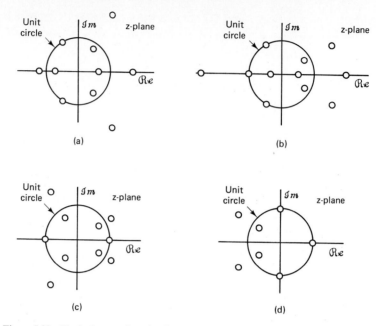

Figure 5.38 Typical zero plots for linear phase systems. (a) Type I. (b) Type II. (c) Type III. (d) Type IV.

Thus, $H(z)$ *must* have a zero at $z = 1$ for both M even and M odd. If $z = -1$, Eq. (5.165) gives

$$H(-1) = (-1)^{-M+1} H(-1). \qquad (5.167)$$

In this case if $(M - 1)$ is odd (M even), $H(-1) = -H(-1)$, so $z = -1$ *must* be a zero of $H(z)$ if M is even. Figures 5.38(c) and 5.38(d) show typical zero locations for type III and IV systems, respectively.

These constraints on the zeros are important in designing FIR linear phase systems since they impose limitations on the types of frequency responses that can be achieved. For example, we note that when approximating a highpass filter using a symmetric impulse response, M should not be odd since the frequency response is constrained to be zero at $\omega = \pi$ ($z = -1$).

5.7.4 Relation of FIR Linear Phase Systems to Minimum-Phase Systems

The previous discussion shows that all FIR linear phase systems with real impulse response have zeros either on the unit circle or at conjugate reciprocal locations. Thus it is easily shown that the system function of any FIR linear phase system can be factored into a minimum-phase term $H_{min}(z)$, a maximum-phase term $H_{max}(z)$, and a term $H_{uc}(z)$ containing only zeros on the unit circle; i.e.,

$$H(z) = H_{min}(z)H_{uc}(z)H_{max}(z), \qquad (5.168a)$$

where

$$H_{max}(z) = H_{min}(z^{-1})z^{-M_i} \qquad (5.168b)$$

and M_i is the number of zeros of $H_{min}(z)$. In Eq. (5.168a), $H_{min}(z)$ has all M_i of its zeros *inside* the unit circle and $H_{uc}(z)$ has all M_o of its zeros *on* the unit circle. $H_{max}(z)$ has all M_i of its zeros *outside* the unit circle and, from Eq. (5.168b), its zeros are the reciprocals of the zeros of $H_{min}(z)$. The order of the system function $H(z)$ is therefore $M = 2M_i + M_o$.

Example 5.17

As a simple example of the use of Eq. (5.168), consider the minimum-phase system function of Eq. (5.109), for which the frequency response is plotted in Fig. 5.25. The system $H_{max}(z)$ obtained by applying Eq. (5.168b) to $H_{min}(z)$ in Eq. (5.109) is

$$H_{max}(z) = (0.9)^2(1 - 1.1111e^{j0.6\pi}z^{-1})(1 - 1.1111e^{-j0.6\pi}z^{-1})$$
$$\times (1 - 1.25^{-j0.8\pi}z^{-1})(1 - 1.25e^{j0.8\pi}z^{-1}).$$

$H_{max}(z)$ has the frequency response shown in Fig. 5.39. Now if these two systems are cascaded, it follows from Eq. (5.168) that the overall system

$$H(z) = H_{min}(z)H_{max}(z)$$

has linear phase. The frequency response of the composite system would be obtained by adding the respective log magnitude, phase, and group delay functions. Therefore,

$$20 \log_{10}|H(e^{j\omega})| = 20 \log_{10}|H_{min}(e^{j\omega})| + 20 \log_{10}|H_{max}(e^{j\omega})|$$
$$= 40 \log_{10}|H_{min}(e^{j\omega})|. \qquad (5.169)$$

Similarly,

$$\measuredangle H(e^{j\omega}) = \measuredangle H_{min}(e^{j\omega}) + \measuredangle H_{max}(e^{j\omega}). \qquad (5.170)$$

From Eq. (5.168b) it follows that

$$\measuredangle H_{max}(e^{j\omega}) = -\omega M_i - \measuredangle H_{min}(e^{j\omega}). \qquad (5.171)$$

and

$$\measuredangle H(e^{j\omega}) = -\omega M_i,$$

where $M_i = 4$ is the number of zeros of $H_{min}(z)$. In like manner, the group delay functions of $H_{min}(e^{j\omega})$ and $H_{max}(e^{j\omega})$ combine to give

$$grd[H(e^{j\omega})] = M_i = 4.$$

The frequency-response plots for the composite system are given in Fig. 5.40. Note that the curves are sums of the corresponding functions in Figs. 5.25 and 5.39.

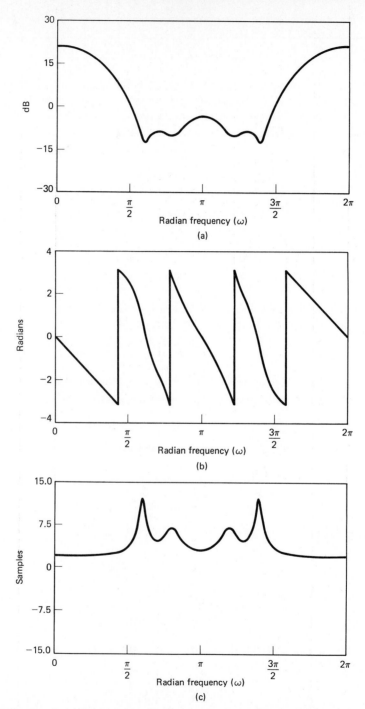

Figure 5.39 Frequency response of maximum-phase system having the same magnitude as the system in Fig. 5.25. (a) Log magnitude. (b) Phase (principal value). (c) Group delay.

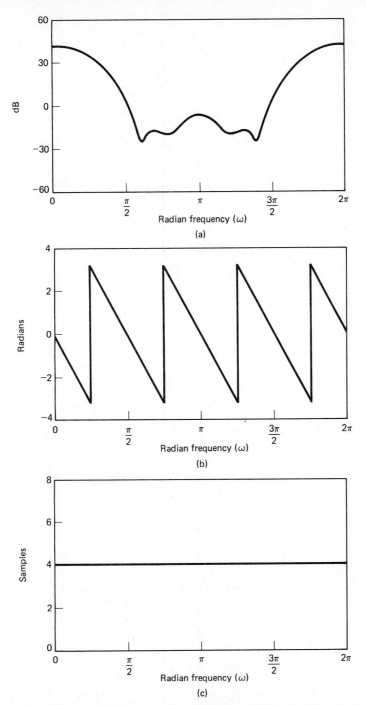

Figure 5.40 Frequency response of cascade of maximum-phase and minimum-phase systems, yielding a linear phase system. (a) Log magnitude. (b) Phase (principal value). (c) Group delay.

5.8 SUMMARY

In this chapter, we developed and explored the representation and analysis of LTI systems using the Fourier and z-transforms. The importance of transform analysis for LTI systems stems directly from the fact that complex exponentials are eigenfunctions of such systems, and the associated eigenvalues correspond to the system function or frequency response.

A particularly important class of LTI systems is that characterized by linear constant-coefficient difference equations. Transform analysis is particularly useful for analyzing these systems, since the Fourier transform or z-transform converts a difference equation to an algebraic equation. In particular, the system function is a ratio of polynomials the coefficients of which correspond directly to the coefficients in the difference equation. The roots of these polynomials provide a useful system representation in terms of the pole-zero plot. Systems characterized by difference equations may have an impulse response that is infinite in duration (IIR) or finite in duration (FIR).

The frequency response of LTI systems is often characterized in terms of magnitude and phase or group delay, which is the negative of the derivative of the phase. Linear phase is often a desirable characteristic of a system frequency response since it is a relatively mild form of phase distortion, corresponding to a time shift. The importance of FIR systems lies in part in the fact that such systems can be easily designed to have exactly linear phase (or generalized linear phase) while for a given set of frequency response magnitude specifications, IIR systems are more efficient. These and other tradeoffs will be discussed in detail in Chapter 7.

While in general for LTI systems the frequency response magnitude and phase are independent, for minimum-phase systems the magnitude uniquely specifies the phase and the phase uniquely specifies the magnitude to within a scale factor. Non-minimum-phase systems can be represented as the cascade combination of a minimum-phase system and an allpass system. Relations between Fourier transform magnitude and phase will be discussed in considerably more detail in Chapter 10.

PROBLEMS

5.1. In the system shown in Fig. P5.1-1, $H(e^{j\omega})$ is an ideal lowpass filter as indicated. Determine whether for some choice of input $x[n]$ and cutoff frequency ω_c, the output $y[n]$ can

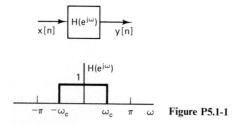

Figure P5.1-1

be the pulse described as follows and shown in Fig. P5.1-2.

$$y[n] = \begin{cases} 1, & 0 \le n \le 10, \\ 0, & \text{otherwise} \end{cases}$$

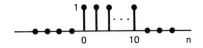

$y[n]$

0   10   n   **Figure P5.1-2**

5.2. Let $h_{lp}[n]$ denote the impulse response of an ideal lowpass filter with unity passband gain and with cutoff frequency $\omega_c = \pi/4$. Figure P5.2 shows five systems, each of which is equivalent to an ideal LTI frequency-selective filter. For each system in Fig. P5.2, sketch the equivalent frequency response, indicating explicitly the band-edge frequencies in terms of ω_c. In each case, specify whether it is a lowpass, highpass, bandpass, bandstop, or multiband filter.

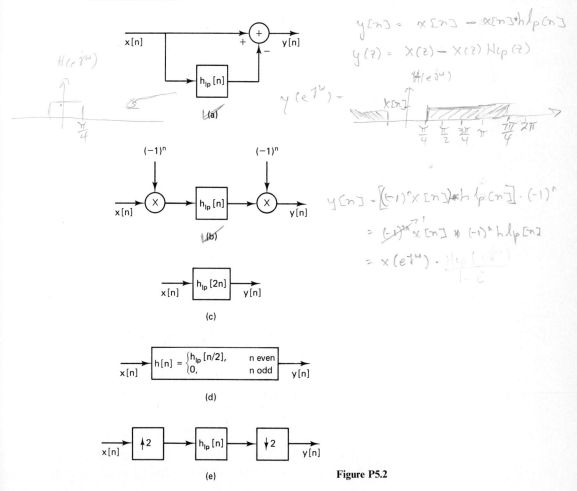

Figure P5.2

5.3. In the system shown in Fig. P5.3-1, assume that the input $x[n]$ can be expressed in the form

$$x[n] = s[n]\cos(\omega_0 n). \tag{P5.3}$$

Also assume that $s[n]$ is lowpass and relatively narrowband, i.e., $S(e^{j\omega}) = 0$ for $|\omega| > \Delta$, with Δ very small so that $X(e^{j\omega})$ is narrowband around $\omega = \pm\omega_0$.

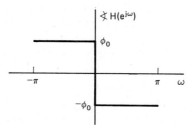

$x[n]$ $H(e^{j\omega})$ $y[n]$ **Figure P5.3-1**

(a) If $|H(e^{j\omega})| = 1$ and $\measuredangle H(e^{j\omega})$ is as shown in Fig. P5.3-2, show that $y[n] = s[n]\cos(\omega_0 n - \phi_0)$.

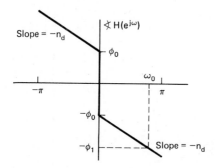

Figure P5.3-2

(b) If $|H(e^{j\omega})| = 1$ and $\measuredangle H(e^{j\omega})$ is as shown in Fig. P5.3-3, show that $y[n]$ can be expressed in the form

$$y[n] = s[n - n_d]\cos(\omega_0 n - \phi_0 - \omega_0 n_d).$$

Also show that $y[n]$ can be equivalently expressed as

$$y[n] = s[n - n_d]\cos(\omega_0 n - \phi_1),$$

where $-\phi_1$ is the phase of $H(e^{j\omega})$ at $\omega = \omega_0$.

Slope = $-n_d$ $\measuredangle H(e^{j\omega})$

ϕ_0

ω_0

$-\pi$ π

$-\phi_0$

$-\phi_1$ Slope = $-n_d$

Figure P5.3-3

(c) The group delay associated with $H(e^{j\omega})$ is defined as

$$\tau_{gr}(\omega) = -\frac{d}{d\omega}\arg[H(e^{j\omega})]$$

and the phase delay is defined as $\tau_{ph}(\omega) = -(1/\omega)\measuredangle H(e^{j\omega})$. Assume that $|H(e^{j\omega})|$ is unity over the bandwidth of $x[n]$. Based on your results in parts (a) and (b) and on the assumption that $x[n]$ is narrowband, show that if $\tau_{gr}(\omega_0)$ and $\tau_{ph}(\omega_0)$ are both integers, then

$$y[n] = s[n - \tau_{gr}(\omega_0)]\cos\{\omega_0[n - \tau_{ph}(\omega_0)]\}.$$

This then shows that for a narrowband signal $x[n]$, $*H(e^{j\omega})$ effectively applies a delay of $\tau_{gr}(\omega_0)$ to the envelope $s[n]$ of $x[n]$ and a delay of $\tau_{ph}(\omega_0)$ to the carrier $\cos\omega_0 n$.

(d) Referring to the discussion in Chapter 3 associated with noninteger delays of a sequence, how would you interpret the effect of group delay and phase delay if $\tau_{gr}(\omega_0)$ and/or $\tau_{ph}(\omega_0)$ are not integers?

5.4. Consider a stable linear time-invariant system with input $x[n]$ and output $y[n]$. The input and output satisfy the difference equation

$$y[n-1] - \tfrac{10}{3}y[n] + y[n+1] = x[n].$$

(a) Plot the poles and zeros in the z-plane.
(b) Find the impulse response $h[n]$.

5.5. Consider a linear time-invariant discrete-time system for which the input $x[n]$ and output $y[n]$ are related by the second-order difference equation

$$y[n-1] + \tfrac{1}{3}y[n-2] = x[n].$$

(handwritten) $\dfrac{Y(z)}{X(z)} = \dfrac{1}{z^{-1} + \frac{1}{3}z^{-2}} = \dfrac{1}{z^{-1}(1 + \frac{1}{3}z^{-1})}$

From the list below choose *two* possible impulse responses for the system.

(a) $(-\tfrac{1}{3})^{n+1}u[n+1]$ *(handwritten)* $|z| > \tfrac{1}{3}$
(b) $3^{n+1}u[n+1]$
(c) $3(-3)^{n+2}u[-n-2]$
(d) $\tfrac{1}{3}(-\tfrac{1}{3})^n u[-n-2]$
(e) $(-\tfrac{1}{3})^{n+1}u[-n-2]$
(f) $(\tfrac{1}{3})^{n+1}u[n+1]$
(g) $(-3)^{n+1}u[n]$
(h) $n^{1/3}u[n]$

(handwritten)
$H(z) = \dfrac{z}{1 + \frac{1}{3}z^{-1}}$

$H(z) = z^{+1}\left(\dfrac{1}{1 + \frac{1}{3}z^{-1}}\right)$

$h(n) = (-\tfrac{1}{3})^n u[n]$

$h(n+1) = (-\tfrac{1}{3})^{n+1} u[n+1]$

5.6. When the input to a linear time-invariant system is

$$x[n] = (\tfrac{1}{2})^n u[n] + (2)^n u[-n-1],$$

the output is

$$y[n] = 6(\tfrac{1}{2})^n u[n] - 6(\tfrac{3}{4})^n u[n].$$

(a) Find the system function $H(z)$ of the system. Plot the poles and zeros of $H(z)$ and indicate the region of convergence.
(b) Find the impulse response $h[n]$ of the system for all values of n.
(c) Write the difference equation that characterizes the system.
(d) Is the system stable? Is it causal?

5.7. Consider a system described by a linear constant-coefficient difference equation with initial-rest conditions. The step response of the system is given by

$$y[n] = (\tfrac{1}{3})^n u[n] + (\tfrac{1}{4})^n u[n] + u[n].$$

(a) Determine the difference equation.
(b) Determine the impulse response of the system.
(c) Determine whether or not the system is stable.

5.8. The following information is known about a linear time-invariant system:
  (i) The system is causal.
  (ii) When the input is

$$x[n] = -\tfrac{1}{3}(\tfrac{1}{2})^n u[n] - \tfrac{4}{3}(2)^n u[-n-1],$$

then the z-transform of the output is

$$Y(z) = \frac{1-z^{-2}}{(1-\tfrac{1}{2}z^{-1})(1-2z^{-1})}.$$

  (a) Find the z-transform of $x[n]$.
  (b) What are the possible choices for the region of convergence of $Y(z)$?
  (c) What are the possible choices for the impulse response of the system?

5.9. A linear time-invariant system with an impulse response

$$h[n] = 2(\tfrac{1}{2})^n u[n]$$

is excited by the input sequence $x[n]$. Use z-transforms to determine the response $y[n]$ of the system to the following input signals.
  (a) $x[n] = 5(\tfrac{3}{4})^n u[n]$
  (b) $x[n] = nu[n]$
In each case, specify the region of convergence of $Y(z)$.

5.10. When the input to a linear time-invariant system is

$$x[n] = 5u[n],$$

the output is

$$y[n] = [2(\tfrac{1}{2})^n + 3(-\tfrac{3}{4})^n]u[n].$$

  (a) Find the system function $H(z)$ of the system. Plot the poles and zeros of $H(z)$ and indicate the region of convergence.
  (b) Find the impulse response of the system for all values of n.
  (c) Write the difference equation that characterizes the system.

5.11. The system function $H(z)$ of a causal linear time-invariant system has the pole-zero configuration shown in Fig. P5.11. It is also known that $H(z) = 6$ when $z = 1$.
  (a) Determine $H(z)$.
  (b) Determine the impulse response $h[n]$ of the system.
  (c) Determine the response of the system to the following input signals:
  (i) $x[n] = u[n] - \tfrac{1}{2}u[n-1]$
  (ii) The sequence $x[n]$ obtained from sampling the continuous-time signal

$$x(t) = 50 + 10\cos 20\pi t + 30\cos 40\pi t$$

at a sampling frequency $\Omega_s = 2\pi(40)$ rad/s.

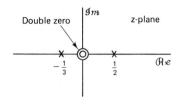

Figure P5.11

5.12. A causal linear time-invariant system is described by the difference equation

$$y[n] = y[n-1] + y[n-2] + x[n-1].$$

(a) Find the system function $H(z) = Y(z)/X(z)$ for this system. Plot the poles and zeros of $H(z)$ and indicate the region of convergence.

(b) Find the impulse response of this system.

(c) You should have found this to be an unstable system. Find a stable (noncausal) impulse response that satisfies the difference equation.

5.13. Consider a linear time-invariant system with input $x[n]$ and output $y[n]$ for which

$$y[n-1] - \tfrac{5}{2}y[n] + y[n+1] = x[n].$$

The system may or may not be stable and/or causal.

By considering the pole-zero pattern associated with this difference equation, determine three possible choices for the impulse response of the system. Show that each choice satisfies the difference equation. Specifically indicate which choice corresponds to a stable system and which choice corresponds to a causal system.

5.14. The system function of a linear time-invariant system is given by

$$H(z) = \frac{21}{(1 - \tfrac{1}{2}z^{-1})(1 - 2z^{-1})(1 - 4z^{-1})}.$$

It is known that the system is not stable and that the impulse response is two-sided.

(a) Determine the impulse response $h[n]$ of the system.

(b) The impulse response found in part (a) can be expressed as the sum of a causal and an anticausal impulse response $h_1[n]$ and $h_2[n]$ respectively. Determine the corresponding system functions $H_1(z)$ and $H_2(z)$.

5.15. A signal $x[n]$ is processed by a linear time-invariant system $H(z)$ and then downsampled by a factor of 2 to yield $y[n]$, as shown in Fig. P5.15-1. The pole-zero plot for $H(z)$ is shown in Fig. P5.15-2.

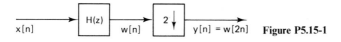

Figure P5.15-1

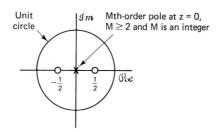

Figure P5.15-2

(a) Determine and sketch $h[n]$, the impulse response of the system $H(z)$.

(b) A second system is shown in Fig. P5.15-3, in which the signal $x[n]$ is first time compressed by a factor of 2 and then passed through an LTI system $G(z)$ to obtain $r[n]$.

Figure P5.15-3

Determine whether $G(z)$ can be chosen so that $y[n] = r[n]$ for any input $x[n]$. If your answer is no, clearly explain. If your answer is yes, specify $G(z)$. If your answer depends on the value of M, clearly explain how. (M is constrained to be an integer greater than or equal to 2.)

5.16. If the system function $H(z)$ of a linear time-invariant system has a pole-zero diagram as shown in Fig. P5.16 and the system is causal, can the inverse system $H_i(z)$, where $H(z)H_i(z) = 1$, be both causal and stable? Clearly justify your answer.

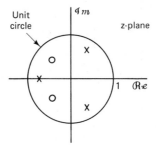

Figure P5.16

5.17. Consider a linear time-invariant system whose system function $H(z)$ is

$$H(z) = \frac{z^{-2}}{(1 - \frac{1}{2}z^{-1})(1 - 3z^{-1})}.$$

(a) Suppose the system is known to be stable. Determine the output $y[n]$ when the input $x[n]$ is the unit step sequence.

(b) Suppose the region of convergence of $H(z)$ includes $z = \infty$. Determine $y[n]$ evaluated at $n = 2$, when $x[n]$ is as shown in Fig. P5.17.

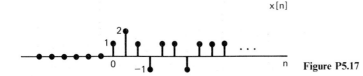

Figure P5.17

(c) Suppose we wish to recover $x[n]$ from $y[n]$ by processing $y[n]$ with an LTI system whose impulse response is given by $h_i[n]$. Determine $h_i[n]$. Does $h_i[n]$ depend on the region of convergence of $H(z)$?

5.18. The Fourier transform of a stable linear time-invariant system is purely real and is shown in Fig. P5.18. Determine whether this system has a stable inverse system.

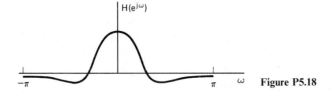

Figure P5.18

5.19. The signal $y[n]$ is the output of a linear time-invariant system with input $x[n]$, which is zero-mean white noise. The system is described by the difference equation

$$y[n] = \sum_{k=1}^{N} a_k y[n-k] + \sum_{k=0}^{M} b_k x[n-k], \qquad b_0 = 1.$$

(a) What is the z-transform $\Phi_{yy}(z)$ of the autocorrelation function $\phi_{yy}[n]$?

Sometimes it is of interest to "whiten" the spectrum of $y[n]$ by processing $y[n]$ with a linear filter; i.e., we wish to find a system such that its output power spectrum will be flat for input $y[n]$. Suppose that we know the autocorrelation function $\phi_{yy}[n]$ and its z-transform $\Phi_{yy}(z)$ but not the a_k's and the b_k's.

(b) Discuss a procedure to find a system function $H_w(z)$ of the whitening filter.

(c) Is the whitening filter unique?

5.20. A sequence $x[n]$ is the output of a linear time-invariant system whose input is $s[n]$. This system is described by the difference equation

$$x[n] = s[n] - e^{8\alpha} s[n-8], \qquad \text{(P5.20-1)}$$

where $0 < \alpha$.

(a) Find the system function

$$H_1(z) = \frac{X(z)}{S(z)}$$

and plot its poles and zeros in the z-plane. Indicate the region of convergence.

(b) We wish to recover $s[n]$ from $x[n]$ with a linear time-invariant system. Find the system function

$$H_2(z) = \frac{Y(z)}{X(z)}$$

such that $y[n] = s[n]$. Find all possible regions of convergence for $H_2(z)$ and, for each, tell whether or not the system is causal and stable.

(c) Find all possible choices for the impulse response $h_2[n]$ such that

$$y[n] = h_2[n] * x[n] = s[n]. \qquad \text{(P5.20-2)}$$

(d) For all choices determined in part (c), demonstrate by explicitly evaluating the convolution in Eq. (P5.20-2) that when $s[n] = \delta[n]$, $y[n] = \delta[n]$.

Note: As discussed in Problem 3.3, Eq. (P5.20-1) represents a simple model for a multipath channel. The system(s) determined in parts (b) and (c), then, correspond to compensation systems to correct for the multipath distortion.

5.21. In many practical situations we are faced with the problem of recovering a signal that has been "blurred" by a convolution process. We can model this blurring process as a linear filtering operation as depicted in Fig. P5.21-1, where the blurring impulse response is as shown in Fig. P5.21-2. The purpose of this problem is to consider ways to recover $x[n]$ from $y[n]$.

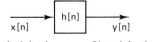

Desired signal   Blurred signal   **Figure P5.21-1**

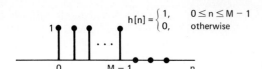

$$h[n] = \begin{cases} 1, & 0 \le n \le M-1 \\ 0, & \text{otherwise} \end{cases}$$

Figure P5.21-2

(a) One approach to recovering $x[n]$ from $y[n]$ is to use an inverse filter; i.e., $y[n]$ is filtered by a system whose frequency response is

$$H_i(e^{j\omega}) = \frac{1}{H(e^{j\omega})},$$

where $H(e^{j\omega})$ is the Fourier transform of $h[n]$. For the impulse response $h[n]$ shown in Fig. P5.21-2, discuss the practical problems involved in implementing the inverse filtering approach. Be complete, but also be brief and to the point.

(b) Because of the difficulties involved in inverse filtering, the following approach is suggested for recovering $x[n]$ from $y[n]$: The blurred signal $y[n]$ is processed by the system shown in Fig. P5.21-3, which produces an output $w[n]$ from which we can extract an improved replica of $x[n]$. The impulse responses $h_1[n]$ and $h_2[n]$ are shown in Fig. P5.21-4. Explain in detail the working of this system. In particular, state precisely the conditions under which we can recover $x[n]$ exactly from $w[n]$. [*Hint:* Consider the impulse response of the overall system from $x[n]$ to $w[n]$.]

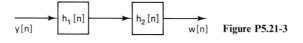

Figure P5.21-3

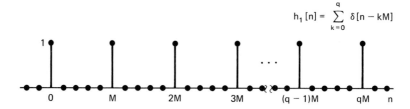

$$h_1[n] = \sum_{k=0}^{q} \delta[n - kM]$$

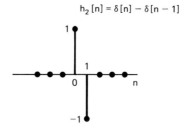

$$h_2[n] = \delta[n] - \delta[n - 1]$$

Figure P5.21-4

(c) Let us now attempt to generalize this approach to arbitrary finite-length blurring impulse responses $h[n]$; i.e., assume only that

$$h[n] = 0, \quad n < 0 \quad \text{and} \quad n \ge M.$$

Further, assume that $h_1[n]$ is the same as in Fig. P5.21-4. How must $H_2(z)$ and $H(z)$ be related for the system to work as in part (b)? What condition must $H(z)$ satisfy in order that it be possible to implement $H_2(z)$ as a causal system?

5.22. Consider a linear time-invariant system whose impulse response $h[n]$ is

$$h[n] = (\tfrac{1}{2})^n u[n] + (\tfrac{1}{3})^n u[n].$$

The input $x[n]$ is zero for $n < 0$ but in general may be nonzero for $0 \leq n \leq \infty$. We would like to compute the output $y[n]$ for $0 \leq n \leq 10^9$ and in particular want to compare the use of an FIR filter and an IIR filter for obtaining $y[n]$ over this interval.

(a) Determine the linear constant-coefficient difference equation for the IIR system relating $x[n]$ and $y[n]$.

(b) Determine the impulse response $h_1[n]$ of the minimum-length LTI FIR filter whose output $y_1[n]$ is identical to the output $y[n]$ for $0 \leq n \leq 10^9$.

(c) Specify the linear constant-coefficient difference equation associated with the FIR filter in part (b).

(d) Compare the number of arithmetic operations (multiplications and additions) required to obtain $y[n]$ for $0 \leq n \leq 10^9$ using the linear constant-coefficient difference equations in part (a) and in part (c).

5.23. Consider a causal linear time-invariant system with system function $H(z)$ and real impulse response. $H(z)$ evaluated for $z = e^{j\omega}$ is shown in Fig. P5.23.

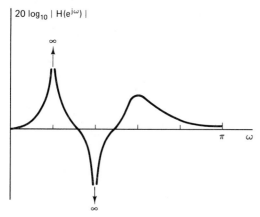

Figure P5.23

(a) Carefully sketch a pole-zero plot for $H(z)$ showing all information about the pole and zero locations that can be inferred from the figure.

(b) What can be said about the length of the impulse response?

(c) Specify whether $\measuredangle H(e^{j\omega})$ is linear.

(d) Specify whether the system is stable.

5.24. A causal linear time-invariant system has system function

$$H(z) = \frac{(1 - 1.5z^{-1} - z^{-2})(1 + 0.9z^{-1})}{(1 - z^{-1})(1 + 0.7jz^{-1})(1 - 0.7jz^{-1})}.$$

(a) Write the difference equation that is satisfied by the input and the output of the system.

(b) Plot the pole-zero diagram and indicate the region of convergence for the system function.

    (c) Sketch $|H(e^{j\omega})|$.

    (d) State whether the following are true or false about the system.

        (i)   The system is stable.

        (ii) The impulse response approaches a constant for large n.

        (iii) The magnitude of the frequency response has a peak at approximately $\omega = \pm\pi/4$.

        (iv) The system has a stable and causal inverse.

5.25. Consider a causal linear time-invariant system with system function

$$H(z) = \frac{1 - a^{-1}z^{-1}}{1 - az^{-1}},$$

where a is real.

    (a) Write the difference equation that relates the input and the output of this system.

    (b) For what range of values of a is this system stable?

    (c) For $a = \frac{1}{2}$, plot the pole-zero diagram and shade the region of convergence.

    (d) Find the impulse response $h[n]$ for this system.

    (e) Show that this system is an allpass system, i.e., that the magnitude of the frequency response is a constant. Also, specify the value of the constant.

5.26. The system function of a linear time-invariant system has the pole-zero plot shown in Fig. P5.26. Specify whether the following statements are true, false, or cannot be determined from the information given.

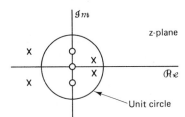

Figure P5.26

    (a) The system is stable.

    (b) The system is causal.

    (c) If the system is causal, then it must be stable.

    (d) If the system is stable, then it must have a two-sided impulse response.

5.27. In this problem, we demonstrate that for a rational z-transform, a factor of the form $(z - z_0)$ and a factor of the form $z/(z - z_0^*)$ contribute the same phase.

    (a) Let $H(z) = z - 1/a$, where a is real and $0 < a < 1$. Sketch the poles and zeros of the system including an indication of those at $z = \infty$. Determine $\angle H(e^{j\omega})$, the phase of the system.

    (b) Let $G(z)$ be specified such that it has poles at the conjugate-reciprocal location of zeros of $H(z)$ and zeros at the conjugate-reciprocal location of poles of $H(z)$, including those at zero and ∞. Sketch the pole-zero diagram of $G(z)$. Determine $\angle G(e^{j\omega})$, the phase of the system, and show that it is identical to $\angle H(e^{j\omega})$.

5.28. Let $h_1[n], h_2[n], \ldots, h_M[n]$ denote M finite-duration sequences, all of duration N, i.e., $h_k[n] = 0$ for $n < 0$ and $n \geq N$. The magnitude of the Fourier transform of each of these sequences is identical. None of the sequences is proportional (with either a real or complex constant of proportionality) to any of the others.

(a) What is the maximum value of M if the sequences are not restricted to be real?

(b) What is the maximum value of M if the sequences are restricted to be real?

5.29. Consider a causal sequence $x[n]$ with the following z-transform:

$$X(z) = \frac{(1 - \frac{1}{2}z^{-1})(1 - \frac{1}{4}z^{-1})(1 - \frac{1}{5}z)}{(1 - \frac{1}{6}z)}.$$

For what values of α is $\alpha^n x[n]$ a real and minimum-phase sequence?

5.30. Prove the validity of the following two statements:

(a) The convolution of two minimum-phase sequences is also minimum phase.

(b) The sum of two minimum-phase sequences is not necessarily minimum phase. Specifically, give an example of both a minimum-phase and a non-minimum-phase sequence that can be formed as the sum of two minimum-phase sequences.

5.31. A sequence $r[n]$ is defined by the relationship

$$r[n] = \sum_{m=-\infty}^{\infty} h[m]h[n+m] = h[n] * h[-n],$$

where $h[n]$ is a minimum-phase sequence, and $r[n]$ is specified as follows:

$$r[n] = \tfrac{4}{3}(\tfrac{1}{2})^n u[n] + \tfrac{4}{3} 2^n u[-n-1].$$

(a) Find $R(z)$ and sketch the pole-zero diagram.

(b) Determine the minimum-phase sequence $h[n]$ to within a scale factor of ± 1. Also determine its z-transform $H(z)$.

5.32. Consider the linear time-invariant system whose system function is

$$H(z) = (1 - 0.9e^{j0.6\pi}z^{-1})(1 - 0.9e^{-j0.6\pi}z^{-1})(1 - 1.25e^{j0.8\pi}z^{-1})(1 - 1.25e^{-j0.8\pi}z^{-1}).$$

(a) Find all causal system functions that result in the same frequency-response magnitude as $H(z)$ and for which the impulse responses are real-valued and of the same length as the impulse response associated with $H(z)$. (There are four different such system functions.) Explicitly identify which one is minimum phase and which, to within a time shift, is maximum phase.

(b) Find the impulse responses for the system functions in part (a).

(c) For each of the sequences in part (b), compute and plot the quantity

$$E[n] = \sum_{m=0}^{n} (h[m])^2$$

for $0 \le n \le 5$. Indicate explicitly which plot corresponds to the minimum-phase system.

5.33. A *maximum-phase* sequence is a stable sequence whose z-transform has all its poles and zeros *outside* the unit circle.

(a) Show that maximum-phase sequences are anticausal, i.e., they are zero for $n > 0$.

FIR maximum-phase sequences can be made causal by including a finite amount of delay. A finite-duration causal maximum-phase sequence having a given Fourier transform magnitude can be obtained by reflecting all the zeros of the z-transform of a minimum-phase sequence to conjugate-reciprocal positions outside the unit circle. That is, we can express the z-transform of a maximum-phase causal finite-duration sequence as

$$H_{max}(z) = H_{min}(z)H_{ap}(z).$$

Obviously, this process ensures that $|H_{max}(e^{j\omega})| = |H_{min}(e^{j\omega})|$. Now, the z-transform of a finite-duration minimum-phase sequence can be expressed as

$$H_{min}(z) = h_{min}[0] \prod_{k=1}^{M} (1 - c_k z^{-1}), \qquad |c_k| < 1.$$

(b) Obtain an expression for the allpass system function required to reflect all the zeros of $H_{min}(z)$ to positions outside the unit circle.

(c) Show that $H_{max}(z)$ can be expressed as

$$H_{max}(z) = z^{-M} H_{min}(z^{-1}).$$

(d) Using the result of part (c), express the maximum-phase sequence $h_{max}[n]$ in terms of $h_{min}[n]$.

5.34. It is not possible to obtain a causal and stable inverse system (perfect compensator) for a non-minimum-phase system. In this problem, we study an approach to compensating only the magnitude of the frequency response of a non-minimum-phase system.

Suppose that a stable non-minimum-phase linear time-invariant discrete-time system with a rational system function $H(z)$ is cascaded with a compensating system $H_c(z)$ as shown in Fig. P5.34.

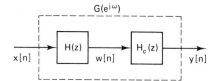

Figure P5.34

(a) How should $H_c(z)$ be chosen so that it is stable and causal and so that the magnitude of the overall effective frequency response is unity? (Recall that $H(z)$ can always be represented as $H(z) = H_{ap}(z)H_{min}(z)$.)

(b) What are the corresponding system functions $H_c(z)$ and $G(z)$?

(c) Assume that

$$H(z) = (1 - 0.8e^{j0.3\pi}z^{-1})(1 - 0.8e^{-j0.3\pi}z^{-1})(1 - 1.2e^{j0.7\pi}z^{-1})(1 - 1.2e^{-j0.7\pi}z^{-1}).$$

Find $H_{min}(z)$, $H_{ap}(z)$, $H_c(z)$, and $G(z)$ for this case, and construct the pole-zero plots for each system function.

5.35. Let $h_{min}[n]$ denote a minimum-phase sequence with z-transform $H_{min}(z)$. If $h[n]$ is a causal non-minimum-phase sequence whose Fourier transform magnitude is equal to $|H_{min}(e^{j\omega})|$, show that

$$|h[0]| < |h_{min}[0]|.$$

(Use the initial value theorem together with Eq. 5.103.)

5.36. One of the interesting and important properties of minimum-phase sequences is the minimum-energy delay property; i.e., of all the causal sequences having the same Fourier transform magnitude function $|H(e^{j\omega})|$, the quantity

$$E[n] = \sum_{m=0}^{n} |h[m]|^2$$

is maximum for all $n \geq 0$ when $h[n]$ is the minimum-phase sequence. This result is proved as follows: Let $h_{min}[n]$ be a minimum-phase sequence with z-transform $H_{min}(z)$. Furthermore, let z_k be a zero of $H_{min}(z)$ so that we can express $H_{min}(z)$ as

$$H_{min}(z) = Q(z)(1 - z_k z^{-1}), \qquad |z_k| < 1,$$

where $Q(z)$ is again minimum phase. Now consider another sequence $h[n]$ with z-transform $H(z)$ such that

$$|H(e^{j\omega})| = |H_{min}(e^{j\omega})|$$

and such that $H(z)$ has a zero at $z = 1/z_k^*$ instead of at z_k.

(a) Express $H(z)$ in terms of $Q(z)$.

(b) Express $h[n]$ and $h_{min}[n]$ in terms of the minimum-phase sequence $q[n]$ that has z-transform $Q(z)$.

(c) To compare the distribution of energy of the two sequences, show that

$$\varepsilon = \sum_{m=0}^{n} |h_{min}[m]|^2 - \sum_{m=0}^{n} |h[m]|^2 = (1 - |z_k|^2)|q[m]|^2.$$

(d) Using the result of part (c), argue that

$$\sum_{m=0}^{n} |h[m]|^2 \le \sum_{m=0}^{n} |h_{min}[m]|^2 \qquad \text{for all } n.$$

5.37. A causal allpass system $H_{ap}(z)$ has input $x[n]$ and output $y[n]$.

(a) If $x[n]$ is a real minimum-phase sequence (which also implies that $x[n] = 0$ for $n < 0$), using Eq. (5.118) show that

$$\sum_{k=0}^{n} |x[k])|^2 \ge \sum_{k=0}^{n} |y[k]|^2. \qquad (P5.37)$$

(b) Show that Eq. (P5.37) holds even if $x[n]$ is not minimum phase but is zero for $n < 0$.

5.38. Shown in Fig. P5.38 are eight different finite-duration sequences. Each sequence is 4 points long. The magnitude of the Fourier transform is the same for all sequences. Which of the sequences has all the zeros of its z-transform *inside* the unit circle?

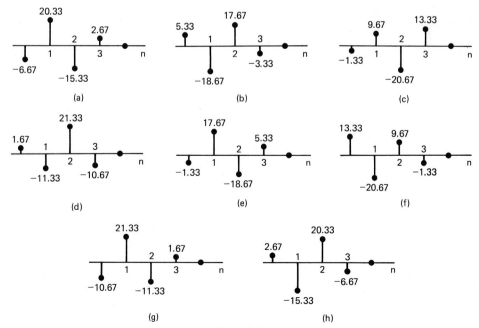

Figure P5.38

5.39. In the design of either continuous-time or discrete-time filters, we often approximate a specified magnitude characteristic without particular regard to the phase. For example, standard design techniques for lowpass and bandpass filters are derived from consideration of the magnitude characteristics only.

   In many filtering problems, we would prefer that the phase characteristics be zero or linear. For causal filters it is impossible to have zero phase. However, for many filtering applications, it is not necessary that the impulse response of the filter be zero for $n < 0$ if the processing is not to be carried out in real time.

   One technique commonly used in discrete-time filtering when the data to be filtered are of finite duration and are stored, for example, in computer memory is to process the data forward and then backward through the same filter.

   Let $h[n]$ be the impulse response of a causal filter with an arbitrary phase characteristic. Assume that $h[n]$ is real and denote its Fourier transform by $H(e^{j\omega})$. Let $x[n]$ be the data that we want to filter.

(a) *Method A*: The filtering operation is performed as shown in Fig. P5.39-1.

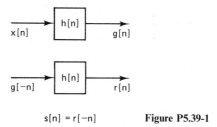

$$s[n] = r[-n] \qquad \text{Figure P5.39-1}$$

1. Determine the overall impulse response $h_1[n]$ that relates $x[n]$ and $s[n]$, and show that it has a zero-phase characteristic.
2. Determine $|H_1(e^{j\omega})|$ and express it in terms of $|H(e^{j\omega})|$ and $\measuredangle H(e^{j\omega})$.

(b) *Method B*: As depicted in Fig. P5.39-2, process $x[n]$ through the filter $h[n]$ to get $g[n]$. Also process $x[n]$ backward through $h[n]$ to get $r[n]$. The output $y[n]$ is then taken as the sum of $g[n]$ and $r[-n]$. This composite set of operations can be represented by a filter, with input $x[n]$, output $y[n]$, and impulse response $h_2[n]$.

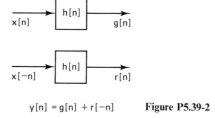

$$y[n] = g[n] + r[-n] \qquad \text{Figure P5.39-2}$$

1. Show that the composite filter $h_2[n]$ has a zero-phase characteristic.
2. Determine $|H_2(e^{j\omega})|$ and express it in terms of $|H(e^{j\omega})|$ and $\measuredangle H(e^{j\omega})$.

(c) Suppose that we are given a sequence of finite duration on which we would like to perform a bandpass zero-phase filtering operation. Furthermore, assume that we are given the bandpass filter $h[n]$, with frequency response as specified in Fig. P5.39-3, which has the magnitude characteristic that we desire but linear phase. To achieve zero phase, we could use either method A or B. Determine and sketch $|H_1(e^{j\omega})|$ and

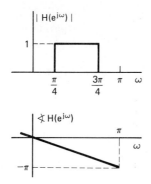

Figure P5.39-3

$|H_2(e^{j\omega})|$. From these results, which method would you use to achieve the desired bandpass filtering operation? Explain why. More generally, if $h[n]$ has the desired magnitude but a nonlinear phase characteristic, which method is preferable to achieve a zero-phase characteristic?

5.40. Consider the class of discrete-time filters whose frequency response has the form

$$H(e^{j\omega}) = |H(e^{j\omega})|e^{-j\alpha\omega},$$

where $|H(e^{j\omega})|$ is a real and nonnegative function of ω and α is a real constant. As discussed in Section 5.7.1, this class of filters is referred to as *linear phase* filters.

Also consider the class of discrete-time filters whose frequency response has the form

$$H(e^{j\omega}) = A(e^{j\omega})e^{-j\alpha\omega + j\beta},$$

where $A(e^{j\omega})$ is a real function of ω, α is a real constant, and β is a real constant. As discussed in Section 5.7.2, filters in this class are referred to as *generalized linear phase* filters.

For each of the filters in Fig. P5.40, determine whether it is a generalized linear phase filter. If it is, then find $A(e^{j\omega})$, α, and β and indicate if it is also a linear phase filter.

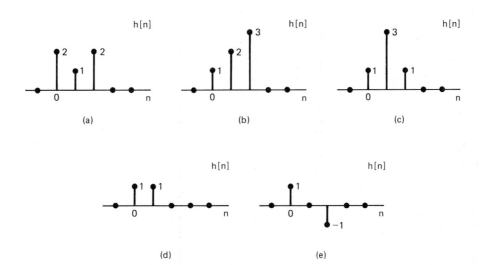

Figure P5.40

5.41. Each of the pole-zero plots in Fig. P5.41, together with the specification of the region of convergence, describes a linear time-invariant system with system function $H(z)$. In each case, determine if any of the following statements are true. Justify your answer with a brief statement or a counterexample.

(i) The system has zero phase or generalized linear phase.

(ii) The system has a stable inverse $H_i(z)$.

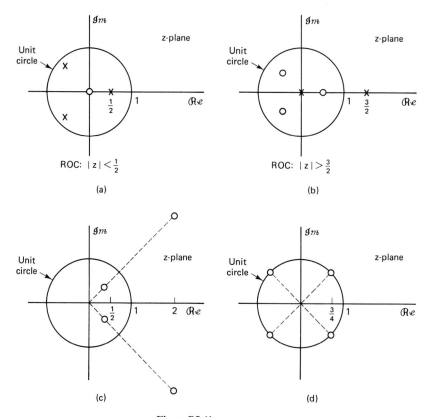

Figure P5.41

5.42. Figure P5.42-1 shows two different interconnections of three systems. The impulse responses $h_1[n]$, $h_2[n]$, and $h_3[n]$ are as shown in Fig. P5.42-2. Determine whether system A and/or system B has generalized linear phase.

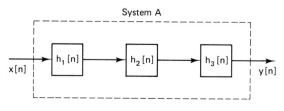

Figure P5.42-1(a)

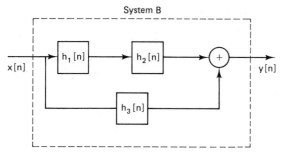

System B

Figure P5.42-1(b)

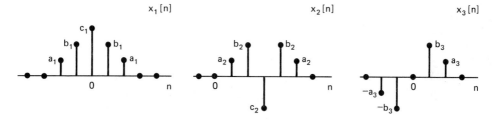

Figure P5.42-2

5.43. Determine whether the following statement is true or false. If it is true, concisely state your reasoning. If it is false, give a counterexample.

Statement: If the system function $H(z)$ has poles anywhere other than at the origin or infinity, then the system cannot have zero phase or generalized linear phase.

5.44. The overall system of Fig. P5.44 is a discrete-time linear time-invariant system with frequency response $H(e^{j\omega})$ and impulse response $h[n]$.

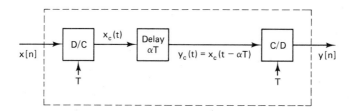

Figure P5.44

(a) $H(e^{j\omega})$ can be expressed in the form

$$H(e^{j\omega}) = A(e^{j\omega})e^{j\phi(\omega)},$$

with $A(e^{j\omega})$ real. Determine and sketch $A(e^{j\omega})$ and $\phi(\omega)$ for $|\omega| < \pi$.

(b) Sketch $h[n]$ for the following.
  (i) $\alpha = 3$
  (ii) $\alpha = 3\frac{1}{2}$
  (iii) $\alpha = 3\frac{1}{4}$

(c) Consider a discrete-time linear time-invariant system for which

$$H(e^{j\omega}) = A(e^{j\omega})e^{j\alpha\omega}, \qquad |\omega| < \pi,$$

with $A(e^{j\omega})$ real. What can be said about the symmetry of $h[n]$ for the following?
(i) $\alpha =$ integer
(ii) $\alpha = M/2$, where M is an odd integer
(iii) General α

5.45. Consider the class of FIR filters that have $h[n]$ real, $h[n] = 0$ for $n < 0$ and $n > M$, and one of the following symmetry properties:

$$\text{Symmetric:} \qquad h[n] = h[M - n]$$

$$\text{Antisymmetric:} \qquad h[n] = -h[M - n]$$

All filters in this class have generalized linear phase, i.e., have frequency response of the form

$$H(e^{j\omega}) = A(e^{j\omega})e^{-j\alpha\omega + j\beta},$$

where $A(e^{j\omega})$ is a real function of ω, α is a real constant, and β is a real constant.
For the following table, show that $A(e^{j\omega})$ has the indicated form, and find the values of α and β.

Type	Symmetry	Filter Length ($M + 1$)	Form of $A(e^{j\omega})$	α	β
I	Symmetric	Odd	$\displaystyle\sum_{n=0}^{M/2} a[n]\cos \omega n$		
II	Symmetric	Even	$\displaystyle\sum_{n=1}^{(M+1)/2} b[n]\cos \omega(n - 1/2)$		
III	Antisymmetric	Odd	$\displaystyle\sum_{n=1}^{M/2} c[n]\sin \omega n$		
IV	Antisymmetric	Even	$\displaystyle\sum_{n=1}^{(M+1)/2} d[n]\sin \omega(n - 1/2)$		

Here are several helpful suggestions.

- For type I filters, first show that $H(e^{j\omega})$ can be written in the form

$$H(e^{j\omega}) = \sum_{n=0}^{(M-2)/2} h[n]e^{-j\omega n} + \sum_{n=0}^{(M-2)/2} h[M - n]e^{-j\omega[M-n]} + h[M/2]e^{-j\omega(M/2)}.$$

- The analysis for type III filters is very similar to that for type I, with the exception of a sign change and removal of one of the above terms.
- For type II filters, first write $H(e^{j\omega})$ in the form

$$H(e^{j\omega}) = \sum_{n=0}^{(M-1)/2} h[n]e^{-j\omega n} + \sum_{n=0}^{(M-1)/2} h[M - n]e^{-j\omega[M-n]}$$

and then pull out a common factor of $e^{-j\omega(M/2)}$ from both sums.
- The analysis for type IV filters is very similar to that for type II filters.

5.46. **(a)** For each of the four types of causal linear phase FIR filters discussed in Section 5.7.3, determine whether the associated symmetry imposes any constraint on the frequency response at $\omega = 0$ and/or $\omega = \pi$.

(b) For each kind of desired filter listed below, indicate which of the four FIR filter types would be useful to consider in approximating the desired filter.

Lowpass

Bandpass

Highpass

Bandstop

Differentiator

5.47. Let $h_{lp}[n]$ denote the impulse response of an FIR generalized linear phase lowpass filter. The impulse response $h_{hp}[n]$ of an FIR generalized linear phase highpass filter can be obtained by the transformation

$$h_{hp}[n] = (-1)^n h_{lp}[n].$$

If we decide to design a highpass filter using this transformation and we wish the resulting highpass filter to be symmetric, which of the four types of generalized linear phase FIR filters can we use for the design of the lowpass filter? Your answer should consider *all* the possible types.

5.48. A causal, linear, time-invariant discrete-time system has system function

$$H(z) = \frac{(1 - 0.5z^{-1})(1 + 4z^{-2})}{(1 - 0.64z^{-2})}.$$

(a) Find expressions for a minimum-phase system $H_1(z)$ and an allpass system $H_{ap}(z)$ such that

$$H(z) = H_1(z)H_{ap}(z).$$

(b) Find expressions for a different minimum-phase system $H_2(z)$ and a generalized linear phase FIR system $H_{lin}(z)$ such that

$$H(z) = H_2(z)H_{lin}(z).$$

5.49. **(a)** A minimum-phase system with system function $H_{min}(z)$ is such that

$$H_{min}(z)H_{ap}(z) = H_{lin}(z),$$

where $H_{ap}(z)$ is an allpass system function and $H_{lin}(z)$ is a causal generalized linear phase system. What does this tell you about the poles and zeros of $H_{min}(z)$?

(b) A generalized linear phase FIR system has an impulse response with real values and $h[n] = 0$ for $n < 0$ and for $n \geq 8$. The system function of this system has a zero at $z = 0.8e^{j\pi/4}$ and another zero at $z = -2$. What is $H(z)$?

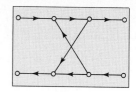

Structures for Discrete-Time Systems

6

6.0 INTRODUCTION

As we saw in Chapter 5, a linear time-invariant system with a rational system function has the property that the input and output sequences satisfy a linear constant-coefficient difference equation. Since the system function is the z-transform of the impulse response and since the difference equation satisfied by the input and output can be determined by inspection from the system function, it follows that the difference equation, the impulse response, and the system function are equivalent characterizations of the input/output relation of a linear time-invariant discrete-time system. When such systems are implemented with discrete-time analog or digital hardware, the difference equation or the system function representation must be converted to an algorithm or structure that can be realized in the desired technology. As we will see in this chapter, systems described by linear constant-coefficient difference equations can be represented by structures consisting of an interconnection of the basic operations of addition, multiplication by a constant, and delay, the exact implementation of which is dictated by the technology to be used.

As an illustration of the computation associated with a difference equation, consider the system described by the system function

$$H(z) = \frac{b_0 + b_1 z^{-1}}{1 - a z^{-1}}, \qquad |z| > |a|. \tag{6.1}$$

The impulse response of this system is

$$h[n] = b_0 a^n u[n] + b_1 a^{n-1} u[n-1] \tag{6.2}$$

290

and the first-order difference equation that is satisfied by the input and output sequences is

$$y[n] - ay[n-1] = b_0 x[n] + b_1 x[n-1]. \qquad (6.3)$$

Since the system has an infinite-duration impulse response, it is not possible to implement the system by discrete convolution. However, rewriting Eq. (6.3) in the form

$$y[n] = ay[n-1] + b_0 x[n] + b_1 x[n-1] \qquad (6.4)$$

provides the basis for an algorithm for recursive computation of the output at any time n in terms of the previous output $y[n-1]$, the current input sample $x[n]$, and previous input sample $x[n-1]$. As discussed in Section 2.5, if we further assume initial-rest conditions (if $x[n] = 0$ for $n < 0$, then $y[n] = 0$ for $n < 0$) and if we use Eq. (6.4) as a recurrence formula for computing the output from past values of the output and present and past values of the input, the system will be linear and time-invariant. A similar procedure can be applied to the more general case of an Nth-order difference equation. However, the algorithm suggested by Eq. (6.4) and its generalization for higher-order difference equations is not the only computational algorithm for implementing a particular system, and often it is not the most preferable. As we will see, an unlimited variety of computational structures result in the same relation between the input sequence $x[n]$ and the output sequence $y[n]$.

In the remainder of this chapter, we consider the important issues in the implementation of linear time-invariant discrete-time systems. We first present the block diagram and signal flow graph description of computational structures or networks for linear constant-coefficient difference equations representing linear, time-invariant, causal systems. Using a combination of algebraic manipulations and manipulations of block diagram representations, we derive a number of basic equivalent structures for implementing a causal, linear, time-invariant system. Although two structures may be equivalent with regard to their input/output characteristics for infinite-precision representations of coefficients and variables, they may have vastly different behavior when the numerical precision is limited. This is the major reason that it is of interest to study different implementation structures. The effects of finite-precision representation of the system coefficients and the effects of truncation or rounding of intermediate computations are considered in the latter sections of the chapter.

6.1 BLOCK DIAGRAM REPRESENTATION OF LINEAR CONSTANT-COEFFICIENT DIFFERENCE EQUATIONS

The implementation of a linear time-invariant discrete-time system by iteratively evaluating a recurrence formula obtained from a difference equation requires that delayed values of the output, input, and intermediate sequences be available. The delay of sequence values implies the need for storage of past sequence values. Also, we must provide means for multiplication of the delayed sequence values by the

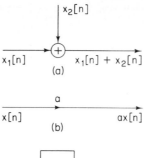

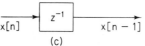

Figure 6.1 Block diagram symbols.
(a) Addition of two sequences.
(b) Multiplication of a sequence by a
constant. (c) Unit delay.

coefficients as well as means for adding the resulting products. Therefore, the basic
elements required for implementation of a linear time-invariant discrete-time system
are adders, multipliers, and memory for storing delayed sequence values. The
interconnection of these basic elements is conveniently depicted by block diagrams
composed of the basic pictorial symbols shown in Fig. 6.1. Figure 6.1(a) represents
means for adding two sequences. In general block diagram notation, an adder may
have any number of inputs. However, in almost all practical implementations, adders
have only two inputs. In all the diagrams of this chapter we indicate this explicitly by
limiting the number of inputs as in Fig. 6.1(a). Figure 6.1(b) depicts multiplication of
a sequence by a constant, and Fig. 6.1(c) depicts delaying a sequence by one
sample. In digital implementations, the delay operation can be implemented by
providing a storage register for each unit delay that is required. In analog discrete-
time implementations such as switched-capacitor filters, the delays are implemented
by charge storage devices. The unit delay system is represented in Fig. 6.1(c) by its
system function, z^{-1}. Delays of more than one sample can be denoted as in Fig 6.1(c)
with a system function of z^{-M}, where M is the number of samples of delay; however,
actual implementation of M samples of delay would generally be done by cascading
M unit delays. In a real-time implementation, these unit delays might form a shift
register that is clocked at the sampling rate of the input signal.

Example 6.1

As an example of the representation of a difference equation in terms of the elements in
Fig. 6.1, consider the second-order difference equation

$$y[n] = a_1 y[n-1] + a_2 y[n-2] + bx[n]. \qquad (6.5)$$

The corresponding system function is

$$H(z) = \frac{b}{1 - a_1 z^{-1} - a_2 z^{-2}}. \qquad (6.6)$$

The block diagram representation of the system realization based on Eq. (6.5) is shown in
Fig. 6.2. Such diagrams give a pictorial representation of a computation algorithm for
implementing the system. When the system is implemented on either a general-purpose
computer or a special-purpose microcomputer, network structures such as that shown in

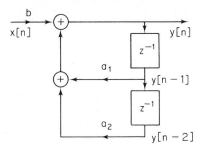

Figure 6.2 Example of a block diagram representation of a difference equation.

Fig. 6.2 serve as the basis for a program that implements the system. If the system is implemented with discrete components or as a complete system with VLSI technology, the block diagram is the basis for determining a hardware architecture for the system. In both cases, diagrams such as Fig. 6.2 show explicitly that we must provide storage for the delayed variables (in this case $y[n-1]$ and $y[n-2]$) and also the coefficients of the difference equation (in this case a_1, a_2, and b). Furthermore, we see from Fig. 6.2 that an output sequence value $y[n]$ is computed by first forming the products $a_1 y[n-1]$ and $a_2 y[n-2]$, adding them, and then adding the result to $bx[n]$. Thus, Fig. 6.2 conveniently depicts the complexity of the associated computational algorithm, the steps of the algorithm, and the amount of hardware required to realize the system.

Example 6.1 can be generalized to higher-order difference equations of the form†

$$y[n] - \sum_{k=1}^{N} a_k y[n-k] = \sum_{k=0}^{M} b_k x[n-k], \tag{6.7}$$

with the corresponding system function

$$H(z) = \frac{\sum_{k=0}^{M} b_k z^{-k}}{1 - \sum_{k=1}^{N} a_k z^{-k}}. \tag{6.8}$$

Rewriting Eq. (6.7) as a recurrence formula for $y[n]$ in terms of a linear combination of past values of the output sequence and current and past values of the input sequence leads to the relation

$$y[n] = \sum_{k=1}^{N} a_k y[n-k] + \sum_{k=0}^{M} b_k x[n-k]. \tag{6.9}$$

---

† The form used in previous chapters for a general Nth-order difference equation was

$$\sum_{k=0}^{N} a_k y[n-k] = \sum_{k=0}^{M} b_k x[n-k].$$

In the remainder of the book, it will be more convenient to use the form in Eq. (6.7), where the coefficient of $y[n]$ is normalized to 1 and the coefficients associated with the delayed output appear with a positive sign after they have been moved to the right-hand side of the equation as in Eq. (6.9).

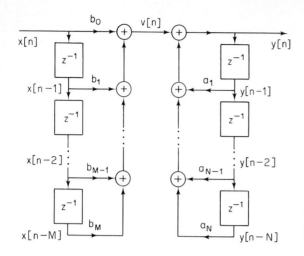

Figure 6.3 Block diagram representation for a general Nth-order difference equation.

The block diagram of Fig. 6.3 is an explicit pictorial representation of Eq. (6.9). More precisely, it represents the pair of difference equations

$$v[n] = \sum_{k=0}^{M} b_k x[n-k], \tag{6.10a}$$

$$y[n] = \sum_{k=1}^{N} a_k y[n-k] + v[n]. \tag{6.10b}$$

The assumption of a two-input adder implies that the additions are done in a specified order. That is, Fig. 6.3 shows that the products $a_N y[n-N]$ and $a_{N-1} y[n-N+1]$ must be computed, then added, and the resulting sum added to $a_{N-2} y[n-N+2]$, and so on. After $y[n]$ has been computed, the delay variables must be updated by moving $y[n-N+1]$ into the register holding $y[n-N]$, and so on.

The block diagram of Fig. 6.3 can be rearranged or modified in a variety of ways without changing the overall system function. Each appropriate rearrangement of the block diagram represents a *different* computational algorithm for implementing the *same* system. For example, the block diagram of Fig. 6.3 can be viewed as a cascade of two systems, the first representing the computation of $v[n]$ from $x[n]$ and the second representing the computation of $y[n]$ from $v[n]$. Since each of the two systems is a linear time-invariant system (assuming initial-rest conditions for the delay registers), the order in which the two systems are cascaded can be reversed, as shown in Fig. 6.4, without affecting the overall system function. In Fig. 6.4, for convenience we have assumed that $M = N$. Clearly, there is no loss of generality since if $M \neq N$, some of the coefficients a_k or b_k in Fig. 6.4 would be zero, and the diagram could be simplified accordingly.

In terms of the system function $H(z)$ in Eq. (6.8), Fig. 6.3 can be viewed as an implementation of $H(z)$ through the decomposition

$$H(z) = H_2(z)H_1(z) = \left(\frac{1}{1 - \sum_{k=1}^{N} a_k z^{-k}} \right) \left(\sum_{k=0}^{M} b_k z^{-k} \right) \tag{6.11}$$

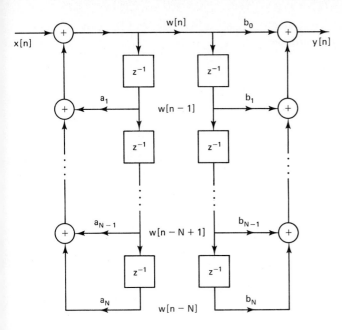

Figure 6.4   Rearrangement of block diagram of Fig. 6.3. We assume for convenience that $N = M$. If $N \neq M$, some of the coefficients will be zero.

or equivalently through the pair of equations

$$V(z) = H_1(z)X(z) = \left(\sum_{k=0}^{M} b_k z^{-k} \right) X(z), \tag{6.12a}$$

$$Y(z) = H_2(z)V(z) = \left(\frac{1}{1 - \displaystyle\sum_{k=1}^{N} a_k z^{-k}} \right) V(z). \tag{6.12b}$$

Figure 6.4, on the other hand, represents $H(z)$ as

$$H(z) = H_1(z)H_2(z) = \left(\sum_{k=0}^{M} b_k z^{-k} \right) \left(\frac{1}{1 - \displaystyle\sum_{k=1}^{N} a_k z^{-k}} \right) \tag{6.13}$$

or equivalently through the equations

$$W(z) = H_2(z)X(z) = \left(\frac{1}{1 - \displaystyle\sum_{k=1}^{N} a_k z^{-k}} \right) X(z), \tag{6.14a}$$

$$Y(z) = H_1(z)W(z) = \left(\sum_{k=0}^{M} b_k z^{-k} \right) W(z). \tag{6.14b}$$

In the time domain, Fig. 6.4 and, equivalently, Eqs. (6.14a) and (6.14b) can be represented by the pair of difference equations

$$w[n] = \sum_{k=1}^{N} a_k w[n-k] + x[n], \qquad (6.15a)$$

$$y[n] = \sum_{k=0}^{M} b_k w[n-k]. \qquad (6.15b)$$

The block diagrams of Figs. 6.3 and 6.4 have several important differences. In Fig. 6.3 the zeros of $H(z)$, represented by $H_1(z)$, are implemented first, followed by the poles, represented by $H_2(z)$. In Fig. 6.4, the poles are implemented first followed by the zeros. Theoretically, the order of implementation does not affect the overall system function. However, as we will see, when a difference equation is implemented with finite-precision arithmetic, there can be a significant difference between two systems that are theoretically equivalent. Another important point concerns the number of delay elements in the two systems. As drawn, the systems in Figs. 6.3 and 6.4 both have a total of $(N + M)$ delay elements. However, the block diagram of Fig. 6.4 can be redrawn by noting that exactly the same signal, $w[n]$, is stored in the two chains of delay elements in Fig. 6.4. Consequently, the two can be collapsed into one chain as indicated in Fig. 6.5.

The total number of delay elements in Fig. 6.5 is less than it is in either Fig. 6.3 or Fig. 6.4, and in fact it is the minimum number required to implement a system with system function given by Eq. (6.8). Specifically, the minimum number of delays required in general is $\max(N, M)$. An implementation with the minimum number of delay elements is commonly referred to as a *canonic form* implementation. The noncanonic block diagram in Fig. 6.3 is referred to as the *direct form I* implementation

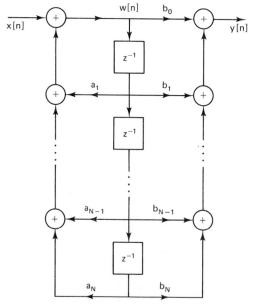

Figure 6.5 Combination of delays in Fig. 6.4.

of the general Nth-order system because it is a direct realization of the difference equation satisfied by the input $x[n]$ and the output $y[n]$, which in turn can be written directly from the system function by inspection. Figure 6.5 is often referred to as the *direct form II or canonic direct form* implementation. Knowing that Fig. 6.5 is an appropriate realization structure for $H(z)$ given by Eq. (6.8), we can go directly back and forth in a straightforward manner between the system function and the block diagram (or the equivalent difference equation).

In the preceding discussion we developed two equivalent block diagrams for implementing a linear time-invariant system with system function given by Eq. (6.8). These block diagrams, which represent different computational algorithms for implementing the system, were obtained by manipulations based on the linearity of the system and the algebraic properties of the system function. Indeed, since the basic difference equations that represent a linear time-invariant system are linear, equivalent sets of difference equations can be obtained simply by linear transformations of the variables of the difference equations. Thus, there are an unlimited number of equivalent realizations of any given system. In Section 6.3, using an approach similar to that used in this section, we will develop a number of other important and useful equivalent structures for implementing a system with system function as in Eq. (6.8). Before discussing these other forms, however, it is convenient to introduce signal flow graphs as an alternative to block diagrams for representing difference equations.

6.2 SIGNAL FLOW GRAPH REPRESENTATION OF LINEAR CONSTANT-COEFFICIENT DIFFERENCE EQUATIONS

A signal flow graph representation of a difference equation is essentially the same as a block diagram representation except for a few notational differences. Formally, a signal flow graph is a network of directed branches that connect at nodes. Associated with each node is a variable or node value. The value associated with node k might be denoted w_k, or, since node variables for digital filters are generally sequences, we often indicate this explicitly with the notation $w_k[n]$. Branch (j, k) denotes a branch originating at node j and terminating at node k, with the direction from j to k being indicated by an arrowhead on the branch. This is shown in Fig. 6.6. Each branch has an input signal and an output signal. The input signal from node j to branch (j, k) is the node value $w_j[n]$. In a linear signal flow graph, which is the only class we will consider, the output of a branch is a linear transformation of the input to the branch. The simplest example is a constant gain, i.e., when the output of the branch is

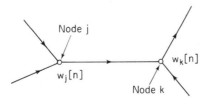

Figure 6.6   Example of nodes and branches in a signal flow graph.

simply a constant multiple of the input to the branch. The linear operation represented by the branch is typically indicated next to the arrowhead showing the direction of the branch. For the case of a constant multiplier, the constant is simply shown next to the arrowhead. When an explicit indication of the branch operation is omitted, this indicates a branch transmittance of unity, or the identity transformation. By definition, the node value at each node in a graph is the sum of the outputs of all the branches entering the node.

To complete the definition of signal flow graph notation, we define two special types of nodes. *Source nodes* are nodes that have no entering branches. Source nodes are used to represent the injection of external inputs or signal sources into a graph. *Sink nodes* are nodes that have only entering branches. Sink nodes are used to extract outputs from a graph. Source nodes, sink nodes, and simple branch gains are illustrated in the signal flow graph of Fig. 6.7. The linear equations represented by Fig. 6.7 are as follows:

$$w_1[n] = x[n] + aw_2[n] + bw_2[n],$$

$$w_2[n] = cw_1[n], \tag{6.16}$$

$$y[n] = dx[n] + ew_2[n].$$

Addition, multiplication by a constant, and delay are the basic operations required to implement a linear constant-coefficient difference equation. Since these are all linear operations, it is possible to use signal flow graph notation to depict algorithms for implementing linear time-invariant discrete-time systems. As an example of how the flow graph concepts just discussed can be applied to the representation of a difference equation, consider the block diagram in Fig. 6.8(a), which is the direct form II realization of the system whose system function is given by Eq. (6.1). A signal flow graph corresponding to this system is shown in Fig. 6.8(b). In the representation of difference equations, the node variables are sequences. In Fig. 6.8(b), node 0 is a source node whose value is determined by the input sequence $x[n]$, and node 5 is a sink node whose value is denoted $y[n]$. Notice that the source and sink nodes are connected to the rest of the graph by unity-gain branches to clearly denote the input and output of the system. Obviously, nodes 3 and 5 have identical values. In Fig. 6.8(b) all branches except one (the delay branch (2, 4)) can be represented by a simple branch gain; i.e., the output signal is a constant multiple of the branch input. Delay cannot be represented in the time domain by a branch gain. However, the z-transform representation of a unit delay is multiplication by the factor z^{-1}. If we represented the difference equations by their corresponding z-transform equations, all the branches would be characterized by their system functions. In this case, each branch gain would be a function of z; e.g., a unit delay branch would have a gain of z^{-1}. By convention, we normally represent the variables

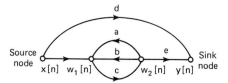

Figure 6.7   Example of a signal flow graph showing source and sink nodes.

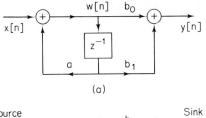

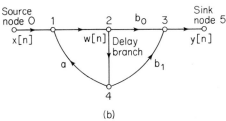

Figure 6.8   (a) Block diagram representation of a first-order digital filter. (b) Structure of the signal flow graph corresponding to the block diagram in (a).

in a signal flow graph as sequences rather than as z-transforms of sequences. However, to simplify the notation, we normally indicate a delay branch by showing its branch gain as z^{-1}, but it is understood that the output of such a branch is the branch input delayed by one sequence value. The graph of Fig. 6.8(b) is shown in Fig. 6.9 with this convention. The equations represented by Fig. 6.9 are as follows:

$$w_1[n] = aw_4[n] + x[n], \tag{6.17a}$$

$$w_2[n] = w_1[n], \tag{6.17b}$$

$$w_3[n] = b_0w_2[n] + b_1w_4[n], \tag{6.17c}$$

$$w_4[n] = w_2[n-1], \tag{6.17d}$$

$$y[n] = w_3[n]. \tag{6.17e}$$

Comparison of Fig. 6.8(a) and Fig. 6.9 shows that there is direct correspondence between branches in the block diagram and branches in the flow graph. In fact, the important difference between the two is that nodes in the flow graph represent both branching points and adders, whereas in the block diagram a special symbol is used for adders. A branching point in the block diagram is represented in the flow graph by a node that has only one incoming branch and one or more outgoing branches. An adder in the block diagram is represented in the signal flow graph by a node that has two (or more) incoming branches. Signal flow graphs are therefore totally equivalent to block diagrams as pictorial representations of difference equations, but they are simpler to draw. Like block diagrams, they can be manipulated graphically to gain

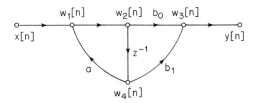

Figure 6.9   Signal flow graph of Fig. 6.8(b) with the delay branch indicated by z^{-1}.

insight into the properties of a given system. A large body of signal flow graph theory exists that can be directly applied to discrete-time systems when they are represented in this form (Mason and Zimmermann, 1960; Chow and Cassignol, 1962). Although we will use flow graphs primarily for their pictorial value, we will make use of certain theorems relating to signal flow graphs in examining alternate structures for implementing linear systems.

Equations (6.17) define a multistep algorithm for computing the output of the linear time-invariant system from the input sequence $x[n]$. This example illustrates the kind of data precedence relations that generally arise in the implementation of IIR systems. Equations (6.17) cannot be computed in arbitrary order. Equations (6.17a) and (6.17c) require multiplications and additions, but Eqs.(6.17b) and (6.17e) simply rename variables. Equation (6.17d) represents the "updating" of the memory of the system. It would be implemented simply by replacing the contents of the memory register representing $w_4[n]$ by the value of $w_2[n]$, but it would have to be done consistently either *before* or *after* the evaluation of all the other equations. Initial-rest conditions would be imposed in this case by defining $w_2[-1] = 0$ or $w_4[0] = 0$. Clearly, Eqs. (6.17) must be computed in the order given, except that the last two could be interchanged or Eq. (6.17d) could be consistently evaluated first.

The flow graph represents a set of difference equations, with one equation being written at each node of the network. In the case of the flow graph of Fig. 6.9, we can eliminate some of the variables rather easily to obtain the pair of equations

$$w_2[n] = aw_2[n-1] + x[n], \tag{6.18a}$$

$$y[n] = b_0 w_2[n] + b_1 w_2[n-1], \tag{6.18b}$$

which are in the form of Eqs. (6.15a) and (6.15b). Often, the manipulation of the difference equations of a flow graph is difficult when dealing with the time-domain variables due to feedback of delayed variables. In such cases, it is always possible to work with the z-transform representation where all branches are simple gains. Problems 6.1–6.12 illustrate the utility of z-transform analysis of flow graphs for obtaining equivalent sets of difference equations.

6.3 BASIC STRUCTURES FOR IIR SYSTEMS

In Section 6.1 we introduced two alternative structures for implementing a linear time-invariant system with system function as in Eq. (6.8). In this section we present the signal flow graph representations of those systems, and we also develop several other commonly used equivalent flow graph network structures. Our discussion will make it clear that for any given rational system function, a wide variety of equivalent sets of difference equations or network structures exists. One consideration in the choice among these different structures is computational complexity. For example, in some digital implementations, structures with the fewest constant multipliers and the fewest delay branches are often most desirable. This is because multiplication is generally a time-consuming and costly operation in digital hardware and because each delay element corresponds to a memory register. Consequently, the reduction of the number of constant multipliers means an increase of speed, and a reduction in the number of delay elements means a reduction in memory requirements.

Other, more subtle tradeoffs arise in VLSI implementations, where chip area is often an important measure of efficiency. Modularity and simplicity of data transfer on the chip are also often very desirable in such implementations. In multiprocessor implementations, the most important considerations are often related to partitioning of the algorithm and communication requirements between processors. Another major consideration is the effects of finite register length and finite-precision arithmetic. These effects depend on the way in which the computations are organized, i.e., on the structure of the signal flow graph. Sometimes it is desirable to use a structure that does not have the minimum number of multipliers and delay elements if that structure is less sensitive to finite register length effects.

In this section we develop several of the most commonly used forms for implementing a linear time-invariant IIR system and obtain their flow graph representations.

6.3.1 Direct Forms

In Section 6.1, we obtained block diagram representations of the direct form I (Fig. 6.3) and direct form II or canonic direct form (Fig. 6.5) structures for a linear time-invariant system whose input and output satisfy a difference equation of the form

$$y[n] - \sum_{k=1}^{N} a_k y[n-k] = \sum_{k=0}^{M} b_k x[n-k], \tag{6.19}$$

with the corresponding rational system function

$$H(z) = \frac{\sum_{k=0}^{M} b_k z^{-k}}{1 - \sum_{k=1}^{N} a_k z^{-k}}. \tag{6.20}$$

In Fig. 6.10, the direct form I structure of Fig. 6.3 is shown using signal flow graph conventions, and Fig. 6.11 shows the signal flow graph representation of the direct form II structure of Fig. 6.5. Again, we have assumed for convenience that $N = M$. It should be noted that we have drawn the flow graph so that each node has no more than two inputs. A node in a signal flow graph may have any number of inputs, but, as indicated earlier, this two-input convention results in a graph that is more closely related to programs and architectures for implementing the computation of the difference equations represented by the graph.

Example 6.2

To illustrate the direct form I and direct form II structures, consider the system function

$$H(z) = \frac{1 + 2z^{-1} + z^{-2}}{1 - 0.75z^{-1} + 0.125z^{-2}}. \tag{6.21}$$

Since the coefficients in the direct form structures correspond directly to the coefficients of the numerator and denominator polynomials (taking into account the minus sign in the denominator of Eq. 6.20), we can draw these structures by inspection with reference to Figs. 6.10 and 6.11. The direct form I and direct form II structures for this example are shown in Figs. 6.12 and 6.13, respectively.

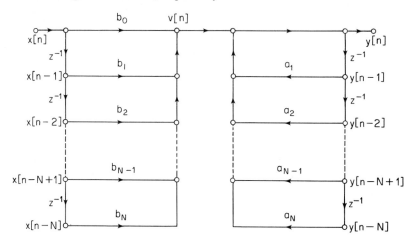

Figure 6.10   Signal flow graph of direct form I structure for an Nth-order system.

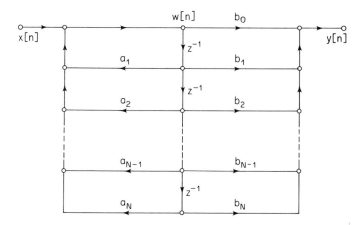

Figure 6.11   Signal flow graph of direct form II structure for an Nth-order system.

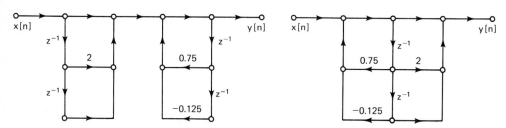

Figure 6.12   Direct form I structure for Example 6.2.

Figure 6.13   Direct form II structure for Example 6.2.

6.3.2 Cascade Form

The direct form structures were obtained directly from the system function $H(z)$ written as a ratio of polynomials in the variable z^{-1} as in Eq. (6.20). If we factor the numerator and denominator polynomials, we can express $H(z)$ in the form

$$H(z) = A \frac{\prod_{k=1}^{M_1} (1 - g_k z^{-1}) \prod_{k=1}^{M_2} (1 - h_k z^{-1})(1 - h_k^* z^{-1})}{\prod_{k=1}^{N_1} (1 - c_k z^{-1}) \prod_{k=1}^{N_2} (1 - d_k z^{-1})(1 - d_k^* z^{-1})}, \tag{6.22}$$

where $M = M_1 + 2M_2$ and $N = N_1 + 2N_2$. In this expression, the first-order factors represent real zeros at g_k and real poles at c_k, and the second-order factors represent complex conjugate pairs of zeros at h_k and h_k^* and complex conjugate pairs of poles at d_k and d_k^*. This represents the most general distribution of poles and zeros when all the coefficients in Eq. (6.20) are real. Equation (6.22) suggests a class of structures consisting of a cascade of first- and second-order systems. There is considerable freedom in the choice of composition of the subsystems and in the order in which the subsystems are cascaded. In practice, however, it is often desirable to implement the cascade realization using a minimum of storage and computation. A modular structure that is advantageous for many types of implementations is obtained by combining pairs of real factors and complex conjugate pairs into second-order factors so that Eq. (6.22) can be expressed as

$$H(z) = \prod_{k=1}^{N_s} \frac{b_{0k} + b_{1k} z^{-1} + b_{2k} z^{-2}}{1 - a_{1k} z^{-1} - a_{2k} z^{-2}}, \tag{6.23}$$

where $N_s = \lfloor (N+1)/2 \rfloor$ is the largest integer contained in $(N+1)/2$. In writing $H(z)$ in this form, we have assumed that $M \leq N$ and that the real poles and zeros have been combined in pairs. If there are an odd number of real zeros, one of the coefficients b_{2k} will be zero. Likewise, if there are an odd number of real poles, one of the coefficients a_{2k} will be zero. The individual second-order sections can be implemented using either of the direct form structures; however, the previous discussion shows that we can implement a cascade structure with a minimum number of multiplications and a minimum number of delay elements if we use the direct form II structure for each second-order section. A cascade structure for a sixth-order system using three direct form II second-order sections is shown in Fig. 6.14. The difference equations represented by a general cascade of direct form II second-order sections are of the form

$$y_0[n] = x[n], \tag{6.24a}$$

$$w_k[n] = a_{1k} w_k[n-1] + a_{2k} w_k[n-2] + y_{k-1}[n], \qquad k = 1, 2, \ldots, N_s, \tag{6.24b}$$

$$y_k[n] = b_{0k} w_k[n] + b_{1k} w_k[n-1] + b_{2k} w_k[n-2], \qquad k = 1, 2, \ldots, N_s, \tag{6.24c}$$

$$y[n] = y_{N_s}[n]. \tag{6.24d}$$

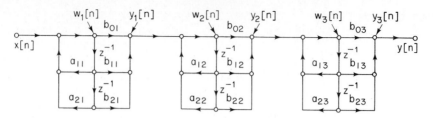

Figure 6.14 Cascade structure for a sixth-order system with a direct form II realization of each second-order subsystem.

It is easy to see that a variety of theoretically equivalent systems can be obtained by simply pairing the poles and zeros in different ways and by ordering the second-order sections in different ways. Indeed, if there are N_s second-order sections, there are N_s factorial pairings of the poles with zeros and N_s factorial orderings of the resulting second-order sections, or a total of $(N_s$ factorial$)^2$ different pairings and orderings. Although these all have the same overall system function and corresponding input/output relation when infinite-precision arithmetic is used, their behavior with finite-precision arithmetic can be quite different, as we will see in Section 6.9.

Example 6.3

Let us again consider the system function of Eq. (6.21). Since this is a second-order system, a cascade structure with direct form II second-order sections reduces to the structure of Fig. 6.13. Alternatively, to illustrate the cascade structure, we can use first-order systems by expressing $H(z)$ as a product of first-order factors, as in

$$H(z) = \frac{1 + 2z^{-1} + z^{-2}}{1 - 0.75z^{-1} + 0.125z^{-2}} = \frac{(1 + z^{-1})(1 + z^{-1})}{(1 - 0.5z^{-1})(1 - 0.25z^{-1})}. \tag{6.25}$$

Since all of the poles and zeros are real, a cascade structure with first-order sections has real coefficients. If the poles and/or zeros were complex, only a second-order section would have real coefficients. Figure 6.15 shows two equivalent cascade structures both of

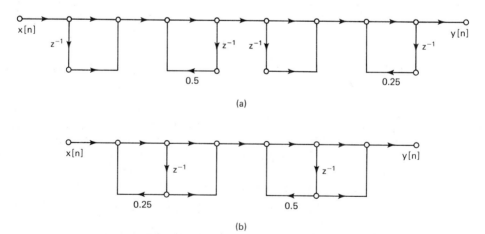

Figure 6.15 Cascade structures for Example 6.3. (a) Direct form I subsections. (b) Direct form II subsections.

which have the system function in Eq. (6.25). The difference equations represented by the flow graphs of Fig. 6.15 can be written down easily. Problem 6.5 is concerned with finding other equivalent system configurations.

A final comment should be made about our definition of the system function for the cascade form. As defined in Eq. (6.23), each second-order section has five constant multipliers. For comparison, let us assume that $M = N$ in $H(z)$ as given by Eq. (6.20) and furthermore assume that N is an even integer, so that $N_s = N/2$. Then the direct form I and II structures have $2N + 1$ constant multipliers, while the cascade form structure suggested by Eq. (6.23) has $5N/2$ constant multipliers. For the sixth-order system in Fig. 6.14, we require a total of 15 multipliers, while the equivalent direct forms would require a total of 13 multipliers. Another definition of the cascade form is

$$H(z) = b_0 \prod_{k=1}^{N_s} \frac{1 + \tilde{b}_{1k}z^{-1} + \tilde{b}_{2k}z^{-2}}{1 - a_{1k}z^{-1} - a_{2k}z^{-2}}, \tag{6.26}$$

where b_0 is the leading coefficient in the numerator polynomial of Eq. (6.20) and $\tilde{b}_{ik} = b_{ik}/b_{0k}$ for $i = 1, 2$ and $k = 1, 2, \ldots, N_s$. This form for $H(z)$ suggests a cascade of four-multiplier second-order sections, with a single overall gain constant b_0. This cascade form has the same number of constant multipliers as the direct form structures. As discussed in Section 6.9, the five-multiplier second-order sections are commonly used when implemented with fixed-point arithmetic because they make it possible to distribute the gain of the system and thereby control the size of signals at various critical points in the system. When floating-point arithmetic is used and dynamic range is not a problem, then the four-multiplier second-order sections can be used to decrease the amount of computation. Further simplification results for zeros on the unit circle. In this case, $\tilde{b}_{2k} = 1$, and we require only three multipliers per second-order section.

6.3.3 Parallel Form

As an alternative to factoring the numerator and denominator polynomials of $H(z)$, we can express a rational system function as a partial fraction expansion in the form

$$H(z) = \sum_{k=0}^{N_p} C_k z^{-k} + \sum_{k=1}^{N_1} \frac{A_k}{1 - c_k z^{-1}} + \sum_{k=1}^{N_2} \frac{B_k(1 - e_k z^{-1})}{(1 - d_k z^{-1})(1 - d_k^* z^{-1})}, \tag{6.27}$$

where $N = N_1 + 2N_2$. If $M \geq N$, then $N_p = M - N$; otherwise, the first summation in Eq. (6.27) is not included. If the coefficients a_k and b_k are real in Eq. (6.20), then the quantities A_k, B_k, C_k, c_k, and e_k are all real. In this form the system function can be interpreted as representing a parallel combination of first- and second-order IIR systems, with possibly N_p simple scaled delay paths. Alternatively, we may group the real poles in pairs so that $H(z)$ can be expressed as

$$H(z) = \sum_{k=0}^{N_p} C_k z^{-k} + \sum_{k=1}^{N_s} \frac{e_{0k} + e_{1k}z^{-1}}{1 - a_{1k}z^{-1} - a_{2k}z^{-2}}, \tag{6.28}$$

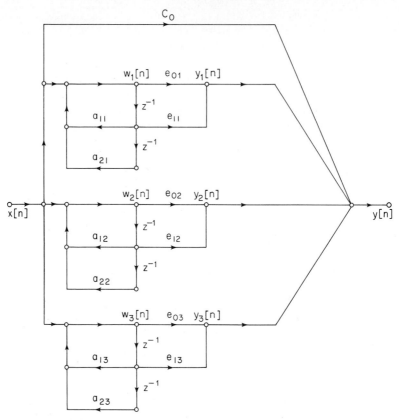

Figure 6.16   Parallel form structure for sixth-order system with the real and complex poles grouped in pairs.

where, as in the cascade form, $N_s = \lfloor (N + 1)/2 \rfloor$ is the largest integer contained in $(N + 1)/2$, and if $N_p = M - N$ is negative, the first sum is not present in Eq. (6.28). A typical example for $N = M = 6$ is shown in Fig. 6.16. The general difference equations for the parallel form with second-order direct form II sections are

$$w_k[n] = a_{1k}w_k[n - 1] + a_{2k}w_k[n - 2] + x[n] \qquad \text{for } k = 1, 2, \ldots, N_s, \qquad (6.29a)$$

$$y_k[n] = e_{01}w_k[n] + e_{1k}w_k[n - 1] \qquad \text{for } k = 1, 2, \ldots, N_s, \qquad (6.29b)$$

$$y[n] = \sum_{k=0}^{N_p} C_k x[n - k] + \sum_{k=1}^{N_s} y_k[n]. \qquad (6.29c)$$

If $M < N$, then the first summation in Eq. (6.29c) is not included.

Example 6.4

Again consider the system function used in Examples 6.2 and 6.3. For the parallel form, we must express $H(z)$ in the form of either Eq. (6.27) or Eq. (6.28). If we use second-order sections, $H(z)$ is

$$H(z) = \frac{1 + 2z^{-1} + z^{-2}}{1 - 0.75z^{-1} + 0.125z^{-2}} = 8 + \frac{-7 + 8z^{-1}}{1 - 0.75z^{-1} + 0.125z^{-2}}. \qquad (6.30)$$

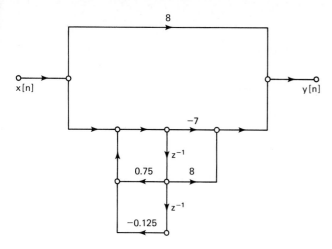

Figure 6.17 Parallel form structure for Example 6.4 using a second-order system.

The parallel form realization for this example with a second-order section is shown in Fig. 6.17.

Since all the poles are real, we can obtain an alternative parallel form realization by expanding $H(z)$ as

$$H(z) = 8 + \frac{18}{1 - 0.5z^{-1}} - \frac{25}{1 - 0.25z^{-1}}. \tag{6.31}$$

The resulting parallel form with first-order sections is shown in Fig. 6.18. As in the general case, the difference equations represented by both Figs. 6.17 and 6.18 can be written down by inspection.

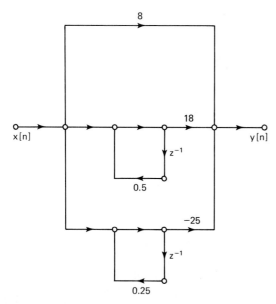

Figure 6.18 Parallel form structure for Example 6.4 using first-order systems.

6.3.4 Feedback in IIR Systems

All the flow graphs of this section have feedback loops; i.e., they have closed paths that begin at a node and return to that node by traversing branches only in the direction of their arrowheads. Such a structure in the flow graph implies that a node variable in a loop depends directly or indirectly on itself. A simple example is shown in Fig. 6.19(a), which represents the difference equation

$$y[n] = ay[n-1] + x[n]. \tag{6.32}$$

Such loops are necessary (but not sufficient) to generate infinitely long impulse responses. This can be seen if we consider a network with no loops. In this case, any path from the input to the output can pass through each delay element only once. Therefore, the longest delay between the input and output would occur for a path that passes through all of the delay elements in the network. Thus, for a network with no loops, the impulse response is no longer than the total number of delay elements in the network. From this we conclude that if a network has no loops, then the system function has only zeros (except for poles at $z = 0$) and the number of zeros can be no more than the number of delay elements in the network.

Returning to the simple example of Fig. 6.19(a), we see that when the input is the impulse sequence, the single-input sample continually recirculates in the feedback loop with either increasing ($|a| > 1$) or decreasing ($|a| < 1$) amplitude due to multiplication by the constant a so that the impulse response is $h[n] = a^n u[n]$. This is the way that feedback can create an infinitely long impulse response.

If a system function has poles, a corresponding block diagram or signal flow graph will have feedback loops. On the other hand, neither poles in the system

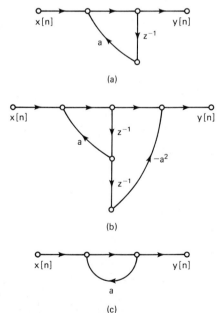

(a)

(b)

(c)

Figure 6.19 (a) System with feedback loop. (b) FIR system with feedback loop. (c) Noncomputable system.

function nor loops in the network are sufficient for the impulse response to be infinitely long. Figure 6.19(b) shows a network with a loop, but with an impulse response of finite length. This is because the pole of the system function cancels with a zero; i.e., for Fig. 6.19(b),

$$H(z) = \frac{1 - a^2 z^{-2}}{1 - az^{-1}} = \frac{(1 - az^{-1})(1 + az^{-1})}{1 - az^{-1}} = 1 + az^{-1}. \tag{6.33}$$

The impulse response of this system is $x[n] = \delta[n] + a\delta[n-1]$. This system is a simple example of a general class of FIR systems called *frequency sampling systems*. This class of systems is considered in more detail in Problems 6.16 and 6.17.

Loops in a network pose special problems in implementing the computations implied by the network. As we have discussed, it must be possible to compute the node variables in a network in sequence such that all necessary values are available when needed. In some cases, there is no way to order the computations so that the node variables of a flow graph can be computed in sequence. Such a network is called *noncomputable* (Crochiere and Oppenheim, 1975). A simple noncomputable network is shown in Fig. 6.19(c). The difference equation for this network is

$$y[n] = ay[n] + x[n]. \tag{6.34}$$

In this form, we cannot compute $y[n]$ because the right-hand side of the equation involves the quantity we wish to compute. The fact that a flow graph is noncomputable does *not* mean the equations represented by the flow graph cannot be solved; indeed, the solution to Eq. (6.34) is $y[n] = x[n]/(1 - a)$. It simply means that the flow graph does not represent a set of difference equations that can be solved successively for the node variables. The key to computability of a flow graph is that all loops must contain at least one unit delay element. Thus, in manipulating flow graphs representing implementations of linear time-invariant systems, we must be careful not to create delay-free loops. Problem 6.6 deals with a system having a delay-free loop. Problem 7.21 shows how a delay-free loop can be introduced.

6.4 TRANSPOSED FORMS

The theory of linear signal flow graphs provides a variety of procedures for transforming signal flow graphs into different forms while leaving the overall system function between input and output unchanged. One of these procedures, called *flow graph reversal* or *transposition*, leads to a set of transposed system structures that provide some useful alternatives to the structures discussed in the previous section.

Transposition of a flow graph is accomplished by reversing the directions of all branches in the network while keeping the branch transmittances as they were and reversing the roles of the input and output so that source nodes become sink nodes and vice versa. For single-input/single-output systems, the resulting flow graph has the same system function as the original graph if the input and output nodes are interchanged. Although we will not formally prove this result here,† we will demonstrate that it is valid with two examples.

Example 6.5

The first-order system corresponding to the flow graph in Fig. 6.20 has system function

$$H(z) = \frac{1}{1 - az^{-1}}. \tag{6.35}$$

To obtain the transposed form for this system, we reverse the directions of all the branch arrows, taking the output where the input was and injecting the input where the output was. The result is shown in Fig. 6.21. It is usually convenient to draw the transposed network with the input on the left and the output on the right as shown in Fig. 6.22. Comparing Figs. 6.20 and 6.22, we note that the only difference is that in Fig. 6.20, we multiply the *delayed* output sequence $y[n-1]$ by the coefficient a, whereas in Fig. 6.22 we multiply the output $y[n]$ by the coefficient a and then delay the resulting product. Since the two operations can be interchanged, we can conclude by inspection that the original system in Fig. 6.20 and the corresponding transposed system in Fig. 6.22 have the same system function.

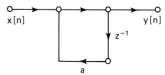

Figure 6.20   Flow graph of simple first-order system.

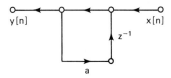

Figure 6.21   Transposed form of Fig. 6.20.

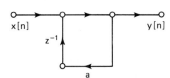

Figure 6.22   Structure of Fig. 6.21 redrawn with input on left.

In Example 6.5, it is straightforward to see that the original system and its transpose have the same system function. However, for more complicated graphs, the result is often not so obvious. This is illustrated by the next example.

---

† For a detailed discussion and proof of this theorem, see Crochiere and Oppenheim (1975). The theorem also follows directly from Mason's gain formula of signal flow graph theory (Mason and Zimmermann, 1960; Chow and Cassignol, 1962).

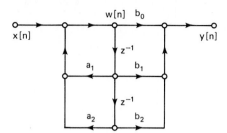

Figure 6.23   Direct form II structure for Example 6.6.

Example 6.6

Consider the basic second-order section depicted in Fig. 6.23. The corresponding difference equations for this system are

$$w[n] = a_1 w[n-1] + a_2 w[n-2] + x[n], \tag{6.36a}$$

$$y[n] = b_0 w[n] + b_1 w[n-1] + b_2 w[n-2]. \tag{6.36b}$$

The transposed flow graph is shown in Fig. 6.24; its corresponding difference equations are

$$v_0[n] = b_0 x[n] + v_1[n-1], \tag{6.37a}$$

$$y[n] = v_0[n], \tag{6.37b}$$

$$v_1[n] = a_1 y[n] + b_1 x[n] + v_2[n-1], \tag{6.37c}$$

$$v_2[n] = a_2 y[n] + b_2 x[n]. \tag{6.37d}$$

Equations (6.36) and (6.37) are different ways to organize the computation of the output samples $y[n]$ from the input samples $x[n]$, and it is not immediately clear that the two sets of difference equations are equivalent. One way to show this equivalence is to use the z-transform representations of both sets of equations, solve for the ratio $Y(z)/X(z) = H(z)$ in both cases, and compare the result. Another way is to substitute Eq. (6.37d) into Eq. (6.37c), substitute the result into Eq. (6.37a), and finally substitute that result into Eq. (6.37b). The result is

$$y[n] = a_1 y[n-1] + a_2 y[n-2] + b_0 x[n] + b_1 x[n-1] + b_2 x[n-2]. \tag{6.38}$$

Since the network of Fig. 6.23 is a direct form II structure, it is easily seen that the input and output of the system in Fig. 6.23 also satisfies the difference equation (6.38). Therefore, for initial-rest conditions, the systems in Figs. 6.23 and 6.24 are equivalent.

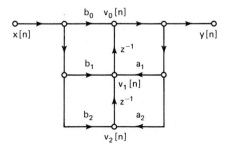

Figure 6.24   Transposed direct form II structure for Example 6.6.

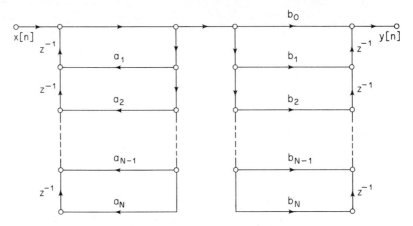

Figure 6.25 General flow graph resulting from applying the transposition theorem to the direct form I structure of Fig. 6.10.

The transposition theorem can be applied to any of the structures that we have discussed so far. For example, the result of applying this theorem to the direct form I structure of Fig. 6.10 is shown in Fig. 6.25, and, similarly, the structure obtained by transposing the direct form II structure of Fig. 6.11 is shown in Fig. 6.26. Clearly, if a signal flow graph configuration is transposed, the number of delay branches and the number of coefficients remain the same. Thus, the transposed direct form II structure is also a canonic structure.

An important point is evident by comparing Figs. 6.11 and 6.26. Whereas the direct form II structure implements the poles first and then the zeros, the transposed direct form II structure implements the zeros first and then the poles. These differences can become important in the presence of quantization in finite-precision digital implementations or in the presence of noise in discrete-time analog implementations.

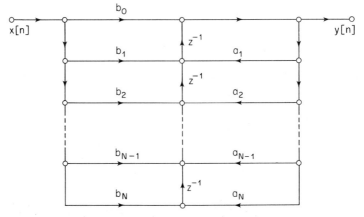

Figure 6.26 General flow graph resulting from applying the transposition theorem to the direct form II structure of Fig. 6.11.

When the transposition theorem is applied to cascade or parallel structures, the individual second-order systems are replaced by transposed structures. For example, applying the transposition theorem to Fig. 6.14 results in a cascade of three transposed direct form II sections (like the one in Example 6.6) with the same coefficients as in Fig. 6.14, but with the order of the cascade reversed. A similar statement can be made about the transposition of Fig. 6.16.

The transposition theorem further emphasizes that an infinite variety of implementation structures exists for any given rational system function. The transposition theorem provides a simple procedure for generating new structures. We will discuss one more class of IIR structures in Section 6.6, but the problems of implementing systems with finite-precision arithmetic have motivated the development of many more classes of equivalent structures than we can discuss here. Thus we concentrate only on the most commonly used structures.

6.5 BASIC NETWORK STRUCTURES FOR FIR SYSTEMS

The direct, cascade, and parallel form structures discussed in Sections 6.3 and 6.4 are the most common basic structures for IIR systems. These structures were developed under the assumption that the system function had both poles and zeros. Although the direct and cascade forms for IIR systems include FIR systems as a special case, there are additional specific forms for FIR systems.

6.5.1 Direct Form

For causal FIR systems, the system function has only zeros (except for poles at $z = 0$), and, since the coefficients a_k are all zero, the difference equation of Eq. (6.9) reduces to

$$y[n] = \sum_{k=0}^{M} b_k x[n - k], \tag{6.39}$$

which we recognize as the discrete convolution of $x[n]$ with the impulse response

$$h[n] = \begin{cases} b_n & \text{for } n = 0, 1, \dots, M, \\ 0 & \text{otherwise.} \end{cases} \tag{6.40}$$

The direct form I and direct form II structures in Figs. 6.10 and 6.11 both reduce to the direct form FIR structure shown in Fig. 6.27. Because of the chain of delay elements

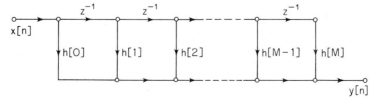

Figure 6.27   Direct form realization of an FIR system.

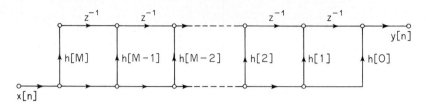

Figure 6.28 Transposition of the network of Fig. 6.27.

across the top of the diagram, this structure is also referred to as a *tapped delay line* structure or a *transversal filter* structure. As seen from Fig. 6.27, the signal at each tap along this chain is weighted by the appropriate coefficient (impulse response value) and the resulting products are summed to form the output $y[n]$.

The transposed direct form for the FIR case is obtained by applying the transposition theorem to Fig. 6.27 or, equivalently, by setting the coefficients a_k to zero in Fig. 6.25 or 6.26. The result is shown in Fig. 6.28.

6.5.2 Cascade Form

The cascade form for FIR systems is obtained by factoring the polynomial system function. That is, we represent $H(z)$ as

$$H(z) = \sum_{n=0}^{M} h[n]z^{-n} = \prod_{k=1}^{M_s} (b_{0k} + b_{1k}z^{-1} + b_{2k}z^{-2}), \tag{6.41}$$

where $M_s = \lfloor (M + 1)/2 \rfloor$ is the largest integer contained in $(M + 1)/2$. If M is odd, one of the coefficients b_{2k} will be zero, since $H(z)$ in that case would have an odd number of real zeros. The flow graph representing Eq. (6.41) is shown in Fig. 6.29, which is identical in form to Fig. 6.14, with the coefficients a_{1k} and a_{2k} all zero. Each of the second-order sections in Fig. 6.29 uses the direct form shown in Fig. 6.27. Another alternative is to use transposed direct form second-order sections or, equivalently, to apply the transposition theorem to Fig. 6.29.

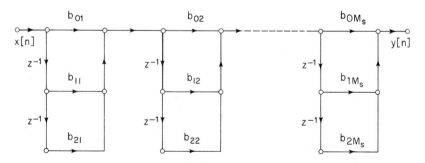

Figure 6.29 Cascade form realization of an FIR system.

6.5.3 Structures for Linear Phase FIR Systems

In Chapter 5, we showed that causal FIR systems have generalized linear phase if the impulse response satisfies the symmetry conditions

$$h[M - n] = h[n] \qquad \text{for } n = 0, 1, \ldots, M \qquad (6.42a)$$

or

$$h[M - n] = -h[n] \qquad \text{for } n = 0, 1, \ldots, M. \qquad (6.42b)$$

With either of these conditions, the number of coefficient multipliers can be essentially halved. To see this, consider the following manipulations of the discrete convolution equation, assuming that M is an even integer corresponding to type I or type III systems:

$$y[n] = \sum_{k=0}^{M} h[k]x[n-k]$$

$$= \sum_{k=0}^{M/2-1} h[k]x[n-k] + h[M/2]x[n-M/2] + \sum_{k=M/2+1}^{M} h[k]x[n-k]$$

$$= \sum_{k=0}^{M/2-1} h[k]x[n-k] + h[M/2]x[n-M/2] + \sum_{k=0}^{M/2-1} h[M-k]x[n-M+k].$$

For type I systems, we use Eq. (6.42a) to obtain

$$y[n] = \sum_{k=0}^{M/2-1} h[k](x[n-k] + x[n-M+k]) + h[M/2]x[n-M/2]. \quad (6.43)$$

For type III systems, we use Eq. (6.42b) to obtain

$$y[n] = \sum_{k=0}^{M/2-1} h[k](x[n-k] - x[n-M+k]). \qquad (6.44)$$

For the case of M an odd integer, the corresponding equations are, for type II systems,

$$y[n] = \sum_{k=0}^{(M-1)/2} h[k](x[n-k] + x[n-M+k]) \qquad (6.45)$$

and, for type IV systems,

$$y[n] = \sum_{k=0}^{(M-1)/2} h[k](x[n-k] - x[n-M+k]). \qquad (6.46)$$

Equations (6.43)–(6.46) imply structures with either $M/2 + 1$, $M/2$, or $(M+1)/2$ coefficient multipliers rather than the M coefficient multipliers of the general direct form structure of Fig. 6.27. Figure 6.30 shows the structure implied by Eq. (6.43) and Fig. 6.31 shows the structure implied by Eq. (6.45).

In our discussion of linear phase systems in Section 5.7.3, we showed that the symmetry condition of Eq. (6.42) causes the zeros of $H(z)$ to occur in mirror-image pairs. That is, if z_0 is a zero of $H(z)$, then $1/z_0$ is also a zero of $H(z)$. Furthermore, if

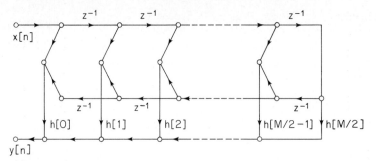

Figure 6.30 Direct form structure for an FIR linear phase system when M is an even integer.

$h[n]$ is real, then the zeros of $H(z)$ occur in complex conjugate pairs. As a consequence, real zeros not on the unit circle occur in reciprocal pairs. Complex zeros not on the unit circle occur in groups of four, corresponding to the complex conjugates and reciprocals. If a zero is on the unit circle, its reciprocal is also its conjugate. Consequently, complex zeros on the unit circle are conveniently grouped in pairs. Zeros at $z = \pm 1$ are their own reciprocal and complex conjugate. The four cases are summarized in Fig. 6.32, where the zeros at z_1, z_1^*, $1/z_1$, and $1/z_1^*$ are considered as a group of four. The zeros at z_2 and $1/z_2$ are considered as a group of two, as are the zeros at z_3 and z_3^*. The zero at z_4 is considered singly. If $H(z)$ has the zeros shown in Fig. 6.32 it can be factored into a product of first-, second-, and fourth-order factors. Each of these factors is a polynomial whose coefficients have the same symmetry as the coefficients of $H(z)$; i.e., each factor is a linear phase polynomial in z^{-1}. Therefore, the system can be implemented as a cascade of first-, second-, and fourth-order systems. For example, the system function corresponding to the zeros of Fig. 6.32 can be expressed as

$$H(z) = h[0](1 + z^{-1})(1 + az^{-1} + z^{-2})(1 + bz^{-1} + z^{-2})$$
$$\times (1 + cz^{-1} + dz^{-2} + cz^{-3} + z^{-4}), \tag{6.47}$$

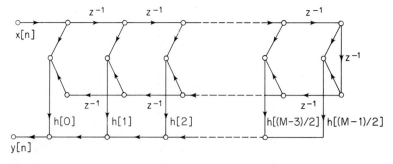

Figure 6.31 Direct form structure for an FIR linear phase system when M is an odd integer.

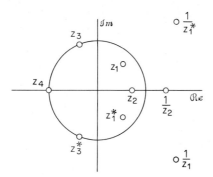

Figure 6.32 Symmetry of zeros for a linear phase FIR filter.

where

$$a = (z_2 + 1/z_2), \quad b = 2\mathcal{R}e\{z_3\}, \quad c = -2\mathcal{R}e\{z_1 + 1/z_1\}, \quad d = 2 + |z_1 + 1/z_1|^2.$$

This representation suggests a cascade structure consisting of linear phase elements. It can be seen that the order of this system function polynomial is $M = 9$, and the number of different coefficient multipliers is five. This is the same number $((M + 1)/2 = 5)$ of constant multipliers required for implementing the system in the linear phase direct form of Fig. 6.30. Thus, with no additional multiplications, we obtain a modular structure in terms of a cascade of short linear phase FIR systems.

6.6 LATTICE STRUCTURES

All of the basic structures for IIR and FIR systems that we have discussed so far have been developed either directly from the system function or from the input/output difference equation. Many other classes of structures with special properties have also been developed. One approach is based on state-variable representations and linear transformations (Crochiere and Oppenheim, 1975; Roberts and Mullis, 1987). Another useful approach is to develop digital filter structures that are analogous to analog filter structures that have certain desirable properties. The *wave digital filters* studied extensively by Fettweis (1986) are an example of this approach. Another interesting class of structures, called *lattice structures*, is motivated by the theory of autoregressive signal modeling (Markel and Gray, 1976).

Most of these other structures are difficult to discuss because of the extensive background necessary to understand their motivation and their properties. However, because of the many applications of lattice structures in signal modeling, spectrum estimation, and adaptive filtering, we will provide a brief introduction to that class of structures. Our approach will be to simply define the structure as a flow graph and then use flow graph manipulations and z-transform analysis to illuminate the properties of the structure.

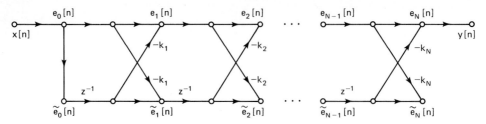

Figure 6.33   Signal flow graph of an FIR lattice system.

6.6.1 FIR Lattice

Figure 6.33 shows the signal flow graph of an Nth-order FIR lattice system. The difference equations represented by this network are

$$e_0[n] = \tilde{e}_0[n] = x[n], \tag{6.48a}$$

$$e_i[n] = e_{i-1}[n] - k_i\tilde{e}_{i-1}[n-1], \qquad i = 1, 2, \ldots, N, \tag{6.48b}$$

$$\tilde{e}_i[n] = -k_ie_{i-1}[n] + \tilde{e}_{i-1}[n-1], \qquad i = 1, 2, \ldots, N, \tag{6.48c}$$

$$y[n] = e_N[n]. \tag{6.48d}$$

To see that this is a useful and interesting structure, we must understand its properties. First, we recognize this as an FIR system. Specifically, in Fig. 6.33, all of the signal flow is from left to right. Since there are no loops, the impulse response has finite length. If the input is $x[n] = \delta[n]$, this impulse propagates immediately to the output along the top row of horizontal branches, with unity gain. All other paths from the input to the output pass through at least one delay element. Thus, $h[0] = 1$. An impulse at the input also propagates to the output through the bottom row of horizontal branches, encountering N delay elements and a final gain of $-k_N$. Therefore, $h[N] = -k_N$. All other paths from the input to the output zigzag between the top row of horizontal branches and the bottom row. Therefore, each of these paths passes through at least one unit delay element and at most $(N - 1)$ delay elements. Therefore, impulse response values for $0 < n < N$ are determined by sums of products of the k_i's. Thus, we conclude that the system function relating $Y(z)$ to $X(z)$ is of the form†

$$H(z) = \frac{Y(z)}{X(z)} = A(z) = \left[1 - \sum_{m=1}^{N} a_m z^{-m}\right]. \tag{6.49}$$

That is, $H(z)$ is a polynomial of order N. The leading coefficient is 1 because the only path from the input to the output that has no overall delay is the row of branches along the top of the graph, and the branch gains along that path are all unity. The network of Fig. 6.33 is called a lattice network because of its highly regular structure formed by cascading elementary sections of the general form shown in Fig. 6.34. The coefficients $\{k_i\}$ in Fig. 6.33 are referred to as the *k-parameters*.

---

† Our notation is suggestive of the notation used for the denominator of the system function of an IIR system. This is to facilitate the later use of the results developed in this section.

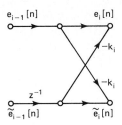

Figure 6.34 Basic lattice section for the FIR lattice system in Fig. 6.33.

The normal method of defining an FIR system is by its impulse response coefficients, which in the notation of Eq. (6.49) are

$$h[n] = \begin{cases} 1 & \text{for } n = 0, \\ -a_n & \text{for } n = 1, 2, \ldots, N, \\ 0 & \text{otherwise.} \end{cases} \qquad (6.50)$$

If this is to be a useful structure, we must be able to obtain a relation between the k-parameters and the impulse response coefficients. Equivalently, if we can find the system function for the system of Fig. 6.33, we will have found its impulse response as well. This can be done by exploiting the repetitive nature of the difference equations in Eq. (6.48) to obtain a recurrence formula for computing $A(z)$ in terms of the intermediate system functions

$$A_i(z) = \frac{E_i(z)}{E_0(z)} = \left[1 - \sum_{m=1}^{i} a_m^{(i)} z^{-m} \right] \qquad (6.51a)$$

and

$$\tilde{A}_i(z) = \frac{\tilde{E}_i(z)}{\tilde{E}_0(z)}. \qquad (6.51b)$$

By definition, $A_0(z) = \tilde{A}_0(z) = 1$. Also notice that since $X(z) = E_0(z) = \tilde{E}_0(z)$, then $A(z) = A_N(z)$.

To derive the desired recurrence formula, we begin by obtaining the z-transforms of Eqs. (6.48); i.e.,

$$E_0(z) = \tilde{E}_0(z) = X(z), \qquad (6.52a)$$

$$E_i(z) = E_{i-1}(z) - k_i z^{-1} \tilde{E}_{i-1}(z), \qquad i = 1, 2, \ldots, N, \qquad (6.52b)$$

$$\tilde{E}_i(z) = -k_i E_{i-1}(z) + z^{-1} \tilde{E}_{i-1}(z), \qquad i = 1, 2, \ldots, N, \qquad (6.52c)$$

$$Y(z) = E_N(z). \qquad (6.52d)$$

Now if Eqs. (6.51) are substituted into Eqs. (6.52), it can be shown by induction that $A_i(z)$ satisfies the recurrence formula

$$A_i(z) = A_{i-1}(z) - k_i z^{-i} A_{i-1}(z^{-1}) \qquad (6.53a)$$

and

$$\tilde{A}_i(z) = z^{-i} A_i(z^{-1}). \qquad (6.53b)$$

The recurrence formula for the polynomial coefficients that is implied by Eqs. (6.53) can be obtained by substituting Eq. (6.51a) into Eq. (6.53a) and collecting terms for the coefficient of z^{-m}. The result is

$$a_i^{(i)} = k_i, \tag{6.54a}$$

$$a_m^{(i)} = a_m^{(i-1)} - k_i a_{i-m}^{(i-1)}, \qquad m = 1, 2, \ldots, (i-1). \tag{6.54b}$$

This recursion is repeated for $i = 1, 2, \ldots, N$, and the final set of coefficients for $A(z) = A_N(z)$ is

$$a_m = a_m^{(N)}, \qquad m = 1, 2, \ldots, N. \tag{6.54c}$$

Equations (6.54) arise in the theory of autoregressive signal modeling (Markel and Gray, 1976; Makhoul, 1975; Rabiner and Schafer, 1978). In that context, the k-parameters are called *reflection coefficients* or *PARCOR coefficients*. When used in signal modeling, the k-parameters are estimated from a data signal, and Eqs. (6.54) are used to obtain the coefficients of the polynomial $A(z)$.

Given a set of k-parameters, we can find the system function and therefore the impulse response. On the other hand, suppose that we have the system function in polynomial form and we wish to find the k-parameters for the structure of Fig. 6.33. That is, we have

$$a_m^{(N)} = a_m, \qquad m = 1, 2, \ldots, N. \tag{6.55}$$

To find the k-parameters given the polynomial coefficients, we reverse the recursion to obtain successively the polynomials $A_{i-1}(z)$ in terms of $A_i(z)$. The k-parameters are obtained as a by-product of this recursion, which starts by observing from Eq. (6.54a) that

$$k_N = a_N^{(N)}. \tag{6.56a}$$

Then Eqs. (6.54) can be manipulated into the reverse recursion

$$k_i = a_i^{(i)}, \tag{6.56b}$$

$$a_m^{(i-1)} = \frac{a_m^{(i)} + k_i a_{i-m}^{(i)}}{1 - k_i^2}, \qquad m = 1, 2, \ldots, (i-1). \tag{6.56c}$$

The recursion is repeated for $i = N, (N-1), \ldots, 1$ until all the k_i's have been computed. The following simple example illustrates the use of Eq. (6.56).

Example 6.7

Consider the FIR system shown in Fig. 6.35(a). Its system function is

$$A(z) = (1 - 0.8jz^{-1})(1 + 0.8jz^{-1})(1 - 0.9z^{-1})$$

$$= 1 - 0.9z^{-1} + 0.64z^{-2} - 0.576z^{-3}. \tag{6.57}$$

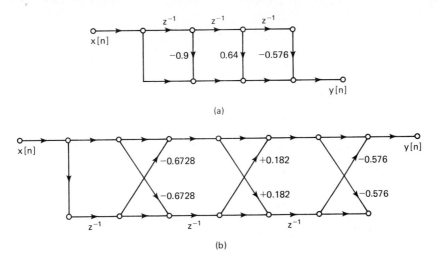

Figure 6.35   Flow graphs for Example 6.7. (a) Direct form. (b) Lattice form (coefficients rounded).

From Eq. (6.57) we see that the polynomial coefficients are $a_1^{(3)} = 0.9$, $a_2^{(3)} = -0.64$, and $a_3^{(3)} = 0.576$. Using Eqs. (6.56), we compute the k-parameters as follows:

$$k_3 = a_3^{(3)} = 0.576,$$

$$a_1^{(2)} = \frac{a_1^{(3)} + k_3 a_2^{(3)}}{1 - k_3^2} = 0.79518245,$$

$$a_2^{(2)} = \frac{a_2^{(3)} + k_3 a_1^{(3)}}{1 - k_3^2} = -0.18197491,$$

$$k_2 = a_2^{(2)} = -0.18197491,$$

$$a_1^{(1)} = \frac{a_1^{(2)} + k_2 a_1^{(2)}}{1 - k_2^2} = 0.67275747,$$

$$k_1 = a_1^{(1)} = 0.67275747.$$

The corresponding FIR lattice system is therefore as shown in Fig. 6.35(b). In Problem 6.19 the coefficients of the polynomial $A(z)$ are computed by using Eqs. (6.54) with the above values of the k-parameters and by tracing an impulse input through all paths in the networks of Fig. 6.35.

We have shown that the lattice structure of Fig. 6.33 is an FIR system. We have also shown how to compute its system function and how to obtain the parameters of the flow graph from the system function. It should be noted that the flow graph of Fig. 6.33 has $2N$ constant multipliers, although its impulse response has only N nonunity coefficients. Thus, such structures are not as efficient as the direct form FIR structure. As a result, the structure of Fig. 6.33 is not generally used for implementing FIR systems with constant coefficients. However, the structure has some desirable properties for implementing adaptive filters (Haykin, 1986) and it plays a central role in the theory of autoregressive signal modeling (Markel and Gray, 1976; Makhoul,

1975; Rabiner and Schafer, 1978). Most important, the FIR lattice system also suggests a lattice structure for IIR systems, which we consider next.

6.6.2 All-Pole Lattice

A lattice system with an all-pole system function $H(z) = 1/A(z)$ can be developed from the FIR lattice of the previous section. Suppose that we are given the output of Fig. 6.33, $e_N[n]$, and we wish to compute the input, $e_0[n]$. This can be done by solving Eqs. (6.48) backward, i.e., computing $e_{i-1}[n]$ from $e_i[n]$, and so on. Specifically, if we solve Eq. (6.48b) for $e_{i-1}[n]$ in terms of $e_i[n]$ and $\tilde{e}_{i-1}[n-1]$ we obtain the set of difference equations†

$$e_N[n] = x[n], \tag{6.58a}$$

$$e_{i-1}[n] = e_i[n] + k_i \tilde{e}_{i-1}[n-1], \qquad i = N, (N-1), \ldots, 1, \tag{6.58b}$$

$$\tilde{e}_i[n] = -k_i e_{i-1}[n] + \tilde{e}_{i-1}[n-1], \qquad i = N, (N-1), \ldots, 1, \tag{6.58c}$$

$$y[n] = e_0[n] = \tilde{e}_0[n]. \tag{6.58d}$$

Equations (6.58b) and (6.58c) are represented by the lattice section flow graph shown in Fig. 6.36, and the end-to-end connection of such sections with the input $x[n] = e_N[n]$ gives the output $y[n] = e_0[n] = \tilde{e}_0[n]$, as shown in Fig. 6.37; i.e., the system of Fig. 6.37 is the *inverse* system to the FIR lattice of Fig. 6.33. As a result, it follows that the system function for the system of Fig. 6.37 is $H(z) = 1/A(z)$, where $A(z)$ is the system function of the FIR lattice in Fig. 6.33. Thus, since $A(z)$ is a polynomial of order N, the system function of the system in Fig. 6.37 has only poles (except for zeros at $z = 0$).

Now all the analysis of Section 6.6.1 applies to the *denominator* of the system function. Given the k-parameters for Fig. 6.37, we can find the coefficients of the

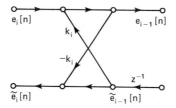

Figure 6.36 Basic lattice section for IIR lattice system.

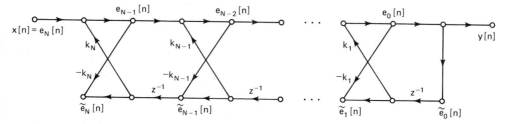

Figure 6.37 Flow graph for general all-pole IIR lattice system (inverse of Fig. 6.33).

† Note that by convention we label the input to the system as $x[n]$ and the output as $y[n]$.

denominator polynomial using the recursion of Eqs. (6.54), and given an all-pole system function $1/A(z)$, we can find the corresponding k-parameters for Fig. 6.37 using Eqs. (6.56).

Example 6.8

As an example of an IIR lattice system, consider the system function

$$H(z) = \frac{1}{1 - 0.9z^{-1} + 0.64z^{-2} - 0.576z^{-3}}, \qquad (6.59)$$

which is the reciprocal of the system function of the FIR system in Example 6.7. Figure 6.38(a) shows the direct form realization of this system, while Fig. 6.38(b) shows the equivalent IIR lattice system using the k-parameters obtained by the analysis of Example 6.7.

Since Fig. 6.37 is an IIR system, we must be concerned about the location of the roots of the polynomial $A(z)$. It can be shown that a necessary and sufficient condition for all the roots of $A(z)$ to be *inside* the unit circle is $|k_i| < 1$, $i = 1, \ldots, N$ (Markel and Gray, 1976). Therefore, if we are given an arbitrary system function with denominator polynomial $A(z)$, we can test for stability of the system by using Eqs. (6.56) to find the corresponding k-parameters, and if any of the k_i have magnitudes greater than or equal to unity, the system is unstable. In fact, even if we do not seek a lattice structure implementation of the system, the recursion of Eqs. (6.56) can be used as a stability test.

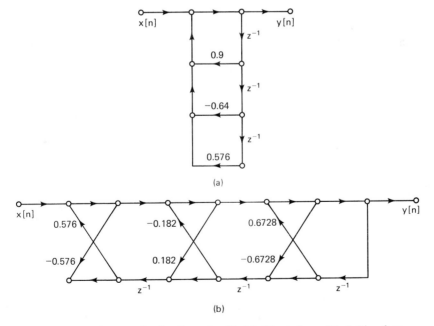

Figure 6.38 Flow graphs for Example 6.8. (a) Direct form. (b) Lattice form (coefficients rounded).

6.6.3 Normalized Lattice

Other useful all-pole lattice systems can be derived from the system of Fig. 6.37. Figure 6.37 consists of a chain of modules like Fig. 6.36, each of which computes the rightward directed output $e_{i-1}[n]$ as a function of $e_i[n]$ and $\tilde{e}_{i-1}[n-1]$ (Eq. 6.58b) and the leftward directed output $\tilde{e}_i[n]$ as a function of $e_{i-1}[n]$ and $\tilde{e}_{i-1}[n-1]$ (Eq. 6.58c). A new modular block can be found by solving for the required outputs $e_{i-1}[n]$ and $\tilde{e}_i[n]$ both as functions of $e_i[n]$ and $\tilde{e}_{i-1}[n-1]$. This is achieved by substituting Eq. (6.58b) into Eq. (6.58c) and by retaining Eq. (6.58b) and the resulting new equation as the difference equations for the ith modular block of the lattice. The resulting difference equations are

$$e_N[n] = x[n], \tag{6.60a}$$

$$e_{i-1}[n] = e_i[n] + k_i \tilde{e}_{i-1}[n-1], \qquad i = N, (N-1), \ldots, 1, \tag{6.60b}$$

$$\tilde{e}_i[n] = -k_i e_i[n] + (1 - k_i^2)\tilde{e}_{i-1}[n-1], \qquad i = N, (N-1), \ldots, 1, \tag{6.60c}$$

$$y[n] = e_0[n] = \tilde{e}_0[n]. \tag{6.60d}$$

These equations can be represented by the block diagram of Fig. 6.39 where, for the system of Eqs. (6.60), the ith section is represented by the signal flow graph of Fig. 6.40(a).

Still another lattice system can be obtained by observing that when the lattice sections of Fig. 6.40(a) are used in Fig. 6.39, all of the feedback loops that start at some node in the graph and return to that node consist of some portion of the leftward-directed bottom path and an equal portion of the rightward-directed upper path. The total loop gain remains the same if the multipliers on the bottom, $(1 - k_i^2)$, are factored into two parts and one part is moved to the opposite branch along the top of the graph. One factorization is suggested by the fact that for stable systems, $|k_i| < 1$. Thus using the Pythagorean theorem, we can define

$$\sin \theta_i = k_i, \tag{6.61a}$$

$$\cos \theta_i = \sqrt{1 - k_i^2}. \tag{6.61b}$$

This leads to the lattice section depicted in Fig. 6.40(b). When sections of this form are used in Fig. 6.39, the resulting system is called a *normalized lattice* system (Gray and Markel, 1975). Because the quantities $\cos \theta_i$ have been moved up to the top row

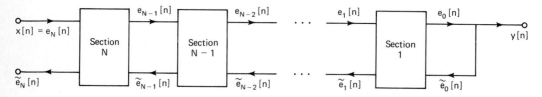

Figure 6.39 General representation of all-pole IIR lattice system.

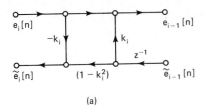

(a)

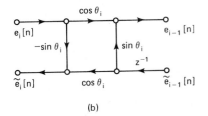

$$\sin \theta_i = k_i$$
$$\cos \theta_i = (1 - k_i^2)^{1/2}$$

(b)

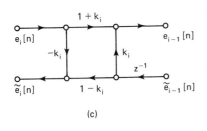

Figure 6.40 Basic lattice sections for use in Fig. 6.39. (a) Three-multiplier form. (b) Four-multiplier, normalized form. (c) Four-multiplier, Kelly-Lochbaum form.

(c)

of branches, the overall gain of the system is changed by the product of all these terms, but the poles remain the same. Therefore, the system function of the normalized lattice is

$$H(z) = \frac{\prod\limits_{i=1}^{N} \cos \theta_i}{A(z)}, \qquad (6.62)$$

where $A(z)$ is related to the k-parameters as described before. An interesting feature of the normalized lattice sections is that the two outputs $e_{i-1}[n]$ and $\tilde{e}_i[n]$ are related to the two inputs $e_i[n]$ and $\tilde{e}_{i-1}[n]$ by a complex multiplication of the form

$$e_{i-1}[n] + j\tilde{e}_i[n] = e^{-j\theta_i}(e_i[n] + j\tilde{e}_{i-1}[n-1]). \qquad (6.63)$$

This property reflects the gain normalization of the sections of Fig. 6.40(b) since it follows from Eq. (6.63) that $|e_{i-1}[n] + j\tilde{e}_i[n]| = |e_i[n] + j\tilde{e}_{i-1}[n-1]|$ for all sections of the system.

Another manipulation of the basic section of Fig. 6.40(a) is based on the factorization $(1 - k_i^2) = (1 + k_i)(1 - k_i)$. This leads to a basic section of the form shown in Fig. 6.40(c). When these sections are used in Fig. 6.39, the resulting lattice system is called the *Kelly-Lochbaum structure*. This structure was first derived as an acoustic tube model for speech synthesis (Kelly and Lochbaum, 1962). As in the case of the normalized lattice, it follows that the modification of moving the factor $(1 + k_i)$ to the top row leads to a system function of the form

$$H(z) = \frac{\prod_{i=1}^{N}(1 + k_i)}{A(z)}. \tag{6.64}$$

6.6.4 Lattice Systems with Poles and Zeros

The general IIR lattice system of Fig. 6.39 is restricted to an all-pole system function. However, both poles and zeros can be realized by the system of Fig. 6.41 (Gray and Markel, 1973, 1975). To see this, we recall that the IIR lattice was constructed in Section 6.6.2 so as to be the inverse of the FIR lattice, and since the notation for the IIR lattice is consistent with the notation for the FIR lattice, all the relationships derived in Section 6.6.1 hold for the IIR lattice as well. Thus, it follows from Eqs. (6.51b) and (6.53b) that

$$\tilde{E}_i(z) = z^{-i}A_i(z^{-1})\tilde{E}_0(z) = z^{-i}A_i(z^{-1})Y(z), \tag{6.65}$$

so that the system function $\tilde{H}_i(z) = \tilde{E}_i(z)/X(z)$ is the cascade of $z^{-i}A_i(z^{-1})$ with $1/A(z)$; i.e.,

$$\tilde{H}_i(z) = \frac{\tilde{E}_i(z)}{X(z)} = \frac{z^{-i}A_i(z^{-1})}{A(z)}. \tag{6.66}$$

(It is worth mentioning that the system function $\tilde{H}_N(z)$ relating the output $\tilde{E}_N(z)$ to the input $X(z)$ is an allpass system function. Thus, as a side benefit, we have shown that the IIR lattice structure can be used to implement allpass systems.)

The system function for the system in Fig. 6.41 is therefore

$$H(z) = \frac{Y(z)}{X(z)} = \sum_{i=0}^{N} \frac{c_i z^{-i}A_i(z^{-1})}{A(z)} = \frac{B(z)}{A(z)}. \tag{6.67}$$

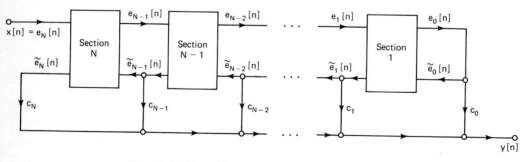

Figure 6.41   General lattice structure for realizing both poles and zeros.

From the properties of the ith-order polynomials, $A_i(z^{-1})$, it follows that $B(z)$ is an Nth-order polynomial of the form

$$B(z) = \sum_{m=0}^{N} b_m z^{-m}, \tag{6.68}$$

where the coefficients $\{b_m\}$ satisfy the equations

$$b_m = c_m - \sum_{i=m+1}^{N} c_i a_{i-m}^{(i)}. \tag{6.69}$$

Thus to obtain a lattice realization for a general rational system function we must obtain the k-parameters from $A(z)$ using the recursion of Eqs. (6.56), and Eq. (6.69) must be solved for the c-parameters required to realize the numerator $B(z)$. The preceding discussion assumes that the lattice sections of Fig. 6.36 or Fig. 6.40(a) are used in Fig. 6.41. If the modified sections of Fig. 6.40(b) or 6.40(c) are used, the c-parameters must be appropriately modified.

Example 6.9

Suppose that we wish to obtain a lattice structure for the system function

$$H(z) = \frac{1 + 3z^{-1} + 3z^{-2} + z^{-3}}{1 - 0.9z^{-1} + 0.64z^{-2} - 0.576z^{-3}}. \tag{6.70}$$

The k-parameters are obtained as in Examples 6.7 and 6.8. The c-parameters are obtained using Eq. (6.69) as follows:

$$c_3 = b_3 = 1,$$
$$c_2 = b_2 + c_3 a_1^{(3)} = 3.9,$$
$$c_1 = b_1 + c_2 a_1^{(2)} + c_3 a_2^{(3)} = 5.46121156,$$
$$c_0 = b_0 + c_1 a_1^{(1)} + c_2 a_2^{(2)} + c_3 a_3^{(3)} = 4.54036872.$$

Figure 6.42(a) shows the direct form II structure for this system, and Fig. 6.42(b) shows the equivalent lattice structure.

All of the lattice structures discussed in this section require more computation than equivalent direct or cascade form implementations. The lattice section in Fig. 6.36 requires two multipliers per section, that in Fig. 6.40(a) requires three multipliers per section, and those in Figs. 6.40(b) and 6.40(c) require four multipliers per section for a total of $2N$, $3N$, and $4N$ multiplications, respectively, to realize the N poles of the system function. Direct and cascade forms would require only N multiplications to realize N poles. What is gained in the lattice structures is decreased sensitivity to quantization effects in the implementation of the system. The remaining sections of this chapter explore these issues.

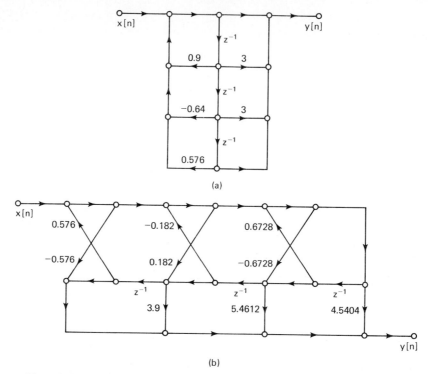

Figure 6.42 Flow graphs for Example 6.9. (a) Direct form II. (b) Lattice form (coefficients rounded).

6.7 OVERVIEW OF FINITE-PRECISION NUMERICAL EFFECTS

We have seen that a particular linear time-invariant discrete-time system can be implemented by a variety of computational structures. One motivation for considering alternatives to the simple direct form structures is that different structures that are theoretically equivalent may behave differently when implemented with finite numerical precision. In this section we give a brief introduction to the major numerical problems in implementing discrete-time systems. More detailed analysis of these finite-wordlength effects is given in Sections 6.8–6.10.

6.7.1 Number Representations

In theoretical analysis of discrete-time systems, we generally assume that signal values and system coefficients are represented in the real number system. With analog discrete-time systems, the limited precision of circuit components makes it difficult to realize coefficients exactly. When implementing digital signal processing systems, we must represent signals and coefficients in some digital number system that must always be of finite precision. Most digital computers or special-purpose hardware use a binary number system.

The problem of finite numerical precision has already been discussed in Section 3.7.2 in the context of A/D conversion. We showed there that the output samples from an A/D converter are quantized and thus can be represented by fixed-point binary numbers. For compactness and simplicity in implementing arithmetic, one of the bits of the binary number is assumed to indicate the algebraic sign of the number. Formats such as *sign and magnitude*, *one's complement*, and *two's complement* are possible, but two's complement is most common.† A real number can be represented with infinite precision in two's-complement form as

$$x = X_m \left(-b_0 + \sum_{i=1}^{\infty} b_i 2^{-i} \right), \tag{6.71}$$

where X_m is an arbitrary scale factor and the b_i's are either 0 or 1. The quantity b_0 is referred to as the *sign bit*. If $b_0 = 0$, then $0 \leq x \leq X_m$, and if $b_0 = 1$, then $-X_m \leq x < 0$. Thus, any real number whose magnitude is less than or equal to X_m can be represented by Eq. (6.71). An arbitrary real number x would require an infinite number of bits for its exact binary representation. As we saw in the case of A/D conversion, if we use only a finite number of bits $(B + 1)$, then the representation of Eq. (6.71) must be modified to

$$\hat{x} = Q_B[x] = X_m \left(-b_0 + \sum_{i=1}^{B} b_i 2^{-i} \right) = X_m \hat{x}_B. \tag{6.72}$$

The resulting binary representation is quantized so that the smallest difference between numbers is

$$\Delta = X_m 2^{-B}. \tag{6.73}$$

In this case, the quantized numbers are in the range $-X_m \leq \hat{x} < X_m$. The fractional part of $\hat{x}$ can be represented with the positional notation

$$\hat{x}_B = b_{0\diamond}b_1 b_2 b_3 \cdots b_B, \tag{6.74}$$

where $\diamond$ represents the binary point.

The operation of quantizing a number to $(B + 1)$ bits can be implemented by rounding or by truncation, but in either case quantization is a nonlinear memoryless operation. Figures 6.43(a) and 6.43(b) show the input/output relation for two's-complement rounding and truncation, respectively, for the case $B = 2$. In considering the effects of quantization, we often define the *quantization error* as

$$e = Q_B[x] - x. \tag{6.75}$$

For the case of two's-complement rounding, $-\Delta/2 < e \leq \Delta/2$, and for two's-complement truncation, $-\Delta < e \leq 0$.‡

If a number is larger than X_m (overflow), we must implement some method of determining the quantized result. In the two's-complement arithmetic system, this

---

† A detailed description of binary number systems and corresponding arithmetic is given by Knuth (1981).

‡ Note that Eq. (6.72) also represents the result of rounding or truncating any $(B_1 + 1)$-bit binary representation, where $B_1 > B$. In this case Δ would be replaced by $(\Delta - X_m 2^{-B_1})$ in the bounds on the size of the quantization error.

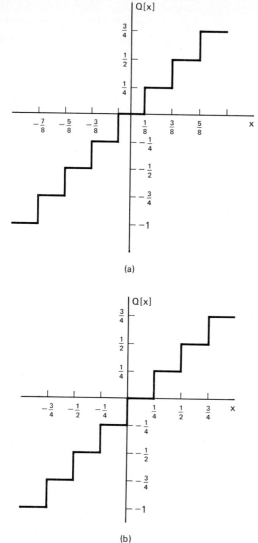

(a)

(b)

Figure 6.43   Nonlinear relationships representing two's-complement (a) rounding and (b) truncation for $B = 2$.

need arises when we add two numbers whose sum is greater than X_m. For example, consider the 4-bit two's-complement number 0111, which in decimal form is 7. If we add the number 0001, the carry propagates all the way to the sign bit so that the result is 1000, which in decimal form is -8. Thus, the resulting error can be very large when overflow occurs. Figure 6.44(a) shows the two's-complement quantizer, including the effect of regular two's-complement arithmetic overflow. An alternative, which is called *saturation overflow* or *clipping*, is shown in Fig. 6.44(b). This method of handling overflow is generally implemented for A/D conversion, and it sometimes is implemented for addition of two's-complement numbers. With this approach, the size of the error does not increase abruptly when overflow occurs; however, a disadvantage of this method of handling overflow is that it voids the following interesting and

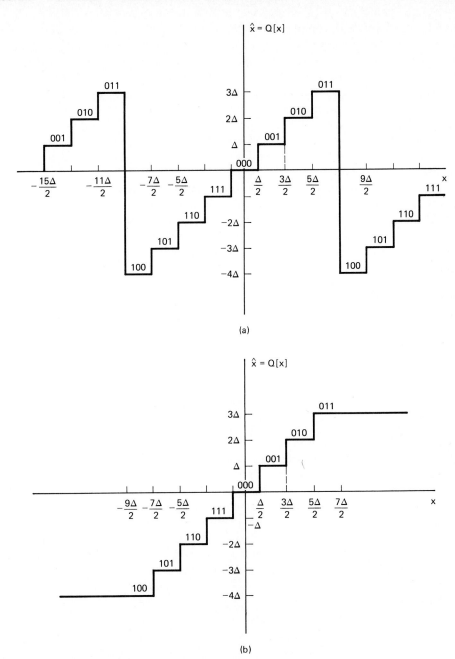

Figure 6.44 Two's-complement rounding. (a) Natural overflow. (b) Saturation.

useful property of two's-complement arithmetic: *If several two's-complement numbers whose sum would not overflow are added, then the result of two's-complement accumulation of these numbers is correct even though intermediate sums might overflow.* The properties of two's-complement arithmetic are reviewed in Problem 3.34.

As we discussed in Section 3.7.2, both quantization and overflow introduce errors in digital representations of numbers. Unfortunately, to minimize overflow while keeping the number of bits the same, we must increase X_m and thus increase the size of quantization errors proportionately. Thus, to simultaneously achieve wider dynamic range and lower quantization error, we must increase the number of bits in the binary representation.

So far, we have simply stated that the quantity X_m is an arbitrary scale factor; however, this factor has several useful interpretations. In A/D conversion, we considered X_m to be the full-scale amplitude of the A/D converter. In this case we think of it as a real number representing a voltage in the system. Thus, X_m serves as a calibration constant for relating binary numbers in the range $-1 \le \hat{x}_B < 1$ to analog signal amplitudes.

In signal processing implementations, it is common to assume that all signal variables and all coefficients are binary fractions. Thus, if we multiply a $(B + 1)$-bit signal variable by a $(B + 1)$-bit coefficient, the result is a $(2B + 1)$-bit fraction that can be conveniently reduced to $(B + 1)$ bits by rounding or truncating the least significant bits. With this convention, the quantity X_m can be thought of as a scale factor that allows the representation of numbers that are greater than 1 in magnitude. For example, in fixed-point computations, it is common to assume that each binary number has a scale factor of the form $X_m = 2^c$. Thus, a value $c = 2$ implies that the binary point is actually located between b_2 and b_3 of the binary word in Eq. (6.74). Often this scale factor is not explicitly represented. Instead, it is implicit in the implementation program or hardware architecture.

Still another way of thinking about the scale factor X_m leads to the *floating-point representations*, where the exponent c of the scale factor is called the *characteristic* and the fractional part $\hat{x}_B$ is called the *mantissa*. Both the characteristic and the mantissa are represented explicitly as binary numbers in floating-point arithmetic systems. Floating-point representations provide a convenient means for maintaining both a wide dynamic range and small quantization noise; however, quantization error manifests itself in a somewhat different way. We will briefly discuss the implications of floating-point quantization error in Section 6.9.5.

6.7.2 Quantization in Implementing Systems

Numerical quantization affects the implementation of linear time-invariant discrete-time systems in several ways. As a simple illustration, consider Fig. 6.45(a), which shows a block diagram for a system in which a bandlimited continuous-time signal $x_c(t)$ is sampled to obtain the sequence $x[n]$, which is the input to a linear time-invariant system whose system function is

$$H(z) = \frac{1}{1 - az^{-1}}. \tag{6.76}$$

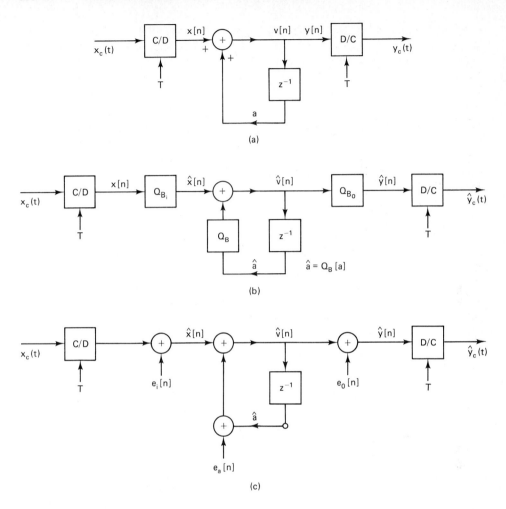

Figure 6.45   Implementation of discrete-time filtering of an analog signal. (a) Ideal system. (b) Nonlinear model. (c) Linearized model.

The output of this system, $y[n]$, is converted by ideal bandlimited interpolation to the bandlimited signal $y_c(t)$.

A more realistic model is shown in Fig. 6.45(b). In a practical setting, sampling would be done with an A/D converter with finite precision of $(B_i + 1)$ bits. The system would be implemented with binary arithmetic of $(B + 1)$ bits. The coefficient a in Fig. 6.45(a) would be represented with $(B + 1)$ bits of precision. Also, the delayed variable $\hat{v}[n - 1]$ would be stored in a $(B + 1)$-bit register, and when the $(B + 1)$-bit number $\hat{v}[n - 1]$ is multiplied by the $(B + 1)$-bit number $\hat{a}$, the resulting product will be $(2B + 1)$ bits in length. If we assume that a $(B + 1)$-bit adder is used, the product $\hat{a}\hat{v}[n - 1]$ must be quantized (i.e., rounded or truncated) to $(B + 1)$ bits before it can be added to the $(B_i + 1)$-bit input sample $\hat{x}[n]$. (Since it is generally true that $B_i < B$, the $(B_i + 1)$ bits of the input samples can be placed anywhere in the $(B + 1)$-bit binary word with appropriate sign extension.) The coefficient a has been quantized, so leaving aside the other quantization errors, the system response cannot in general be the same as in Fig. 6.45(a). Finally, the $(B + 1)$-bit samples $\hat{y}[n]$, computed by iterating the difference equation represented by the block diagram, would be converted to an analog signal by a $(B_o + 1)$-bit D/A converter. When $B_o < B$, the output samples must be quantized before D/A conversion.

Although the model of Fig. 6.45(b) is accurate, it is difficult to analyze. The system in Fig. 6.45(b) is nonlinear due to the quantizers and the possibility of overflow at the adder. Also, quantization errors are introduced at many points in the system. The effects of these errors are impossible to analyze precisely, since they depend on the input signal, which we generally consider to be unknown. Thus, we are forced to adopt several different approximate approaches to simplify the analysis of such systems.

The effect of quantizing the system parameters, such as the coefficient a in Fig. 6.45(a), is generally determined separately from the effects of quantization in data conversion or in implementing difference equations. That is, the ideal coefficients of a system function are replaced by their quantized values, and the resulting response functions are tested to see if quantization has degraded performance to unacceptable levels. For the example of Fig. 6.45, if the real number a is quantized to $(B + 1)$ bits, we must consider whether the resulting system with system function

$$\hat{H}(z) = \frac{1}{1 - \hat{a}z^{-1}} \tag{6.77}$$

is close enough to the desired system function $H(z)$ given by Eq. (6.76). Since there are only 2^{B+1} different $(B + 1)$-bit binary numbers, the pole of $H(z)$ can occur only at 2^{B+1} locations on the real axis of the z-plane. This type of analysis is discussed in more general terms in Section 6.8.

The nonlinearity of the system of Fig. 6.45(b) causes behavior that cannot occur in a linear system. Specifically, systems such as this can exhibit *zero-input limit cycles* whereby the output oscillates periodically when the input is zero. Limit cycles are caused both by quantization and by overflow. Although the analysis of such phenomena is difficult, some useful approximate results have been developed. Limit cycles are discussed in Section 6.10.

If care is taken in the design of a digital implementation, we can ensure that overflow occurs only rarely and quantization errors are small. Under these conditions, the system of Fig. 6.45(b) behaves very much like a linear system (with quantized coefficients) in which quantization errors are injected at the input and output and at internal points in the network where rounding or truncation occurs. Therefore, we can replace the model of Fig. 6.45(b) by the linearized model of Fig. 6.45(c), where the quantizers are replaced by additive noise sources (Gold and Rader, 1969; Jackson, 1970a, 1970b). Figure 6.45(c) is exactly equivalent to Fig. 6.45(b) if we know each of the noise sources exactly. However, as discussed in Section 3.7.3, useful results are obtained if we assume a random noise model for the quantization noise in A/D conversion. This same approach can be used in analyzing the effects of arithmetic quantization in digital implementations of linear systems. As seen in Fig. 6.45(c), each noise source injects a signal that is processed by different parts of the system, but since we assume that all parts of the system are linear, we can compute the overall effect by superposition. In Section 6.9, we illustrate this style of analysis for several important systems.

In the simple example of Fig. 6.45, there is little flexibility in the choice of structure. However, for higher-order systems, we have seen that there is a wide variety of choices. Some of the structures are less sensitive to coefficient quantization than others. Similarly, because different structures have different quantization noise sources and because these noise sources are filtered in different ways by the system, we will find that structures that are theoretically equivalent sometimes have vastly different performance when finite-precision arithmetic is used to implement them.

6.8 THE EFFECTS OF COEFFICIENT QUANTIZATION

Linear time-invariant discrete-time systems are generally used to perform a filtering operation. Methods for design of both FIR and IIR filters are discussed in Chapter 7. These design methods typically assume a particular form for the system function, and they compute the coefficients of the system with high (generally at least 32-bit floating-point) accuracy. The result of the filter design process is a system function for which we must choose an implementation structure (a set of difference equations) from infinitely many theoretically equivalent implementations. Although we are almost always interested in implementations that require the least hardware or software complexity, it is not always possible to base the choice of implementation structure on this criterion alone. As we will see in Section 6.9, the implementation structure determines the quantization noise generated internally in the system. Also, some structures are more sensitive than others to perturbations of the coefficients. As we pointed out in Section 6.7, the standard approach to the study of these issues is to treat them independently. In this section we consider the effects of quantizing the system parameters.

6.8.1 Effects of Coefficient Quantization in IIR Systems

When the parameters of a rational system function or corresponding difference equation are quantized, the poles and zeros of the system function move to new positions in the z-plane. Equivalently, the frequency response is perturbed from the original frequency response. If the system implementation structure is highly sensitive to perturbations of the coefficients, the resulting system may no longer meet the original design specifications, or an IIR system might even become unstable.

A detailed sensitivity analysis for the general case is complex and usually of limited value in specific cases. However, even though simulation of the system is generally necessary in specific cases, we can learn a great deal from a simple analysis of how the zeros and poles of the direct form structures are affected by quantization of

the coefficients of the difference equation. The system function representation corresponding to both direct forms is the ratio of polynomials

$$H(z) = \frac{\displaystyle\sum_{k=0}^{M} b_k z^{-k}}{1 - \displaystyle\sum_{k=1}^{N} a_k z^{-k}}. \tag{6.78}$$

The sets of coefficients $\{a_k\}$ and $\{b_k\}$ are the ideal infinite-precision coefficients in both
direct form implementation structures. If we quantize these coefficients, we obtain the
system function

$$\hat{H}(z) = \frac{\displaystyle\sum_{k=0}^{M} \hat{b}_k z^{-k}}{1 - \displaystyle\sum_{k=1}^{N} \hat{a}_k z^{-k}}, \tag{6.79}$$

where $\hat{a}_k = a_k + \Delta a_k$ and $\hat{b}_k = b_k + \Delta b_k$ are the quantized coefficients that differ from
the original coefficients by the quantization errors Δa_k and Δb_k.

Now consider the effect of quantizing the denominator coefficients on the
locations of the poles of the system. First, the quantization error in a given coefficient
affects *all* the poles of the system function. To quantify this effect, suppose that the
poles are all first order and that they are located at $z = z_i$ for $i = 1, 2, \ldots, N$ so that the
denominator of Eq. (6.78) can be expressed as

$$A(z) = 1 - \sum_{k=1}^{N} a_k z^{-k} = \prod_{j=1}^{N} (1 - z_j z^{-1}). \tag{6.80}$$

Furthermore, let us define the poles of $\hat{H}(z)$ as $z_i + \Delta z_i$ for $i = 1, 2, \ldots, N$. The error in
the location of the ith pole can be expressed in terms of the errors in the coefficients as

$$\Delta z_i = \sum_{k=1}^{N} \frac{\partial z_i}{\partial a_k} \Delta a_k, \qquad i = 1, 2, \ldots, N. \tag{6.81}$$

Using the two forms for $A(z)$ in Eq. (6.80) and the fact that

$$\left(\frac{\partial A(z)}{\partial z_i} \right)_{z=z_i} \frac{\partial z_i}{\partial a_k} = \left(\frac{\partial A(z)}{\partial a_k} \right)_{z=z_i}, \tag{6.82}$$

it follows that

$$\frac{\partial z_i}{\partial a_k} = \frac{z_i^{N-k}}{\displaystyle\prod_{j=1, j \neq i}^{N} (z_i - z_j)} \tag{6.83}$$

for $i = 1, 2, \ldots, N$ and $k = 1, 2, \ldots, N$. Equation (6.83) can be substituted into Eq.
(6.81) to determine the total change in a particular pole z_i due to all the changes in the
denominator coefficients. Equation (6.83) is also a measure of the sensitivity of the ith
pole to the quantization error in the kth coefficient in the denominator of $H(z)$. Since

the zeros of $H(z)$ depend only on the numerator coefficients, an analogous result can be derived for the sensitivity of the zeros to the quantization errors in the b_k's.

A result of this form was first used by Kaiser (1966) to show that if the poles (or zeros) are tightly clustered, it is possible that small errors in the denominator (numerator) coefficients can cause large shifts of the poles (zeros) for the direct form structures. This can be seen by considering the denominator of Eq. (6.83). Each factor $(z_i - z_j)$ may be represented as a vector in the z-plane, as shown in Fig. 6.46. The magnitude of the denominator of Eq. (6.83) is equal to the product of the lengths of the vectors from all the remaining poles to the pole location z_i. Thus, if poles (zeros) are tightly clustered as in Fig. 6.46(a), corresponding to a narrow

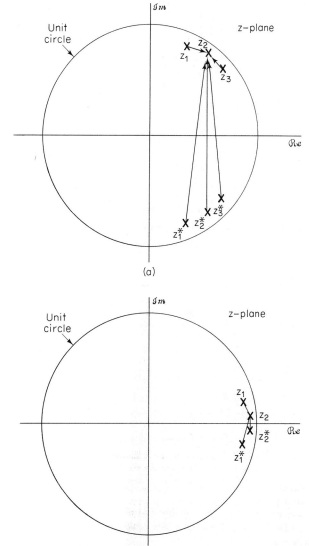

(a)

(b)

Figure 6.46 Representation of the factors of Eq. (6.83) as vectors in the z-plane for determining the sensitivity of the pole at z_2. (a) Narrow bandpass filter. (b) Narrow bandwidth lowpass filter.

bandpass filter, or as in Fig. 6.46(b), corresponding to a narrow-bandwidth lowpass filter, then we can expect the poles of the direct form structure to be quite sensitive to quantization errors in the coefficients. Furthermore, it is evident that the larger the number of clustered poles (zeros), the greater the sensitivity.

The cascade and parallel form system functions, which are given by Eq. (6.23) and (6.28) respectively, consist of combinations of second-order direct form systems. However, in both cases, each pair of complex conjugate poles is realized independently of all the other poles. Thus, the error in a particular pole pair is independent of its distance from the other poles of the system function. For the cascade form, the same argument holds for the zeros, since they are realized as independent second-order factors. Thus, the cascade form is generally much less sensitive to coefficient quantization than the equivalent direct form realization.

As seen in Eq. (6.28), the zeros of the parallel form system function are realized implicitly. They result implicitly from combining the quantized second-order sections to obtain a common denominator. Thus, a particular zero is affected by quantization errors in the numerator and denominator coefficients of *all* the second-order sections. However, for most practical filter designs, the parallel form is also found to be much less sensitive to coefficient quantization than the equivalent direct forms because the second-order subsystems are not extremely sensitive to quantization. In many practical filters, the zeros are often widely distributed around the unit circle or in some cases they may all be located at $z = \pm 1$. In the latter case, the zeros mainly provide much higher attenuation around frequencies $\omega = 0$ and $\omega = \pi$ than is specified, and thus movements of zeros away from $z = \pm 1$ do not significantly degrade the system performance.

Example 6.10

As an illustration of the effect of coefficient quantization, consider the example of a bandpass IIR elliptic filter. The filter was designed to meet the following specifications:

$$0.99 \le |H(e^{j\omega})| \le 1.01, \qquad 0.3\pi \le \omega \le 0.4\pi,$$

$$|H(e^{j\omega})| \le 0.01 \text{ (i.e., } -40 \text{ dB}), \qquad \omega \le 0.29\pi,$$

$$|H(e^{j\omega})| \le 0.01 \text{ (i.e., } -40 \text{ dB}), \qquad 0.41\pi \le \omega \le \pi.$$

A filter of order 12 is required to meet these very sharp cutoff narrowband specifications. In the filter design process, the coefficients of the cascade form were computed with 32-bit floating-point accuracy. These are defined as "unquantized." Table 6.1 gives the

TABLE 6.1 UNQUANTIZED CASCADE FORM COEFFICIENTS FOR A 12TH-ORDER ELLIPTIC FILTER

k	a_{1k}	a_{2k}	b_{0k}	b_{1k}	b_{2k}
1	0.738409	−0.850835	0.135843	0.026265	0.135843
2	0.960374	−0.860000	0.278901	−0.444500	0.278901
3	0.629449	−0.931460	0.535773	−0.249249	0.535773
4	1.116458	−0.940429	0.697447	−0.891543	0.697447
5	0.605182	−0.983693	0.773093	−0.425920	0.773093
6	1.173078	−0.986166	0.917937	−1.122226	0.917937

TABLE 6.2   16-BIT QUANTIZED CASCADE FORM COEFFICIENTS FOR A 12TH-ORDER ELLIPTIC FILTER

k	a_{1k}	a_{2k}	b_{0k}	b_{1k}	b_{2k}
1	24196×2^{-15}	-27880×2^{-15}	17805×2^{-17}	3443×2^{-17}	17805×2^{-17}
2	31470×2^{-15}	-28180×2^{-15}	18278×2^{-16}	-29131×2^{-16}	18278×2^{-16}
3	20626×2^{-15}	-30522×2^{-15}	17556×2^{-15}	-8167×2^{-15}	17556×2^{-15}
4	18292×2^{-14}	-30816×2^{-15}	22854×2^{-15}	-29214×2^{-15}	22854×2^{-15}
5	19831×2^{-15}	-32234×2^{-15}	25333×2^{-15}	-13957×2^{-15}	25333×2^{-15}
6	19220×2^{-14}	-32315×2^{-15}	15039×2^{-14}	-18387×2^{-14}	15039×2^{-14}

coefficients of the six second-order sections (as defined in Eq. 6.23). Figure 6.47(a) shows the log magnitude in dB and Fig. 6.47(b) shows the magnitude response in the passband (i.e., $0.3\pi \le \omega \le 0.4\pi$) for the unquantized system.

To illustrate the effect of coefficient quantization, the coefficients in Table 6.1 were quantized to 16-bit accuracy. The resulting coefficients are shown in Table 6.2. The fixed-point coefficients are shown as a decimal integer times a power of 2 scale factor. The binary representation would be obtained by converting the decimal integer to a binary number. In a fixed-point implementation, the scale factor would be represented only implicitly in the data shifts that would be necessary to line up the binary points of products prior to addition to other products. Notice that binary points of the coefficients are not all in the same location. For example, all the coefficients with scale factor 2^{-15} have their binary points between the sign bit, b_0, and the highest fractional bit, b_1, as shown in Eq. (6.74). However, numbers whose magnitudes do not exceed 0.5, such as the coefficient b_{02}, can be shifted left by one or more bit positions.† Thus, the binary point for b_{02} is actually to the left of the sign bit. On the other hand, numbers whose magnitudes exceed 1 but are less than 2, such as a_{16}, must have their binary points moved one position to the right, i.e., between b_1 and b_2 in Eq. (6.74).

Figure 6.47(c) shows the magnitude response in the passband for the quantized cascade form implementation. The frequency response is only slightly degraded in the passband region.

To obtain other equivalent structures, the cascade form system function must be rearranged into a different form. For example, if a parallel form structure is determined (by partial fraction expansion of the unquantized system function) and the resulting coefficients quantized to 16 bits as before, the frequency response in the passband is as shown in Fig. 6.47(d).

If the second-order factors in the numerator and denominator of the unquantized cascade form are multiplied together to form 12th-order numerator and denominator polynomials, and if the resulting coefficients are quantized to 16 bits and used in a direct form structure, the log magnitude of the frequency response is as shown in Fig. 6.47(e). In this case the passband response is virtually destroyed by the quantization. Comparing Fig. 6.47(e) with Fig. 6.47(a) shows that the widely separated zeros on either side of the passband have not moved off the unit circle. However, the unquantized system has a pair of zeros on the unit circle just at both stopband edges. These zeros, being clustered, move away from the unit circle with quantization and thus contribute to the overall degradation of the frequency response.

---

† The use of different binary point locations retains greater accuracy in the coefficients, but it complicates the programming or system architecture.

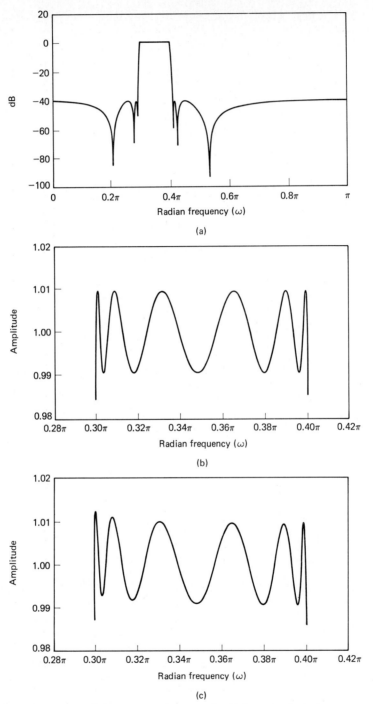

Figure 6.47 IIR coefficient quantization example. (a) Log magnitude for unquantized elliptic bandpass filter. (b) Passband for unquantized cascade case. (c) Passband for cascade structure with 16-bit coefficients.

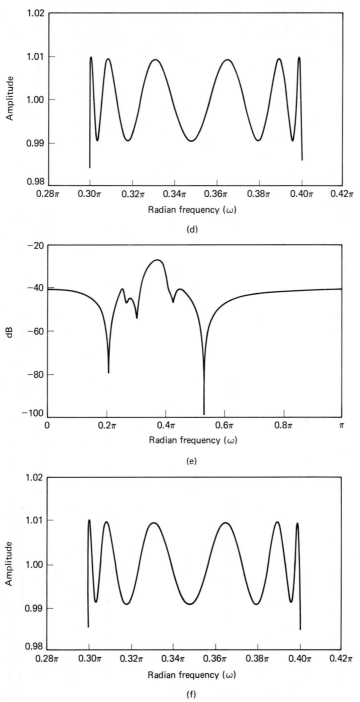

Figure 6.47 (*continued*) (d) Passband for parallel structure with 16-bit coefficients. (e) Log magnitude for direct form with 16-bit coefficients. (f) Passband for normalized lattice with 16-bit coefficients.

A final example is obtained by using Eqs. (6.56), (6.61), and (6.69) to compute a normalized lattice implementation for the system. The resulting magnitude response in the passband, plotted in Fig. 6.47(f), shows small degradation on the same order as the cascade and parallel forms.

Example 6.10 illustrates the robustness of the cascade, parallel, and normalized lattice forms to the effects of coefficient quantization, and it also illustrates the extreme sensitivity of the direct forms for high-order filters. Because of this sensitivity, the direct forms are rarely used for implementing anything other than second-order systems.† Since the cascade and parallel forms can be configured to require the same amount of memory and the same or only slightly more computation as the canonic direct form, these modular structures are the most commonly used. More complex structures like the normalized lattice may be more robust for very short wordlengths, but they require significantly more computation for systems of the same order.

Even for the second-order systems that are used to implement the cascade and parallel forms, there remains some flexibility to improve the robustness to coefficient quantization. Consider a complex conjugate pole pair implemented using the direct form, as in Fig. 6.48. With infinite-precision coefficients, this network has poles at $z = re^{j\theta}$ and $z = re^{-j\theta}$. However, if the coefficients $2r \cos \theta$ and $-r^2$ are quantized, only a finite number of different pole locations is possible. The poles must lie on a grid in the z-plane defined by the intersection of concentric circles (corresponding to the quantization of r^2) and vertical lines (corresponding to the quantization of $2r \cos \theta$). Such a grid is illustrated in Fig. 6.49(a) for 4-bit quantization (3 bits plus sign); i.e., r^2 is restricted to seven positive values and zero, while $2r \cos \theta$ is restricted to seven positive values, eight negative values, and zero. Figure 6.49(b) shows a denser grid obtained with 7-bit quantization (6 bits plus sign). The plots of Fig. 6.49 are, of course, symmetrically mirrored into each of the other quadrants of the z-plane. Notice that for the direct form, the grid is rather sparse around the real axis. Thus, poles located around $\theta = 0$ or $\theta = \pi$ may be shifted more than those around $\theta = \pi/2$. Of course, it is always possible that the infinite-precision pole location is very close to one of the allowed quantized poles. In this case, quantization causes no problem whatsoever, but in general, quantization can be expected to degrade system performance.

An alternative second-order structure for realizing poles at $z = re^{j\theta}$ and $z = re^{-j\theta}$ is shown in Fig. 6.50. This structure has been called the *coupled form* for the

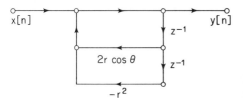

x[n]

y[n]

z^{-1}

$2r \cos \theta$

z^{-1}

$-r^2$

Figure 6.48   Direct form implementation of a complex-conjugate pole pair.

---

† An exception is in speech synthesis, where systems of 10th order and higher are routinely implemented using the direct form. This is possible because in speech synthesis the poles of the system function are widely separated (Rabiner and Schafer, 1978).

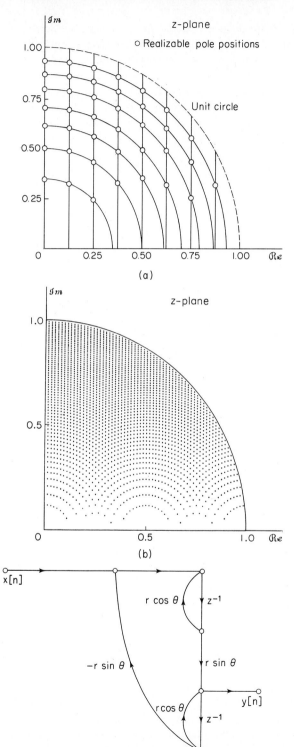

Figure 6.49 Pole locations for the second-order IIR direct form system of Fig. 6.48. (a) 4-bit quantization of coefficients. (b) 7-bit quantization.

Figure 6.50 Coupled form implementation of a complex-conjugate pole pair.

second-order system (Rader and Gold, 1967). It is easily verified that the systems of Figs. 6.48 and 6.50 have the same poles for infinite-precision coefficients. To implement Fig. 6.50 we must quantize $r \cos \theta$ and $r \sin \theta$. Since these quantities are the real and imaginary parts, respectively, of the pole locations, the quantized pole locations are at intersections of evenly spaced horizontal and vertical lines in the z-plane. Figure 6.51(a) and 6.51(b) show the possible pole locations for 4-bit and 7-bit quantization, respectively. In this case the density of pole locations is uniform throughout the interior of the unit circle. Twice as many constant multipliers are required to achieve this more uniform density. In some situations, this extra computation might be justified to achieve more accurate pole location with reduced wordlength.

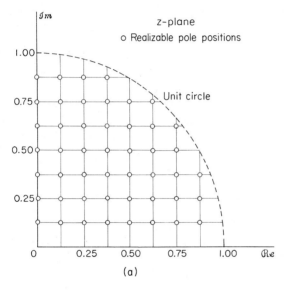

(a)

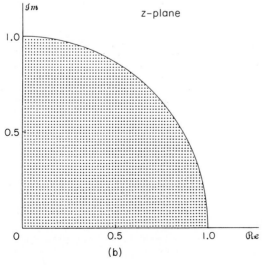

(b)

Figure 6.51 Pole locations for coupled form second-order IIR system of Fig. 6.50. (a) 4-bit quantization of coefficients. (b) 7-bit quantization.

6.8.2 Effects of Coefficient Quantization in FIR Systems

For FIR systems, we must be concerned only with the locations of the zeros of the system function since for causal FIR systems all the poles are at $z = 0$. Although we have just seen that the direct form structure should be avoided for high-order IIR systems, it turns out that the direct form structure is commonly used for FIR systems. To understand why this is so, we express the system function for a direct form FIR system in the form

$$H(z) = \sum_{n=0}^{M} h[n]z^{-n}. \tag{6.84}$$

Now suppose that the coefficients $\{h[n]\}$ are quantized, resulting in a new set of coefficients $\{\hat{h}[n] = h[n] + \Delta h[n]\}$. The system function for the quantized system is

$$\hat{H}(z) = \sum_{n=0}^{M} \hat{h}[n]z^{-n} = H(z) + \Delta H(z), \tag{6.85}$$

where

$$\Delta H(z) = \sum_{n=0}^{M} \Delta h[n]z^{-n}. \tag{6.86}$$

Thus, the system function (and therefore also the frequency response) of the quantized system is linearly related to the quantization errors in the impulse response coefficients. For this reason, the quantized system can be represented as in Fig. 6.52, which shows the unquantized system in parallel with an error system, whose impulse response is the sequence of quantization error samples $\{\Delta h[n]\}$ and whose system function is the corresponding z-transform, $\Delta H(z)$.

One approach to studying the sensitivity of the direct form FIR structure is to examine the sensitivity of the zeros to quantization errors in the impulse response coefficients. The results obtained for the poles in Eqs. (6.81)–(6.83) apply here to the zeros if the coefficients $\{a_k\}$ in those equations are replaced by the impulse response coefficients $\{h[k]\}$. The same type of conclusions can be drawn for the FIR case. If the zeros of $H(z)$ are tightly clustered, then their locations will be highly sensitive to quantization errors in the impulse response coefficients. The reason that the direct form FIR system is widely used is that for most linear phase FIR filters, the zeros are more or less uniformly spread in the z-plane. We demonstrate this by example.

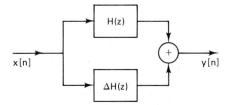

Figure 6.52   Representation of coefficient quantization in FIR systems.

Example 6.11

As an example of the effect of coefficient quantization in the FIR case, consider a linear phase lowpass filter designed to meet the following specifications:

$$0.99 \le |H(e^{j\omega})| \le 1.01, \qquad 0 \le \omega \le 0.4\pi,$$

$$|H(e^{j\omega})| \le 0.001 \text{ (i.e., } -60\text{dB)}, \qquad 0.6\pi \le \omega \le \pi.$$

This filter was designed using the Parks-McClellan design technique, which will be discussed in Section 7.6.3. The details of the design of this particular system are considered in Section 7.7.1.

Table 6.3 shows the unquantized impulse response coefficients for the system along with quantized coefficients for 16-, 14-, 13-, and 8-bit quantization.

Figure 6.53 gives a comparison of the frequency responses of the various systems. Figure 6.53(a) shows the log magnitude in dB of the frequency response for unquantized coefficients. Figures 6.53(b), (c), (d), (e), and (f) show the passband and stopband approximation errors (errors in approximating unity in the passband and zero in the stopband) for the unquantized, 16-, 14-, 13-, and 8-bit quantized cases, respectively. From Fig. 6.53 we see that the system meets the specifications for the unquantized case and both the 16-bit and 14-bit quantized cases. However, with 13-bit quantization the stopband approximation error becomes greater than 0.001, and with 8-bit quantization the stopband approximation error is over 10 times as large as specified. Thus, we see that at least 14-bit coefficients are required for direct form implementation of this system. However, this is not a serious limitation since 16-bit or 14-bit coefficients are well matched to many of the technologies that might be used to implement such a filter.

It is worth noting that for this example, all of the unquantized coefficients have magnitudes less than 0.5. Consequently, if all of the coefficients (and therefore the impulse response) are doubled prior to quantization, more efficient use of the available bits will result, corresponding in effect to increasing B by 1. In Table 6.3 and Fig. 6.53, we did not take this potential for increased accuracy into account.

TABLE 6.3   UNQUANTIZED AND QUANTIZED COEFFICIENTS FOR AN OPTIMUM FIR LOWPASS FILTER $(M = 27)$

Coefficient	Unquantized	16 bits	14 bits	13 bits	8 bits
$h[0] = h[27]$	1.359657×10^{-3}	45×2^{-15}	11×2^{-13}	6×2^{-12}	0×2^{-1}
$h[1] = h[26]$	-1.616993×10^{-3}	-53×2^{-15}	-13×2^{-13}	-7×2^{-12}	0×2^{-1}
$h[2] = h[25]$	-7.738032×10^{-3}	-254×2^{-15}	-63×2^{-13}	-32×2^{-12}	-1×2^{-1}
$h[3] = h[24]$	-2.686841×10^{-3}	-88×2^{-15}	-22×2^{-13}	-11×2^{-12}	0×2^{-1}
$h[4] = h[23]$	1.255246×10^{-2}	411×2^{-15}	103×2^{-13}	51×2^{-12}	2×2^{-1}
$h[5] = h[22]$	6.591530×10^{-3}	216×2^{-15}	54×2^{-13}	27×2^{-12}	1×2^{-1}
$h[6] = h[21]$	-2.217952×10^{-2}	-727×2^{-15}	-182×2^{-13}	-91×2^{-12}	-3×2^{-1}
$h[7] = h[20]$	-1.524663×10^{-2}	-500×2^{-15}	-125×2^{-13}	-62×2^{-12}	-2×2^{-1}
$h[8] = h[19]$	3.720668×10^{-2}	1219×2^{-15}	305×2^{-13}	152×2^{-12}	5×2^{-1}
$h[9] = h[18]$	3.233332×10^{-2}	1059×2^{-15}	265×2^{-13}	132×2^{-12}	4×2^{-1}
$h[10] = h[17]$	-6.537057×10^{-2}	-2142×2^{-15}	-536×2^{-13}	-268×2^{-12}	-8×2^{-1}
$h[11] = h[16]$	-7.528754×10^{-2}	-2467×2^{-15}	-617×2^{-13}	-308×2^{-12}	-10×2^{-1}
$h[12] = h[15]$	1.560970×10^{-1}	5115×2^{-15}	1279×2^{-13}	639×2^{-12}	20×2^{-1}
$h[13] = h[14]$	4.394094×10^{-1}	14399×2^{-15}	3600×2^{-13}	1800×2^{-12}	56×2^{-1}

To gain a better understanding of results of Example 6.11, let us recall that according to Eqs. (6.85) and (6.86) the frequency response of an FIR system with quantized coefficients is

$$\hat{H}(e^{j\omega}) = H(e^{j\omega}) + \Delta H(e^{j\omega}), \tag{6.87}$$

where

$$\Delta H(e^{j\omega}) = \sum_{n=0}^{M} \Delta h[n]e^{-j\omega n}. \tag{6.88}$$

To simplify the analysis, let us assume that all coefficients are less than 1 and that they are all scaled with the binary point between the sign bit and the most significant fractional bit. Then it follows from our discussion in Section 6.7.1 that if the impulse response coefficients are rounded to $(B + 1)$ bits (including sign), then

$$-2^{-(B+1)} < \Delta h[n] \le 2^{-(B+1)}. \tag{6.89}$$

A bound on the size of the frequency-response error is obtained by the following simple manipulations:

$$|\Delta H(e^{j\omega})| = \left| \sum_{n=0}^{M} \Delta h[n]e^{-j\omega n} \right| \le \sum_{n=0}^{M} |\Delta h[n]||e^{-j\omega n}| \le (M+1)2^{-(B+1)}, \tag{6.90}$$

where we have replaced the coefficient errors $\Delta h[n]$ and the complex exponentials $e^{-j\omega n}$ by their maximum magnitudes to obtain the final bound on the right in Eq. (6.90).

Let us apply the bound to the system in Example 6.11. For $M + 1 = 28$, the error bounds for the 16-, 14-, 13-, and 8-bit quantized systems are 0.000427, 0.001709, 0.003418, and 0.109375, respectively. A careful comparison of Figs. 6.53(c)–(f) with Fig. 6.53(b) shows that the deviations introduced by the quantization errors are less than the corresponding bounds in all cases.

The bound of Eq. (6.90) is a pessimistic bound, since all of the quantization errors would have to be of the same sign and equal to the maximum value for the error to be as large as the bound. Also, the bound of Eq. (6.90) suggests that the error increases linearly with the length of the impulse response. Chan and Rabiner (1973c) assumed a random variable model for the coefficient errors and derived a sharper bound on the variance of the frequency-response error. In this bound, the error variance increases as the square root of the length of the impulse response.

So far we have not made any assumptions about the phase response of the FIR system. However, the possibility of generalized linear phase is one of the major advantages of an FIR system. Recall that a linear phase FIR system has either a symmetric $(h[M - n] = h[n])$ or antisymmetric $(h[M - n] = -h[n])$ impulse response. These linear phase conditions are easily preserved for the direct form quantized system. Thus, all the systems in Example 6.11 have precisely linear phase regardless of the coarseness of the quantization. In situations where quantization is very coarse, or for high-order systems with closely spaced zeros, it may be worthwhile to realize smaller sets of zeros independently with a cascade form FIR system. To maintain linear phase, each of the sections in the cascade must also have linear

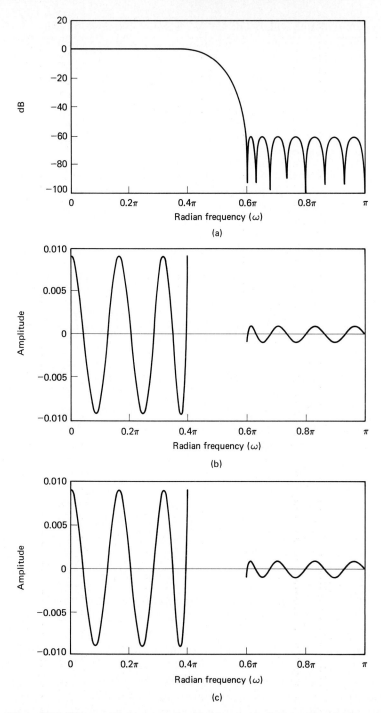

Figure 6.53 FIR quantization example. (a) Log magnitude for unquantized case. (b) Approximation error for unquantized case (error not defined in transition band). (c) Approximation error for 16-bit quantization.

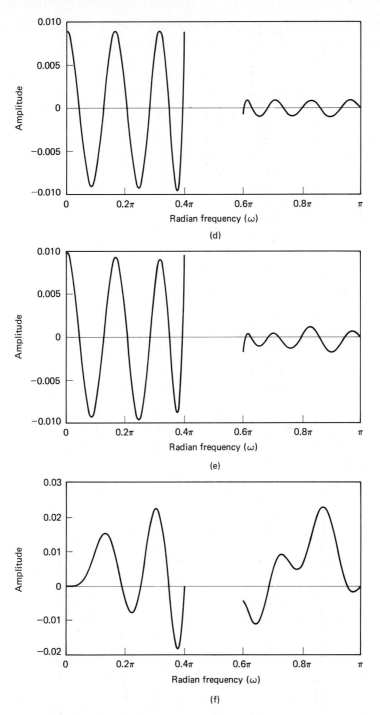

Figure 6.53   (*continued*) (d) Approximation error for 14-bit quantization. (e) Approximation error for 13-bit quantization. (f) Approximation error for 8-bit quantization.

phase. Recall that the zeros of a linear phase system must occur as illustrated in Fig. 6.32. For example, if we use second-order sections of the form $(1 + az^{-1} + z^{-2})$ for each complex conjugate pair of zeros on the unit circle, the zero can move only on the unit circle when the coefficient a is quantized. This prevents zeros from moving away from the unit circle, thereby lessening their attenuating effect. Similarly, real zeros inside the unit circle and at the reciprocal location outside the unit circle would remain real. Also, zeros at $z = \pm 1$ can be realized exactly by first-order systems. If a pair of complex conjugate zeros inside the unit circle is realized by a second-order system rather than a fourth-order system, then we must ensure that for each complex zero inside the unit circle there is a conjugate reciprocal zero outside the unit circle. This can be done by expressing the fourth-order factor corresponding to zeros at $z = re^{j\theta}$ and $z = r^{-1}e^{-j\theta}$ as

$$1 + cz^{-1} + dz^{-2} + cz^{-3} + z^{-4}$$

$$= (1 - 2r\cos\theta z^{-1} + r^2 z^{-2})\frac{1}{r^2}(r^2 - 2r\cos\theta z^{-1} + z^{-2}). \qquad (6.91)$$

This corresponds to the subsystem shown in Fig. 6.54. This system uses the same coefficients, $-2r\cos\theta$ and r^2, to realize both the zeros inside the unit circle and the conjugate reciprocal zeros outside the unit circle. Thus the linear phase condition is preserved under quantization. Notice that the factor $(1 - 2r\cos\theta z^{-1} + r^2 z^{-2})$ is identical to the denominator of the second-order direct form IIR system of Fig. 6.48. Therefore, the set of quantized zeros is as depicted in Fig. 6.49. More detail on cascade realizations of FIR systems is given by Herrmann and Schüssler (1970b).

6.8.3 Other Approaches to Sensitivity Analysis

The purpose of our discussion in this section has been to indicate the important issues that arise due to quantization of coefficients in difference equations. Our theoretical analysis has been limited in scope, but the results given are suggestive of conditions that lead to high sensitivity of the system response to coefficient quantization. Our discussion has mainly illustrated the commonly used approach of designing a system, picking a realization form, quantizing the coefficients, and checking the response of the resulting quantized system against the original specifications. More general approaches to the analysis of coefficient quantization errors have been developed

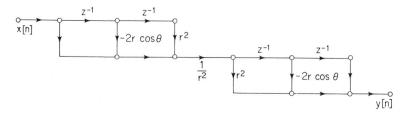

Figure 6.54 Subnetwork to implement fourth-order factors in a linear phase FIR system such that linear phase is maintained independent of parameter quantization.

using matrix formulations (Crochiere, 1972; Roberts and Mullis, 1987). Also available is Dehner's quantization analysis program for IIR systems (DSP Committee, 1979).

6.9 EFFECTS OF ROUNDOFF NOISE IN DIGITAL FILTERS

Difference equations realized with finite-precision arithmetic are nonlinear systems. Although it is important in general to understand how this nonlinearity affects the performance of discrete-time systems, a precise analysis of arithmetic quantization effects is generally not required in practical applications where we are concerned with the performance of a specific system. Indeed, just as with coefficient quantization, the most effective approach is often to simulate the system and measure its performance. For example, a common objective in quantization error analysis is to choose the digital wordlength such that the digital system is a sufficiently accurate realization of the desired linear system and at the same time requires a minimum of hardware or software complexity. The digital wordlength can, of course, be changed only in steps of one bit, and as we have already seen in Section 3.7.2, the addition of one bit to the wordlength reduces the size of quantization errors by a factor of 2. Thus, the choice of digital wordlength is insensitive to inaccuracies in the quantization error analysis; an analysis that is correct to within 30–40% is often adequate. For this reason, many of the important effects of quantization can be studied using linear models. We develop such models in this section and illustrate their use with several examples. An exception is the phenomenon of zero-input limit cycles, which are strictly nonlinear phenomena. We limit our study of nonlinear models for digital filters to a brief introduction to zero-input limit cycles in Section 6.10.

6.9.1 Analysis of the Direct Form IIR Structures

To introduce the basic ideas, let us consider the direct form structure for a linear time-invariant discrete-time system. The flow graph of a direct form I second-order system is shown in Fig. 6.55(a). The general Nth-order difference equation for the direct form I structure is

$$y[n] = \sum_{k=1}^{N} a_k y[n-k] + \sum_{k=0}^{M} b_k x[n-k] \tag{6.92}$$

and the system function is

$$H(z) = \frac{\sum_{k=0}^{M} b_k z^{-k}}{1 - \sum_{k=1}^{N} a_k z^{-k}} = \frac{B(z)}{A(z)}. \tag{6.93}$$

Let us assume that all signal values and coefficients are represented by $(B + 1)$-bit fixed-point binary numbers. Then in implementing Eq. (6.92) with a $(B + 1)$-bit adder, it would be necessary to reduce the length of the $(2B + 1)$-bit products to $(B + 1)$ bits. Since all numbers are treated as fractions, we would discard the least significant B bits by either rounding or truncation. This is represented by replacing each constant multiplier branch in Fig. 6.55(a) by a constant multiplier followed by a quantizer as in the nonlinear model of Fig. 6.55(b). The difference equation corresponding to Fig. 6.55(b) is the nonlinear equation

$$\hat{y}[n] = \sum_{k=1}^{N} Q[a_k \hat{y}[n-k]] + \sum_{k=0}^{M} Q[b_k x[n-k]]. \tag{6.94}$$

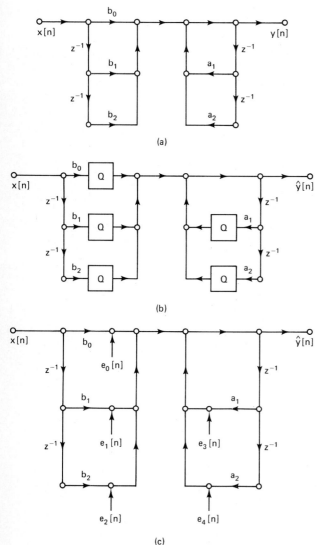

(a)

(b)

(c)

Figure 6.55 Models for direct form I system. (a) Infinite-precision model. (b) Nonlinear quantized model. (c) Linear noise model.

Figure 6.55(c) shows an alternative representation where the quantizers are replaced by noise sources that are equal to the quantization error at the output of each quantizer. For example, rounding or truncation of a product $bx[n]$ is represented by a noise source of the form

$$e[n] = Q[bx[n]] - bx[n]. \tag{6.95}$$

If the noise sources are known exactly, then Fig. 6.55(c) is exactly equivalent to Fig. 6.55(b). However, Fig. 6.55(c) is most useful when we assume that each quantization noise source has the following properties:

1. Each quantization noise source $e[n]$ is a wide-sense stationary white-noise process.
2. Each noise source has a uniform distribution of amplitudes over one quantization interval.
3. Each quantization noise source is *uncorrelated* with the input to the corresponding quantizer, all other quantization noise sources, and the input to the system.

These are identical to the assumptions made in the analysis of A/D conversion in Section 3.7. Strictly speaking, these assumptions cannot be valid since the quantization error depends directly on the input to the quantizer. This is readily apparent for constant and sinusoidal signals. However, experimental and theoretical analyses have shown (Bennett, 1948; Widrow, 1956, 1961) that in many situations, the model just described leads to accurate predictions of statistical averages such as mean, variance, and correlation function. This is true when the input signal is a complicated wideband signal such as speech, where the signal fluctuates rapidly among all the quantization levels and traverses many quantization levels in going from sample to sample (Gold and Rader, 1969). The simple linear noise model presented here allows us to characterize the noise generated in the system by averages such as mean and variance and to determine how these averages are modified by the system.

For $(B + 1)$-bit quantization, we showed in Section 6.7 that for rounding,

$$-\frac{1}{2}2^{-B} < e[n] \le \frac{1}{2}2^{-B} \tag{6.96a}$$

and for two's-complement truncation,

$$-2^{-B} < e[n] \le 0. \tag{6.96b}$$

Thus, according to our second assumption above, the probability density functions for the random variables representing quantization error are the uniform densities shown in Fig. 6.56(a) for rounding and in Fig. 6.56(b) for truncation. The mean and variance for rounding are

$$m_e = 0, \tag{6.97a}$$

$$\sigma_e^2 = \frac{2^{-2B}}{12}. \tag{6.97b}$$

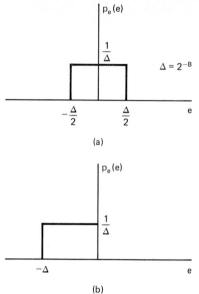

Figure 6.56 Probability density function for quantization errors. (a) Rounding. (b) Truncation.

For two's-complement truncation, the mean and variance are

$$m_e = -\frac{2^{-B}}{2},$$ (6.98a)

$$\sigma_e^2 = \frac{2^{-2B}}{12}.$$ (6.98b)

With rounding, the autocorrelation sequence of a quantization noise source is, according to the first assumption,

$$\phi_{ee}[n] = \sigma_e^2 \delta[n].$$ (6.99)

With this model for each of the noise sources in Fig. 6.55(c), we can now proceed to determine the effect of the quantization noise on the output of the system. To aid in doing this it is helpful to observe that all of the noise sources in Fig. 6.55(c) are effectively injected between the part of the system that implements the zeros and the part that implements the poles. Thus, Fig. 6.57 is equivalent to Fig. 6.55(c) if $e[n]$ in Fig. 6.57 is

$$e[n] = e_0[n] + e_1[n] + e_2[n] + e_3[n] + e_4[n].$$ (6.100)

Since we assume that all the noise sources are independent of the input and independent of each other, the variance of the combined noise sources for the second-order direct form I case is

$$\sigma_e^2 = \sigma_{e_0}^2 + \sigma_{e_1}^2 + \sigma_{e_2}^2 + \sigma_{e_3}^2 + \sigma_{e_4}^2 = 5 \cdot \frac{2^{-2B}}{12},$$ (6.101)

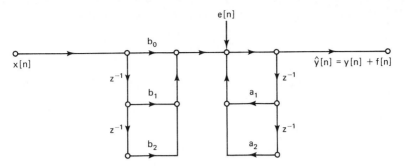

Figure 6.57   Linear noise model for direct form I with noise sources combined.

and for the general direct form I case,

$$\sigma_e^2 = (M + 1 + N)\frac{2^{-2B}}{12}. \tag{6.102}$$

Now to obtain an expression for the output noise, we note from Fig. 6.57 that the system has two inputs, $x[n]$ and $e[n]$, and, since the system is now assumed to be linear, the output $\hat{y}[n]$ can be represented as $\hat{y}[n] = y[n] + f[n]$, where $y[n]$ is the response of the ideal unquantized system to the input sequence $x[n]$, and $f[n]$ is the response of the system to the input $e[n]$. The output $y[n]$ is given by the difference equation in Eq. (6.92), but since $e[n]$ is injected after the zeros and before the poles, the output noise satisfies the difference equation

$$f[n] = \sum_{k=1}^{N} a_k f[n-k] + e[n]. \tag{6.103}$$

To determine the mean and variance of the output noise sequence, we can use some results from Section 2.10. Consider a linear system with system function $H_{ef}(z)$ with a white noise input $e[n]$ and corresponding output $f[n]$. Then from Eqs. (2.167) and (2.168), the mean of the output is

$$m_f = m_e \sum_{n=-\infty}^{\infty} h_{ef}[n] = m_e H_{ef}(e^{j0}) \tag{6.104}$$

and from Eq. (2.173) it follows that since $e[n]$ is a white-noise sequence, the power spectrum of the output noise is simply

$$P_{ff}(\omega) = \sigma_e^2 |H_{ef}(e^{j\omega})|^2 + 2\pi m_f^2 \delta(\omega). \tag{6.105}$$

Therefore, the variance of the output can be found using Eq. (2.174) to be

$$\sigma_f^2 = \sigma_e^2 \frac{1}{2\pi} \int_{-\pi}^{\pi} |H_{ef}(e^{j\omega})|^2 \, d\omega. \tag{6.106}$$

Using Parseval's theorem in Eq. (2.146), we can also express σ_f^2 as

$$\sigma_f^2 = \sigma_e^2 \sum_{n=-\infty}^{\infty} |h_{ef}[n]|^2, \tag{6.107}$$

or using the z-transform version of Parseval's theorem (Eq. 4.83),

$$\sigma_f^2 = \sigma_e^2 \frac{1}{2\pi j} \oint_C H_{ef}(z) H_{ef}(z^{-1}) z^{-1} \, dz, \tag{6.108}$$

which can be evaluated conveniently using Cauchy's residue theorem as discussed in Section 4.7.

We will use the results summarized in Eqs. (6.104)–(6.108) often in our analysis of quantization noise in linear systems. For example, for the direct form I system of Fig. 6.57, $H_{ef}(z) = 1/A(z)$; i.e., the system function from the point where all the noise sources are injected to the output consists only of the poles of the system function $H(z)$ in Eq. (6.93). Thus, we conclude that in general the total output variance due to internal roundoff or truncation is

$$\sigma_f^2 = (M + 1 + N) \frac{2^{-2B}}{12} \frac{1}{2\pi j} \oint_C \frac{z^{-1} \, dz}{A(z) A(z^{-1})}$$

$$= (M + 1 + N) \frac{2^{-2B}}{12} \sum_{n=-\infty}^{\infty} |h_{ef}[n]|^2, \tag{6.109}$$

where $h_{ef}[n]$ is the impulse response corresponding to $H_{ef}(z) = 1/A(z)$. The use of the preceding results is illustrated by the following examples.

Example 6.12

Suppose that we wish to implement a stable system having the system function

$$H(z) = \frac{b}{1 - az^{-1}}. \tag{6.110}$$

Figure 6.58 shows the flow graph of the linear noise model for the implementation in which products are quantized before addition. Each noise source is filtered by the system whose impulse response is $h_{ef}[n] = a^n u[n]$. From Eq. (6.105), the power spectrum of the output noise is

$$P_{ff}(\omega) = 2 \frac{2^{-2B}}{12} \cdot \frac{1}{1 + a^2 - 2a \cos \omega} + \frac{2\pi m_e^2 b}{1 - a} \delta(\omega), \tag{6.111}$$

where m_e is zero for rounding and $-2^{-B}/2$ for two's-complement truncation. The total noise variance at the output is

$$\sigma_f^2 = 2 \frac{2^{-2B}}{12} \sum_{n=0}^{\infty} a^{2n} = 2 \frac{2^{-2B}}{12} \frac{1}{1 - a^2}. \tag{6.112}$$

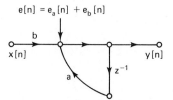

Figure 6.58   First-order linear noise model.

From Eq. (6.112) we see that the output noise variance increases as the pole at $z = a$ approaches the unit circle. Thus, to maintain the noise variance below a specified level as $|a|$ approaches unity, we must use longer wordlengths. The following example also illustrates this point.

Example 6.13

Consider a stable second-order direct form I system with system function

$$H(z) = \frac{b_0 + b_1 z^{-1} + b_2 z^{-2}}{(1 - re^{j\theta}z^{-1})(1 - re^{-j\theta}z^{-1})}. \tag{6.113}$$

The linear noise model for this system is shown in Fig. 6.55(c) or, equivalently Fig. 6.57, with $a_1 = 2r \cos \theta$ and $a_2 = -r^2$. In this case, the total output noise power can be expressed in the form

$$\sigma_f^2 = 5 \frac{2^{-2B}}{12} \frac{1}{2\pi j} \oint_C \frac{z \, dz}{(z - re^{j\theta})(z - re^{-j\theta})(1 - re^{j\theta}z)(1 - re^{-j\theta}z)}. \tag{6.114}$$

From Cauchy's residue theorem it follows that

$$\sigma_f^2 = 5 \frac{2^{-2B}}{12} \left[\frac{re^{j\theta}}{(re^{j\theta} - re^{-j\theta})(1 - r^2 e^{j2\theta})(1 - r^2)} + \frac{re^{-j\theta}}{(re^{-j\theta} - re^{j\theta})(1 - r^2)(1 - r^2 e^{-j2\theta})} \right]$$

$$= 5 \frac{2^{-2B}}{12} \left(\frac{1 + r^2}{1 - r^2} \right) \frac{1}{r^4 + 1 - 2r^2 \cos 2\theta} \tag{6.115}$$

As in Example 6.12, we see that as the complex conjugate poles approach the unit circle ($r \to 1$), the total output noise variance increases, thus requiring longer word-lengths to maintain the variance below a prescribed level.

The techniques of analysis developed so far for the direct form I structure can also be applied to the direct form II structure. The nonlinear difference equations for the direct form II structure are of the form

$$\hat{w}[n] = \sum_{k=1}^{N} Q[a_k \hat{w}[n - k]] + x[n], \tag{6.116a}$$

$$\hat{y}[n] = \sum_{k=0}^{M} Q[b_k \hat{w}[n - k]]. \tag{6.116b}$$

Figure 6.59(a) shows the linear noise model for a second-order direct form II system. A noise source has been introduced after each multiplication, indicating that the products are quantized to $(B + 1)$ bits before addition. Figure 6.59(b) shows an equivalent linear model where we have moved the noise sources resulting from implementation of the poles and combined them into a single noise source $e_a[n] = e_3[n] + e_4[n]$ at the input. Likewise, the noise sources due to implementation of the zeros are combined into the single noise source $e_b[n] = e_0[n] + e_1[n] + e_2[n]$ that is added directly to the output. From this equivalent model it follows immediately that for rounding ($m_e = 0$), the power spectrum of the output noise is

$$P_{ff}(\omega) = N \frac{2^{-2B}}{12} |H(e^{j\omega})|^2 + (M + 1) \frac{2^{-2B}}{12} \tag{6.117}$$

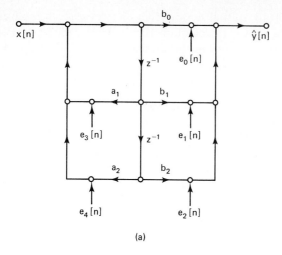

(a)

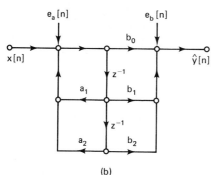

(b)

Figure 6.59 Linear noise models for direct form II. (a) Showing quantization of individual products. (b) With noise sources combined.

and the output noise variance is

$$\sigma_f^2 = N \frac{2^{-2B}}{12} \frac{1}{2\pi j} \oint_C H(z)H(z^{-1})z^{-1} \, dz + (M+1)\frac{2^{-2B}}{12}$$

$$= N \frac{2^{-2B}}{12} \sum_{n=-\infty}^{\infty} |h[n]|^2 + (M+1)\frac{2^{-2B}}{12}. \qquad (6.118)$$

That is, the white noise produced in implementing the poles is filtered by the entire system, while the white noise produced in implementing the zeros is added directly to the output of the system. In writing Eq. (6.118), we have assumed that the N noise sources at the input are independent, so that their sum has N times the variance of a single quantization noise source. The same assumption was made about the $(M+1)$ noise sources at the output.

A comparison of Eq. (6.109) with Eq. (6.118) shows that the direct form I and direct form II structures are affected differently by the quantization of products in implementing the corresponding difference equations. In general, other equivalent structures such as cascade, parallel, lattice, and transposed forms will have different total output noise variance than either of the direct form structures. We are interested in the system with the smallest output noise variance (subject to dynamic range constraints, to be discussed next). However, even though Eqs. (6.109) and (6.118) are different, we cannot say which system will have the smaller output noise variance unless we know specific values for the coefficients of the system. In other words, it is not possible to state that a particular structural form will always produce the least output noise.

It is possible to improve the noise performance of the direct form systems (and therefore cascade and parallel forms as well) by using a $(2B + 1)$-bit adder to accumulate the sum of products required in both direct form systems. For example, for the direct form I implementation, we could use a difference equation of the form

$$\hat{y}[n] = Q\left[\sum_{k=1}^{N} a_k \hat{y}[n-k] + \sum_{k=0}^{M} b_k x[n-k]\right], \tag{6.119}$$

i.e., the sums of products are accumulated with $(2B + 1)$- or $(2B + 2)$-bit accuracy, and the result is quantized to $(B + 1)$ bits for output and storage in the delay memory. In the direct form I case, this means that the quantization noise is still filtered by the poles, but the factor $(M + 1 + N)$ in Eq. (6.109) is replaced by unity. Similarly, for the direct form II realization, the difference equations of Eq. (6.116) can be replaced by

$$\hat{w}[n] = Q\left[\sum_{k=1}^{N} a_k \hat{w}[n-k] + x[n]\right], \tag{6.120a}$$

$$\hat{y}[n] = Q\left[\sum_{k=0}^{M} b_k \hat{w}[n-k]\right]. \tag{6.120b}$$

This implies a single noise source at both the input and output, so the factors N and $(M + 1)$ in Eq. (6.118) are replaced by unity. Thus, the use of a double-length accumulator word can significantly reduce quantization noise in direct form systems.

These results are easily modified for two's-complement truncation. Recall from Eqs. (6.97) and (6.98) that the variance of a truncation noise source is the same as a rounding noise source, but the mean of a truncation noise source is not zero. Consequently, the formulas in Eqs. (6.109) and (6.118) for the total output noise variance also hold for truncation. However, the output noise will have a nonzero average value that can be computed using Eq. (6.104).

6.9.2 Scaling in Fixed-Point Implementations of IIR Systems

The possibility of overflow is another important consideration in the implementation of IIR systems using fixed-point arithmetic. If we follow the convention that each fixed-point number represents a fraction (with possible assumed scale factor), each node in the network must be constrained to have a magnitude less than 1 to avoid overflow. If $w_k[n]$ denotes the value of the kth node variable and $h_k[n]$ denotes the impulse response from the input $x[n]$ to the node variable $w_k[n]$, then

$$|w_k[n]| = \left|\sum_{m=-\infty}^{\infty} x[n-m]h_k[m]\right|. \tag{6.121}$$

The bound

$$|w_k[n]| \le x_{max} \sum_{m=-\infty}^{\infty} |h_k[m]| \tag{6.122}$$

is obtained by replacing $x[n-m]$ by its maximum value x_{max} and using the fact that the magnitude of a sum is less than or equal to the sum of the magnitudes. Therefore, a sufficient condition that $|w_k[n]| < 1$ is

$$x_{max} < \frac{1}{\displaystyle\sum_{m=-\infty}^{\infty} |h_k[m]|} \tag{6.123}$$

for all nodes in the network. If x_{max} does not satisfy Eq. (6.123), then we can multiply $x[n]$ by a scaling multiplier s at the input to the system so that sx_{max} satisfies Eq. (6.123) for all nodes in the network; i.e.,

$$sx_{max} < \frac{1}{\displaystyle\max_k \left[\sum_{m=-\infty}^{\infty} |h_k[m]| \right]}. \tag{6.124}$$

Scaling the input in this way guarantees that overflow never occurs at any of the nodes in the network. Equation (6.123) is necessary as well as sufficient, since an input always exists such that Eq. (6.122) is satisfied with equality. (See Eq. 2.55 in the discussion of stability in Section 2.4.) However, Eq. (6.123) leads to a very conservative scaling of the input for most signals.

Another approach to scaling is to assume that the input is a narrowband signal, modeled as $x[n] = x_{max} \cos \omega_0 n$. In this case, the node variables will be

$$w_k[n] = |H_k(e^{j\omega_0})| x_{max} \cos(\omega_0 n + \sphericalangle H_k(e^{j\omega_0})). \tag{6.125}$$

Therefore, overflow is avoided for *all* sinusoidal signals if

$$\max_{k, |\omega| \le \pi} |H_k(e^{j\omega})| x_{max} < 1 \tag{6.126}$$

or if the input is scaled by the scale factor

$$sx_{max} < \frac{1}{\displaystyle\max_{k, |\omega| \le \pi} |H_k(e^{j\omega})|}. \tag{6.127}$$

Still another approach is to scale the input so that the total energy of each node variable sequence is less than or equal to the total energy of the input sequence. We can derive the appropriate scale factor by applying the Schwarz inequality and Parseval's theorem. The total energy of the sequence $w_k[n]$ is bounded by

$$\sum_{n=-\infty}^{\infty} |w_k[n]|^2 = \frac{1}{2\pi} \int_{-\pi}^{\pi} |H_k(e^{j\omega})X(e^{j\omega})|^2 \, d\omega \le \sum_{n=-\infty}^{\infty} |x[n]|^2 \frac{1}{2\pi} \int_{-\pi}^{\pi} |H_k(e^{j\omega})|^2 \, d\omega. \tag{6.128}$$

Therefore, to ensure that $\sum|w_k[n]|^2 \leq \sum|x[n]|^2$, we can multiply the sequence $x[n]$ by a scale factor s such that

$$s^2 \leq \frac{1}{\dfrac{1}{2\pi}\displaystyle\int_{-\pi}^{\pi}|H_k(e^{j\omega})|^2 d\omega} = \frac{1}{\displaystyle\sum_{n=-\infty}^{\infty}|h_k[n]|^2}. \tag{6.129}$$

It can be shown that

$$\left\{\sum_{n=-\infty}^{\infty}|h_k[n]|^2\right\}^{1/2} \leq \max_{k,\omega}|H_k(e^{j\omega})| \leq \sum_{n=-\infty}^{\infty}|h_k[n]|. \tag{6.130}$$

Therefore, Eqs. (6.124), (6.127), and (6.129) give three decreasingly conservative ways of obtaining scaling multipliers for the input to a digital filter. Of the three, Eq. (6.129) is generally the easiest to evaluate analytically because the integral in its complex form can be evaluated using Cauchy's residue theorem. On the other hand, Eq. (6.124) is difficult to evaluate analytically except for the simplest systems. Of course, for specific systems with known coefficients, the scale factors can be estimated by computing the impulse response or frequency response numerically.

If the input must be scaled down ($s < 1$), the signal-to-noise ratio at the output of the system will be reduced. Figure 6.60 shows second-order direct form I and direct form II systems with scaling multipliers at the input. In determining the scaling multiplier for the direct form systems, it is not necessary to examine each node in the network. Some nodes do not represent addition and thus cannot overflow. Other

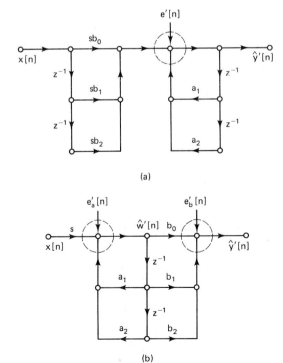

Figure 6.60   Scaling of direct form systems. (a) Direct form I. (b) Direct form II.

nodes represent partial sums. If we use nonsaturation two's-complement arithmetic, such nodes are permitted to overflow if certain key nodes do not. For example, in Fig. 6.60(a) we can focus on the node enclosed by the dashed circle. In Fig. 6.60(a) the scaling multiplier is shown combined with the b_k's so that the noise source is the same as in Fig. 6.57, i.e., it has five times the power of a single quantization noise source.†
Since the noise source is again filtered only by the poles, the output noise power is the same in Figs. 6.57 and 6.60(a). However, the overall system function of the system in Fig. 6.60(a) is $sH(z)$ instead of $H(z)$, so the unquantized component of the output is $y'[n] = sy[n]$ instead of $y[n]$. Since the noise is injected after the scaling, the ratio of signal power to noise power in the scaled system is s^2 times the signal-to-noise ratio for Fig. 6.57. Since $s < 1$ if scaling is required to avoid overflow, the signal-to-noise ratio is reduced by scaling.

The same is true for the direct form II system of Fig. 6.60(b). In this case we must determine the scaling multiplier to avoid overflow at both of the circled nodes in Fig. 6.60(b). Again the overall gain of the system is s times the gain of the system in Fig. 6.59(b), but it may be necessary to implement the scaling multiplier explicitly in this case to avoid overflow at the node on the left. This scaling multiplier adds an additional noise component to $e_a[n]$, so the noise power at the input is in general $(N + 1)2^{-2B}/12$. Otherwise, the noise sources are filtered by the system in exactly the same way in both Fig. 6.59(b) and 6.60(b). Therefore, the signal power is multiplied by s^2 and the noise at the output is again given by Eq. (6.118) with N replaced by $(N + 1)$. The signal-to-noise ratio is again reduced if scaling is required to avoid overflow.

Example 6.14

To illustrate the interaction of scaling and roundoff noise, consider the system of Example 6.12 with system function given by Eq. (6.110). If the scaling multiplier is combined with the coefficient b, we obtain the flow graph of Fig. 6.61 for the scaled system. Suppose that the input is a white random signal with amplitudes uniformly distributed between -1 and $+1$. Then the total signal variance is $\sigma_x^2 = 1/3$. To guarantee no overflow in computing $\hat{y}'[n]$, we use Eq. (6.124) to compute the scale factor

$$s = \frac{1}{\sum_{n=0}^{\infty} |b||a|^n} = \frac{1 - |a|}{|b|}. \tag{6.131}$$

The output noise variance was determined in Example 6.12 to be

$$\sigma_{f'}^2 = 2 \frac{2^{-2B}}{12} \frac{1}{1 - a^2} = \sigma_f^2 \tag{6.132}$$

and the variance of the output $y'[n]$ is

$$\sigma_{y'}^2 = \left(\frac{1}{3}\right) \frac{s^2 b^2}{1 - a^2} = s^2 \sigma_y^2. \tag{6.133}$$

---

† This eliminates a separate scaling multiplication and quantization noise source. However, scaling (and quantizing) the b_k's can change the frequency response of the system. If a separate input scaling multiplier precedes the implementation of the zeros in Fig. 6.60(a), then a single quantization noise source would contribute to the output noise through the system $H(z)$.

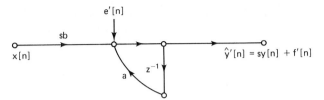

Figure 6.61   Scaled first-order system.

Therefore, the signal-to-noise ratio is

$$\frac{\sigma_{\hat{y}'}^2}{\sigma_{f'}^2} = s^2 \frac{\sigma_y^2}{\sigma_f^2} = \left(\frac{1 - |a|}{|b|}\right)^2 \frac{\sigma_y^2}{\sigma_f^2}. \tag{6.134}$$

As the pole of the system approaches the unit circle, the signal-to-noise ratio decreases because the quantization noise is amplified by the system and because the high gain of the system forces the input to be scaled down to avoid overflow. Again we see that overflow and quantization noise work in opposition to decrease the performance of the system.

6.9.3 Analysis of Cascade and Parallel IIR Structures

The previous results of this section can be applied directly to the analysis of parallel and cascade structures using second-order direct form subsystems. The interaction of scaling and quantization is particularly interesting in the cascade form. Our general comments on cascade systems will be interwoven with a specific example.

An elliptic lowpass filter was designed to meet the following specifications:

$$0.99 \le |H(e^{j\omega})| \le 1.01, \qquad |\omega| \le 0.5\pi,$$

$$|H(e^{j\omega})| \le 0.01, \qquad\qquad 0.56\pi \le |\omega| \le \pi.$$

The system function of the resulting system is

$$H(z) = 0.079459 \prod_{k=1}^{3} \left(\frac{1 + b_{1k}z^{-1} + z^{-2}}{1 - a_{1k}z^{-1} - a_{2k}z^{-2}}\right) = 0.079459 \prod_{k=1}^{3} H_k(z), \tag{6.135}$$

where the coefficients in Eq. (6.135) are given in Table 6.4. Notice that all the zeros of $H(z)$ are on the unit circle in this example; however, that need not be the case in general.

Figure 6.62(a) shows a flow graph of a possible implementation of this system as a cascade of second-order transposed direct form II subsystems. The gain constant, 0.079459, is such that the overall gain of the system is approximately unity in the

TABLE 6.4   COEFFICIENTS FOR ELLIPTIC LOWPASS FILTER IN CASCADE FORM

k	a_{1k}	a_{2k}	b_{1k}
1	0.478882	−0.172150	1.719454
2	0.137787	−0.610077	0.781109
3	−0.054779	−0.902374	0.411452

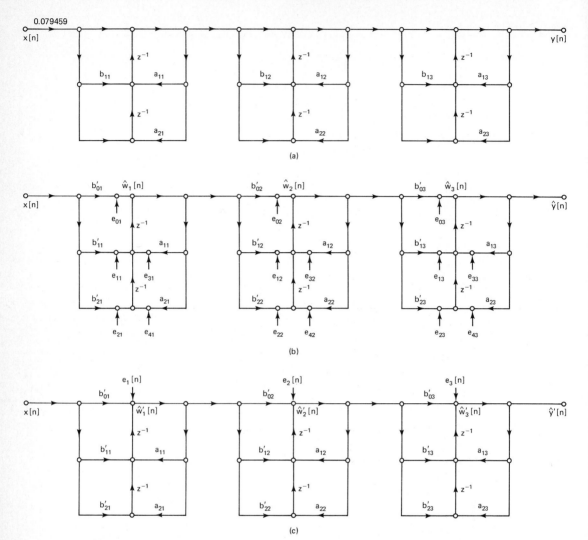

Figure 6.62 Models for sixth-order cascade system with transposed direct form II subsystems. (a) Infinite-precision model. (b) Linear noise model for scaled system showing quantization of individual multiplications. (c) Linear noise model with noise sources combined.

passband, and it is assumed that this guarantees no overflow at the output of the system. Figure 6.62(a) shows the gain constant located at the input to the system. This approach reduces the amplitude of the signal immediately, with the result that the subsequent filter sections must have high gain to produce an overall gain of unity. Since the quantization noise sources are introduced after the gain of 0.079459, but are likewise amplified by the rest of the system, this is not a good approach. Ideally, the overall gain constant, being less than unity, should be placed at the very end of the cascade so that the signal and noise will be attenuated by the same amount. However, this creates the possibility of overflow along the cascade. Therefore, a better approach is to distribute the gain among the three stages of the system so that overflow is just avoided at each stage of the cascade. This distribution is represented by

$$H(z) = s_1 H_1(z) s_2 H_2(z) s_3 H_3(z), \tag{6.136}$$

where $s_1 s_2 s_3 = 0.079459$. The scaling multipliers can be incorporated into the coefficients of the numerators of the individual system functions $H'_k(z) = s_k H_k(z)$ as in

$$H(z) = \prod_{k=1}^{3} \left(\frac{b'_{0k} + b'_{1k} z^{-1} + b'_{2k} z^{-2}}{1 - a_{1k} z^{-1} - a_{2k} z^{-2}} \right) = \prod_{k=1}^{3} H'_k(z), \tag{6.137}$$

where $b'_{0k} = b'_{2k} = s_k$ and $b'_{1k} = s_k b_{1k}$. The resulting scaled system is depicted in Fig. 6.62(b).

Also shown in Fig. 6.62(b) are quantization noise sources representing the quantization of the products before addition. Figure 6.62(c) shows an equivalent noise model, where it is recognized that all the noise sources in a particular section are filtered only by the poles of that section (and the subsequent subsystems). Figure 6.62(c) also uses the fact that delayed white-noise sources are still white noise and are independent of all the other noise sources, so that all five sources in a subsection can be combined into a single noise source having five times the variance of a single quantization noise source.† Since the noise sources are assumed independent, the variance of the output noise is the sum of the variances due to the three noise sources in Fig. 6.62(c). Therefore, for rounding, the power spectrum of the output noise is

$$P_{f'f'}(\omega) = 5 \frac{2^{-2B}}{12} \left[\frac{s_2^2 |H_2(e^{j\omega})|^2 s_3^2 |H_3(e^{j\omega})|^2}{|A_1(e^{j\omega})|^2} + \frac{s_3^2 |H_3(e^{j\omega})|^2}{|A_2(e^{j\omega})|^2} + \frac{1}{|A_3(e^{j\omega})|^2} \right] \tag{6.138}$$

and the total output noise variance is

$$\sigma_{f'}^2 = 5 \frac{2^{-2B}}{12} \left[\frac{1}{2\pi j} \oint_C \frac{s_2^2 H_2(z) H_2(z^{-1}) s_3^2 H_3(z) H_3(z^{-1}) z^{-1} \, dz}{A_1(z) A_1(z^{-1})} \right.$$
$$\left. + \frac{1}{2\pi j} \oint_C \frac{s_3^2 H_3(z) H_3(z^{-1}) z^{-1} \, dz}{A_2(z) A_2(z^{-1})} + \frac{1}{2\pi j} \oint_C \frac{z^{-1} \, dz}{A_3(z) A_3(z^{-1})} \right]. \tag{6.139}$$

If a double-length accumulator is available, it would be necessary to quantize only the sums that are the inputs to the delay elements in Fig. 6.62(b). In this case the factor of 5 in Eqs. (6.138) and (6.139) would be changed to 3. Furthermore, if a double-length register were used to implement the delay elements in Fig. 6.62(b), only the variables $\hat{w}_k[n]$ would have to be quantized, and there would be only one quantization noise source per subsystem. In this case the factor of 5 in Eqs. (6.138) and (6.139) would be changed to 1.

---

† This discussion can be generalized to show that the transposed direct form II has the same noise behavior as the direct form I system.

The scale factors s_k are chosen to avoid overflow at points along the cascade system. In this case, we will use the scaling convention of Eq. (6.127). Therefore, the scaling constants are chosen to satisfy

$$s_1 \max_{|\omega| \le \pi} |H_1(e^{j\omega})| < 1, \tag{6.140a}$$

$$s_1 s_2 \max_{|\omega| \le \pi} |H_1(e^{j\omega})H_2(e^{j\omega})| < 1, \tag{6.140b}$$

$$s_1 s_2 s_3 = 0.079459. \tag{6.140c}$$

The last condition was assumed to ensure no overflow at the output of the system. For the coefficients of Table 6.4, the resulting scale factors are $s_1 = 0.186447$, $s_2 = 0.529236$, and $s_3 = 0.805267$.

Equations (6.138) and (6.139) show that the shape of the output noise power spectrum and the total output noise variance depends on the way that zeros and poles are paired to form the second-order sections and on the order of the second-order sections in the cascade form realization. Indeed, it is easily seen that for N_s sections, there are N_s factorial ways to pair the poles and zeros and there are likewise N_s factorial ways to order the resulting second-order sections, a total of $(N_s!)^2$ different systems. In addition, we can choose either direct form I or direct form II (or their transposes) for the implementation of the second-order sections. In our example, this implies that there are 144 different cascade systems to consider if we wish to determine the system with the lowest output noise variance. For five cascaded sections, there are 57,600 different systems! Clearly, the complete analysis of even low-order systems is a tedious task, since an expression like Eq. (6.139) must be evaluated for each pairing and ordering. Hwang (1974) used dynamic programming and Liu and Peled (1975) used a heuristic approach to reduce the amount of computation. Dehner's program for design and analysis of cascade form IIR filters includes pairing and ordering optimization (DSP Committee, 1979).

In spite of the difficulty of finding the optimum pairing and ordering, Jackson (1970a, 1970b, 1986) found that good results are almost always obtained by applying simple rules of the following form:

1. The pole that is closest to the unit circle should be paired with the zero that is closest to it in the z-plane.
2. Rule 1 should be repeatedly applied until all the poles and zeros have been paired.
3. The resulting second-order sections should be ordered according to the closeness of the poles to the unit circle, either in order of increasing closeness to the unit circle or in order of decreasing closeness to the unit circle.

The pairing rules are based on the observation that subsystems with high peak gain are undesirable because they can cause overflow and because they can amplify quantization noise. Pairing a pole that is close to the unit circle with an adjacent zero tends to reduce the peak gain of that section.

One motivation for rule 3 is suggested by Eq. (6.138). We see that the frequency responses of some of the subsystems appear more than once in the equation for the power spectrum of the output noise. If we do not want the output noise variance spectrum to be highly peaked around a pole that is close to the unit circle, then it is advantageous to have the frequency-response component due to that pole not appear frequently in Eq. (6.138). This suggests moving such "high Q" poles to the beginning of the cascade. On the other hand, the frequency response from the input to a particular node in the network will involve a product of the frequency responses of the subsystems that precede that node. Thus, to avoid excessive reduction of the signal level in the early stages of the cascade, we should place the poles that are close to the unit circle last in order. Thus, the question of ordering hinges on a variety of considerations, including total output noise variance and the shape of the output noise spectrum. Jackson (1970a, 1970b) used L_p norms to quantify the analysis of the pairing and ordering problem and gave a much more detailed set of "rules of thumb" for obtaining good results without exhaustive evaluation of all possibilities.

The pole-zero plot for our example system is shown in Fig. 6.63. The paired poles and zeros are circled. In this case we have chosen to order the sections from least peaked to most peaked frequency response, thereby accepting a somewhat more peaked output noise spectrum in exchange for larger scale factors in the early stages of the cascade. Figure 6.64 illustrates how the frequency responses of the individual sections combine to form the overall frequency response. Figures 6.64(a)–(c) show the frequency responses of the individual unscaled subsystems. Figures 6.64(d)–(f) show how the overall frequency response is built up. Notice that the scaling equations of Eq. (6.140) ensure that the maximum gain from the input to the output of any subsystem is less than unity.

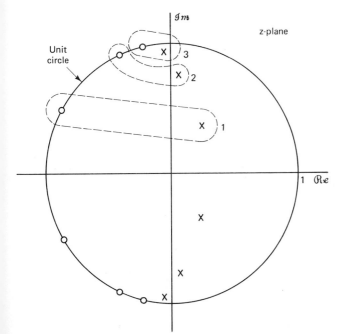

Figure 6.63   Pole-zero plot for sixth-order example system of Fig. 6.62 showing pairing of poles and zeros.

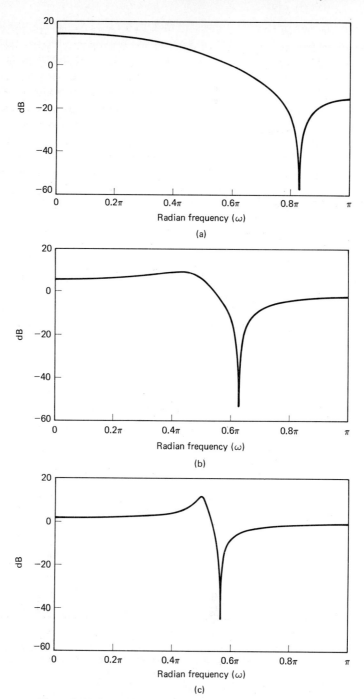

Figure 6.64 Frequency response functions for example system.
(a) $20 \log_{10} |H_1(e^{j\omega})|$. (b) $20 \log_{10} |H_2(e^{j\omega})|$. (c) $20 \log_{10} |H_3(e^{j\omega})|$.

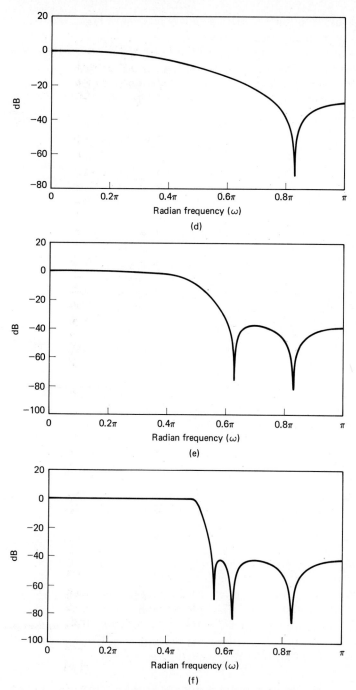

Figure 6.64   (*continued*) (d) $20 \log_{10} |H'_1(e^{j\omega})|$. (e) $20 \log_{10} |H'_1(e^{j\omega})H'_2(e^{j\omega})|$.
(f) $20 \log_{10} |H'_1(e^{j\omega})H'_2(e^{j\omega})H'_3(e^{j\omega})| = 20 \log_{10} |H'(e^{j\omega})|$.

The example of this subsection shows the complexity of the issues that arise in fixed-point implementations of cascade IIR systems. The parallel form is somewhat simpler because the issue of pairing and ordering does not arise. However, scaling is still required to avoid overflow in individual second-order subsystems and when the outputs of the subsystems are summed to produce the overall output. The techniques that we have developed must therefore be applied for the parallel form as well. Jackson (1986) discusses the analysis of the parallel form in detail and concludes that the total output noise power of the parallel form is typically comparable to that of the best pairing and orderings of the cascade form. Even so, the cascade form is more commonly used, because for commonly used IIR filters where the zeros of the system function are on the unit circle, the cascade form can be implemented with fewer multipliers with more control over the locations of the zeros.

6.9.4 Analysis of Direct Form FIR Systems

Since the direct form I and direct form II IIR systems include the direct form FIR system as a special case (i.e., all a_k's zero in Figs. 6.10 and 6.11), the results and analysis techniques of Sections 6.9.1 and 6.9.2 apply to FIR systems if we eliminate all reference to the poles of the system function and eliminate the feedback paths in all the signal flow graphs. Therefore, we only review these results in the context of FIR systems.

The direct form FIR system is simply the discrete convolution

$$y[n] = \sum_{k=0}^{M} h[k]x[n-k]. \tag{6.141}$$

Figure 6.65(a) shows the ideal unquantized direct form FIR system, and Fig. 6.65(b) shows the linear noise model for the system, assuming that all products are quantized before additions are performed. The effect is to inject $(M + 1)$ white-noise sources directly at the output of the system so that the total output noise variance is

$$\sigma_f^2 = (M + 1)\frac{2^{-2B}}{12}. \tag{6.142}$$

This is exactly the result we would obtain by setting $N = 0$ and $h_{ef}[n] = \delta[n]$ in Eqs. (6.109) and (6.118). When a double-length accumulator is available, we would need only to quantize the output. Therefore, the factor $(M + 1)$ in Eq. (6.142) would be replaced by unity. This makes the double-length accumulator a very attractive hardware feature for implementing FIR systems.

Overflow is also a problem for fixed-point realizations of FIR systems in direct form. For two's-complement arithmetic, we need to be concerned only about the size of the output, since all the other sums in Fig. 6.65(b) are partial sums. Thus, the impulse response coefficients can be scaled to reduce the possibility of overflow. Scaling multipliers can be determined using any of the alternatives discussed in Section 6.9.2. Of course, scaling the impulse response reduces the gain of the system, and therefore the signal-to-noise ratio at the output is reduced as discussed in Section 6.9.2.

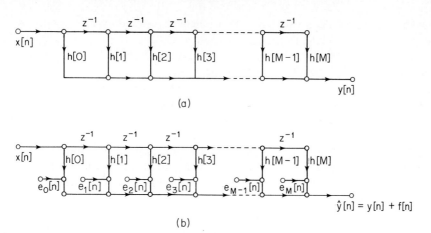

Figure 6.65 Direct form realization of an FIR system. (a) Infinite-precision model. (b) Linear noise model.

Example 6.15

Consider the system in Example 6.11. The impulse response coefficients for the system are given in Table 6.3. Simple arithmetic shows that

$$\sum_{n=0}^{27} |h[n]| = 1.751352,$$

$$\left(\sum_{n=0}^{27} |h[n]|^2\right)^{1/2} = 0.679442,$$

and, from Fig. 6.53(b), we see that

$$\max_{|\omega| \le \pi} |H(e^{j\omega})| \approx 1.009.$$

These numbers satisfy the relationship in Eq. (6.130). We see that the system as given is scaled so that overflow is theoretically possible for a sinusoidal signal whose amplitude is greater than $1/1.009 = 0.9911$, but even so, overflow is unlikely for most signals. Indeed, since the filter has linear phase, we can argue that for wideband signals, since the gain in the passband is approximately unity and the gain elsewhere is less than unity, then the output signal should be smaller than the input signal.

In Section 6.5.3 we showed that linear phase systems like the one in Example 6.15 can be implemented with about half the multiplications of the general FIR system. This is evident from the signal flow graphs of Figs. 6.30 and 6.31. In these cases it should be clear that the output noise variance would be cut in half if products were quantized before addition.

In Section 6.5.3 we discussed cascade realizations of FIR systems. The results and analysis techniques of Section 6.9.3 apply to cascade realizations of FIR systems; but for FIR systems with no poles, the pairing and ordering problem reduces to just an ordering problem. As in the case of IIR cascade systems, the analysis of all possible orderings can be very difficult if the system is composed of many subsystems. Chan and Rabiner (1973b, 1973c) studied this problem and found experimentally that the noise performance is relatively insensitive to the ordering. Their results suggest that a good ordering is one for which the frequency response from each noise source to the output is relatively flat and for which the peak gain is small.

6.9.5 Floating-Point Realizations of Discrete-Time Systems

From the preceding discussion it is clear that the limited dynamic range of fixed-point arithmetic makes it necessary to carefully scale the input and intermediate signal levels in fixed-point digital realizations of discrete-time systems. The need for such scaling can be essentially eliminated by using floating-point numerical representations and floating-point arithmetic.

In floating-point representations, a real number x is represented by the binary number $2^c \hat{x}_M$, where the exponent c of the scale factor is called the *characteristic* and $\hat{x}_M$ is a fractional part called the *mantissa*. Both the characteristic and the mantissa are represented explicitly as fixed-point binary numbers in floating-point arithmetic systems. Floating-point representations provide a convenient means for maintaining both a wide dynamic range and small quantization noise; however, quantization error manifests itself in a somewhat different way. Floating-point arithmetic generally maintains high accuracy and wide dynamic range by adjusting the characteristic and normalizing the mantissa so that $0.5 \le \hat{x}_M < 1$. When floating-point numbers are multiplied, their characteristics are added and their mantissas are multiplied. Thus, the mantissa must be quantized. When two floating-point numbers are added, their characteristics must be adjusted to be the same by moving the binary point of the mantissa of the smaller number. Thus, addition results in quantization too. If we assume that the range of the characteristic is sufficient so that no numbers become larger than 2^c, then quantization affects only the mantissa, but the error in the mantissa is also scaled by 2^c. Thus, a quantized floating-point number is conveniently represented as

$$\hat{x} = x(1 + \varepsilon) = x + \varepsilon x. \tag{6.143}$$

By representing the quantization error as a fraction ε of x we automatically represent the fact that the quantization error is scaled up and down with the signal level.

The above mentioned properties of floating-point arithmetic complicate the quantization error analysis of floating-point implementations of discrete-time systems. First, noise sources must be inserted both after each multiplication and after each addition. An important consequence is that, in contrast to fixed-point arithmetic, the *order* in which multiplications and additions are performed can sometimes make a big difference. More important for analysis, we can no longer justify the assumption that the quantization noise sources are white noise and are independent of the signal. In fact, in Eq. (6.143) the noise is expressed explicitly in terms of the signal. Therefore, we can no longer analyze the noise without making assumptions about the nature of the input signal. If the input is assumed to be known (e.g., white noise), a reasonable assumption is that the relative error ε is independent of x and uniformly distributed white noise.

With these types of assumptions, useful results have been obtained by Sandberg (1967), Liu and Kaneko (1969), Weinstein and Oppenheim (1969), and Kan and Aggarwal (1971). In particular, Weinstein and Oppenheim compared floating-point and fixed-point realizations of first- and second-order IIR systems. They showed that if the number of bits representing the floating-point mantissa is equal to the length of the fixed-point word, then floating-point arithmetic leads to higher signal-to-noise ratio at the output. The difference was found to be greater for poles close to the unit circle. However, additional bits are required to represent the characteristic, and the greater the desired dynamic range, the more bits are required for the characteristic. Also, the hardware to implement floating-point arithmetic is much more complex than that for fixed-point arithmetic. Therefore, the use of floating-point arithmetic entails increased wordlength and increased complexity in the arithmetic unit. Its major advantage is that it essentially eliminates the problems of overflow, and if a sufficiently long mantissa is used, quantization also becomes much less a problem. This translates into greater simplicity in system design and implementation.

6.10 ZERO-INPUT LIMIT CYCLES IN FIXED-POINT REALIZATIONS OF IIR DIGITAL FILTERS

For stable IIR discrete-time systems implemented with infinite-precision arithmetic, if the excitation becomes zero and remains zero for n greater than some value n_0, the output for $n > n_0$ will decay asymptotically toward zero. For the same system, implemented with finite-register-length arithmetic, the output may continue to oscillate indefinitely while the input remains equal to zero. This effect is often referred to as *zero-input limit cycle behavior* and is a consequence either of the nonlinear quantizers in the feedback loop of the system or of overflow of additions.

6.10.1 Limit Cycles Due to Roundoff and Truncation

The limit cycle behavior of a digital filter is complex and difficult to analyze, and we will not attempt to treat the topic in any general sense. For simple first- and second-order filters, however, it is possible to understand the effect and to develop an interpretation of the oscillations in terms of the effective filter poles moving onto the unit circle. The effect is best illustrated by an example.

Example 6.16

Consider the first-order system characterized by the difference equation

$$y[n] = ay[n-1] + x[n], \qquad |a| < 1. \tag{6.144}$$

The signal flow graph of this system is shown in Fig. 6.66(a). Let us assume that the register length for storing the coefficient a, the input $x[n]$, and the filter node variable $y[n-1]$ is 4 bits (i.e., a sign bit to the left of the binary point and 3 bits to the right of the binary point). Because of the finite-length registers, the product $ay[n-1]$ must be rounded or truncated to 4 bits before addition to $x[n]$. The flow graph representing the

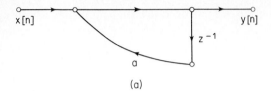

(a)

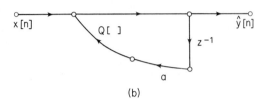

(b)

Figure 6.66 First-order IIR system.
(a) Infinite-precision linear system.
(b) Nonlinear system due to
quantization.

actual realization based on Eq. (6.144) is shown in Fig. 6.66(b). Assuming rounding of
the product, the actual output $\hat{y}[n]$ satisfies the nonlinear difference equation

$$\hat{y}[n] = Q[a\hat{y}[n-1]] + x[n], \tag{6.145}$$

where $Q[\cdot]$ represents the rounding operation. Let us assume that $a = 1/2 = 0_\diamond 100$ and
that the input is $x[n] = (7/8)\delta[n] = (0_\diamond 111)\delta[n]$. Using Eq. (6.145), we see that for
$n = 0$, $\hat{y}[0] = 7/8 = 0_\diamond 111$. To obtain $\hat{y}[1]$, we multiply $\hat{y}[0]$ by a, obtaining the result
$a\hat{y}[0] = 0_\diamond 011|100$, a 7-bit number that must be rounded to 4 bits. This number, 7/16, is
exactly halfway between the two 4-bit quantization levels 4/8 and 3/8. If we choose
always to round upward in such cases, then $0_\diamond 011100$ rounded to 4 bits is $0_\diamond 100 =$
1/2. Since $x[1] = 0$, then $\hat{y}[1] = 0_\diamond 100 = 1/2$. Continuing to iterate the difference
equation gives $\hat{y}[2] = Q[a\hat{y}[1]] = 0_\diamond 010 = 1/4$ and $\hat{y}[3] = 0_\diamond 001 = 1/8$. In both these
cases no rounding is necessary. However, to obtain $\hat{y}[4]$, we must round the 7-bit
number $a\hat{y}[3] = 0_\diamond 000100$ to $0_\diamond 001$. The same result is obtained for all values of
$n \geq 3$. The output sequence for this example is shown in Fig. 6.67(a). If $a = -1/2$ we can
carry out the preceding computation again to show that the output is as shown in Fig.
6.67(b). Thus, because of rounding of the product $a\hat{y}[n-1]$, the output reaches a
constant value of 1/8 when $a = 1/2$ and a periodic steady-state oscillation between $+1/8$
and $-1/8$ when $a = -1/2$. These are periodic outputs similar to those that would be
obtained from a first-order pole at $z = \pm 1$ instead of at $z = \pm 1/2$. When $a = +1/2$,
the period of the oscillation is 1, and when $a = -1/2$, the period of oscillation is 2. Such
steady-state periodic outputs are called *limit cycles*, and their existence was first noted by
Blackman (1965), who referred to the amplitude intervals to which such limit cycles are
confined as *dead bands*. In this case the dead band is $-2^{-B} \leq \hat{y}[n] \leq +2^{-B}$, where
$B = 3$.

The possible existence of a zero-input limit cycle is important in applications
where a digital filter is to be in continuous operation, since it is generally desired that
the output approach zero when the input is zero. For example, suppose that a speech
signal is sampled, filtered by a digital filter, and then converted back to an acoustic
signal using a D/A converter. In such a situation it would be very undesirable for the
filter to enter a periodic limit cycle whenever the input is zero, since the limit cycle
would produce an audible tone.

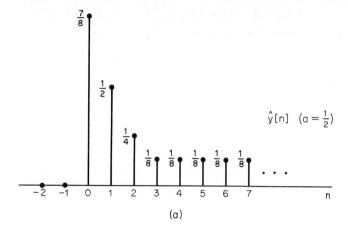

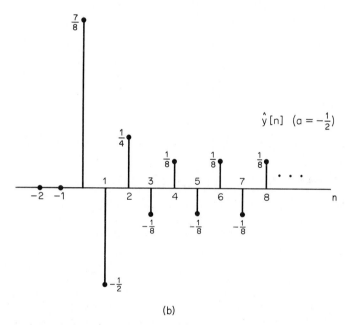

Figure 6.67 Response of the first-order system of Fig. 6.66 to an impulse.
(a) $a = \frac{1}{2}$. (b) $a = -\frac{1}{2}$.

Jackson (1969) has considered limit cycle behavior in first- and second-order systems with an analysis based on the observation that in the limit cycle, the system behaves as if the poles of the system were on the unit circle. Specifically, consider the first-order filter of Example 6.16. By the definition of rounding,

$$|Q[a\hat{y}[n-1]] - a\hat{y}[n-1]| \le \tfrac{1}{2}(2^{-B}). \tag{6.146}$$

Furthermore, for values of n in the limit cycle,

$$|Q[a\hat{y}[n-1]]| = |\hat{y}[n-1]|, \tag{6.147}$$

i.e., the effective value of a is 1, corresponding to the pole of the filter being on the unit circle. The range of values for which this condition is met is

$$|\hat{y}[n-1]| - |a\hat{y}[n-1]| \le \tfrac{1}{2}(2^{-B}) \tag{6.148}$$

or, solving for $|\hat{y}[n-1]|$,

$$|\hat{y}[n-1]| \le \frac{\tfrac{1}{2}(2^{-B})}{1-|a|}. \tag{6.149}$$

Equation (6.149) defines the dead band for the first-order filter. As a result of rounding, values within the dead band are quantized in steps of 2^{-B}. (Note that Eq. 6.149 gives the correct value for the dead band when $|a| = 1/2$.) Whenever the node variable $\hat{y}[n-1]$ falls within the dead band when the input is zero, the filter enters into a limit cycle and remains in this mode until an input is applied to carry the output out of the dead band.

A second-order filter demonstrates a larger variety of limit cycle behavior. Consider the second-order difference equation

$$y[n] = x[n] + a_1 y[n-1] + a_2 y[n-2]. \tag{6.150}$$

With $a_1^2 < -4a_2$, the filter poles are complex conjugates, and with $a_2 = -1$, the poles are on the unit circle. If the products are rounded before accumulation in implementing Eq. (6.150), then

$$\hat{y}[n] = x[n] + Q[a_1 \hat{y}[n-1]] + Q[a_2 \hat{y}[n-2]], \tag{6.151}$$

where $Q[\cdot]$ again represents rounding of the indicated products. As before, by the definition of rounding,

$$|Q[a_2 \hat{y}[n-2]] - a_2 \hat{y}[n-2]| \le \tfrac{1}{2}(2^{-B}). \tag{6.152}$$

With $x[n] = 0$, the poles of the system will appear to be on the unit circle if

$$Q[a_2 \hat{y}[n-2]] = -\hat{y}[n-2]. \tag{6.153}$$

Substituting this equation into Eq. (6.152), we obtain

$$|\hat{y}[n-2]| - |1+a_2| \le \tfrac{1}{2}(2^{-B}) \tag{6.154}$$

or, solving for $|\hat{y}[n-2]|$,

$$|\hat{y}[n-2]| \le \frac{\tfrac{1}{2}(2^{-B})}{|1+a_2|}. \tag{6.155}$$

Thus, if $\hat{y}[n-2]$ falls in this range when the input is zero, the effective value of a_2 is such that the poles of the system are apparently on the unit circle. Under these conditions, the value of a_1 controls the frequency of oscillation.

In a second mode of limit cycle behavior that can occur in second-order filters, the effect of rounding is to place the effective poles at $z = +1$ and $z = -1$. The dead band corresponding to this mode is bounded by $1/(1 - |a_1| - a_2)$. The amplitude of a limit cycle in this band is of course quantized in steps of 2^{-B} (Jackson, 1969).

This discussion has been concerned only with zero-input limit cycles that result from rounding in first- and second-order IIR systems. Although the analysis was somewhat heuristic, the simple formulas that result have been found to be consistent with experimental results and are useful in predicting limit cycle behavior in IIR digital filters. A similar style of analysis can be used for truncation (see, for example, Problem 6.33). In the case of parallel realizations of higher-order systems, the outputs of the individual second-order systems are independent when the input is zero. Thus, the previous analysis can be applied directly. In the case of cascade realizations, only the first section has zero input. Succeeding sections may exhibit their own characteristic limit cycle behavior or they may appear to be simply filtering the limit cycle output of a previous section. For high-order systems realized by other filter structures, the limit cycle behavior becomes more complex, as does its analysis. When the input is nonzero, the quantization effects depend on the input, and the style of analysis of this section is completely inappropriate except for simple inputs such as an impulse, a unit step, or a sinusoid. In other cases the complexity of the quantization phenomena force us to a statistical model.

In addition to giving an understanding of limit cycle effects in digital filters, the preceding results are useful when the zero-input limit cycle response of a system is the desired output. This is the case, for example, when one is concerned with digital sine-wave oscillators for signal generation and for generation of coefficients for discrete Fourier transformations.

6.10.2 Limit Cycles Due to Overflow

In addition to the classes of limit cycles discussed in the preceding section, a more severe type of limit cycle can occur due to overflow. The effect of overflow is to insert a gross error in the output, and in some cases the filter output thereafter oscillates between large-amplitude limits. Such limit cycles have been referred to as *overflow oscillations*. The problem of oscillations caused by overflow is discussed in detail by Ebert et al. (1969). Overflow oscillations are illustrated by the following example.

Example 6.17

Consider a second-order system realized by the following difference equation:

$$\hat{y}[n] = x[n] + Q[a_1 \hat{y}[n-1]] + Q[a_2 \hat{y}[n-2]], \tag{6.156}$$

where $Q[\cdot]$ represents two's-complement rounding with a wordlength of 3 bits plus sign. Overflow can occur with two's-complement addition of the rounded products. Suppose that $a_1 = 3/4 = 0_\diamond 110$, and $a_2 = -3/4 = 1_\diamond 010$, and assume that $x[n]$ remains equal to zero for $n \geq 0$. Furthermore, assume that $\hat{y}[-1] = 3/4 = 0_\diamond 110$, and $\hat{y}[-2] = -3/4 = 1_\diamond 010$. Now the output at sample $n = 0$ is

$$\hat{y}[0] = 0_\diamond 110 \times 0_\diamond 110 + 1_\diamond 010 \times 1_\diamond 010.$$

If we evaluate the products using two's-complement arithmetic, we obtain

$$\hat{y}[0] = 0_\diamond 100100 + 0_\diamond 100100$$

and if we choose to round upward when a number is halfway between two quantization levels, the result of two's-complement addition is

$$\hat{y}[0] = 0_\diamond 101 + 0_\diamond 101 = 1_\diamond 010 = -\tfrac{3}{4}.$$

In this case the binary carry overflows into the sign bit, thus changing the positive sum into a negative number. Repeating the process gives

$$\hat{y}[1] = 1_\diamond 011 + 1_\diamond 011 = 0_\diamond 110 = \tfrac{3}{4}.$$

In this case, the binary carry resulting from the sum of the sign bits is lost and the negative sum is mapped into a positive number. Clearly, $\hat{y}[n]$ will continue to oscillate between $+3/4$ and $-3/4$ until an input is applied. Thus $\hat{y}[n]$ has entered a periodic limit cycle with a period of 2 and an amplitude of almost the full-scale amplitude of the implementation.

This example illustrates how overflow oscillations occur. Much more complex behavior can be exhibited by higher-order systems, and other frequencies can occur. Some results similar to those of Section 6.10.1 are available for predicting when overflow oscillations can be supported by a difference equation (Ebert et al., 1969). Overflow oscillations can be avoided by using the saturation overflow characteristic of Fig. 6.44(b) (Ebert et al., 1969).

One approach to the general problem of limit cycles is to seek structures that do not support limit cycle oscillations. Such structures have been derived using state-space representations (Barnes and Fam, 1977; Mills et al., 1978) and concepts analogous to passivity in analog systems (Rao and Kailath, 1984; Fettweis, 1986). However, these structures generally require more computation than an equivalent cascade or parallel form implementation. By adding more bits to the computational wordlength, we can generally avoid overflow. Similarly, since round-off limit cycles usually are limited to the least significant bits of the binary word, additional bits can be used to reduce the effective amplitude of the limit cycle. Also, Claasen et al. (1973) showed that if a double-length accumulator is used so that quantization occurs after accumulation of products, then limit cycles due to roundoff are much less likely to occur in second-order systems. Thus, the tradeoff between wordlength and computational algorithm complexity arises for limit cycles just as it does for coefficient quantization and roundoff noise.

Finally, it is important to point out that zero-input limit cycles due to both overflow and roundoff are a phenomenon unique to IIR systems. FIR systems cannot support zero-input limit cycles because they have no feedback paths. The output of an FIR system will be zero no later than $(M + 1)$ samples after the input goes to zero and remains there. This is a major advantage of FIR systems in applications where limit cycle oscillations cannot be tolerated.

6.11 SUMMARY

In this chapter we have considered many aspects of the problem of implementing a linear time-invariant discrete-time system. The first half of the chapter was devoted to basic implementation structures. After introducing block diagram and signal flow graphs as pictorial representations of difference equations, we discussed a number of basic structures for IIR and FIR discrete-time systems. These included the direct form I, direct form II, cascade form, parallel form, lattice form, and transposed form. It was shown that these forms are all equivalent when implemented with infinite-precision arithmetic. However, the different structures are most significant in the context of finite-precision implementations. Therefore, the remainder of the chapter addressed problems associated with finite precision or quantization in digital implementations of the basic structures.

We began with a brief review of digital number representation and an overview that showed that quantization effects are important in sampling (discussed in Chapter 3), in representing the coefficients of a discrete-time system, and in implementing systems using finite-precision arithmetic. We illustrated the effect of quantization of the coefficients of a difference equation through simple sensitivity analyses and through several examples of the effects of coefficient quantization. This issue was treated independently of the effects of finite-precision arithmetic, which we showed introduces nonlinearity into the system. We showed that in some cases this non-linearity was responsible for limit cycles that may persist after the input to a system has become zero. We also showed that quantization effects can be modeled in terms of independent random white-noise sources that are injected internally in the network. Such linear noise models were developed for the direct form structures and for the cascade structure. In all of our discussion of quantization effects, the underlying theme was the conflict between the desire for fine quantization and the need to maintain a wide range of possible signal amplitudes. We saw that in fixed-point implementations, one can be improved at the expense of the other, but to improve one while leaving the other unaffected requires that we increase the number of bits used to represent coefficients and signal amplitudes. This can be done either by increasing the fixed-point wordlength or by adopting a floating-point representation.

Our discussion of quantization effects serves two purposes. First, we developed several results that can be useful in guiding the design of practical implementations. We found that quantization effects depend greatly on the structure used and on the specific parameters of the system to be implemented, and even though simulation of the system is generally necessary to evaluate performance, many of the results discussed are useful in making intelligent decisions in the design process. A second, equally important purpose of this part of the chapter was to illustrate a style of analysis that can be applied in studying quantization effects in a variety of digital signal processing algorithms. The examples of this chapter indicate the types of assumptions and approximations that are commonly made in studying quantization effects. We will apply the analysis techniques developed here to the study of quantization in the computation of the discrete Fourier transform in Chapter 9.

PROBLEMS

6.1. Determine the system function of the two networks in Fig. P6.1 and show that they have the same poles.

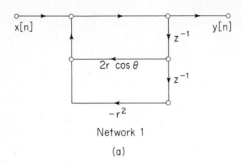

Network 1

(a)

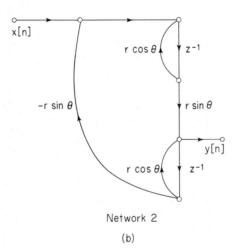

Network 2

(b)

Figure P6.1

6.2. The signal flow graph of Fig. P6.2 represents a linear difference equation with constant coefficients. Determine the difference equation that relates the output $y[n]$ to the input $x[n]$.

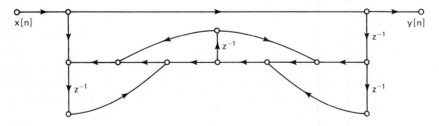

Figure P6.2

6.3. Consider a general flow graph (denoted network A) consisting of coefficient multipliers and delay elements, as shown in Fig. P6.3-1. If the system is initially at rest, its behavior is completely specified by its impulse response $h[n]$. We wish to modify the system to create a new flow graph (denoted network A_1) with impulse response $h_1[n] = (-1)^n h[n]$.

   (a) If $H(e^{j\omega})$ is as given in Fig. P6.3-2, sketch $H_1(e^{j\omega})$.

   (b) Explain how to modify network A by simple modifications of the coefficient multipliers and/or the delay branches of network A to form the new network A_1, such that the impulse response of A_1 is $h_1[n]$.

   (c) If network A is as given in Fig. P6.3-3, show how to modify it by simple modifications to *only the coefficient multipliers* so that the resulting network A_1 has impulse response $h_1[n]$.

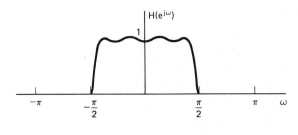

Network A   **Figure P6.3-1**

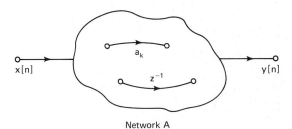

Figure P6.3-2

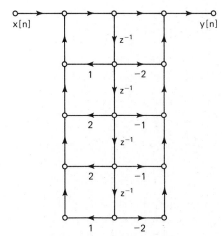

Figure P6.3-3

6.4. For many applications it is useful to have a system that will generate a sinusoidal sequence. One possible way to do this is with a system whose impulse response is $h[n] = e^{j\omega_0 n}u[n]$. The real and imaginary parts of $h[n]$ are therefore $h_r[n] = (\cos \omega_0 n)u[n]$ and $h_i[n] = (\sin \omega_0 n)u[n]$.

In implementing a system with a complex impulse response, the real and imaginary parts are distinguished as separate outputs. By first writing the complex difference equation required to produce the desired impulse response and then separating it into its real and imaginary parts, draw a flow graph that will implement this system. The flow graph that you draw can have only real coefficients. This implementation is sometimes called the *coupled form oscillator* since when the input is excited by an impulse, the outputs are sinusoidal.

6.5. For the system function

$$H(z) = \frac{1 + 2z^{-1} + z^{-2}}{1 - \frac{3}{4}z^{-1} + \frac{1}{8}z^{-2}},$$

draw the flow graphs of all possible realizations for this system as cascades of first-order systems.

6.6. The flow graph shown in Fig. P6.6 is noncomputable; i.e., it is not possible to compute the output using the difference equations represented by the flow graph because it contains a closed loop having no delay elements.

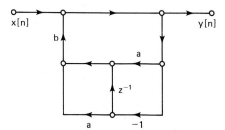

Figure P6.6

(a) Write the difference equations for the system of Fig. P6.6 and from them find the system function of the network.

(b) From the system function, obtain a flow graph that is computable.

6.7. Consider a causal linear time-invariant system whose system function is

$$H(z) = \frac{1 + \frac{1}{5}z^{-1}}{(1 - \frac{1}{2}z^{-1} + \frac{1}{3}z^{-2})(1 + \frac{1}{4}z^{-1})}.$$

(a) Draw the signal flow graphs for implementations of the system in each of the following forms:
  (i) Direct form I
  (ii) Direct form II
  (iii) Cascade form using first- and second-order direct form II sections
  (iv) Parallel form using first- and second-order direct form II sections
  (v) Transposed direct form II

(b) Write the difference equations for the flow graph of part (v) in (a) and show that this system has the correct system function.

6.8. Several flow graphs are shown in Fig. P6.8. Determine the transpose of each flow graph and verify that in each case the original and transpose flow graphs have the same system function.

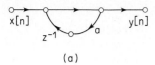

(a)

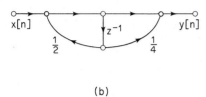

(b)

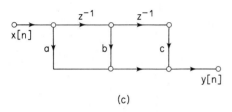

(c)

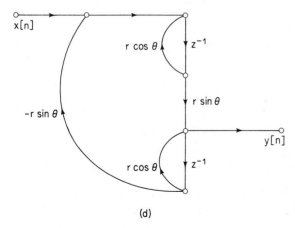

(d)   **Figure P6.8**

6.9. Figure P6.9(a)–(f) shows six systems. Determine which one of the last five, (b)–(f), has the same system function as (a). You should be able to eliminate some of the possibilities by inspection.

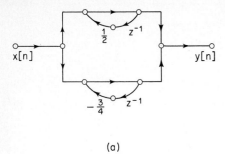

(a)

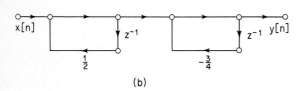

(b)

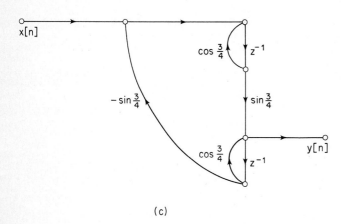

(c)

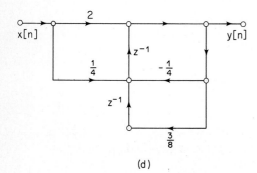

(d)

Figure P6.9

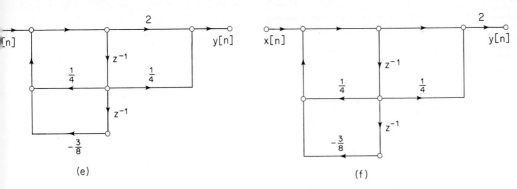

Figure P6.9 (continued)

6.10. Consider the system in Fig. P6.10.
   (a) Find the system function relating the z-transforms of the input and output.
   (b) Write the difference equation that is satisfied by the input sequence $x[n]$ and the output sequence $y[n]$.
   (c) Draw a signal flow graph that has the same input/output relationship as the system in Fig. P6.10 but that has the smallest possible number of delay elements.

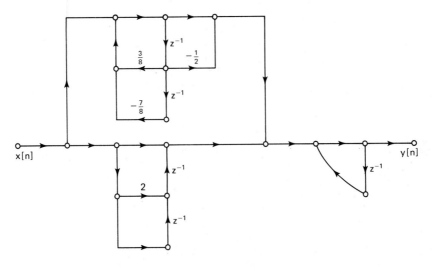

Figure P6.10

6.11. A linear time-invariant system is realized by the flow graph shown in Fig. P6.11.
   (a) Write the difference equation(s) represented by this flow graph.
   (b) What is the system function of the system?
   (c) In the realization of Fig. P6.11, how many real multiplications and real additions are required to compute each sample of the output? (Assume that $x[n]$ is real and assume that multiplication by 1 does not count in the total.)

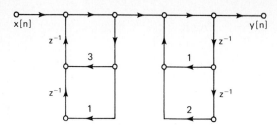

Figure P6.11

(d) The realization of Fig. P6.11 requires four storage registers (delay elements). Is it possible to reduce the number of storage registers by using a different structure? If so, draw the flow graph; if not, explain why the number of storage registers cannot be reduced.

6.12. A linear time-invariant system with system function

$$H(z) = \frac{0.2(1 + z^{-1})^6}{(1 - 2z^{-1} + \frac{7}{8}z^{-2})(1 + z^{-1} + \frac{1}{2}z^{-2})(1 - \frac{1}{2}z^{-1} + z^{-2})}$$

is to be implemented using a flow graph of the form shown in Fig. P6.12.

(a) Fill in all the coefficients in the diagram of Fig. P6.12. Is your solution unique?

(b) Define appropriate node variables in Fig. P6.12 and write the set of difference equations that is represented by the flow graph.

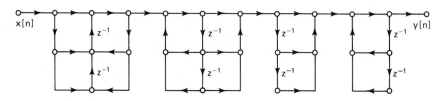

Figure P6.12

6.13. The following simple FORTRAN program implements a second-order discrete-time filter. The input sequence resides in the array X of length L and the output sequence is placed in the array Y.

```
DIMENSION X(L),Y(L)
F = 0.
G = 0.
DO 1 N = 1, L
   Y(N) = X(N) + F
   F = - X(N) + 0.75 * Y(N) + G
   G = - 0.125 * Y(N)
1  CONTINUE
STOP
END
```

Draw the flow graph representation of this system.

6.14. Determine the impulse response of each of the systems in Fig. P6.14.

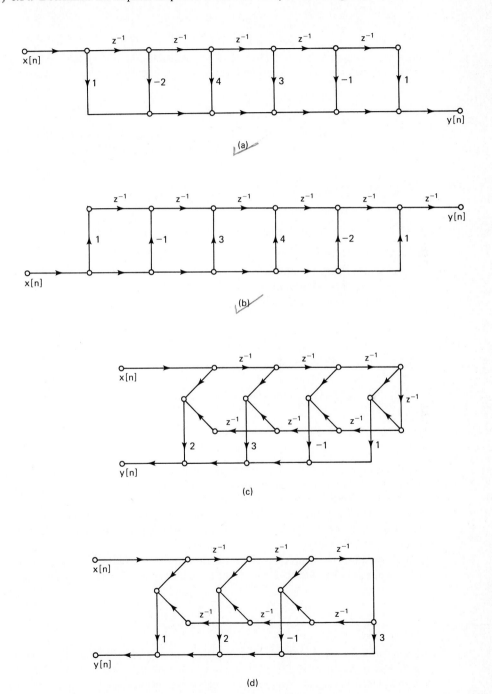

Figure P6.14

6.15. The impulse response of a linear time-invariant system is

$$h[n] = \begin{cases} a^n, & 0 \le n \le 7, \\ 0, & \text{otherwise.} \end{cases}$$

(a) Draw the flow graph of a direct form FIR implementation of the system.
(b) Show that the corresponding system function can be expressed as

$$H(z) = \frac{1 - a^8 z^{-8}}{1 - az^{-1}}, \qquad |z| > |a|.$$

(c) Draw the flow graph of another system having the same system function and consisting of a cascade of an FIR system with an IIR system (assume $|a| < 1$).
(d) Which implementation of the system requires
  (i) the most storage (delay elements)?
  (ii) the most arithmetic (multiplications and additions per output sample)?

6.16. Consider an FIR system whose impulse response is

$$h[n] = \begin{cases} \frac{1}{15}(1 + \cos[(2\pi/15)(n - n_0)]), & 0 \le n \le 14, \\ 0, & \text{otherwise.} \end{cases}$$

This system is an example of a frequency sampling filter.
(a) Sketch the impulse response of this system for the case $n_0 = \frac{15}{2}$.
(b) Show that the system function of this system can be expressed as

$$H(z) = (1 - z^{-15}) \cdot \frac{1}{15}\left[\frac{1}{1 - z^{-1}} + \frac{\frac{1}{2}e^{-j2\pi n_0/15}}{1 - e^{j2\pi/15}z^{-1}} + \frac{\frac{1}{2}e^{j2\pi n_0/15}}{1 - e^{-j2\pi/15}z^{-1}} \right].$$

(c) Show that if $n_0 = \frac{15}{2}$, the frequency response of the system can be expressed as

$$H(e^{j\omega}) = \frac{1}{15}e^{-j\omega 7}\left\{ \frac{\sin(\omega 15/2)}{\sin(\omega/2)} + \frac{1}{2}\frac{\sin[(\omega - 2\pi/15)15/2]}{\sin[(\omega - 2\pi/15)/2]} \right.$$

$$\left. + \frac{1}{2}\frac{\sin[(\omega + 2\pi/15)15/2]}{\sin[(\omega + 2\pi/15)/2]} \right\}.$$

Use this expression to sketch the magnitude of the frequency response of the system for $n_0 = \frac{15}{2}$. Obtain a similar expression for $n_0 = 0$. Does this system have generalized linear phase? Sketch the magnitude response for $n_0 = 0$.
(d) Draw a signal flow graph of an implementation of this system as a cascade of an FIR system whose system function is $1 - z^{-15}$ and a parallel combination of a first- and second-order IIR system.

6.17. In this problem we will develop some of the properties of a class of discrete-time systems called frequency sampling filters. This class of filters has system functions of the form

$$H(z) = (1 - z^{-N}) \cdot \sum_{k=0}^{N-1} \frac{\tilde{H}[k]/N}{1 - z_k z^{-1}},$$

where $z_k = e^{j(2\pi/N)k}$ for $k = 0, 1, \ldots, N - 1$.
(a) System functions such as $H(z)$ can be implemented as a cascade of an FIR system whose system function is $(1 - z^{-N})$ with a parallel combination of first-order IIR systems. Draw the signal flow graph of such an implementation.
(b) Show that $H(z)$ as defined above is an $(N - 1)$st degree polynomial in z^{-1}. To do this it is necessary to show that $H(z)$ has no poles other than $z = 0$ and that it has no

powers of z^{-1} higher than $(N-1)$. What does this imply about the length of the impulse response of the system?

(c) Show that the impulse response is given by the following expression:

$$h[n] = \left(\frac{1}{N}\sum_{k=0}^{N-1}\tilde{H}[k]e^{j(2\pi/N)kn}\right)(u[n] - u[n-N]).$$

[*Hint*: Find the impulse responses of the FIR and the IIR parts of the system and convolve them to find the overall impulse response.]

(d) Use l'Hôpital's rule to show that

$$H(z_m) = H(e^{j(2\pi/N)m}) = \tilde{H}[m], \qquad m = 0, 1, \ldots, N-1,$$

i.e., show that the constants $\tilde{H}[m]$ are samples of the frequency response of the system at equally spaced frequencies $\omega_m = (2\pi/N)m$ for $m = 0, 1, \ldots, N-1$. It is this property that accounts for the name of this class of FIR systems.

(e) In general both the poles z_k of the IIR part and the samples of the frequency response $\tilde{H}[k]$ will be complex. However, if $h[n]$ is real, we can find an implementation involving only real quantities. Specifically, show that if $h[n]$ is real and N is an even integer, then $H(z)$ can be expressed as

$$H(z) = (1 - z^{-N})\left\{\frac{H(1)/N}{1 - z^{-1}} + \frac{H(-1)/N}{1 + z^{-1}}\right.$$

$$\left. + \sum_{k=1}^{(N/2)-1}\frac{2|H(e^{j(2\pi/N)k})|}{N}\cdot\frac{\cos[\theta(2\pi k/N)] - z^{-1}\cos[\theta(2\pi k/N) - 2\pi k/N]}{1 - 2\cos(2\pi k/N)z^{-1} + z^{-2}}\right\},$$

where $H(e^{j\omega}) = |H(e^{j\omega})|e^{j\theta(\omega)}$. Draw the signal flow graph representation of such a system when $N = 16$ and $H(e^{j\omega_k}) = 0$ for $k = 3, 4, \ldots, 14$.

6.18. In Chapter 3 we showed that in general the sampling rate of a discrete-time signal can be reduced by a combination of linear filtering and time compression. Figure P6.18 shows a block diagram of an M-to-1 decimator that can be used to reduce the sampling rate by an integer factor M. According to the model, the linear filter operates at the high sampling rate. However, if M is large, most of the output samples of the filter will be discarded by the compressor. In some cases more efficient implementations are possible.

(a) Assume that the filter is an FIR system with impulse response such that $h[n] = 0$ for $n < 0$ and for $n > 10$. Draw the flow graph for the complete system of Fig. P6.18 for this filter.

(b) Note that some of the branch operations can be commuted with the compression operation. Using this fact, draw the flow graph of a more efficient realization of the system of part (a). By what factor has the total computation required in obtaining the output $y[n]$ been decreased?

(c) Now suppose that the filter in Fig. P6.18 has system function

$$H(z) = \frac{1}{1 - \frac{1}{2}z^{-1}}, \qquad |z| > \frac{1}{2}.$$

Draw the flow graph of the direct form realization of the complete system in Fig. P6.18. With this system for the linear filter, can the total computation per output sample be reduced? If so, by what factor?

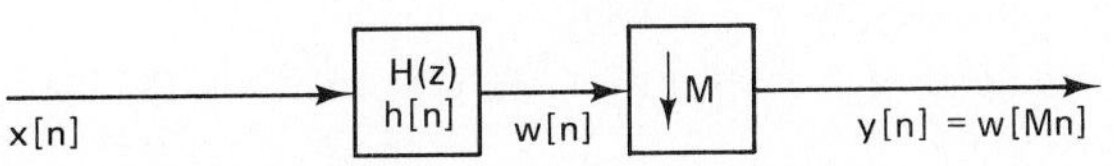

Figure P6.18

(d) Finally, suppose that the filter in Fig. P6.18 has system function

$$H(z) = \frac{1 + \frac{7}{8}z^{-1}}{1 - \frac{1}{2}z^{-1}}, \qquad |z| > \frac{1}{2}.$$

Draw the flow graph for the complete system of Fig. P6.18 using each of the following forms for the linear filter:

(i) direct form I

(ii) direct form II

(iii) transposed direct form I

(iv) transposed direct form II

For which of the four forms can the system of Fig. P6.18 be more efficiently implemented by commuting operations with the compressor?

6.19. Figure P6.19 shows the direct form and lattice form flow graphs for the FIR system discussed in Example 6.7. We wish to verify that the two flow graphs have the same system function.

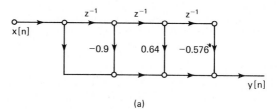

(a)

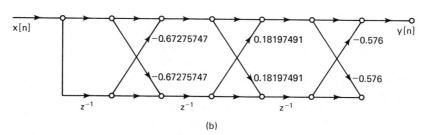

(b)

Figure P6.19

(a) Use Eqs. (6.54) to find the coefficients of the system function polynomial $A(z)$ for the system of Fig. P6.19(b) from the k-parameters and compare $A(z)$ to the system function of the system in Fig. P6.19(a).

(b) Compute the impulse responses of the two systems in Fig. P6.19 by simply tracing an impulse input through all paths in the flow graphs and summing the impulses that arrive at the output with the same delay.

6.20. The lattice network shown in Fig. P6.20-1 has two inputs $x_1[n]$ and $x_2[n]$ and two outputs $y_1[n]$ and $y_2[n]$.

(a) Write the two difference equations relating the two outputs to the two inputs.

If the output $y_1[n]$ in Fig. P6.20-1 is connected to the input $x_2[n]$, we obtain the network of Fig. P6.20-2.

(b) Determine the system function $H_{11}(z) = Y_1(z)/X_1(z)$ for the network of Fig. P6.20-2. For what values of K will the system be stable?

(c) Determine the system function $H_{21}(z) = Y_2(z)/X_1(z)$ for the network of Fig. P6.20-2. Plot $|H_{21}(e^{j\omega})|$ for $-\pi < \omega < \pi$.

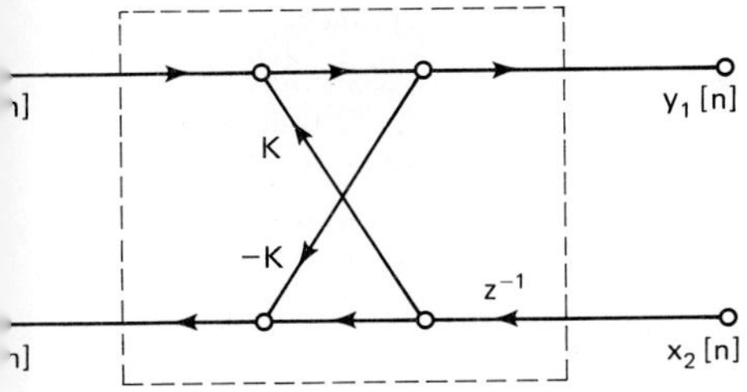

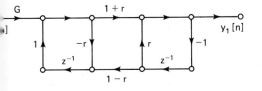

Figure P6.20-1

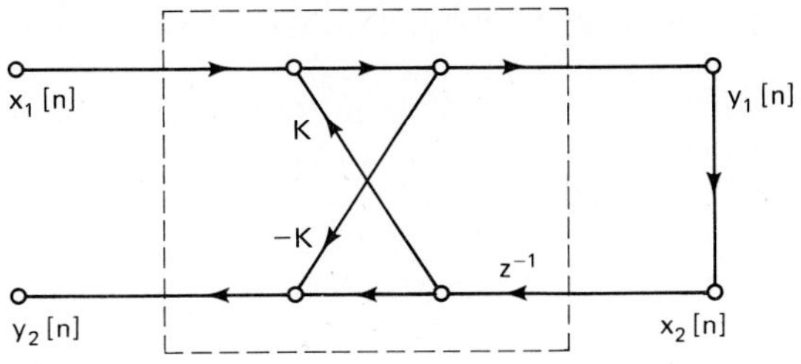

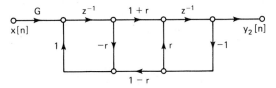

Figure P6.20-2

6.21. Consider the discrete-time system depicted in Fig. P6.21-1.
   (a) Write the set of difference equations represented by the flow graph of Fig. P6.21-1.
   (b) Determine the system function $H_1(z) = Y_1(z)/X(z)$ of the system in Fig. P6.21-1, and determine the magnitudes and angles of the poles of $H_1(z)$ as a function of r for $-1 < r < 1$.
   (c) Figure P6.21-2 shows a different flow graph obtained from the flow graph of Fig. P6.21-1 by moving the delay elements to the opposite top branch. How is the system function $H_2(z) = Y_2(z)/X(z)$ related to $H_1(z)$?

Figure P6.21-1   Figure P6.21-2

6.22. Speech production can be modeled by a linear system representing the vocal cavity, which is excited by puffs of air released through the vibrating vocal cords. One approach to synthesizing speech involves representing the vocal cavity as a connection of cylindrical acoustic tubes with equal length but with varying cross-sectional areas, as depicted in Fig. P6.22. Let us assume that we want to simulate this system in terms of the volume velocity representing airflow. The input is coupled into the vocal tract through a small constriction, the vocal cords. We will assume that the input is represented by a change in volume velocity at the left end but that the boundary condition for traveling waves at the left end is that the net volume velocity must be zero. (This is analogous to an electrical transmission line driven by a current source.) The output is considered to be the volume velocity at the right end. We assume that each section is a lossless acoustic transmission line.

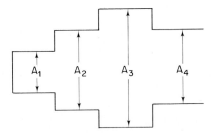

Figure P6.22

At each interface between sections, a forward-traveling wave f^+ is transmitted to the next section with one coefficient and reflected as a backward-traveling wave f^- with a different coefficient. Similarly, a backward-traveling wave f^- arriving at an interface is transmitted with one coefficient and reflected with a different coefficient. Specifically, if we consider a forward-traveling wave f^+ in a tube with cross-sectional area A_1 arriving at the interface with a tube of cross-sectional area A_2, then the forward-traveling wave transmitted is $(1 + r)f^+$ and the reflected wave is is rf^+, where

$$r = \frac{A_2 - A_1}{A_2 + A_1}.$$

Consider the length of each section to be 3.4 cm, with the velocity of sound in air equal to 34,000 cm/s. Draw a flow graph that will implement the four-section model in Fig. P6.22, with the output sampled at 20,000 samples/s.

In spite of the lengthy introduction, this a reasonably straightforward problem. If you find it difficult to think in terms of acoustic tubes, think in terms of transmission-line sections with different characteristic impedances. Just as with transmission lines, it is difficult to express the impulse response in closed form. Therefore, draw the network directly from physical considerations, in terms of forward- and backward-traveling pulses in each section.

6.23. The three networks in Fig. P6.23 are all equivalent implementations of the same two-input/two-output linear time-invariant system.

(a) Write the difference equations for network A.

(b) Find values of a, b, c, and d for network B in terms of r in network A such that the two systems are equivalent.

(c) Find values of e and f for network C in terms of r in network A such that the two systems are equivalent.

(d) Why might networks B or C be preferred over network A? What possible advantage could network A have over network B or C?

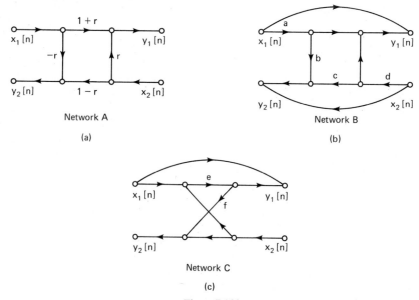

Network A

(a)

Network B

(b)

Network C

(c)

Figure P6.23

6.24. In modeling the effects of roundoff and truncation in digital filter implementations, quantized variables are represented as

$$\hat{x}[n] = Q[x[n]] = x[n] + e[n],$$

where $Q[\cdot]$ denotes either rounding or truncation to $(B + 1)$ bits and $e[n]$ is the *quantization error*. We assume that the quantization noise sequence is a stationary white-noise sequence such that

$$\mathcal{E}\{(e[n] - m_e)(e[n + m] - m_e)\} = \sigma_e^2 \delta[m]$$

and that the amplitudes of the noise sequence values are uniformly distributed over the quantization step $\Delta = 2^{-B}$. The first-order probability densities for rounding and truncation are shown in Figs. P6.24(a) and (b), respectively.

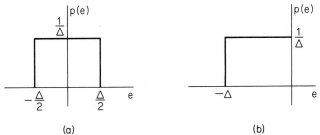

(a)   (b)   **Figure P6.24**

(a) Find the mean m_e and the variance σ_e^2 for the noise due to rounding.
(b) Find the mean m_e and the variance σ_e^2 for the noise due to truncation.

6.25. Consider a linear time-invariant system with two inputs as depicted in Fig. P6.25. Let $h_1[n]$ and $h_2[n]$ be the impulse responses from nodes 1 and 2, respectively, to the output, node 3. Show that if $x_1[n]$ and $x_2[n]$ are uncorrelated, then their corresponding outputs $y_1[n]$ and $y_2[n]$ are also uncorrelated.

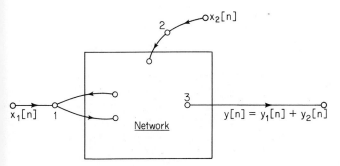

Figure P6.25

6.26. Consider an allpass system with system function

$$H(z) = \frac{z^{-1} - 0.54}{1 - 0.54z^{-1}}.$$

A flow graph for an implementation of the system is shown in Fig. P6.26.
(a) Determine the coefficients b, c, and d such that the flow graph in Fig. P6.26 is a realization of $H(z)$.

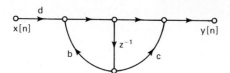

Figure P6.26

(b) In a practical implementation of the network in Fig. P6.26, the coefficients b, c, and d might be quantized by rounding the exact value to the nearest tenth (e.g., 0.54 will be rounded to 0.5 and 1.8518... will be rounded to 1.9). Would the resulting system still be an allpass system?

The difference equation relating input and output of the allpass system with system function $H(z)$ can be expressed as

$$y[n] = 0.54(y[n-1] - x[n]) + x[n-1].$$

(c) Draw the flow graph of a network that requires two delay elements but requires only one multiplication by a constant other than ± 1.

(d) With quantized coefficients, would the network of part (c) be an allpass system?

The primary disadvantage of the implementation in part (c) compared with the implementation in part (a) is that it requires two delay elements. However, for higher-order systems it is necessary to implement a cascade of allpass systems. For N allpass sections in cascade, it is possible to use allpass sections in the form determined in part (c) while only requiring $(N+1)$ delay elements. This is accomplished by sharing a delay element between sections.

(e) Consider the allpass system with system function

$$H(z) = \left(\frac{z^{-1} - a}{1 - az^{-1}}\right)\left(\frac{z^{-1} - b}{1 - bz^{-1}}\right).$$

Draw the flow graph of a "cascade" realization composed of two sections of the form obtained in part (c) with one delay element shared between the two sections. The resulting network should have only three delay elements.

(f) With quantized coefficients a and b, would the network in part (e) be an allpass system?

6.27. The networks in Fig. P6.27 all have the same system function. Assume that the systems in Fig. P6.27 are all implemented using $(B+1)$-bit fixed-point arithmetic in all the

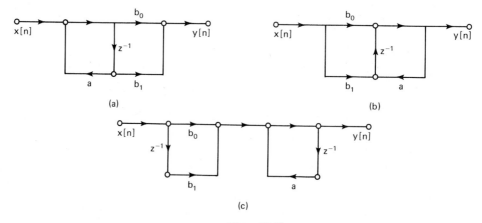

(a)

(b)

(c)

Figure P6.27

computations. Assume also that all products are rounded to $(B + 1)$ bits *before* additions are performed.

(a) Draw linear noise models for each of the systems in Fig. P6.27.

(b) Two of the networks in Fig. P6.27 have the *same* total output noise power due to arithmetic roundoff. Without explicitly computing the output noise power, determine which two have the same output noise power.

(c) Determine the output noise power for each of the networks in Fig. P6.27. Express your answer in terms of σ_B^2, the power of a single roundoff noise source.

6.28. The flow graph of Fig. P6.28 depicts an implementation of a first-order system.

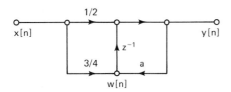

Figure P6.28

(a) Find the system function that describes the relationship between the input $x[n]$ and the output $y[n]$; i.e., $H(z) = Y(z)/X(z)$. Also find the impulse response $h[n]$.

(b) Find the system function that describes the relationships between the input $x[n]$ and the node variable $w[n]$; i.e., $H_{xw}(z) = W(z)/X(z)$. Also find the corresponding impulse response, $h_{xw}[n]$.

(c) What is the maximum value of $x[n]$ such that it is guaranteed that $|y[n]| < 1$ *and* $|w[n]| < 1$ for all n? How does the maximum value of $x[n]$ vary with $|a| < 1$?

(d) The products in Fig. P6.28 are quantized to $(B + 1)$ bits *before* any additions are carried out. Draw a linear noise model for the quantized system. Simplify your flow graph by combining as many sources as possible. Indicate the power of all noise sources in the resulting flow graph.

(e) Determine an expression for the total noise power at the output of the system. Obtain a closed-form expression for all sums or integrals. What happens to the noise power as $|a|$ approaches 1?

(f) Consider the cases (i) $a = 0.9$ and (ii) $a = 0.98$ with $x[n]$ a uniformly distributed white-noise input such that $|x[n]| < x_{max}$. Find x_{max} in each case such that no overflow can occur and compute the signal-to-noise ratio at the output for each case.

6.29. The three systems depicted in Fig. P6.29 all have system function

$$H(z) = \frac{0.2(1 + z^{-1})}{1 - 0.5z^{-1}}$$

for infinite-precision arithmetic. Assume that the systems in Fig. P6.29 are implemented with finite-precision arithmetic, with products rounded to $(B + 1)$ bits *before* addition.

(a) For each system in Fig. P6.29, determine the maximum value x_{max} of the input such that no overflow can occur anywhere in the system.

(b) For each system in Fig. P6.29, draw the linear noise model for implementation with $(B + 1)$-bit arithmetic combining independent sources where possible. Determine the total noise power at the output of each system.

(c) Assume that in each system in Fig. P6.29 the input $x[n]$ is a white-noise signal with a uniform distribution of amplitudes with maximum value x_{max} as determined in part (a). For each case, determine the signal-to-noise ratio at the output.

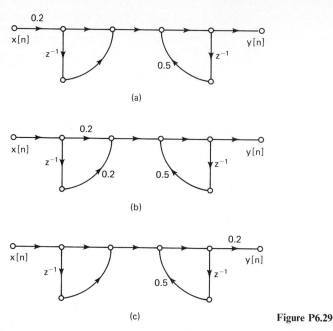

(a)

(b)

(c)

Figure P6.29

6.30. Consider the system shown in Fig. P6.30. Assume that $X_c(j\Omega) = 0$ for $|\Omega| \geq \pi/T$, and assume that the A/D converter is a 16-bit converter configured so that the maximum signal level is unity. The system represented by the flow graph part of Fig. P6.30 is implemented with 16-bit two's-complement arithmetic $(B + 1 = 16)$. Each 16-bit number represents a fraction less than unity, and each product is rounded *before* addition.

(a) Draw a linear noise model for the complete system in Fig. P6.30, i.e., include the effects of the A/D converter and the arithmetic quantization in the flow graph. Indicate the power of each noise source in terms of σ_{16}^2, the power of a single 16-bit rounding operation.

(b) How large can the input signal be if we want to be absolutely sure that $y[n] < 1$? Does your answer imply that all 16 bits of the A/D word can be used?

(c) Compute the total noise power at the output of the system due to all the noise sources in the complete system found in part (a).

Now assume that the A/D converter in Fig. P6.30 has 12-bit output samples, with the A/D converter again configured so that the maximum signal amplitude is unity. Assume as before that 16-bit two's-complement arithmetic with rounding is used to implement the computations represented by the flow graph.

(d) How would the linear noise model of part (a) change? Again find the total noise power at the output of the system due to all the noise sources in the complete system. Again express your answer in terms of σ_{16}^2, the total noise power of a single 16-bit rounding operation.

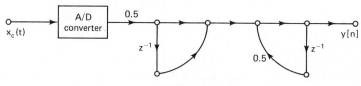

Figure P6.30

6.31. Consider the system function

$$H(z) = \frac{1 - 0.25z^{-1}}{1 - 0.25z^{-2}}.$$

The flow graphs in Fig. P6.31 represent four possible implementations of this system function.

(a) For each of the networks in Fig. P6.31, draw the flow graph of the linear noise model for implementation with $(B + 1)$-bit arithmetic with products rounded before addition.

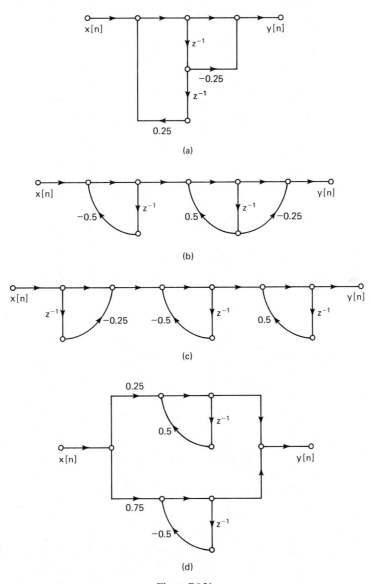

Figure P6.31

(b) For each network, find the noise power at the output of the system.

(c) For each network in Fig. P6.31, circle the critical nodes at which overflow must be avoided. Assume two's-complement arithmetic. For each network, determine x_{max} such that $|y[n]| < 1$ if $|x[n]| < x_{max}$.

(d) Assume that the input is a white-noise signal with uniform amplitude distribution. In each case assume that the maximum value of the input is the value determined in part (c). Compute the signal-to-noise power ratio for each of the cases in Fig. P6.31. Which network has the largest signal-to-noise ratio? Why?

6.32. This problem considers an alternative to fixed-point arithmetic called *block-floating-point* arithmetic. This approach is discussed in detail in Oppenheim (1970). To illustrate the procedure, consider a first-order system defined by the difference equation

$$y[n] = \alpha y[n-1] + x[n],$$

corresponding to the flow graph in Fig. P6.32-1. If the multiplication and addition in the system are implemented in fixed point, the full register length is not utilized on every iteration since the input must be scaled so that the *largest* output value is unity. Block-floating-point arithmetic corresponds to multiplying $x[n]$ and $y[n-1]$ by a gain $A[n]$ to jointly normalize them and then dividing $y[n]$ by this gain; i.e.,

$$y[n] = \frac{1}{A[n]} (\alpha y[n-1]A[n] + x[n]A[n]).$$

The normalizing coefficient $A[n]$ changes from iteration to iteration. The roundoff noise is therefore reduced by the factor $1/A[n]$ at each iteration of the difference equation.

Throughout this problem assume that $x[n]$ is a white-noise process with zero mean and uniform amplitude distribution between $-x_0$ and $+x_0$ such that $|y[n]| < 1$.

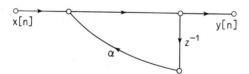

Figure P6.32-1

(a) Show that the network in Fig. P6.32-2 is equivalent to the network in Fig. P6.32-1 if quantization effects are neglected.

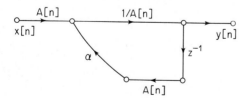

Figure P6.32-2

(b) The multiplication and addition are carried out with fixed-point arithmetic. The gain $A[n]$ is chosen as

$$A[n] = 2^{c[n]}, \qquad c[n] \geq 0,$$

so that it corresponds to a left shift of the registers containing $x[n]$ and $y[n-1]$. Since $A[n]$ is chosen to jointly normalize $x[n]$ and $y[n-1]$, we have

$$\tfrac{1}{2} \leq A[n] \max\{|x[n]|, |y[n-1]|\} < 1,$$

where $\max\{|[n]|, |y[n-1]|\}$ is the largest value of $|x[n]|$ and $|y[n-1]|$ for each n. We must therefore require that

$$\frac{\frac{1}{2}}{\max\{|x[n]|, |y[n-1]|\}} \le A[n] < \frac{1}{\max\{|x[n]|, |y[n-1]|\}}. \qquad \text{(P6.32)}$$

Since multiplication by a positive power of 2 introduces no roundoff noise, the noise sources introduced due to rounding of products are shown in Fig. P6.32-3.

Assuming that $e_1[n]$ and $e_2[n]$ are uniformly distributed between $-\frac{1}{2}(2^{-B})$ and $+\frac{1}{2}(2^{-B})$ and that they are uncorrelated with each other and with $x[n]$ (and consequently uncorrelated with $A[n]$), determine the variance of the output noise $f[n]$ for the network in Fig. P6.32-3. Express your answer in terms of $k^2 = \mathcal{E}[(1/A[n])^2]$. Show that the output noise variance due to arithmetic roundoff for the network of Fig. P6.32-3 is always *greater* than would be obtained if the first-order system were implemented directly as in Fig. P6.32-1 using fixed-point arithmetic.

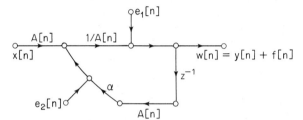

Figure P6.32-3

(c) The network in Fig. P6.32-2 can be modified to the equivalent form shown in Fig. P6.32-4. Let $e_1[n]$, $e_2[n]$, and $e_3[n]$ represent the noise introduced by rounding the multiplications by $1/A[n]$, $\Delta[n]$, and α, respectively. Again assume that $e_1[n]$, $e_2[n]$, and $e_3[n]$ are uniformly distributed between $-\frac{1}{2}(2^{-B})$ and $+\frac{1}{2}(2^{-B})$ and are independent of each other and independent of $x[n]$ and $y[n]$. Determine the variance of the total output noise due to these three noise sources. Express your answer in terms of $k^2 = \mathcal{E}[(1/A[n])^2]$.

(d) Assume that $\alpha = 1 - \delta$, with $\delta \ll 1$ (high-gain case). For this value of α, the absolute value of the output $|y[n-1]|$ can be assumed to be greater than the absolute value of the input $|x[n]|$; i.e., the normalizing coefficient $A[n]$ is determined solely by the output. With this assumption and using Eq. (P6.32-1), show that an upper bound on the quantity $k^2 = E[(1/A[n])^2]$ is $k^2 \le 4\sigma_y^2$.

(e) To determine the noise-to-signal ratio at the output, we must require that the maximum value of $x[n]$ be small enough to ensure that $|y[n]| < 1$. Let $x[n]$ be a uniformly distributed white-noise process, and $\alpha = 1 - \delta$, with $\delta \ll 1$. With these assumptions and the results of part (d), determine the noise-to-signal power ratio for the network of Fig. P6.32-4. Compare this noise-to-signal ratio with the corresponding result for fixed-point arithmetic.

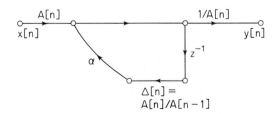

Figure P6.32-4

6.33. Consider a first-order system of the form

$$y[n] = ay[n-1] + x[n].$$

Assume that all variables and coefficients are represented in sign-and-magnitude form with the results of multiplications being truncated before additions are performed. Thus, the nonlinear difference equation implemented is

$$\hat{y}[n] = Q[a\hat{y}[n-1]] + x[n],$$

where $Q[\cdot]$ represents sign-and-magnitude truncation. (See Problem 3.33 for a consideration of sign-and-magnitude number representation.)

Consider the possibility of a zero-input limit cycle of the form $|\hat{y}[n]| = |\hat{y}[n-1]|$ for all n. Show that if the ideal system is stable, then no zero-input limit cycle can exist. Is the same result true for two's-complement truncation?

6.34. Consider the first-order system shown in Fig. P6.34. The quantizer $Q[\cdot]$ represents rounding of the product $ay[n-1]$ to $(B+1)$ bits. All numbers are fixed-point fractions with a wordlength of B bits plus sign. The input is zero, but the system is started with initial condition $y[-1] = A$. Because of the quantizer there is a range of values of A referred to as the *dead band*, for which the effective value of the coefficient a is ± 1; i.e., $|Q[aA]| = A$. Once the output falls into the dead band, it will either oscillate or remain constant, depending on whether the effective coefficient is positive or negative.

(a) Determine in terms of a and B the range of values of A corresponding to the dead band.

(b) For $B = 6$ bits and $A = \frac{1}{16}$, sketch $y[n]$ for the cases $a = +\frac{15}{16}$ and $a = -\frac{15}{16}$.

(c) For $B = 6$ bits and $A = \frac{1}{2}$, sketch $y[n]$ for $a = -\frac{15}{16}$.

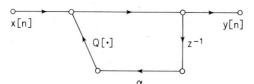

Figure P6.34

6.35. The flow graph of a first-order system is shown in Fig. P6.35-1.

(a) Assuming infinite-precision arithmetic, find the response of the system to the input

$$x[n] = \begin{cases} \frac{1}{2}, & n \geq 0 \\ 0, & n < 0. \end{cases}$$

What is the response of the system for large n?

Now suppose that the system is implemented with fixed-point arithmetic. The coefficient and all variables in the network are represented in sign-and-magnitude notation with 5-bit registers. That is, all numbers are to be considered signed fractions represented as

$$b_0 b_1 b_2 b_3 b_4,$$

where b_0, b_1, b_2, b_3, and b_4 are either 0 or 1 and

$$|\text{Register value}| = b_1 2^{-1} + b_2 2^{-2} + b_3 2^{-3} + b_4 2^{-4}.$$

If $b_0 = 0$, the fraction is positive, and if $b_0 = 1$, the fraction is negative. The result of a multiplication of a sequence value by a coefficient is truncated before additions occur, i.e., only the sign bit and the most significant four bits are retained.

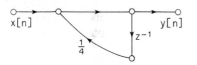

Figure P6.35-1

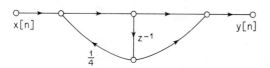

Figure P6.35-2

(b) Compute the response of the quantized system to the input of part (a) and plot the responses of both the quantized and unquantized systems for $0 \leq n \leq 5$. How do the responses compare for large n?

(c) Now consider the system depicted in Fig. P6.35-2, where

$$x[n] = \begin{cases} \frac{1}{2}(-1)^n, & n \geq 0, \\ 0, & n < 0. \end{cases}$$

Repeat parts (a) and (b) for this system and input.

6.36. Consider the second-order system shown in Fig. P6.36. The system is realized with fixed-point arithmetic, and the results of all multiplications are rounded. All numbers are fixed-point fractions with a wordlength of $(B + 1)$ bits including sign. There is a dead band region for $y[n]$ for which the "effective" value of the coefficient $-r^2$ is -1. When this range of $y[n]$ is reached, the effective pole locations can be considered to be on the unit circle and the angular positions of the poles will change. Assume that $x[n] = 0$ for all n, $y[-1] = A \neq 0$, and $y[-2] = 0$.

(a) Find the dead band for $y[n]$, i.e., the values of A for which $Q[-r^2 A] = -A$.

(b) By obtaining a lower bound on A, find the range of values of r for which it is possible to have a dead band.

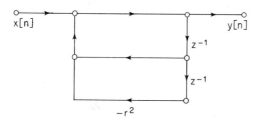

Figure P6.36

6.37. Consider the second-order nonlinear difference equation depicted in Fig. P6.37-1. The branch operation $f[\cdot]$ is defined by the function of Fig. P6.37-2, which was shown in Problem 3.34 to represent two's-complement overflow. (Figure P6.37-2 neglects quantization.) In other words, the function $f[\cdot]$ accounts for possible overflow in forming the sum

$$ay[n-1] + by[n-2] + x[n].$$

A simple example will illustrate the possibility of the existence of zero-input limit cycles.

(a) First, determine the range of values of a and b for stability of the system when no overflow occurs; i.e., assume that the magnitude of the sum is less than unity so that the system behaves linearly. Plot the region of the $(a - b)$-plane corresponding to stability under linear conditions.

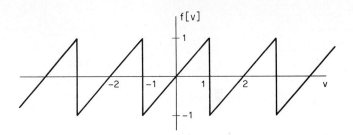

Figure P6.37-1

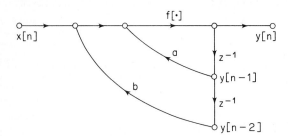

Figure P6.37-2

(b) Suppose that $x[n] = 0$. What conditions on a and b ensure that no overflow occurs, i.e., that $|y[n]| < 1$? Shade the corresponding region in the $(a - b)$-plane.

(c) Consider the possibility that with $x[n] = 0$, $y[n] = y_0 > 0$ for all n. What is the value of y_0 when a and b satisfy the requirements for linear stability?

(d) What values of a and b are consistent with the stability constraint and $0 < y_0 < 1$?

(e) Now consider the possibility of a zero-input limit cycle of period 2, i.e., $y[n] = (-1)^n y_0$ for all n. What value of $0 < y_0 < 1$ is possible for a and b satisfying the linear stability constraint?

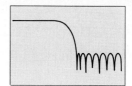

Filter Design Techniques

7

7.0 INTRODUCTION

Filters are a particularly important class of linear time-invariant systems. Strictly speaking, the term *frequency-selective filter* suggests a system that passes certain frequency components and totally rejects all others, but in a broader context any system that modifies certain frequencies relative to others is also called a filter. While the primary emphasis in this chapter is on the design of frequency-selective filters, some of the techniques are more broadly applicable. Also, we concentrate in this chapter on the design of causal filters, although in many contexts filters need not be restricted to causal designs. Very often, noncausal filters are designed and implemented by modifying causal designs.

The design of filters involves the following stages: (1) the specification of the desired properties of the system; (2) the approximation of the specifications using a causal discrete-time system; and (3) the realization of the system. Although these three steps are certainly not independent, we focus our attention in this chapter primarily on the second step, the first being highly dependent on the application and the third dependent on the technology to be used for the implementation. In a practical setting, the desired filter is often implemented with digital computation and used to filter a signal that is derived from a continuous-time signal by means of periodic sampling followed by analog-to-digital conversion. For this reason, it has become common to refer to discrete-time filters as *digital filters* even though the underlying design techniques most often relate only to the discrete-time nature of the signals and systems.

When a discrete-time filter is to be used for discrete-time processing of continuous-time signals in the configuration of Fig. 7.1, the specifications for both the discrete-time filter and the effective continuous-time filter are typically (but not always) given in the frequency domain. This is especially common for frequency-selective filters such as lowpass, bandpass, and highpass filters. As shown in Section

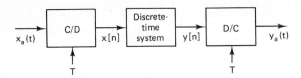

Figure 7.1 Basic system for discrete-time filtering of continuous-time signals.

3.4, if a discrete-time system is used as in Fig. 7.1 and if the input is bandlimited and the sampling frequency is high enough to avoid aliasing, then the overall system behaves as a linear time-invariant continuous-time system with frequency response

$$H_{eff}(j\Omega) = \begin{cases} H(e^{j\Omega T}), & |\Omega| < \pi/T, \\ 0, & |\Omega| > \pi/T. \end{cases} \tag{7.1a}$$

In such cases, it is straightforward to convert from specifications on the effective continuous-time filter to specifications on the discrete-time filter through the relation $\omega = \Omega T$. That is, $H(e^{j\omega})$ is specified over one period by the equation

$$H(e^{j\omega}) = H_{eff}\left(j\frac{\omega}{T}\right), \qquad |\omega| < \pi. \tag{7.1b}$$

This is illustrated in Example 7.1.

Example 7.1

Consider a discrete-time filter that is to be used to lowpass filter a continuous-time signal using the basic configuration of Fig. 7.1. Specifically, we want the overall system of Fig. 7.1 to have the following properties when the sampling rate is 10^4 samples/s ($T = 10^{-4}$ s):

1. The gain $|H_{eff}(j\Omega)|$ should be within ± 0.01 (0.086 dB) of unity (zero dB) in the frequency band $0 \le \Omega \le 2\pi(2000)$.
2. The gain should be no greater than 0.001 (-60 dB) in the frequency band $2\pi(3000) \le \Omega$.

Such a set of lowpass specifications on $|H_{eff}(j\Omega)|$ can be depicted as in Fig. 7.2(a), where the limits of tolerable approximation error are indicated by the shaded horizontal lines. For this specific example, the parameters of Fig. 7.2(a) would be

$$\delta_1 = 0.01 \ (20\log_{10}(1 + \delta_1) = 0.086 \text{ dB}),$$

$$\delta_2 = 0.001 \ (20\log_{10}\delta_2 = -60 \text{ dB}),$$

$$\Omega_p = 2\pi(2000),$$

$$\Omega_s = 2\pi(3000).$$

Since the sampling rate is 10^4 samples/s, the gain of the ideal system is identically zero above $\Omega = 2\pi(5000)$ due to the ideal discrete-to-continuous (D/C) converter.

The tolerance scheme for the digital filter is shown in Fig. 7.2(b). It is the same as Fig. 7.2(a) except that it is plotted as a function of normalized frequency ($\omega = \Omega T$) and it need only be plotted in the range $0 \le \omega \le \pi$, since the remainder can be inferred from symmetry properties (assuming $h[n]$ is real) and periodicity of $H(e^{j\omega})$. From Eq. (7.1b) it follows that the *passband*, within which the magnitude of the frequency response must approximate unity with an error of $\pm\delta_1$, is

$$(1 - \delta_1) \le |H(e^{j\omega})| \le (1 + \delta_1), \qquad |\omega| \le \omega_p, \tag{7.2}$$

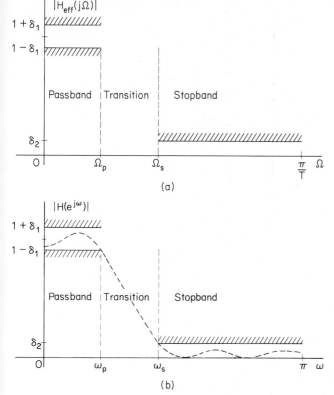

Figure 7.2 (a) Specifications for effective frequency response of overall system in Fig. 7.1 for the case of lowpass filter. (b) Corresponding specifications for the discrete-time system in Fig. 7.1.

where in this example $\delta_1 = 0.01$ and $\omega_p = 2\pi(2000) \cdot 10^{-4} = 0.4\pi$ radians. The other approximation band is the *stopband*, in which the magnitude response must approximate zero with an error less than δ_2; i.e.,

$$|H(e^{j\omega})| \leq \delta_2, \qquad \omega_s \leq |\omega| \leq \pi. \tag{7.3}$$

In this example, $\delta_2 = 0.001$ and $\omega_s = 2\pi(3000) \cdot 10^{-4} = 0.6\pi$ radians. The passband cutoff frequency ω_p and the stopband cutoff frequency ω_s are given in terms of z-plane angle. To approximate the ideal lowpass filter in this way with a realizable system, we must provide a transition band of nonzero width $(\omega_s - \omega_p)$ in which the magnitude response changes smoothly from passband to stopband. The dashed curve in Fig. 7.2(b) is the magnitude response of a system that meets the prescribed specification.

There are many applications in which a discrete-time signal to be filtered is not derived from a continuous-time signal, and there are a variety of means besides periodic sampling for representing continuous-time signals in terms of sequences. (See, for example, Steiglitz, 1965, and Oppenheim and Johnson, 1972.) Also, in most of the design techniques that we discuss, the sampling period plays no role whatsoever in the approximation procedure. For these reasons, we take the point of view that the filter design problem begins from a set of desired specifications in terms of the discrete-time frequency variable ω. Depending on the specific application or context, these specifications may or may not have been obtained from a consideration of filtering in the framework of Fig. 7.1.

Many of the filters used in practice are specified by a tolerance scheme similar to that in Example 7.1, with no constraints on the phase response other than those imposed implicitly by stability and causality requirements. For example, the poles of the system function for a causal and stable infinite impulse response (IIR) filter must lie inside the unit circle. Similarly, in designing finite impulse response (FIR) filters, we often impose the constraint of linear phase. This again removes phase from consideration in the design process.

Given a set of specifications in the form of Fig. 7.2, we must determine the system function of a discrete-time linear system whose frequency response falls within the prescribed tolerances. This is a problem in functional approximation. Design of IIR filters implies approximation by a rational function of z, while design of FIR filters implies polynomial approximation. Our discussion distinguishes between design techniques that are appropriate for IIR filters and those that are appropriate for FIR filters. We discuss a variety of design techniques for both types of filters, ranging from closed-form procedures, which involve only substitution of design specifications into design formulas, to algorithmic techniques, where a solution is obtained by an iterative procedure.

7.1 DESIGN OF DISCRETE-TIME IIR FILTERS FROM CONTINUOUS-TIME FILTERS

The traditional approach to the design of discrete-time IIR filters involves the transformation of a continuous-time filter into a discrete-time filter meeting prescribed specifications. This is a reasonable approach for several reasons:

- The art of continuous-time IIR filter design is highly advanced and, since useful results can be achieved, it is advantageous to use the design procedures already developed for continuous-time filters.
- Many useful continuous-time IIR design methods have relatively simple closed-form design formulas. Therefore, discrete-time IIR filter design methods based on such standard continuous-time design formulas are rather simple to carry out.
- The standard approximation methods that work well for continuous-time IIR filters do not lead to simple closed-form design formulas when these methods are applied directly to the discrete-time IIR case.

The fact that continuous-time filter designs can be mapped to discrete-time filter designs is totally unrelated to and independent of whether the discrete-time filter is to be used in the configuration of Fig. 7.1 for processing continuous-time signals. We emphasize again that the design procedure for the discrete-time system begins from a set of *discrete-time* specifications. Henceforth, we assume that these specifications have been determined by an analysis like that of Example 7.1 or by some other method. We will use continuous-time filter approximation methods only as a convenience in determining the discrete-time filter that meets the desired specifications. Indeed, the continuous-time filter on which the approximation is based may have a frequency response that is vastly different from the effective frequency response when the discrete-time filter is used in the configuration of Fig. 7.1.

In designing a discrete-time filter by transforming a prototype continuous-time filter, the specifications on the continuous-time filter are obtained by a transformation of the specifications for the desired discrete-time filter. The system function $H_c(s)$ or impulse response $h_c(t)$ of the continuous-time filter is then obtained through one of the established approximation methods used for continuous-time filter design, examples of which are discussed in Appendix B. Next, the system function $H(z)$ or impulse response $h[n]$ for the discrete-time filter is obtained by applying to $H_c(s)$ or $h_c(t)$ a transformation of the type discussed in this section.

In such transformations we generally require that the essential properties of the continuous-time frequency response be preserved in the frequency response of the resulting discrete-time filter. Specifically, this implies that we want the imaginary axis of the s-plane to map onto the unit circle of the z-plane. A second condition is that a stable continuous-time filter should be transformed to a stable discrete-time filter. This means that if the continuous-time system has poles only in the left half of the s-plane, then the discrete-time filter must have poles only inside the unit circle. These constraints are basic to all the techniques discussed in this section.

7.1.1 Filter Design by Impulse Invariance

In Section 3.4.2 we discussed the concept of *impulse invariance*, where a discrete-time system is defined by sampling the impulse response of a continuous-time system. We showed that impulse invariance provides a direct means of computing samples of the output of a bandlimited continuous-time system for bandlimited input signals. Alternatively, in the context of filter design we can think of impulse invariance as a method for obtaining a discrete-time system whose frequency response is determined by the frequency response of a continuous-time system.

In the impulse invariance design procedure for transforming continuous-time filters into discrete-time, the impulse response of the discrete-time filter is chosen as equally spaced samples of the impulse response of the continuous-time filter; i.e.,

$$h[n] = T_d h_c(nT_d), \tag{7.4}$$

where T_d represents a sampling interval. As we will see, because we begin the design problem with the discrete-time filter specifications, the parameter T_d in Eq. (7.4) in fact has no role whatsoever in the design process or the resulting discrete-time filter. However, since it is customary to include this parameter in defining the procedure, we include it in the following discussion. As we will see, even if the filter is used in the basic configuration of Fig. 7.1, the design sampling period T_d need not be the same as the sampling period T associated with the C/D and D/C conversion.

When impulse invariance is used as a means for designing a discrete-time filter with a specified frequency response, we are especially interested in the relationship between the frequency responses of the discrete-time and continuous-time filters.

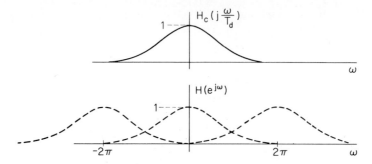

Figure 7.3   Illustration of aliasing in the impulse invariance design technique.

From the discussion of sampling in Chapter 3, it follows that the frequency response of the discrete-time filter obtained through Eq. (7.4) is related to the frequency response of the continuous-time filter by

$$H(e^{j\omega}) = \sum_{k=-\infty}^{\infty} H_c\left(j\frac{\omega}{T_d} + j\frac{2\pi}{T_d}k\right). \tag{7.5}$$

If the continuous-time filter is bandlimited, so that

$$H_c(j\Omega) = 0, \qquad |\Omega| \geq \pi/T_d, \tag{7.6}$$

then

$$H(e^{j\omega}) = H_c\left(j\frac{\omega}{T_d}\right), \qquad |\omega| \leq \pi, \tag{7.7}$$

i.e., the discrete-time and continuous-time frequency responses are related by a linear scaling of the frequency axis, i.e., $\omega = \Omega T_d$. Unfortunately, any practical continuous-time filter cannot be exactly bandlimited, and consequently, interference between successive terms in Eq. (7.5) occurs, causing aliasing, as illustrated in Fig. 7.3. However, if the continuous-time filter approaches zero at high frequencies, the aliasing may be negligibly small and a useful discrete-time filter can result from the sampling of the impulse response of a continuous-time filter.

In the impulse invariance design procedure, the discrete-time filter specifications are first transformed to continuous-time filter specifications through the use of Eq. (7.7). We illustrate this in Example 7.2.

Example 7.2

To illustrate the impulse invariance design procedure and the transformation implied by Eqs. (7.4) and (7.5), let us assume that the specifications for the desired discrete-time filter are as shown in Fig. 7.4(a) with $\delta_1 = 0.10875$, $\delta_2 = 0.17783$, $\omega_p = 0.2\pi$, and $\omega_s = 0.3\pi$. This corresponds to a maximum gain of -15 dB ($20 \log_{10} 0.17783$) in the stopband and a maximum deviation of 1 dB *below* 0 dB gain in the passband (i.e., $20 \log_{10}(1) - 20 \log_{10}(1 - 0.10875) = -1$ dB). Note that in this case the passband tolerance is between $1 - \delta_1$ and 1 rather than between $1 - \delta_1$ and $1 + \delta_1$ as in Fig. 7.2(b). Historically, most continuous-time filter approximation methods were developed

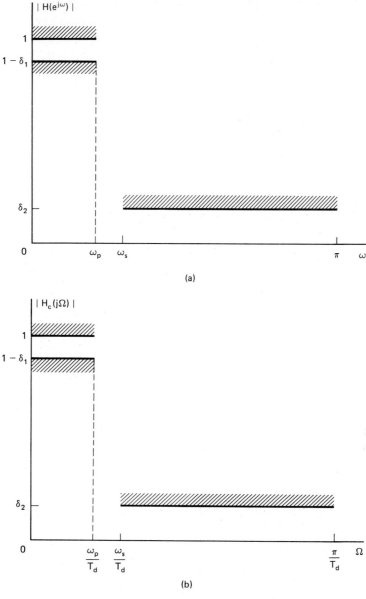

Figure 7.4   Specifications for Example 7.2. (a) Specifications for the desired digital filter. (b) Specifications for the analog filter to be transformed by impulse invariance.

for design of passive systems (having gain less than unity). Therefore, the design specifications for continuous-time filters have typically been formulated so that the passband is less than or equal to unity as in this example. Conversion between specifications in the form of Fig. 7.4(a) and specifications in the form of Fig. 7.2(b) is easily accomplished by multiplying the system function by an appropriate constant. See Problem 7.5 for a consideration of the conversion process.

Assuming that the aliasing involved in the transformation from $H_c(j\Omega)$ to $H(e^{j\omega})$ will be negligible, we obtain the specifications on $H_c(j\Omega)$ by applying the relation

$$\Omega = \omega/T_d \tag{7.8}$$

to obtain the continuous-time filter specifications shown in Fig. 7.4(b). After obtaining a suitable continuous-time filter based on these specifications, the continuous-time filter with system function $H_c(s)$ is transformed to the desired discrete-time filter with system function $H(z)$. We develop the algebraic details of the transformation from $H_c(s)$ to $H(z)$ below. Note, however, that in the transformation back to discrete-time frequency, $H(e^{j\omega})$ will be related to $H_c(j\Omega)$ through Eq. (7.7), which again applies the transformation of Eq. (7.8) to the frequency axis. As a consequence, the "sampling" parameter T_d cannot be used to control aliasing. Since the basic specifications are in terms of discrete-time frequency, if the sampling rate is increased (T_d is made smaller), then the cutoff frequency of the continuous-time filter must increase in proportion. In practice, to compensate for aliasing that might occur in the transformation from $H_c(s)$ to $H(z)$, the continuous-time filter may be somewhat overdesigned, i.e., designed to exceed the specifications, particularly in the stopband.

While the impulse invariance transformation from continuous time to discrete time is defined in terms of time-domain sampling, it is easy to carry out as a transformation on the system function. To develop this transformation, let us consider the system function of the continuous-time filter expressed in terms of a partial fraction expansion, so that†

$$H_c(s) = \sum_{k=1}^{N} \frac{A_k}{s - s_k}. \tag{7.9}$$

The corresponding impulse response is

$$h_c(t) = \begin{cases} \sum_{k=1}^{N} A_k e^{s_k t}, & t \geq 0, \\ \\ 0, & t < 0. \end{cases} \tag{7.10}$$

The impulse response of the discrete-time filter obtained by sampling $T_d h_c(t)$ is

$$\begin{aligned} h[n] = T_d h_c(nT_d) &= \sum_{k=1}^{N} T_d A_k e^{s_k nT_d} u[n] \\ &= \sum_{k=1}^{N} T_d A_k (e^{s_k T_d})^n u[n]. \end{aligned} \tag{7.11}$$

The system function $H(z)$ of the discrete-time filter is therefore given by

$$H(z) = \sum_{k=1}^{N} \frac{T_d A_k}{1 - e^{s_k T_d} z^{-1}}. \tag{7.12}$$

---

†For simplicity, we assume in the discussion that all poles of $H(s)$ are single order. In Problem 7.2 we consider the modifications required for multiple-order poles.

In comparing Eqs. (7.9) and (7.12) we observe that a pole at $s = s_k$ in the s-plane transforms to a pole at $e^{s_k T_d}$ in the z-plane and the coefficients in the partial fraction expansions of $H_c(s)$ and $H(z)$ are equal except for the scaling multiplier T_d. If the continuous-time filter is stable, corresponding to the real part of s_k less than zero, then the magnitude of $e^{s_k T_d}$ will be less than unity, so that the corresponding pole in the discrete-time filter is inside the unit circle. Therefore, the causal discrete-time filter is also stable. While the poles in the s-plane map to poles in the z-plane according to the relationship $z_k = e^{s_k T_d}$, it is important to recognize that the impulse invariance design procedure does not correspond to a simple mapping of the s-plane to the z-plane by that relationship. In particular, the zeros in the discrete-time system function are a function of the poles and the coefficients $T_d A_k$ in the partial fraction expansion, and they will not in general be mapped in the same way the poles are mapped. We illustrate the impulse invariance design procedure with the following example.

Example 7.3

Let us consider the design of a lowpass discrete-time filter by applying impulse invariance to an appropriate Butterworth continuous-time filter.† The specifications for the discrete-time filter are those used in Example 7.2 and shown in Fig. 7.4. Specifically,

$$0.89125 \le |H(e^{j\omega})| \le 1, \qquad 0 \le |\omega| \le 0.2\pi, \tag{7.13a}$$

$$|H(e^{j\omega})| \le 0.17783, \qquad 0.3\pi \le |\omega| \le \pi. \tag{7.13b}$$

Since the parameter T_d cancels in the impulse invariance procedure, we can choose $T_d = 1$ so that $\omega = \Omega$. In Problem 7.4 this same example is considered with the parameter T_d explicitly included to illustrate how and where it cancels.

In designing the filter using impulse invariance on a continuous-time Butterworth filter, we must first transform the discrete-time specifications to specifications on the continuous-time filter. We recall that impulse invariance corresponds to a linear mapping between Ω and ω in the absence of aliasing. For this example, we will assume that the effect of aliasing is negligible. After the design is complete, the performance of the resulting filter can then be evaluated.

Because of the above considerations, we want to design a continuous-time Butterworth filter with magnitude function $|H_c(j\Omega)|$ for which

$$0.89125 \le |H_c(j\Omega)| \le 1, \qquad 0 \le |\Omega| \le 0.2\pi, \tag{7.14a}$$

$$|H_c(j\Omega)| \le 0.17783, \qquad 0.3\pi \le |\Omega| \le \pi. \tag{7.14b}$$

Since the magnitude response of an analog Butterworth filter is a monotonic function of frequency, Eqs. (7.14a) and (7.14b) will be satisfied if

$$|H_c(j0.2\pi)| \ge 0.89125 \tag{7.15a}$$

and

$$|H_c(j0.3\pi)| \le 0.17783. \tag{7.15b}$$

Specifically, the magnitude-squared function of a Butterworth filter is of the form

$$|H_c(j\Omega)|^2 = \frac{1}{1 + (\Omega/\Omega_c)^{2N}}, \tag{7.16}$$

---

† Continuous-time Butterworth and Chebyshev filters are discussed in Appendix B.

so that the filter design process consists of determining the parameters N and Ω_c to meet the desired specifications. Using Eqs. (7.16) in Eqs. (7.15) with equality leads to the equations

$$1 + \left(\frac{0.2\pi}{\Omega_c}\right)^{2N} = \left(\frac{1}{0.89125}\right)^2 \tag{7.17a}$$

and

$$1 + \left(\frac{0.3\pi}{\Omega_c}\right)^{2N} = \left(\frac{1}{0.17783}\right)^2. \tag{7.17b}$$

The solution of these two equations is $N = 5.8858$ and $\Omega_c = 0.70474$. The parameter N, however, must be an integer; consequently, so that the specifications are met or exceeded, we round N up to the nearest integer, $N = 6$. Now both passband and stopband specifications cannot be met exactly. As the value of Ω_c varies, there is a tradeoff in the amount by which the stopband and passband specifications are exceeded. If we substitute $N = 6$ into Eq. (7.17a) we obtain $\Omega_c = 0.7032$. With this value, the passband specifications (of the continuous-time filter) will be met exactly and the stopband specifications (of the continuous-time filter) will be exceeded. This allows some margin for aliasing in the discrete-time filter. With $\Omega_c = 0.7032$ and with $N = 6$, the 12 poles of the magnitude-squared function $H_c(s)H_c(-s) = 1/[1 + (s/j\Omega_c)^{2N}]$ are uniformly distributed in angle on a circle of radius $\Omega_c = 0.7032$, as indicated in Fig. 7.5. Consequently, the poles of $H_c(s)$ are the three pole pairs in the left half of the s-plane with coordinates

Pole pair 1: $-0.182 \pm j(0.679)$

Pole pair 2: $-0.497 \pm j(0.497)$

Pole pair 3: $-0.679 \pm j(0.182)$

so that

$$H_c(s) = \frac{0.12093}{(s^2 + 0.365s + 0.495)(s^2 + 0.995s + 0.495)(s^2 + 1.359s + 0.495)}. \tag{7.18}$$

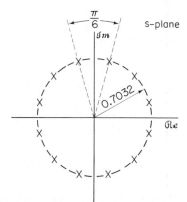

Figure 7.5 s-plane locations for poles of $H_c(s)H_c(-s)$ for sixth-order Butterworth filter in Example 7.3.

If we express $H_c(s)$ as a partial fraction expansion, perform the transformation of Eq. (7.12), and then combine complex conjugate terms, the resulting system function of the discrete-time filter is

$$H(z) = \frac{0.287 - 0.447z^{-1}}{1 - 1.297z^{-1} + 0.695z^{-2}} + \frac{-2.143 + 1.145z^{-1}}{1 - 1.069z^{-1} + 0.370z^{-2}}$$
$$+ \frac{1.856 - 0.630z^{-1}}{1 - 0.997z^{-1} + 0.257z^{-2}}. \tag{7.19}$$

As is evident from Eq. (7.19), the system function resulting from the impulse invariance design procedure may be realized directly in parallel form. If cascade or direct form is desired, the separate second-order terms must be combined in an appropriate way. Parallel, cascade, and direct-form structures for this example are considered in Problem 7.11.

The frequency-response functions of the discrete-time system are shown in Fig. 7.6. We recall that the prototype continuous-time filter was designed to meet the specifications exactly at the passband edge and to exceed the specifications at the stopband edge, and this turns out to be true for the resulting discrete-time filter. This is an indication that the continuous-time filter was sufficiently bandlimited so that aliasing presented no problem. Indeed, the difference between $20 \log_{10} |H(e^{j\omega})|$ and $20 \log_{10} |H_c(j\omega)|$ is not visible on this plotting scale except for a slight deviation around $\omega = \pi$. (Recall that $T_d = 1$, so $\Omega = \omega$.) Sometimes aliasing is much more of a problem. If the resulting discrete-time filter fails to meet the specifications because of aliasing, there is no alternative with impulse invariance but to try again with a higher-order filter or a different adjustment of the filter parameters, holding the order fixed.

The basis for impulse invariance is to choose an impulse response for the discrete-time filter that is similar in some sense to the impulse response of the continuous-time filter. The use of this procedure is often motivated not so much by a desire to maintain the impulse response shape as by the knowledge that if the continuous-time filter is bandlimited, then the discrete-time filter frequency response will closely approximate the continuous-time frequency response. However, in some filter design problems, a primary objective may be to control some aspect of the time response, such as the impulse response or the step response. In such cases a natural approach might be to design the discrete-time filter by impulse invariance or by step invariance. In the latter case, the response of the discrete-time filter to a sampled unit step function is defined to be the sequence obtained by sampling the continuous-time step response. If the continuous-time filter has good step response characteristics, such as small rise time and low peak overshoot, these characteristics will be preserved in the discrete-time filter. Clearly, this concept of waveform invariance can be extended to the preservation of the output waveshape for a variety of inputs, as illustrated in Problem 7.1. This problem illustrates that transforming the same continuous-time filter by impulse invariance and also by step invariance (or other waveform invariance criteria) does not lead to the same discrete-time filter in the two cases.

In the impulse invariance design procedure, the relationship between continuous-time and discrete-time frequency is linear; consequently, except for aliasing, the shape of the frequency response is preserved. This is in contrast to the procedure

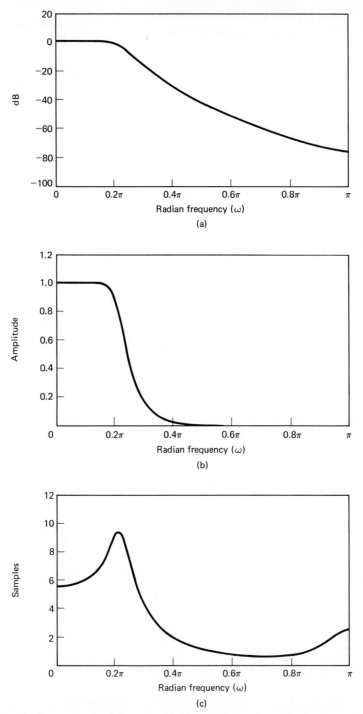

Figure 7.6 Frequency response of sixth-order Butterworth filter transformed by impulse invariance. (a) Log magnitude in dB. (b) Magnitude. (c) Group delay.

discussed next, which is based on an algebraic transformation. We note in conclusion that the impulse invariance technique is appropriate only for bandlimited filters. Highpass or bandstop continuous-time filters, for example, would require additional bandlimiting to avoid severe aliasing distortion if impulse invariance design is used.

7.1.2 Bilinear Transformation

The technique discussed in this section avoids the problem of aliasing by using the bilinear transformation, an algebraic transformation between the variables s and z that maps the entire $j\Omega$-axis in the s-plane to one revolution of the unit circle in the z-plane. Since $-\infty \leq \Omega \leq \infty$ maps onto $-\pi \leq \omega \leq \pi$, the transformation between the continuous-time and discrete-time frequency variables must be nonlinear. Therefore, the use of this technique is restricted to situations where the corresponding warping of the frequency axis is acceptable.

With $H_c(s)$ denoting the continuous-time system function and $H(z)$ the discrete-time system function, the bilinear transformation corresponds to replacing s by

$$s = \frac{2}{T_d}\left(\frac{1 - z^{-1}}{1 + z^{-1}}\right), \tag{7.20}$$

that is,

$$H(z) = H_c\left[\frac{2}{T_d}\left(\frac{1 - z^{-1}}{1 + z^{-1}}\right)\right]. \tag{7.21}$$

As in impulse invariance, a "sampling" parameter T_d is included in the definition of the bilinear transformation. Historically, this parameter has been included because the difference equation corresponding to $H(z)$ can be obtained by applying the trapezoidal integration rule to the differential equation corresponding to $H_c(s)$, with T_d representing the numerical integration step size. (See Kaiser, 1966 and Problem 7.14.) However, in filter design, our use of the bilinear transformation is based on the properties of the algebraic transformation of Eq. (7.20). As with impulse invariance, the parameter T_d is of no consequence in the design procedure since we assume that the design problem always begins with specifications on the discrete-time filter $H(e^{j\omega})$. When these specifications are mapped to continuous-time specifications and then the continuous-time filter is mapped back to a discrete-time filter, the effect of T_d will cancel. Although we will retain the parameter T_d in our discussion, in specific problems and examples any convenient value of T_d can be chosen.

To develop the properties of the algebraic transformation of Eq. (7.20), we solve for z to obtain

$$z = \frac{1 + (T_d/2)s}{1 - (T_d/2)s} \tag{7.22}$$

and, substituting $s = \sigma + j\Omega$ into Eq. (7.22), we obtain

$$z = \frac{1 + \sigma T_d/2 + j\Omega T_d/2}{1 - \sigma T_d/2 - j\Omega T_d/2}. \tag{7.23}$$

If $\sigma < 0$, then from Eq. (7.23) it follows that $|z| < 1$ for any value of Ω. Similarly, if $\sigma > 0$, then $|z| > 1$ for all Ω. That is, if a pole of $H_c(s)$ is in the left-half s-plane, its image in the z-plane will be inside the unit circle. Therefore causal stable continuous-time filters map into causal stable discrete-time filters.

Next, to show that the $j\Omega$-axis of the s-plane maps onto the unit circle, we substitute $s = j\Omega$ into Eq. (7.22), obtaining

$$z = \frac{1 + j\Omega T_d/2}{1 - j\Omega T_d/2}. \tag{7.24}$$

From Eq. (7.24) it is clear that $|z| = 1$ for all values of s on the $j\Omega$-axis. That is, the $j\Omega$-axis maps onto the unit circle, so Eq. (7.24) takes the form

$$e^{j\omega} = \frac{1 + j\Omega T_d/2}{1 - j\Omega T_d/2}. \tag{7.25}$$

To derive a relationship between the variables ω and Ω, it is useful to return to Eq. (7.20) and substitute $z = e^{j\omega}$:

$$s = \frac{2}{T_d}\left(\frac{1 - e^{-j\omega}}{1 + e^{-j\omega}}\right), \tag{7.26}$$

or, equivalently,

$$s = \sigma + j\Omega = \frac{2}{T_d}\left[\frac{2e^{-j\omega/2}(j\sin\omega/2)}{2e^{-j\omega/2}(\cos\omega/2)}\right] = \frac{2j}{T_d}\tan(\omega/2). \tag{7.27}$$

Equating real and imaginary parts on both sides of Eq. (7.27) leads to the relations $\sigma = 0$ and

$$\Omega = \frac{2}{T_d}\tan(\omega/2), \tag{7.28a}$$

or

$$\omega = 2\arctan(\Omega T_d/2). \tag{7.28b}$$

These properties of the bilinear transformation as a mapping from the s-plane to the z-plane are summarized in Figs. 7.7 and 7.8. From Eq. (7.28b) and Fig. 7.8 we see that the range of frequencies $0 \leq \Omega \leq \infty$ maps to $0 \leq \omega \leq \pi$ while the range $-\infty \leq \Omega \leq 0$ maps to $-\pi \leq \omega \leq 0$. The bilinear transformation avoids the problem

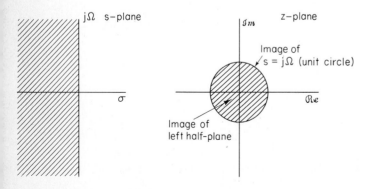

Figure 7.7 Mapping of the s-plane onto the z-plane using the bilinear transformation.

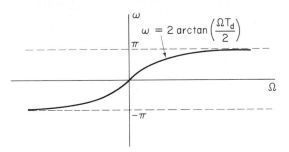

Figure 7.8 Mapping of the continuous-time frequency axis onto the unit circle by bilinear transformation.

of aliasing encountered with the use of impulse invariance because it maps the entire imaginary axis of the s-plane onto the unit circle in the z-plane. The price paid for this, however, is the nonlinear compression of the frequency axis depicted in Fig. 7.8. Consequently, the design of discrete-time filters using the bilinear transformation is useful only when this compression can be tolerated or compensated for, as in the case of filters that approximate ideal piecewise-constant magnitude response characteristics. For example, to design a lowpass filter, we require an approximation to the ideal lowpass characteristic shown in Fig. 7.9. If we were able to design an ideal lowpass continuous-time filter with a cutoff frequency $\Omega_c = (2/T_d)\tan(\omega_c/2)$ and mapped it to the z-plane by means of the bilinear transformation, the ideal frequency response of Fig. 7.9 would result. Of course, neither in the continuous-time case nor in the discrete-time case are we able to realize an ideal filter of this type. In general, we would have to approximate such a frequency response by allowing some deviation from unity in the passband and some deviation from zero in the stopband with a transition band of nonzero width. Figure 7.10 depicts the mapping of a continuous-time frequency response and tolerance scheme to a corresponding discrete-time frequency response and tolerance scheme. If the critical frequencies (such as the passband and stopband edge frequencies) of the continuous-time filter are prewarped according to the equation

$$\Omega = \frac{2}{T_d}\tan(\omega/2), \tag{7.29}$$

then when the continuous-time filter is transformed to the discrete-time filter using Eq. (7.21), the discrete-time filter will meet the desired specifications.

Typical frequency-selective continuous-time approximations are Butterworth, Chebyshev, and elliptic filters. The closed-form design formulas of these continuous-time approximation methods make the design procedure rather straightforward. A

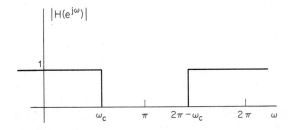

Figure 7.9 Frequency response of an ideal lowpass filter.

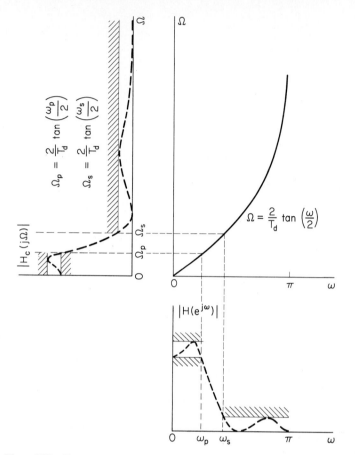

Figure 7.10 Frequency warping inherent in the bilinear transformation of a continuous-time lowpass filter into a discrete-time lowpass filter. To achieve the desired discrete-time cutoff frequencies, the continuous-time cutoff frequencies must be prewarped as indicated.

Butterworth continuous-time filter is monotonic in the passband and in the stopband. A type I Chebyshev filter has an equiripple characteristic in the passband and it varies monotonically in the stopband or vice versa. An elliptic filter is equiripple in both the passband and the stopband, and a type II Chebyshev filter is monotonic in the passband and equiripple in the stopband. Clearly, these properties will be preserved when the filter is mapped to a digital filter with the bilinear transformation. This is illustrated by the dashed elliptic-type approximation shown in Fig. 7.10.

Although the bilinear transformation can be used effectively in mapping a piecewise-constant magnitude response characteristic from the s-plane to the z-plane, the distortion in the frequency axis also manifests itself as a warping of the phase response of the filter. For example, Fig. 7.11 shows the result of applying the bilinear transformation to an ideal linear phase factor $e^{-s\alpha}$. If we substitute Eq. (7.20) for s and evaluate the result on the unit circle, the phase angle is $-(2\alpha/T_d)\tan(\omega/2)$. The dotted lines in Fig. 7.11 show the periodic linear phase function $-(\omega\alpha/T_d)$, while the solid

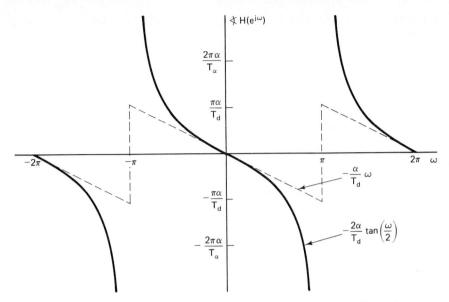

Figure 7.11   Illustration of the effect of the bilinear transformation on a linear phase characteristic. (Dashed line is linear phase and solid line is phase resulting from bilinear transformation.)

curve shows the function $-(2\alpha/T_d)\tan(\omega/2)$. From this it should be evident that if we were interested in a discrete-time lowpass filter with a linear phase characteristic, we could not obtain such a filter by applying the bilinear transformation to a continuous-time lowpass filter with a linear phase characteristic.

The following example illustrates the use of the bilinear transformation in filter design.

Example 7.4

Consider the discrete-time filter specifications of Example 7.3, where we illustrated the impulse invariance technique for the design of a discrete-time filter. The specifications on the discrete-time filter are

$$0.89125 \leq |H(e^{j\omega})| \leq 1, \qquad 0 \leq \omega \leq 0.2\pi, \tag{7.30a}$$

$$|H(e^{j\omega})| \leq 0.17783, \qquad 0.3\pi \leq \omega \leq \pi. \tag{7.30b}$$

In carrying out the design using the bilinear transformation, the critical frequencies of the discrete-time filter must be prewarped to the corresponding continuous-time frequencies using Eq. (7.29) so that the frequency distortion inherent in the bilinear transformation will map them back to the correct discrete-time critical frequencies. For this specific filter, with $|H_c(j\Omega)|$ representing the magnitude response function of the continuous-time filter, we require that

$$0.89125 \leq |H_c(j\Omega)| \leq 1, \qquad 0 \leq \Omega \leq \frac{2}{T_d}\tan\left(\frac{0.2\pi}{2}\right), \tag{7.31a}$$

$$|H_c(j\Omega)| \leq 0.17783, \qquad \frac{2}{T_d}\tan\left(\frac{0.3\pi}{2}\right) \leq \Omega < \infty. \tag{7.31b}$$

For convenience we choose $T_d = 1$. Also, as with Example 7.3, since a continuous-time Butterworth filter has a monotonic magnitude response, we can equivalently require that

$$|H_c(j2\tan(0.1\pi))| \geq 0.89125 \qquad (7.32a)$$

and

$$|H_c(j2\tan(0.15\pi))| \leq 0.17783. \qquad (7.32b)$$

The form of the magnitude-squared function for the Butterworth filter is

$$|H_c(j\Omega)|^2 = \frac{1}{1 + (\Omega/\Omega_c)^{2N}}. \qquad (7.33)$$

Solving for N and Ω_c with the equality sign in Eqs. (7.32a) and (7.32b), we obtain

$$1 + \left(\frac{2\tan(0.1\pi)}{\Omega_c}\right)^{2N} = \left(\frac{1}{0.89}\right)^2 \qquad (7.34a)$$

and

$$1 + \left(\frac{2\tan(0.15\pi)}{\Omega_c}\right)^{2N} = \left(\frac{1}{0.178}\right)^2, \qquad (7.34b)$$

and solving for N in Eqs. (7.34a) and (7.34b) gives

$$N = \frac{\log[((\frac{1}{0.178})^2 - 1)/((\frac{1}{0.89})^2 - 1)]}{2\log[\tan(0.15\pi)/\tan(0.1\pi)]} \qquad (7.35)$$

$$= 5.30466.$$

Since N must be an integer, we choose $N = 6$. Substituting $N = 6$ into Eq. (7.34b), we obtain $\Omega_c = 0.76622$. For this value of Ω_c, the passband specifications are exceeded and the stopband specifications are met exactly. This is reasonable for the bilinear transformation since we do not have to be concerned with aliasing. That is, with proper prewarping, we can be certain that the resulting discrete-time filter will meet the specifications exactly at the desired stopband edge.

In the s-plane, the 12 poles of the magnitude-squared function are uniformly distributed in angle on a circle of radius 0.76622, as shown in Fig. 7.12. The system

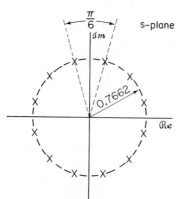

Figure 7.12 s-plane locations for poles of $H_c(s)H_c(-s)$ for sixth-order Butterworth filter in Example 7.4.

function of the continuous-time filter obtained by selecting the left half-plane poles is

$$H_c(s) = \frac{0.20238}{(s^2 + 0.396s + 0.5871)(s^2 + 1.083s + 0.5871)(s^2 + 1.4802s + 0.5871)}. \quad (7.36)$$

The system function $H(z)$ for the discrete-time filter is then obtained by applying the bilinear transformation to $H_c(s)$ with $T_d = 1$, resulting in

$$H(z) = \frac{0.0007378(1 + z^{-1})^6}{(1 - 1.2686z^{-1} + 0.7051z^{-2})(1 - 1.0106z^{-1} + 0.3583z^{-2})}$$

$$\times \frac{1}{(1 - 0.9044z^{-1} + 0.2155z^{-2})}. \quad (7.37)$$

The magnitude, log magnitude, and group delay of the frequency response of the discrete-time filter are shown in Fig. 7.13. At $\omega = 0.2\pi$ the magnitude is 0.56 dB below 0 dB, and at $\omega = 0.3\pi$ the log magnitude is exactly -15 dB.

Since the bilinear transformation maps the entire $j\Omega$-axis of the s-plane onto the unit circle, the magnitude response of the discrete-time filter falls off much more rapidly than the original continuous-time filter. In particular, the behavior of $H(e^{j\omega})$ at $\omega = \pi$ corresponds to the behavior of $H_c(j\Omega)$ at $\Omega = \infty$. Therefore, since the continuous-time Butterworth filter has a sixth-order zero at $s = \infty$, the resulting discrete-time filter has a sixth-order zero at $z = -1$. In Problem 7.12 the cascade, parallel, and direct form implementations of the filter are considered.

It it interesting to note that since the general form of the Nth-order Butterworth continuous-time filter is as given by Eq. (7.33) and since ω and Ω are related by Eq. (7.29), it follows that the general Nth-order Butterworth discrete-time filter has magnitude-squared function

$$|H(e^{j\omega})|^2 = \frac{1}{1 + \left(\dfrac{\tan(\omega/2)}{\tan(\omega_c/2)}\right)^{2N}}, \quad (7.38)$$

where $\tan(\omega_c/2) = \Omega_c T_d/2$.

The frequency-response function of Eq. (7.38) has the same properties as the continuous-time Butterworth response; i.e., it is maximally flat† and $|H(e^{j\omega_c})|^2 = 0.5$. However, the function in Eq. (7.38) is periodic with period 2π and it falls off more sharply than the continuous-time Butterworth response.

We do not design discrete-time Butterworth filters directly by starting with Eq. (7.38) because it is not straightforward to determine the z-plane locations of the poles (all the zeros are at $z = -1$) of the magnitude-squared function of Eq. (7.38). It is necessary to determine the poles so as to factor the magnitude-squared function into $H(z)H(z^{-1})$ and thereby determine $H(z)$. It is much easier to find the s-plane pole locations (all the zeros are at infinity), factor the continuous-time system function, and then transform the left half-plane poles by the bilinear transformation as we did in Example 7.4.

Equations of the form of Eq. (7.38) may also be obtained for discrete-time Chebyshev filters, but the same difficulties arise in their use. Thus, the two-step

---

† The first $(2N - 1)$ derivatives of $|H(e^{j\omega})|^2$ are zero at $\omega = 0$.

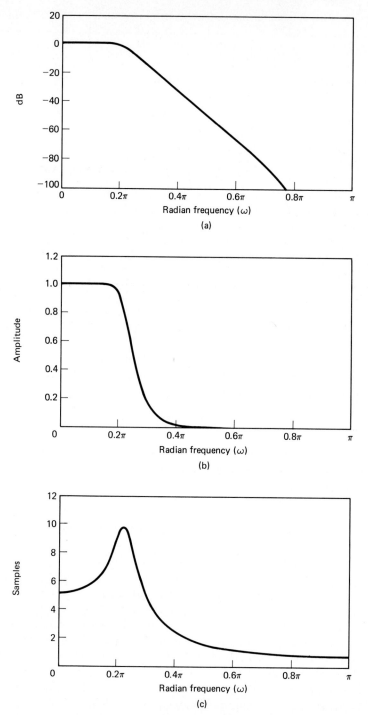

Figure 7.13 Frequency response of sixth-order Butterworth filter transformed by bilinear transform. (a) Log magnitude in dB. (b) Magnitude. (c) Group delay.

approach described above has become the established method of designing IIR frequency-selective filters.

7.1.3 Bilinear Transformation Design Examples

The major approximation methods for frequency-selective IIR analog filters are the Butterworth, Chebyshev, and elliptic function approximation methods. The details of these methods can be found in Guillemin (1957), Daniels (1974), Weinberg (1975), and Lam (1979). These methods are generally explained and developed in terms of lowpass filter approximations. This is the approach followed in Appendix B, where we summarize the essential features of some of these methods.

   To illustrate the properties of the basic IIR approximation methods, we present in this section some examples of realizations of a set of filter specifications by each of the basic methods. The details of the design computations are not presented since they are tedious and lengthy and are best carried out by computer programs that incorporate the appropriate closed-form design equations.

   Suppose we wish to design a lowpass digital filter to meet the specifications of Example 7.1; i.e.,

$$0.99 \leq |H(e^{j\omega})| \leq 1.01, \qquad |\omega| \leq 0.4\pi \tag{7.39a}$$

and

$$|H(e^{j\omega})| \leq 0.001, \qquad 0.6\pi \leq |\omega| \leq \pi. \tag{7.39b}$$

   In terms of the tolerance scheme of Fig. 7.2(b), $\delta_1 = 0.01$, $\delta_2 = 0.001$, $\omega_p = 0.4\pi$, and $\omega_s = 0.6\pi$.

   These specifications are sufficient to determine the input parameters to the Butterworth, Chebyshev, and elliptic design formulas. Note that the specifications are only on the magnitudes of the frequency response. The phase is implicitly determined by the nature of the approximating functions.

Example 7.5   Butterworth Approximation

   For the specification of Eqs. (7.39), the Butterworth approximation method requires a system of 14th order. The frequency response of the discrete-time filter that results from the bilinear transformation of the appropriate prewarped Butterworth filter is shown in Fig. 7.14. Figure 7.14(a) shows the log magnitude in dB; Fig. 7.14(b) shows the magnitude of $|H(e^{j\omega})|$ in the passband only; and Fig. 7.14(c) shows the group delay of the filter. From these plots, we see that the Butterworth frequency response decreases monotonically with frequency, and the gain of the filter becomes very small above about $\omega = 0.7\pi$. Note from Fig. 7.14(b) that in this example the Butterworth frequency response has been normalized so that it has gain greater than unity in the passband.

   In the Butterworth example, the specifications are exceeded at the passband and stopband edges because of rounding the order up to the next integer. However, we note that the specifications are far exceeded in the stopband. The reason for this is

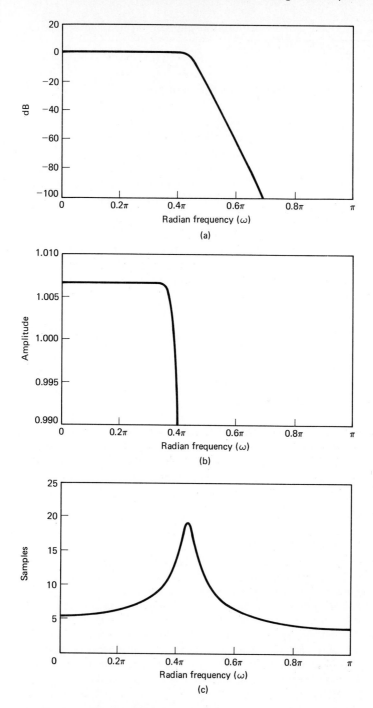

Figure 7.14   Frequency reponse of 14th-order Butterworth filter in Example 7.5. (a) Log magnitude in dB. (b) Detailed plot of magnitude in passband. (c) Group delay.

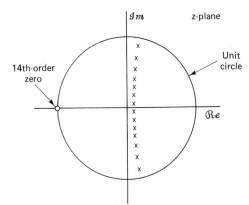

Figure 7.15 Pole-zero plot of 14th-order Butterworth filter in Example 7.5.

evident from Fig. 7.15, which shows the pole-zero plot for the 14th-order Butterworth filter. Because the continuous-time Butterworth filter has 14 zeros at $s = \infty$, the bilinear transformation creates 14 zeros at $z = -1$ for the digital filter. It is reasonable to expect that a lower-order filter might still satisfy the specifications even if it did not exceed them so greatly in the stopband. This expectation motivates the use of Chebyshev or equiripple approximation.

Example 7.6   Chebyshev Approximation

This method has two forms. Chebyshev type I approximations have equiripple behavior in the passband, and Chebyshev type II approximations have equiripple behavior in the stopband. Both methods lead to the same order for a given set of specifications. For the specifications of Eqs. (7.39), the required order is 8 rather than 14 as for the Butterworth approximation. Figure 7.16 shows log magnitude, passband magnitude, and group delay for the type I approximation to the specifications of Eqs. (7.39). Note that the frequency response oscillates with equal maximum error on either side of the desired gain of unity in the passband.

Figure 7.17 shows the frequency-response functions for the Chebyshev type II approximation to the specifications of Eqs. (7.39). In this case the equiripple approximation behavior is in the stopband. The pole-zero plots for the Chebyshev filters are shown in Fig. 7.18. Note that the Chebyshev type I system is similar to the Butterworth system in that it has all eight of its zeros at $z = -1$. On the other hand, the type II system has its zeros arrayed on the unit circle. These zeros are positioned by the design equations so as to achieve the equiripple behavior in the stopband.

In both cases of Chebyshev approximation, the monotonic behavior in either the stopband or the passband suggests that perhaps a lower-order system might be obtained if equiripple approximation were used in both the passband and the stopband. Indeed, it can be shown (see Papoulis, 1957) that for fixed values of δ_1, δ_2, ω_p, and ω_s in the tolerance scheme of Fig. 7.2(b), the lowest-order filter is obtained when the approximation error ripples equally between the extremes of the two approximation bands. Since this equiripple behavior is achieved with a rational

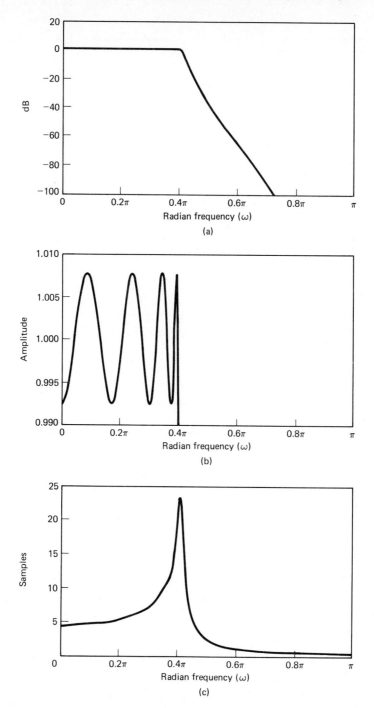

Figure 7.16 Frequency response of eighth-order Chebyshev type I filter in Example 7.6. (a) Log magnitude in dB. (b) Detailed plot of magnitude in passband. (c) Group delay.

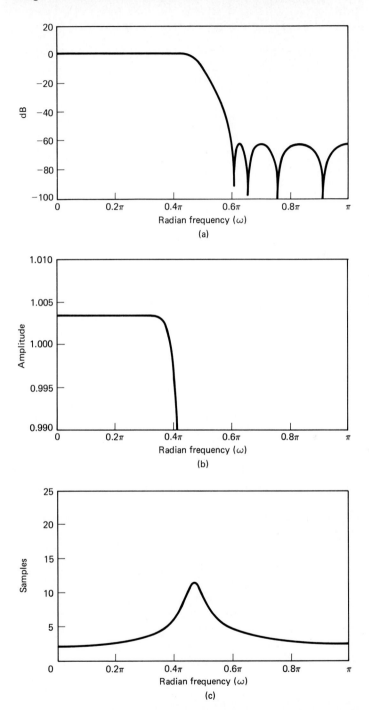

Figure 7.17   Frequency response of eighth-order Chebyshev type II filter in Example 7.6. (a) Log magnitude in dB. (b) Detailed plot of magnitude in passband. (c) Group delay.

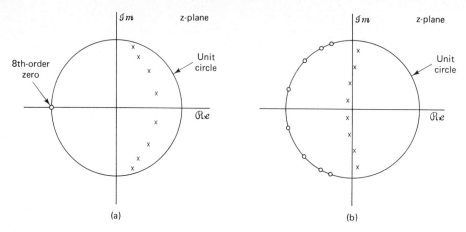

Figure 7.18 Pole-zero plot of eighth-order Chebyshev filters in Example 7.6. (a) Type I. (b) Type II.

function that involves elliptic functions, such systems are generally called elliptic filters.

Example 7.7   Elliptic Approximation†

The specifications of Eqs. (7.39) are met by an elliptic filter of order six. This is the *lowest-order* rational function approximation to the specifications. Figure 7.19 clearly shows the equiripple behavior in both approximation bands. Figure 7.20 shows that the elliptic filter, like the Chebyshev type II, has its zeros arrayed in the stopband region of the unit circle.

Bilinear transformation of analog filters designed by Butterworth, Chebyshev, or elliptic approximation methods is a standard method of design of IIR digital filters. The previous examples illustrate several important general features of such filters. In all cases, the resulting system function $H(z)$ has all its zeros on the unit circle and (for stability) all its poles inside the unit circle. As a result, all the approximation methods yield digital filters with nonconstant group delay or, equivalently, nonlinear phase. The greatest deviation from constant group delay occurs in all cases at the edge of the passband or in the transition band. In general, the Chebyshev type II approximation method yields the smallest delay in the passband and the widest region of the passband over which the group delay is approximately constant. However, if phase linearity is not an issue, then elliptic approximation yields the lowest-order system function, and therefore elliptic filters will generally require the least computation to implement a given filter specification.

---

† The design equations for elliptic filters are too involved to be appropriately summarized in Appendix B. They can be found in Storer (1957), Weinberg (1975), and Parks and Burrus (1987). A program for elliptic filter design was given by Gray and Markel (1976), and extensive tables for elliptic filter designs are available in Zverev (1967).

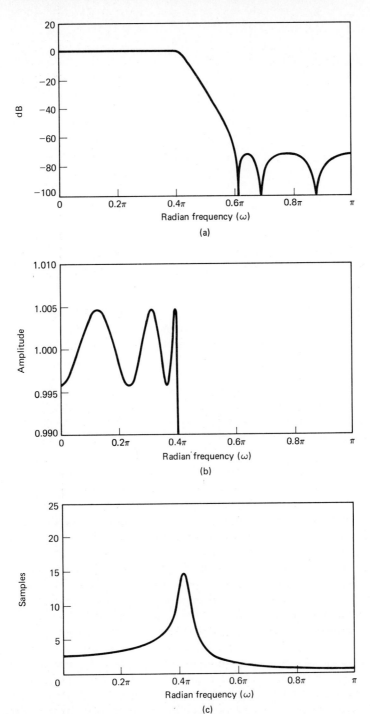

Figure 7.19   Frequency response of sixth-order elliptic filter in Example 7.7. (a) Log magnitude in dB. (b) Detailed plot of magnitude in passband. (c) Group delay.

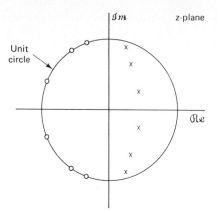

Figure 7.20   Pole-zero plot of sixth-order elliptic filter in Example 7.7.

7.2 FREQUENCY TRANSFORMATIONS OF LOWPASS IIR FILTERS

The previous examples have illustrated the use of the impulse invariance and the bilinear transformation methods for the design of IIR digital filters from continuous-time system functions having *lowpass* frequency-selective properties. Other types of frequency-selective filters such as highpass, bandpass, and bandstop filters are also of interest. Typical tolerance schemes for the four most commonly used types of frequency-selective discrete-time filters are shown in Fig. 7.21. Parts (a), (b), (c), and (d) depict approximation schemes for lowpass, highpass, bandpass, and bandstop filters, respectively. The traditional approach to the design of such frequency-selective continuous-time filters is first to design a frequency-normalized prototype lowpass filter and then, using an algebraic transformation, derive the desired lowpass, highpass, bandpass, or bandstop filter from the prototype lowpass filter (see Daniels, 1974). In the case of discrete-time frequency-selective filters, we could design a continuous-time frequency-selective filter of the desired type and then transform it to a discrete-time filter. This procedure would be acceptable with the bilinear transformation, but impulse invariance clearly could not be used to transform highpass and bandstop continuous-time filters into corresponding discrete-time filters because of the aliasing that results from sampling. An alternative procedure that works with either the bilinear transformation or impulse invariance is to design a discrete-time prototype lowpass filter and then perform an algebraic transformation on it to obtain the desired frequency-selective discrete-time filter.

Frequency-selective filters of the lowpass, highpass, bandpass, and bandstop types can be obtained from a lowpass discrete-time filter by use of transformations very similar to the bilinear transformation used to transform continuous-time system functions into discrete-time system functions. To see how this is done, assume that we are given a lowpass system function $H_{lp}(Z)$ that we wish to transform to a new system function $H(z)$, which has either lowpass, highpass, bandpass, or bandstop characteristics when evaluated on the unit circle. Note that we associate the complex variable Z

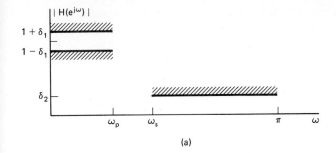

(a)

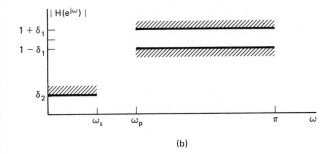

(b)

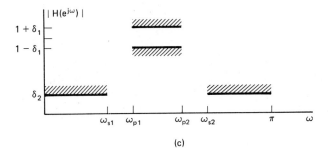

(c)

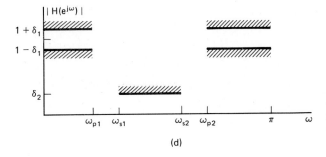

(d)

Figure 7.21   Tolerance schemes for frequency-selective digital filters. (a) Lowpass. (b) Highpass. (c) Bandpass. (d) Bandstop.

with the prototype lowpass filter and the complex variable z with the transformed filter. Then we define a mapping from the Z-plane to the z-plane of the form

$$Z^{-1} = G(z^{-1}) \tag{7.40}$$

such that

$$H(z) = H_{lp}(Z)\Big|_{Z^{-1}=G(z^{-1})}. \tag{7.41}$$

Instead of expressing Z as a function of z, we have assumed in Eq. (7.40) that Z^{-1} is expressed as a function of z^{-1}. Thus, according to Eq. (7.41), in obtaining $H(z)$ from $H_{lp}(Z)$ we simply replace Z^{-1} everywhere in $H_{lp}(Z)$ by the function $G(z^{-1})$. This is a convenient representation because $H_{lp}(Z)$ is normally expressed as a rational function of Z^{-1}.

If $H_{lp}(Z)$ is the rational system function of a causal and stable system, we naturally require that the transformed system function $H(z)$ be a rational function of z^{-1} and that the system also be causal and stable. This places the following constraints on the transformation $Z^{-1} = G(z^{-1})$:

1. $G(z^{-1})$ must be a rational function of z^{-1}.
2. The inside of the unit circle of the Z-plane must map to the inside of the unit circle of the z-plane.
3. The unit circle of the Z-plane must map onto the unit circle of the z-plane.

Let θ and ω be the frequency variables (angles) in the Z-plane and z-plane, respectively, i.e., on the respective unit circles $Z = e^{j\theta}$ and $z = e^{j\omega}$. Then for condition 3 to hold, it must be true that

$$e^{-j\theta} = |G(e^{-j\omega})|e^{j\sphericalangle G(e^{-j\omega})} \tag{7.42}$$

and, thus,

$$|G(e^{-j\omega})| = 1. \tag{7.43}$$

Therefore, the relationship between the frequency variables is

$$-\theta = \sphericalangle G(e^{-j\omega}). \tag{7.44}$$

It has been shown by Constantinides (1970) that the most general form of the function $G(z^{-1})$ that satisfies all the above requirements is

$$Z^{-1} = G(z^{-1}) = \pm \prod_{k=1}^{N} \frac{z^{-1} - \alpha_k}{1 - \alpha_k z^{-1}}. \tag{7.45}$$

From our discussion of allpass systems in Chapter 5, it should be clear that $G(z^{-1})$ as given in Eq. (7.45) satisfies Eq. (7.43), and it is easily shown that Eq. (7.45) maps the inside of the unit circle of the Z-plane to the inside of the unit circle of the z-plane if and only if $|\alpha_k| < 1$. By choosing appropriate values for N and the constants α_k, a variety of mappings can be obtained. The simplest is one that transforms a lowpass filter into another lowpass filter. For this case,

$$Z^{-1} = G(z^{-1}) = \frac{z^{-1} - \alpha}{1 - \alpha z^{-1}}. \tag{7.46}$$

If we substitute $Z = e^{j\theta}$ and $z = e^{j\omega}$, we obtain

$$e^{-j\theta} = \frac{e^{-j\omega} - \alpha}{1 - \alpha e^{-j\omega}}, \tag{7.47}$$

from which it follows that

$$\omega = \arctan\left[\frac{(1 - \alpha^2)\sin\theta}{2\alpha + (1 + \alpha^2)\cos\theta}\right]. \tag{7.48}$$

This relationship is plotted in Fig. 7.22 for different values of α. Although a warping of the frequency scale is evident in Fig. 7.22 (except in the case $\alpha = 0$, which corresponds to $Z^{-1} = z^{-1}$), if the original system has a piecewise-constant lowpass frequency response with cutoff frequency θ_p, then the transformed system will likewise have a similar lowpass response with cutoff frequency ω_p determined by the choice of α. Solving for α in terms of θ_p and ω_p, we obtain

$$\alpha = \frac{\sin[(\theta_p - \omega_p)/2]}{\sin[(\theta_p + \omega_p)/2]}. \tag{7.49}$$

Thus, to use these results to obtain a lowpass filter $H(z)$ with cutoff frequency ω_p from an already available lowpass filter $H_{lp}(Z)$ with cutoff frequency θ_p, we would use Eq. (7.49) to determine α in the expression

$$H(z) = H_{lp}(Z)\Big|_{Z^{-1} = (z^{-1} - \alpha)/(1 - \alpha z^{-1})}. \tag{7.50}$$

(Problem 7.21 explores how the lowpass-lowpass transformation can be used to obtain a network structure for a variable cutoff frequency filter where the cutoff frequency is determined by a single parameter α.)

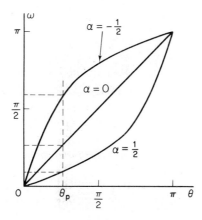

Figure 7.22 Warping of the frequency scale in lowpass-to-lowpass transformation.

TABLE 7.1   TRANSFORMATIONS FROM A LOWPASS DIGITAL FILTER PROTOTYPE OF CUTOFF FREQUENCY θ_p

Filter Type	Transformation	Associated Design Formulas
Lowpass	$Z^{-1} = \dfrac{z^{-1} - \alpha}{1 - \alpha z^{-1}}$	$\alpha = \dfrac{\sin\left(\dfrac{\theta_p - \omega_p}{2}\right)}{\sin\left(\dfrac{\theta_p + \omega_p}{2}\right)}$ $\omega_p = $ desired cutoff frequency
Highpass	$Z^{-1} = -\dfrac{z^{-1} + \alpha}{1 + \alpha z^{-1}}$	$\alpha = -\dfrac{\cos\left(\dfrac{\theta_p + \omega_p}{2}\right)}{\cos\left(\dfrac{\theta_p - \omega_p}{2}\right)}$ $\omega_p = $ desired cutoff frequency
Bandpass	$Z^{-1} = -\dfrac{z^{-2} - \dfrac{2\alpha k}{k+1}z^{-1} + \dfrac{k-1}{k+1}}{\dfrac{k-1}{k+1}z^{-2} - \dfrac{2\alpha k}{k+1}z^{-1} + 1}$	$\alpha = \dfrac{\cos\left(\dfrac{\omega_{p2} + \omega_{p1}}{2}\right)}{\cos\left(\dfrac{\omega_{p2} - \omega_{p1}}{2}\right)}$ $k = \cot\left(\dfrac{\omega_{p2} - \omega_{p1}}{2}\right)\tan\left(\dfrac{\theta_p}{2}\right)$ $\omega_{p1} = $ desired lower cutoff frequency $\omega_{p2} = $ desired upper cutoff frequency
Bandstop	$Z^{-1} = \dfrac{z^{-2} - \dfrac{2\alpha}{1+k}z^{-1} + \dfrac{1-k}{1+k}}{\dfrac{1-k}{1+k}z^{-2} - \dfrac{2\alpha}{1+k}z^{-1} + 1}$	$\alpha = \dfrac{\cos\left(\dfrac{\omega_{p2} + \omega_{p1}}{2}\right)}{\cos\left(\dfrac{\omega_{p2} - \omega_{p1}}{2}\right)}$ $k = \tan\left(\dfrac{\omega_{p2} - \omega_{p1}}{2}\right)\tan\left(\dfrac{\theta_p}{2}\right)$ $\omega_{p1} = $ desired lower cutoff frequency $\omega_{p2} = $ desired upper cutoff frequency

Transformations from a lowpass filter to highpass, bandpass, and bandstop filters can be derived in a similar manner. These transformations are summarized in Table 7.1. In the design formulas, all of the cutoff frequencies are assumed to be between zero and π radians. The following example illustrates the use of such transformations.

Example 7.8

Consider a type I Chebyshev lowpass filter whose system function is

$$H_{lp}(Z) = \frac{0.001836(1 + Z^{-1})^4}{(1 - 1.5548Z^{-1} + 0.6493Z^{-2})(1 - 1.4996Z^{-1} + 0.8482Z^{-2})}. \quad (7.51)$$

This fourth-order system was designed using the bilinear transformation to meet the specifications of Examples 7.3 and 7.4, i.e.,

$$0.89125 \leq |H_{lp}(e^{j\theta})| \leq 1, \qquad 0 \leq \theta \leq 0.2\pi, \tag{7.52a}$$

$$|H_{lp}(e^{j\theta})| \leq 0.17783, \qquad 0.3\pi \leq \theta \leq \pi. \tag{7.52b}$$

Thus the passband cutoff frequency is $\theta_p = 0.2\pi$. The frequency response of this filter is shown in Fig. 7.23.

Suppose that we want a highpass filter with passband cutoff frequency $\omega_p = 0.6\pi$. From Table 7.1 we obtain

$$\alpha = -\frac{\cos[(0.2\pi + 0.6\pi)/2]}{\cos[(0.2\pi - 0.6\pi)/2]} = -0.38197. \tag{7.53}$$

Thus, using the lowpass-highpass transformation indicated in Table 7.1, we obtain

$$H(z) = H_{lp}(Z)\Big|_{Z^{-1} = -[(z^{-1} - 0.38197)/(1 - 0.38197z^{-1})]}$$

$$= \frac{0.02426(1 - z^{-1})^4}{(1 + 1.0416z^{-1} + 0.4019z^{-2})(1 + 0.5661z^{-1} + 0.7657z^{-2})}. \tag{7.54}$$

The frequency response of this system is shown in Fig. 7.24. Note that except for some distortion of the frequency scale, the highpass frequency response appears very much as if the lowpass response were shifted in frequency by π. Also note that the fourth-order zero at $Z = -1$ for the lowpass filter now appears at $z = 1$ for the highpass filter. This example also verifies the obvious fact that equiripple passband (and stopband) behavior is preserved by frequency transformations of this type. Finally, note that the group delay in Fig. 7.24(c) is not simply a stretched and shifted version of Fig. 7.23(c). This is because the phase variations are stretched and shifted, so that the *derivative* of the phase is smaller for the highpass filter.

The use of frequency transformations in the design of a discrete-time filter is often more complicated than Example 7.8 suggests. Generally the prototype lowpass filter would not be given; rather it would be necessary to *determine* a lowpass filter such that after transformation, the resulting highpass, bandpass, or bandstop filter would meet a given set of specifications. Thus, for a highpass filter, the passband and stopband frequencies ω_p and ω_s would be specified. Obviously, if we specify ω_p and ω_s for the highpass filter, we cannot choose both θ_p and θ_s arbitrarily since only one pair of frequencies (θ_p and ω_p or θ_s and ω_s) enters into the formula for determining the transformation parameter α. One way of using the design formulas is to fix α and then obtain an equation relating θ to ω. This equation can be used to determine prewarped values of θ_p and θ_s that would transform to the desired passband and stopband cutoff frequencies ω_p and ω_s. Then a lowpass filter can be designed with the specified passband and stopband approximation errors δ_1 and δ_2 and the computed cutoff frequencies θ_p and θ_s.

Generally the relationship between θ, α, and ω would be nonlinear and similar in form to Eq. (7.48). However, the choice $\alpha = 0$ is particularly simple for the lowpass-to-highpass transformation since

$$Z^{-1} = -z^{-1}, \tag{7.55}$$

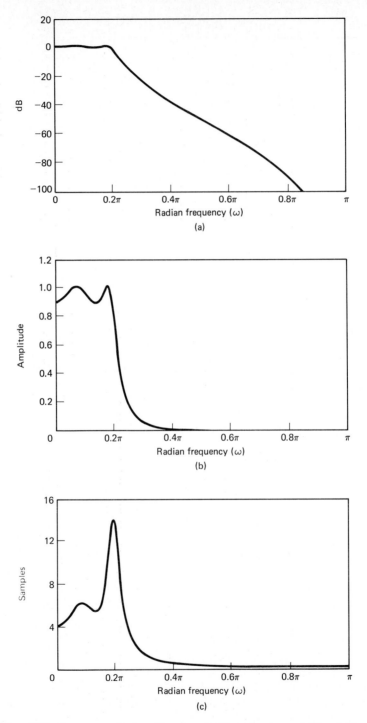

Figure 7.23   Frequency response of fourth-order Chebyshev lowpass filter. (a) Log magnitude in dB. (b) Magnitude. (c) Group delay.

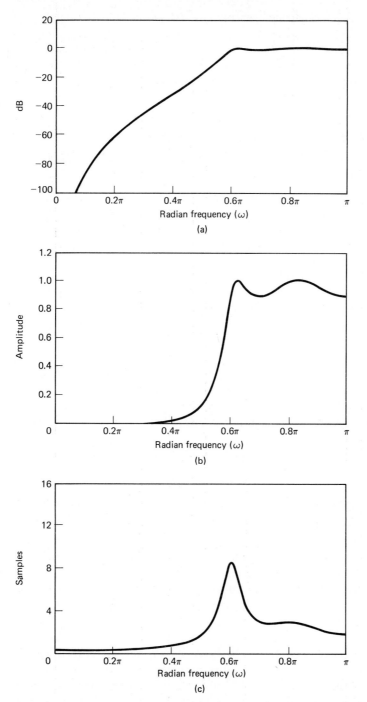

Figure 7.24   Frequency response of fourth-order Chebyshev highpass filter obtained by frequency transformation. (a) Log magnitude in dB. (b) Magnitude. (c) Group delay.

in which case the frequency variables are related by

$$\theta = \pi + \omega. \tag{7.56}$$

In the bandpass or bandstop case, the problem is somewhat more involved since there are more free parameters.

Sk·p ## 7.3 COMPUTER-AIDED DESIGN OF DISCRETE-TIME IIR FILTERS

In Section 7.1 we showed that discrete-time filters can be designed by designing an appropriate continuous-time filter and then transforming it into a discrete-time filter. This approach is reasonable when we can take advantage of continuous-time designs that have closed-form formulas or that can use extensive existing design tables; filters for which this is possible are frequency-selective filters such as Butterworth, Chebyshev, or elliptic filters. In general, however, analytical formulas do not exist for the design of either continuous-time or discrete-time filters to match arbitrary frequency-response specifications or other types of specifications. In these more general cases, design procedures have been developed that are algorithmic, generally relying on the use of a computer to solve sets of linear or nonlinear equations. Often the computer-aided design techniques apply equally well to the design of either continuous-time or discrete-time filters with only minor modification. Therefore, nothing is gained by first obtaining a continuous-time filter design and then transforming this design to a discrete-time filter.

Most algorithmic design procedures for IIR filters take the following form:

1. $H(z)$ is assumed to be a rational function. It can be represented as a ratio of polynomials in z (or z^{-1}), as a product of numerator and denominator factors (zeros and poles), or as a product of second-order factors.
2. The orders of the numerator and denominator of $H(z)$ are fixed.
3. An ideal desired frequency response and a corresponding approximation error criterion is chosen.
4. By a suitable optimization algorithm, the free parameters (numerator and denominator coefficients, zeros and poles, etc.) are varied in a systematic way to minimize the approximation error according to the assumed error criterion.
5. The set of parameters that minimizes the approximation error determines the system function of the desired system.

In such a procedure, the designer must either be satisfied with the filter that minimizes the approximation error or change the problem and recompute the solution. Generally this means changing the order of the numerator and denominator of the system function. Thus, algorithmic methods are not as elegant as methods based on closed-form design formulas, but the latter are available principally for frequency-selective filters, while algorithmic procedures generally are not so limited.

For nonstandard response properties, algorithmic methods may be the only possible approach.

7.3.1   Deczky's Method

From the description of the general approach, it is clear that many possibilities exist for defining algorithmic design procedures. Many error criteria are available, and there are many optimization algorithms. To illustrate the algorithmic approach, we will discuss a method due to Deczky (1972). We focus on this approach because it is effective, flexible, and easy to modify and, most important, a FORTRAN program for its implementation is readily available. (See DSP Committee, 1979.)

In Deczky's method, the system function of the filter is represented in terms of its poles and zeros. For simplicity of notation we assume that all the poles and zeros occur in complex conjugate pairs so that

$$H(z) = K_o \prod_{k=1}^{N_s} \frac{(1 - z_k z^{-1})(1 - z_k^* z^{-1})}{(1 - p_k z^{-1})(1 - p_k^* z^{-1})}. \tag{7.57}$$

Modifications to include a fixed number of real poles and/or zeros or to restrict some of the zeros to lie on the unit circle are easily made. Other more specialized system function forms are also possible. For example, to design an allpass system to compensate for the phase or group delay of a known system, the form of $H(z)$ would be

$$H(z) = \prod_{k=1}^{N_s} \frac{(p_k - z^{-1})(p_k^* - z^{-1})}{(1 - p_k z^{-1})(1 - p_k^* z^{-1})}. \tag{7.58}$$

Another interesting example of the flexibility of this approach is the work of Richards (1982) who showed how to modify the Deczky program to design recursive decimation filters where the denominator of $H(z)$ is a polynomial in z^{-R} (R is the decimation factor) rather than z^{-1} as in Eq. (7.57).

To further simplify certain expressions required by the optimization algorithms, it is particularly convenient if the poles and zeros are expressed in their polar representations. That is,

$$z_k = r_{zk} e^{j\theta_{zk}}, \tag{7.59a}$$

$$p_k = r_{pk} e^{j\theta_{pk}}. \tag{7.59b}$$

The order of both the numerator and the denominator is $2N_s$, although it is not necessary that the orders be equal.

As originally formulated by Deczky, the method permits minimization of approximation error for both the frequency-response magnitude and the group delay. If we define

$$A(\omega) = |H(e^{j\omega})| \tag{7.60a}$$

and

$$\tau(\omega) = -\frac{d}{d\omega} \arg[H(e^{j\omega})], \tag{7.60b}$$

then the approximation error for the magnitude is defined as

$$E_A = \sum_{i=1}^{L} W_A(\omega_i)|A(\omega_i) - A_d(\omega_i)|^{2m} \tag{7.61}$$

and the approximation error for the group delay is

$$E_\tau = \sum_{i=1}^{L} W_\tau(\omega_i)|\tau(\omega_i) - \tau_d(\omega_i) - \tau_o|^{2q}, \tag{7.62}$$

where m and q are integer parameters specified by the designer, $A_d(\omega)$ and $\tau_d(\omega)$ are the desired ideal magnitude and group delay functions, respectively, and τ_o is a variable parameter that is adjusted in the optimization process to account for constant delays that cannot be achieved by the assumed form of $H(z)$. Formulas for $A(\omega)$ and $\tau(\omega)$ as required in Eqs. (7.61) and (7.62) are easily derived by setting $z = e^{j\omega}$ in Eq. (7.57) or (7.58). These formulas are given by Deczky (1972).

The functions $W_A(\omega)$ and $W_\tau(\omega)$ are error weighting functions that can be used to distribute the approximation error as desired. Notice that the approximation errors defined in Eqs. (7.61) and (7.62) are sums of weighted errors at a discrete set of frequencies ω_i, $i = 1, 2, \ldots, L$. The choice of these frequencies is part of the initialization of the algorithm. One possibility is to distribute them evenly across the interval $0 \le \omega \le \pi$, or they may be concentrated in regions of special interest in the frequency response. The total approximation error is defined as

$$E = \alpha E_A + (1 - \alpha)E_\tau. \tag{7.63}$$

This is the error that is minimized by the Deczky program given in DSP Committee (1979). Note that $\alpha = 1$ and $\alpha = 0$ correspond respectively to approximation of magnitude only and group delay only. Values of α between zero and one correspond to joint approximation of magnitude and group delay.

For a particular choice of m and q in Eqs. (7.61) and (7.62), there are $4N_s + 2$ variable parameters in this formulation, given by the parameter vector

$$\boldsymbol{\phi} = \{r_{z1}, \ldots, r_{zN_s}, \theta_{z1}, \ldots, \theta_{zN_s}, r_{p1}, \ldots, r_{pN_s}, \theta_{p1}, \ldots, \theta_{pN_s}, K_o, \tau_o\}. \tag{7.64}$$

These parameters must be systematically varied to minimize the approximation error E.

The error E is a nonlinear function of the $4N_s + 2$ variables. Conceptually, finding the minimum of such a function involves the solution of the set of $4N_s + 2$ nonlinear equations

$$\frac{\partial E}{\partial \phi_i} = \alpha \frac{\partial E_A}{\partial \phi_i} + (1 - \alpha)\frac{\partial E_\tau}{\partial \phi_i} = 0, \qquad i = 1, 2, \ldots, 4N_s + 2, \tag{7.65}$$

where ϕ_i represents the ith component of the parameter vector. This solution is not easily found, and in general there is no guarantee that a solution is a global minimum since the error function will have many local minima. Deczky used the iterative algorithm of Fletcher and Powell (1963) for finding a local minimum of the approximation error. In its original form, the Fletcher-Powell algorithm performs an unconstrained minimization. However, in the IIR filter design problem, we want to restrict our search for a minimum to the region of parameter space corresponding to stable systems, i.e., $0 \le r_{pk} < 1$ for all k. To apply this constraint, Deczky suggested that the basic Fletcher-Powell algorithm be modified so that if at a given stage in the iterative solution, one of the r_{pk}'s is to be changed so that the new value of r_{pk} will be greater than one, then the amount of change is successively halved until the new value of r_{pk} will be less than one. Deczky showed that with this constraint, starting with an initial stable approximation, a bounded $A_d(\omega)$ and $\tau_d(\omega)$, and positive weighting functions $W_A(\omega)$ and $W_\tau(\omega)$, a local minimum of E exists for $2m \ge 2$ and $2q \ge 2$ such that the optimum parameters correspond to a stable system.

The types of solutions obtained by such optimization depend strongly on the integer parameters m and q. Raising the magnitude and group delay errors to the $2m$ and $2q$ powers, respectively, ensures that all individual errors have a positive contribution to the total error and at the same time leads to an easily differentiable error function. This is important since the Fletcher-Powell algorithm requires an equation for the gradient of E as a function of the free parameters. Formulas for $\partial E_A / \partial \phi_i$ and $\partial E_\tau / \partial \phi_i$ for $H(z)$ as in Eq. (7.57) are given by Deczky (1972). If $m = q = 1$, the mean-square error is minimized. If m and q are large, then small errors become much less important than large errors, so that most of the effort goes into minimizing the maximum error. In fact, it can be shown that as $m, q \to \infty$, the solution approaches the minimax solution.

7.3.2 Group Delay Equalization

As an example of the utility of Deczky's method, consider the design of an allpass filter to compensate for phase nonlinearity of a linear system. This design problem, depicted in Fig. 7.25, is common in communications systems, where it is often necessary to "equalize" the group delay of a communications channel.

The overall effective frequency response of the system in Fig. 7.25 is

$$G(e^{j\omega}) = H_f(e^{j\omega})H_{ap}(e^{j\omega}), \tag{7.66}$$

where $H_f(e^{j\omega})$ represents the fixed frequency response of the system to be compensated and $H_{ap}(e^{j\omega})$ represents the frequency response of the compensating allpass system to be designed. Obviously, if $H_{ap}(e^{j\omega})$ is an allpass system, then $|G(e^{j\omega})| = |H_f(e^{j\omega})|$. The group delay of the overall system is

$$\mathrm{grd}[G(e^{j\omega})] = \mathrm{grd}[H_f(e^{j\omega})] + \mathrm{grd}[H_{ap}(e^{j\omega})]. \tag{7.67}$$

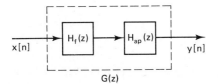

Figure 7.25 Block diagram of allpass group delay compensation of a linear system.

The group delay will be perfectly equalized if the group delay of the allpass system is

$$\tau_d(\omega) = \text{grd}[H_{ap}(e^{j\omega})] = -\text{grd}[H_f(e^{j\omega})]. \qquad (7.68)$$

However, recall from our discussion of allpass systems in Chapter 5 that

$$\text{grd}[H_{ap}(e^{j\omega})] \geq 0, \qquad 0 \leq \omega \leq \pi. \qquad (7.69)$$

Therefore, only negative group delay can be canceled by an allpass system. This is the reason that the free parameter τ_o is included in the group delay approximation error definition of Eq. (7.62).

Consequently, we seek an allpass system such that the *variations* of its group delay are the negative of the variations of the group delay of the fixed system. It is explicitly recognized that the allpass system will also introduce some amount of additional positive delay. The following example given by Deczky (1972) illustrates the design of an allpass compensator.

Example 7.9

Figure 7.26(a) shows the attenuation or loss function $(-20 \log_{10}|H(e^{j\omega})|)$ for a fourth-order elliptic lowpass digital filter. The group delay function for this filter is the solid curve in Figs. 7.26(b) and 7.26(c). To illustrate how this design method can be used to compensate for group delay distortion, the desired group delay was set equal to the negative of the group delay of the elliptic filter, and the system function of the filter to be designed was assumed to be allpass as in Eq. (7.58). Figure 7.26(b) shows the results for $\alpha = 0$, $N_s = 1$, and $2q = 2$. The weighting function $W_{\tau}(\omega)$ was chosen to emphasize the group delay error in the frequency band corresponding to the passband of the elliptic filter (approximately $0 \leq \omega \leq 0.5\pi$). The dashed curve in Fig. 7.26(b) shows the overall group delay of the cascade of the elliptic filter with the allpass equalizing system. (Recall that the group delay functions add.) The amount of variation of the overall group delay is seen to be somewhat less in the filter passband than for the elliptic filter alone. If this is not satisfactory performance, more allpass sections can be used. With four allpass sections and $2q = 10$, the resulting overall group delay is the dashed curve in Fig. 7.26(c). In this case, the group delay is constant to within a ripple of about 1.3 samples. It can be seen that the value $2q = 10$ yields very nearly equiripple or minimax behavior in the passband. Notice that the allpass equalizer significantly increases the overall delay of the compensated system.

The previous discussion and the example of Deczky's method of IIR filter design points out some general facts about algorithmic design methods. First, the choice of filter order and structure must be guided by experience in designing filters of a given type. Second, designing for a set of prescribed specifications usually requires some amount of trial and error. Third, the initial choice of parameters can influence the rate of convergence of the iteration and the ultimate solution, since convergence to a local minimum is often all that can be guaranteed.

In spite of these restrictions, an experienced filter designer can solve a wide range of filtering problems using the flexibility of the algorithmic approach. This approach has led to the development of other optimization approaches such as the use of linear programming (Rabiner, Graham, and Helms, 1974) and other methods (Kaiser and Dolan in DSP Committee, 1979; Martinez and Parks, 1978).

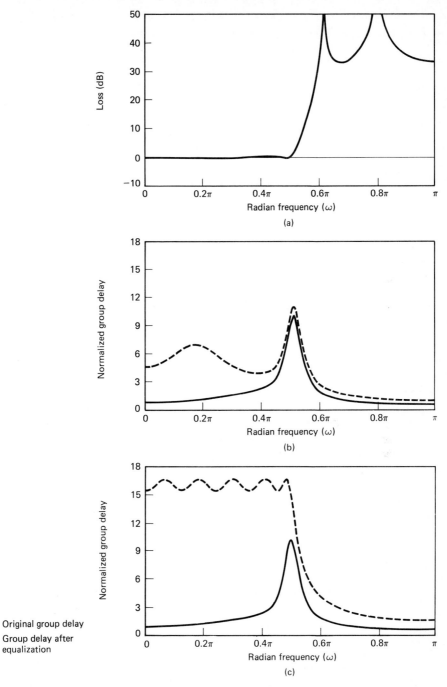

Figure 7.26   Allpass group delay compensation of an elliptic filter. (a) Loss characteristic of a fourth-order elliptic lowpass filter. (b) Group delay equalization by a two-section allpass filter with $2q = 2$. (c) Group delay equalization by a four-section allpass filter with $2q = 10$. (After Deczky, 1972.)

7.4 DESIGN OF FIR FILTERS BY WINDOWING

As discussed in Section 7.1, the most commonly used techniques for design of IIR filters are based on transformations of continuous-time IIR systems into discrete-time IIR systems. This is partly because continuous-time filter design was a highly advanced art before discrete-time filters were of interest and partly because of the difficulty of implementing a noniterative direct design method for IIR filters.

In contrast, FIR filters are almost entirely restricted to discrete-time implementations. Consequently the design techniques for FIR filters are based on directly approximating the desired frequency response of the discrete-time system. Furthermore, most techniques for approximating the magnitude response of an FIR system assume a linear phase constraint, thereby avoiding the problem of spectrum factorization that complicates the direct design of IIR filters.

The simplest method of FIR filter design is called the *window method*. This method generally begins with an ideal desired frequency response that can be represented as

$$H_d(e^{j\omega}) = \sum_{n=-\infty}^{\infty} h_d[n]e^{-j\omega n}, \tag{7.70}$$

where $h_d[n]$ is the corresponding impulse response sequence, which can be expressed in terms of $H_d(e^{j\omega})$ as

$$h_d[n] = \frac{1}{2\pi} \int_{-\pi}^{\pi} H_d(e^{j\omega})e^{j\omega n} \, d\omega. \tag{7.71}$$

Many idealized systems are defined by piecewise-constant or piecewise-functional frequency responses with discontinuities at the boundaries between bands. As a result, they have impulse responses that are noncausal and infinitely long. The most straightforward approach to obtaining a causal FIR approximation to such systems is to truncate the ideal response. Equation (7.70) can be thought of as a Fourier series representation of the periodic frequency response $H_d(e^{j\omega})$, with the sequence $h_d[n]$ playing the role of the Fourier coefficients. Thus, the approximation of an ideal filter by truncation of the ideal impulse response is identical to the issue of the convergence of Fourier series, a subject that has received a great deal of study. A particularly important concept from this theory is the Gibbs phenomenon, which was discussed in Example 2.17. In the following discussion we will see how this nonuniform convergence phenomenon manifests itself in the design of FIR filters.

The simplest way to obtain a causal FIR filter from $h_d[n]$ is to define a new system with impulse response $h[n]$ given by†

$$h[n] = \begin{cases} h_d[n], & 0 \le n \le M, \\ 0, & \text{otherwise.} \end{cases} \tag{7.72}$$

---

† The notation for FIR systems was established in Chapter 5. That is, M is the order of the system function polynomial. Thus, $(M + 1)$ is the length or duration of the impulse response. Often in the literature, N is used for the length of the impulse response of an FIR filter; however, we have used N to denote the order of the denominator polynomial in the system function of an IIR filter. Thus, to avoid confusion and maintain consistency throughout this book, we will always consider the length of the impulse response of an FIR filter to be $(M + 1)$.

More generally, we can represent $h[n]$ as the product of the desired impulse response and a finite-duration "window" $w[n]$; i.e.,

$$h[n] = h_d[n]w[n], \tag{7.73}$$

where for simple truncation as in Eq. (7.72) the window is the *rectangular window*

$$w[n] = \begin{cases} 1, & 0 \le n \le M, \\ 0, & \text{otherwise.} \end{cases} \tag{7.74}$$

It follows from the modulation or windowing theorem (Section 2.9.7) that

$$H(e^{j\omega}) = \frac{1}{2\pi} \int_{-\pi}^{\pi} H_d(e^{j\theta})W(e^{j(\omega - \theta)}) \, d\theta. \tag{7.75}$$

That is, $H(e^{j\omega})$ is the periodic convolution of the desired ideal frequency response with the Fourier transform of the window. Thus, the frequency response $H(e^{j\omega})$ will be a "smeared" version of the desired response $H_d(e^{j\omega})$. Figure 7.27(a) depicts typical functions $H_d(e^{j\theta})$ and $W(e^{j(\omega - \theta)})$ as required in Eq. (7.75).

If $w[n] = 1$ for all n (i.e., we do not truncate at all), $W(e^{j\omega})$ is a periodic impulse train with period 2π, and therefore $H(e^{j\omega}) = H_d(e^{j\omega})$. This interpretation suggests that if $w[n]$ is chosen so that $W(e^{j\omega})$ is concentrated in a narrow band of frequencies around $\omega = 0$, then $H(e^{j\omega})$ will "look like" $H_d(e^{j\omega})$ except where $H_d(e^{j\omega})$ changes very

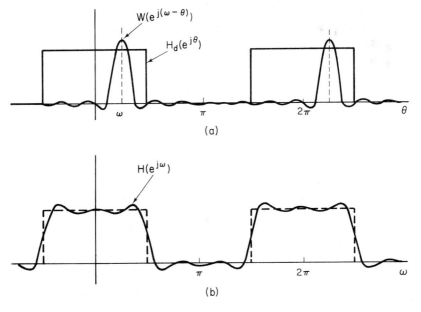

(a)

(b)

Figure 7.27 (a) Convolution process implied by truncation of the ideal impulse response. (b) Typical approximation resulting from windowing the ideal impulse response.

abruptly. Thus, the choice of window is governed by the desire to have $w[n]$ as short as possible in duration so as to minimize computation in the implementation of the filter while having $W(e^{j\omega})$ approximate an impulse; that is, we want $W(e^{j\omega})$ to be highly concentrated in frequency so that the convolution of Eq. (7.75) faithfully reproduces the desired frequency response. These are conflicting requirements, as can be seen in the case of the rectangular window of Eq. (7.74), where

$$W(e^{j\omega}) = \sum_{n=0}^{M} e^{-j\omega n} = \frac{1 - e^{-j\omega(M+1)}}{1 - e^{-j\omega}} = e^{-j\omega M/2} \frac{\sin[\omega(M+1)/2]}{\sin(\omega/2)}. \tag{7.76}$$

The magnitude of the function $\sin[\omega(M+1)/2]/\sin(\omega/2)$ is plotted in Fig. 7.28 for the case $M = 7$. Note that $W(e^{j\omega})$ for the rectangular window has generalized linear phase. As M increases, the width of the "main lobe" decreases. The main lobe is usually defined as the region between the first zero crossings on either side of the origin. For the rectangular window, the mainlobe width is $\Delta\omega_m = 4\pi/(M+1)$. However, for the rectangular window, the sidelobes are significantly large and, in fact, as M increases, the peak amplitudes of the mainlobe and the sidelobes grow in a manner such that the area under each lobe is a constant while the width of each lobe decreases with M. Consequently, as $W(e^{j(\omega-\theta)})$ "slides by" a discontinuity of $H_d(e^{j\theta})$ with increasing ω, the integral of $W(e^{j(\omega-\theta)})H_d(e^{j\theta})$ will oscillate as each sidelobe of $W(e^{j(\omega-\theta)})$ moves past the discontinuity. This result is depicted in Fig. 7.27(b). Since the area under each lobe remains constant with increasing M, the oscillations occur more rapidly but do not decrease in amplitude as M increases.

In the theory of Fourier series, it is well known that this nonuniform convergence, the *Gibbs phenomenon*, can be moderated through the use of a less abrupt truncation of the Fourier series. By tapering the window smoothly to zero at each end, the height of the sidelobes can be diminished; however, this is achieved at the expense of a wider mainlobe and thus a wider transition at the discontinuity.

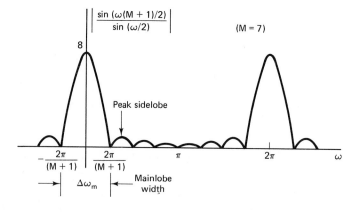

Figure 7.28   Magnitude of the Fourier transform of a rectangular window ($M = 7$).

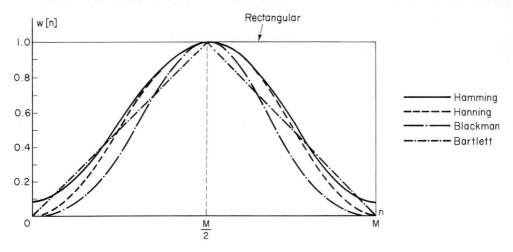

Figure 7.29   Commonly used windows.

7.4.1 Properties of Commonly Used Windows

Some commonly used windows are shown in Fig. 7.29.† These windows are defined by the following equations:

Rectangular

$$w[n] = \begin{cases} 1, & 0 \leq n \leq M, \\ 0, & \text{otherwise} \end{cases} \qquad (7.77a)$$

Bartlett (*triangular*)

$$w[n] = \begin{cases} 2n/M, & 0 \leq n \leq M/2, \\ 2 - 2n/M, & M/2 < n \leq M, \\ 0, & \text{otherwise} \end{cases} \qquad (7.77b)$$

Hanning

$$w[n] = \begin{cases} 0.5 - 0.5 \cos(2\pi n/M), & 0 \leq n \leq M, \\ 0, & \text{otherwise} \end{cases} \qquad (7.77c)$$

Hamming

$$w[n] = \begin{cases} 0.54 - 0.46 \cos(2\pi n/M), & 0 \leq n \leq M, \\ 0, & \text{otherwise} \end{cases} \qquad (7.77d)$$

† The Bartlett, Hanning, Hamming, and Blackman windows are all named after their originators. The Hanning window is associated with Julius von Hann, an Austrian meteorologist, and is sometimes referred to as the Hann window. The term "hanning" was used by Blackman and Tukey (1958) to describe the operation of applying this window to a signal and has since become the most widely used name for the window, with varying preferences for the choice of "Hanning" or "hanning."

Blackman

$$w[n] = \begin{cases} 0.42 - 0.5\cos(2\pi n/M) + 0.08\cos(4\pi n/M), & 0 \le n \le M, \\ 0, & \text{otherwise} \end{cases} \quad (7.77e)$$

(For convenience, Fig. 7.29 shows these windows plotted as functions of a continuous variable; however, as specified in Eqs. 7.77, the window sequence is defined only at integer values of n.)

As will be discussed in Chapter 11, the windows defined in Eqs. (7.77) are commonly used for spectrum analysis as well as FIR filter design. They have the desired property that their Fourier transforms are concentrated around $\omega = 0$, and they have a simple functional form that allows them to be computed easily. The Fourier transform of the Bartlett window can be expressed as a product of Fourier transforms of rectangular windows, and the Fourier transforms of the other windows can be expressed as sums of frequency-shifted Fourier transforms of the rectangular window as given by Eq. (7.76). (See Problem 7.25.)

The function $20\log_{10}|W(e^{j\omega})|$ is plotted in Fig. 7.30 for each of these windows with $M = 50$. The rectangular window clearly has the narrowest mainlobe and thus, for a given length, it should yield the sharpest transitions of $H(e^{j\omega})$ at a discontinuity of $H_d(e^{j\omega})$. However, the first sidelobe is only about 13 dB below the main peak, resulting in oscillations of $H(e^{j\omega})$ of considerable size around discontinuities of $H_d(e^{j\omega})$. Table 7.2, which compares the windows of Eqs. (7.77), clearly shows that by

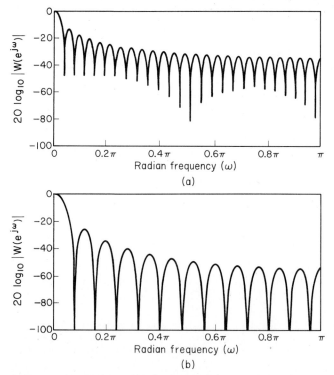

Figure 7.30 Fourier transforms (log magnitude) of windows of Fig. 7.29. (a) Rectangular. (b) Bartlett.

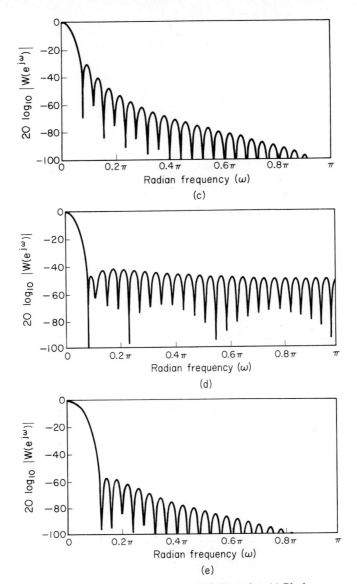

Figure 7.30 *(continued)* (c) Hanning. (d) Hamming. (e) Blackman.

tapering the window smoothly to zero as with the Hamming, Hanning, and Blackman windows, the sidelobes (second column) are greatly reduced; however, the price paid is a much wider mainlobe (third column) and thus wider transitions at discontinuities of $H_d(e^{j\omega})$. The other columns of Table 7.2 will be discussed later.

7.4.2 Incorporation of Generalized Linear Phase

In designing many types of FIR filters, it is desirable to obtain causal systems with generalized linear phase response. All the windows of Eqs. (7.77) have been defined in

TABLE 7.2 COMPARISON OF COMMONLY USED WINDOWS

Window Type	Peak Sidelobe Amplitude (Relative)	Approximate Width of Mainlobe	Peak Approximation Error $20 \log_{10} \delta$ (dB)	Equivalent Kaiser Window β	Transition Width of Equivalent Kaiser Window
Rectangular	-13	$4\pi/(M + 1)$	-21	0	$1.81\pi/M$
Bartlett	-25	$8\pi/M$	-25	1.33	$2.37\pi/M$
Hanning	-31	$8\pi/M$	-44	3.86	$5.01\pi/M$
Hamming	-41	$8\pi/M$	-53	4.86	$6.27\pi/M$
Blackman	-57	$12\pi/M$	-74	7.04	$9.19\pi/M$

anticipation of this need. Specifically, note that all the windows have the property

$$w[n] = \begin{cases} w[M - n], & 0 \le n \le M, \\ 0, & \text{otherwise,} \end{cases} \tag{7.78}$$

i.e., they are symmetric about the point $M/2$. As a result, their Fourier transforms are of the form

$$W(e^{j\omega}) = W_e(e^{j\omega})e^{-j\omega M/2}, \tag{7.79}$$

where $W_e(e^{j\omega})$ is a real even function of ω. This is illustrated by Eq. (7.76). The convention of Eq. (7.78) leads to causal filters in general and, if the desired impulse response is also symmetric about $M/2$, i.e., $h_d[M - n] = h_d[n]$, then the windowed impulse response will also have that symmetry and the resulting frequency response will have generalized linear phase, that is,

$$H(e^{j\omega}) = A_e(e^{j\omega})e^{-j\omega M/2}, \tag{7.80}$$

where $A_e(e^{j\omega})$ is real and is an even function of ω. Similarly, if the desired impulse response is antisymmetric about $M/2$, i.e., $h_d[M - n] = -h_d[n]$, then the windowed impulse response will also be antisymmetric about $M/2$ and the resulting frequency response will have generalized linear phase with $90°$ constant phase shift; i.e.,

$$H(e^{j\omega}) = jA_o(e^{j\omega})e^{-j\omega M/2}, \tag{7.81}$$

where $A_o(e^{j\omega})$ is real and is an odd function of ω.

Although the preceding statements are obvious if we consider the product of the symmetric window with the symmetric (or antisymmetric) desired impulse response, it is useful to consider the frequency-domain representation. Suppose $h_d[M - n] = h_d[n]$. Then

$$H_d(e^{j\omega}) = H_e(e^{j\omega})e^{-j\omega M/2}, \tag{7.82}$$

where $H_e(e^{j\omega})$ is real and even.

If the window is symmetric, we can substitute Eqs. (7.79) and (7.82) into Eq. (7.75) to obtain

$$H(e^{j\omega}) = \frac{1}{2\pi} \int_{-\pi}^{\pi} H_e(e^{j\theta})e^{-j\theta M/2} W_e(e^{j(\omega - \theta)})e^{-j(\omega - \theta)M/2} \, d\theta. \tag{7.83}$$

A simple manipulation of the phase factors leads to

$$H(e^{j\omega}) = A_e(e^{j\omega})e^{-j\omega M/2}, \qquad (7.84)$$

where

$$A_e(e^{j\omega}) = \frac{1}{2\pi} \int_{-\pi}^{\pi} H_e(e^{j\theta}) W_e(e^{j(\omega - \theta)}) \, d\theta. \qquad (7.85)$$

Thus, we see that the resulting system has generalized linear phase and, moreover, the real function $A_e(e^{j\omega})$ is the result of the periodic convolution of the real functions $H_e(e^{j\omega})$ and $W_e(e^{j\omega})$.

The detailed behavior of the convolution of Eq. (7.85) determines the magnitude response of the filter that results from windowing. The following example illustrates this for a linear phase lowpass filter.

Example 7.10

The desired frequency response is defined as

$$H_{lp}(e^{j\omega}) = \begin{cases} e^{-j\omega M/2}, & |\omega| < \omega_c, \\ 0, & \omega_c < |\omega| \le \pi, \end{cases} \qquad (7.86)$$

where the generalized linear phase factor has been incorporated into the definition of the ideal lowpass filter. The corresponding ideal impulse response is

$$h_{lp}[n] = \frac{1}{2\pi} \int_{-\omega_c}^{\omega_c} e^{-j\omega M/2} e^{j\omega n} \, d\omega = \frac{\sin[\omega_c(n - M/2)]}{\pi(n - M/2)}, \qquad (7.87)$$

for $-\infty < n < \infty$. It is easily shown that $h_{lp}[M - n] = h_{lp}[n]$, so if we use a symmetric window in the equation

$$h[n] = \frac{\sin[\omega_c(n - M/2)]}{\pi(n - M/2)} w[n], \qquad (7.88)$$

then a linear phase system will result.

The upper part of Fig. 7.31 depicts the character of the amplitude response that would result for all the windows of Eqs. (7.77) except the Bartlett window, which is rarely used for filter design. (The Bartlett window would produce a monotonic function $A_e(e^{j\omega})$ because $W_e(e^{j\omega})$ is a positive function.) Figure 7.31 displays the important properties of window method approximations to desired frequency responses that have step discontinuities. It applies accurately when ω_c is not close to zero or to π and when the width of the mainlobe is smaller than $2\omega_c$. At the bottom of the figure is a typical Fourier transform for a symmetric window. This function should be visualized in different positions as an aid in understanding the shape of the approximation $A_e(e^{j\omega})$ in the vicinity of ω_c.

When $\omega = \omega_c$, the symmetric function $W_e(e^{j(\omega - \theta)})$ is centered on the discontinuity, and about one-half its area contributes to $A_e(e^{j\omega})$. Similarly, we can see that the peak overshoot occurs when $W_e(e^{j(\omega - \theta)})$ is shifted such that the first negative sidelobe on the right is just to the right of ω_c. Similarly, the peak negative undershoot occurs when the first negative sidelobe on the left is just to the left of ω_c. This means that the distance between the peak ripples on either side of the discontinuity is approximately the mainlobe width $\Delta\omega_m$ as shown. The transition width $\Delta\omega$ as defined in Fig. 7.31 is therefore somewhat less than the mainlobe width. Finally, due to the symmetry of

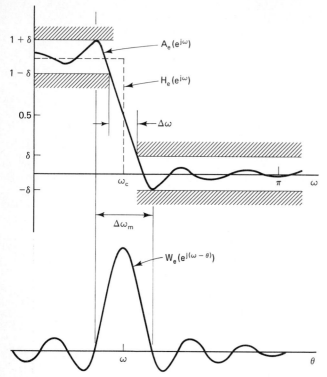

Figure 7.31 Illustration of type of approximation obtained at a discontinuity of the ideal frequency response.

$W_e(e^{j(\omega-\theta)})$, the approximation tends to be symmetric around ω_c; i.e., the approximation overshoots by δ in the passband and undershoots by the same amount in the stopband.

The fourth column of Table 7.2 shows the peak approximation error (in dB) for the windows of Eqs. (7.77). Clearly the windows with the smaller sidelobes yield better approximations at a discontinuity of the ideal response. Also the third column, which shows mainlobe width, suggests that narrower transition regions can be achieved by increasing M. Thus, through the choice of the window shape and duration, we can control the properties of the resulting FIR filter. However, trying different windows and adjusting lengths by trial and error is not a very satisfactory way to design filters. Fortunately, a simple formalization of the window method has been developed by Kaiser (1974).

7.4.3 The Kaiser Window Filter Design Method

The tradeoff between the mainlobe width and sidelobe area can be quantified by seeking the window function that is maximally concentrated around $\omega = 0$ in the frequency domain. This issue was considered in depth in a series of classic papers by Slepian et al. (1961). The solution found in this work involves prolate spheroidal wave functions, which are difficult to compute and therefore unattractive for filter design. However, Kaiser (1966, 1974) found that a near-optimal window could be formed using the zeroth-order modified Bessel function of the first kind, a function that is much easier to compute. The Kaiser window is defined as

$$
w[n] = \begin{cases} \dfrac{I_0[\beta(1 - [(n - \alpha)/\alpha]^2)^{1/2}]}{I_0(\beta)}, & 0 \le n \le M, \\[3mm] 0, & \text{otherwise,} \end{cases}
\tag{7.89}
$$

where $\alpha = M/2$, and $I_0(\,\cdot\,)$ represents the zeroth-order modified Bessel function of the first kind. In contrast to the other windows in Eqs. (7.77), the Kaiser window has two parameters: the length $(M + 1)$ and a shape parameter β. By varying $(M + 1)$ and β, the window length and shape can be adjusted to trade sidelobe amplitude for mainlobe width. Figure 7.32(a) shows continuous envelopes of Kaiser windows of length $M + 1 = 21$ for $\beta = 0$, 3, and 6. Notice from Eq. (7.89) that the case $\beta = 0$ reduces to the rectangular window. Figure 7.32(b) shows the corresponding Fourier transforms of the Kaiser windows in Fig. 7.32(a). Figure 7.32(c) shows Fourier transforms of Kaiser windows with $\beta = 6$ and $M = 10, 20$, and 40. These plots clearly show that the desired tradeoff can be achieved. If the window is tapered more, the sidelobes of the Fourier transform become smaller, but the mainlobe becomes wider. Figure 7.32(c) shows that increasing M while holding β constant causes the mainlobe to decrease in width but does not affect the amplitude of the sidelobes. In fact, through extensive numerical experimentation, Kaiser obtained a pair of formulas that permit the filter designer to predict in advance the values of M and β needed to meet a given frequency-selective filter specification. Figure 7.31 is also typical of approximations obtained using the Kaiser window, and Kaiser (1974) found that over a usefully wide range of conditions, the peak approximation error (δ in Fig. 7.31) is determined by the choice of β. Given that δ is fixed, the passband cutoff frequency ω_p of the lowpass filter is defined to be the highest frequency such that $|H(e^{j\omega})| \ge 1 - \delta$. The stopband cutoff frequency ω_s is defined to be the lowest frequency such that $|H(e^{j\omega})| \le \delta$. Therefore, the transition region has width

$$
\Delta\omega = \omega_s - \omega_p
\tag{7.90}
$$

for the lowpass filter approximation. Defining

$$
A = -20 \log_{10} \delta,
\tag{7.91}
$$

Kaiser determined empirically that the value of β needed to achieve a specified value of A is given by

$$
\beta = \begin{cases} 0.1102(A - 8.7), & A > 50, \\ 0.5842(A - 21)^{0.4} + 0.07886(A - 21), & 21 \le A \le 50, \\ 0.0, & A < 21. \end{cases}
\tag{7.92}
$$

(Recall that the case $\beta = 0$ is the rectangular window for which $A = 21$.) Furthermore, to achieve prescribed values of A and $\Delta\omega$, M must satisfy

$$
M = \frac{A - 8}{2.285\Delta\omega}.
\tag{7.93}
$$

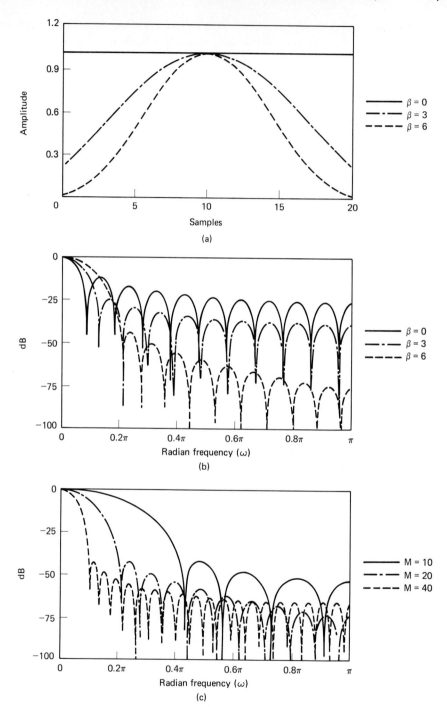

Figure 7.32 (a) Kaiser windows for $\beta = 0$, 3, and 6 and $M = 20$. (b) Fourier transforms corresponding to windows in (a). (c) Fourier transforms of Kaiser windows with $\beta = 6$ and $M = 10$, 20, and 40.

Equation (7.93) predicts M to within ± 2 over a wide range of values of $\Delta\omega$ and A. Thus, with these formulas, the Kaiser window design method requires almost no iteration or trial and error.

Example 7.11

Using the design formulas for the Kaiser window, it is straightforward to design an FIR lowpass filter to meet prescribed specifications. The procedure is as follows:

1. First the specifications must be established. This means selecting the desired ω_p, ω_s, and the maximum tolerable approximation error. For window design, the resulting filter will have the same peak error δ in both the passband and the stopband. For this example, we use the same specifications as in Examples 7.5, 7.6, and 7.7, i.e., $\omega_p = 0.4\pi$, $\omega_s = 0.6\pi$, $\delta_1 = 0.01$, and $\delta_2 = 0.001$. Since filters designed by the window method inherently have $\delta_1 = \delta_2$, we must set $\delta = 0.001$.

2. The cutoff frequency of the underlying ideal lowpass filter must be found. Due to the symmetry of the approximation at the discontinuity of $H_d(e^{j\omega})$, we would set

$$\omega_c = \frac{\omega_p + \omega_s}{2} = 0.5\pi.$$

3. To determine the parameters of the Kaiser window, we first compute

$$\Delta\omega = \omega_s - \omega_p = 0.2\pi, \qquad A = -20\log_{10}\delta = 60.$$

We substitute these two quantities into Eqs. (7.92) and (7.93) to obtain the required values of β and M. For this example the formulas predict

$$\beta = 5.653, \qquad M = 37.$$

4. The impulse response of the filter is computed using Eqs. (7.88) and (7.89):

$$h[n] = \begin{cases} \dfrac{\sin\omega_c(n-\alpha)}{\pi(n-\alpha)} \cdot \dfrac{I_0[\beta(1-[(n-\alpha)/\alpha]^2)^{1/2}]}{I_0(\beta)}, & 0 \le n \le M, \\[2ex] 0, & \text{otherwise,} \end{cases}$$

where $\alpha = M/2 = 37/2 = 18.5$. Since $M = 37$ is an odd integer, the resulting linear phase system would be of type II. (See Section 5.7.3 for the definitions of the four types of FIR systems with generalized linear phase.) The response characteristics of the filter are shown in Fig. 7.33. Figure 7.33(a), which shows the impulse response, displays the characteristic symmetry of a type II system. Figure 7.33(b), which shows the log magnitude response in dB, indicates that $H(e^{j\omega})$ is zero at $\omega = \pi$ or, equivalently, that $H(z)$ has a zero at $z = -1$ as required for a type II FIR system. Figure 7.33(c) shows the approximation error in the passband and stopbands. This error function is defined as

$$E_A(\omega) = \begin{cases} 1 - A_e(e^{j\omega}), & 0 \le \omega \le \omega_p, \\ 0 - A_e(e^{j\omega}), & \omega_s \le \omega \le \pi. \end{cases} \tag{7.94}$$

(The error is not defined in the transition region, $0.4\pi < \omega < 0.6\pi$.) Note the symmetry of the approximation error and also note that the peak approximation error is slightly greater than $\delta = 0.001$. Increasing M to 38 results in a type I filter for which $\delta = 0.0008$.

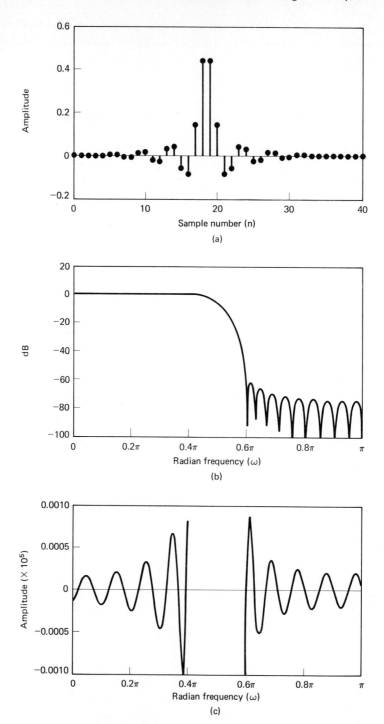

Figure 7.33 Response functions for Example 7.11. (a) Impulse response ($M = 37$). (b) Log magnitude. (c) Approximation error.

Finally, note that it is not necessary to plot either the phase or the group delay since we know that the phase is precisely linear and the delay is $M/2 = 18.5$ samples.

One of the reasons that windows such as the Hanning or Hamming window are widely used is that they are expressed in terms of familiar functions that are easily computed. Although the modified Bessel function in the Kaiser window is not a generally familiar function, it nevertheless can be easily computed using the following FORTRAN program due to Kaiser (1974):

```
      REAL FUNCTION I0(X)
      DS=1.0
      D=0.0
1     D=D+2.0
      DS=DS * X * X/(D * D)
      S=S+DS
      IF(DS-.2E-8 * S) 2, 1, 1
2     I0=S
      RETURN
      END
```

7.4.4 Relationship of the Kaiser Window to Other Windows

The basic principle of the window design method is to truncate the ideal impulse response with a finite-length window such as one of those discussed in this section. The corresponding effect in the frequency domain is that the ideal frequency response is convolved with the Fourier transform of the window. If the ideal filter is a lowpass filter, the discontinuity in the ideal filter frequency response is smeared as the mainlobe of the Fourier transform of the window moves across the discontinuity in the convolution process. To a first approximation, the width of the resulting transition band is determined by the width of the mainlobe of the Fourier transform of the window, and the passband and stopband ripples are determined by the sidelobes of the Fourier transform of the window. Because the passband and stopband ripples are produced by integration of the symmetric window sidelobes, the ripples in the passband and the stopband are approximately the same. Furthermore, to a very good approximation, the maximum passband and stopband deviation is not dependent on M and can be changed only by changing the shape of the window used. This is illustrated by Kaiser's formula, Eq. (7.92), for the window shape parameter, which is independent of M. The inherent tradeoff in the choice of window is that to achieve less passband and stopband ripple, a window with a greater mainlobe width or large M is generally required. The last two columns of Table 7.2 compare the Kaiser window to the windows of Eqs. (7.77). The fifth column gives the Kaiser window shape parameter that yields the same peak approximation error (δ) as the window indicated in the first column. The sixth column shows the corresponding transition width (from Eq. 7.93) for filters designed with the Kaiser window. This formula would be a much better predictor of transition width for the other windows than would the mainlobe width given in the third column.

7.5 EXAMPLES OF FIR FILTER DESIGN BY THE KAISER WINDOW METHOD

The use of the window method is, of course, not restricted to lowpass filters. Windows can be used to truncate any ideal impulse response to obtain a causal FIR approximation. In this section, we give several examples that illustrate the technique using the Kaiser window. These examples also serve as a convenient vehicle for pointing out some important properties of FIR systems.

7.5.1 Highpass Filter

The ideal highpass filter with generalized linear phase has the frequency response

$$H_{hp}(e^{j\omega}) = \begin{cases} 0, & 0 \le |\omega| < \omega_c \\ e^{-j\omega M/2}, & \omega_c < |\omega| \le \pi. \end{cases} \tag{7.95}$$

The corresponding impulse response can be found by evaluating the inverse transform of $H_{hp}(e^{j\omega})$ or we can observe that

$$H_{hp}(e^{j\omega}) = e^{-j\omega M/2} - H_{lp}(e^{j\omega}), \tag{7.96}$$

where $H_{lp}(e^{j\omega})$ is given by Eq. (7.86). Thus, $h_{hp}[n]$ is

$$h_{hp}[n] = \frac{\sin \pi(n - M/2)}{\pi(n - M/2)} - \frac{\sin \omega_c(n - M/2)}{\pi(n - M/2)}, \qquad -\infty < n < \infty. \tag{7.97}$$

To design an FIR approximation to the highpass filter we can proceed in a similar manner to Example 7.11.

Suppose that we wish to design a highpass filter to meet the specifications $\omega_s = 0.35\pi$, $\omega_p = 0.5\pi$, and $\delta_1 = \delta_2 = \delta = 0.021$. Since the ideal response also has a discontinuity, we can apply Kaiser's formulas in Eqs. (7.92) and (7.93) to estimate the required values of $\beta = 2.6$ and $M = 24$. Figure 7.34 shows the response characteristics that result when a Kaiser window with these parameters is applied to $h_{hp}[n]$ with $\omega_c = (0.35\pi + 0.5\pi)/2$. Note that since M is an even integer, the filter is a type I FIR system with linear phase and the delay is precisely $M/2 = 12$ samples. In this case the actual peak approximation error is $\delta = 0.0213$ rather than 0.021 as specified. In a practical setting this would probably be good enough, but if it is not, we might naturally increase M to 25, keeping β the same, and design another filter. Such a filter is shown in Fig. 7.35. Note that this type II system is highly unsatisfactory because of the zero of $H(z)$ that is forced by the linear phase constraint to be at $z = -1$ ($\omega = \pi$). Thus, increasing the filter order leads to a worse result. Of course, $M = 26$ would lead to a type I system that would exceed the specifications. Type II FIR linear phase systems are generally not appropriate approximations for either highpass or bandstop filters because the linear phase contraint forces a zero at $z = -1$.

The previous discussion of highpass filter design can be generalized to the case of multiple passbands and stopbands. Figure 7.36 shows an ideal multiband frequency-selective frequency response. This generalized multiband filter includes lowpass, highpass, bandpass, and bandstop filters as special cases. If such a magnitude

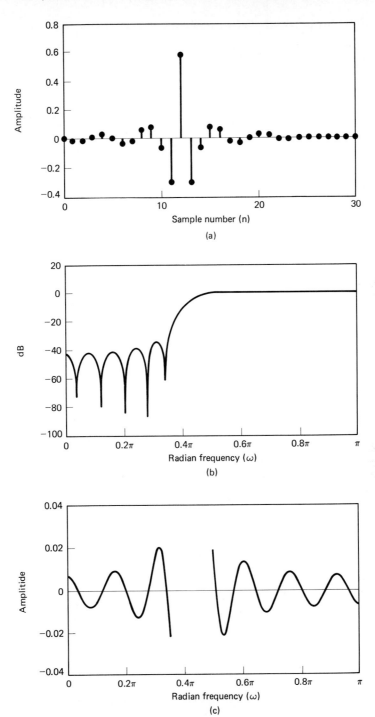

Figure 7.34   Response functions for type I FIR highpass filter. (a) Impulse response ($M = 24$). (b) Log magnitude. (c) Approximation error.

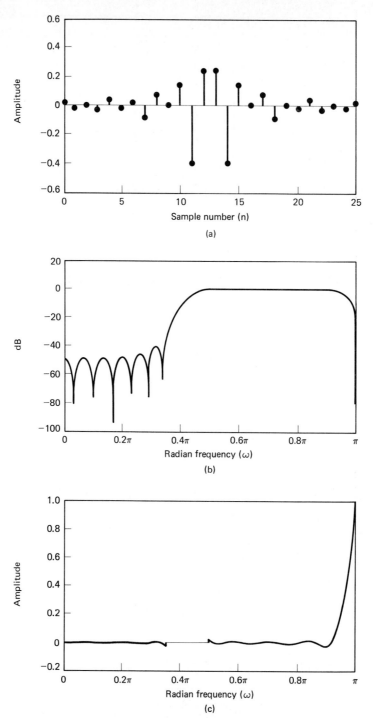

Figure 7.35 Response functions for type II FIR highpass filter. (a) Impulse response ($M = 25$). (b) Log magnitude. (c) Approximation error.

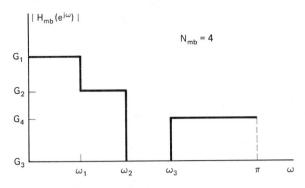

Figure 7.36 Ideal frequency response for multiband filter.

function is multiplied by a linear phase factor $e^{-j\omega M/2}$, the corresponding ideal impulse response is

$$h_{mb}[n] = \sum_{k=1}^{N_{mb}} (G_k - G_{k+1}) \frac{\sin \omega_k (n - M/2)}{\pi(n - M/2)}, \qquad (7.98)$$

where $G_{N_{mb}+1} = 0$. If $h_{mb}[n]$ is multiplied by a Kaiser window, the type of approximations that we have observed at the single discontinuity of the lowpass and highpass systems will occur at *each* of the discontinuities. The behavior will be the same at each discontinuity provided the discontinuities are far enough apart. Thus, Kaiser's formulas for the window parameters can be applied to this case to predict approximation error and transition widths. Note that the approximation errors will be scaled by the size of the jump that produces them. That is, if a discontinuity of unity produces a peak error of δ, then a discontinuity of one-half will have a peak error of $\delta/2$.

7.5.2 Discrete-Time Differentiators

Sometimes it is necessary to obtain samples of the derivative of a bandlimited signal from the samples of that signal. Since the Fourier transform of the derivative of a continuous-time signal is $j\Omega$ times the Fourier transform of the signal, it follows that for bandlimited signals a discrete-time system with frequency response $(j\omega/T)$ for $-\pi < \omega < \pi$ (and periodic, with period 2π) will yield output samples that are equal to samples of the derivative of the continuous-time signal. This ideal system was considered in Example 3.2.

For an ideal "discrete-time differentiator" with linear phase, the appropriate frequency response is

$$H_{diff}(e^{j\omega}) = (j\omega)e^{-j\omega M/2}, \qquad -\pi < \omega < \pi. \qquad (7.99)$$

(We have omitted the factor $1/T$.) The corresponding ideal impulse response is

$$h_{diff}[n] = \frac{\cos \pi(n - M/2)}{(n - M/2)} - \frac{\sin \pi(n - M/2)}{\pi(n - M/2)}, \qquad -\infty < n < \infty. \qquad (7.100)$$

If $h_{diff}[n]$ is multiplied by a symmetric window of length $(M + 1)$, then it is easily shown that $h[n] = -h[M - n]$. Thus, the resulting system is either a type III or a type IV generalized linear phase system.

Since Kaiser's formulas were developed for frequency responses with simple magnitude discontinuities, it is not straightforward to apply them to differentiators where the discontinuity in the ideal frequency response is introduced by the phase. Nevertheless, the window method is very effective in designing such systems. To illustrate the window design of a differentiator, suppose $M = 10$ and $\beta = 2.4$. The resulting response characteristics are shown in Fig. 7.37. Figure 7.37(a) shows the antisymmetric impulse response. Since M is even, the system is a type III linear phase system, which implies that $H(z)$ has zeros at *both* $z = 0$ ($\omega = 0$) and $z = -1$ ($\omega = \pi$). This is clearly displayed in the magnitude response shown in Fig. 7.37(b). The phase is exact, since type III systems have $\pi/2$ radians constant phase shift plus a linear phase corresponding in this case to $M/2 = 5$ samples delay. Figure 7.37(c) shows the amplitude approximation error

$$E_{diff}(\omega) = \omega - A_o(e^{j\omega}), \qquad 0 \le \omega \le 0.8\pi, \tag{7.101}$$

where $A_o(e^{j\omega})$ is the amplitude of the approximation. (Note that the error is large around $\omega = \pi$ and is not plotted for frequencies above $\omega = 0.8\pi$.) Clearly, the linearly increasing magnitude is not achieved over the whole band, and, obviously, the relative error (i.e., $E_{diff}(\omega)/\omega$) is very large for low frequencies or high frequencies (around $\omega = \pi$).

Type IV linear phase systems do not constrain $H(z)$ to have a zero at $z = -1$. This type of system leads to much better approximations to the amplitude function, as shown in Fig. 7.38, where $M = 5$ and $\beta = 2.4$. In this case, the amplitude approximation error is very small up to and beyond $\omega = 0.8\pi$. The phase for this system is again $\pi/2$ radians constant phase shift plus linear phase corresponding to a delay of $M/2 = 2.5$ samples. This noninteger delay is the price paid for the exceedingly good amplitude approximation. Instead of obtaining samples of the derivative of the continuous-time signal at the original sampling times $t = nT$, we obtain samples of the derivative at times $t = (n - 2.5)T$. However, in many applications this noninteger delay may not cause a problem or it could be compensated for by other noninteger delays in a more complex system involving other linear phase filters.

7.6 OPTIMUM APPROXIMATIONS OF FIR FILTERS

The design of FIR filters by windowing is straightforward to apply and it is quite general even though it has a number of limitations. However, we often wish to design a filter that is the "best" that can be achieved for a given value of M. It is meaningless to discuss this question in the absence of a definition of an approximation criterion. For example, in the case of the window design method, it follows from the theory of Fourier series that the rectangular window provides the best mean-square approximation to a desired frequency response for a given value of M. That is,

$$h[n] = \begin{cases} h_d[n], & 0 \le n \le M, \\ 0, & \text{otherwise}, \end{cases} \tag{7.102}$$

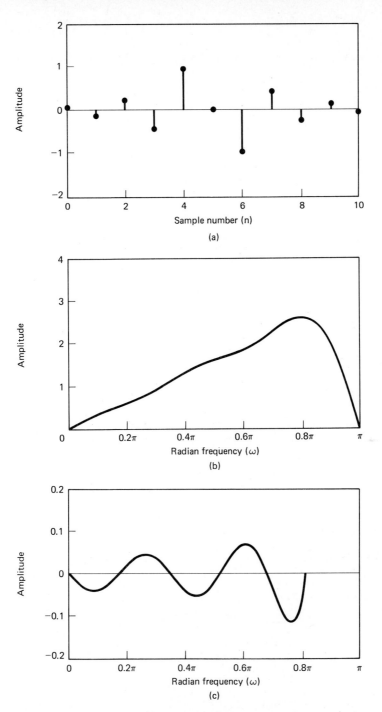

Figure 7.37 Response functions for type III FIR discrete-time differentiator. (a) Impulse response ($M = 10$). (b) Magnitude. (c) Approximation error.

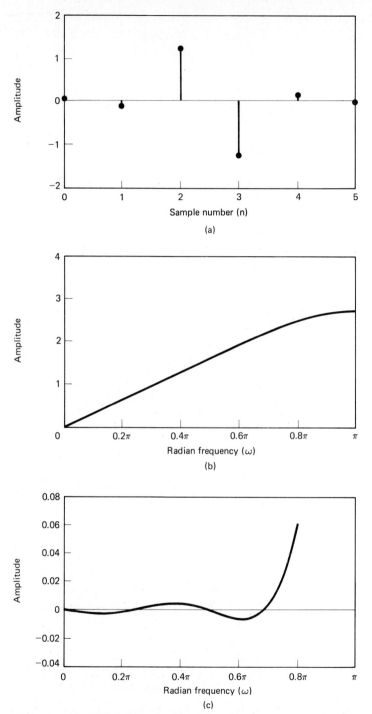

Figure 7.38 Response functions for type IV FIR discrete-time differentiator. (a) Impulse response ($M = 5$). (b) Magnitude. (c) Approximation error.

minimizes the expression

$$\varepsilon^2 = \frac{1}{2\pi} \int_{-\pi}^{\pi} |H_d(e^{j\omega}) - H(e^{j\omega})|^2 d\omega \qquad (7.103)$$

(see Problem 7.23). However, as we have seen, this approximation criterion leads to adverse behavior at discontinuities of $H_d(e^{j\omega})$. Furthermore, the window method does not permit individual control over the approximation errors in different bands. For many applications, better filters result from minimization of maximum error or a frequency-weighted error criterion. Such designs can be achieved using algorithmic techniques.

As the previous examples show, frequency-selective filters designed by windowing often have the property that the error is greatest on either side of a discontinuity of the ideal frequency response, and the error becomes smaller for frequencies away from the discontinuity. Furthermore, as suggested by Fig. 7.31, such filters typically result in approximately equal error in the passband and stopband. (See Figs. 7.33 and 7.34, for example.) We have already seen that for IIR filters, if the approximation error is spread out uniformly in frequency and if the passband and stopband ripples are adjusted separately, a given design specification can be met with a lower-order filter than if the approximation just meets the specification at one frequency and far exceeds it at others. This intuitive notion is confirmed for FIR systems by a theorem to be discussed later in this section.

In the following discussion we consider a particularly effective and widely used algorithmic procedure for the design of FIR filters with generalized linear phase. Although we consider only type I filters in detail, where appropriate we indicate how the results apply to types II, III, and IV generalized linear phase filters.

In designing a causal type I linear phase FIR filter, it is convenient first to consider the design of a zero-phase filter, i.e., one for which

$$h_e[n] = h_e[-n], \qquad (7.104)$$

and then to insert sufficient delay to make it causal. Consequently, we consider $h_e[n]$ satisfying the condition of Eq. (7.104). The corresponding frequency response is given by

$$A_e(e^{j\omega}) = \sum_{n=-L}^{L} h_e[n]e^{-j\omega n}, \qquad (7.105)$$

with $L = M/2$ an integer or, because of Eq. (7.104),

$$A_e(e^{j\omega}) = h_e[0] + \sum_{n=1}^{L} 2h_e[n]\cos(\omega n). \qquad (7.106)$$

Note that $A_e(e^{j\omega})$ is a real, even, and periodic function of ω. A causal system can be obtained from $h_e[n]$ by delaying it by $L = M/2$ samples. The resulting system has impulse response

$$h[n] = h_e[n - M/2] = h[M - n] \qquad (7.107)$$

and frequency response

$$H(e^{j\omega}) = A_e(e^{j\omega})e^{-j\omega M/2}. \qquad (7.108)$$

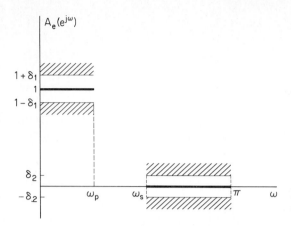

Figure 7.39 Tolerance scheme and ideal response for lowpass filter.

Figure 7.39 shows a tolerance scheme for an approximation to a lowpass filter with a real function such as $A_e(e^{j\omega})$. Unity is to be approximated in the band $0 \leq |\omega| \leq \omega_p$ with maximum absolute error δ_1, and zero is to be approximated in the band $\omega_s \leq |\omega| \leq \pi$ with maximum absolute error δ_2. An algorithmic technique for designing a filter to meet these specifications must in effect systematically vary the $(L + 1)$ unconstrained impulse response values $h_e[n]$, where $0 \leq n \leq L$. Design algorithms have been developed in which some of the parameters $L, \delta_1, \delta_2, \omega_p$, and ω_s are fixed and an iterative procedure is used to obtain optimum adjustments of the remaining parameters. Two distinct approaches have been developed. Herrmann (1969) and Herrmann and Schüssler (1970a) and Hofstetter et al. (1971) developed procedures in which L, δ_1, and δ_2 are fixed and ω_p and ω_s are variable. Parks and McClellan (1972a, 1972b), McClellan and Parks (1973), and Rabiner (1972a, 1972b) developed procedures in which L, ω_p, ω_s and the ratio δ_1/δ_2 are fixed and δ_1 (or δ_2) is variable. In the time period since these different approaches were developed, the Parks-McClellan algorithm has become the dominant method for optimum design of FIR filters. This is because it is the most flexible and the most computationally efficient. Furthermore, a well-tested FORTRAN program is available in DSP Committee (1979) and McClellan et al. (1973). Thus, we will discuss only the Parks-McClellan algorithm here.

The Parks-McClellan algorithm is based on reformulating the filter design problem as a problem in polynomial approximation. Specifically, the terms $\cos(\omega n)$ in Eq. (7.106) can be expressed as a sum of powers of $\cos \omega$ in the form

$$\cos(\omega n) = T_n(\cos \omega), \tag{7.109}$$

where $T_n(x)$ is an nth-order polynomial.† Consequently, Eq. (7.106) can be rewritten as an Lth-order polynomial in $\cos \omega$. Specifically,

$$A_e(e^{j\omega}) = \sum_{k=0}^{L} a_k(\cos \omega)^k, \tag{7.110}$$

---

† More specifically, $T_n(x)$ is the nth-order Chebyshev polynomial, defined as $T_n(x) = \cos(n \cos^{-1} x)$.

where the a_k's are constants that are related to $h_e[n]$, the values of the impulse response. With the substitution $x = \cos \omega$, we can express Eq. (7.110) as

$$A_e(e^{j\omega}) = P(x)\big|_{x = \cos \omega}, \tag{7.111}$$

where $P(x)$ is the Lth-order polynomial

$$P(x) = \sum_{k=0}^{L} a_k x^k. \tag{7.112}$$

We will see that it is not necessary to know the relationship between the a_k's and $h_e[n]$ (although a formula can be obtained); it is enough to know that $A_e(e^{j\omega})$ can be expressed as the Lth-degree trigonometric polynomial of Eq. (7.110).

The key to gaining control over ω_p and ω_s is to fix them at their desired values and let δ_1 and δ_2 vary. Parks and McClellan (1972a, 1972b) showed that with L, ω_p, and ω_s fixed, the frequency-selective filter design problem becomes a problem in Chebyshev approximation over disjoint sets, an important problem in approximation theory and one for which several useful theorems and procedures have been developed (see Cheney, 1966). To formalize the approximation problem in this case, let us define an approximation error function

$$E(\omega) = W(\omega)[H_d(e^{j\omega}) - A_e(e^{j\omega})], \tag{7.113}$$

where the weighting function, $W(\omega)$, incorporates the approximation error parameters into the design process. In this design method, the error function $E(\omega)$, the weighting function $W(\omega)$, and the desired frequency response $H_d(e^{j\omega})$ are defined only over closed subintervals of $0 \le \omega \le \pi$. For example, for approximating a lowpass filter, these functions are defined for $0 \le \omega \le \omega_p$ and $\omega_s \le \omega \le \pi$. The approximating function $A_e(e^{j\omega})$ is not constrained in the transition region(s) (e.g., $\omega_p < \omega < \omega_s$) and it may take any shape necessary to achieve the desired response in the other subintervals.

For example, suppose that we wish to obtain an approximation as in Fig. 7.39, where L, ω_p, and ω_s are fixed design parameters. For this case,

$$H_d(e^{j\omega}) = \begin{cases} 1, & 0 \le \omega \le \omega_p, \\ 0, & \omega_s \le \omega \le \pi. \end{cases} \tag{7.114}$$

The weighting function $W(\omega)$ provides for weighting the approximation errors differently in the different approximation intervals. For the lowpass filter approximation problem, the weighting function is

$$W(\omega) = \begin{cases} \dfrac{1}{K}, & 0 \le \omega \le \omega_p, \\ 1, & \omega_s \le \omega \le \pi, \end{cases} \tag{7.115}$$

where $K = \delta_1/\delta_2$. If $A_e(e^{j\omega})$ is as shown in Fig. 7.40, the weighted approximation error, $E(\omega)$ in Eq. (7.113), would be as indicated in Fig. 7.41. Note that with this weighting, the maximum weighted absolute approximation error is $\delta = \delta_2$ in both bands.

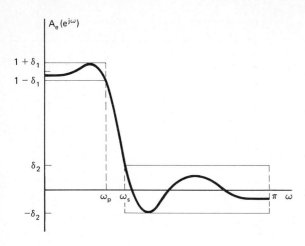

Figure 7.40 Typical frequency response meeting the specifications of Fig. 7.39.

The particular criterion used in this design procedure is the so-called minimax or Chebyshev criterion where within the frequency intervals of interest (the passband and stopband for a lowpass filter) we seek a frequency response $A_e(e^{j\omega})$ that *minimizes* the *maximum* weighted approximation error of Eq. (7.113). Stated more compactly, the best approximation is to be found in the sense of

$$\min_{\{h_e[n]:\ 0\le n\le L\}} \left(\max_{\omega\in F} |E(\omega)| \right),$$

where F is the closed subset of $0 \le \omega \le \pi$ such that $0 \le \omega \le \omega_p$ or $\omega_s \le \omega \le \pi$. Thus, we seek the set of impulse response values that minimize δ in Fig. 7.41.

Parks and McClellan (1972a, 1972b) applied the following theorem of approximation theory to this filter design problem.

Alternation Theorem   Let F_P denote the closed subset consisting of the disjoint union of closed subsets of the real axis x. $P(x)$ denotes an rth-order polynomial

$$P(x) = \sum_{k=0}^{r} a_k x^k.$$

Also, $D_P(x)$ denotes a given desired function of x that is continuous on F_P; $W_P(x)$ is a positive function, continuous on F_P, and $E_P(x)$ denotes the weighted error

$$E_P(x) = W_P(x)[D_P(x) - P(x)].$$

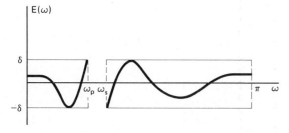

Figure 7.41 Weighted error for the approximation of Fig. 7.40.

The maximum error $\|E\|$ is defined as

$$\|E\| = \max_{x \in F_P} |E_P(x)|.$$

A necessary and sufficient condition that $P(x)$ is the unique rth-order polynomial that minimizes $\|E\|$ is that $E_P(x)$ exhibit *at least* $(r + 2)$ alternations, i.e., there must exist at least $(r + 2)$ values x_i in F_P such that $x_1 < x_2 < \cdots < x_{r+2}$ and such that $E_P(x_i) = -E_P(x_{i+1}) = \pm \|E\|$ for $i = 1, 2, \ldots, (r + 1)$.

At first glance, it may seem to be difficult to relate this formal theorem to the problem of filter design. However, in the following discussion, all of the elements of the theorem will be shown to be important in developing the design algorithm. To aid in understanding the alternation theorem, we will interpret it specifically for the design of a type I lowpass filter.

7.6.1 Optimal Type I Lowpass Filters

For type I filters, the polynomial $P(x)$ is the cosine polynomial $A_e(e^{j\omega})$ in Eq. (7.110), with the transformation of variable $x = \cos \omega$ and $r = L$:

$$P(\cos \omega) = \sum_{k=0}^{L} a_k (\cos \omega)^k. \tag{7.116}$$

$D_P(x)$ is the desired lowpass filter frequency response in Eq. (7.114), with $x = \cos \omega$:

$$D_P(\cos \omega) = \begin{cases} 1, & \cos \omega_p \leq \cos \omega \leq 1, \\ 0, & -1 \leq \cos \omega \leq \cos \omega_s \end{cases}. \tag{7.117}$$

$W_P(\cos \omega)$ is given by Eq. (7.115), rephrased in terms of $\cos \omega$:

$$W_P(\cos \omega) = \begin{cases} \dfrac{1}{K}, & \cos \omega_p \leq \cos \omega \leq 1, \\ 1, & -1 \leq \cos \omega \leq \cos \omega_s, \end{cases} \tag{7.118}$$

and the weighted approximation error is

$$E_P(\cos \omega) = W_P(\cos \omega)[D_P(\cos \omega) - P(\cos \omega)]. \tag{7.119}$$

The closed subset F_P is made up of the union of the intervals $0 \leq \omega \leq \omega_p$ and $\omega_s \leq \omega \leq \pi$ or, in terms of $\cos \omega$, of the intervals $\cos \omega_p \leq \cos \omega \leq 1$ and $-1 \leq \cos \omega \leq \cos \omega_s$. The alternation theorem then states that a set of coefficients a_k in Eq. (7.116) will correspond to the filter representing the unique best approximation to the ideal lowpass filter with the ratio δ_1/δ_2 fixed at K and with passband and stopband edges ω_p and ω_s if and only if $E_P(\cos \omega)$ exhibits at least $(L + 2)$ alternations on F_P, i.e., it alternately equals plus and minus its maximum value. Such approximations are called *equiripple approximations*.

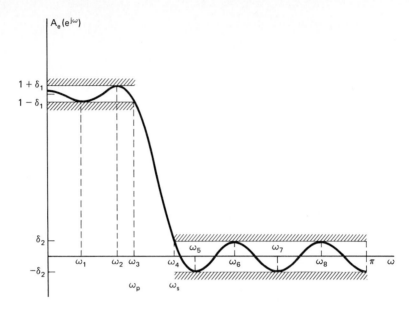

Figure 7.42 Typical example of a lowpass filter approximation that is optimal according to the alternation theorem for $L = 7$.

Figure 7.42 shows a filter frequency response that is optimum according to the alternation theorem for $L = 7$. In this figure, $A_e(e^{j\omega})$ is plotted against ω. To formally test the alternation theorem, we should first redraw it as a function of $x = \cos \omega$. Furthermore, we want to explicitly examine the alternations of $E_P(x)$. Consequently, in Fig. 7.43(a), (b), (c), we show $P(x)$, $W_P(x)$, and $E_P(x)$, respectively, as a function of $x = \cos \omega$. For this example, where $L = 7$, we see that there are a total of nine alternations of the error. Consequently, the alternation theorem is satisfied. An important point is that in counting alternations we include the points $\cos \omega_p$ and $\cos \omega_s$ since, according to the alternation theorem, the subsets (or subintervals) included in F_P are closed, i.e., the endpoints of the intervals are counted. While this might seem to be a small issue, it is in fact very significant, as we will see.

Comparing Figs. 7.42 and 7.43 suggests that when the desired filter is a lowpass filter (or any piecewise-constant filter) we could in fact count the alternations by direct examination of the frequency response, keeping in mind that the maximum error is different (in the ratio $K = \delta_1/\delta_2$) in the passband and stopband.

The alternation theorem states that the optimum filter must have a minimum of $(L + 2)$ alternations but does not exclude the possibility of more than $(L + 2)$ alternations. In fact, we will show that for a lowpass filter, the maximum possible number of alternations is $(L + 3)$. First, however, we illustrate this in Fig. 7.44 for $L = 7$. Figure 7.44(a) has $L + 3 = 10$ alternations, while Figs. 7.44(b), (c), and (d) each have $L + 2 = 9$ alternations. The case of $L + 3$ alternations (Fig. 7.44a) is often referred to as the *extraripple case*. Note that for the extraripple filter, there are alternations at $\omega = 0$ and π as well as at $\omega = \omega_p$ and $\omega = \omega_s$, i.e., at all the band edges. For Figs. 7.44(b) and (c) there are again alternations at ω_p and ω_s but not at

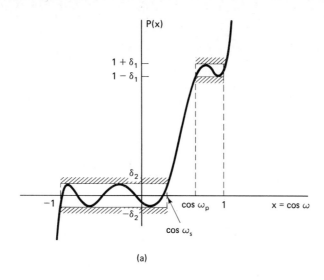

(a)

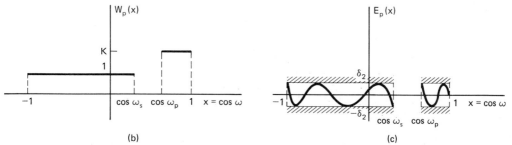

(b)   (c)

Figure 7.43 Equivalent polynomial approximation functions as a function of $x = \cos \omega$. (a) Approximating polynomial. (b) Weighting function. (c) Approximation error.

both $\omega = 0$ and $\omega = \pi$. In Fig. 7.44(d), there are alternations at 0, π, ω_p, and ω_s, but there is one less extremum (point of zero slope) inside the stopband. We also observe that all of these cases are equiripple inside the passband and stopband, i.e., all points of zero slope inside the interval $0 < \omega < \pi$ are frequencies at which the magnitude of the weighted error is maximum. It should also be stressed that since all of the filters in Fig. 7.44 satisfy the alternation theorem for $L = 7$ and for the same value of $K = \delta_1/\delta_2$, then ω_p and/or ω_s must be different for each since the alternation theorem states that the optimum filter under the conditions of the theorem is unique.

The properties referred to above for the filters in Fig. 7.44 result from the alternation theorem. Specifically we will show that for type I lowpass filters:

- The maximum possible number of alternations of the error is $(L + 3)$.
- Alternations will always occur at ω_p and ω_s.
- All points of zero slope *inside* the passband and the stopband (for $0 < \omega < \omega_p$ and $\omega_s < \omega < \pi$) will correspond to alternations; i.e., the filter will be equiripple except possibly at $\omega = 0$ and $\omega = \pi$.

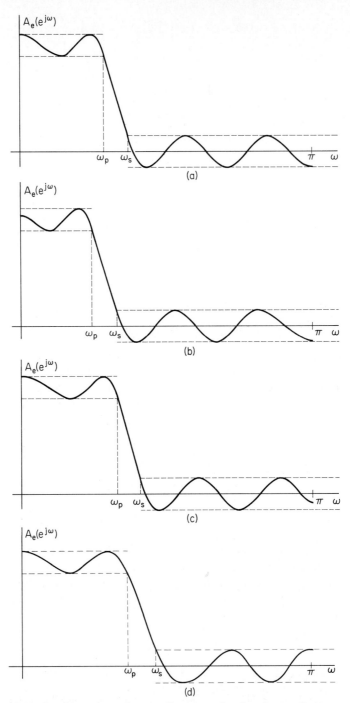

Figure 7.44 Possible optimum lowpass filter approximations for $L = 7$. (a) $L + 3$ alternations (extraripple case). (b) $M + 2$ alternations (extremum at $\omega = \pi$). (c) $M + 2$ alternations (extremum at $\omega = 0$). (d) $M + 2$ alternations (extremum at both $\omega = 0$ and $\omega = \pi$).

The maximum possible number of alternations is (L + 3).   Reference to Fig. 7.42 or 7.44 suggests that the maximum possible number of locations for extrema or alternations are the four band edges ($\omega = 0, \pi, \omega_p, \omega_s$) and the local extrema, i.e., the frequencies at which $A_e(e^{j\omega})$ has zero slope. Since an Lth-degree polynomial can have at most $(L - 1)$ points of zero slope in an open interval, the maximum possible number of locations for alternations are the $(L - 1)$ extrema of the polynomial plus the four band edges, a total of $(L + 3)$. In considering extrema and points of zero slope for trigonometric polynomials, it is important to observe that the trigonometric polynomial

$$P(\cos \omega) = \sum_{k=0}^{L} a_k(\cos \omega)^k, \qquad (7.120)$$

when considered as a function of ω, will always have zero slope at $\omega = 0$ and $\omega = \pi$ even though $P(x)$ considered as a function of x may not have zero slope at the corresponding points $x = 1$ and $x = -1$. This is because

$$\frac{dP(\cos \omega)}{d\omega} = -\sin \omega \left(\sum_{k=0}^{L} k a_k(\cos \omega)^{k-1} \right)$$

$$= -\sin \omega \left(\sum_{k=0}^{L-1} (k + 1)a_{k+1}(\cos \omega)^k \right), \qquad (7.121)$$

which is always zero at $\omega = 0$ and π as well as at the $(L-1)$ roots of the $(L - 1)$st-order polynomial represented by the sum. This behavior at $\omega = 0$ and $\omega = \pi$ is evident in Fig. 7.44. In Fig. 7.44(d), it happens that both $\sin \omega$ and

$$\sum_{k=0}^{L} k a_k(\cos \omega)^{k-1}$$

are zero in Equation (7.121), i.e., the polynomial $P(x)$ also has zero slope at $x = -1 = \cos \pi$.

Alternations always occur at ω_p and ω_s.   For all of the frequency responses in Fig. 7.44, $A_e(e^{j\omega})$ is exactly equal to $1 - \delta_1$ at the passband edge ω_p and exactly equal to $+\delta_2$ at the stopband edge. To suggest why this must always be the case, let us consider whether the filter in Fig. 7.44(a) could also be optimum if we decreased ω_p as indicated in Fig. 7.45. The frequencies at which the magnitude of the maximum weighted error are equal are the frequencies $\omega = 0, \omega_1, \omega_2, \omega_s, \omega_3, \omega_4, \omega_5, \omega_6$, and $\omega = \pi$, for a total of $(L + 2) = 9$. However, not all of the frequencies are alternations, since to be counted in the alternation theorem the error must *alternate* between $\delta = \pm \|E\|$ at these frequencies. Therefore, since the error is negative at both ω_2 and ω_s, the frequencies counted in the alternation theorem are $\omega = 0, \omega_1, \omega_2, \omega_3, \omega_4, \omega_5, \omega_6$, and π, for a total of 8. Since $(L + 2) = 9$, the conditions of the alternation theorem are not satisfied and the frequency response of Fig. 7.45 is not optimum with ω_p and ω_s as indicated. In other words, if ω_p is removed as an alternation frequency it removes two alternations. Since the maximum number is $(L + 3)$, this leaves at most $(L + 1)$, which is not a sufficient number. An identical argument would hold if ω_s

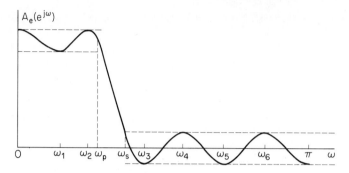

Figure 7.45 Illustration that the passband edge ω_p must be an alternation frequency.

were removed as an alternation frequency. A similar argument can be constructed for highpass filters, but as considered in Problem 7.31, this is not necessarily the case for bandpass or multiband filters.

The filter will be equiripple except possibly at $\omega = 0$ or $\omega = \pi$. The argument here is very similar to the one used to show that both ω_p and ω_s must be alternations. Suppose, for example, that the filter in Fig. 7.44(a) was modified as indicated in Fig. 7.46 so that one point of zero slope did not achieve the maximum error. While the maximum error occurs at nine frequencies, only eight of these can be counted as alternations. Consequently, eliminating one ripple as a point of maximum error reduces the number of alternations by two, leaving $(L + 1)$ as the maximum possible number.

These represent only a few of many properties that can be inferred from the alternation theorem. A variety of others are discussed in Rabiner and Gold (1975). Furthermore, we have considered only type I lowpass filters. While a much broader and detailed discussion of type II, III, and IV filters or more general desired frequency responses is beyond the scope of this book, we briefly consider type II lowpass filters to further emphasize a number of aspects of the alternation theorem.

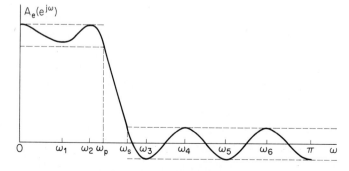

Figure 7.46 Illustration that the frequency response must be equiripple in the approximation bands.

7.6.2 Optimal Type II Lowpass Filters

A type II causal filter is one for which $h[n] = 0$ outside the range $0 \le n \le M$, with the filter length $(M + 1)$ even, i.e., M odd, and with the symmetry property

$$h[n] = h[M - n]. \tag{7.122}$$

Consequently, the frequency response $H(e^{j\omega})$ can be expressed in the form

$$H(e^{j\omega}) = e^{-j\omega M/2} \sum_{n=0}^{(M-1)/2} 2h[n] \cos\left[\omega\left(\frac{M}{2} - n\right)\right]. \tag{7.123}$$

Letting $b[n] = 2h[(M + 1)/2 - n]$, $n = 1, 2, \ldots, (M + 1)/2$, Eq. (7.123) can be rewritten as

$$H(e^{j\omega}) = e^{-j\omega M/2} \left\{ \sum_{n=1}^{(M+1)/2} b[n] \cos\left[\omega\left(n - \frac{1}{2}\right)\right] \right\}. \tag{7.124}$$

To apply the alternation theorem to the design of type II filters, we must be able to identify the problem as one of polynomial approximation. To accomplish this, we express the summation in Eq. (7.124) in the form

$$\sum_{n=1}^{(M+1)/2} b[n] \cos\left[\omega\left(n - \frac{1}{2}\right)\right] = \cos(\omega/2)\left[\sum_{n=0}^{(M-1)/2} \tilde{b}[n] \cos(\omega n)\right]. \tag{7.125}$$

(See Problem 7.35.) The summation on the right-hand side of Eq. (7.125) can now be represented as a trigonometric polynomial $P(\cos \omega)$ so that

$$H(e^{j\omega}) = e^{-j\omega M/2} \cos(\omega/2)P(\cos \omega), \tag{7.126a}$$

where

$$P(\cos \omega) = \sum_{k=0}^{L} a_k (\cos \omega)^k \tag{7.126b}$$

and $L = (M - 1)/2$. The coefficients a_k in Eq. (7.126b) are related to the coefficients $\tilde{b}[n]$ in Eq. (7.125), which in turn are related to the coefficients $b[n] = 2h[(M + 1)/2 - n]$ in Eq. (7.124). As in the type I case, it is not necessary to obtain an explicit relationship between the impulse response and the a_k's. We now can apply the alternation theorem to the weighted error between $P(\cos \omega)$ and the desired frequency response. For a desired type I lowpass filter with a specified ratio K of passband to stopband ripple, the desired function is given by Eq. (7.114) and the weighting function for the error is given by Eq. (7.115). For type II lowpass filters, because of the presence of the factor $\cos(\omega/2)$ in Eq. (7.126a), the desired function to be approximated by the polynomial $P(\cos \omega)$ is defined as

$$H_d(e^{j\omega}) = D_P(\cos \omega) = \begin{cases} \dfrac{1}{\cos(\omega/2)}, & 0 \le \omega \le \omega_p, \\[2mm] 0, & \omega_s \le \omega \le \pi, \end{cases} \tag{7.127}$$

and the weighting function to be applied to the error is

$$
W(\omega) = W_P(\cos \omega) = \begin{cases} \dfrac{\cos(\omega/2)}{K}, & 0 \le \omega \le \omega_p, \\[2ex] \cos(\omega/2), & \omega_s \le \omega \le \pi. \end{cases} \tag{7.128}
$$

Consequently, type II filter design is a different polynomial approximation problem than type I filter design.

In this section we have only outlined the design of type II filters, principally to highlight the requirement that the design problem first be formulated as a polynomial approximation problem. A similar set of issues arises in the design of type III and type IV linear phase FIR filters. Specifically, these classes also can be formulated as polynomial approximation problems, but in each class the weighting function applied to the error has a trigonometric form just as it does for type II filters (see Problem 7.35). A detailed discussion of the design and properties of these classes of filters can be found in Rabiner and Gold (1975).

The details of the problem formulation for type I and II linear phase systems have been illustrated for the lowpass filter case. However, the discussion of the type II case in particular should suggest that there is great flexibility in the choice of both the desired response function $H_d(e^{j\omega})$ and the weighting function $W(\omega)$. For example, the weighting function can be defined in terms of the desired function so as to yield equiripple percentage error approximation. This approach is valuable in designing type III and IV differentiator systems.

7.6.3 The Parks-McClellan Algorithm

The alternation theorem states necessary and sufficient conditions on the error for optimality in the Chebyshev sense. Although the theorem does not state explicitly how to find the optimum filter, the conditions given serve as the basis for an efficient algorithm for finding it. While our discussion is phrased in terms of type I lowpass filters, the algorithm easily generalizes.

From the alternation theorem, we know that the optimum filter $A_e(e^{j\omega})$ will satisfy the following set of equations:

$$
W(\omega_i)[H_d(e^{j\omega_i}) - A_e(e^{j\omega_i})] = (-1)^{i+1}\delta, \qquad i = 1, 2, \ldots, (L+2), \tag{7.129}
$$

where δ is the optimum error and $A_e(e^{j\omega})$ is given by either Eq. (7.106) or Eq. (7.110). Using Eq. (7.110) for $A_e(e^{j\omega})$, these equations can be written as

$$
\begin{bmatrix}
1 & x_1 & x_1^2 & \cdots & x_1^L & \dfrac{1}{W(\omega_1)} \\[1.5ex]
1 & x_2 & x_2^2 & \cdots & x_2^L & \dfrac{-1}{W(\omega_2)} \\[1.5ex]
\vdots & \vdots & \vdots & & \vdots & \vdots \\[1.5ex]
1 & x_{L+2} & x_{L+2}^2 & \cdots & x_{L+2}^L & \dfrac{(-1)^{L+2}}{W(\omega_{L+2})}
\end{bmatrix}
\begin{bmatrix}
a_0 \\[1.5ex] a_1 \\[1.5ex] \vdots \\[1.5ex] \delta
\end{bmatrix}
=
\begin{bmatrix}
H_d(e^{j\omega_1}) \\[1.5ex]
H_d(e^{j\omega_2}) \\[1.5ex]
\vdots \\[1.5ex]
H_d(e^{j\omega_{L+2}})
\end{bmatrix}, \tag{7.130}
$$

where $x_i = \cos \omega_i$. This set of equations serves as the basis for an iterative algorithm for finding the optimum $A_e(e^{j\omega})$. The procedure begins by guessing a set of alternation frequencies $\omega_i, i = 1, 2, \ldots, (L + 2)$. Note that ω_p and ω_s are fixed and, based on our discussion in Section 7.6.1, are necessarily members of the set of alternation frequencies. Specifically if $\omega_\ell = \omega_p$, then $\omega_{\ell+1} = \omega_s$. The set of equations (7.130) could be solved for the set of coefficients a_k and δ. However, a more efficient alternative is to use polynomial interpolation. Specifically, Parks and McClellan (1972a, 1972b) found that for the given set of the extremal frequencies, δ is given by the formula

$$\delta = \frac{\sum\limits_{k=1}^{L+2} b_k H_d(e^{j\omega_k})}{\sum\limits_{k=1}^{L+2} \dfrac{b_k(-1)^{k+1}}{W(\omega_k)}}, \tag{7.131}$$

where

$$b_k = \prod_{\substack{i=1 \\ i \neq k}}^{L+2} \frac{1}{(x_k - x_i)} \tag{7.132}$$

and, as above, $x_i = \cos \omega_i$. That is, if $A_e(e^{j\omega})$ is determined by the set of coefficients a_k that satisfy Eq. (7.130), with δ given by Eq. (7.131), then the error function goes through $\pm\delta$ at the $(L + 2)$ frequencies ω_i or, equivalently, $A_e(e^{j\omega})$ has values $1 \pm K\delta$ if $0 \leq \omega_i \leq \omega_p$ and $\pm\delta$ if $\omega_s \leq \omega_i \leq \pi$. Now since $A_e(e^{j\omega})$ is known to be an Lth-order trigonometric polynomial, we can interpolate a trigonometric polynomial through $(L + 1)$ of the $(L + 2)$ known values $E(\omega_i)$ (or equivalently $A_e(e^{j\omega_i})$). Parks and McClellan used the Lagrange interpolation formula to obtain

$$A_e(e^{j\omega}) = P(\cos \omega) = \frac{\sum\limits_{k=1}^{L+1} [d_k/(x - x_k)]C_k}{\sum\limits_{k=1}^{L+1} [d_k/(x - x_k)]}, \tag{7.133a}$$

where $x = \cos \omega$, $x_i = \cos \omega_i$,

$$C_k = H_d(e^{j\omega_k}) - \frac{(-1)^{k+1}\delta}{W(\omega_k)}, \tag{7.133b}$$

and

$$d_k = \prod_{\substack{i=1 \\ i \neq k}}^{L+1} \frac{1}{(x_k - x_i)} = \frac{b_k}{(x_k - x_{L+2})}. \tag{7.133c}$$

Although only the frequencies $\omega_1, \omega_2, \ldots, \omega_{L+1}$ are used in fitting the Lth-degree polynomial, we can be assured that the polynomial also takes on the correct value at ω_{L+2} because Eqs. (7.130) are satisfied by the resulting $A_e(e^{j\omega})$.

Now, $A_e(e^{j\omega})$ is available at any desired frequency without the need to solve the set of equations (7.130) for the coefficients a_k. The polynomial of Eq. (7.133a) can be used to evaluate $A_e(e^{j\omega})$ and also $E(\omega)$ on a dense set of frequencies in the passband

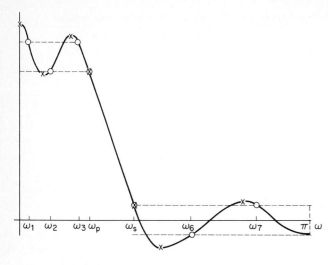

Figure 7.47 Illustration of the Parks–McClellan algorithm for equiripple approximation.

and stopband. If $|E(\omega)| \leq \delta$ for all ω in the passband and stopband, then the optimum approximation has been found. Otherwise we must find a new set of extremal frequencies.

Figure 7.47 shows a typical example for a type I lowpass filter before the optimum has been found. Clearly the set of frequencies ω_i used to find δ (as represented by open circles in Fig. 7.47) was such that δ was too small. Adopting the philosophy of the Remez method, the extremal frequencies are exchanged for a completely new set defined by the $(L + 2)$ largest peaks of the error curve. The points marked with $\times$ would be the new set of frequencies for the example shown in Fig. 7.47. As before, ω_p and ω_s must be selected as extremal frequencies. Recall that there are at most $(L - 1)$ local minima and maxima in the open intervals $0 < \omega < \omega_p$ and $\omega_s < \omega < \pi$. The remaining extrema can be at either $\omega = 0$ or $\omega = \pi$. If there is a maximum of the error function at both 0 and π, then the frequency at which the greatest error occurs is taken as the new estimate of extremum frequency. The cycle—computation of δ, fitting a polynomial to the assumed error peaks, and then locating the actual error peaks—is repeated until δ does not change from its previous value by more than a prescribed small amount. This value of δ is then the desired minimum maximum weighted approximation error. A flow chart (Rabiner and Gold, 1975) for the algorithm is shown in Fig. 7.48.

In this algorithm all the impulse response values $h_e[n]$ are implicitly varied on each iteration to obtain the desired optimum approximation, but the values of $h_e[n]$ are never explicitly computed. After the algorithm has converged, the impulse response can be computed from samples of the polynomial representation using the discrete Fourier transform, as will be discussed in Chapter 8.

7.6.4 Characteristics of Optimum FIR Filters

The optimum FIR filters have the smallest maximum weighted approximation error δ for a prescribed transition band $(\omega_s - \omega_p)$. For the weighting function of Eq. (7.115),

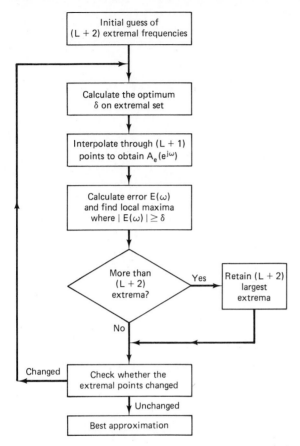

Figure 7.48   Flowchart of Parks–McClellan algorithm.

the resulting maximum stopband approximation error is $\delta_2 = \delta$, and the maximum passband approximation error is $\delta_1 = K\delta$. If given values of δ_1 and δ_2 are desired, the algorithm presented in the preceding section can be employed iteratively to determine a filter with prescribed values of δ_1 and δ_2 by fixing ω_p and M and varying ω_s until the desired δ_1 and δ_2 are obtained. A summary of results of this type (see Rabiner and Gold, 1975) is given in Fig. 7.49, where $\Delta\omega/2\pi = (\omega_s - \omega_p)/2\pi$ is plotted against $\omega_p/2\pi$ for $\delta_1 = \delta_2 = 0.1$ and several values of M. The curves show that as ω_p increases, the transition width $\Delta\omega$ attains local minima. These points on the curves correspond to the extraripple ($L + 3$ extrema) filters. All points between the minima correspond to filters that are optimum according to the alternation theorem. The filters for $M = 8$ and $M = 10$ are type I filters, while $M = 9$ corresponds to a type II filter. It is interesting to note that for some choices of parameters, a shorter filter ($M = 9$) may be better (smaller transition width) than a longer filter ($M = 10$). This may at first seem surprising and even contradictory. However, the cases $M = 9$ and $M = 10$ represent fundamentally different types of filters. Interpreted another way, filters for $M = 9$ cannot be considered to be special cases of $M = 10$ with one point set to zero, since this would violate the linear phase symmetry requirement. On the other hand, $M = 8$

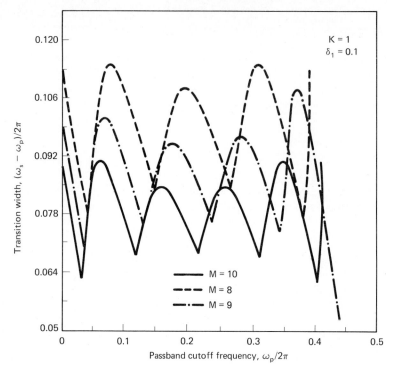

Figure 7.49 Dependence of transition width on cutoff frequency for optimum approximations of a lowpass filter.

could always be thought of as a special case of $M = 10$ with the first and last samples set to zero. For this reason, an optimal filter for $M = 8$ cannot be better than one for $M = 10$. This restriction can be seen in Fig. 7.49, where the curve for $M = 8$ is always above or equal to the one for $M = 10$. The points at which they touch correspond to identical impulse responses, with the $M = 10$ filter having the first and last points equal to zero.

Herrmann et al. (1973) did an extensive computational study of the relationships among the parameters M, δ_1, δ_2, ω_p, and ω_s for equiripple lowpass approximations, and Kaiser (1974) subsequently obtained the following simplified formula as a fit to their data:

$$M = \frac{-10\log_{10}(\delta_1\delta_2) - 13}{2.324\Delta\omega}, \tag{7.134}$$

where $\Delta\omega = \omega_s - \omega_p$. We can see by comparing Eq. (7.134) to the design formula Eq. (7.93) for the Kaiser window method that for the comparable case ($\delta_1 = \delta_2 = \delta$), the optimal approximations provide about 5 dB better approximation error for a given value of M. Another important advantage of the equiripple filters is that δ_1 and δ_2 need not be equal as required for the window method.

sky

7.7 EXAMPLES OF FIR EQUIRIPPLE APPROXIMATION

The Parks-McClellan algorithm for optimum equiripple approximation of FIR filters can be used to design a wide variety of FIR filters. In this section we give several examples that illustrate some of the properties of the optimum approximation and suggest the great flexibility that is afforded by the design method.

7.7.1 Lowpass Filter

For the lowpass filter case, we again approximate the set of specifications used in Examples 7.5, 7.6, 7.7, and 7.11 so that we can compare all the major design methods on the same lowpass filter specifications. These specifications call for $\omega_p = 0.4\pi$, $\omega_s = 0.6\pi$, $\delta_1 = 0.01$, and $\delta_2 = 0.001$. In contrast to the window method, the Parks-McClellan algorithm can accommodate the different approximation error in the passband and stopband by fixing the weighting function parameter at $K = \delta_1/\delta_2 = 10$.

Substituting the above specifications into Eq. (7.134) and rounding up yields the estimate $M = 26$ for the value of M that is necessary to achieve the specifications. Figures 7.50(a), (b), and (c) show the impulse response, log magnitude, and approximation error, respectively, for the optimum filter with $M = 26$ and $\omega_p = 0.4\pi$ and $\omega_s = 0.6\pi$. Figure 7.50(c) shows the *unweighted* approximation error

$$E_A(\omega) = \frac{E(\omega)}{W(\omega)} = \begin{cases} 1 - A_e(e^{j\omega}), & 0 \le \omega \le \omega_p, \\ 0 - A_e(e^{j\omega}), & \omega_s \le \omega \le \pi, \end{cases} \tag{7.135}$$

rather than the weighted error used in the formulation of the design algorithm. The weighted error would be identical to Fig. 7.50(c) except that the error would be divided by 10 in the passband.† The alternations of the approximation error are clearly in evidence in Fig. 7.50(c). There are seven alternations in the passband and eight in the stopband, for a total of 15 alternations. Since $L = M/2$ for type I (M even) systems, and $M = 26$, the minimum number of alternations is $(L + 2) = (26/2 + 2) = 15$. Thus, the filter of Fig. 7.50 is the optimum filter for $M = 26$ and $\omega_p = 0.4\pi$ and $\omega_p = 0.6\pi$. However, Fig. 7.50(c) shows that the filter fails to meet the original specifications on passband and stopband error. (The maximum errors in the passband and stopband are 0.0116 and 0.00116, respectively.) Thus, to meet or exceed the specifications, we must increase M to 27.

The filter response functions for the case $M = 27$ are shown in Fig. 7.51. Now the passband and stopband approximation errors are slightly less than the specified values. (The maximum errors in the passband and stopband are 0.0092 and 0.00092, respectively.) In this case there are again seven alternations in the passband and eight alternations in the stopband, for a total of 15. Note that since $M = 27$, this is a type II

---

† For frequency selective filters, the unweighted approximation error also conveniently displays the passband and stopband behavior since $A_e(e^{j\omega}) = 1 - E(\omega)$ in the passband, and $A_e(e^{j\omega}) = -E(\omega)$ in the stopband.

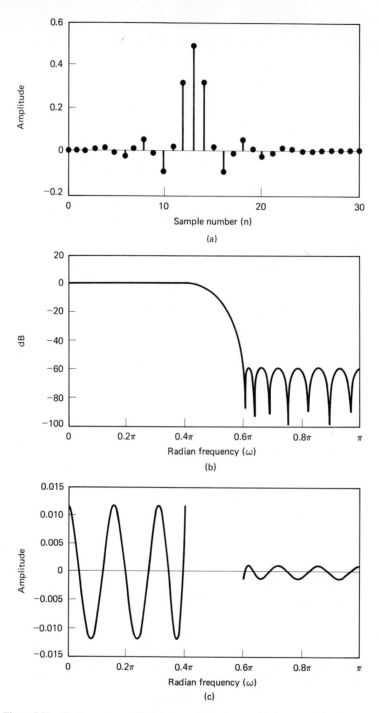

Figure 7.50   Optimum type I FIR lowpass filter for $\omega_p = 0.4\pi$, $\omega_s = 0.6\pi$, $K = 10$, and $M = 26$. (a) Impulse response. (b) Log magnitude. (c) Approximation error (unweighted).

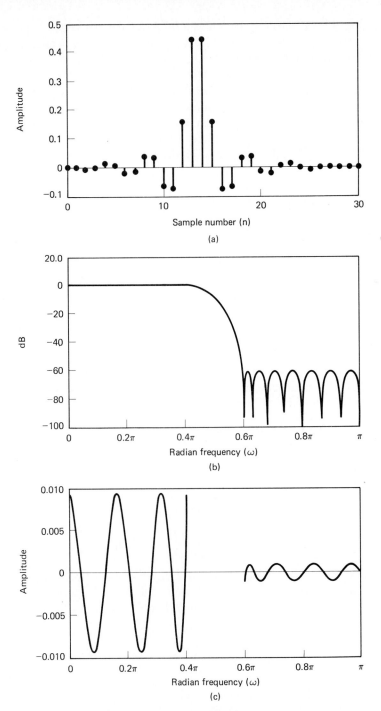

Figure 7.51   Optimum type II FIR lowpass filter for $\omega_p = 0.4\pi$, $\omega_s = 0.6\pi$, $K = 10$, and $M = 27$. (a) Impulse response. (b) Log magnitude. (c) Approximation error (unweighted).

system and for type II systems, the order of the implicit approximating polynomial is $L = (M - 1)/2 = (27 - 1)/2 = 13$. Thus, the minimum number of alternations is still 15. Notice also that in the type II case, the system is constrained to have a zero of its system function at $z = -1$ or $\omega = \pi$. This is clearly shown in Figs. 7.51(b) and (c).

If we compare the results of this example to the results of Example 7.11, we find that the Kaiser window method requires a value $M = 38$ to meet or exceed the specifications, while the Parks-McClellan method requires $M = 27$. This disparity is accentuated because the window method produces approximately equal maximum errors in the passband and stopband while the Parks-McClellan method can weight the errors differently.

7.7.2 Compensation for Zero-Order Hold

In many cases, a discrete-time filter is designed to be used in a system such as depicted in Fig. 7.52; i.e., the filter is used to process a sequence of samples $x[n]$ to obtain a sequence $y[n]$, which is then the input to a D/A converter and continuous-time lowpass filter (as an approximation to the ideal D/C converter) for reconstruction of a continuous-time signal $y_c(t)$. Such a system arises as part of a system for discrete-time filtering of a continuous-time signal, as discussed in Section 3.7. In Section 3.7.4, we showed that if the D/A converter holds its output constant for the entire sampling period T, the Fourier transform of the output $y_c(t)$ is

$$Y_c(j\Omega) = H_r(j\Omega)H_o(j\Omega)H(e^{j\Omega T})X(e^{j\Omega T}), \tag{7.136}$$

where $H_r(j\Omega)$ is the frequency response of the lowpass reconstruction filter with unity gain and cutoff frequency π/T, and

$$H_o(j\Omega) = \frac{\sin(\Omega T/2)}{\Omega/2} e^{-j\Omega T/2} \tag{7.137}$$

is the frequency response of the zero-order hold of the D/A converter. (Note that $H_r(j\Omega)$ can have unity gain because $H_o(j\Omega)$ has gain of T at $\Omega = 0$.) In Section 3.7.4 we suggested that compensation for $H_o(j\Omega)$ could be incorporated into the continuous-time reconstruction filter; i.e., $H_r(j\Omega)$ could be replaced by

$$\tilde{H}_r(j\Omega) = \frac{\Omega T/2}{\sin(\Omega T/2)} H_r(j\Omega) \tag{7.138}$$

so that the effect of the discrete-time filter $H(e^{j\Omega T})$ would be undistorted by the zero-order hold. Another approach is to build the compensation into the discrete-time filter by designing a filter $\tilde{H}(e^{j\Omega T})$ such that

$$\tilde{H}(e^{j\Omega T}) = \frac{\Omega T/2}{\sin(\Omega T/2)} H(e^{j\Omega T}). \tag{7.139}$$

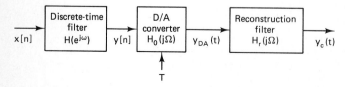

Figure 7.52 Precompensation of a discrete-time filter for the effects of a D/A converter.

A D/A-compensated lowpass filter can be readily designed by the Parks-McClellan algorithm if we simply define the desired response as

$$\tilde{H}_d(e^{j\omega}) = \begin{cases} \dfrac{\omega/2}{\sin(\omega/2)}, & 0 \le \omega \le \omega_p, \\ 0, & \omega_s \le \omega \le \pi. \end{cases} \tag{7.140}$$

Figure 7.53 shows the response functions for such a filter, where the specifications are again $\omega_p = 0.4\pi$, $\omega_p = 0.6\pi$, $\delta_1 = 0.01$, and $\delta_2 = 0.001$. In this case, the specifications are met with $M = 28$ rather than $M = 27$ as in the previous constant-gain case. Thus, for essentially no penalty we have incorporated compensation for the D/A converter so that the effective passband of the filter will be flat. (To emphasize the sloping nature of the passband, Fig. 7.53c shows the magnitude response in the passband rather than the approximation error as in the frequency response plots for the other FIR examples.)

7.7.3 Bandpass Filter

Section 7.6 focused entirely on the lowpass optimal FIR case for which there are only two approximation bands. However, bandpass and bandstop filters require three approximation bands. To design such filters, it is necessary to generalize the discussion of Section 7.6 to the multiband case. This requires that we explore the implications of the alternation theorem and the properties of the approximating polynomial in this more general context. First, recall that the alternation theorem as stated does not assume any limit on the number of disjoint approximation intervals. Therefore the *minimum* number of alternations for the optimum approximation is still $(L + 2)$. However, multiband filters can have more than $(L + 3)$ alternations because there are more band edges. (Problem 7.31 explores this issue.) This means that some of the statements proved in Section 7.6.1 are not true in the multiband case. For example, it is *not* necessary for all the local maxima or minima of $A_e(e^{j\omega})$ to lie inside the approximation intervals. Thus, local extrema can occur in the transition regions, and the approximation need not be equiripple in the approximation regions.

To illustrate this, consider the desired response

$$H_d(e^{j\omega}) = \begin{cases} 0, & 0 \le \omega \le 0.3\pi, \\ 1, & 0.35\pi \le \omega \le 0.6\pi, \\ 0, & 0.7\pi \le \omega \le \pi, \end{cases} \tag{7.141}$$

and the error weighting function

$$W(\omega) = \begin{cases} 1, & 0 \le \omega \le 0.3\pi, \\ 1, & 0.35\pi \le \omega \le 0.6\pi, \\ 0.2, & 0.7\pi \le \omega \le \pi. \end{cases} \tag{7.142}$$

A value of $M + 1 = 75$ was chosen for the length of the impulse response of the filter. Figure 7.54 shows the response functions for the resulting filter. First note that

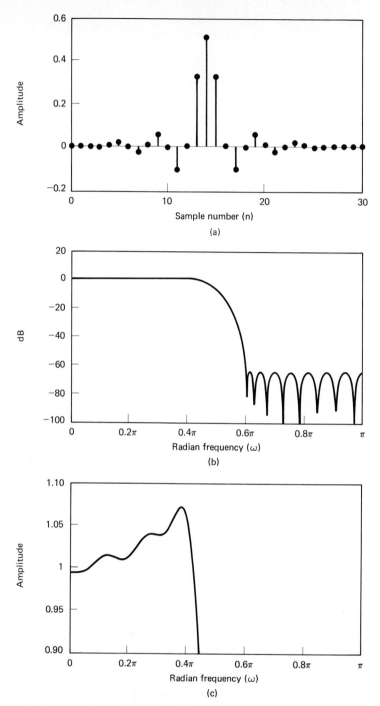

Figure 7.53 Optimum D/A-compensated lowpass filter for $\omega_p = 0.4\pi$, $\omega_s = 0.6\pi$, $K = 10$, and $M = 28$. (a) Impulse response. (b) Log magnitude. (c) Magnitude response in passband.

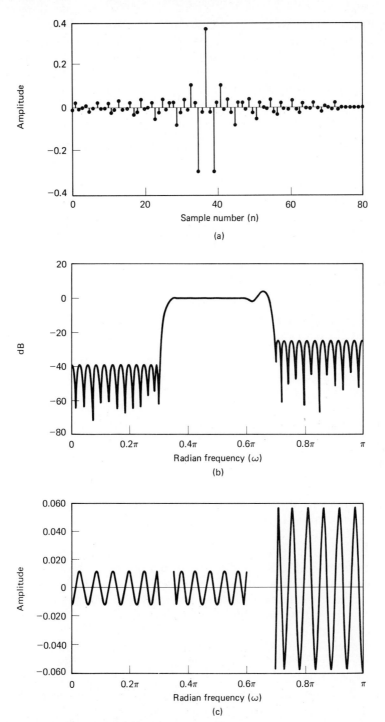

Figure 7.54 Optimum FIR bandpass filter for $M = 74$. (a) Impulse response. (b) Log magnitude. (c) Approximation error (unweighted).

the transition region from the second approximation band to the third is no longer monotonic. However, the use of two local extrema in this unconstrained region does not violate the alternation theorem. Since $M = 74$, the filter is a type I system and the order of the implicit approximating polynomial is $L = M/2 = 74/2 = 37$. Thus, the alternation theorem requires at least $L + 2 = 39$ alternations. It can be readily seen in Fig. 7.54(c), which shows the unweighted approximation error, that there are 13 alternations in each band, for a total of 39.

Such approximations as shown in Fig. 7.54 are optimum in the sense of the alternation theorem, but they would probably be unacceptable in a filtering application. In general, there is no guarantee that the transition regions of a multiband filter will be monotonic, because the Parks-McClellan algorithm leaves these regions completely unconstrained. When such a response results for a particular choice of the filter parameters, acceptable transition regions can usually be obtained by systematically changing one or more of the band edge frequencies, the impulse response length, or the error weighting function and redesigning the filter.

7.8 COMMENTS ON IIR AND FIR DIGITAL FILTERS

This chapter has been concerned with design methods for linear time-invariant discrete-time systems. We have discussed a wide range of methods of designing both infinite-duration and finite-duration impulse response filters. Questions that naturally arise are: What type of system is best, IIR or FIR? Why give so many different design methods? Which method yields the best results? As in any engineering design problem, it is generally not possible to give a precise answer as to what is best. We have discussed a wide variety of methods for both IIR and FIR filters because no single type of filter and no single design method is best for all circumstances.

The choice between an FIR filter and an IIR filter depends on the importance to the design problem of the advantages of each type. IIR filters, for example, have the advantage that a variety of frequency-selective filters can be designed using closed-form design formulas. That is, once the problem has been specified in terms appropriate for a given approximation method (e.g., Butterworth, Chebyshev, or elliptic), then the order of the filter that will meet the specifications can be computed and the coefficients (or poles and zeros) of the digital filter can be obtained by straightforward substitution into a set of design equations. This kind of simplicity of the design procedure makes it feasible to design IIR filters by manual computation if necessary and it leads to straightforward noniterative computer programs for IIR filter design.

These methods are limited to frequency-selective filters and they permit only the magnitude response to be specified. If other magnitude shapes are desired, or if it is necessary to approximate a prescribed phase or group delay response, an algorithmic procedure will be required. The procedure developed by Deczky is representative of algorithmic design methods that have been developed for IIR systems.

Closed-form design equations do not exist for FIR filters. Although the window method is straightforward to apply, some iteration may be necessary to meet a prescribed specification. The Parks-McClellan algorithm leads to lower-order filters than the window method, and both methods can be implemented on a personal computer or workstation. Although it is possible in principle to design frequency-selective IIR discrete-time filters using only a hand calculator and tables of continuous-time filter design parameters, the price paid for simplicity of the design process is loss of flexibility in the attainable filter response. The closed-form IIR designs are limited to a few classes such as lowpass, highpass, bandpass, and bandstop filters. Furthermore, these approximations generally disregard the phase response of the filter; i.e., the underlying continuous-time filters assume minimum phase. Thus, although we might obtain an elliptic lowpass filter with excellent magnitude response by a relatively simple computational procedure, its group delay response will be very poor, especially at the band edge.

In contrast, FIR filters can have precisely (generalized) linear phase. Also, the window method and most of the algorithmic methods afford the possibility of approximating rather arbitrary frequency-response characteristics with little more difficulty than is encountered in the design of lowpass filters. In addition, the design problem for FIR filters is much more under control than the IIR design problem because of the existence of an optimality theorem for FIR filters that is meaningful in a wide range of practical situations. Design techniques for FIR filters without linear phase have been given by Chen and Parks (1987), Parks and Burrus (1987), and Schüssler and Steffen (1988).

Finally, questions of economics arise in implementing a discrete-time filter. Economic concerns are usually measured in terms of hardware complexity, chip area, or computational speed. These factors are more or less directly related to the order of the filter required to meet a given specification. If we put aside phase considerations, it is generally true that a given magnitude response specification can be met most efficiently with an IIR filter. However, in many cases, the linear phase available with an FIR filter may be well worth the extra cost, and in applications such as decimation or interpolation, FIR filters may be just as efficient as IIR filters. A detailed consideration of questions of computational efficiency is given in Rabiner, Kaiser, Herrmann, and Dolan (1974).

Thus, a multitude of tradeoffs must be considered in designing a digital filter. Clearly, the final choice will most often be made by engineering judgment on such questions as the formulation of the specifications, the method of implementation, and computational facilities and design software available for carrying out the design.

7.9 SUMMARY

In this chapter we have considered a variety of design techiques for both infinite-duration and finite-duration impulse response discrete-time filters. Our emphasis was on frequency-domain specification of the desired filter characteristics since this is most common in practice. Our objective was to give a general picture of the wide range of possibilities available for discrete-time filter design while also giving sufficient detail about some of the techniques so that they may be applied directly without further consultation of the extensive literature on discrete-time filter design. Consequently, we gave considerable attention to the well-established methods of impulse invariance and bilinear transformation and a briefer discussion of algorithmic design methods for IIR filters, since the latter are used much less in practice. In the FIR case, considerable detail was given on both the windowing method and the Parks-McClellan algorithmic methods of filter design.

The chapter concludes with some remarks on the choice between the two classes of digital filters. The main point of this discussion is that the choice is not always clear-cut and may depend on a multitude of factors that are often difficult to quantify or discuss in general terms. However, it should be clear from this chapter and Chapter 6 that digital filters are characterized by great flexibility in design and implementation. This flexibility makes it possible to implement rather sophisticated signal processing schemes that in many cases would be difficult, if not impossible, to implement by analog means.

PROBLEMS

7.1. Consider a continuous-time system with impulse response $h_c(t)$ and system function

$$H_c(s) = \frac{s + a}{(s + a)^2 + b^2}.$$

(a) Use impulse invariance to determine $H_1(z)$ for a discrete-time system such that $h_1[n] = h_c(nT)$.

(b) Use step invariance to determine $H_2(z)$ for a discrete-time system such that $s_2[n] = s_c(nT)$, where

$$s_2[n] = \sum_{k=-\infty}^{n} h_2[k] \quad \text{and} \quad s_c(t) = \int_{-\infty}^{t} h_c(\tau)d\tau.$$

(c) Determine the step response $s_1[n]$ of system 1 and the impulse response $h_2[n]$ of system 2. Is it true that $h_2[n] = h_1[n] = h_c(nT)$? Is it true that $s_1[n] = s_2[n] = s_c(nT)$?

7.2. Assume that $H_c(s)$ has an rth-order pole at $s = s_0$, so that $H_c(s)$ can be expressed as

$$H_c(s) = \sum_{k=1}^{r} \frac{A_k}{(s - s_0)^k} + G_c(s),$$

where $G_c(s)$ has only first-order poles.

(a) Give a formula for determining the constants A_k from $H_c(s)$.

(b) Obtain an expression for the impulse response $h_c(t)$ in terms of s_0 and $g_c(t)$, the inverse Laplace transform of $G_c(s)$.

(c) Suppose that we define $h[n] = T_d h_c(nT_d)$ to be the impulse response of a discrete-time filter. Using the result of part (b), write an expression for the system function $H(z)$.

(d) Describe a direct procedure for obtaining $H(z)$ from $H_c(s)$.

7.3. Suppose that we are given a continuous-time lowpass filter with frequency response $H_c(j\omega)$ such that

$$1 - \delta_1 \leq |H_c(j\Omega)| \leq 1 + \delta_1, \qquad |\Omega| \leq \Omega_p,$$

$$|H_c(j\Omega)| \leq \delta_2, \qquad |\Omega| \geq \Omega_s.$$

A set of discrete-time lowpass filters can be obtained from $H_c(s)$ by using the bilinear transformation, i.e.,

$$H(z) = H_c(s)\Big|_{s=(2/T_d)[(1-z^{-1})/(1+z^{-1})]},$$

with T_d variable.

(a) Assuming that Ω_p is fixed, find the value of T_d such that the corresponding passband cutoff frequency for the discrete-time system is $\omega_p = \pi/2$.

(b) With Ω_p fixed, sketch ω_p as a function of $0 < T_d < \infty$.

(c) With both Ω_p and Ω_s fixed, sketch the transition region $\Delta\omega = (\omega_s - \omega_p)$ as a function of $0 < T_d < \infty$.

7.4. A discrete-time lowpass filter is to be designed by applying the impulse invariance method to a continuous-time Butterworth filter having magnitude-squared function

$$|H_c(j\Omega)|^2 = \frac{1}{1 + (\Omega/\Omega_c)^{2N}}.$$

The specifications for the discrete-time system are those of Example 7.3, i.e.,

$$0.89125 \le |H(e^{j\omega})| \le 1, \qquad 0 \le |\omega| \le 0.2\pi,$$

$$|H(e^{j\omega})| \le 0.17783, \qquad 0.3\pi \le |\omega| \le \pi.$$

Assume as in Example 7.3 that aliasing will not be a problem; i.e., design the continuous-time Butterworth filter to meet passband and stopband specifications as determined by the desired discrete-time filter.

(a) Sketch the tolerance bounds on the magnitude of the frequency response $|H_c(j\Omega)|$ of the continuous-time Butterworth filter such that after applying the impulse invariance method (i.e., $h[n] = T_d h_c(nT_d)$), the resulting discrete-time filter will satisfy the given design specifications. Do not assume that $T_d = 1$ as in Example 7.3.

(b) Determine the integer order N and the quantity $T_d\Omega_c$ such that the continuous-time Butterworth filter exactly meets the specifications determined in part (a) at the passband edge.

(c) Note that if $T_d = 1$, your answer in part (b) should give the values of N and Ω_c obtained in Example 7.3. Use this observation to determine the system function $H_c(s)$ for $T_d \ne 1$ and to argue that the system function $H(z)$ that results from impulse invariance design with $T_d \ne 1$ is the same as the result for $T_d = 1$ given by Eq. (7.19).

7.5. We wish to use impulse invariance or the bilinear transformation to design a discrete-time filter that meets specifications of the following form:

$$1 - \delta_1 \le |H(e^{j\omega})| \le 1 + \delta_1, \qquad 0 \le \omega \le \omega_p,$$

$$|H(e^{j\omega})| \le \delta_2, \qquad \omega_s \le \omega \le \pi. \tag{P7.5-1}$$

For historical reasons most of the design formulas, tables, or charts for continuous-time filters are normally specified with a peak gain of 1 in the passband; i.e.,

$$1 - \hat{\delta}_1 \le |H_c(j\Omega)| \le 1, \qquad 0 \le \Omega \le \Omega_p,$$

$$|H_c(\Omega)| \le \hat{\delta}_2, \qquad \Omega_s \le \Omega. \tag{P7.5-2}$$

Useful design charts for continuous-time filters specified in this form were given by Rabiner, Kaiser, Herrmann, and Dolan (1974).

(a) To use such tables and charts to design discrete-time systems with peak gain of $(1 + \delta_1)$, it is necessary to convert the discrete-time specifications into specifications of the form of Eq. (P7.5-2). This can be done by dividing the discrete-time specifications by $(1 + \delta_1)$. Use this approach to obtain an expression for $\hat{\delta}_1$ and $\hat{\delta}_2$ in terms of δ_1 and δ_2.

(b) In Example 7.3, we designed a discrete-time filter with maximum passband gain of unity. This filter can be converted to a filter satisfying a set of specifications such as those in Eq. (P7.5-1) by multiplying by a constant of the form $(1 + \delta_1)$. Find the required value of δ_1 and the corresponding value of δ_2 for this example, and use Eq. (7.19) to determine the coefficients of the system function of the new filter.

(c) Repeat part (b) for the filter in Example 7.4.

7.6. Consider a continuous-time system with system function

$$H_c(s) = \frac{1}{s}.$$

This system is called an *integrator* since the output $y_c(t)$ is related to the input $x_c(t)$ by

$$y_c(t) = \int_{-\infty}^{t} x_c(\tau)d\tau.$$

Suppose a discrete-time system is obtained by applying the bilinear transformation to $H_c(s)$.

(a) What is the system function $H(z)$ of the resulting discrete-time system? What is the impulse response $h[n]$?

(b) If $x[n]$ is the input and $y[n]$ is the output of the resulting discrete-time system, write the difference equation that is satisfied by the input and output. What problems do you anticipate in implementing the discrete-time system using this difference equation?

(c) Obtain an expression for the frequency response $H(e^{j\omega})$ of the system. Sketch the magnitude and phase of the discrete-time system for $0 \le |\omega| \le \pi$. Compare them to the magnitude and phase of the frequency response $H_c(j\Omega)$ of the continuous-time integrator. Under what conditions could the discrete-time "integrator" be considered a good approximation to the continuous-time integrator?

Now consider a continuous-time system with system function

$$G_c(s) = s.$$

This system is called a *differentiator* since the output is the derivative of the input. Suppose a discrete-time system is obtained by applying the bilinear transformation to $G_c(s)$.

(d) What is the system function $G(z)$ of the resulting discrete-time system? What is the impulse response $g[n]$?

(e) Obtain an expression for the frequency response $G(e^{j\omega})$ of the system. Sketch the magnitude and phase of the discrete-time system for $0 \le |\omega| \le \pi$. Compare them to the magnitude and phase of the frequency response $G_c(j\Omega)$ of the continuous-time differentiator. Under what conditions could the discrete-time "differentiator" be considered a good approximation to the continuous-time differentiator?

(f) The continuous-time integrator and differentiator are exact inverses of one another. Is the same true of the discrete-time approximations?

7.7. A continuous-time filter with impulse response $h_c(t)$ and frequency-response magnitude

$$|H_c(j\Omega)| = \begin{cases} |\Omega|, & |\Omega| < 10\pi, \\ 0, & |\Omega| > 10\pi, \end{cases}$$

is to be used as the prototype for design of a discrete-time filter. The resulting discrete-time system is to be used in the configuration of Fig. P7.7 to filter the continuous-time signal $x_c(t)$.

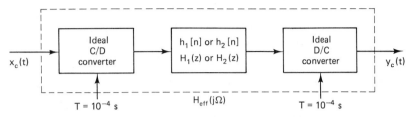

Figure P7.7

(a) A discrete-time system with impulse response $h_1[n]$ and system function $H_1(z)$ is obtained from the prototype continuous-time system by impulse invariance with $T_d = 0.01$; i.e., $h_1[n] = 0.01h_c(0.01n)$. Plot the magnitude of the overall effective frequency response $H_{eff}(j\Omega) = Y_c(j\Omega)/X_c(j\Omega)$ when this discrete-time system is used in Fig. P7.7.

(b) Alternatively, suppose that a discrete-time system with impulse response $h_2[n]$ and system function $H_2(z)$ is obtained from the prototype continuous-time system by the bilinear transformation with $T_d = 2$; i.e.,

$$H_2(z) = H_c(s)\Big|_{s = (1 - z^{-1})/(1 + z^{-1})}.$$

Plot the magnitude of the overall effective frequency response $H_{eff}(j\Omega)$ when this discrete-time system is used in Fig. P7.7.

7.8. The system function of a discrete-time system is

$$H(z) = \frac{2}{1 - e^{-0.2}z^{-1}} - \frac{1}{1 - e^{-0.4}z^{-1}}.$$

(a) Assume that this discrete-time filter was designed by the impulse invariance method with $T_d = 2$; i.e., $h[n] = 2h_c(2n)$, where $h_c(t)$ is real. Find the system function $H_c(s)$ of a continuous-time filter that could have been the basis for the design. Is your answer unique? If not, find another system function $H_c(s)$.

(b) Assume that $H(z)$ was obtained by the bilinear transform method with $T_d = 2$. Find the system function $H_c(s)$ that could have been the basis for the design. Is your answer unique? If not, find another $H_c(s)$.

7.9. *Impulse invariance* and the *bilinear transformation* are two methods for designing discrete-time filters. Both methods transform a continuous-time system function $H_a(s)$ into a discrete-time system function $H(z)$. Answer the following questions by indicating which method(s) will yield the desired result.

(a) A minimum-phase continuous-time system has all its poles and zeros in the left-half s-plane. If a minimum-phase continuous-time system is transformed into a discrete-time system, which method(s) will result in a minimum-phase discrete-time system?

(b) If the continuous-time system is an allpass system, its poles will be at locations s_k in the left-half s-plane and its zeros will be at corresponding locations $-s_k$ in the right-half s-plane. Which design method(s) will result in an allpass discrete-time system?

(c) Which design method(s) will guarantee that

$$H(e^{j\omega})\Big|_{\omega=0} = H_c(j\Omega)\Big|_{\Omega=0} ?$$

(d) If the continuous-time system is a bandstop filter, which method(s) will result in a discrete-time bandstop filter?

(e) Suppose that $H_1(z)$, $H_2(z)$, and $H(z)$ are transformed versions of $H_{c1}(s)$, $H_{c2}(s)$, and $H_c(s)$, respectively. Which design method(s) will guarantee that $H(z) = H_1(z)H_2(z)$ whenever $H_c(s) = H_{c1}(s)H_{c2}(s)$?

(f) Suppose that $H_1(z)$, $H_2(z)$, and $H(z)$ are transformed versions of $H_{c1}(s)$, $H_{c2}(s)$, and $H_c(s)$, respectively. Which design method(s) will guarantee that $H(z) = H_1(z) + H_2(z)$ whenever $H_c(s) = H_{c1}(s) + H_{c2}(s)$?

(g) Assume that two continuous-time system functions satisfy the condition

$$\frac{H_{c1}(j\Omega)}{H_{c2}(j\Omega)} = \begin{cases} e^{-j\pi/2}, & \Omega > 0, \\ e^{j\pi/2}, & \Omega < 0. \end{cases}$$

If $H_1(z)$ and $H_2(z)$ are transformed versions of $H_{c1}(s)$ and $H_{c2}(s)$, respectively, which design method(s) will result in discrete-time systems such that

$$\frac{H_1(e^{j\omega})}{H_2(e^{j\omega})} = \begin{cases} e^{-j\pi/2}, & 0 < \omega < \pi, \\ e^{j\pi/2}, & -\pi < \omega < 0? \end{cases}$$

(Such systems are called "90-degree phase splitters.")

7.10. A discrete-time filter with system function $H(z)$ is designed by transforming a continuous-time filter with system function $H_c(s)$. It is desired that

$$H(e^{j\omega})\Big|_{\omega=0} = H_c(j\Omega)\Big|_{\Omega=0} .$$

(a) Is it possible that the above condition could hold for a filter designed by impulse invariance? If so, what condition(s), if any, would $H_c(j\Omega)$ have to satisfy?

(b) Is it possible that the above condition could hold for a filter designed using the bilinear transformation? If so, what conditions, if any, would $H_c(j\Omega)$ have to satisfy?

7.11. In Example 7.3 a lowpass filter was designed by the impulse invariance method. The resulting system function was

$$H(z) = \frac{0.287 - 0.447z^{-1}}{1 - 1.297z^{-1} + 0.695z^{-2}} + \frac{-2.143 + 1.145z^{-1}}{1 - 1.069z^{-1} + 0.370z^{-2}}$$

$$+ \frac{1.856 - 0.630z^{-1}}{1 - 0.997z^{-1} + 0.257z^{-2}}.$$

(a) Draw the signal flow graph of an implementation of the system in the parallel form using second-order sections.

(b) Draw the signal flow graphs of implementations of the system in direct form I and direct form II.

(c) Draw the signal flow graph of an implementation of the system as a cascade of second-order sections. (*Note*: This part requires access to a polynomial-root-finding computer program.)

7.12. In Example 7.4 a lowpass filter was designed by the bilinear transformation method. The resulting system function was

$$H(z) = \frac{0.0007378(1 + z^{-1})^6}{(1 - 1.2686z^{-1} + 0.7051z^{-2})(1 - 1.0106z^{-1} + 0.3583z^{-2})}$$

$$\times \frac{1}{(1 - 0.9044z^{-1} + 0.2155z^{-2})}.$$

(a) Draw the signal flow graph of an implementation of the system as a cascade of second-order sections.

(b) Draw the signal flow graphs of implementations of the system in direct form I and direct form II.

(c) Draw the signal flow graph of an implementation of the system in the parallel form using second-order sections. (*Note*: This part requires lengthy computations to find the partial fraction expansion and should be done with the aid of a computer program.)

7.13. If a linear time-invariant continuous-time system has a rational system function, then its input and output satisfy an ordinary linear differential equation with constant coefficients. A standard procedure in simulation of such systems is to use finite-difference approximations to the derivatives in the differential equations. In particular, since for continuous differentiable functions $y_c(t)$,

$$\frac{dy_c(t)}{dt} = \lim_{T \to 0} \left[\frac{y_c(t) - y_c(t - T)}{T} \right],$$

it seems plausible that if T is "small enough," we should obtain a good approximation if we replace $dy_c(t)/dt$ by $[y_c(t) - y_c(t - T)]/T$.

While this simple approach may be useful in simulation of continuous-time systems, it is *not* generally a useful method of designing discrete-time systems for filtering applications. To understand the effect of approximating differential equations by difference equations, it is helpful to consider a specific example. Assume that the system function of a continuous-time system is

$$H_c(s) = \frac{A}{s + c},$$

where A and c are constants.

(a) Show that the input $x_c(t)$ and the output $y_c(t)$ of the system satisfy the differential equation

$$\frac{dy_c(t)}{dt} + cy_c(t) = Ax_c(t).$$

(b) Evaluate the differential equation at $t = nT$, and substitute

$$\left. \frac{dy_c(t)}{dt} \right|_{t = nT} \approx \frac{y_c(nT) - y_c(nT - T)}{T},$$

i.e., replace the first derivative by the *first backward difference*.

(c) Define $x[n] = x_c(nT)$ and $y[n] = y_c(nT)$. With this notation and the result of part (b), obtain a difference equation relating $x[n]$ and $y[n]$, and determine the system function $H(z) = Y(z)/X(z)$ of the resulting discrete-time system.

(d) Show that for this example,

$$H(z) = H_c(s)\Big|_{s=(1-z^{-1})/T},$$

i.e., show that $H(z)$ can be obtained directly from $H_c(s)$ by the mapping

$$s = \frac{1-z^{-1}}{T}.$$

(It can be shown that if higher-order derivatives are approximated by repeated application of the first backward difference, then the result of part (d) holds for higher-order systems as well.)

(e) For the mapping of part (d), determine the contour in the z-plane to which the Ω-axis of the s-plane maps. Also determine the region of the z-plane that corresponds to the left half of the s-plane. If the continuous-time system with system function $H_c(s)$ is stable, will the discrete-time system obtained by first backward difference approximation also be stable? Will the frequency response of the discrete-time system be a faithful reproduction of the frequency response of the continuous-time system? How will the stability and frequency response be affected by the choice of T?

(f) Assume that the first derivative is approximated by the *first forward difference*; i.e.,

$$\frac{dy_c(t)}{dt}\bigg|_{t=nT} \approx \frac{y_c(nT+T) - y_c(nT)}{T}.$$

Determine the corresponding mapping from the s-plane to the z-plane and repeat part (e) for this mapping.

7.14. Consider a linear time-invariant continuous-time system with rational system function $H_c(s)$. The input $x_c(t)$ and the output $y_c(t)$ satisfy an ordinary linear differential equation with constant coefficients. One approach to simulating such systems is to use numerical techniques to integrate the differential equation. In this problem we demonstrate that if the trapezoidal integration formula is used, this approach is equivalent to transforming the continuous-time system function $H_c(s)$ to a discrete-time system function $H(z)$ using the bilinear transformation.

To demonstrate this, consider the continuous-time system function

$$H_c(s) = \frac{A}{s+c},$$

where A and c are constants. The corresponding differential equation is

$$\dot{y}_c(t) + cy_c(t) = Ax_c(t),$$

where

$$\dot{y}_c(t) = \frac{dy_c(t)}{dt}.$$

(a) Show that $y_c(nT)$ can be expressed in terms of $\dot{y}_c(t)$ as

$$y_c(nT) = \int_{(nT-T)}^{nT} \dot{y}_c(\tau)d\tau + y_c(nT-T).$$

The definite integral in this equation represents the area beneath the function $\dot{y}_c(t)$ for the interval from $(nT-T)$ to nT. Figure P7.14 shows a function $\dot{y}_c(t)$ and a shaded trapezoid-shaped region whose area approximates the area beneath the curve. This approximation to the integral is known as the *trapezoidal approximation*. Clearly, as

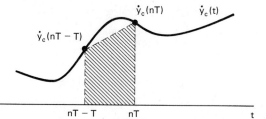

Figure P7.14

T approaches zero, the approximation improves. Use this approximation to obtain an expression for $y_c(nT)$ in terms of $y_c(nT - T)$, $\dot{y}_c(nT)$, and $\dot{y}_c(nT - T)$.

(b) Use the differential equation to obtain an expression for $\dot{y}_c(nT)$ and substitute this expression into the expression obtained in part (a).

(c) Define $x[n] = x_c(nT)$ and $y[n] = y_c(nT)$. With this notation and the result of part (b), obtain a difference equation relating $x[n]$ and $y[n]$ and determine the system function $H(z) = Y(z)/X(z)$ of the resulting discrete-time system.

(d) Show that for this example

$$H(z) = H_c(s)\Big|_{s = (2/T)[(1 - z^{-1})/(1 + z^{-1})]},$$

i.e., show that $H(z)$ can be obtained directly from $H_c(s)$ by the bilinear transformation. (For higher-order differential equations, repeated trapezoidal integration applied to the highest-order derivative of the output will result in the same conclusion for a general continuous-time system with rational system function.)

7.15. In this problem we consider a method of filter design that might be called *autocorrelation invariance*. Consider a stable continuous-time system with impulse response $h_c(t)$ and system function $H_c(s)$. The autocorrelation function of the system is defined as

$$\phi_c(\tau) = \int_{-\infty}^{\infty} h_c(t)h_c(t + \tau)d\tau,$$

and for a real impulse response, it is easily shown that the Laplace transform of $\phi_c(\tau)$ is $\Phi_c(s) = H_c(s)H_c(-s)$. Similarly, consider a discrete-time system with impulse response $h[n]$ and system function $H(z)$. The autocorrelation function of a discrete-time system is defined as

$$\phi[m] = \sum_{n=-\infty}^{\infty} h[n]h[n + m],$$

and for a real impulse response, $\Phi(z) = H(z)H(z^{-1})$.

   Autocorrelation invariance implies that a discrete-time filter is defined by equating the autocorrelation function of the discrete-time system to the sampled autocorrelation function of a continuous-time system; i.e.,

$$\phi[m] = T_d\phi_c(mT_d), \qquad -\infty < m < \infty.$$

The following design procedure is proposed for autocorrelation invariance when $H_c(s)$ is a rational function having N first-order poles at s_k, $k = 1, 2, \ldots, N$, and $M < N$ zeros:

1. Obtain a partial fraction expansion of $\Phi_c(s)$ in the form

$$\Phi_c(s) = \sum_{k=1}^{N} \left(\frac{A_k}{s - s_k} + \frac{B_k}{s + s_k} \right).$$

2. Form the z-transform

$$\Phi(z) = \sum_{k=1}^{N} \left(\frac{T_d A_k}{1 - e^{s_k T_d} z^{-1}} + \frac{T_d B_k}{1 - e^{s_k T_d} z^{-1}} \right).$$

3. Find the poles and zeros of $\Phi(z)$ and form a minimum-phase system function $H(z)$ from the poles and zeros of $\Phi(z)$ that are *inside* the unit circle.

(a) Justify each step in the proposed design procedure; i.e., show that the autocorrelation function of the resulting discrete-time system is a sampled version of the autocorrelation function of the continuous-time system. To verify the procedure, it may be helpful to try it out on the first-order system with impulse response

$$h_c(t) = e^{-\alpha t} u(t)$$

and system function

$$H_c(s) = \frac{1}{s + \alpha}.$$

(b) What is the relationship between $|H(e^{j\omega})|^2$ and $|H_c(j\Omega)|^2$? What types of frequency-response functions would be appropriate for autocorrelation invariance design?

(c) Is the system function obtained in step 3 unique? If not, describe how to obtain additional autocorrelation invariant discrete-time systems.

7.16. Suppose that we are given an ideal lowpass discrete-time filter with frequency response

$$H(e^{j\omega}) = \begin{cases} 1, & |\omega| < \pi/4, \\ 0, & \pi/4 < |\omega| \leq \pi. \end{cases}$$

We wish to derive new filters from this prototype by manipulations of the impulse response $h[n]$.

(a) Plot the frequency response $H_1(e^{j\omega})$ for the system whose impulse response is $h_1[n] = h[2n]$.

(b) Plot the frequency response $H_2(e^{j\omega})$ for the system whose impulse response is

$$h_2[n] = \begin{cases} h[n/2], & n = 0, \pm 2, \pm 4, \ldots, \\ 0, & \text{otherwise.} \end{cases}$$

(c) Plot the frequency response $H_3(e^{j\omega})$ for the system whose impulse response is $h_3[n] = e^{j\pi n} h[n] = (-1)^n h[n]$.

7.17. Consider a continuous-time lowpass filter $H_c(s)$ with passband and stopband specifications

$$1 - \delta_1 \leq |H_c(j\Omega)| \leq 1 + \delta_1, \qquad |\Omega| \leq \Omega_p,$$

$$|H_c(j\Omega)| \leq \delta_2, \qquad \Omega_s \leq |\Omega| \leq \pi.$$

This filter is transformed to a lowpass discrete-time filter $H_1(z)$ by the transformation

$$H_1(z) = H_c(s) \Big|_{s = (1 - z^{-1})/(1 + z^{-1})},$$

and the same continuous-time filter is transformed to a highpass discrete-time filter by the transformation

$$H_2(z) = H_c(s) \Big|_{s = (1 + z^{-1})/(1 - z^{-1})}.$$

(a) Determine a relationship between the passband cutoff frequency Ω_p of the continuous-time lowpass filter and the passband cutoff frequency ω_{p1} of the discrete-time lowpass filter.

(b) Determine a relationship between the passband cutoff frequency Ω_p of the continuous-time lowpass filter and the passband cutoff frequency ω_{p2} of the discrete-time highpass filter.

(c) Determine a relationship beween the passband cutoff frequency ω_{p1} of the discrete-time lowpass filter and the passband cutoff frequency ω_{p2} of the discrete-time highpass filter.

(d) The network in Fig. P7.17 depicts an implementation of the discrete-time lowpass filter with system function $H_1(z)$. The coefficients A, B, C, and D are real. How should these coefficients be modified to obtain a network that implements the discrete-time highpass filter with system function $H_2(z)$?

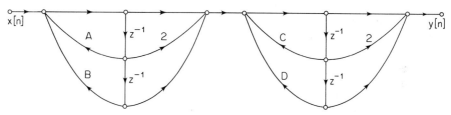

Figure P7.17

7.18. A discrete-time system with system function $H(Z)$ and impulse response $h[n]$ has frequency response

$$H(e^{j\theta}) = \begin{cases} A, & |\theta| < \theta_c, \\ 0, & \theta_c < |\theta| \le \pi, \end{cases}$$

where $0 < \theta_c < \pi$. This filter is transformed into a new filter by the transformation $Z = -z^2$; i.e.,

$$H_1(z) = H(Z)\Big|_{Z=-z^2} = H(-z^2).$$

(a) Obtain a relationship between the frequency variable θ for the original lowpass system $H(Z)$ and the frequency variable ω for the new system $H_1(z)$.

(b) Sketch and carefully label the frequency response $H_1(e^{j\omega})$ for the new filter.

(c) Obtain a relationship expressing $h_1[n]$ in terms of $h[n]$.

(d) Assume that $H(Z)$ can be realized by the set of difference equations

$$g[n] = x[n] - a_1 g[n-1] - b_1 f[n-2],$$

$$f[n] = a_2 g[n-1] + b_2 f[n-1],$$

$$y[n] = c_1 f[n] - c_2 g[n-1],$$

where $x[n]$ is the input and $y[n]$ is the output of the system. Determine a set of difference equations that will realize the transformed system $H_1(z) = H(-z^2)$.

7.19. Consider designing a discrete-time filter with system function $H(z)$ from a continuous-time filter with rational system function $H_c(s)$ by the following transformation:

$$H(z) = H_c(s)\Big|_{s=\beta[(1-z^{-\alpha})/(1+z^{-\alpha})]},$$

where α is a nonzero integer and β is real.

(a) If $\alpha > 0$, for what values of β does a stable, causal, continuous-time filter with rational $H_c(s)$ always lead to a stable, causal, discrete-time filter with rational $H(z)$?

(b) If $\alpha < 0$, for what values of β does a stable, causal, continuous-time filter with rational $H_c(s)$ always lead to a stable, causal, discrete-time filter with rational $H(z)$?

(c) For $\alpha = 2$ and $\beta = 1$, determine to what contour in the z-plane the $j\Omega$-axis of the s-plane maps.

(d) Suppose that the continuous-time filter is a stable lowpass filter with passband frequency response such that

$$1 - \delta_1 \leq |H_c(j\Omega)| \leq 1 + \delta_1 \qquad \text{for} \quad |\Omega| \leq 1.$$

If the discrete-time system $H(z)$ is obtained by the above transformation with $\alpha = 2$ and $\beta = 1$, determine the values of ω in the interval $|\omega| \leq \pi$ for which

$$1 - \delta_1 \leq |H(e^{j\omega})| \leq 1 + \delta_1.$$

7.20. A discrete-time highpass filter can be obtained from a continuous-time lowpass filter by the following transformation:

$$H(z) = H_c(s)\Big|_{s = [(1 + z^{-1})/(1 - z^{-1})]},$$

(a) Show that the above transformation maps the $j\Omega$-axis of the s-plane onto the unit circle of the z-plane.

(b) Show that if $H_c(s)$ is a rational function with all its poles inside the left-half s-plane, then $H(z)$ will be a rational function with all its poles inside the unit circle of the z-plane.

(c) Suppose a desired highpass discrete-time filter has specifications

$$|H(e^{j\omega})| \leq 0.01, \qquad |\omega| \leq \pi/3,$$

$$0.95 \leq |H(e^{j\omega})| \leq 1.05, \qquad \pi/2 \leq |\omega| \leq \pi.$$

Determine the specifications on the continuous-time lowpass filter so that the desired highpass discrete-time filter results from the above transformation.

7.21. Let $H_{lp}(Z)$ denote the system function for a discrete-time lowpass filter. The implementations of such a system can be represented by linear signal flow graphs consisting of adders, gains, and unit delay elements as in Fig. P7.21(a). We want to implement a lowpass filter for which the cutoff frequency can be varied by changing a single parameter. The proposed strategy is to replace each unit delay element in a flow graph representing $H_{lp}(Z)$ by the network shown in Fig. P7.21(b), where α is real and $|\alpha| < 1$.

(a) Let $H(z)$ denote the system function for the filter resulting when the network of Fig. P7.21(b) is substituted for each unit delay branch in the network that implements $H_{lp}(Z)$. Show that $H(z)$ and $H_{lp}(Z)$ are related by a mapping of the Z-plane into the z-plane.

(b) If $H(e^{j\omega})$ and $H_{lp}(e^{j\theta})$ are the frequency responses of the two systems, determine the relationship between the frequency variables ω and θ. Sketch ω as a function of θ for $\alpha = 0, \pm 0.5$, and show that $H(e^{j\omega})$ is a lowpass filter. Also, if θ_p is the passband cutoff frequency for the original lowpass filter $H_{lp}(Z)$, obtain an equation for ω_p, the cutoff frequency of the new filter $H(z)$, as a function of α and θ_p.

(c) Assume that the original lowpass filter has system function

$$H_{lp}(Z) = \frac{1}{1 - 0.9Z^{-1}}.$$

(a)

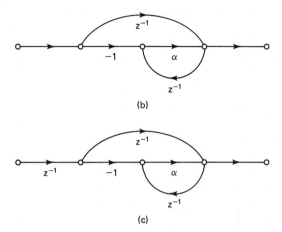

(b)

(c)   **Figure P7.21**

Draw the flow graph of an implementation of $H_{lp}(Z)$ and also draw the flow graph of the implementation of $H(z)$ obtained by replacing the unit delay elements in the first flow graph by the network in Fig. P7.21(b). Does the resulting network correspond to a computable difference equation?

(d) If $H_{lp}(Z)$ corresponds to an FIR system implemented in direct form, would the flow graph manipulation lead to a computable difference equation? If the FIR system $H_{lp}(Z)$ was linear phase, would the resulting system $H(z)$ also be linear phase? If the FIR system has impulse response length $(M + 1)$ samples, what would be the length of the impulse response of the transformed system?

(e) To avoid the difficulties that arose in part (c), it is suggested that the network of Fig. P7.21(b) be cascaded with a unit delay element as depicted in Fig. P7.21(c). Repeat the analysis of part (a) when the network of Fig. P7.21(c) is substituted for each unit delay element. Determine an equation that expresses θ as a function of ω, and show that if $H_{lp}(e^{j\theta})$ is a lowpass filter, then $H(e^{j\omega})$ is *not* a lowpass filter.

7.22. If we are given a basic filter module (hardware or computer subroutine), it is sometimes possible to use it repetitively to implement a new filter with sharper frequency-response characteristics. One approach is to cascade the filter with itself two or more times, but it can easily be shown that while stopband errors are squared (thereby reducing them if they are less than 1), this approach will increase the passband approximation error. Another approach, suggested by Tukey (1977), is shown in the block diagram of Fig. P7.22-1. Tukey called this approach "twicing."

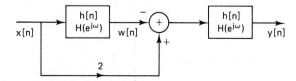

Figure P7.22-1

(a) Assume that the basic system has a symmetric finite-duration impulse response, i.e.,

$$h[n] = \begin{cases} h[-n], & -L \le n \le L, \\ 0 & \text{otherwise.} \end{cases}$$

Determine whether the overall impulse response $g[n]$ is (i) FIR and (ii) symmetric.

(b) Suppose that $H(e^{j\omega})$ satisfies the following approximation error specifications:

$$(1 - \delta_1) \le H(e^{j\omega}) \le (1 + \delta_1), \qquad 0 \le \omega \le \omega_p,$$

$$-\delta_2 \le H(e^{j\omega}) \le \delta_2, \qquad \omega_s \le \omega \le \pi.$$

If the basic system has these specifications, it can be shown that the overall frequency response $G(e^{j\omega})$ satisfies specifications of the form

$$A \le G(e^{j\omega}) \le B, \qquad 0 \le \omega \le \omega_p,$$

$$C \le G(e^{j\omega}) \le D, \qquad \omega_s \le \omega \le \pi.$$

Determine A, B, C, and D in terms of δ_1 and δ_2. If $\delta_1 \ll 1$ and $\delta_2 \ll 1$, what are the approximate maximum passband and stopband approximation errors for $G(e^{j\omega})$?

(c) As determined in part (b), Tukey's twicing method improves the passband approximation error, but it increases the stopband error. Kaiser and Hamming (1977) generalized the twicing method so as to improve *both* the passband and the stopband. They called their approach "sharpening." The simplest sharpening system that improves both passband and stopband is shown in Fig. P7.22-2. Again assume that the impulse response of the basic system is as given in part (a). Repeat part (b) for the system of Fig. P7.22-2.

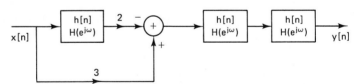

Figure P7.22-2

(d) The basic system was assumed to be noncausal. If the impulse response of the basic system is a causal linear phase FIR system such that

$$h[n] = \begin{cases} h[M - n], & 0 \le n \le M, \\ 0, & \text{otherwise,} \end{cases}$$

how should the systems of Figs. P7.22-1 and P7.22-2 be modified? What type(s) (I, II, III, or IV) of causal linear phase FIR system(s) can be used? What are the lengths of the impulse responses $g[n]$ for the systems in Figs. P7.22-1 and P7.22-2 (in terms of L)?

7.23. Let $h_d[n]$ denote the impulse response of an ideal desired system with corresponding frequency response $H_d(e^{j\omega})$, and let $h[n]$ and $H(e^{j\omega})$ denote the impulse response and frequency response, respectively, of an FIR approximation to the ideal system. Assume that $h[n] = 0$ for $n < 0$ and $n > M$. We wish to choose the $(M + 1)$ samples of the impulse response so as to minimize the mean-square error of the frequency response defined as

$$\epsilon^2 = \frac{1}{2\pi} \int_{-\pi}^{\pi} |H_d(e^{j\omega}) - H(e^{j\omega})|^2 d\omega.$$

(a) Use Parseval's relation to express the error function in terms of the sequences $h_d[n]$ and $h[n]$.

(b) Using the result of part (a), determine the values of $h[n]$ for $0 \leq n \leq M$ that minimize ϵ^2.

(c) The FIR filter determined in part (b) could have been obtained by a windowing operation. That is, $h[n]$ could have been obtained by multiplying the desired infinite-length sequence $h_d[n]$ by a certain finite-length sequence $w[n]$. Determine the necessary window $w[n]$ such that the optimal impulse response is $h[n] = w[n]h_d[n]$.

7.24. An *ideal discrete-time Hilbert transformer* is a system that introduces -90 degrees ($-\pi/2$ radians) of phase shift for $0 < \omega < \pi$ and $+90$ degrees ($+\pi/2$ radians) of phase shift for $+\pi < \omega < 0$. The magnitude of the frequency response is constant (unity) for $0 < \omega < \pi$ and for $-\pi < \omega < 0$. Such systems are also called *ideal 90-degree phase shifters*.

(a) Give an equation for the ideal frequency response $H_d(e^{j\omega})$ of an ideal discrete-time Hilbert transformer with constant (nonzero) group delay. Plot the phase response of this system for $-\pi < \omega < \pi$.

(b) What type(s) of FIR linear phase systems (I, II, III, or IV) can be used to approximate the ideal Hilbert transformer in part (a)?

(c) Suppose that we wish to use the window method to design a linear phase approximation to the ideal Hilbert transformer. Use $H_d(e^{j\omega})$ given in part (a) to determine the ideal impulse response $h_d[n]$ if the FIR system is to be such that $h_d[n] = 0$ for $n < 0$ and $n > M$.

(d) What is the delay of the system if $M = 21$? Sketch the magnitude of the frequency response of the FIR approximation for this case assuming a rectangular window.

(e) What is the delay of the system if $M = 20$? Sketch the magnitude of the frequency response of the FIR approximation for this case assuming a rectangular window.

7.25. The commonly used windows presented in Section 7.4.1 can all be expressed in terms of rectangular windows. This fact can be used to obtain expressions for the Fourier transforms of the Bartlett window and the raised cosine family of windows, which includes the Hanning, Hamming, and Blackman windows.

(a) Show that the $(M + 1)$-point Bartlett window, defined by Eq. (7.77b), can be expressed as the convolution of two smaller rectangular windows. Use this fact to show that the Fourier transform of the $(M + 1)$-point Bartlett window is

$$W_B(e^{j\omega}) = e^{-j\omega M/2}\left(\frac{\sin(\omega M/4)}{\sin(\omega/2)}\right)^2 \quad \text{for } M \text{ even,}$$

or

$$W_B(e^{j\omega}) = e^{-j\omega M/2}\left\{\frac{\sin[\omega(M + 1)/4]}{\sin(\omega/2)}\right\}\left\{\frac{\sin[\omega(M - 1)/4]}{\sin(\omega/2)}\right\} \quad \text{for } M \text{ odd.}$$

(b) It can easily be seen that the $(M + 1)$-point raised cosine windows defined by Eqs. (7.77c)–(7.77e) can all be expresed in the form

$$w[n] = [A + B \cos(2\pi n/M) + C \cos(4\pi n/M)]w_R[n],$$

where $w_R[n]$ is an $(M + 1)$-point rectangular window. Use this relation to find the Fourier transform of the general raised cosine window.

(c) Using appropriate choices for A, B, and C and the result determined in part (b), sketch the magnitude of the Fourier transform of the Hanning window.

7.26. We wish to use the Kaiser window method to design a discrete-time filter with generalized linear phase that meets specifications of the following form:

$$|H(e^{j\omega})| \le 0.01, \qquad 0 \le \omega \le 0.25\pi,$$

$$0.95 \le |H(e^{j\omega})| \le 1.05, \qquad 0.35\pi \le \omega \le 0.6\pi,$$

$$|H(e^{j\omega})| \le 0.01, \qquad 0.65\pi \le \omega \le \pi.$$

(a) Determine the minimum length $(M + 1)$ of the impulse response and the value of the Kaiser window parameter β for a filter that meets the above specifications.
(b) What is the delay of the filter?
(c) Determine the ideal impulse response $h_d[n]$ to which the Kaiser window should be applied.

7.27. Consider the following ideal frequency response for a multiband filter:

$$H_d(e^{j\omega}) = \begin{cases} e^{-j\omega M/2}, & 0 \le |\omega| < 0.3\pi, \\ 0, & 0.3\pi < |\omega| < 0.6\pi, \\ 0.5e^{-j\omega M/2}, & 0.6\pi < |\omega| \le \pi. \end{cases}$$

The impulse response $h_d[n]$ is multiplied by a Kaiser window with $M = 48$ and $\beta = 3.68$, resulting in a linear phase FIR system with impulse response $h[n]$.
(a) What is the delay of the filter?
(b) Determine the ideal impulse response $h_d[n]$.
(c) Determine the set of approximation error specifications that is satisfied by the FIR filter; i.e., determine the parameters $\delta_1, \delta_2, \delta_3, A, B, \omega_{p1}, \omega_{s1}, \omega_{s2},$ and ω_{p2} in

$$1 - \delta_1 \le |H(e^{j\omega})| \le 1 + \delta_1, \qquad 0 \le \omega \le \omega_{p1},$$

$$|H(e^{j\omega})| \le \delta_2, \qquad \omega_{s1} \le \omega \le \omega_{s2},$$

$$1 - \delta_3 \le |H(e^{j\omega})| \le 1 + \delta_3, \qquad \omega_{p2} \le \omega \le \pi.$$

7.28. Consider the design of a lowpass linear phase FIR filter using the Parks-McClellan algorithm. Use the alternation theorem to argue that the approximation must decrease monotonically in the "don't care" region between the passband and the stopband approximation intervals. [*Hint*: Show that all the local maxima and minima of the trigonometric polynomial must be in either the passband or the stopband to satisfy the alternation theorem.]

7.29. Figure P7.29 shows the frequency response $A_e(e^{j\omega})$ of a discrete-time FIR system for which the impulse response is

$$h_e[n] = \begin{cases} h_e[-n], & -L \le n \le L, \\ 0, & \text{otherwise.} \end{cases}$$

(a) Show that $A_e(e^{j\omega})$ *cannot* correspond to an FIR filter generated by the Parks-McClellan algorithm with a passband edge frequency of $\pi/3$, a stopband edge frequency of $2\pi/3$, and an error weighting function of unity in the passband and stopband. Clearly explain your reasoning. [*Hint*: The alternation theorem states that the best approximation is unique.]
(b) Based on Fig. P7.29 and the statement that $A_e(e^{j\omega})$ cannot correspond to an optimal filter, what can be concluded about the value of L?

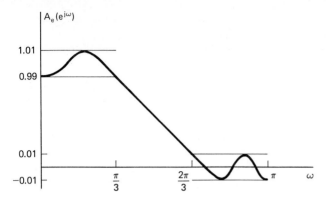

Figure P7.29

7.30. An optimal equiripple FIR linear phase filter was designed by the Parks-McClellan algorithm. The magnitude of its frequency response is shown in Fig. P7.30. The maximum approximation error in the passband is $\delta_1 = 0.0531$, and the maximum approximation error in the stopband is $\delta_2 = 0.085$. The passband and stopband cutoff frequencies are $\omega_p = 0.4\pi$ and $\omega_s = 0.58\pi$.

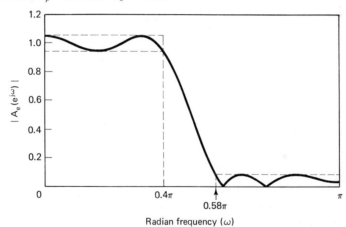

Radian frequency (ω)

Figure P7.30

(a) What was the error weighting function $W(\omega)$ that was used in the optimization?

(b) Carefully sketch the weighted approximation error; i.e., sketch

$$E(\omega) = W(\omega)[H_d(e^{j\omega}) - A_e(e^{j\omega})].$$

(Note that Fig. P7.30 shows $|A_e(e^{j\omega})|$.)

(c) What type (I, II, III, or IV) of linear phase system is this? Explain how you can tell.

(d) What is the length of the impulse response of the system?

(e) If this system is causal, what is the smallest delay that it can have?

(f) Plot the zeros of the system function $H(z)$ as accurately as you can in the z-plane.

7.31. Consider the design of a type I bandpass linear phase FIR filter using the Parks-McClellan algorithm. The impulse response length is $M + 1 = 2L + 1$. Recall that for type I systems, the frequency response is of the form $H(e^{j\omega}) = A_e(e^{j\omega})e^{-j\omega M/2}$, and the Parks-McClellan algorithm finds the function $A_e(e^{j\omega})$ that minimizes the maximum value of the error function

$$E(\omega) = W(\omega)[H_d(e^{j\omega}) - A_e(e^{j\omega})], \qquad \omega \in F,$$

where F is a closed subset of the interval $0 \leq \omega \leq \pi$, $W(\omega)$ is a weighting function, and $H_d(e^{j\omega})$ defines the desired frequency response in the approximation intervals F. The tolerance scheme for a bandpass filter is shown in Fig. P7.31.

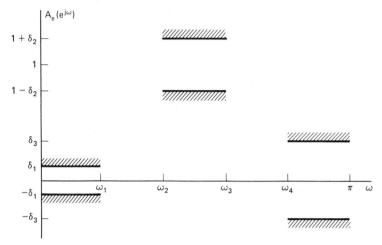

Figure P7.31

(a) Give the equation for the desired response $H_d(e^{j\omega})$ for the tolerance scheme in Fig. P7.31.

(b) Give the equation for the weighting function $W(\omega)$ for the tolerance scheme in Fig. P7.31.

(c) What is the *minimum* number of alternations of the error function for the optimum filter?

(d) What is the *maximum* number of alternations of the error function for the optimum filter?

(e) Sketch a "typical" weighted error function $E(\omega)$ that could possibly be the error function for an optimum bandpass filter if $M = 14$. Assume the *maximum* number of alternations.

(f) Now suppose that M, ω_1, ω_2, ω_3, the weighting function, and the desired function are kept the same, but ω_4 is *increased* so that the transition band $(\omega_4 - \omega_3)$ is increased. Will the optimum filter for these new specifications *necessarily* have a *smaller* value of the maximum approximation error than the optimum filter associated with the original specifications? Clearly show your reasoning.

(g) In the lowpass filter case, all local minima and maxima of $A_e(e^{j\omega})$ must occur in the approximation bands $\omega \in F$. They *cannot* occur in the "don't care" bands. Also, in the lowpass case, the local minima and maxima that occur in the approximation bands must be alternations of the error. Show that this is not necessarily true in the bandpass filter case. Specifically, use the alternation theorem to show (i) that local maxima and minima of $A_e(e^{j\omega})$ are not restricted to the approximation bands and (ii) that local maxima and minima in the approximation bands need not be alternations.

7.32. Consider the system in Fig. P7.32.

1. Assume that $X_c(j\Omega) = 0$ for $|\Omega| \geq \pi/T$ and that $H_r(j\Omega)$ denotes the ideal lowpass reconstruction filter such that

$$H_r(j\Omega) = \begin{cases} 1, & |\Omega| < \pi/T, \\ 0, & |\Omega| > \pi/T. \end{cases}$$

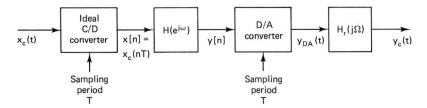

Figure P7.32

2. The D/A converter has a built-in zero-order hold circuit so that

$$y_{DA}(t) = \sum_{n=-\infty}^{\infty} y[n]h_0(t - nT),$$

where $h_0(t)$ is

$$h_0(t) = \begin{cases} 1, & 0 \le t < T, \\ 0, & \text{otherwise.} \end{cases}$$

(We neglect quantization in the D/A converter.)

3. The second system in Fig. P7.32 is a linear phase FIR discrete-time system with frequency response $H(e^{j\omega})$.

We wish to design the FIR system using the Parks-McClellan algorithm so as to compensate for the effects of the zero-order hold system.

(a) The Fourier transform of the output is $Y_c(j\Omega) = H_{eff}(j\Omega)X_c(j\Omega)$. Determine an expression for $H_{eff}(j\Omega)$ in terms of $H(e^{j\Omega T})$ and T.

(b) If the linear phase FIR system is such that $h[n] = 0$ for $n < 0$ and $n > 51$, and $T = 10^{-4}$ s, what is the overall time delay (in ms) between $x_c(t)$ and $y_c(t)$?

(c) Suppose that when $T = 10^{-4}$ s we want the effective frequency response to be equiripple (in both the passband and the stopband) within the following tolerances:

$$0.99 \le |H_{eff}(j\Omega)| \le 1.01, \qquad |\Omega| \le 2\pi(1000),$$

$$|H_{eff}(j\Omega)| \le 0.01, \qquad 2\pi(2000) \le |\Omega| \le 2\pi(5000).$$

We want to achieve this by designing an optimum linear phase filter (using the Parks-McClellan algorithm) that includes compensation for the zero-order hold. Give an equation for the ideal response $H_d(e^{j\omega})$ that should be used. Find and sketch the weighting function $W(\omega)$ that should be used. Sketch a "typical" frequency response $H(e^{j\omega})$ that might result.

(d) How would you modify your results in part (c) to include magnitude compensation for a reconstruction filter $H_r(j\Omega)$ with zero gain above $\Omega = 2\pi(5000)$ but with sloping passband?

7.33. After a discrete-time signal is lowpass filtered, it is often downsampled or decimated as depicted in Fig. P7.33-1. Linear phase FIR filters are often desirable in such applications, but if the lowpass filter in Fig. P7.31-1 has a narrow transition band, an FIR system will have a long impulse response and thus will require a large number of multiplications and additions per output sample.

In this problem, we will study the merits of a multistage implementation of the system in Fig. P7.33-1. Such implementations are particularly useful when ω_s is small and the decimation factor M is large. A general multistage implementation is depicted in Fig. P7.33-2. The strategy is to use a wider transition band in the lowpass filters of the earlier

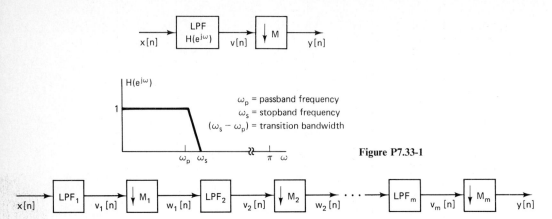

Figure P7.33-1

Figure P7.33-2

stages, thereby reducing the length of the required filter impulse responses in those stages. As decimation occurs, the number of signal samples is reduced, and we can progressively reduce the widths of the transition bands of the filters that operate on the decimated signal. In this manner, the overall number of computations required to implement the decimator may be reduced.

(a) If no aliasing is to occur as a result of the decimation in Fig. P7.33-1, what is the maximum allowable decimation factor M in terms of ω_s?

(b) Let $M = 100$, $\omega_s = \pi/100$, and $\omega_p = 0.9\pi/100$ in the system in Fig. P7.33-1. If $x[n] = \delta[n]$, sketch $V(e^{j\omega})$ and $Y(e^{j\omega})$.

Now consider a two-stage implementation of the decimator for $M = 100$, as depicted in Fig. P7.33-3, where $M_1 = 50$, $M_2 = 2$, $\omega_{p_1} = 0.9\pi/100$, $\omega_{p_2} = 0.9\pi/2$, and $\omega_{s_2} = \pi/2$. We must choose ω_{s_1}, or equivalently the transition band of LPF_1, $(\omega_{s_1} - \omega_{p_1})$, such that the two-stage implementation yields the same equivalent passband and stopband frequencies as the single-stage decimator. (We are not concerned about the detailed shape of the

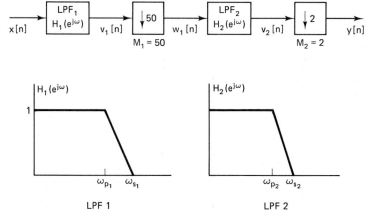

Figure P7.33-3

frequency response in the transition band, except that both systems should have monotonically decreasing response in the transition band.)

(c) For an arbitrary ω_{s_1} and the input $x[n] = \delta[n]$, sketch $V_1(e^{j\omega})$, $W_1(e^{j\omega})$, $V_2(e^{j\omega})$, and $Y(e^{j\omega})$ for the two-stage decimator of Fig. P7.33-3.

(d) Find the *largest* value of ω_{s_1} such that the two-stage decimator yields the same equivalent passband and stopband cutoff frequencies as the single-stage system in part (b).

In addition to a nonzero transition bandwidth, the lowpass filters must also differ from ideal by passband and stopband approximation errors of δ_p and δ_s, respectively. Assume that linear phase equiripple FIR approximations are used. It follows from Eq. (7.134) that for optimum lowpass filters,

$$N \simeq \frac{-10 \log_{10}(\delta_p \delta_s) - 13}{2.324 \Delta \omega} + 1, \qquad \text{(P7.33)}$$

where N is the length of the impulse response and $\Delta \omega = \omega_s - \omega_p$ is the transition band of the lowpass filter. Equation (P7.33) provides the basis for comparison of the two implementations of the decimator. (Equation 7.93 could be used in place of Eq. P7.33 to estimate the impulse response length when the filters are designed by the Kaiser window method.)

(e) Assume that $\delta_p = 0.01$ and $\delta_s = 0.001$ for the lowpass filter in the single-stage implementation. Compute the length of its impulse response N, and determine the number of multiplications required to compute each sample of the output. Take advantage of the symmetry of the impulse response of the linear phase FIR system. (Note that in this decimation application, only every Mth sample of the output need be computed; i.e., the compressor commutes with the multiplications of the FIR system.)

(f) Using the value of ω_{s_1} found in part (d), compute the impulse response lengths N_1 and N_2 of LPF_1 and LPF_2, respectively, in the two-stage decimator of Fig. P7.33-3. Determine the total number of multiplications required to compute each sample of the output in the two-stage decimator.

(g) If the approximation error specifications $\delta_p = 0.01$ and $\delta_s = 0.001$ are used for both filters in the two-stage decimator, the overall passband ripple may be greater than 0.01 since the passband ripples of the two stages can possibly reinforce; e.g., $(1 + \delta_p)(1 + \delta_p) > (1 + \delta_p)$. To compensate for this, the filters in the two-stage implementation can each be designed to have only one-half the passband ripple of the single-stage implementation. Therefore, assume that $\delta_p = 0.005$ and $\delta_s = 0.001$ for each filter in the two-stage decimator. Compute the impulse response lengths N_1 and N_2 of LPF_1 and LPF_2, respectively, and determine the total number of multiplications required to compute each sample of the output.

(h) Should we also reduce the specification on the stopband approximation error for the filters in the two-stage decimator?

(i) *Optional.* The combination of $M_1 = 50$ and $M_2 = 2$ may not yield the smallest total number of multiplications per output sample. Other integer choices for M_1 and M_2 are possible such that $M_1 M_2 = 100$. Determine the values of M_1 and M_2 that minimize the number of multiplications per output sample.

7.34. In this problem we develop a technique for designing discrete-time filters with minimum phase. Such filters have all their poles and zeros inside (or on) the unit circle. (In this problem we allow zeros on the unit circle.) Let us first consider the problem of converting

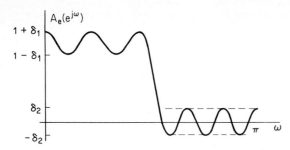

Figure P7.34-1

a type I linear phase FIR equiripple lowpass filter to a minimum-phase system. If $H(e^{j\omega})$ is the frequency response of a type I linear phase filter, then

1. The corresponding impulse response $h[n]$ is real and

$$h[n] = \begin{cases} h[M - n], & 0 \le n \le M, \\ 0, & \text{otherwise,} \end{cases}$$

   where M is an even integer.
2. It follows from (1) that $H(e^{j\omega}) = A_e(e^{j\omega})e^{-j\omega n_0}$, where $A_e(e^{j\omega})$ is real and $n_0 = M/2$ is an integer.
3. The passband ripple is δ_1; i.e., in the passband $A_e(e^{j\omega})$ oscillates between $(1 + \delta_1)$ and $(1 - \delta_1)$ (see Fig. P7.34-1).
4. The stopband ripple is δ_2; i.e., in the stopband $-\delta_2 \le A_e(e^{j\omega}) \le \delta_2$ and $A_e(e^{j\omega})$ oscillates between $-\delta_2$ and $+\delta_2$ (see Fig. P7.34-1).

The following technique was proposed by Herrmann and Schüssler (1970a) for converting this linear phase system into a minimum-phase system that has a system function $H_{min}(z)$ and unit sample response $h_{min}[n]$. (In this problem, we assume that minimum-phase systems can have zeros *on* the unit circle.)

   Step A. Create a new sequence

$$h_1[n] = \begin{cases} h[n], & n \ne n_0, \\ h[n_0] + \delta_2, & n = n_0. \end{cases}$$

   Step B. Recognize that $H_1(z)$ can be expressed in the form

$$H_1(z) = z^{-n_0} H_2(z) H_2(1/z) = z^{-n_0} H_3(z)$$

   for some $H_2(z)$, where $H_2(z)$ has all its poles and zeros inside or on the unit circle and $h_2[n]$ is real.

   Step C. Define

$$H_{min}(z) = \frac{H_2(z)}{\sqrt{1 + \delta_2}}.$$

   The denominator constant $\sqrt{1 + \delta_2}$ normalizes the passband so that the resulting frequency response $H_{min}(e^{j\omega})$ will oscillate about a value of unity.

   (a) Show that if $h_1[n]$ is chosen as in step A, then $H_1(e^{j\omega})$ can be written as

$$H_1(e^{j\omega}) = e^{-j\omega n_0} H_3(e^{j\omega}),$$

where $H_3(e^{j\omega})$ is real and nonnegative for all values of ω.

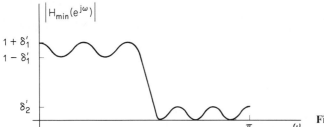

Figure P7.34-2

(b) If $H_3(e^{j\omega}) \geq 0$, as was shown in part (a), show that there exists an $H_2(z)$ such that

$$H_3(z) = H_2(z)H_2(1/z),$$

where $H_2(z)$ is minimum phase and $h_2[n]$ is real (i.e., justify step B).

(c) Demonstrate that the new filter $H_{min}(e^{j\omega})$ is an equiripple lowpass filter (i.e., that its magnitude characteristic is of the form shown in Fig. P7.34-2) by evaluating δ_1' and δ_2'. What is the length of the new impulse response $h_{min}[n]$?

(d) In parts (a), (b), and (c) we assumed that we started with a type I FIR linear phase filter. Will this technique work if we remove the linear phase constraint? Will it work if we use a type II FIR linear phase system?

7.35. Suppose that we have a program that finds the set of coefficients $a[n]$, $n = 0, 1, \ldots, L$ that minimizes

$$\max_{\omega \in F}\left\{\left|W(\omega)[H_d(e^{j\omega}) - \sum_{n=0}^{L} a[n] \cos \omega n]\right|\right\},$$

given L, F, $W(\omega)$, and $H_d(e^{j\omega})$. We have shown that the solution to this optimization problem implies a noncausal FIR zero-phase system with impulse response satisfying $h_e[n] = h_e[-n]$. By delaying $h_e[n]$ by L samples we obtain a causal type I FIR linear phase system with frequency response

$$H(e^{j\omega}) = e^{-j\omega M/2} \sum_{n=0}^{L} a[n] \cos \omega n = \sum_{n=0}^{2L} h[n]e^{-j\omega n},$$

where the impulse response is related to the coefficients $a[n]$ by

$$a[n] = \begin{cases} 2h[M/2 - n] & \text{for } 1 \leq n \leq L, \\ h[M/2] & \text{for } n = 0, \end{cases}$$

and $M = 2L$ is the order of the system function polynomial (the length of the impulse response is $M + 1$).

The other three types (II, III, and IV) of linear phase FIR filters can be designed by the available program if we make suitable modifications to the weighting function $W(\omega)$ and the desired frequency response $H_d(e^{j\omega})$. To see how to do this, it is necessary to manipulate the expressions for the frequency response into the standard form assumed by the program.

(a) Assume that we wish to design a causal type II FIR linear phase system such that $h[n] = h[M - n]$ for $n = 0, 1, \ldots, M$, where M is an odd integer. Show that the frequency response of this type of system can be expressed as

$$H(e^{j\omega}) = e^{-j\omega M/2} \sum_{n=1}^{(M+1)/2} b[n] \cos \omega(n - \tfrac{1}{2}),$$

and determine the relationship between the coefficients $b[n]$ and $h[n]$.

(b) Show that the summation

$$\sum_{n=1}^{(M+1)/2} b[n] \cos \omega(n - \tfrac{1}{2})$$

can be written as

$$\cos(\omega/2) \sum_{n=0}^{(M-1)/2} \tilde{b}[n] \cos \omega n$$

by obtaining an expression for $b[n]$ for $n = 1, 2, \ldots, (M+1)/2$ in terms of $\tilde{b}[n]$ for $n = 0, 1, \ldots, (M-1)/2$. [*Hint*: Note carefully that $b[n]$ is to be expressed in terms of $\tilde{b}[n]$. Also, use the trigonometric identity $\cos \alpha \cos \beta = \tfrac{1}{2} \cos(\alpha + \beta) + \tfrac{1}{2} \cos(\alpha - \beta)$.]

(c) If we wish to use the given program to design type II systems (M even) for a given F, $W(\omega)$, and $H_d(e^{j\omega})$, show how to obtain $\tilde{L}$, $\tilde{F}$, $\tilde{W}(\omega)$, and $\tilde{H}_d(e^{j\omega})$ in terms of M, F, $W(\omega)$, and $H_d(e^{j\omega})$ such that if we run the program using $\tilde{L}$, $\tilde{F}$, $\tilde{W}(\omega)$, and $\tilde{H}_d(e^{j\omega})$ we may use the resulting set of coefficients to determine the impulse response of the desired type II system.

(d) Parts (a)–(c) can be repeated for types III and IV causal linear phase FIR systems where $h[n] = -h[M - n]$. For these cases you must show that for type III systems (M even), the frequency response can be expressed as

$$H(e^{j\omega}) = e^{-j\omega M/2} \sum_{n=1}^{M/2} c[n] \sin \omega n$$

$$= e^{-j\omega M/2} \sin \omega \sum_{n=0}^{(M-2)/2} \tilde{c}[n] \cos \omega n,$$

and for type IV systems (M odd),

$$H(e^{j\omega}) = e^{-j\omega M/2} \sum_{n=1}^{(M+1)/2} d[n] \sin \omega(n - \tfrac{1}{2})$$

$$= e^{-j\omega M/2} \sin(\omega/2) \sum_{n=0}^{(M-1)/2} \tilde{d}[n] \cos \omega n.$$

As in part (b), it is necessary to express $c[n]$ in terms of $\tilde{c}[n]$ and $d[n]$ in terms of $\tilde{d}[n]$ using the trigonometric identity, $\sin \alpha \cos \beta = \tfrac{1}{2} \sin(\alpha + \beta) + \tfrac{1}{2} \sin(\alpha - \beta)$. McClellan and Parks (1973) and Rabiner and Gold (1975) give more detail on issues raised in this problem.

7.36. In this problem we consider a method of obtaining an implementation of a variable-cutoff linear phase filter. Assume that we are given a zero-phase filter designed by the Parks-McClellan method. Its frequency response can be represented as

$$A_e(e^{j\theta}) = \sum_{k=0}^{L} a_k(\cos \theta)^k,$$

and its system function can therefore be represented as

$$A_e(Z) = \sum_{k=0}^{L} a_k \left(\frac{Z + Z^{-1}}{2} \right)^k,$$

with $e^{j\theta} = Z$. (We use Z for the original system and z for the system to be obtained by transformation of the original system.)

(a) Using the above expression for the system function, draw a block diagram or flow graph of an implementation of the system using multiplications by the coefficients a_k, additions, and elemental systems having system function $(Z + Z^{-1})/2$.

(b) What is the length of the impulse response of the system? The overall system can be made causal by cascading the system with a delay of L samples. Distribute this delay as unit delays so that all parts of the network will be causal.

(c) Suppose that we obtain a new system function from $A_e(Z)$ by the substitution

$$B_e(z) = A_e(Z)\Big|_{(Z+Z^{-1})/2 = \alpha_0 + \alpha_1[(z+z^{-1})/2]} .$$

Using the flow graph obtained in part (b), draw the flow graph of a system that implements the system function $B_e(z)$. What is the length of the impulse response of this system? Modify the network as in part (b) to make the overall system and all parts of the network causal.

(d) If $A_e(e^{j\theta})$ is the frequency response of the original filter and $B_e(e^{j\omega})$ is the frequency response of the transformed filter, determine the relationship between θ and ω.

(e) The frequency response of the original optimal filter is shown in Fig. P7.36. For the case $\alpha_1 = 1 - \alpha_0$ and $0 \le \alpha_0 < 1$, describe how the frequency response $B_e(e^{j\omega})$ changes as α_0 varies. [*Hint*: Plot $A_e(e^{j\theta})$ and $B_e(e^{j\omega})$ as functions of $\cos \theta$ and $\cos \omega$. Are the resulting transformed filters also optimal in the sense of having the minimum-maximum weighted approximation errors in the transformed passband and stopband?]

(f) *Optional*. Repeat part (e) for the case $\alpha_1 = 1 + \alpha_0$ and $-1 < \alpha_0 \le 0$.

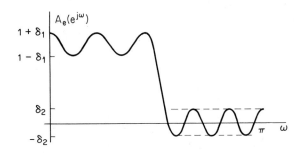

Figure P7.36

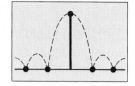

The Discrete Fourier Transform

8

8.0 INTRODUCTION

In Chapters 2 and 4 we discussed the representation of sequences and linear time-invariant systems in terms of the Fourier and z-transforms. For finite-duration sequences, it is possible to develop an alternative Fourier representation, referred to as the *discrete Fourier transform* (DFT). The DFT is itself a sequence rather than a function of a continuous variable, and it corresponds to samples, equally spaced in frequency, of the Fourier transform of the signal. In addition to its theoretical importance as a Fourier representation of sequences, the DFT plays a central role in the implementation of a variety of digital signal processing algorithms. This is because efficient algorithms exist for the computation of the DFT. These algorithms will be discussed in detail in Chapter 9. A variety of applications of the DFT will be described in Chapters 11 and 12.

Although several points of view can be taken toward the derivation and interpretation of the DFT representation of a finite-duration sequence, we have chosen to base our presentation on the relationship between periodic sequences and finite-length sequences. We will begin by considering the Fourier series representation of periodic sequences. While this representation is important in its own right, we are most often interested in the application of Fourier series results to the representation of finite-length sequences. We accomplish this by constructing a periodic sequence for which each period is identical to the finite-length sequence. As we will see, the Fourier series representation of the periodic sequence corresponds to the DFT of the finite-length sequence. Thus our approach is to define the Fourier series representation for periodic sequences and to study the properties of such representations. Then we repeat essentially the same derivations assuming that the sequence to be represented is a finite-length sequence. This approach to the DFT emphasizes the fundamental inherent periodicity of the DFT representation and ensures that this periodicity is not overlooked in applications of the DFT.

8.1 REPRESENTATION OF PERIODIC SEQUENCES: THE DISCRETE FOURIER SERIES

Consider a sequence $\tilde{x}[n]$ that is periodic† with period N so that $\tilde{x}[n] = \tilde{x}[n + rN]$ for any integer value of r. As with continuous-time periodic signals, such a sequence can be represented by a Fourier series corresponding to a sum of harmonically related complex exponential sequences, i.e., complex exponentials with frequencies that are integer multiples of the fundamental frequency $(2\pi/N)$ associated with the periodic sequence $\tilde{x}[n]$. These periodic complex exponentials $e_k[n]$ are of the form

$$e_k[n] = e^{j(2\pi/N)kn} = e_k[n + rN], \tag{8.1}$$

where k is an integer, and the Fourier series representation then has the form‡

$$\tilde{x}[n] = \frac{1}{N} \sum_k \tilde{X}[k] e^{j(2\pi/N)kn}. \tag{8.2}$$

The Fourier series representation of a continuous-time periodic signal generally requires infinitely many harmonically related complex exponentials, whereas the Fourier series for any discrete-time signal with period N requires only N harmonically related complex exponentials. To see this, note that the harmonically related complex exponentials $e_k[n]$ in Eq. (8.1) are identical for values of k separated by N, i.e., $e_0[n] = e_N[n]$, $e_1[n] = e_{N+1}[n]$, and, in general,

$$e_k[n] = e_{k+\ell N}[n], \tag{8.3}$$

where ℓ is an integer. Consequently, the set of N periodic complex exponentials $e_0[n]$, $e_1[n]$, ..., $e_{N-1}[n]$ defines all the distinct periodic complex exponentials with frequencies that are integer multiples of $(2\pi/N)$. Thus the Fourier series representation of a periodic sequence $\tilde{x}[n]$ need contain only N of these complex exponentials and hence it has the form

$$\tilde{x}[n] = \frac{1}{N} \sum_{k=0}^{N-1} \tilde{X}[k] e^{j(2\pi/N)kn}. \tag{8.4}$$

To obtain the sequence of Fourier series coefficients $\tilde{X}[k]$ from the periodic sequence $\tilde{x}[n]$, we exploit the orthogonality of the set of complex exponential

---

† Henceforth we will use the tilde (˜) to denote periodic sequences whenever it is important to clearly distinguish periodic and aperiodic sequences.

‡ The multiplying constant $1/N$ is included in Eq. (8.2) for convenience. It could also be absorbed into the definition of $\tilde{X}[k]$.

sequences. After multiplying both sides of Eq. (8.4) by $e^{-j(2\pi/N)rn}$ and summing from $n = 0$ to $n = N - 1$, we obtain

$$\sum_{n=0}^{N-1} \tilde{x}[n] e^{-j(2\pi/N)rn} = \sum_{n=0}^{N-1} \frac{1}{N} \sum_{k=0}^{N-1} \tilde{X}[k] e^{j(2\pi/N)(k-r)n}. \qquad (8.5)$$

After interchanging the order of summation on the right-hand side, Eq. (8.5) becomes

$$\sum_{n=0}^{N-1} \tilde{x}[n] e^{-j(2\pi/N)rn} = \sum_{k=0}^{N-1} \tilde{X}[k] \left[\frac{1}{N} \sum_{n=0}^{N-1} e^{j(2\pi/N)(k-r)n} \right]. \qquad (8.6)$$

The following identity expresses the orthogonality of the complex exponentials:

$$\frac{1}{N} \sum_{n=0}^{N-1} e^{j(2\pi/N)(k-r)n} = \begin{cases} 1, & k - r = mN, \quad m \text{ an integer,} \\ 0, & \text{otherwise.} \end{cases} \qquad (8.7)$$

The identity can be proved easily (see Problem 8.3), and when applied to the summation in brackets in Eq. (8.6), the result is

$$\sum_{n=0}^{N-1} \tilde{x}[n] e^{-j(2\pi/N)rn} = \tilde{X}[r]. \qquad (8.8)$$

Thus the Fourier series coefficients $\tilde{X}[k]$ in Eq. (8.4) are obtained from $\tilde{x}[n]$ by the relation

$$\tilde{X}[k] = \sum_{n=0}^{N-1} \tilde{x}[n] e^{-j(2\pi/N)kn}. \qquad (8.9)$$

Note that the sequence $\tilde{X}[k]$ as given by Eq. (8.9) is periodic with period N; i.e., $\tilde{X}[0] = \tilde{X}[N]$, $\tilde{X}[1] = \tilde{X}[N+1]$, and, more generally, $\tilde{X}[k] = \tilde{X}[N+k]$ for any integer k.

The Fourier series coefficients can be interpreted to be a sequence of finite length, given by Eq. (8.9) for $k = 0, \ldots, (N-1)$, and zero otherwise or as a periodic sequence defined for all k by Eq. (8.9). Clearly both of these interpretations are acceptable since in Eq. (8.4) we use only the values of $\tilde{X}[k]$ for $0 \le k \le (N-1)$. An advantage of interpreting the Fourier series coefficients $\tilde{X}[k]$ as a periodic sequence is that there is then a duality between the time and frequency domains for the Fourier series representation of periodic sequences. Equations (8.9) and (8.4) together are an analysis/synthesis pair and will be referred to as the *discrete Fourier series* (DFS) representation of a periodic sequence. For convenience in notation, these equations are often written in terms of the complex quantity W_N, defined as

$$W_N = e^{-j(2\pi/N)}. \qquad (8.10)$$

With this notation, the DFS analysis/synthesis pair is expressed as follows:

Analysis equation: $$\tilde{X}[k] = \sum_{n=0}^{N-1} \tilde{x}[n] W_N^{kn}, \qquad (8.11)$$

Synthesis equation: $$\tilde{x}[n] = \frac{1}{N} \sum_{k=0}^{N-1} \tilde{X}[k] W_N^{-kn}, \qquad (8.12)$$

where both $\tilde{X}[k]$ and $\tilde{x}[n]$ are periodic sequences. We will sometimes find it convenient to use the notation

$$\tilde{x}[n] \overset{\mathcal{DFS}}{\longleftrightarrow} \tilde{X}[k] \tag{8.13}$$

to signify the relationships of Eqs. (8.11) and (8.12). The following examples illustrate the use of Eqs. (8.11) and (8.12).

Example 8.1

Let $\tilde{x}[n]$ be the periodic impulse train

$$\tilde{x}[n] = \sum_{r=-\infty}^{\infty} \delta[n + rN]. \tag{8.14}$$

Since $\tilde{x}[n] = \delta[n]$ for $0 \le n \le N-1$, the DFS coefficients for $\tilde{x}[n]$ are found, using Eq. (8.11), to be

$$\tilde{X}[k] = \sum_{n=0}^{N-1} \delta[n] W_N^{kn} = 1. \tag{8.15}$$

In this case $\tilde{X}[k]$ is the same for all k. Thus, substituting Eq. (8.15) into Eq. (8.12) leads to the representation

$$\tilde{x}[n] = \sum_{r=-\infty}^{\infty} \delta[n + rN] = \frac{1}{N} \sum_{k=0}^{N-1} W_N^{-kn} = \frac{1}{N} \sum_{k=0}^{N-1} e^{j(2\pi/N)kn}, \tag{8.16}$$

which is the orthogonality relation of Eq. (8.7) in a slightly different form.

Example 8.1 produced a useful representation of the periodic impulse train in terms of a sum of complex exponentials, where all the complex exponentials have the same magnitude and phase and they add up to unity at integer multiples of N and to zero for all other integers (i.e., the periodic impulse train of Eq. 8.14). The following example is slightly more complicated.

Example 8.2

For this example, $\tilde{x}[n]$ is the sequence shown in Fig. 8.1, where the period is $N = 10$. From Eq. (8.11),

$$\tilde{X}[k] = \sum_{n=0}^{4} W_{10}^{kn} = \sum_{n=0}^{4} e^{-j(2\pi/10)kn}. \tag{8.17}$$

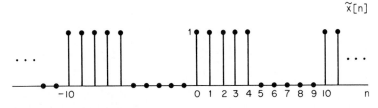

Figure 8.1 Periodic sequence with period $N = 10$ for which the Fourier series representation is to be computed.

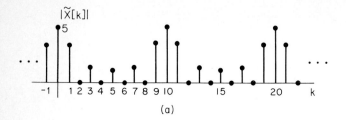

(a)

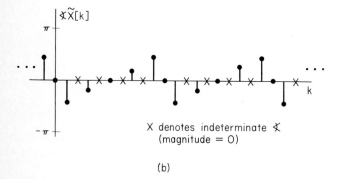

X denotes indeterminate ∢
(magnitude = 0)

(b)

Figure 8.2 Magnitude and phase of the Fourier series coefficients of the sequence of Fig. 8.1.

This finite sum has the closed form

$$\tilde{X}[k] = \frac{1 - W_{10}^{5k}}{1 - W_{10}^{k}} = e^{-j(4\pi k/10)}\frac{\sin(\pi k/2)}{\sin(\pi k/10)}. \tag{8.18}$$

The magnitude and phase of the periodic sequence $\tilde{X}[k]$ are sketched in Fig. 8.2.

The periodic sequence $\tilde{X}[k]$ in Eq. (8.11) has a convenient interpretation as equally spaced samples of the Fourier transform of one period of $\tilde{x}[n]$. To obtain this relationship, let $x[n]$ represent one period of $\tilde{x}[n]$; i.e.,

$$x[n] = \begin{cases} \tilde{x}[n], & 0 \le n \le N - 1, \\ 0, & \text{otherwise,} \end{cases} \tag{8.19}$$

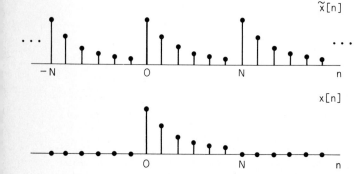

Figure 8.3 Finite-length sequence $x[n]$ that is equal to the periodic sequence $\tilde{x}[n]$ over one period and is zero otherwise.

as illustrated in Fig. 8.3. Then, since $x[n] = \tilde{x}[n]$ for $0 \le n \le N - 1$ and $x[n] = 0$ otherwise, $X(e^{j\omega})$ is

$$X(e^{j\omega}) = \sum_{n=0}^{N-1} x[n]e^{-j\omega n} = \sum_{n=0}^{N-1} \tilde{x}[n]e^{-j\omega n}. \qquad (8.20)$$

Comparing Eq. (8.20) and Eq. (8.11), we see that

$$\tilde{X}[k] = X(e^{j\omega})|_{\omega = 2\pi k/N}. \qquad (8.21)$$

This corresponds to sampling the Fourier transform at N equally spaced frequencies between $\omega = 0$ and $\omega = 2\pi$ with a frequency spacing of $2\pi/N$.

Example 8.3

As an illustration of the relationship between the Fourier series coefficients and the Fourier transform of one period, let us again consider the sequence $\tilde{x}[n]$ shown in Fig. 8.1. One period of $\tilde{x}[n]$ for the sequence in Fig. 8.1 is

$$x[n] = \begin{cases} 1, & 0 \le n \le 4, \\ 0, & \text{otherwise.} \end{cases} \qquad (8.22)$$

The Fourier transform of one period of $\tilde{x}[n]$ is given by

$$X(e^{j\omega}) = \sum_{n=0}^{4} e^{-j\omega n} = e^{-j2\omega} \frac{\sin(5\omega/2)}{\sin(\omega/2)}. \qquad (8.23)$$

Equation (8.21) can be shown to be satisfied for this example by substituting $\omega = 2\pi k/10$ into Eq. (8.23) and comparing the result with Eq. (8.18). The magnitude and phase of $X(e^{j\omega})$ are sketched in Fig. 8.4. Note that the phase is discontinuous at the frequencies where $X(e^{j\omega}) = 0$. That the sequences in Figs. 8.2(a) and (b) correspond to samples of Figs. 8.4(a) and (b), respectively, is demonstrated in Fig. 8.5, where Figs. 8.2 and 8.4 have been superimposed.

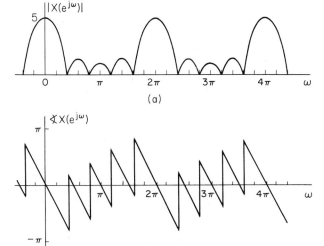

Figure 8.4   Magnitude and phase of the Fourier transform of one period of the sequence in Fig. 8.1.

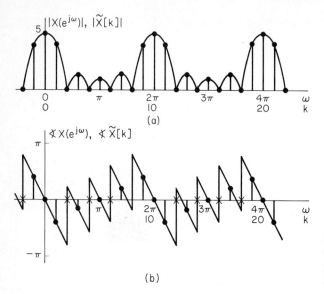

(a)

(b)

Figure 8.5 Overlay of Figs. 8.2 and 8.4 illustrating the DFS coefficients of a periodic sequence as samples of the Fourier transform of one period.

8.2 PROPERTIES OF THE DISCRETE FOURIER SERIES

Just as with Fourier series and Fourier and Laplace transforms for continuous-time signals and with the Fourier and z-transforms for discrete-time aperiodic sequences, certain properties of the discrete Fourier series are of fundamental importance to its successful use in signal processing problems. In this section we summarize these important properties. It is not surprising that many of the basic properties are analogous to properties of the z-transform and Fourier transform. However, we will be careful to point out where the periodicity of both $\tilde{x}[n]$ and $\tilde{X}[k]$ results in some important distinctions. Furthermore, an exact duality exists between the time and frequency domains in the DFS representation that does not exist in the Fourier transform and z-transform representation of sequences.

8.2.1 Linearity

Consider two periodic sequences $\tilde{x}_1[n]$ and $\tilde{x}_2[n]$, both with period N, such that

$$\tilde{x}_1[n] \overset{\mathcal{DFS}}{\longleftrightarrow} \tilde{X}_1[k] \tag{8.24a}$$

and

$$\tilde{x}_2[n] \overset{\mathcal{DFS}}{\longleftrightarrow} \tilde{X}_2[k]. \tag{8.24b}$$

Then

$$a\tilde{x}_1[n] + b\tilde{x}_2[n] \overset{\mathcal{DFS}}{\longleftrightarrow} a\tilde{X}[k] + b\tilde{X}_2[k]. \tag{8.25}$$

This linearity property follows immediately from the form of Eqs. (8.11) and (8.12).

8.2.2 Shift of a Sequence

If a periodic sequence $\tilde{x}[n]$ has Fourier coefficients $\tilde{X}[k]$, then $\tilde{x}[n-m]$ is a shifted version of $\tilde{x}[n]$ and

$$\tilde{x}[n-m] \overset{\mathcal{DFS}}{\longleftrightarrow} W_N^{km}\tilde{X}[k]. \tag{8.26}$$

The proof of this property is considered in Problem 8.4. Any shift that is greater than the period (i.e., $m \geq N$) cannot be distinguished in the time domain from a shorter shift m_1 such that $m = m_1 + m_2 N$, where m_1 and m_2 are integers and $0 \leq m_1 \leq N-1$. (Another way of stating this is that $m_1 = m$ modulo N or, equivalently, m_1 is the remainder when m is divided by N.) It is easily shown that with this representation of m, $W_N^{km} = W_N^{km_1}$; i.e., as it must be, the ambiguity of the shift in the time domain is also manifest in the frequency-domain representation.

Because the sequence of Fourier series coefficients of a periodic sequence is a periodic sequence, a similar result applies to a shift in the Fourier coefficients by an integer ℓ. Specifically,

$$W_N^{-n\ell}\tilde{x}[n] \overset{\mathcal{DFS}}{\longleftrightarrow} \tilde{X}[k-\ell]. \tag{8.27}$$

Note the difference in the sign of the exponents in Eqs. (8.26) and (8.27).

8.2.3 Duality

Because of the strong similarity between the Fourier analysis and synthesis equations in continuous time, there is a duality between the time domain and frequency domain. For the discrete-time Fourier transform of aperiodic signals, a similar duality does not exist since aperiodic signals and their Fourier transforms are very different kinds of functions; aperiodic discrete-time signals are, of course, aperiodic sequences, while their Fourier transforms are always periodic functions of a continuous frequency variable.

From Eqs. (8.11) and (8.12), we see that the DFS analysis and synthesis equations differ only in a factor of $1/N$ and in the sign of the exponent of W_N. Furthermore, a periodic sequence and its DFS coefficients are the same kinds of functions; they are both periodic sequences. Specifically, taking account of the factor $1/N$ and the difference in sign in the exponent between Eqs. (8.11) and (8.12), it follows from Eq. (8.12) that

$$N\tilde{x}[-n] = \sum_{k=0}^{N-1} \tilde{X}[k]W_N^{kn} \tag{8.28}$$

or, interchanging the roles of n and k,

$$N\tilde{x}[-k] = \sum_{n=0}^{N-1} \tilde{X}[n]W_N^{nk}. \tag{8.29}$$

We see that Eq. (8.29) is similar to Eq. (8.11). In other words, the sequence of DFS coefficients of the periodic sequence $\tilde{X}[n]$ is $N\tilde{x}[-k]$, i.e., the original periodic sequence order-reversed and multiplied by N. This duality property is summarized as follows: If

$$\tilde{x}[n] \overset{\mathcal{DFS}}{\longleftrightarrow} \tilde{X}[k], \tag{8.30a}$$

then

$$\tilde{X}[n] \overset{\mathcal{DFS}}{\longleftrightarrow} N\tilde{x}[-k]. \tag{8.30b}$$

8.2.4 Symmetry Properties

As we discussed for the Fourier transform in Section 2.8, the DFS representation of a periodic sequence has a number of important symmetry properties. The derivation of these properties, which is similar in style to the derivations in Chapter 2, is left as an exercise (see Problem 8.5). The resulting properties are summarized as properties 9–17 in Table 8.1 in Section 8.3.

8.2.5 Periodic Convolution

Let $\tilde{x}_1[n]$ and $\tilde{x}_2[n]$ be two periodic sequences each with period N and with discrete Fourier series coefficients denoted by $\tilde{X}_1[k]$ and $\tilde{X}_2[k]$, respectively. If we form the product

$$\tilde{X}_3[k] = \tilde{X}_1[k]\tilde{X}_2[k], \tag{8.31}$$

then the periodic sequence $\tilde{x}_3[n]$ with Fourier series coefficients $\tilde{X}_3[k]$ is

$$\tilde{x}_3[n] = \sum_{m=0}^{N-1} \tilde{x}_1[m]\tilde{x}_2[n-m]. \tag{8.32}$$

This result is not surprising since our previous experience with transforms suggests that multiplication of frequency-domain functions corresponds to convolution of time-domain functions, and Eq. (8.32) looks very much like a convolution sum. Equation (8.32) involves the summation of the product of $\tilde{x}_1[m]$ and $\tilde{x}_2[n-m]$, which is a time-reversed and time-shifted version of $\tilde{x}_2[n]$ just as in aperiodic discrete convolution. However, the sequences in Eq. (8.32) are all periodic with period N, and the summation is over only one period. A convolution in the form of Eq. (8.32) is referred to as a *periodic convolution*. Just as with aperiodic convolution, periodic convolution is commutative; i.e.,

$$\tilde{x}_3[n] = \sum_{m=0}^{N-1} \tilde{x}_2[m]\tilde{x}_1[n-m]. \tag{8.33}$$

To demonstrate that $\tilde{X}_3[k]$ given by Eq. (8.31) is the sequence of Fourier coefficients corresponding to $\tilde{x}_3[n]$ given by Eq. (8.32), let us first apply the DFS analysis equation in Eq. (8.11) to Eq. (8.32) to obtain

$$\tilde{X}_3[k] = \sum_{n=0}^{N-1} W_N^{kn}\left(\sum_{m=0}^{N-1} \tilde{x}_1[m]\tilde{x}_2[n-m]\right), \tag{8.34}$$

which, after interchanging the order of summation, becomes

$$\tilde{X}_3[k] = \sum_{m=0}^{N-1} \tilde{x}_1[m] \sum_{n=0}^{N-1} \tilde{x}_2[n - m] W_N^{kn}. \tag{8.35}$$

The inner sum on the index n is the DFS for the shifted sequence $x_2[n - m]$. Therefore from the shifting property of Section 8.2.2, we obtain

$$\sum_{n=0}^{N-1} \tilde{x}_2[n - m] W_N^{kn} = W_N^{km} \tilde{X}_2[k],$$

which can be substituted into Eq. (8.35) to yield

$$\tilde{X}_3[k] = \sum_{m=0}^{N-1} \tilde{x}_1[m] W_N^{km} \tilde{X}_2[k] = \left(\sum_{m=0}^{N-1} \tilde{x}_1[m] W_N^{km} \right) \tilde{X}_2[k] = \tilde{X}_1[k]\tilde{X}_2[k]. \tag{8.36}$$

In summary,

$$\sum_{m=0}^{N-1} \tilde{x}_1[m]\tilde{x}_2[n - m] \xleftrightarrow{\;\mathcal{DFS}\;} \tilde{X}_1[k]\tilde{X}_2[k]. \tag{8.37}$$

The periodic convolution of periodic sequences thus corresponds to multiplication of the corresponding sequences of Fourier series coefficients.

Since periodic convolutions are somewhat different from aperiodic convolutions, it is worthwhile to consider the mechanics of evaluating Eq. (8.32). First note that Eq. (8.32) calls for the product of sequences $\tilde{x}_1[m]$ and $\tilde{x}_2[n - m] = \tilde{x}_2[-(m - n)]$ viewed as functions of m with n fixed. This is the same as for an aperiodic convolution with the following two major differences:

1. The sum is over the finite interval $0 \le m \le N - 1$.
2. The values of $\tilde{x}_2[n - m]$ in the interval $0 \le m \le N - 1$ repeat periodically for m outside that interval.

These details are illustrated by the following example.

Example 8.4

An illustration of the procedure for forming the periodic convolution of two periodic sequences corresponding to Eq. (8.32) is given in Fig. 8.6, where we have illustrated the sequences $\tilde{x}_2[m]$, $\tilde{x}_1[m]$, $\tilde{x}_2[-m]$, $\tilde{x}_2[1 - m] = \tilde{x}_2[-(m - 1)]$, and $\tilde{x}_2[2 - m] = \tilde{x}_2[-(m - 2)]$. To evaluate $\tilde{x}_3[n]$ in Eq. (8.32) for $n = 2$, for example, we multiply $\tilde{x}_1[m]$ and $\tilde{x}_2[2 - m]$ and then sum the product terms $\tilde{x}_1[m]\tilde{x}_2[2 - m]$ for $0 \le m \le N - 1$ to obtain $\tilde{x}_3[2]$. As n changes, the sequence $\tilde{x}_2[n - m]$ shifts appropriately, and Eq. (8.32) is evaluated for each value of $0 \le n \le N - 1$. Note that as the sequence $\tilde{x}_2[n - m]$ shifts to the right or left, values that leave the interval between the dotted lines at one end reappear at the other end because of the periodicity. Because of the periodicity of $\tilde{x}_3[n]$, there is no need to continue to evaluate Eq. (8.32) outside the interval $0 \le n \le N - 1$.

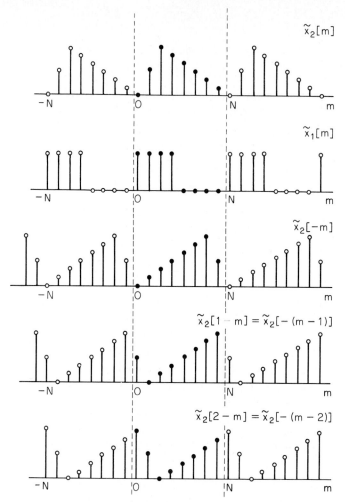

Figure 8.6 Procedure for forming the periodic convolution of two periodic sequences.

The duality theorem (Section 8.2.3) suggests that if the roles of time and frequency are interchanged, we will obtain a result almost identical to the previous result. That is, the periodic sequence

$$\tilde{x}_3[n] = \tilde{x}_1[n]\tilde{x}_2[n], \tag{8.38}$$

where $\tilde{x}_1[n]$ and $\tilde{x}_2[n]$ are periodic sequences each with period N, has the discrete Fourier series coefficients given by

$$\tilde{X}_3[k] = \frac{1}{N}\sum_{\ell=0}^{N-1}\tilde{X}_1[\ell]\tilde{X}_2[k-\ell], \tag{8.39}$$

corresponding to $1/N$ times the periodic convolution of $\tilde{X}_1[k]$ and $\tilde{X}_2[k]$. This result can also be verified by substituting $\tilde{X}_3[k]$ given by Eq. (8.39) into the Fourier series relation of Eq. (8.12) to obtain $\tilde{x}_3[n]$.

8.3 SUMMARY OF PROPERTIES OF THE DFS REPRESENTATION OF PERIODIC SEQUENCES

The properties of the discrete Fourier series representation discussed in Section 8.2 are summarized in Table 8.1.

TABLE 8.1

Periodic Sequence (Period N)	DFS Coefficients (Period N)
1. $\tilde{x}[n]$	$\tilde{X}[k]$ periodic with period N
2. $\tilde{x}_1[n], \tilde{x}_2[n]$	$\tilde{X}_1[k], \tilde{X}_2[k]$ periodic with period N
3. $a\tilde{x}_1[n] + b\tilde{x}_2[n]$	$a\tilde{X}_1[k] + b\tilde{X}_2[k]$
4. $\tilde{X}[n]$	$N\tilde{x}[-k]$
5. $\tilde{x}[n-m]$	$W_N^{km}\tilde{X}[k]$
6. $W_N^{-\ell n}\tilde{x}[n]$	$\tilde{X}[k-\ell]$
7. $\displaystyle\sum_{m=0}^{N-1}\tilde{x}_1[m]\tilde{x}_2[n-m]$ (periodic convolution)	$\tilde{X}_1[k]\tilde{X}_2[k]$
8. $\tilde{x}_1[n]\tilde{x}_2[n]$	$\dfrac{1}{N}\displaystyle\sum_{\ell=0}^{N-1}\tilde{X}_1[\ell]\tilde{X}_2[k-\ell]$ (periodic convolution)
9. $\tilde{x}^*[n]$	$\tilde{X}^*[-k]$
10. $\tilde{x}^*[-n]$	$\tilde{X}^*[k]$
11. $\mathcal{R}e\{\tilde{x}[n]\}$	$\tilde{X}_e[k] = \frac{1}{2}(\tilde{X}[k] + \tilde{X}^*[-k])$
12. $j\mathcal{I}m\{\tilde{x}[n]\}$	$\tilde{X}_o[k] = \frac{1}{2}(\tilde{X}[k] - \tilde{X}^*[-k])$
13. $\tilde{x}_e[n] = \frac{1}{2}(\tilde{x}[n] + \tilde{x}^*[-n])$	$\mathcal{R}e\{\tilde{X}[k]\}$
14. $\tilde{x}_o[n] = \frac{1}{2}(\tilde{x}[n] - \tilde{x}^*[-n])$	$j\mathcal{I}m\{\tilde{X}[k]\}$

Properties 15–17 apply only when $x[n]$ is real.

15. Symmetry properties	$\begin{cases} \tilde{X}[k] = \tilde{X}^*[-k] \\ \mathcal{R}e\{\tilde{X}[k]\} = \mathcal{R}e\{\tilde{X}[-k]\} \\ \mathcal{I}m\{\tilde{X}[k]\} = -\mathcal{I}m\{\tilde{X}[-k]\} \\ |\tilde{X}[k]| = |\tilde{X}[-k]| \\ \sphericalangle\tilde{X}[k] = -\sphericalangle\tilde{X}[-k] \end{cases}$

16. $\tilde{x}_e[n] = \frac{1}{2}(\tilde{x}[n] + \tilde{x}[-n])$	$\mathcal{R}e\{\tilde{X}[k]\}$
17. $\tilde{x}_o[n] = \frac{1}{2}(\tilde{x}[n] - \tilde{x}[-n])$	$j\mathcal{I}m\{\tilde{X}[k]\}$

8.4 THE FOURIER TRANSFORM OF PERIODIC SIGNALS

As discussed in Section 2.7, uniform convergence of the Fourier transform of a sequence requires that the sequence be absolutely summable, and mean-square convergence requires that the sequence be square summable. Periodic sequences satisfy neither condition because they do not approach zero as n approaches $\pm\infty$. However, as we discussed briefly in Section 2.7, sequences that can be expressed as a sum of complex exponentials can be considered to have a Fourier transform representation in the form of Eq. (2.131), i.e., as a train of impulses. Similarly, it is often useful to incorporate the discrete Fourier series representation of periodic signals within the framework of the Fourier transform. This can be done by interpreting the Fourier transform of a periodic signal to be an impulse train in the frequency domain with the impulse values proportional to the DFS coefficients for the sequence. Specifically, if $\tilde{x}[n]$ is periodic with period N and the corresponding discrete Fourier series coefficients are $\tilde{X}[k]$, then the Fourier transform of $\tilde{x}[n]$ is defined to be the impulse train

$$\tilde{X}(e^{j\omega}) \triangleq \sum_{k=-\infty}^{\infty} \frac{2\pi}{N} \tilde{X}[k]\delta\left(\omega - \frac{2\pi k}{N}\right). \tag{8.40}$$

Note that $\tilde{X}(e^{j\omega})$ has the necessary periodicity with period 2π since $\tilde{X}[k+N] = \tilde{X}[k]$ and the impulses are spaced at integer multiples of $2\pi/N$, where N is an integer. To show that $\tilde{X}(e^{j\omega})$ as defined in Eq. (8.40) is a Fourier transform representation of the periodic sequence $\tilde{x}[n]$, we substitute Eq. (8.40) into the inverse Fourier transform equation (2.112); i.e.,

$$\frac{1}{2\pi}\int_{0-}^{2\pi-} \tilde{X}(e^{j\omega})e^{j\omega n}\,d\omega = \frac{1}{2\pi}\int_{0-}^{2\pi-} \sum_{k=-\infty}^{\infty} \frac{2\pi}{N} \tilde{X}[k]\delta\left(\omega - \frac{2\pi k}{N}\right)e^{j\omega n}\,d\omega. \tag{8.41}$$

Recall that in evaluating the inverse Fourier transform, we can integrate over *any* interval of length 2π since the integrand $\tilde{X}(e^{j\omega})e^{j\omega n}$ is periodic with period 2π. In Eq. (8.41) the integration limits are denoted $0-$ and $2\pi-$, which means that the integration is from just before $\omega = 0$ to just before $\omega = 2\pi$. These limits are convenient because they include the impulse at $\omega = 0$ and exclude the impulse at $\omega = 2\pi$. Interchanging the order of integration and summation leads to

$$\frac{1}{2\pi}\int_{0-}^{2\pi-} \tilde{X}(e^{j\omega})e^{j\omega n}\,d\omega = \frac{1}{N}\sum_{k=-\infty}^{\infty} \tilde{X}[k]\int_{0-}^{2\pi-} \delta\left(\omega - \frac{2\pi k}{N}\right)e^{j\omega n}\,d\omega$$
$$= \frac{1}{N}\sum_{k=0}^{N-1} \tilde{X}[k]e^{j(2\pi/N)kn}. \tag{8.42}$$

The final form of Eq. (8.42) results because only the impulses corresponding to $k = 0, 1, \ldots, (N-1)$ are included in the interval between $\omega = 0-$ and $\omega = 2\pi-$.

Comparing Eq. (8.42) and Eq. (8.12) we see that the right-hand side of Eq. (8.42) is exactly equal to the Fourier series representation for $\tilde{x}[n]$ as specified by Eq. (8.12). Consequently the inverse Fourier transform of the impulse train in Eq. (8.40) is the periodic signal $\tilde{x}[n]$ as desired.

Although the Fourier transform of a periodic sequence does not converge in the normal sense, the introduction of impulses permits us to include periodic sequences within the framework of Fourier transform analysis. This approach was also used in Chapter 2 to obtain a Fourier transform representation of other nonsummable sequences such as the two-sided constant sequence or the general two-sided sine or cosine sequence. Although the discrete Fourier series representation is adequate for most purposes, the Fourier transform representation of Eq. (8.40) sometimes leads to simpler or more compact expressions and simplified analysis.

8.5 SAMPLING THE FOURIER TRANSFORM

In this section we discuss the general relationship between an aperiodic sequence with Fourier transform $X(e^{j\omega})$ and the periodic sequence for which the DFS coefficients correspond to samples of $X(e^{j\omega})$ equally spaced in frequency. We will find this relationship to be particularly important when we discuss the discrete Fourier transform and its properties later in this chapter.

Consider an aperiodic sequence $x[n]$ with Fourier transform $X(e^{j\omega})$, and assume that a sequence $\tilde{X}[k]$ is obtained by sampling $X(e^{j\omega})$ at frequencies $\omega_k = 2\pi k/N$; i.e.,

$$\tilde{X}[k] = X(e^{j\omega})|_{\omega=(2\pi/N)k} = X(e^{j(2\pi/N)k}). \tag{8.43}$$

Since the Fourier transform is periodic in ω with period 2π, the resulting sequence is periodic in k with period N. Also, since the Fourier transform is equal to the z-transform evaluated on the unit circle, it follows that $\tilde{X}[k]$ can also be obtained by sampling $X(z)$ at N equally spaced points on the unit circle. Thus,

$$\tilde{X}[k] = X(z)|_{z=e^{j(2\pi/N)k}} = X(e^{j(2\pi/N)k}). \tag{8.44}$$

These sampling points are depicted in Fig. 8.7 for $N = 8$.

We see that the sequence of samples $\tilde{X}[k]$, being periodic with period N, *could* be the sequence of discrete Fourier series coefficients of a sequence $\tilde{x}[n]$. To obtain that sequence, we can simply substitute $\tilde{X}[k]$ into Eq. (8.12):

$$\tilde{x}[n] = \frac{1}{N} \sum_{k=0}^{N-1} \tilde{X}[k] W_N^{-kn}. \tag{8.45}$$

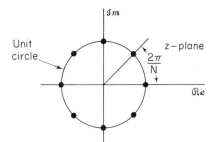

Figure 8.7 Points on the unit circle at which $X(z)$ is sampled to obtain the periodic sequence $\tilde{X}[k]$ ($N = 8$).

Since we have made no assumption about $x[n]$ other than that the Fourier transform exists, $X(e^{j\omega})$ is

$$X(e^{j\omega}) = \sum_{m=-\infty}^{\infty} x[m]e^{-j\omega m}. \tag{8.46}$$

Substituting Eq. (8.46) into Eq. (8.43) and then substituting the resulting expression for $\tilde{X}[k]$ into Eq. (8.45) gives

$$\tilde{x}[n] - \frac{1}{N} \sum_{k=0}^{N-1} \left[\sum_{m=-\infty}^{\infty} x[m]e^{-j(2\pi/N)km} \right] W_N^{-kn}, \tag{8.47}$$

which becomes, after interchanging the order of summation,

$$\tilde{x}[n] = \sum_{m=-\infty}^{\infty} x[m] \left[\frac{1}{N} \sum_{k=0}^{N-1} W_N^{-k(n-m)} \right]. \tag{8.48}$$

The term in brackets in Eq. (8.48) can be seen from either Eq. (8.7) or Eq. (8.16) to be the Fourier series representation of the periodic impulse train of Example 8.1. Specifically,

$$\frac{1}{N} \sum_{k=0}^{N-1} W_N^{-k(n-m)} = \sum_{r=-\infty}^{\infty} \delta[n - m + rN] \tag{8.49}$$

and, therefore,

$$\tilde{x}[n] = x[n] * \sum_{r=-\infty}^{\infty} \delta[n + rN] = \sum_{r=-\infty}^{\infty} x[n + rN], \tag{8.50}$$

where $*$ denotes aperiodic convolution. That is, $\tilde{x}[n]$ is the periodic sequence that results from the aperiodic convolution of $x[n]$ with a periodic unit-impulse train. Thus the periodic sequence $\tilde{x}[n]$, corresponding to $\tilde{X}[k]$ obtained by sampling $X(e^{j\omega})$, is formed from $x[n]$ by adding together an infinite number of shifted replicas of $x[n]$. The shifts are positive and negative integer multiples of N, the period of the sequence $\tilde{X}[k]$. This is illustrated in Fig. 8.8. In the example of Fig. 8.8, the sequence $x[n]$ is of length 9 and the value of N in Eq. (8.50) is $N = 12$. Consequently, the delayed replications of $x[n]$ do not overlap, and one period of the periodic sequence $\tilde{x}[n]$ is recognizable as $x[n]$. This is consistent with the discussion in Section 8.1 where we showed that the Fourier series coefficients for a periodic sequence are samples of the Fourier transform of one period. In Fig. 8.9 the same sequence $x[n]$ is used but the value of N is now $N = 7$. In this case the replicas of $x[n]$ overlap and one period of $\tilde{x}[n]$ is no longer identical to $x[n]$. In both cases, however, Eq. (8.43) still holds, i.e., in both cases the DFS coefficients of $\tilde{x}[n]$ are samples of the Fourier transform of $x[n]$ spaced in frequency at integer multiples of $2\pi/N$. This discussion should be reminiscent of our discussion of sampling in Chapter 3. The difference is that here we are sampling in the frequency domain rather than in the time domain. However, the general outlines of the mathematical representations are very similar.

For the example in Fig. 8.8, the original sequence $x[n]$ can be recovered from $\tilde{x}[n]$ by extracting one period. Equivalently, the Fourier transform $X(e^{j\omega})$ can be

x[n]

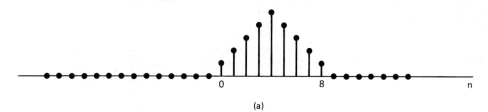

(a)

$$\widetilde{x}[n] = \sum_{r=-\infty}^{\infty} x[n + r/2]$$

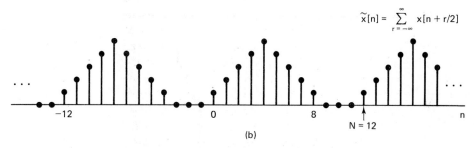

N = 12

(b)

Figure 8.8   (a) Finite-length sequence $x[n]$. (b) Periodic sequence $\widetilde{x}[n]$ corresponding to sampling the Fourier transform of $x[n]$ with $N = 12$.

recovered from the samples spaced in frequency by $2\pi/12$. In contrast, in Fig. 8.9, $x[n]$ cannot be recovered by extracting one period of $\widetilde{x}[n]$, and, equivalently, $X(e^{j\omega})$ cannot be recovered from its samples if the sample spacing is only $2\pi/7$. In effect, for the case illustrated in Fig. 8.8, the Fourier transform of $x[n]$ has been sampled at a sufficiently small spacing (in frequency) to be able to recover it from these samples, whereas Fig. 8.9 represents a case for which the Fourier transform has been undersampled. The relationship between $x[n]$ and one period of $\widetilde{x}[n]$ in the undersampled case of Fig. 8.9 can be thought of as a form of aliasing in the time domain, essentially identical to the frequency-domain aliasing (discussed in Chapter 3) that results from undersampling in the time domain. Obviously, time-domain aliasing can be avoided only if $x[n]$ has finite length, just as frequency-domain aliasing can be avoided only for signals that have bandlimited Fourier transforms.

$$\widetilde{x}[n] = \sum_{r=-\infty}^{\infty} x[n + r7]$$

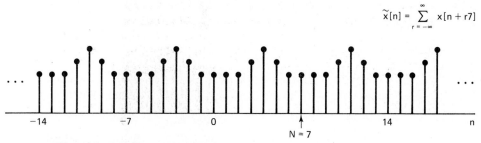

N = 7

Figure 8.9   Periodic sequence $\widetilde{x}[n]$ corresponding to sampling the Fourier transform of $x[n]$ in Fig. 8.8(a) with $N = 7$.

This discussion highlights several important concepts that will play a central role in the remainder of this chapter. We have seen that samples of the Fourier transform of an aperiodic sequence $x[n]$ can be thought of as DFS coefficients of a periodic sequence $\tilde{x}[n]$ obtained through periodic replication of $x[n]$. If $x[n]$ has finite length and we take a sufficient number of equally spaced samples of its Fourier transform (specifically, a number greater than or equal to the length of $x[n]$), then the Fourier transform is recoverable from these samples, and, equivalently, $x[n]$ is recoverable from $\tilde{x}[n]$ through the relation

$$x[n] = \begin{cases} \tilde{x}[n], & 0 \le n \le N - 1, \\ 0, & \text{otherwise.} \end{cases} \tag{8.51}$$

A direct relationship between $X(e^{j\omega})$ and its samples $\tilde{X}[k]$—i.e., an interpolation formula for $X(e^{j\omega})$—can be derived (see Problem 8.8). However, the essence of our previous discussion is that it is not necessary to know $X(e^{j\omega})$ at all frequencies if $x[n]$ has finite length. Given a finite-length sequence $x[n]$, we can form a periodic sequence using Eq. (8.50), which in turn can be represented by a Fourier series. Alternatively, given the sequence of Fourier coefficients $\tilde{X}[k]$, we can find $\tilde{x}[n]$ and then use Eq. (8.51) to obtain $x[n]$. When the Fourier series is used in this way to represent finite-length sequences, it is called the discrete Fourier transform (DFT). In developing, discussing, and applying the DFT, it is always important to remember that the representation through samples of the Fourier transform is in effect a representation of the finite-duration sequence by a periodic sequence, one period of which is the finite-duration sequence that we wish to represent.

8.6 FOURIER REPRESENTATION OF FINITE-DURATION SEQUENCES: THE DISCRETE FOURIER TRANSFORM

In this section we formalize the point of view suggested at the end of the previous section. We begin by considering a finite-length sequence $x[n]$ of length N samples such that $x[n] = 0$ outside the range $0 \le n \le N - 1$. In many instances we will want to assume that a sequence has length N even if its length is $M \le N$. In such cases we simply recognize that the last $(N - M)$ samples have zero amplitude. To each finite-length sequence of length N we can always associate a periodic sequence $\tilde{x}[n]$ given by

$$\tilde{x}[n] = \sum_{r=-\infty}^{\infty} x[n + rN]. \tag{8.52a}$$

The finite-length sequence $x[n]$ can be recovered from $\tilde{x}[n]$ through Eq. (8.51), i.e.,

$$x[n] = \begin{cases} \tilde{x}[n], & 0 \le n \le N - 1, \\ 0, & \text{otherwise.} \end{cases} \tag{8.52b}$$

Recall from Section 8.4 that the DFS coefficients of $\tilde{x}[n]$ are samples (spaced in frequency by $2\pi/N$) of the Fourier transform of $x[n]$. Since $x[n]$ is assumed to have finite length N, there is no overlap between the terms $x[n + rN]$ for different values of r. Thus, Eq. (8.52a) can alternatively be written as

$$\tilde{x}[n] = x[(n \text{ modulo } N)].\tag{8.53}$$

For convenience we will use the notation $((n))_N$ to denote $(n \text{ modulo } N)$; with this notation Eq. (8.53) is expressed as

$$\tilde{x}[n] = x[((n))_N].\tag{8.54}$$

Note that Eq. (8.54) is equivalent to Eq. (8.52) *only* when $x[n]$ has length less than or equal to N. The finite-duration sequence $x[n]$ is obtained from $\tilde{x}[n]$ by extracting one period as in Eq. (8.52b).

One informal and useful way of visualizing Eq. (8.53) is to think of wrapping the finite-duration sequence $x[n]$ around a cylinder with a circumference equal to the sequence length. As we repeatedly traverse the circumference of the cylinder we see the finite-length sequence periodically repeated. With this interpretation, representation of the finite-length sequence by a periodic sequence corresponds to wrapping the sequence around the cylinder; recovering the finite-length sequence from the periodic sequence using Eq. (8.51) can be visualized as unwrapping the cylinder and laying it flat so that the sequence is displayed on a linear time axis rather than a circular (modulo N) time axis.

As defined in Section 8.1, the sequence of discrete Fourier series coefficients $\tilde{X}[k]$ of the periodic sequence $\tilde{x}[n]$ is itself a periodic sequence with period N. To maintain a duality between the time and frequency domains, we will choose the Fourier coefficients that we associate with a finite-duration sequence to be a finite-duration sequence corresponding to one period of $\tilde{X}[k]$. This finite-duration sequence, $X[k]$, will be referred to as the discrete Fourier transform (DFT). Thus the DFT, $X[k]$, is related to the DFS coefficients, $\tilde{X}[k]$, by

$$X[k] = \begin{cases} \tilde{X}[k], & 0 \le k \le N - 1, \\ 0, & \text{otherwise,} \end{cases}\tag{8.55}$$

and

$$\tilde{X}[k] = X[(k \text{ modulo } N)] = X[((k))_N].\tag{8.56}$$

From Section 8.1, $\tilde{X}[k]$ and $\tilde{x}[n]$ are related by

$$\tilde{X}[k] = \sum_{n=0}^{N-1} \tilde{x}[n]W_N^{kn},\tag{8.57}$$

$$\tilde{x}[n] = \frac{1}{N}\sum_{k=0}^{N-1} \tilde{X}[k]W_N^{-kn}.\tag{8.58}$$

Since the summations in Eqs. (8.57) and (8.58) involve only the interval between zero and $(N-1)$, it follows from Eqs. (8.51)–(8.58) that

$$X[k] = \begin{cases} \displaystyle\sum_{n=0}^{N-1} x[n]W_N^{kn}, & 0 \le k \le N-1, \\[2mm] 0, & \text{otherwise,} \end{cases} \tag{8.59}$$

$$x[n] = \begin{cases} \displaystyle\frac{1}{N}\sum_{k=0}^{N-1} X[k]W_N^{-kn}, & 0 \le n \le N-1, \\[2mm] 0, & \text{otherwise.} \end{cases} \tag{8.60}$$

Generally the DFT analysis and synthesis equations are written as

$$\text{Analysis equation:} \qquad X[k] = \sum_{n=0}^{N-1} x[n]W_N^{kn}, \tag{8.61}$$

$$\text{Synthesis equation:} \qquad x[n] = \frac{1}{N}\sum_{k=0}^{N-1} X[k]W_N^{-kn}. \tag{8.62}$$

That is, the fact that $X[k] = 0$ for k outside the interval $0 \le k \le N-1$ and that $x[n] = 0$ for n outside the interval $0 \le n \le N-1$ is implied but not always stated explicitly. Equation (8.61) is called the *analysis equation* and Eq. (8.62) the *synthesis equation*. The relationship between $x[n]$ and $X[k]$ implied by Eqs. (8.61) and (8.62) will sometimes be denoted as

$$x[n] \xleftrightarrow{\;\mathcal{DFT}\;} X[k]. \tag{8.63}$$

In recasting Eqs. (8.11) and (8.12) in the form of Eqs. (8.61) and (8.62) for finite-duration sequences, we have not eliminated the inherent periodicity. As with the DFS, the DFT $X[k]$ is equal to samples of the periodic Fourier transform $X(e^{j\omega})$, and if Eq. (8.62) is evaluated for values of n outside the interval $0 \le n \le N-1$, the result will not be zero but rather a periodic extension of $x[n]$. The inherent periodicity is always present. Sometimes it causes us difficulty and sometimes we can exploit it, but to totally ignore it is to invite trouble. In defining the DFT representation we are simply recognizing that we are *interested* in values of $x[n]$ only in the interval $0 \le n \le N-1$ because $x[n]$ is really zero outside that interval, and we are *interested* in values of $X[k]$ only in the interval $0 \le k \le N-1$ because these are the only values needed in Eq. (8.62).

Example 8.5

To illustrate the DFT of a finite-duration sequence, consider $x[n]$ shown in Fig. 8.10(a). In determining the DFT, we can consider $x[n]$ as a finite-duration sequence with any length greater than or equal to $N = 5$. Considered as a sequence of length $N = 5$, the periodic sequence $\tilde{x}[n]$ whose DFS corresponds to the DFT of $x[n]$ is shown in Figure

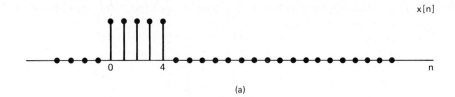

(a)

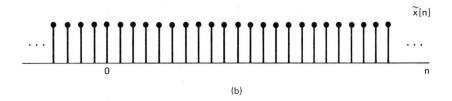

(b)

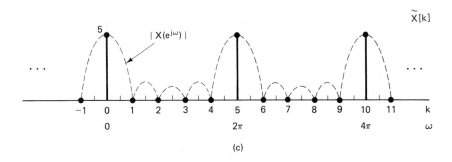

(c)

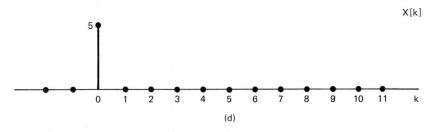

(d)

Figure 8.10 Illustration of the DFT. (a) Finite-length sequence $x[n]$. (b) Periodic sequence $\tilde{x}[n]$ formed from $x[n]$ with period $N = 5$. (c) Fourier series coefficients $\tilde{X}[k]$ for $\tilde{x}[n]$. To emphasize that the Fourier series coefficients are samples of the Fourier transform, $|X(e^{j\omega})|$ is also shown. (d) DFT of $x[n]$.

8.10(b). Since the sequence in Fig. 8.10(b) is constant over the interval $0 \le n \le N - 1$, it follows that

$$\tilde{X}[k] = \sum_{n=0}^{N-1} e^{-j(2\pi k/N)n} = \frac{1 - e^{-j2\pi k}}{1 - e^{-j(2\pi k/N)}}$$

$$= \begin{cases} N, & k = 0, \ \pm N, \ \pm 2N, \ldots, \\ 0, & \text{otherwise,} \end{cases}$$

(8.64)

i.e., the only nonzero DFS coefficients $\tilde{X}[k]$ are at $k = 0$ and integer multiples of $k = N$ (all of which represent the same complex exponential frequency). The DFS coefficients are shown in Fig. 8.10(c). Also shown in Fig. 8.10(c) is the magnitude of the Fourier transform, $|X(e^{j\omega})|$. Clearly, $\tilde{X}[k]$ is a sequence of samples of $X(e^{j\omega})$ at frequencies $\omega_k = 2\pi k/N$. According to Eq. (8.55), the 5-point DFT of $x[n]$ corresponds to the finite-length sequence obtained by extracting one period of $\tilde{X}[k]$. Consequently the 5-point DFT of $x[n]$ is shown in Fig. 8.10(d).

   If instead we consider $x[n]$ to be of length $N = 10$, then the underlying periodic sequence is that shown in Fig. 8.11(b), which is the periodic sequence considered in Example 8.2. Consequently $\tilde{X}[k]$ is as shown in Figs. 8.2 and 8.5 and the 10-point DFT $X[k]$ shown in Figs. 8.11(c) and 8.11(d) is one period of $\tilde{X}[k]$.

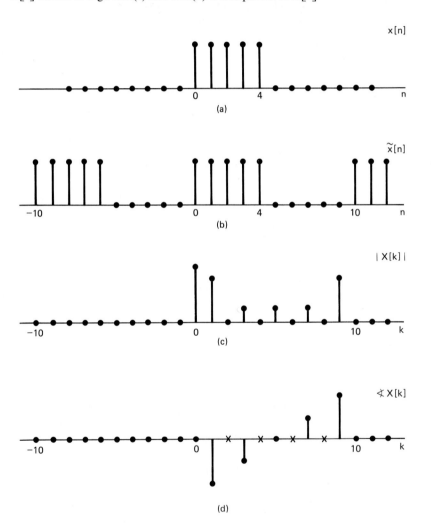

Figure 8.11 Illustration of the DFT. (a) Finite-length sequence $x[n]$. (b) Periodic sequence $\tilde{x}[n]$ formed from $x[n]$ with period $N = 10$. (c) DFT magnitude. (d) DFT phase. (x's indicate indeterminate value.)

The distinction between the finite-duration sequence $x[n]$ and the periodic sequence $\tilde{x}[n]$ related through Eqs. (8.51) and (8.54) may seem minor, since by using these equations it is straightforward to construct one from the other. However, the distinction becomes important in considering properties of the DFT and in considering the effect on $x[n]$ of modifications to $X[k]$. This will become evident in the next section, where we discuss the properties of the DFT representation.

8.7 PROPERTIES OF THE DISCRETE FOURIER TRANSFORM

In this section we consider a number of properties of the DFT for finite-duration sequences. Our discussion parallels the discussion of Section 8.2 for periodic sequences. However, particular attention is paid to the interaction of the finite-length assumption and the implicit periodicity of the DFT representation of finite-length sequences.

8.7.1 Linearity

If two finite-duration sequences $x_1[n]$ and $x_2[n]$ are linearly combined, i.e.,

$$x_3[n] = ax_1[n] + bx_2[n], \tag{8.65}$$

then the DFT of $x_3[n]$ is

$$X_3[k] = aX_1[k] + bX_2[k] \tag{8.66}$$

Clearly if $x_1[n]$ has length N_1 and $x_2[n]$ has length N_2, then the maximum length of $x_3[n]$ will be $N_3 = \max[N_1, N_2]$. Thus the DFTs must be computed with $N \geq N_3$. If, for example, $N_1 < N_2$, then $X_1[k]$ is the DFT of the sequence $x_1[n]$ augmented by $(N_2 - N_1)$ zeros. That is, its N_2-point DFT is

$$X_1[k] = \sum_{n=0}^{N_2-1} x_1[n]W_{N_2}^{kn}, \qquad 0 \leq k \leq N_2 - 1, \tag{8.67}$$

and the N_2-point DFT of $x_2[n]$ is

$$X_2[k] = \sum_{n=0}^{N_2-1} x_2[n]W_{N_2}^{kn}, \qquad 0 \leq k \leq N_2 - 1. \tag{8.68}$$

In summary, if

$$x_1[n] \overset{\mathscr{DFT}}{\longleftrightarrow} X_1[k] \tag{8.69a}$$

and

$$x_2[n] \overset{\mathscr{DFT}}{\longleftrightarrow} X_2[k], \tag{8.69b}$$

then

$$ax_1[n] + bx_2[n] \overset{\mathscr{DFT}}{\longleftrightarrow} aX_1[k] + bX_2[k], \tag{8.70}$$

where the lengths of the sequences and their discrete Fourier transforms are all equal to the maximum of the lengths of $x_1[n]$ and $x_2[n]$. Of course, DFTs of greater length can be computed by augmenting *both* sequences with zero-valued samples.

8.7.2 Circular Shift of a Sequence

According to Section 2.9.2 and property 2 in Table 2.2, if $X(e^{j\omega})$ is the Fourier transform of $x[n]$, then $e^{-j\omega m}X(e^{j\omega})$ is the Fourier transform of the time-shifted sequence $x[n-m]$. In other words, a shift in the time domain by m points (m positive corresponding to a time delay and m negative to a time advance) corresponds in the frequency domain to multiplication of the Fourier transform by the linear phase factor $e^{-j\omega m}$. In Section 8.2.2 we discussed the corresponding property for the DFS coefficients of a periodic sequence; specifically, if a periodic sequence $\tilde{x}[n]$ has Fourier series coefficients $\tilde{X}[k]$, then the shifted sequence $\tilde{x}[n-m]$ has Fourier series coefficients $e^{-j(2\pi k/N)m}\tilde{X}[k]$. Now we will consider the operation in the time domain that corresponds to multiplying the DFT coefficients of a finite-length sequence $x[n]$ by the linear phase factor $e^{-j(2\pi k/N)m}$. Specifically, let $x_1[n]$ denote the finite-length sequence for which the DFT is $e^{-j(2\pi k/N)m}X[k]$, i.e.,

$$x[n] \overset{\mathcal{DFT}}{\longleftrightarrow} X[k] \tag{8.71}$$

and

$$x_1[n] \overset{\mathcal{DFT}}{\longleftrightarrow} X_1[k] = e^{-j(2\pi k/N)m}X[k]. \tag{8.72}$$

Since the N-point DFT represents a finite-duration sequence of length N, both $x[n]$ and $x_1[n]$ must be zero outside the interval $0 \le n \le N-1$, and consequently $x_1[n]$ cannot result from a simple time shift of $x[n]$. The correct result follows directly from the result of Section 8.2.2 and the interpretation of the DFT as the Fourier series coefficients of the periodic sequence $x_1[((n))_N]$. In particular, from Eqs. (8.53) and (8.56) it follows that

$$\tilde{x}[n] = x[((n))_N] \overset{\mathcal{DFS}}{\longleftrightarrow} \tilde{X}[k] = X[((k))_N] \tag{8.73}$$

and, similarly, we can define a periodic sequence $\tilde{x}_1[n]$ such that

$$\tilde{x}_1[n] = x_1[((n))_N] \overset{\mathcal{DFS}}{\longleftrightarrow} \tilde{X}_1[k] = X_1[((k))_N], \tag{8.74}$$

where, by assumption,

$$X_1[k] = e^{-j(2\pi k/N)m}X[k]. \tag{8.75}$$

Therefore the discrete Fourier series coefficients of $\tilde{x}_1[n]$ are

$$\tilde{X}_1[k] = e^{-j[2\pi((k))_N/N]m}X[((k))_N]. \tag{8.76}$$

Note that

$$e^{-j[2\pi((k))_N/N]m} = e^{-j(2\pi k/N)m}. \tag{8.77}$$

That is, since $e^{-j(2\pi k/N)m}$ is periodic with period N in both k and m, we can drop the notation $((k))_N$. Therefore, Eq. (8.76) becomes

$$\tilde{X}_1[k] = e^{-j(2\pi k/N)m}\tilde{X}[k], \tag{8.78}$$

so that it follows from Section 8.2.2 that

$$\tilde{x}_1[n] = \tilde{x}[n-m] = x[((n-m))_N]. \tag{8.79}$$

Thus the finite-length sequence $x_1[n]$ whose DFT is given by Eq. (8.75) is

$$x_1[n] = \begin{cases} \tilde{x}_1[n] = x[((n-m))_N], & 0 \le n \le N-1, \\ 0, & \text{otherwise.} \end{cases} \tag{8.80}$$

Equations (8.79) and (8.80) tell us how to construct $x_1[n]$. The procedure is illustrated in Fig. 8.12 for $m = -2$. Specifically, from $x[n]$ we construct the periodic

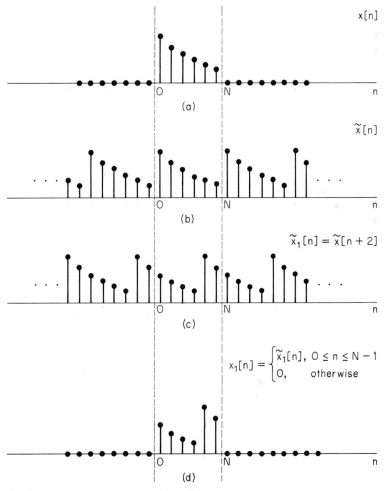

Figure 8.12 Circular shift of a finite-length sequence; i.e., the effect in the time domain of multiplying the DFT of the sequence by a linear phase factor.

sequence $\tilde{x}[n] = x[((n))_N]$ as indicated in Fig. 8.12(b). According to Eq. (8.79), we then shift $\tilde{x}[n]$ by m to obtain $\tilde{x}_1[n] = \tilde{x}[n - m]$ as in Fig. 8.12(c). Finally, using Eq. (8.80), we extract one period of $\tilde{x}_1[n]$ to obtain $x_1[n]$ as indicated in Fig. 8.12(d).

Comparison of Fig. 8.12(a) and (d) indicates clearly that $x_1[n]$ does *not* correspond to a linear shift of $x[n]$, and in fact both sequences are confined to the interval between 0 and $(N - 1)$. By reference to Fig. 8.12 we see that $x_1[n]$ can be formed by shifting $x[n]$ so that as a sequence value leaves the interval 0 to $(N - 1)$ at one end it enters at the other end. In Section 8.6 we suggested the interpretation of forming the periodic sequence $\tilde{x}[n]$ from the finite-length sequence $x[n]$ by displaying $x[n]$ around the circumference of a cylinder with a circumference of exactly N points. As we repeatedly traverse the circumference of the cylinder, the sequence that we see is the periodic sequence $\tilde{x}[n]$. A linear shift of the periodic sequence $\tilde{x}[n]$ corresponds then to a *rotation* of the cylinder. In the context of finite-length sequences and the DFT, such a shift is called a *circular* shift or a *rotation* of the sequence in the interval $0 \leq n \leq N - 1$.

In summary, the circular shift property of the DFT is

$$x[((n - m))_N], \quad 0 \leq n \leq N - 1 \quad \overset{\mathcal{DFT}}{\longleftrightarrow} \quad e^{-j(2\pi k/N)m} X[k]. \tag{8.81}$$

8.7.3 Duality

Since the DFT is so closely associated with the DFS, we would expect the DFT to exhibit a duality property similar to that which we discussed for the DFS in Section 8.2.3. In fact, from an examination of Eqs. (8.61) and (8.62), we see that the analysis and synthesis equations differ only in the factor $1/N$ and the sign of the exponent of the powers of W_N.

The DFT duality property can be easily derived by exploiting the relationship between the DFT and DFS as in our derivation of the circular shift property. Toward this end, consider $x[n]$ and its DFT $X[k]$ and construct the periodic sequences

$$\tilde{x}[n] = x[((n))_N], \tag{8.82a}$$

$$\tilde{X}[k] = X[((k))_N], \tag{8.82b}$$

so that

$$\tilde{x}[n] \overset{\mathcal{DFS}}{\longleftrightarrow} \tilde{X}[k]. \tag{8.83}$$

From the duality property given in Eqs. (8.30),

$$\tilde{X}[n] \overset{\mathcal{DFS}}{\longleftrightarrow} N\tilde{x}[-k]. \tag{8.84}$$

If we define the periodic sequence $\tilde{x}_1[n] = \tilde{X}[n]$, one period of which is the finite-length sequence $x_1[n] = X[n]$, then the DFS coefficients of $\tilde{x}_1[n]$ are $\tilde{X}_1[k] = N\tilde{x}[-k]$. Therefore the DFT of $x_1[n]$ is

$$X_1[k] = \begin{cases} N\tilde{x}[-k], & 0 \leq k \leq N - 1, \\ 0, & \text{otherwise,} \end{cases} \tag{8.85}$$

or, equivalently,

$$X_1[k] = \begin{cases} Nx[((-k))_N], & 0 \le k \le N-1, \\ 0, & \text{otherwise.} \end{cases} \tag{8.86}$$

Consequently, the duality property for the DFT can be expressed as follows: If

$$x[n] \overset{\mathcal{DFS}}{\longleftrightarrow} X[k], \tag{8.87a}$$

then

$$X[n] \overset{\mathcal{DFS}}{\longleftrightarrow} Nx[((-k))_N], \qquad 0 \le k \le N-1. \tag{8.87b}$$

The sequence $Nx[((-k))_N]$ is $Nx[k]$ index-reversed modulo N. As in the case of shifting modulo N, the process of index reversing modulo N is usually best visualized in terms of the underlying periodic sequences.

Example 8.6

To illustrate the duality relationship in Eqs. (8.87), let us consider the sequence $x[n]$ of Example 8.5. Figure 8.13(a) shows the finite-length sequence $x[n]$, and Figs. 8.13(b) and

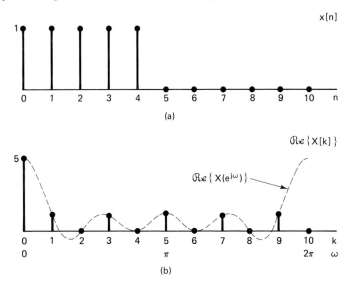

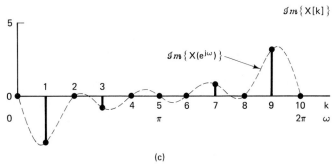

Figure 8.13 (continued) (d) and (e) The real and imaginary parts of the dual sequence $x_1[n] = X[n]$. (f) The DFT of $x_1[n]$.

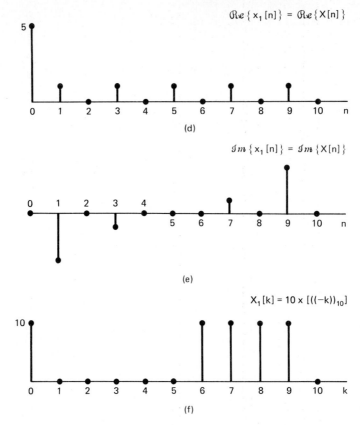

Figure 8.13 (*continued*) (d) and (e) The real and imaginary parts of the dual sequence $x_1[n] = X[n]$. (f) The DFT of $x_1[n]$.

8.13(c) are the real and imaginary parts of the corresponding 10-point DFT $X[k]$. By simply relabeling the horizontal axis, we obtain $x_1[n] = X[n]$, as shown in Figs. 8.13(d) and 8.13(e). According to the duality relation in Eq. (8.87), the 10-point DFT of the (complex-valued) sequence $X[n]$ is the sequence shown in Fig. 8.13(f).

8.7.4 Symmetry Properties

Since the DFT of $x[n]$ is identical to the DFS coefficients of the periodic sequence $\tilde{x}[n] = x[((n))_N]$, symmetry properties associated with the DFT can be inferred from the symmetry properties of the DFS summarized in Table 8.1 in Section 8.3. Specifically, using Eqs. (8.82) together with properties 9 and 10 in Table 8.1, we have

$$x^*[n] \overset{\mathcal{DFT}}{\longleftrightarrow} X^*[((-k))_N], \qquad 0 \le n \le N-1, \tag{8.88}$$

and

$$x^*[((-n))_N] \overset{\mathcal{DFT}}{\longleftrightarrow} X^*[k], \qquad 0 \le n \le N-1. \tag{8.89}$$

Properties 11–14 in Table 8.1 refer to the decomposition of a periodic sequence into the sum of a conjugate-symmetric and a conjugate-antisymmetric sequence. This suggests the decomposition of the finite-duration sequence $x[n]$ into the two finite-duration sequences of duration N corresponding to one period of the conjugate-symmetric and one period of the conjugate-antisymmetric components of $\tilde{x}[n]$. We will denote these components of $x[n]$ as $x_{ep}[n]$ and $x_{op}[n]$. Thus with

$$\tilde{x}[n] = x[((n))_N] \tag{8.90}$$

and

$$\tilde{x}_e[n] = \tfrac{1}{2}\{\tilde{x}[n] + \tilde{x}^*[-n]\} \tag{8.91}$$

and

$$\tilde{x}_o[n] = \tfrac{1}{2}\{\tilde{x}[n] - \tilde{x}^*[-n]\}, \tag{8.92}$$

we define $x_{ep}[n]$ and $x_{op}[n]$ as

$$x_{ep}[n] = \tilde{x}_e[n], \qquad 0 \le n \le N - 1, \tag{8.93}$$

$$x_{op}[n] = \tilde{x}_o[n], \qquad 0 \le n \le N - 1 \tag{8.94}$$

or, equivalently,

$$x_{ep}[n] = \tfrac{1}{2}\{x[((n))_N] + x^*[((-n))_N]\}, \qquad 0 \le n \le N - 1, \tag{8.95a}$$

$$x_{op}[n] = \tfrac{1}{2}\{x[((n))_N] - x^*[((-n))_N]\}, \qquad 0 \le n \le N - 1, \tag{8.95b}$$

with both $x_{ep}[n]$ and $x_{op}[n]$ finite-length sequences, i.e., both zero outside the interval $0 \le n \le N - 1$. Since $((-n))_N = (N - n)$ and $((n))_N = n$ for $0 \le n \le N - 1$, we can also express Eqs. (8.95) as

$$x_{ep}[n] = \tfrac{1}{2}\{x[n] + x^*[N - n]\}, \qquad 1 \le n \le N - 1, \tag{8.96a}$$

$$x_{ep}[0] = \mathcal{R}e\{x[0]\}, \tag{8.96b}$$

$$x_{op}[n] = \tfrac{1}{2}\{x[n] - x^*[N - n]\}, \qquad 1 \le n \le N - 1, \tag{8.96c}$$

$$x_{op}[0] = \mathcal{I}m\{x[0]\}. \tag{8.96d}$$

This form of the equations is convenient since it avoids the modulo N computation of indices.

Clearly $x_{ep}[n]$ and $x_{op}[n]$ are not equivalent to $x_e[n]$ and $x_o[n]$ as defined by Eq. (2.133). However, it can be shown (see Problem 8.28) that

$$x_{ep}[n] = \{x_e[n] + x_e[n - N]\}, \qquad 0 \le n \le N - 1 \tag{8.97}$$

and

$$x_{op}[n] = \{x_o[n] + x_o[n - N]\}, \qquad 0 \le n \le N - 1. \tag{8.98}$$

In other words, $x_{ep}[n]$ and $x_{op}[n]$ can be generated by aliasing $x_e[n]$ and $x_o[n]$ into the interval $0 \le n \le N - 1$. The sequences $x_{ep}[n]$ and $x_{op}[n]$ will be referred to as the *periodic conjugate-symmetric* and *periodic conjugate-antisymmetric components*, respectively, of $x[n]$. When $x_{ep}[n]$ and $x_{op}[n]$ are real, they will be referred to as the periodic even and periodic odd components, respectively. Note that the sequences

$x_{ep}[n]$ and $x_{op}[n]$ are *not* periodic sequences; they are, however, finite-length sequences that are equal to one period of the periodic sequences $\tilde{x}_e[n]$ and $\tilde{x}_o[n]$, respectively.

Equations (8.95) and (8.96) define $x_{ep}[n]$ and $x_{op}[n]$ in terms of $x[n]$. The inverse relation, expressing $x[n]$ in terms of $x_{ep}[n]$ and $x_{op}[n]$, can be obtained by using Eqs. (8.91) and (8.92) to express $\tilde{x}[n]$ as

$$\tilde{x}[n] = \tilde{x}_e[n] + \tilde{x}_o[n]. \tag{8.99}$$

Thus

$$x[n] = \tilde{x}[n] = \tilde{x}_e[n] + \tilde{x}_o[n], \qquad 0 \le n \le N - 1. \tag{8.100}$$

Combining Eqs. (8.100) with Eqs. (8.93) and (8.94), we obtain

$$x[n] = x_{ep}[n] + x_{op}[n]. \tag{8.101}$$

Alternatively, Eqs. (8.96), when added, lead to the result of Eq. (8.101). The symmetry properties of the DFT associated with properties 11–14 in Table 8.1 now follow in a straightforward way. Specifically,

$$\mathcal{R}e\{x[n]\} \overset{\mathcal{DFT}}{\longleftrightarrow} X_{ep}[k], \tag{8.102}$$

$$j\mathcal{I}m\{x[n]\} \overset{\mathcal{DFT}}{\longleftrightarrow} X_{op}[k], \tag{8.103}$$

$$x_{ep}[n] \overset{\mathcal{DFT}}{\longleftrightarrow} \mathcal{R}e\{X[k]\}, \tag{8.104}$$

$$x_{op}[n] \overset{\mathcal{DFT}}{\longleftrightarrow} j\mathcal{I}m\{X[k]\}. \tag{8.105}$$

✦ 8.7.5 Circular Convolution

In Section 8.2.5 we showed that multiplication of the DFS coefficients of two sequences corresponds to a periodic convolution of the sequences. Here we consider two *finite-duration* sequences $x_1[n]$ and $x_2[n]$, both of length N, with DFTs $X_1[k]$ and $X_2[k]$, and we wish to determine the sequence $x_3[n]$ for which the DFT is $X_3[k] = X_1[k]X_2[k]$. To determine $x_3[n]$ we can apply the results of Section 8.2.5. Specifically, $x_3[n]$ corresponds to one period of $\tilde{x}_3[n]$, which is given by Eq. (8.32). Thus

$$x_3[n] = \sum_{m=0}^{N-1} \tilde{x}_1[m]\tilde{x}_2[n - m], \qquad 0 \le n \le N - 1 \tag{8.106}$$

or, equivalently,

$$x_3[n] = \sum_{m=0}^{N-1} x_1[((m))_N]x_2[((n - m))_N], \qquad 0 \le n \le N - 1. \tag{8.107}$$

Since $((m))_N = m$ for $0 \le m \le N - 1$, Eq. (8.107) can be written

$$x_3[n] = \sum_{m=0}^{N-1} x_1[m]x_2[((n - m))_N], \qquad 0 \le n \le N - 1. \tag{8.108}$$

Equations (8.106) and (8.108) differ from a linear convolution of $x_1[n]$ and $x_2[n]$ as defined by Eq. (2.39) in some important respects. In linear convolution, the computation of the sequence value $x_3[n]$ involves multiplying one sequence by a time-reversed and linearly shifted version of the other and then summing the values of the product $x_1[m]x_2[n-m]$ over all m. To obtain successive values of the sequence representing the convolution, the two sequences are successively shifted relative to each other. In contrast, for the convolution as given by Eq. (8.108), the second sequence is circularly time reversed and circularly shifted with respect to the first. For this reason the operation of combining two finite-length sequences according to Eq. (8.108) is called *circular convolution*. More specifically, we refer to Eq. (8.108) as an N-N-*point circular convolution*, explicitly identifying the fact that both sequences have length N (or less) and that the sequences are shifted modulo N. Sometimes the operation of forming a sequence $x_3[n]$ for $0 \le n \le N-1$ using Eq. (8.108) will be denoted

$$x_3[n] = x_1[n] \ⓃⓃ \ x_2[n]. \qquad (8.109)$$

Since the DFT of $x_3[n]$ is $X_3[k] = X_1[k]X_2[k]$ and since $X_1[k]X_2[k] = X_2[k]X_1[k]$, it follows with no further analysis that

$$x_3[n] = x_2[n] \Ⓝ \ x_1[n] \qquad (8.110)$$

or, more specifically,

$$x_3[n] = \sum_{m=0}^{N-1} x_2[m]x_1[((n-m))_N]. \qquad (8.111)$$

That is, circular convolution, like linear convolution, is a commutative operation.

Since circular convolution is really just periodic convolution, Example 8.4 and Fig. 8.6 are also illustrative of circular convolution. However, if we utilize the notion of circular shifting, it is not necessary to construct the underlying periodic sequences. This is illustrated in the following examples.

Example 8.7

An example of circular convolution is provided by the result of Section 8.7.2. Let $x_2[n]$ be a finite-duration sequence of length N, and

$$x_1[n] = \delta[n - n_0], \qquad (8.112)$$

where $0 < n_0 < N$. Clearly $x_1[n]$ can be considered as the finite-duration sequence

$$x_1[n] = \begin{cases} 0, & 0 \le n < n_0, \\ 1, & n = n_0, \\ 0, & n_0 < n \le N - 1. \end{cases} \qquad (8.113)$$

The DFT of $x_1[n]$ is

$$X_1[k] = W_N^{kn_0}. \qquad (8.114)$$

If we form the product

$$X_3[k] = W_N^{kn_0} X_2[k], \qquad (8.115)$$

we see from Section 8.7.2 that the finite-duration sequence corresponding to $X_3[k]$ is the sequence $x_2[n]$ rotated to the right by n_0 samples in the interval $0 \leq n \leq N - 1$. That is, the circular convolution of a sequence $x_2[n]$ with a single delayed unit impulse results in a rotation of the sequence $x_2[n]$ in the interval $0 \leq n \leq N - 1$. This example is illustrated in Fig. 8.14 for $N = 5$ and $n_0 = 1$. Here we show the sequences $x_2[m]$ and $x_1[m]$ and then $x_2[((0 - m))_N]$ and $x_2[((1 - m))_N]$. It is clear from these two cases that the result of circular convolution of $x_2[n]$ with a single shifted unit impulse will be to circularly shift $x_2[n]$. The last sequence shown is $x_3[n]$, the result of the circular convolution of $x_1[n]$ and $x_2[n]$.

Example 8.8

As another example of circular convolution, let

$$x_1[n] = x_2[n] = \begin{cases} 1, & 0 \leq n \leq L - 1, \\ 0, & \text{otherwise.} \end{cases} \qquad (8.116)$$

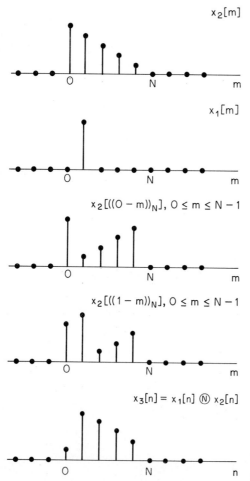

Figure 8.14   Circular convolution of a finite-length sequence $x_2[n]$ with a single delayed impulse.

If $N = L$, the N-point DFTs are

$$X_1[k] = X_2[k] = \sum_{n=0}^{N-1} W_N^{kn}$$

$$= \begin{cases} N, & k = 0, \\ 0, & \text{otherwise.} \end{cases} \tag{8.117}$$

If we explicitly multiply $X_1[k]$ and $X_2[k]$ we obtain

$$X_3[k] = X_1[k]X_2[k] = \begin{cases} N^2, & k = 0, \\ 0, & \text{otherwise,} \end{cases} \tag{8.118}$$

from which it follows that

$$x_3[n] = N, \qquad 0 \le n \le N - 1. \tag{8.119}$$

This result is depicted in Fig. 8.15. Clearly, as the sequence $x_2[((n - m))_N]$ is rotated with respect to $x_1[m]$, the sum of products $x_1[m]x_2[((n - m))_N]$ will always be equal to N.

It is, of course, possible to consider $x_1[n]$ and $x_2[n]$ as $2L$-point sequences by augmenting them with L zeros. If we then perform a $2L$-point circular convolution of the augmented sequences, we obtain the sequence in Fig. 8.16, which can be seen to be identical to the linear convolution of the finite-duration sequences $x_1[n]$ and $x_2[n]$. This important observation will be discussed in much more detail in Section 8.9.

The circular convolution property is represented as

$$x_1[n] \; \textcircled{N} \; x_2[n] \overset{\mathcal{DFS}}{\longleftrightarrow} X_1[k]X_2[k]. \tag{8.120}$$

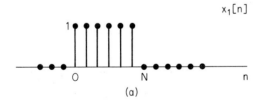

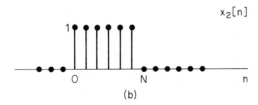

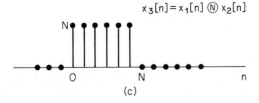

Figure 8.15 N-point circular convolution of two constant sequences of length N.

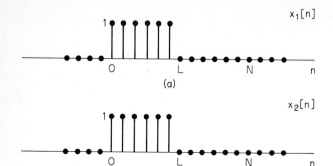

(a)

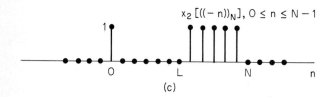

(b)

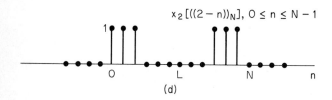

(c)

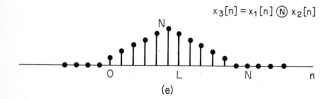

(d)

$x_3[n] = x_1[n] \, \textcircled{N} \, x_2[n]$

Figure 8.16 $2L$-point circular convolution of two constant sequences of length L.

(e)

In view of the duality of the DFT relations, it is not surprising that the DFT of a product of two N-point sequences is the circular convolution of their respective discrete Fourier transforms. Specifically, if $x_3[n] = x_1[n]x_2[n]$, then $X_3[k]$ is

$$X_3[k] = \frac{1}{N} \sum_{\ell=0}^{N-1} X_1[\ell]X_2[((k-\ell))_N] \tag{8.121}$$

or

$$x_1[n]x_2[n] \overset{\mathcal{DFT}}{\longleftrightarrow} X_1[k] \; \text{N} \; X_2[k]. \tag{8.122}$$

8.8 SUMMARY OF PROPERTIES OF THE DISCRETE FOURIER TRANSFORM

The properties of the discrete Fourier transform discussed in Section 8.7 are summarized in Table 8.2. Note that for all of the properties, the expressions given specify $x[n]$ for $0 \leq n \leq N-1$ and $X[k]$ for $0 \leq k \leq N-1$. Both $x[n]$ and $X[k]$ are equal to zero outside those ranges.

TABLE 8.2

Finite-Length Sequence (Length N)	N-Point DFT (Length N)
1. $x[n]$	$X[k]$
2. $x_1[n], x_2[n]$	$X_1[k], X_2[k]$
3. $ax_1[n] + bx_2[n]$	$aX_1[k] + bX_2[k]$
4. $X[n]$	$Nx[((-k))_N]$
5. $x[((n-m))_N]$	$W_N^{km}X[k]$
6. $W_N^{-\ell n}x[n]$	$X[((k-\ell))_N]$
7. $\displaystyle\sum_{m=0}^{N-1} x_1(m)x_2[((n-m))_N]$	$X_1[k]X_2[k]$
8. $x_1[n]x_2[n]$	$\displaystyle\frac{1}{N}\sum_{\ell=0}^{N-1} X_1(\ell)X_2[((k-\ell))_N]$
9. $x^*[n]$	$X^*[((-k))_N]$
10. $x^*[((-n))_N]$	$X^*[k]$
11. $\mathcal{R}e\{x[n]\}$	$X_{ep}[k] = \frac{1}{2}\{X[((k))_N] + X^*[((-k))_N]\}$
12. $j\mathcal{I}m\{x[n]\}$	$X_{op}[k] = \frac{1}{2}\{X[((k))_N] - X^*[((-k))_N]\}$
13. $x_{ep}[n] = \frac{1}{2}\{x[n] + x^*[((-n))_N]\}$	$\mathcal{R}e\{X[k]\}$
14. $x_{op}[n] = \frac{1}{2}\{x[n] - x^*[((-n))_N]\}$	$j\mathcal{I}m\{X[k]\}$

Properties 15–17 apply only when $x[n]$ is real.

15. Symmetry properties	$\begin{cases} X[k] = X^*[((-k))_N] \\ \mathcal{R}e\{X[k]\} = \mathcal{R}e\{X[((-k))_N]\} \\ \mathcal{I}m\{X[k]\} = -\mathcal{I}m\{X[((-k))_N]\} \\ |X[k]| = |X[((-k))_N]| \\ \measuredangle\{X[k]\} = -\measuredangle\{X[((-k))_N]\} \end{cases}$
16. $x_{ep}[n] = \frac{1}{2}\{x[n] + x[((-n))_N]\}$	$\mathcal{R}e\{X[k]\}$
17. $x_{op}[n] = \frac{1}{2}\{x[n] - x[((-n))_N]\}$	$j\mathcal{I}m\{X[k]\}$

8.9 LINEAR CONVOLUTION USING THE DISCRETE FOURIER TRANSFORM

We will show in Chapter 9 that efficient algorithms are available for computing the discrete Fourier transform of a finite-duration sequence. These are known collectively as *fast Fourier transform* (FFT) algorithms. Because these algorithms are available, it is computationally efficient to consider implementing a convolution of two sequences by the following procedure:

1. Compute the N-point discrete Fourier transforms $X_1[k]$ and $X_2[k]$ of the two sequences $x_1[n]$ and $x_2[n]$.
2. Compute the product $X_3[k] = X_1[k]X_2[k]$ for $0 \le k \le N - 1$.
3. Compute the sequence $x_3[n] = x_1[n] \, \mathrm{(N)} \, x_2[n]$ as the inverse DFT of $X_3[k]$.

In most applications, we are interested in implementing a linear convolution of two sequences; i.e., we wish to implement a linear time-invariant system. This is certainly true, for example, in filtering a sequence such as a speech waveform or a radar signal or in computing the autocorrelation function of such signals. As we saw in Section 8.8, the multiplication of discrete Fourier transforms corresponds to a circular convolution of the sequences. To obtain a linear convolution we must ensure that circular convolution has the effect of linear convolution. The discussion at the end of Example 8.8 hints at how this might be done. We now present a more detailed analysis.

8.9.1 Linear Convolution of Two Finite-Length Sequences

Consider a sequence $x_1[n]$ whose length is L points and a sequence $x_2[n]$ whose length is P points, and suppose that we wish to combine these two sequences by linear convolution to obtain a third sequence

$$x_3[n] = \sum_{m=-\infty}^{\infty} x_1[m]x_2[n - m]. \tag{8.123}$$

Figure 8.17(a) shows a typical sequence $x_1[m]$ and Fig. 8.17(b) shows a typical sequence $x_2[n - m]$ for several values of n. Clearly, the product $x_1[m]x_2[n - m]$ is zero for all m whenever $n < 0$ and $n > L + P - 2$. Therefore, $(L + P - 1)$ is the maximum length of the sequence $x_3[n]$ resulting from the linear convolution of a sequence of length L with a sequence of length P.

8.9.2 Circular Convolution as Linear Convolution with Aliasing

As we saw from Examples 8.7 and 8.8, whether a circular convolution corresponding to the product of two N-point DFTs is the same as the linear convolution of the corresponding finite-length sequences depends on the length of the DFT in relation to the length of the finite-length sequences. An extremely useful interpretation of the relationship between circular convolution and linear convolution is in terms of time

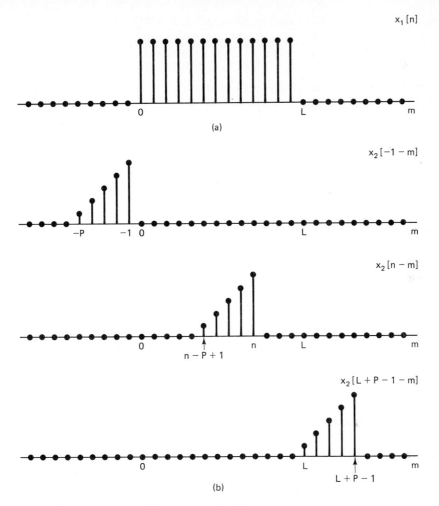

$x_1[n]$

0   L   m

(a)

$x_2[-1-m]$

$-P$ -1 0   L   m

$x_2[n-m]$

0   n   L   m
   $n-P+1$

$x_2[L+P-1-m]$

0   L   m
                    $L+P-1$

(b)

Figure 8.17 Example of linear convolution of two finite-length sequences showing that the result is such that $x_3[n] = 0$ for $n \leq -1$ and for $n \geq L + P - 1$. (a) Finite-length sequence $x_1[n]$. (b) $x_2[n - m]$ for several values of n.

aliasing. Since this interpretation is so important and useful in understanding circular convolution, we will develop it in several ways.

In Section 8.5 we observed that if the Fourier transform $X(e^{j\omega})$ of a sequence $x[n]$ is sampled at frequencies $\omega_k = 2\pi k/N$, then the resulting sequence corresponds to the DFS coefficients of the periodic sequence

$$\tilde{x}[n] = \sum_{r=-\infty}^{\infty} x[n + rN]. \qquad (8.124)$$

From our discussion of the DFT, it follows that the finite-length sequence

$$X[k] = \begin{cases} X(e^{j(2\pi k/N)}), & 0 \leq k \leq N - 1, \\ 0, & \text{otherwise}, \end{cases} \qquad (8.125)$$

is the DFT of one period of $\tilde{x}[n]$ as given by Eq. (8.124), i.e.,

$$x_p[n] = \begin{cases} \tilde{x}[n], & 0 \le n \le N-1, \\ 0, & \text{otherwise.} \end{cases} \tag{8.126}$$

Obviously, if $x[n]$ has length less than or equal to N, no time aliasing occurs and $x_p[n] = x[n]$. However, if the length of $x[n]$ is greater than N, $x_p[n]$ may not be equal to $x[n]$ for some or all values of n. We will henceforth use the subscript p to denote that a sequence is one period of a periodic sequence resulting from an inverse DFT of a sampled Fourier transform. The subscript can be dropped if it is clear that time aliasing is avoided.

The sequence $x_3[n]$ in Eq. (8.123) has Fourier transform

$$X_3(e^{j\omega}) = X_1(e^{j\omega})X_2(e^{j\omega}). \tag{8.127}$$

If we define a DFT

$$X_3[k] = X_3(e^{j(2\pi k/N)}), \qquad 0 \le k \le N-1, \tag{8.128}$$

then it is clear from Eqs. (8.127) and (8.128) that $X_3[k]$ is also equal to

$$X_3[k] = X_1(e^{j(2\pi k/N)})X_2(e^{j(2\pi k/N)}), \qquad 0 \le k \le N-1, \tag{8.129}$$

and therefore

$$X_3[k] = X_1[k]X_2[k]. \tag{8.130}$$

That is, the sequence $x_{3p}[n]$ resulting as the inverse DFT of $X_3[k]$ is

$$x_{3p}[n] = \begin{cases} \displaystyle\sum_{r=-\infty}^{\infty} x_3[n+rN], & 0 \le n \le N-1, \\ 0, & \text{otherwise,} \end{cases} \tag{8.131}$$

and from Eq. (8.130) it follows that

$$x_{3p}[n] = x_1[n] \, \textcircled{N} \, x_2[n]. \tag{8.132}$$

Thus the circular convolution of two finite-length sequences is equivalent to linear convolution of the two sequences followed by time aliasing according to Eq. (8.131).

Note that if N is greater than either L or P, $X_1[k]$ and $X_2[k]$ represent $x_1[n]$ and $x_2[n]$ exactly, but $x_{3p}[n] = x_3[n]$ for all n only if N is greater than or equal to the length of the sequence $x_3[n]$. As we showed in Section 8.9.1, if $x_1[n]$ has length L and $x_2[n]$ has length P, then $x_3[n]$ has maximum length $(L+P-1)$. Therefore, the circular convolution corresponding to $X_1[k]X_2[k]$ is identical to the linear convolution corresponding to $X_1(e^{j\omega})X_2(e^{j\omega})$ if N, the length of the DFTs, satisfies $N \ge L + P - 1$.

Example 8.9

The results of Example 8.8 are easily understood in light of the interpretation just discussed. Note that $x_1[n]$ and $x_2[n]$ are identical constant sequences of length $L = P = 6$ as shown in Fig. 8.18(a). The linear convolution of $x_1[n]$ and $x_2[n]$ is of length $L + P - 1 = 11$ and it has the triangular shape shown in Fig. 8.18(b). In Figs. 8.18(c) and (d) are shown $x_3[n-N]$ and $x_3[n+N]$ for $N = 6$. The N-point circular

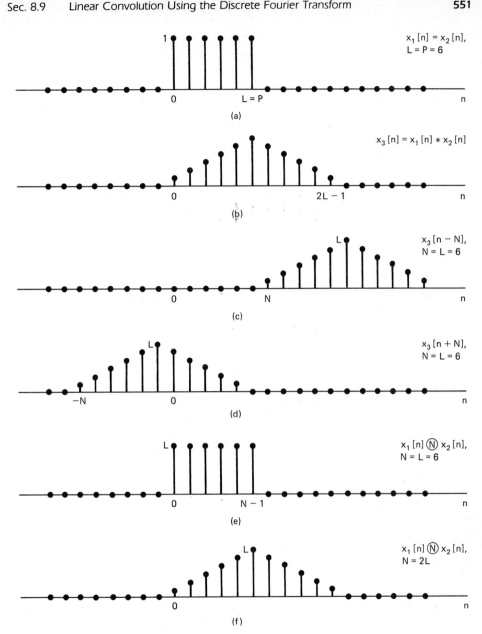

Figure 8.18   Illustration that circular convolution is equivalent to linear convolution followed by aliasing. (a) The sequences $x_1[n]$ and $x_2[n]$ to be convolved. (b) The linear convolution of $x_1[n]$ and $x_2[n]$. (c) $x_3[n-N]$ for $N = 6$. (d) $x_3[n+N]$ for $N = 6$. (e) $x_1[n] \, ⑥ \, x_2[n]$, which is equal to the sum of (b), (c), and (d) in the interval $0 \leq n \leq 5$. (f) $x_1[n] \, ⑫ \, x_2[n]$.

convolution of $x_1[n]$ and $x_2[n]$ can be formed by using Eq. (8.131). This is shown in Fig. 8.18(e) for $N = L = 6$ and in Fig. 8.18(f) for $N = 2L = 12$. Note that for $N = L = 6$, only $x_3[n]$ and $x_3[n + N]$ contribute to the result. For $N = 2L = 12$, only $x_3[n]$ contributes to the result. Since the length of the linear convolution is $(2L - 1)$, the result of the circular convolution for $N = 2L$ is identical to the result of linear convolution for all $0 \le n \le N - 1$. In fact, this would be true for $N = 2L - 1 = 11$ as well.

As Example 8.9 points out, time aliasing in the circular convolution of two finite-length sequences can be avoided if $N \ge L + P - 1$. Also it is clear that if $N = L = P$, all of the sequence values of the circular convolution may be different from those of the linear convolution. However, if $P < L$, some of the sequence values in an L-point circular convolution will be equal to corresponding sequence values of the linear convolution. The time-aliasing interpretation is useful for showing this.

Consider two finite-duration sequences $x_1[n]$ and $x_2[n]$, with $x_1[n]$ of length L and $x_2[n]$ of length P, where $P < L$ as indicated in Figs. 8.19(a) and 8.19(b) respectively. Let us first consider the L-point circular convolution of $x_1[n]$ and $x_2[n]$

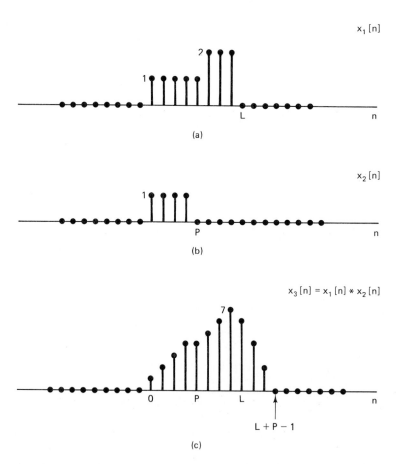

Figure 8.19   An example of linear convolution of two finite-length sequences.

and inquire as to which sequence values in the circular convolution are identical to values that would be obtained from a linear convolution and which are not. The linear convolution of $x_1[n]$ with $x_2[n]$ will be a finite-length sequence of length $(L + P - 1)$ as indicated in Fig. 8.19(c). To determine the L-point circular convolution, we use Eqs. (8.131) and (8.132) so that

$$x_{3p}[n] = \begin{cases} x_1[n] \; \textcircled{L} \; x_2[n] = \displaystyle\sum_{r=-\infty}^{\infty} x_3[n + rL], & 0 \le n \le L - 1, \\ 0, & \text{otherwise.} \end{cases} \quad (8.133)$$

Figure 8.20(a) shows the term in Eq. (8.133) for $r = 0$, and Figs. 8.20(b) and 8.20(c) show the terms for $r = +1$ and $r = -1$ respectively. From Fig. 8.20, it should be clear that in the interval $0 \le n \le L - 1$, $x_{3p}[n]$ is influenced only by $x_3[n]$ and $x_3[n + L]$.

In general whenever $P < L$, only the term $x_3[n + L]$ will alias into the interval $0 \le n \le L - 1$. More specifically, when these terms are summed, the last $(P - 1)$ points of $x_3[n + L]$, which extend from $n = 0$ to $n = P - 2$, will be added to the first $(P - 1)$ points of $x_3[n]$, and the last $(P - 1)$ points of $x_3[n]$, extending from $n = L$ to $n = L + P - 2$, will be discarded. Then $x_{3p}[n]$ is formed by extracting the portion for $0 \le n \le L - 1$. Since the last $(P - 1)$ points of $x_3[n + L]$ and the last $(P - 1)$ points of $x_3[n]$ are identical, we can alternatively view the process of forming the circular convolution $x_{3p}[n]$ through linear convolution plus aliasing as taking the $(P - 1)$ values of $x_3[n]$ from $n = L$ to $n = L + P - 2$ and adding them to the first $(P - 1)$ values of $x_3[n]$. This process is illustrated in Fig. 8.21 for the case $P = 4$ and $L = 8$. Figure 8.21(a) shows the linear convolution $x_3[n]$, with the points for $n \ge L$ denoted by open square symbols. Note that only $(P - 1)$ points for $n \ge L$ are nonzero. Figure 8.21(b) shows the formation of $x_{3p}[n]$ by "wrapping $x_3[n]$ around on itself." The first $(P - 1)$ points are corrupted by the time aliasing, and the remaining points from $n = P - 1$ to $n = L - 1$ (i.e., the last $L - P + 1$ points) are *not* corrupted, i.e., they are identical to what would be obtained with a linear covolution.

From this discussion, it should be clear that if the circular convolution is of sufficient length relative to the lengths of the sequences $x_1[n]$ and $x_2[n]$, then aliasing with nonzero values can be avoided, in which case the circular convolution and linear convolution will be identical. Specifically, if for the case just considered, $x_3[n]$ is replicated with period $N \ge L + P - 1$, then no nonzero overlap will occur. Figures 8.21(c) and 8.21(d) illustrate this case, again for $P = 4$ and $L = 8$, with $N = 11$.

8.9.3 Implementing Linear Time-Invariant Systems Using the DFT

The previous discussion focused on ways of obtaining a linear convolution from a circular convolution. Since linear time-invariant systems can be implemented by convolution, this implies that circular convolution (implemented by the procedure suggested at the beginning of Section 8.9) can be used to implement these systems. To see how this can be done, let us first consider an L-point input sequence $x[n]$ and a P-point impulse response $h[n]$. The linear convolution of these two sequences, which

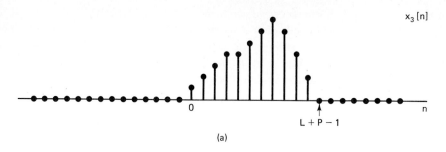

(a)

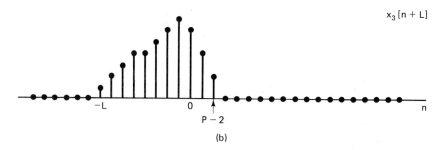

(b)

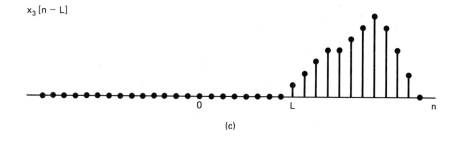

(c)

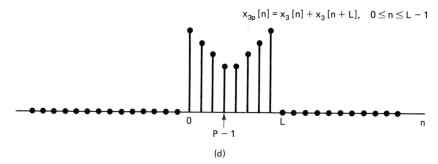

(d)

Figure 8.20　Interpretation of circular convolution as linear convolution followed by aliasing for the circular convolution of the two sequences $x_1[n]$ and $x_2[n]$ in Fig. 8.19.

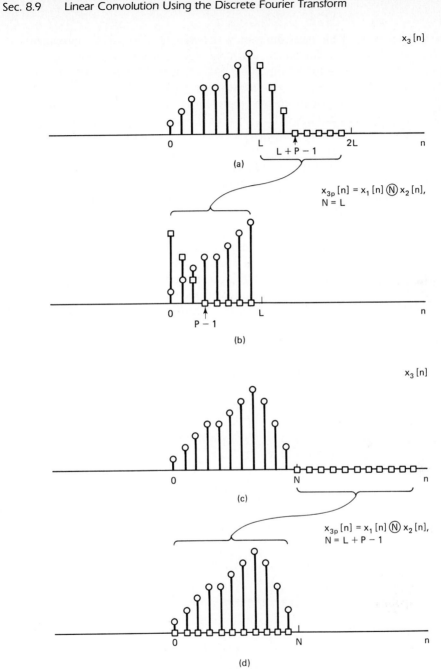

Figure 8.21   Illustration of how the result of a circular convolution "wraps around." (a) and (b) $N = L$, so the aliased "tail" overlaps the first $(P - 1)$ points. (c) and (d) $N = (L + P - 1)$, so no overlap occurs.

will be denoted by $y[n]$, has finite duration with length $(L + P - 1)$. Consequently, as discussed in Section 8.9.2, for the circular convolution and linear convolution to be identical, the circular convolution must have a length of at least $(L + P - 1)$ points. The circular convolution can be achieved by multiplying the DFTs of $x[n]$ and $h[n]$. Since we want the product to represent the DFT of the linear convolution of $x[n]$ and $h[n]$, which has length $(L + P - 1)$, the DFTs that we compute must also be of that length, i.e., both $x[n]$ and $h[n]$ must be augmented with sequence values of zero amplitude. This process is often referred to as *zero padding*.

This procedure permits the computation of the linear convolution of two finite-length sequences using the discrete Fourier transform; i.e., the output of an FIR system whose input also has finite length can be computed with the DFT. In many applications, such as filtering a speech waveform, the input signal is for practical purposes of infinite duration. While theoretically we might be able to store the entire waveform and then implement the procedure discussed above using a DFT for a large number of points, such a DFT is generally impractical to compute. Another consideration is that for this method of filtering no filtered samples can be computed until all the input samples have been collected. Generally we would like to avoid such a large delay in processing. The solution to both of these problems is to use *block convolution*, in which the signal to be filtered is segmented into sections of length L. Each section can then be convolved with the finite-length impulse response and the filtered sections fitted together in an appropriate way. The linear filtering of each block can then be implemented using the DFT.

To illustrate the procedure and to develop the procedure for fitting the filtered sections together, consider the impulse response $h[n]$ of length P and the signal $x[n]$ depicted in Fig. 8.22. Henceforth we will assume that $x[n] = 0$ for $n < 0$ and that the length of $x[n]$ is much greater than P. The sequence $x[n]$ can be represented as a sum of shifted finite-length segments of length L, i.e.,

$$x[n] = \sum_{r=0}^{\infty} x_r[n - rL], \tag{8.134}$$

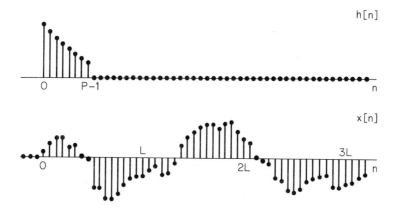

Figure 8.22   Finite-length impulse response $h[n]$ and indefinite-length signal $x[n]$ to be filtered.

where

$$x_r[n] = \begin{cases} x[n + rL], & 0 \le n \le L - 1, \\ 0, & \text{otherwise.} \end{cases} \qquad (8.135)$$

Figure 8.23(a) illustrates this segmentation for $x[n]$ in Fig. 8.22. Note that within each segment the first sample is at $n = 0$; however, the zeroth sample of $x_r[n]$ is the rLth sample of the sequence $x[n]$. This is shown in Fig. 8.23(a) by plotting the segments in their shifted positions.

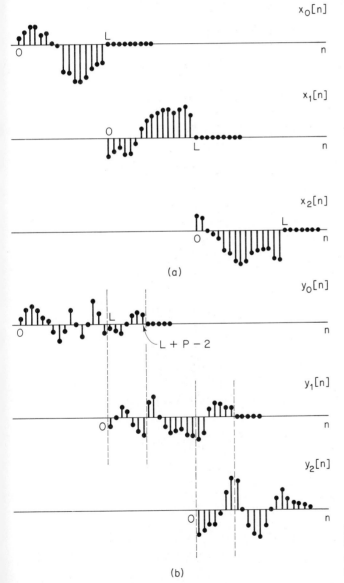

Figure 8.23 (a) Decomposition of $x[n]$ in Fig. 8.22 into nonoverlapping sections of length L. (b) Result of convolving each section with $h[n]$.

Because convolution is a linear time-invariant operation, it follows that

$$y[n] = x[n] * h[n] = \sum_{r=0}^{\infty} y_r[n - rL], \qquad (8.136)$$

where

$$y_r[n] = x_r[n] * h[n]. \qquad (8.137)$$

Since the sequences $x_r[n]$ have only L nonzero points and $h[n]$ is of length P, each of the terms $y_r[n] = x_r[n] * h[n]$ has length $(L + P - 1)$. Thus the linear convolution $x_r[n] * h[n]$ can be obtained by the procedure described above using N-point DFTs where $N \geq L + P - 1$. Since the beginning of each input section is separated from its neighbors by L points and each filtered section has length $(L + P - 1)$, the nonzero points in the filtered sections will overlap by $(P - 1)$ points and these overlap samples must be added in carrying out the sum required by Eq. (8.136). This is illustrated in Fig. 8.23(b), which shows the filtered sections, $y_r[n] = x_r[n] * h[n]$. Just as the input waveform is reconstructed by adding the delayed waveforms in Fig. 8.23(a), the filtered result $x[n] * h[n]$ is constructed by adding the delayed filtered sections depicted in Fig. 8.23(b). This procedure for constructing the filtered output from filtered sections is often referred to as the *overlap-add method* because the filtered sections are overlapped and added to construct the output. The overlapping occurs because the linear convolution of each section with the impulse response is in general longer than the section length. The overlap-add method of block convolution is *not* tied to the DFT and circular convolution. Clearly, all that is required is that the smaller convolutions be computed and the results combined appropriately.

An alternative block convolution procedure, commonly called the *overlap-save method*, corresponds to implementing an L-point circular convolution of a P-point impulse response $h[n]$ with an L-point segment $x_r[n]$ and identifying the part of the circular convolution that corresponds to a linear convolution. The resulting output segments are then "patched together" to form the output. Specifically, we showed that if an L-point sequence is circularly convolved with a P-point sequence $(P < L)$, then the first $(P - 1)$ points of the result are incorrect while the remaining points are identical to those that would be obtained had we implemented a linear convolution. Therefore, we can divide $x[n]$ into sections of length L so that each input section overlaps the preceding section by $(P - 1)$ points. That is, we define the sections $x_r[n]$ as

$$x_r[n] = x[n + r(L - P + 1) - P + 1], \qquad 0 \leq n \leq L - 1, \qquad (8.138)$$

where, as before, we have defined the time origin for each section to be at the beginning of that section rather than at the origin of $x[n]$. This method of sectioning is depicted in Fig. 8.24(a). The circular convolution of each section with $h[n]$ is denoted $y_{rp}[n]$, the extra subscript p indicating that $y_{rp}[n]$ is the result of a circular convolution in which time aliasing has occurred. These sequences are depicted in Fig. 8.24(b). The portion of each output section in the region $0 \leq n \leq P - 2$ is the part

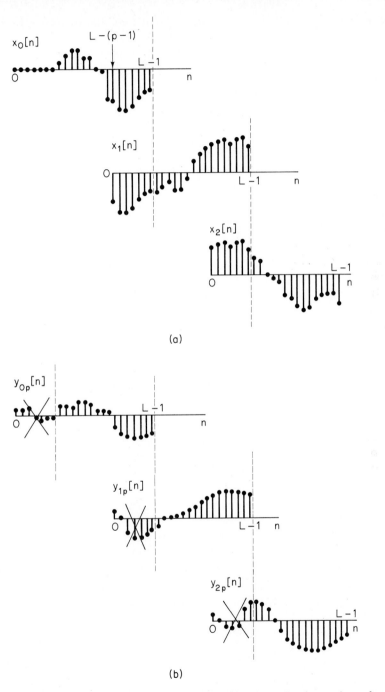

Figure 8.24 (a) Decomposition of $x[n]$ in Fig. 8.22 into overlapping sections of length L. (b) Result of convolving each section with $h[n]$. The portions of each filtered section to be discarded in forming the linear convolution are indicated.

that must be discarded. The remaining samples from successive sections are then abutted to construct the final filtered output. That is,

$$y[n] = \sum_{r=0}^{\infty} y_r[n - r(L - P + 1) + P - 1], \qquad (8.139)$$

where

$$y_r[n] = \begin{cases} y_{rp}[n], & P - 1 \le n \le L - 1, \\ 0, & \text{otherwise}. \end{cases} \qquad (8.140)$$

This procedure is called the overlap-save method because the input segments overlap so that each succeeding input section consists of $(L - P + 1)$ new points and $(P - 1)$ points saved from the previous section.

The utility of the overlap-add and the overlap-save methods of block convolution may not be immediately apparent. In Chapter 9 we consider highly efficient algorithms for computing the DFT. These algorithms, collectively called the fast Fourier transform (FFT), are so efficient that for FIR impulse responses of even modest length (on the order of 25 or 30) it may be more efficient to carry out block convolution using the DFT than to implement the linear convolution directly. The length P at which the DFT method becomes more efficient is, of course, dependent on the hardware and software available to implement the computations. (See Stockham, 1966, and Helms, 1967.)

8.10 SUMMARY

In this chapter we have discussed a Fourier representation of finite-length sequences. This representation, called the discrete Fourier transform (DFT), is based on the discrete Fourier series representation of periodic sequences. By defining a periodic sequence for which each period is identical to the finite-length sequence, the DFT becomes identical to one period of the discrete Fourier series coefficients. Because of the importance of this underlying periodicity, we first examined the properties of discrete Fourier series representations and then interpreted those properties in terms of finite-length sequences. An important result is that the DFT values are equal to samples of the z-transform at equally spaced points on the unit circle. This leads to the notion of time aliasing in the interpretation of DFT properties. This concept was used extensively in the study of circular convolution and its relation to linear convolution. The chapter concluded with a discussion of how the DFT can be used to implement linear convolution of a finite-length impulse response with an indefinitely long input signal.

The DFT is the most widely used example of a general class of finite-length orthogonal transform representations of the form

$$x[n] = \frac{1}{N} \sum_{k=0}^{N-1} A[k]\phi_k[n], \qquad (8.141)$$

$$A[k] = \sum_{n=0}^{N-1} x[n]\phi_k^*[n], \qquad (8.142)$$

where the sequences $\phi_k[n]$ are orthogonal to one another. Other examples include the cosine transform and the Hartley transform, among many others.

PROBLEMS

8.1. Suppose $x_c(t)$ is a periodic continuous-time signal which has period 1 ms and for which the Fourier series is

$$x_c(t) = \sum_{k=-9}^{9} a_k e^{j(2\pi kt/10^{-3})}.$$

The Fourier series coefficients a_k are zero for $|k| > 9$. $x_c(t)$ is sampled with a sample spacing $T = \frac{1}{6} \times 10^{-3}$ s to form $x[n]$. That is,

$$x[n] = x_c\left(\frac{n10^{-3}}{6}\right).$$

(a) Is $x[n]$ periodic and, if so, with what period?
(b) Is the sampling rate above the Nyquist rate, i.e., is T sufficiently small to avoid aliasing?
(c) Find the discrete Fourier series coefficients of $x[n]$ in terms of a_k.

8.2. Suppose $\tilde{x}[n]$ is a periodic sequence with period N. Then $\tilde{x}[n]$ is also periodic with period $3N$. Let $\tilde{X}[k]$ denote the DFS coefficients of $\tilde{x}[n]$ considered as a periodic sequence with period N and let $\tilde{X}_3[k]$ denote the DFS coefficients of $\tilde{x}[n]$ considered as a periodic sequence with period $3N$.
(a) Express $\tilde{X}_3[k]$ in terms of $\tilde{X}[k]$.
(b) By explicitly calculating $\tilde{X}[k]$ and $\tilde{X}_3[k]$, verify your result in part (a) when $\tilde{x}[n]$ is as given in Fig. P8.2.

$$\tilde{x}[n], \quad N = 2$$

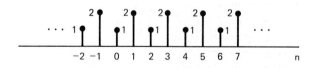

$$-2 \ -1 \quad 0 \quad 1 \quad 2 \quad 3 \quad 4 \quad 5 \quad 6 \quad 7 \qquad n \quad \textbf{Figure P8.2}$$

8.3. In deriving the DFS analysis equation (8.11), we used the identity of Eq. (8.7). To verify this identity, we will consider the two conditions $k - r = mN$ and $k - r \neq mN$ separately.
(a) For $k - r = mN$, show that $e^{j(2\pi/N)(k-r)n} = 1$ and from this that

$$\frac{1}{N} \sum_{n=0}^{N-1} e^{j(2\pi/N)(k-r)n} = 1 \qquad \text{for } k - r = mN.$$

Since k and r are both integers in Eq. (8.7), we can make the substitution $k - r = \ell$ and consider the summation

$$\frac{1}{N} \sum_{n=0}^{N-1} e^{j(2\pi/N)\ell n} = \frac{1}{N} \sum_{n=0}^{N-1} [e^{j(2\pi/N)\ell}]^n.$$

Since this is the sum of a finite number of terms in a geometric series, it can be expressed in closed form as

$$\frac{1}{N} \sum_{n=0}^{N-1} [e^{j(2\pi/N)\ell}]^n = \frac{1 - e^{j(2\pi/N)\ell N}}{1 - e^{j(2\pi/N)\ell}}. \tag{P8.3}$$

(b) For what values of ℓ is the right-hand side of Eq. (P8.3) indeterminate, i.e., are the numerator and denominator both zero?

(c) From the result in part (b), show that if $k - r \neq mN$, then

$$\frac{1}{N} \sum_{n=0}^{N-1} e^{j(2\pi/N)(k-r)n} = 0.$$

8.4. In Section 8.2.2, we stated the property that

$$\text{if} \quad \tilde{x}_1[n] = \tilde{x}[n - m],$$

$$\text{then} \quad \tilde{X}_1[k] = W_N^{km} \tilde{X}[k],$$

where $\tilde{X}[k]$ and $\tilde{X}_1[k]$ are the DFS coefficients of $\tilde{x}[n]$ and $\tilde{x}_1[n]$, respectively. In this problem, we consider the proof of this property.

(a) Using Eq. (8.11) together with an appropriate substitution of variables, show that $\tilde{X}_1[k]$ can be expressed as

$$\tilde{X}_1[k] = W_N^{km} \sum_{r=-m}^{N-1-m} \tilde{x}[r] W_N^{kr}. \tag{P8.4-1}$$

(b) The summation in Eq. (P8.4-1) can be rewritten as

$$\sum_{r=-m}^{N-1-m} \tilde{x}[r] W_N^{kr} = \sum_{r=-m}^{-1} \tilde{x}[r] W_N^{kr} + \sum_{r=0}^{N-1-m} \tilde{x}[r] W_N^{kr}. \tag{P8.4-2}$$

Using the fact that $\tilde{x}[r]$ and W_N^{kr} are both periodic, show that

$$\sum_{r=-m}^{-1} \tilde{x}[r] W_N^{kr} = \sum_{N-m}^{N-1} \tilde{x}[r] W_N^{kr}. \tag{P8.4-3}$$

(c) From your results in parts (a) and (b), show that

$$\tilde{X}_1[k] = W_N^{km} \sum_{r=0}^{N-1} \tilde{x}[r] W_N^{kr} = W_N^{km} \tilde{X}[k].$$

8.5. (a) In Table 8.1 we list a number of symmetry properties of the discrete Fourier series for periodic sequences, several of which we repeat below. Prove that each of these properties is true. In carrying out the proof, you may use the definition of the discrete Fourier series and any previous property in the list (for example, in proving property 3 you may use properties 1 and 2).

Sequence	Discrete Fourier Series
1. $\tilde{x}^*[n]$	$\tilde{X}^*[-k]$
2. $\tilde{x}^*[-n]$	$\tilde{X}^*[k]$
3. $\mathcal{Re}\{\tilde{x}(n)\}$	$\tilde{X}_e[k]$
4. $j\mathcal{Im}\{\tilde{x}(n)\}$	$\tilde{X}_o[k]$

(b) From the properties proved in part (a), show that for a real periodic sequence $\tilde{x}[n]$, the following symmetry properties of the discrete Fourier series hold:

1. $\mathcal{R}e\{\tilde{X}[k]\} = \mathcal{R}e\{\tilde{X}[-k]\}$
2. $\mathcal{I}m\{\tilde{X}[k]\} = -\mathcal{I}m\{\tilde{X}[-k]\}$
3. $|\tilde{X}[k]| = |\tilde{X}[-k]|$
4. $\sphericalangle \tilde{X}[k] = -\sphericalangle \tilde{X}[-k]$

8.6. **(a)** Figure P8.6-1 shows two periodic sequences, $\tilde{x}_1[n]$ and $\tilde{x}_2[n]$, with period $N = 7$. Find a sequence $\tilde{y}_1[n]$ whose DFS is equal to the product of the DFS of $\tilde{x}_1[n]$ and the DFS of $\tilde{x}_2[n]$, i.e.,

$$\tilde{Y}_1[k] = \tilde{X}_1[k]\tilde{X}_2[k].$$

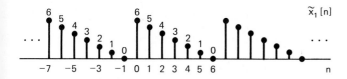

Figure P8.6-1

(b) Figure P8.6-2 shows a periodic sequence $\tilde{x}_3[n]$ with period $N = 7$. Find a sequence $\tilde{y}_2[n]$ whose DFS is equal to the product of the DFS of $\tilde{x}_1[n]$ and the DFS of $\tilde{x}_3[n]$, i.e.,

$$\tilde{Y}_2[k] = \tilde{X}_1[k]\tilde{X}_3[k].$$

Figure P8.6-2

8.7. Figure P8.7 shows several periodic sequences $\tilde{x}[n]$. These sequences can be expressed in a Fourier series as

$$\tilde{x}[n] = \frac{1}{N} \sum_{k=0}^{N-1} \tilde{X}[k]e^{j(2\pi/N)kn}.$$

(a) For which sequences can the time origin be chosen such that all the $\tilde{X}[k]$ are real?
(b) For which sequences can the time origin be chosen such that all the $\tilde{X}[k]$ (except for k an integer multiple of N) are imaginary?
(c) For which sequences does $\tilde{X}[k] = 0$, $k = \pm 2, \pm 4, \pm 6$, etc.?

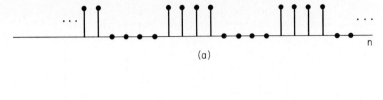

(a)

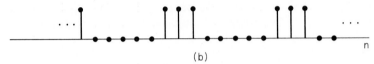

(b)

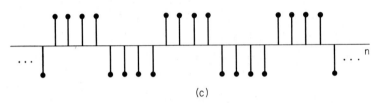

(c)

Figure P8.7

8.8. We stated in Section 8.5 that a direct relationship between $X(e^{j\omega})$ and $\tilde{X}[k]$ can be derived, where $\tilde{X}[k]$ is the DFS coefficients of a periodic sequence and $X(e^{j\omega})$ is the Fourier transform of one period. Since $\tilde{X}[k]$ corresponds to samples of $X(e^{j\omega})$, the relationship then corresponds to an interpolation formula.

One approach to obtaining the desired relationship is to rely on the discussion of Section 8.5, the relationship of Eq. (8.48), and the modulation property of Section 2.9.7. The procedure is as follows:

1. With $\tilde{X}[k]$ denoting the DFS coefficients of $\tilde{x}[n]$, express the Fourier transform $\tilde{X}(e^{j\omega})$ of $\tilde{x}[n]$ as an impulse train.

2. From Eq. (8.51), $x[n]$ can be expressed as $x[n] = \tilde{x}[n]w[n]$, where $w[n]$ is an appropriate finite-length window.

3. Since $x[n] = \tilde{x}[n]w[n]$, from Section 2.9.7, $X(e^{j\omega})$ can be expressed as the (periodic) convolution of $\tilde{X}(e^{j\omega})$ and $W(e^{j\omega})$.

By carrying out the details in this procedure, show that $X(e^{j\omega})$ can be expressed as

$$X(e^{j\omega}) = \frac{1}{N} \sum_k \tilde{X}[k] \frac{\sin[(\omega N - 2\pi k)/2]}{\sin\{[\omega - (2\pi k/N)]/2\}} e^{-j[(N-1)/2](\omega - 2\pi k/N)}.$$

Explicitly specify the limits on the summation.

8.9. Consider the sequence $x[n]$ given by $x[n] = \alpha^n u[n]$. A periodic sequence $\tilde{x}[n]$ is constructed from $x[n]$ in the following way:

$$\tilde{x}[n] = \sum_{r=-\infty}^{\infty} x[n + rN].$$

(a) Determine the Fourier transform $X(e^{j\omega})$ of $x[n]$.

(b) Determine the discrete Fourier series $\tilde{X}[k]$ of $\tilde{x}[n]$.

(c) How is $\tilde{X}[k]$ related to $X(e^{j\omega})$?

8.10. Consider a finite-length sequence $x[n]$ of length N, i.e.,

$$x[n] = 0 \qquad \text{outside} \quad 0 \le n \le N - 1.$$

$X(e^{j\omega})$ denotes the Fourier transform of $x[n]$. $\tilde{X}[k]$ denotes the sequence of 64 equally spaced samples of $X(e^{j\omega})$, i.e.,

$$\tilde{X}[k] = X(e^{j\omega})|_{\omega = 2\pi k/64}.$$

It is known that in the range $0 \le k \le 63$, $\tilde{X}[32] = 1$ and all the other values of $\tilde{X}[k]$ are zero.

(a) If the sequence length is $N = 64$, determine one sequence $x[n]$ consistent with the given information. Indicate whether the answer is unique. If it is, clearly explain why. If it is not, give a second *distinct* choice.

(b) If the sequence length is $N = 192 = 3 \times 64$, determine one sequence $x[n]$ consistent with the constraint that in the range $0 \le k \le 63$, $\tilde{X}[32] = 1$ and all other values in that range are zero. Indicate whether the answer is unique. If it is, clearly explain why. If it is not, give a second *distinct* choice.

8.11. Compute the DFT of each of the following finite-length sequences considered to be of length N (where N is even).

(a) $x[n] = \delta[n]$

(b) $x[n] = \delta[n - n_0], \qquad 0 \le n_0 \le N - 1$

(c) $x[n] = \begin{cases} 1, & n \text{ even}, \quad 0 \le n \le N - 1 \\ 0, & n \text{ odd}, \quad 0 \le n \le N - 1 \end{cases}$

(d) $x[n] = \begin{cases} 1, & 0 \le n \le N/2 - 1 \\ 0, & N/2 \le n \le N - 1 \end{cases}$

(e) $x[n] = \begin{cases} a^n, & 0 \le n \le N - 1 \\ 0, & \text{otherwise} \end{cases}$

8.12. Consider the complex sequence

$$x[n] = \begin{cases} e^{j\omega_0 n}, & 0 \le n \le N - 1, \\ 0, & \text{otherwise.} \end{cases}$$

(a) Find the Fourier transform $X(e^{j\omega})$ of $x[n]$.

(b) Find the N-point DFT $X[k]$ of the finite-length sequence $x[n]$.

(c) Find the DFT of $x[n]$ for the case $\omega_0 = 2\pi k_0/N$ where k_0 is an integer.

8.13. A continuous-time signal is sampled at a sampling rate of 10 kHz and the DFT of 1024 samples computed. Determine the frequency spacing between spectral samples. Justify your answer.

8.14. The DFT of a finite-duration sequence corresponds to samples of its z-transform on the unit circle. For example, the DFT of a 10-point sequence $x[n]$ corresponds to samples of $X(z)$ at the 10 equally spaced points indicated in Fig. P8.14-1. We wish to find the equally spaced samples of $X(z)$ on the contour shown in Fig. P8.14-2; i.e., we wish to obtain

$$X(z)\Big|_{z = 0.5e^{j[(2\pi k/10) + (\pi/10)]}}.$$

Show how to modify $x[n]$ to obtain a sequence $x_1[n]$ such that the DFT of $x_1[n]$ corresponds to the desired samples of $X(z)$.

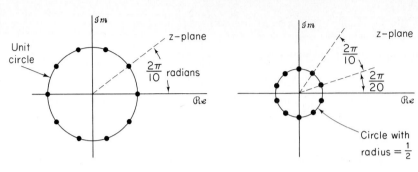

Figure P8.14-1   Figure P8.14-2

8.15. $x[n]$ denotes a finite-length sequence of length N. Show that

$$x[((-n))_N] = x[((N-n))_N].$$

8.16. Figure P8.16 shows a finite-length sequence $x[n]$. Sketch the sequences $x_1[n]$ and $x_2[n]$ specified as

$$x_1[n] = x[((n-2))_4], \qquad 0 \le n \le 3,$$
$$x_2[n] = x[((-n))_4], \qquad 0 \le n \le 3.$$

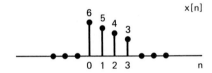

Figure P8.16

8.17. A finite-duration sequence $x[n]$ of length 8 has the 8-point DFT $X[k]$ shown in Fig. P8.17-1. A new sequence $y[n]$ of length 16 is defined by

$$y[n] = \begin{cases} x[n/2], & n \text{ even}, \\ 0, & n \text{ odd}. \end{cases}$$

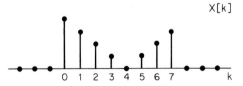

Figure P8.17-1

From the list in Fig. P8.17-2, choose the sketch corresponding to the 16-point DFT of $y[n]$.

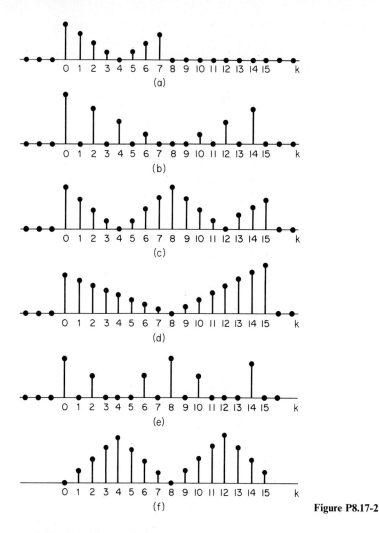

(a)

(b)

(c)

(d)

(e)

(f)   **Figure P8.17-2**

8.18. Consider the finite-length sequence $x[n]$ in Fig. P8.18. Let $X(z)$ be the z-transform of $x[n]$. If we sample $X(z)$ at $z = e^{j(2\pi/4)k}$, $k = 0, 1, 2, 3$, we obtain

$$X[k] = X(z)\Big|_{z = e^{j(2\pi/4)k}}, \qquad k = 0, 1, 2, 3.$$

Sketch the sequence $x_1[n]$ obtained as the inverse DFT of $X[k]$.

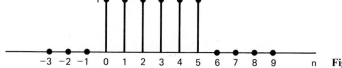

Figure P8.18

8.19. Consider a finite-length sequence $x[n]$ of length N as indicated in Fig. P8.19-1. (The solid line is used to suggest the envelope of the sequence values between 0 and $N-1$.) Two finite-length sequences $x_1[n]$ and $x_2[n]$ of length $2N$ are constructed from $x[n]$ as indicated in Fig. P8.19-2, with $x_1[n]$ and $x_2[n]$ given as follows:

$$x_1[n] = \begin{cases} x[n], & 0 \le n \le N-1, \\ 0, & \text{otherwise,} \end{cases}$$

$$x_2[n] = \begin{cases} x[n], & 0 \le n \le N-1, \\ -x[n-N], & N \le n \le 2N-1, \\ 0, & \text{otherwise.} \end{cases}$$

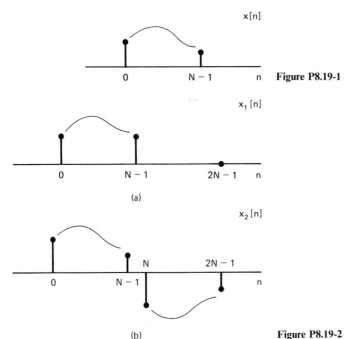

$x[n]$

Figure P8.19-1

$x_1[n]$

(a)

$x_2[n]$

(b)

Figure P8.19-2

The N-point DFT of $x[n]$ is denoted by $X[k]$, and the $2N$-point DFTs of $x_1[n]$ and $x_2[n]$ are denoted by $X_1[k]$ and $X_2[k]$, respectively.

(a) Specify whether $X_2[k]$ can be obtained if $X[k]$ is given. Clearly indicate your reasoning.

(b) Determine the simplest possible relationship to obtain $X[k]$ from $X_1[k]$.

8.20. Let $X(e^{j\omega})$ denote the Fourier transform of the sequence $x[n] = (\frac{1}{2})^n u[n]$. Let $y[n]$ denote a finite-duration sequence of length 10; i.e., $y[n] = 0$, $n < 0$, and $y[n] = 0$, $n \ge 10$. The 10-point DFT of $y[n]$, denoted by $Y[k]$, corresponds to 10 equally spaced samples of $X(e^{j\omega})$; i.e., $Y[k] = X(e^{j2\pi k/10})$. Determine $y[n]$.

8.21. Consider a finite-duration sequence $x[n]$ of length P such that $x[n] = 0$ for $n < 0$ and $n \ge P$. We want to compute samples of the Fourier transform at the N equally spaced frequencies

$$\omega_k = \frac{2\pi k}{N}, \quad k = 0, 1, \dots, N-1.$$

Determine and justify procedures for computing the N samples of the Fourier transform using only one N-point DFT for the following two cases.

(a) $N > P$

(b) $N < P$ [*Hint*: Consider applying time aliasing to $x[n]$.]

8.22. Consider a 20-point finite-duration sequence $x[n]$ such that $x[n] = 0$ outside $0 \le n \le 19$, and let $X(e^{j\omega})$ represent the Fourier transform of $x[n]$.

(a) If it is desired to evaluate $X(e^{j\omega})$ at $\omega = 4\pi/5$ by computing one M-point DFT, determine the smallest possible M and develop a method to obtain $X(e^{j\omega})$ at $\omega = 4\pi/5$ using the smallest M.

(b) If it is desired to evaluate $X(e^{j\omega})$ at $\omega = 10\pi/27$ by computing one L-point DFT, determine the smallest possible L and develop a method to obtain $X(e^{j10\pi/27})$ using the smallest L.

8.23. Consider a real finite-length sequence $x[n]$ with Fourier transform $X(e^{j\omega})$ and DFT $X[k]$. If

$$\mathscr{Im}\{X[k]\} = 0, \qquad k = 0, 1, \ldots, N - 1,$$

can we conclude that

$$\mathscr{Im}\{X(e^{j\omega})\} = 0, \qquad -\pi \le \omega \le \pi?$$

Concisely state your reasoning if your answer is yes. Give a counterexample if your answer is no.

8.24. Consider the finite-length sequence $x[n]$ in Fig. P8.24. The 4-point DFT of $x[n]$ is denoted $X[k]$. Plot the sequence $y[n]$ whose DFT is

$$Y[k] = W_4^{3k} X[k].$$

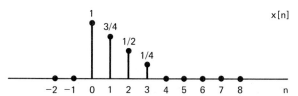

Figure P8.24

8.25. The two 8-point sequences $x_1[n]$ and $x_2[n]$ shown in Fig. P8.25 have DFTs $X_1[k]$ and $X_2[k]$, respectively. Determine the relationship between $X_1[k]$ and $X_2[k]$.

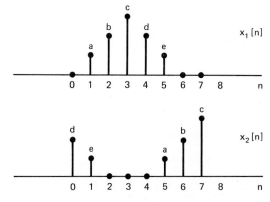

Figure P8.25

8.26. Let $X[k]$ denote the N-point DFT of the N-point sequence $x[n]$.

(a) Show that if $x[n]$ satisfies the relation

$$x[n] = -x[N - 1 - n],$$

then $X[0] = 0$. Consider separately the cases of N even and N odd.

(b) Show that if N is even and if

$$x[n] = x[N - 1 - n],$$

then $X[N/2] = 0$.

8.27. The even part of a real sequence $x[n]$ is defined by

$$x_e[n] = \frac{x[n] + x[-n]}{2}.$$

Suppose that $x[n]$ is a real finite-length sequence defined such that $x[n] = 0$ for $n < 0$ and $n \geq N$. Let $X[k]$ denote the N-point DFT of $x[n]$.

(a) Is $\mathscr{R}e\{X[k]\}$ the DFT of $x_e[n]$?

(b) What is the inverse DFT of $\mathscr{R}e\{X[k]\}$ in terms of $x[n]$?

8.28. In Chapter 2, the conjugate-symmetric and conjugate-antisymmetric components of a sequence $x[n]$ were defined, respectively, as

$$x_e[n] = \tfrac{1}{2}(x[n] + x^*[-n]),$$

$$x_o[n] = \tfrac{1}{2}(x[n] - x^*[-n]).$$

In Section 8.7.4, we found it convenient to define the periodic conjugate-symmetric and periodic conjugate-antisymmetric components of a sequence of finite duration N as

$$x_{ep}[n] = \tfrac{1}{2}\{x[((n))_N] + x^*[((-n))_N]\}, \qquad 0 \leq n \leq N - 1,$$

$$x_{op}[n] = \tfrac{1}{2}\{x[((n))_N] - x^*[((-n))_N]\}, \qquad 0 \leq n \leq N - 1.$$

(a) Show that $x_{ep}[n]$ can be related to $x_e[n]$ and that $x_{op}[n]$ can be related to $x_o[n]$ by the relations

$$x_{ep}[n] = (x_e[n] + x_e[n - N]), \qquad 0 \leq n \leq N - 1,$$

$$x_{op}[n] = (x_o[n] + x_o[n - N]), \qquad 0 \leq n \leq N - 1.$$

(b) $x[n]$ is considered to be a sequence of length N, and in general $x_e[n]$ cannot be recovered from $x_{ep}[n]$, and $x_o[n]$ cannot be recovered from $x_{op}[n]$. Show that with $x[n]$ considered as a sequence of length N, but with $x[n] = 0$, $n > N/2$, $x_e[n]$ can be obtained from $x_{ep}[n]$, and $x_o[n]$ can be obtained from $x_{op}[n]$.

8.29. Consider the real finite-length sequence $x[n]$ shown in Fig. P8.29.

(a) Sketch the finite-length sequence $y[n]$ whose 6-point DFT is

$$Y[k] = W_6^{4k} X[k],$$

where $X[k]$ is the 6-point DFT of $x[n]$.

(b) Sketch the finite-length sequence $w[n]$ whose 6-point DFT is

$$W[k] = \mathscr{R}e\{X[k]\}.$$

(c) Sketch the finite-length sequence $q[n]$ whose 3-point DFT is

$$Q[k] = X[2k], \qquad k = 0, 1, 2.$$

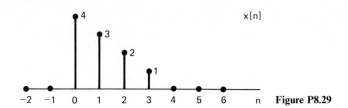

Figure P8.29

8.30. Figure P8.30 shows two finite-length sequences. Sketch their 6-point circular convolution.

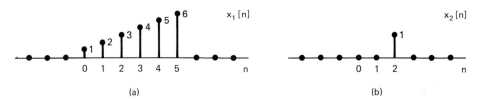

(a)   (b)

Figure P8.30

8.31. Figure P8.31 shows two finite-length sequences. Sketch their N-point circular convolution for $N = 6$ and for $N = 10$.

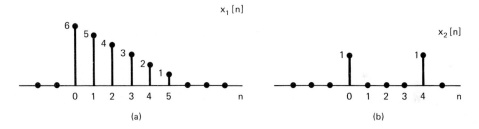

(a)   (b)

Figure P8.31

8.32. Suppose we have two 4-point sequences $x[n]$ and $h[n]$ as follows:

$$x[n] = \cos\left(\frac{\pi n}{2}\right), \qquad n = 0, 1, 2, 3,$$

$$h[n] = 2^n, \qquad n = 0, 1, 2, 3.$$

(a) Calculate the 4-point DFT $X[k]$.
(b) Calculate the 4-point DFT $H[k]$.
(c) Calculate $y[n] = x[n]④h[n]$ by doing the circular convolution directly.
(d) Calculate $y[n]$ of part (c) by multiplying the DFTs of $x[n]$ and $h[n]$ and performing an inverse DFT.

8.33. Determine a sequence $x[n]$ that satisfies all of the following three conditions:
        Condition 1: The Fourier transform of $x[n]$ has the following form:

$$X(e^{j\omega}) = 1 + A_1 \cos \omega + A_2 \cos 2\omega,$$

where A_1 and A_2 are some unknown constants.

Condition 2: The sequence $x[n] * \delta[n - 3]$, evaluated at $n = 2$, is 5.

Condition 3: For the 3-point sequence $w[n]$ as shown in Fig. P8.33, the result of the 8-point circular convolution of $w[n]$ and $x[n - 3]$ is 11 when $n = 2$, i.e.,

$$\left. \sum_{m=0}^{7} w[m]x[((n - 3 - m))_8] \right|_{n=2} = 11.$$

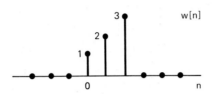

$w[n]$ **Figure P8.33**

8.34. Consider the finite-length sequence

$$x[n] = 2\delta[n] + \delta[n - 1] + \delta[n - 3].$$

We perform the following operation on this sequence:

(i) We compute the 5-point DFT $X[k]$.

(ii) We compute a 5-point inverse DFT of $Y[k] = X[k]^2$ to obtain a sequence $y[n]$.

    (a) Determine the sequence $y[n]$ for $n = 0, 1, 2, 3, 4$.

    (b) If N-point DFTs are used in the two-step procedure, how should we choose N so that $y[n] = x[n] * x[n]$ for $0 \le n \le N - 1$?

8.35. Consider a finite-duration sequence $x[n]$, which is zero for $n < 0$ and $n \ge N$, where N is even. The z-transform of $x[n]$ is denoted by $X(z)$. Table P8.35-1 lists seven sequences obtained from $x[n]$. Table P8.35-2 lists nine sequences obtained from $X(z)$. For each sequence in Table P8.35-1, find its DFT in Table P8.35-2. The size of the transform considered must be greater than or equal to the length of the sequence $g_k[n]$. *For purposes of illustration only* assume that $x[n]$ can be represented by the envelope shown in Fig. P8.35.

TABLE P8.35-1

$g_1[n] = x[N - 1 - n]$

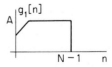

$g_2[n] = (-1)^n x[n]$

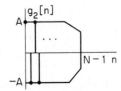

$$g_3[n] = \begin{cases} x[n], & 0 \le n \le N-1, \\ x[n-N], & N \le n \le 2N-1, \\ 0, & \text{otherwise} \end{cases}$$

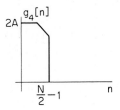

$$g_4[n] = \begin{cases} x[n] + x[n+N/2], & 0 \le n \le N/2-1, \\ 0, & \text{otherwise} \end{cases}$$

$$g_5[n] = \begin{cases} x[n], & 0 \le n \le N-1, \\ 0, & N \le n \le 2N-1, \\ 0, & \text{otherwise} \end{cases}$$

$$g_6(n) = \begin{cases} x[n/2], & n \text{ even}, \\ 0, & n \text{ odd} \end{cases}$$

$$g_7[n] = x(2n)$$

TABLE P8.35-2

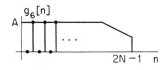

$$H_1[k] = X(e^{j2\pi k/N})$$
$$H_2[k] = X(e^{j2\pi k/2N})$$
$$H_3[k] = \begin{cases} 2X(e^{j2\pi k/2N}), & k \text{ even}, \\ 0, & k \text{ odd} \end{cases}$$
$$H_4[k] = X(e^{j2\pi k/(2N-1)})$$
$$H_5[k] = 0.5\{X(e^{j2\pi k/N}) + X[e^{j2\pi(k+N/2)/N}]\}$$
$$H_6[k] = X(e^{j4\pi k/N})$$
$$H_7[k] = e^{j2\pi k/N}X(e^{-j2\pi k/N})$$
$$H_8[k] = X[e^{j(2\pi/N)(k+N/2)}]$$
$$H_9[k] = X(e^{-j2\pi k/N})$$

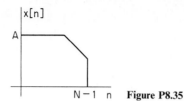

Figure P8.35

8.36. Show from Eqs. (8.59) and (8.60) that with $x[n]$ as an N-point sequence and $X[k]$ as its N-point DFT,

$$\sum_{n=0}^{N-1} |x[n]|^2 = \frac{1}{N} \sum_{k=0}^{N-1} |X[k]|^2.$$

This is commonly referred to as *Parseval's relation* for the DFT.

8.37. $x[n]$ is a real-valued finite-length sequence of length 10 and is nonzero in the interval from 0 to 9, i.e.,

$$x[n] = 0, \quad n < 0, n \geq 10,$$
$$x[n] \neq 0, \quad 0 \leq n \leq 9.$$

$X(e^{j\omega})$ denotes the Fourier transform of $x[n]$, and $X[k]$ denotes the 10-point DFT of $x[n]$.

Determine a choice for $x[n]$ so that $X[k]$ is real-valued for all k and

$$X(e^{j\omega}) = A(\omega)e^{j\alpha\omega}, \quad |\omega| < \pi,$$

where $A(\omega)$ is real and α is a nonzero real constant.

8.38. $x[n]$ is a real-valued, nonnegative, finite-length sequence of length N, i.e., $x[n]$ is real and nonnegative for $0 \leq n \leq N - 1$ and is zero otherwise. The N-point DFT of $x[n]$ is $X[k]$, and the Fourier transform of $x[n]$ is $X(e^{j\omega})$.

Determine whether each of the following statements is true or false. If you indicate it is true, clearly show your reasoning. If false, construct a counterexample.

(a) If $X(e^{j\omega})$ is expressible in the form

$$X(e^{j\omega}) = B(\omega)e^{j\alpha\omega},$$

where $B(\omega)$ is real and α is a real constant, then $X[k]$ can be expressed in the form

$$X[k] = A[k]e^{j\gamma k},$$

where $A[k]$ is real and γ is a real constant.

(b) If $X[k]$ is expressible in the form

$$X[k] = A[k]e^{j\gamma k},$$

where $A[k]$ is real and γ is a real constant, then $X(e^{j\omega})$ can be expressed in the form

$$X(e^{j\omega}) = B(\omega)e^{j\alpha\omega},$$

where $B(\omega)$ is real and α is a real constant.

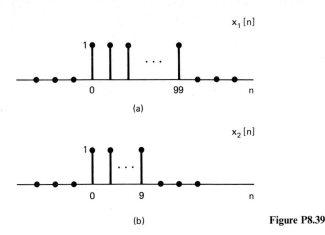

(a)

(b)   **Figure P8.39**

8.39. Figure P8.39 shows two sequences, $x_1[n]$ and $x_2[n]$, where

$$x_1[n] = \begin{cases} 1, & 0 \le n \le 99, \\ 0, & \text{otherwise,} \end{cases}$$

$$x_2[n] = \begin{cases} 1, & 0 \le n \le 9, \\ 0, & \text{otherwise.} \end{cases}$$

(a) Determine and sketch the linear convolution $x_1[n] * x_2[n]$.
(b) Determine and sketch the 100-point circular convolution $x_1[n] \,(100)\, x_2[n]$.

(c) Determine and sketch the 110-point circular convolution $x_1[n] \,(110)\, x_2[n]$.

8.40. Two finite-length sequences $x_1[n]$ and $x_2[n]$, which are zero outside the interval $0 \le n \le 99$, are circularly convolved to form a new sequence $y[n]$, i.e.,

$$y[n] = x_1[n] \,(100)\, x_2[n] = \sum_{k=0}^{99} x_1[k]x_2[((n-k))_{100}], \qquad 0 \le n \le 99.$$

If, in fact, $x_1[n]$ is nonzero only for $10 \le n \le 39$, determine the set of values of n for which $y[n]$ is guaranteed to be identical to the *linear* convolution of $x_1[n]$ and $x_2[n]$.

8.41. Consider two finite-length sequences $x[n]$ and $h[n]$ for which $x[n] = 0$ outside the interval $0 \le n \le 49$ and $h[n] = 0$ outside the interval $0 \le n \le 9$.
(a) What is the maximum possible number of nonzero values in the *linear* convolution of $x[n]$ and $h[n]$?
(b) The 50-point *circular* convolution of $x[n]$ and $h[n]$ is

$$x[n] \,(50)\, h[n] = 10, \qquad 0 \le n \le 49.$$

The first 5 points of the *linear* convolution of $x[n]$ and $h[n]$ are

$$x[n] * h[n] = 5, \qquad 0 \le n \le 4.$$

Determine as many points as possible of the linear convolution of $x[n] * h[n]$.

8.42. Consider two finite-duration sequences $x[n]$ and $y[n]$. $x[n]$ is zero for $n < 0, n \ge 40$, and $9 < n < 30$, and $y[n]$ is zero for $n < 10$ and $n > 19$, as indicated in Fig. P8.42.

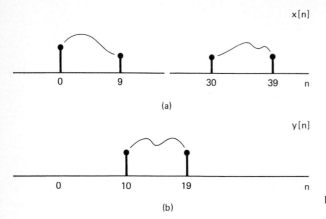

(a)

(b)

Figure P8.42

Let $w[n]$ denote the linear convolution of $x[n]$ and $y[n]$. Let $g[n]$ denote the 40-point circular convolution of $x[n]$ and $y[n]$:

$$w[n] = x[n] * y[n] = \sum_{k=-\infty}^{\infty} x[k]y[n-k],$$

$$g[n] = x[n] \,\textcircled{40}\, y[n] = \sum_{k=0}^{39} x[k]y[((n-k))_{40}].$$

(a) Determine the values of n for which $w[n]$ can be nonzero.

(b) Determine the values of n for which $w[n]$ can be obtained from $g[n]$. Explicitly specify at what index values n in $g[n]$ these values of $w[n]$ appear.

8.43. $x[n]$ and $y[n]$ are two real-valued, positive, finite-length sequences of length 256, i.e.,

$$
\begin{aligned}
x[n] > 0, && 0 \le n \le 255, \\
y[n] > 0, && 0 \le n \le 255, \\
x[n] = y[n] = 0, && \text{otherwise.}
\end{aligned}
$$

$r[n]$ denotes the *linear* convolution of $x[n]$ and $y[n]$. $R(e^{j\omega})$ denotes the Fourier transform of $r[n]$. $R_s[k]$ denotes 128 equally spaced samples of $R(e^{j\omega})$, i.e.,

$$R_s[k] = R(e^{j\omega})|_{\omega = 2\pi k/128}, \qquad k = 0, 1, \ldots, 127.$$

Given $x[n]$ and $y[n]$, we want to obtain $R_s[k]$ as efficiently as possible. The *only* modules available are those shown in Fig. P8.43. The costs associated with each module are as follows:

> Modules I and II are free
>
> Module III: 10
>
> Module IV: 50
>
> Module V: 100

By appropriately connecting one or several of each, construct a system for which the inputs are $x[n]$ and $y[n]$ and the output is $R_s[k]$. The important considerations are (a) whether the system works and (b) how efficient it is. The lower the *total* cost, the more efficient.

Module I

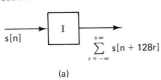

(a)

Module III

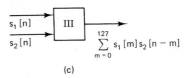

(c)

Module II

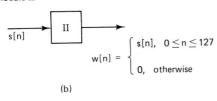

(b)

Module IV

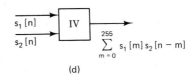

(d)

Module V

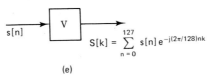

(e)

Figure P8.43

8.44. Two finite-duration sequences $h_1[n]$ and $h_2[n]$ of length 8 are sketched in Fig. P8.44. They are related by a circular shift.
   (a) Specify whether the magnitudes of the 8-point DFTs are equal.
   (b) We wish to implement a lowpass FIR filter and must use either $h_1[n]$ or $h_2[n]$ as the impulse response. Which one of the following statements is correct?
   (i) $h_1[n]$ is a better lowpass filter than $h_2[n]$.
   (ii) $h_2[n]$ is a better lowpass filter than $h_2[n]$.
   (iii) They are both about equally good (or bad) as lowpass filters.
   For which choice of impulse response is the corresponding frequency response a better approximation to a lowpass filter, or are they both equally good?

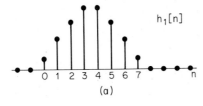

(a)

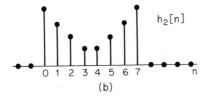

(b)   **Figure P8.44**

8.45. $y[n]$ is the output of a stable LTI system with system function $H(z) = 1/(z - bz^{-1})$, where b is a known constant. We would like to recover the input signal $x[n]$ by operating on $y[n]$.

The following procedure is proposed for recovering part of $x[n]$ from the data $y[n]$:

1. Using $y[n]$, $0 \le n \le N - 1$, calculate $Y[k]$, the N-point DFT of $y[n]$.
2. Form $V[k]$ as

$$V[k] = (W_N^{-k} - bW_N^k)Y[k].$$

3. Calculate the inverse DFT of $V[k]$ to obtain $v[n]$.

For which values of the index n in the range $n = 0, 1, \dots, N - 1$ are we guaranteed that

$$x[n] = v[n]?$$

8.46. A modified discrete Fourier transform (MDFT) was proposed (Vernet, 1971) that computes samples of the z-transform on the unit circle offset from those computed by the DFT. In particular, with $X_M[k]$ denoting the MDFT of $x(n)$,

$$X_M[k] = X(z)\Big|_{z = e^{j[2\pi k/N + \pi/N]}}, \qquad k = 0, 1, 2, \dots, N - 1.$$

Assume that N is even.

(a) The N-point MDFT of a sequence $x[n]$ corresponds to the N-point DFT of a sequence $x_M[n]$, which is easily constructed from $x[n]$. Determine $x_M[n]$ in terms of $x[n]$.

(b) If $x[n]$ is real, all the points in the DFT are not independent since the DFT is conjugate symmetric; i.e., $X[k] = X^*[((-k))_N]$ for $0 \le k \le N - 1$. Similarly, if $x[n]$ is real, all the points in the MDFT are not independent. Determine for $x[n]$ real the relationship between points in $X_M[k]$.

(c) (i) Let $R[k] = X_M[2k]$; that is, $R[k]$ contains the even-numbered points in $X_M[k]$. From your answer in part (b), show that $X_M[k]$ can be recovered from $R[k]$.

(ii) $R[k]$ can be considered to be the $N/2$-point MDFT of an $N/2$-point sequence $r[n]$. Determine a simple expression relating $r[n]$ directly to $x[n]$.

According to parts (b) and (c), the N-point MDFT of a real sequence $x[n]$ can be computed by forming $r[n]$ from $x[n]$ and then computing the $N/2$-point MDFT of $r[n]$. The next two parts are directed at showing that the MDFT can be used to implement a linear convolution.

(d) Consider three sequences $x_1[n]$, $x_2[n]$, and $x_3[n]$ all of length N. Let $X_{1M}[k]$, $X_{2M}[k]$, and $X_{3M}[k]$, respectively, denote the MDFTs of the three sequences. If

$$X_{3M}[k] = X_{1M}[k]X_{2M}[k],$$

express $x_3[n]$ in terms of $x_1[n]$ and $x_2[n]$. Your expression must be of the form of a single summation over a "combination" of $x_1[n]$ and $x_2[n]$ in the same style as (but not identical to) a circular convolution.

(e) It is convenient to refer to the result in part (d) as a modified circular convolution. If the sequences $x_1[n]$ and $x_2[n]$ are both zero for $n \ge N/2$, show that the modified circular convolution of $x_1[n]$ and $x_2[n]$ is identical to the linear convolution of $x_1[n]$ and $x_2[n]$.

8.47. In some applications in coding theory it is necessary to compute a 63-point circular convolution of two 63-point sequences $x[n]$ and $h[n]$. Suppose that the only computational devices available are multipliers, adders, and processors that compute N-point DFTs with N restricted to be a power of 2.

  (a) It is possible to compute the 63-point circular convolution of $x[n]$ and $h[n]$ using a number of 64-point DFTs, inverse DFTs, and the overlap-add method. How many DFTs are needed? [*Hint*: Consider each of the 63-point sequences as the sum of a 32-point sequence and 31-point sequence.]

  (b) Specify an algorithm that computes the 63-point circular convolution of $x[n]$ and $h[n]$ using two 128-point DFTs and one 128-point inverse DFT.

  (c) We could also compute the 63-point circular convolution of $x[n]$ and $h[n]$ by computing their linear convolution in the time domain and then aliasing the results. In terms of multiplications, which of these three methods is most efficient? Which is least efficient? (Assume that one complex multiply requires four real multiplies and that $x[n]$ and $h[n]$ are real.)

8.48. We want to implement the linear convolution of a 10,000-point sequence with an FIR impulse response that is 100 points long. The convolution is to be implemented by using DFTs and inverse DFTs of length 256.

  (a) If the overlap-add method is used, what is the minimum number of 256-point DFTs and the minimum number of 256-point inverse DFTs needed to implement the convolution for the entire 10,000-point sequence? Justify your answer.

  (b) If the overlap-save method is used, what is the minimum number of 256-point DFTs and the minimum number of 256-point inverse DFTs needed to implement the convolution for the entire 10,000-point sequence? Justify your answer.

  (c) We will see in Chapter 9 that when N is a power of 2, an N-point DFT or inverse DFT requires $(N/2)\log_2 N$ complex multiplications and $N \log_2 N$ complex additions. For the same filter and impulse response length considered in parts (a) and (b), compare the number of arithmetic operations (multiplications and additions) required in the overlap-add method, the overlap-save method, and direct convolution.

8.49. We want to filter a very long string of data with an FIR filter whose impulse response is 50 samples long. We wish to implement this filter with a DFT using the overlap-save technique. The procedure is as follows:

  1. The input sections must be overlapped by V samples.

  2. From the output of each section we must extract M samples such that when these samples from each section are abutted, the resulting sequence is the desired filtered output.

  Assume that the input segments are 100 samples long and that the size of the DFT is 128 $(= 2^7)$ points. Further assume that the output sequence from the circular convolution is indexed from point 0 to point 127.

  (a) Determine V.

  (b) Determine M.

  (c) Determine the index of the beginning and the end of the M points extracted; i.e., determine which of the 128 points from the circular convolution are extracted to be abutted with the result from the previous section.

8.50. A problem that often arises in practice is one in which a signal $x[n]$ has been filtered by a linear time-invariant system, the output of which is the distorted signal $y[n]$. We wish to recover the original signal $x[n]$ by processing $y[n]$. In theory, $x[n]$ can be recovered from

$y[n]$ by passing $y[n]$ through an inverse filter having a system function equal to the reciprocal of the system function of the distorting filter.

Suppose that the distortion is caused by an FIR filter with impulse response

$$h[n] = \delta[n] - \frac{1}{2}\,\delta[n - n_0],$$

where n_0 is a positive integer, i.e., the distortion of $x[n]$ takes the form of an echo at delay n_0.

(a) Determine the z-transform $H(z)$ and the N-point DFT $H[k]$ of the impulse response $h[n]$. Assume that $N = 4n_0$.

(b) Let $H_i(z)$ denote the system function of the inverse filter and let $h_i[n]$ be the corresponding impulse response. Determine $h_i[n]$. Is this an FIR or an IIR filter? What is the duration of $h_i[n]$?

(c) Suppose that we use an FIR filter of length N in an attempt to implement the inverse filter, and let the N-point DFT of the FIR filter be

$$G[k] = 1/H[k], \qquad k = 0, 1, \ldots, N - 1.$$

What is the impulse response $g[n]$ of the FIR filter?

(d) It might appear that the FIR filter with DFT $G[k] = 1/H[k]$ implements the inverse filter perfectly. After all, one might argue that the FIR distorting filter has an N-point DFT $H[k]$ and the FIR filter in cascade has an N-point DFT $G[k] = 1/H[k]$, and since $G[k]H[k] = 1$ for all k, we have implemented an allpass, nondistorting filter. Briefly explain the fallacy in this argument.

(e) Perform the convolution of $g[n]$ with $h[n]$ and thus determine how well the FIR filter with N-point DFT $G[k] = 1/H[k]$ implements the inverse filter.

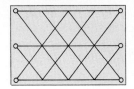

Computation of the Discrete Fourier Transform

9

9.0 INTRODUCTION

The discrete Fourier transform (DFT) plays an important role in the analysis, design, and implementation of discrete-time signal processing algorithms and systems. The basic properties of the Fourier transform and discrete Fourier transform, discussed in Chapters 2 and 8, make it particularly convenient to analyze and design systems in the Fourier domain. Equally important is the fact that efficient algorithms exist for explicitly computing the DFT. As a result, the DFT is an important component in many practical applications of discrete-time systems.

As discussed in Chapter 8, the DFT is identical to samples of the Fourier transform at equally spaced frequencies. Consequently, computation of the N-point DFT corresponds to the computation of N samples of the Fourier transform at N equally spaced frequencies, $\omega_k = 2\pi k/N$, i.e., at N points on the unit circle in the z-plane. In this chapter, we discuss several methods for computing values of the DFT. The major focus of this chapter is a particularly efficient class of algorithms for the digital computation of the N-point DFT. Collectively, these efficient algorithms are called *fast Fourier transform* (FFT) algorithms, and we discuss them in Sections 9.3–9.6. To achieve the highest efficiency, the FFT algorithms must compute all N values of the DFT. When we require values of the DFT over only a portion of the frequency range $0 \le \omega < 2\pi$, other algorithms may be more efficient and flexible even though they are less efficient than the FFT algorithms for computation of all the values of the DFT. Examples of such algorithms are the Goertzel algorithm, discussed in Section 9.2, and the chirp transform algorithm, discussed in Section 9.7.2.

There are many ways to measure complexity and efficiency of an implementation or algorithm, and a final assessment depends on both the available technology and the intended application. We will use the number of arithmetic multiplications and additions as a measure of computational complexity. This measure is simple to apply, and the number of multiplications and additions is directly related to computational speed when algorithms are implemented on general-purpose digital computers or special-purpose microprocessors. However, other measures are sometimes more appropriate. For example, in custom VLSI implementations, chip area is often the most important consideration, and chip area may not be directly related to the number of arithmetic operations.

In terms of multiplications and additions, the class of FFT algorithms can be orders of magnitude more efficient than competing algorithms. The efficiency of FFT algorithms is so high, in fact, that in many cases the most efficient procedure for implementing a convolution is to compute the transform of the sequences to be convolved, multiply their transforms, and then compute the inverse transform of the product of transforms. The details of this technique were discussed in Section 8.8. In seeming contradiction to this, a set of algorithms (mentioned briefly in Section 9.7) for evaluation of the DFT (or a more general set of samples of the Fourier transform) derive their efficiency from a reformulation of the Fourier transform in terms of a convolution and thereby implement the Fourier transform computation by using efficient procedures for evaluation of the associated convolution. This suggests the possibility of implementing a convolution by multiplication of DFTs, where the DFTs have been implemented by first expressing them as convolutions and then taking advantage of efficient procedures for implementing the associated convolutions. While this seems on the surface to be a basic contradiction, we will see in Section 9.7 that in certain cases it is an entirely reasonable approach and is not at all a contradiction.

9.1 EFFICIENT COMPUTATION OF THE DISCRETE FOURIER TRANSFORM

As defined in Chapter 8, the DFT of a finite-length sequence of length N is

$$X[k] = \sum_{n=0}^{N-1} x[n] W_N^{kn}, \qquad k = 0, 1, \ldots, N-1, \tag{9.1}$$

where $W_N = e^{-j(2\pi/N)}$. The inverse discrete Fourier transform is given by

$$x[n] = \frac{1}{N} \sum_{k=0}^{N-1} X[k] W_N^{-kn}, \qquad n = 0, 1, \ldots, N-1. \tag{9.2}$$

In Eqs. (9.1) and (9.2), both $x[n]$ and $X[k]$ may be complex. Since the expressions of Eqs. (9.1) and (9.2) differ only in the sign of the exponent of W_N and in the scale factor

$1/N$, a discussion of computational procedures for Eq. (9.1) applies with straightforward modifications to Eq. (9.2). (See Problem 9.1.)

To create a frame of reference for our discussion of computation of the DFT, let us first consider direct evaluation of the DFT equation in Eq. (9.1). Since $x[n]$ may be complex, N complex multiplications and $(N - 1)$ complex additions are required to compute each value of the DFT if we use Eq. (9.1) directly as a formula for computation. To compute all N values therefore requires a total of N^2 complex multiplications and $N(N - 1)$ complex additions. Expressing Eq. (9.1) in terms of operations on real numbers, we obtain

$$X[k] = \sum_{n=0}^{N-1} [(\mathscr{R}e\{x[n]\}\mathscr{R}e\{W_N^{kn}\} - \mathscr{I}m\{x[n]\}\mathscr{I}m\{W_N^{kn}\})$$

$$+ j(\mathscr{R}e\{x[n]\}\mathscr{I}m\{W_N^{kn}\} + \mathscr{I}m\{x[n]\}\mathscr{R}e\{W_N^{kn}\})], \qquad (9.3)$$

$$k = 0, 1, \ldots, N - 1,$$

which shows that each complex multiplication requires four real multiplications and two real additions, and each complex addition requires two real additions. Therefore, for each value of k, the direct computation of $X[k]$ requires $4N$ real multiplications and $(4N - 2)$ real additions.† Since $X[k]$ must be computed for N different values of k, the direct computation of the discrete Fourier transform of a sequence $x[n]$ requires $4N^2$ real multiplications and $N(4N - 2)$ real additions. Besides the multiplications and additions called for by Eq. (9.3), the digital computation of the DFT on a general-purpose digital computer or with special-purpose hardware also requires provision for storing and accessing the N complex input sequence values $x[n]$ and values of the complex coefficients W_N^{kn}. Since the amount of computation, and thus the computation time, is approximately proportional to N^2, it is evident that the number of arithmetic operations required to compute the DFT by the direct method becomes very large for large values of N. For this reason, we are interested in computational procedures that reduce the number of multiplications and additions.

Most approaches to improving the efficiency of the computation of the DFT exploit the symmetry and periodicity properties of W_N^{kn}; specifically,

1. $W_N^{k[N-n]} = W_N^{-kn} = (W_N^{kn})^*$ complex conjugate symmetry;
2. $W_N^{kn} = W_N^{k(n+N)} = W_N^{(k+N)n}$ periodicity in n and k.

As an illustration, using the first property, i.e., the symmetry of the implicit cosine and sine functions, we can group terms in the summation in Eq. (9.3) for n and $(N - n)$. For example,

$$\mathscr{R}e\{x[n]\}\mathscr{R}e\{W_N^{kn}\} + \mathscr{R}e\{x[N - n]\}\mathscr{R}e\{W_N^{k[N-n]}\}$$

$$= (\mathscr{R}e\{x[n]\} + \mathscr{R}e\{x[N - n]\})\mathscr{R}e\{W_N^{kn}\}$$

---

† Throughout the discussion, the figure for the number of computations is only approximate. Multiplication by W_N^0, for example, does not in fact require a multiplication. Nevertheless, the estimate of computational complexity obtained by including such multiplications is sufficiently accurate to permit comparisons between different classes of algorithms.

and

$$- \mathcal{I}m\{x[n]\}\mathcal{I}m\{W_N^{kn}\} - \mathcal{I}m\{x[N-n]\}\mathcal{I}m\{W_N^{k[N-n]}\}$$

$$= -(\mathcal{I}m\{x[n]\} + \mathcal{I}m\{x[N-n]\})\mathcal{I}m\{W_N^{kn}\}$$

Similar groupings can be used for the other terms in Eq. (9.3). In this way, the number of multiplications can be reduced by approximately a factor of 2. We can also take advantage of the fact that for certain values of the product kn, the sine and cosine functions take on the value 1 or 0, thereby eliminating the need for multiplications. However, reductions of this type still leave us with an amount of computation that is proportional to N^2. Fortunately, the second property, the periodicity of the complex sequence W_N^{kn}, can be exploited in achieving significantly greater reductions of the computation.

Computational algorithms that exploit both the symmetry and the periodicity of the sequence W_N^{kn} were known long before the era of high-speed digital computation. At that time, any scheme that reduced manual computation by even a factor of 2 was welcomed. Heideman et al. (1984) have traced the origins of the FFT back to Gauss as early as 1805. Runge (1905) and later Danielson and Lanczos (1942) described algorithms for which computation was roughly proportional to $N \log N$ rather than N^2.† However, the distinction was not of great importance for the small values of N that were feasible for hand computation. The possibility of greatly reduced computation was generally overlooked until about 1965, when Cooley and Tukey (1965) published an algorithm for the computation of the discrete Fourier transform that is applicable when N is a composite number, i.e., N is the product of two or more integers. The publication of their paper touched off a flurry of activity in the application of the discrete Fourier transform to signal processing and resulted in the discovery of a number of highly efficient computational algorithms. Collectively, the entire set of such algorithms has come to be known as the *fast Fourier transform*, or the FFT.

FFT algorithms are based on the fundamental principle of decomposing the computation of the discrete Fourier transform of a sequence of length N into successively smaller discrete Fourier transforms. The manner in which this principle is implemented leads to a variety of different algorithms, all with comparable improvements in computational speed. In this chapter we are concerned with two basic classes of FFT algorithms. The first, called decimation-in-time, derives its name from the fact that in the process of arranging the computation into smaller transformations, the sequence $x[n]$ (generally thought of as a time sequence) is decomposed into successively smaller subsequences. In the second general class of algorithms, the sequence of discrete Fourier transform coefficients $X[k]$ is decomposed into smaller subsequences; hence its name, decimation-in-frequency.

In this chapter we consider a number of algorithms for computing the discrete Fourier transform. The algorithms vary in efficiency, but all of them require fewer multiplications and additions than direct evaluation of Eq. (9.3). We begin in the next section with a discussion of Goertzel's algorithm (Goertzel, 1958), which requires

---

† See Cooley et al. (1967) for a historical summary of results related to the FFT.

computation proportional to N^2 but with a smaller constant of proportionality than the direct method. One of the principal advantages of Goertzel's algorithm is that it is not restricted to computation of the DFT but is in fact equally valid for the computation of any desired set of frequency samples.

In Sections 9.3–9.6 we present a detailed discussion of FFT algorithms for which computation is proportional to $N \log N$. This class of algorithms is considerably more efficient in terms of arithmetic operations than the Goertzel algorithm but is specifically oriented toward computation of *all* the values of the DFT. We do not attempt to be exhaustive in our coverage of this class of algorithms, but we illustrate the general principles common to all algorithms of this type by considering in detail only a few of the more commonly used schemes.

In Section 9.7 we discuss algorithms that rely on formulating the computation of the DFT in terms of a convolution. Finally, in Section 9.8 we consider the effects of arithmetic quantization in DFT computations.

9.2 THE GOERTZEL ALGORITHM

The Goertzel algorithm (Goertzel, 1958) is an example of how the periodicity of the sequence W_N^{kn} can be used to reduce computation. To derive the algorithm, we begin by noting that

$$W_N^{-kN} = e^{j(2\pi/N)Nk} = e^{j2\pi k} = 1. \tag{9.4}$$

This is, of course, a direct result of the periodicity of W_N^{-kn}. Because of Eq. (9.4) we may multiply the right side of Eq. (9.1) by W_N^{-kN} without affecting the equation. Thus,

$$X[k] = W_N^{-kN} \sum_{r=0}^{N-1} x[r] W_N^{kr} = \sum_{r=0}^{N-1} x[r] W_N^{-k(N-r)}. \tag{9.5}$$

To suggest the final result, let us define the sequence

$$y_k[n] = \sum_{r=-\infty}^{\infty} x[r] W_N^{-k(n-r)} u[n-r]. \tag{9.6}$$

From Eqs. (9.5) and (9.6) and the fact that $x[n] = 0$ for $n < 0$ and $n \geq N$, it follows that

$$X[k] = y_k[n] \Big|_{n=N}. \tag{9.7}$$

Equation (9.6) can be interpreted as a discrete convolution of the finite-duration sequence $x[n], 0 \leq n \leq N - 1$, with the sequence $W_N^{-kn} u[n]$. Consequently, $y_k[n]$ can be viewed as the response of a system with impulse response $W_N^{-kn} u[n]$ to a finite-length input $x[n]$. In particular, $X[k]$ is the value of the output when $n = N$.

A system with impulse response $W_N^{-kn} u[n]$ is depicted in Fig. 9.1, where initial rest conditions are assumed. Since the general input $x[n]$ and the coefficient W_N^{-k} are both complex, the computation of each new value of $y_k[n]$ using the system of Fig. 9.1

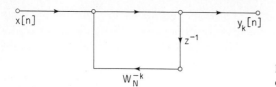

Figure 9.1　Flow graph of first-order complex recursive computation of $X[k]$.

requires 4 real multiplications and 4 real additions. All the intervening values $y_k[1]$, $y_k[2], \ldots, y_k[N-1]$ must be computed in order to compute $y_k[N] = X[k]$, so the use of the system in Fig. 9.1 as a computational algorithm requires $4N$ real multiplications and $4N$ real additions to compute $X[k]$ for a particular value of k. Thus, this procedure is slightly less efficient than the direct method. However, the method of Fig. 9.1 avoids the computation or storage of the coefficients W_N^{kn} since these quantities are implicitly computed by the recursion implied by Fig. 9.1.

It is possible to retain this simplification while reducing the number of multiplications by a factor of 2. To see how this may be done, note that the system function of the system of Fig. 9.1 is

$$H_k[z] = \frac{1}{1 - W_N^{-k} z^{-1}}. \tag{9.8}$$

Multiplying both the numerator and the denominator of $H_k[z]$ by the factor $(1 - W_N^k z^{-1})$, we obtain

$$
\begin{aligned}
H_k[z] &= \frac{1 - W_N^k z^{-1}}{(1 - W_N^{-k} z^{-1})(1 - W_N^k z^{-1})} \\
&= \frac{1 - W_N^k z^{-1}}{1 - 2\cos(2\pi k/N)z^{-1} + z^{-2}}.
\end{aligned} \tag{9.9}
$$

The signal flow graph of Fig. 9.2 corresponds to the system of Eq. (9.9).

If the input is complex, only two real multiplications per sample are required to implement the poles of this system, since the coefficients are real and the factor -1 need not be counted as a multiplication. As in the case of the first-order system, for complex input four real additions per sample are required to implement the poles (if the input is complex). Since we only need to bring the system to a state from which $y_k[N]$ can be computed, the complex multiplication by $-W_N^k$ required to implement the zero of the system function need not be performed at every iteration of the difference equation, but only after the Nth iteration. Thus, the total computation is $2N$ real multiplications and $4N$ real additions for the poles plus 4 real multiplications and 4 real additions for the zero. The total computation is therefore $2(N+2)$ real

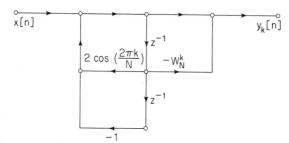

Figure 9.2　Flow graph of second-order recursive computation of $X[k]$ (Goertzel algorithm).

multiplications and $4(N + 1)$ real additions, about half the number of real multiplications required with the direct method. In this more efficient scheme, we still have the advantage that $\cos(2\pi k/N)$ and W_N^k are the only coefficients that must be computed and stored. The coefficients W_N^{kn} are again computed implicitly in the iteration of the recursion formula implied by Fig. 9.2.

As an additional advantage of the use of this network, let us consider the computation of the z-transform of $x[n]$ at conjugate locations on the unit circle, that is, the computation of $X[k]$ and $X[N - k]$. It is straightforward to verify that the network in the form of Fig. 9.2 required to compute $X[N - k]$ has exactly the same poles as that in Fig. 9.2, but the coefficient for the zero is the complex conjugate of that in Fig. 9.2 (see Problem 9.2). Since the zero is implemented only on the final iteration, the $2N$ multiplications and $4N$ additions required for the poles can be used for the computation of two DFT values. Thus, for the computation of all N values of the discrete Fourier transform using the Goertzel algorithm, the number of real multiplications required is approximately N^2 and the number of real additions is approximately $2N^2$. While this is more efficient than the direct computation of the discrete Fourier transform, the amount of computation is still proportional to N^2.

In either the direct method or the Goertzel method we do not need to evaluate $X[k]$ at all N values of k. Indeed, we can evaluate $X[k]$ for any M values of k, with each DFT value being computed by a recursive system of the form of Fig. 9.2 with appropriate coefficients. In this case the total computation is proportional to NM. The Goertzel method and the direct method are attractive when M is small; however, as indicated previously, algorithms are available for which the computation is proportional to $N \log_2 N$ when N is a power of 2. Therefore, when M is less than $\log_2 N$, either the Goertzel algorithm or the direct method may in fact be the most efficient method, but when all N values of $X[k]$ are required, the decimation-in-time algorithms, to be considered next, are roughly $(N/\log_2 N)$ times more efficient than either the direct method or the Goertzel method.

9.3 DECIMATION-IN-TIME FFT ALGORITHMS

When computing the DFT, dramatic efficiency results from decomposing the computation into successively smaller DFT computations. In this process we exploit both the symmetry and the periodicity of the complex exponential $W_N^{kn} = e^{-j(2\pi/N)kn}$. Algorithms in which the decomposition is based on decomposing the sequence $x[n]$ into successively smaller subsequences, are called *decimation-in-time algorithms*.

The principle of the decimation-in-time algorithm is most conveniently illustrated by considering the special case of N an integer power of 2, i.e., $N = 2^v$. Since N is an even integer, we can consider computing $X[k]$ by separating $x[n]$ into two $(N/2)$-point† sequences consisting of the even-numbered points in $x[n]$ and the

---

† When discussing FFT algorithms, we use the words *sample* and *point* interchangeably to mean sequence value. Also, we refer to a sequence of length N as an N-point sequence, and the DFT of a sequence of length N will be called an N-point DFT.

odd-numbered points in $x[n]$. With $X[k]$ given by

$$X[k] = \sum_{n=0}^{N-1} x[n]W_N^{nk}, \qquad k = 0, 1, \ldots, N-1, \tag{9.10}$$

and separating $x[n]$ into its even- and odd-numbered points, we obtain

$$X[k] = \sum_{n \text{ even}} x[n]W_N^{nk} + \sum_{n \text{ odd}} x[n]W_N^{nk}, \tag{9.11}$$

or, with the substitution of variables $n = 2r$ for n even and $n = 2r + 1$ for n odd,

$$
\begin{aligned}
X[k] &= \sum_{r=0}^{(N/2)-1} x[2r]W_N^{2rk} + \sum_{r=0}^{(N/2)-1} x[2r+1]W_N^{(2r+1)k} \\
&= \sum_{r=0}^{(N/2)-1} x[2r](W_N^2)^{rk} + W_N^k \sum_{r=0}^{(N/2)-1} x[2r+1](W_N^2)^{rk}.
\end{aligned}
\tag{9.12}
$$

But $W_N^2 = W_{N/2}$ since

$$W_N^2 = e^{-2j(2\pi/N)} = e^{-j2\pi/(N/2)} = W_{N/2}. \tag{9.13}$$

Consequently, Eq. (9.12) can be rewritten as

$$
\begin{aligned}
X[k] &= \sum_{r=0}^{(N/2)-1} x[2r]W_{N/2}^{rk} + W_N^k \sum_{r=0}^{(N/2)-1} x[2r+1]W_{N/2}^{rk} \\
&= G[k] + W_N^k H[k].
\end{aligned}
\tag{9.14}
$$

Each of the sums in Eq. (9.14) is recognized as an $(N/2)$-point DFT, the first sum being the $(N/2)$-point DFT of the even-numbered points of the original sequence and the second being the $(N/2)$-point DFT of the odd-numbered points of the original sequence. Although the index k ranges over N values, $k = 0, 1, \ldots, N-1$, each of the sums must be computed only for k between 0 and $(N/2) - 1$, since $G[k]$ and $H[k]$ are each periodic in k with period $N/2$. After the two DFTs are computed, they are combined according to Eq. (9.14) to yield the N-point DFT $X[k]$. Figure 9.3 depicts this computation for $N = 8$. In this figure we have used the signal flow graph conventions that were introduced in Chapter 6 for representing difference equations. That is, branches entering a node are summed to produce the node variable. When no coefficient is indicated, the branch transmittance is assumed to be unity. For other branches, the transmittance of a branch is an integer power of W_N.

    In Fig. 9.3 two 4-point DFTs are computed, with $G[k]$ designating the 4-point DFT of the even-numbered points and $H[k]$ designating the 4-point DFT of the odd-numbered points. $X[0]$ is then obtained by multiplying $H[0]$ by W_N^0 and adding the product to $G[0]$. $X[1]$ is obtained by multiplying $H[1]$ by W_N^1 and adding that result to $G[1]$. Equation (9.14) states that to compute $X[4]$ we should multiply $H[4]$ by W_N^4 and add the result of $G[4]$. However, since $G[k]$ and $H[k]$ are both periodic in k with period 4, $H[4] = H[0]$ and $G[4] = G[0]$. Thus $X[4]$ is obtained by multiplying $H[0]$ by W_N^4 and adding the result to $G[0]$. As shown in Fig. 9.3, the values $X[5]$, $X[6]$, and $X[7]$ are obtained similarly.

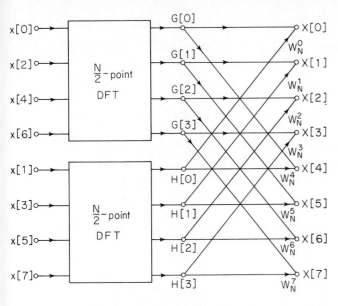

Figure 9.3 Flow graph of the decimation-in-time decomposition of an N-point DFT computation into two $(N/2)$-point DFT computations ($N = 8$).

With the computation restructured according to Eq. (9.14), we can compare the number of multiplications and additions required with those required for a direct computation of the DFT. Previously we saw that for direct computation without exploiting symmetry, N^2 complex multiplications and additions were required.† By comparison, Eq. (9.14) requires the computation of two $(N/2)$-point DFTs, which in turn requires $2(N/2)^2$ complex multiplications and approximately $2(N/2)^2$ complex additions if we do the $(N/2)$-point DFTs by the direct method. Then the two $(N/2)$-point DFTs must be combined, requiring N complex multiplications, corresponding to multiplying the second sum by W_N^k, and N complex additions, corresponding to adding that product to the first sum. Consequently, the computation of Eq. (9.14) for all values of k requires at most $N + 2(N/2)^2$ or $N + N^2/2$ complex multiplications and complex additions. It is easy to verify that for $N > 2$, the total $N + N^2/2$ will be less than N^2.

Equation (9.14) corresponds to breaking the original N-point computation into two $(N/2)$-point DFT computations. If $N/2$ is even, as it is when N is equal to a power of 2, then we can consider computing each of the $(N/2)$-point DFTs in Eq. (9.14) by breaking each of the sums in Eq. (9.14) into two $(N/4)$-point DFTs, which would then be combined to yield the $(N/2)$-point DFTs. Thus, $G[k]$ in Eq. (9.14) would be represented as

$$G[k] = \sum_{r=0}^{(N/2)-1} g[r] W_{N/2}^{rk} = \sum_{\ell=0}^{(N/4)-1} g[2\ell] W_{N/2}^{2\ell k} + \sum_{\ell=0}^{(N/4)-1} g[2\ell+1] W_{N/2}^{(2\ell+1)k} \quad (9.15)$$

or

$$G[k] = \sum_{\ell=0}^{(N/4)-1} g[2\ell] W_{N/4}^{\ell k} + W_{N/2}^{k} \sum_{\ell=0}^{(N/4)-1} g[2\ell+1] W_{N/4}^{\ell k}. \quad (9.16)$$

† For simplicity we shall assume that N is large, so that $(N-1)$ can be approximated by N.

Similarly, $H[k]$ would be represented as

$$H[k] = \sum_{\ell=0}^{(N/4)-1} h[2\ell]W_{N/4}^{\ell k} + W_{N/2}^{k} \sum_{\ell=0}^{(N/4)-1} h[2\ell+1]W_{N/4}^{\ell k}. \qquad (9.17)$$

Consequently, the $(N/2)$-point DFT $G[k]$ can be obtained by combining the $(N/4)$-point DFTs of the sequences $g[2\ell]$ and $g[2\ell+1]$. Similarly, the $(N/2)$-point DFT $H[k]$ can be obtained by combining the $(N/4)$-point DFTs of the sequences $h[2\ell]$ and $h[2\ell+1]$. Thus, if the 4-point DFTs in Fig. 9.3 are computed according to Eqs. (9.16) and (9.17), then that computation would be carried out as indicated in Fig. 9.4. Inserting the computation of Fig. 9.4 into the flow graph of Fig. 9.3, we obtain the complete flow graph of Fig. 9.5, where we have expressed the coefficients in terms of powers of W_N rather than powers of $W_{N/2}$, using the fact that $W_{N/2} = W_N^2$.

For the 8-point DFT that we have been using as an illustration, the computation has been reduced to a computation of 2-point DFTs. The 2-point DFT of, for example, $x[0]$ and $x[4]$ is depicted in Fig. 9.6. With the computation of Fig. 9.6 inserted in the flow graph of Fig. 9.5, we obtain the complete flow graph for computation of the 8-point DFT, as shown in Fig. 9.7.

For the more general case, but with N still a power of 2, we would proceed by decomposing the $(N/4)$-point transforms in Eq. (9.16) and (9.17) into $(N/8)$-point transforms and continue until left with only 2-point transforms. This requires v stages of computation, where $v = \log_2 N$. Previously we found that in the original decomposition of an N-point transform into two $(N/2)$-point transforms, the number of complex multiplications and additions required was $N + 2(N/2)^2$. When the $(N/2)$-point transforms are decomposed into $(N/4)$-point transforms, then the factor of $(N/2)^2$ is replaced by $N/2 + 2(N/4)^2$, so the overall computation then requires $N + N + 4(N/4)^2$ complex multiplications and additions. If $N = 2^v$, this can be done at most $v = \log_2 N$ times, so that after carrying out this decomposition as many times as possible, the number of complex multiplications and additions is equal to $N \log_2 N$.

The flow graph of Fig. 9.7 displays the operations explicitly. By counting branches with transmittances of the form W_N^r, we note that each stage has N complex multiplications and N complex additions. Since there are $\log_2 N$ stages, we have a total of $N \log_2 N$ complex multiplications and additions. This is the substantial computational savings that we have previously indicated was possible. For example, if $N = 2^{10} = 1024$, then $N^2 = 2^{20} = 1,048,576$, and $N \log_2 N = 10,240$, a reduction of more than 2 orders of magnitude!

The computation in the flow graph of Fig. 9.7 can be reduced further by exploiting the symmetry and periodicity of the coefficients W_N^r. We first note that in

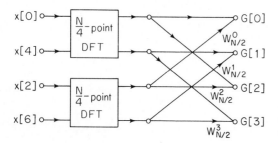

Figure 9.4 Flow graph of the decimation-in-time decomposition of an $(N/2)$-point DFT computation into two $(N/4)$-point DFT computations ($N = 8$).

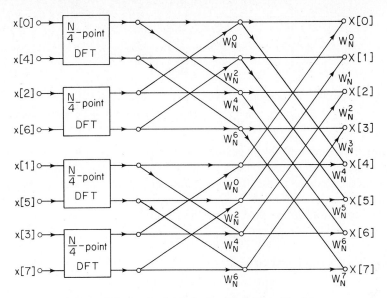

Figure 9.5   Result of substituting Fig. 9.4 into Fig. 9.3.

Figure 9.6   Flow graph of a 2-point DFT.

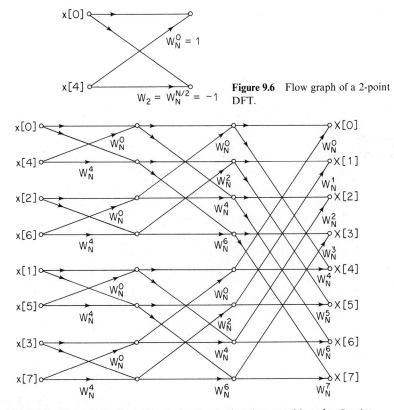

Figure 9.7   Flow graph of complete decimation-in-time decomposition of an 8-point DFT computation.

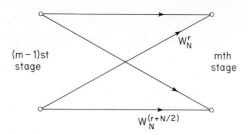

$(m-1)$st
stage

mth
stage

Figure 9.8   Flow graph of basic butterfly computation in Fig. 9.7.

proceeding from one stage to the next in Fig. 9.7, the basic computation is in the form of Fig. 9.8, i.e., it involves obtaining a pair of values in one stage from a pair of values in the preceding stage, where the coefficients are always powers of W_N and the exponents are separated by $N/2$. Because of the shape of the flow graph, this elementary computation is called a *butterfly*. Since

$$W_N^{N/2} = e^{-j(2\pi/N)N/2} = e^{-j\pi} = -1, \tag{9.18}$$

the factor $W_N^{r+N/2}$ can be written as

$$W_N^{r+N/2} = W_N^{N/2} W_N^r = -W_N^r. \tag{9.19}$$

With this observation, the butterfly computation of Fig. 9.8 can be simplified to the form shown in Fig. 9.9, which requires only one complex multiplication instead of two. Using the basic flow graph of Fig. 9.9 as a replacement for butterflies of the form of Fig. 9.8, we obtain from Fig. 9.7 the flow graph of Fig. 9.10. In particular, the number of complex multiplications has been reduced by a factor of 2 over the number in Fig. 9.7.

9.3.1 In-Place Computations

The flow graph of Fig. 9.10 describes an algorithm for the computation of the discrete Fourier transform. Particularly important in the flow graph of Fig. 9.10 are the branches connecting the nodes and the transmittance of each of these branches. No matter how the nodes in the flow graph are rearranged, it will always represent the same computation provided that the connections between the nodes and the transmittances of the connections are maintained. The particular form for the flow graph in Fig. 9.10 arose out of deriving the algorithm by separating the original sequence into the even-numbered and odd-numbered points and then continuing to create smaller and smaller subsequences in the same way. An interesting by-product of this derivation is that this flow graph, in addition to describing an efficient procedure for computing the discrete Fourier transform, also suggests a useful way of

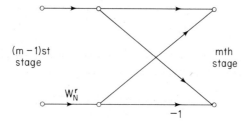

$(m-1)$st
stage

mth
stage

Figure 9.9   Flow graph of simplified butterfly computation requiring only one complex multiplication.

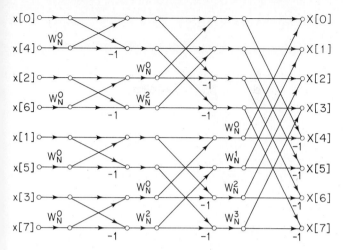

Figure 9.10 Flow graph of 8-point DFT using the butterfly computation of Fig. 9.9.

storing the original data and storing the results of the computation in intermediate arrays.

   To see this, it is useful to note that according to Fig. 9.10, each stage of the computation takes a set of N complex numbers and transforms them into another set of N complex numbers through basic butterfly computations of the form of Fig. 9.9. This process is repeated $v = \log_2 N$ times, resulting in the computation of the desired discrete Fourier transform. When implementing the computations depicted in Fig. 9.10 we can imagine the use of two arrays of (complex) storage registers, one for the array being computed and one for the data being used in the computation. For example, in computing the first array in Fig. 9.10, one set of storage registers would contain the input data and the second set would contain the computed results for the first stage. While the validity of Fig. 9.10 is not tied to the order in which the input data are stored, we can order the set of complex numbers in the same order that they appear in Fig. 9.10 (from top to bottom). We denote the sequence of complex numbers resulting from the mth stage of computation as $X_m[\ell]$, where $\ell = 0, 1, \ldots, N - 1$, and $m = 1, 2, \ldots, v$. Furthermore, for convenience, we define the set of input samples as $X_0[\ell]$. We can think of $X_{m-1}[\ell]$ as the input array and $X_m[\ell]$ as the output array for the mth stage of computations. Thus for the case of $N = 8$ as in Fig. 9.10,

$$X_0[0] = x[0]$$
$$X_0[1] = x[4]$$
$$X_0[2] = x[2]$$
$$X_0[3] = x[6]$$
$$X_0[4] = x[1]$$
$$X_0[5] = x[5]$$
$$X_0[6] = x[3]$$
$$X_0[7] = x[7]$$

(9.20)

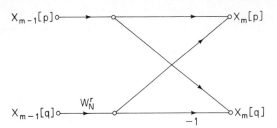

Figure 9.11   Flow graph of Eqs. (9.21).

Using this notation, we can label the input and output of the butterfly computation in Fig. 9.9 as indicated in Fig. 9.11, with the associated equations

$$X_m[p] = X_{m-1}[p] + W_N^r X_{m-1}[q],\qquad(9.21\text{a})$$

$$X_m[q] = X_{m-1}[p] - W_N^r X_{m-1}[q].\qquad(9.21\text{b})$$

In Eqs. (9.21), p, q, and r vary from stage to stage in a manner that is readily inferred from Fig. 9.10 and from Eqs. (9.11), (9.14), (9.16), etc. It is clear from Figs. 9.10 and 9.11 that only the complex numbers in locations p and q of the $(m-1)$st array are required to compute the elements p and q of the mth array. Thus, only one complex array of N storage registers is physically necessary to implement the complete computation if $X_m[p]$ and $X_m[q]$ are stored in the same storage registers as $X_{m-1}[p]$ and $X_{m-1}[q]$, respectively. This kind of computation is commonly referred to as an *in-place* computation. The fact that the flow graph of Fig. 9.10 (or Fig. 9.7) represents an in-place computation is tied to the fact that we have associated nodes in the flow graph that are on the same horizontal line with the same storage location and the fact that the computation between two arrays consists of a butterfly computation in which the input nodes and the output nodes are horizontally adjacent.

In order that the computation may be done in place as discussed above, the input sequence must be stored (or at least accessed) in a nonsequential order, as shown in the flow graph of Fig. 9.10. In fact, the order in which the input data are stored and accessed is referred to as *bit-reversed* order. To see what is meant by this terminology, we note that for the 8-point flow graph that we have been discussing, three binary digits are required to index through the data. Writing the indices in Eq. (9.20) in binary form, we obtain the following set of equations:

$$X_0[000] = x[000]$$

$$X_0[001] = x[100]$$

$$X_0[010] = x[010]$$

$$X_0[011] = x[110]$$

$$X_0[100] = x[001] \qquad (9.22)$$

$$X_0[101] = x[101]$$

$$X_0[110] = x[011]$$

$$X_0[111] = x[111].$$

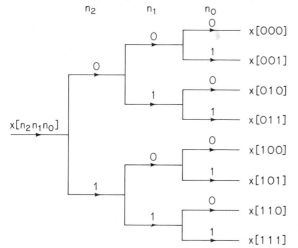

Figure 9.12   Tree diagram depicting normal-order sorting.

If (n_2, n_1, n_0) is the binary representation of the index of the sequence $x[n]$, then the sequence value $x[n_2, n_1, n_0]$ is stored in the array position $X_0[n_0, n_1, n_2]$. That is, in determining the position of $x[n_2, n_1, n_0]$ in the input array, we must reverse the order of the bits of the index n.

Let us first consider the process depicted in Fig. 9.12 for sorting a data sequence in normal order by successive examination of the bits representing the data index. If the most significant bit of the data index is zero, $x[n]$ belongs in the top half of the sorted array; otherwise it belongs in the bottom half. Next, the top half and bottom half subsequences can be sorted by examining the second most significant bit, and so on.

To see why bit-reversed order is necessary for in-place computation, recall the process that resulted in Fig. 9.7 and Fig. 9.10. The sequence $x[n]$ was first divided into the even-numbered samples, with the even-numbered samples occurring in the top half of Fig. 9.3 and the odd-numbered samples occurring in the bottom half. Such a separation of the data can be carried out by examining the least significant bit $[n_0]$ in the index n. If the least significant bit is 0, the sequence value corresponds to an even-numbered sample and therefore will appear in the top half, and if the least significant bit is 1, the sequence value corresponds to an odd-numbered sample and consequently will appear in the bottom half of the array $X_0[\ell]$. Next the even- and odd-indexed subsequences are sorted into their even- and odd-indexed parts, and this can be done by examining the second least significant bit in the index. Considering first the even-indexed subsequence, if the second least significant bit is 0, the sequence value is an even-numbered term in the subsequence, and if the second least significant bit is 1, then the sequence value has an odd-numbered index in this subsequence. The same process is carried out for the subsequence formed from the original odd-indexed sequence values. This process is repeated until N subsequences of length 1 are obtained. This sorting into even- and odd-indexed subsequences is depicted by the tree diagram of Fig. 9.13.

The tree diagrams of Figs. 9.12 and 9.13 are identical except that for normal sorting we examine the bits representing the index from left to right whereas for the

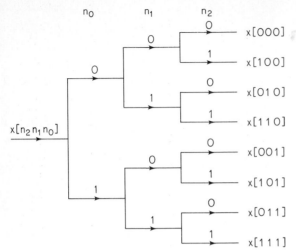

Figure 9.13 Tree diagram depicting bit-reversed sorting.

sorting leading naturally to Figs. 9.7 or 9.10, we examine the bits in reverse order, right to left, resulting in bit-reversed sorting. Thus, the necessity for bit-reversed ordering of the sequence $x[n]$ results from the manner in which the DFT computation is decomposed into successively smaller DFT computations in arriving at Figs. 9.7 and 9.10.

9.3.2 Alternative Forms

Although it is reasonable to store the results of each stage of the computation in the order in which the nodes appear in Fig. 9.10, it is certainly not necessary to do so. No matter how the nodes of Fig. 9.10 are rearranged, the result will always be a valid computation of the discrete Fourier transform of $x[n]$ as long as the branch transmittances are unchanged. Only the order in which data are accessed and stored will change. If we associate the nodes with indexing of an array of complex storage locations, it is clear from our previous discussion that a flow graph corresponding to an in-place computation results only if the rearrangement of nodes is such that the input and output nodes for each butterfly computation are horizontally adjacent. Otherwise two complex storage arrays will be required. Figure 9.10, is, of course, such an arrangement. Another is depicted in Fig. 9.14. In this case the input sequence is in normal order and the sequence of DFT values is in bit-reversed order. Figure 9.14 can be obtained from Fig. 9.10 as follows: All the nodes that are horizontally adjacent to $x[4]$ in Fig. 9.10 are interchanged with all the nodes horizontally adjacent to $x[1]$. Similarly, all the nodes that are horizontally adjacent to $x[6]$ in Fig. 9.10 are interchanged with those that are horizontally adjacent to $x[3]$. The nodes horizontally adjacent to $x[0]$, $x[2]$, $x[5]$, and $x[7]$ are not disturbed. The resulting flow graph in Fig. 9.14 corresponds to the form of the decimation-in-time algorithm originally given by Cooley and Tukey (1965).

The only difference between Figs. 9.10 and 9.14 is in the ordering of the nodes. The branch transmittances (powers of W_N) remain the same. There are, of

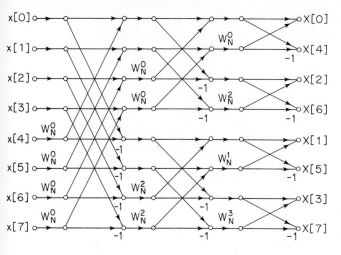

Figure 9.14 Rearrangement of Fig. 9.10 with input in normal order and output in bit-reversed order.

course, a large variety of possible orderings. However, most do not make much sense from a computational viewpoint. As one example, suppose that the nodes are ordered such that the input and output both appear in normal order. A flow graph of this type is shown in Fig. 9.15. In this case, however, the computation cannot be carried out in place because the butterfly structure does not continue past the first stage. Thus, two complex arrays of length N would be required to perform the computation depicted in Fig. 9.15.

In realizing the computations depicted by Figs. 9.10, 9.14, and 9.15, it is clearly necessary to access elements of intermediate arrays in nonsequential order. Thus, for greater computational speed, the complex numbers must be stored in random access memory. For example, in the computation of the first array in Fig. 9.10 from the input array, the inputs to each butterfly computation are adjacent node variables and are thought of as being stored in adjacent storage locations. In the computation of the second intermediate array from the first, the inputs to a butterfly are separated by two

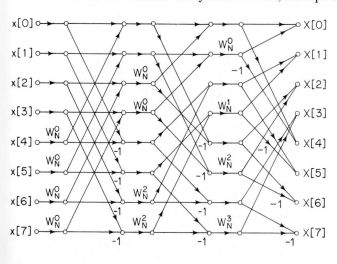

Figure 9.15 Rearrangement of Fig. 9.10 with both input and output in normal order.

storage locations; and in the computation of the third array from the second, the inputs to a butterfly computation are separated by four storage locations. If $N > 8$, the separation between butterfly inputs is 8 for the fourth stage, 16 for the fifth stage, and so on. The separation in the last (vth) stage is $N/2$.

In Fig. 9.14 the situation is similar in that to compute the first array from the input data we use data separated by 4, to compute the second array from the first array we use input data separated by 2, and then finally in computing the last array we use adjacent data. Although it is straightforward to imagine simple algorithms for modifying index registers to access the data in either the flow graph of Fig. 9.10 or Fig. 9.14, the data are not accessed sequentially, so random access memory would be very desirable. In the flow graph of Fig. 9.15, the data are accessed nonsequentially, the computation is not in place, and a scheme for indexing the data is considerably more complicated than in either of the two previous cases. Consequently, this structure has no apparent advantages.

Some forms have advantages even if in-place computation is not possible. A rearrangement of the flow graph in Fig. 9.10 that is particularly useful when random access memory is not available is shown in Fig. 9.16. This flow graph represents the decimation-in-time algorithm originally given by Singleton (1969). (See DSP Committee, 1979, for a program using serial memory.) Note first that in this flow graph the input is in bit-reversed order and the output is in normal order. The important feature of this flow graph is that the geometry is identical for each stage; only the branch transmittances change from stage to stage. This makes it possible to access data sequentially. Suppose that we have four separate disk files (or four sequentially addressed memories such as magnetic tape units), and suppose that the first half of the input sequence (in bit-reversed order) is stored in one file and the second half is stored in a second file. Then the sequence can be accessed sequentially in files 1 and 2 and the results written sequentially on files 3 and 4, with the first half of the new array being written to file 3 and the second half to file 4. Then at the next stage of computation, files 3 and 4 are the input, and the output is written to files 1 and 2. This is repeated for each of the v stages. Such an algorithm could be useful in computing the DFT of very long sequences.

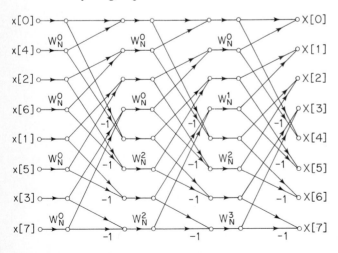

Figure 9.16 Rearrangement of Fig. 9.10 having the same geometry for each stage, thereby permitting sequential data accessing and storage.

9.4 DECIMATION-IN-FREQUENCY FFT ALGORITHMS

The decimation-in-time FFT algorithms were all based on the decomposition of the DFT computation by forming smaller and smaller subsequences of the input sequence $x[n]$. Alternatively we can consider dividing the output sequence $X[k]$ into smaller and smaller subsequences in the same manner. FFT algorithms based on this procedure are commonly called *decimation-in-frequency* algorithms.

To develop these FFT algorithms, let us again restrict the discussion to N a power of 2 and consider computing separately the even-numbered frequency samples and the odd-numbered frequency samples. Since $X[k]$ is

$$X[k] = \sum_{n=0}^{N-1} x[n] W_N^{nk}, \qquad k = 0, 1, \dots, N-1, \tag{9.23}$$

the even-numbered frequency samples are

$$X[2r] = \sum_{n=0}^{N-1} x[n] W_N^{n(2r)}, \qquad r = 0, 1, \dots, (N/2) - 1, \tag{9.24}$$

which can be expressed as

$$X[2r] = \sum_{n=0}^{(N/2)-1} x[n] W_N^{2nr} + \sum_{n-N/2}^{N-1} x[n] W_N^{2nr}. \tag{9.25}$$

With a substitution of variables in the second summation in Eq. (9.25), we obtain

$$X[2r] = \sum_{n=0}^{(N/2)-1} x[n] W_N^{2nr} + \sum_{n=0}^{(N/2)-1} x[n + (N/2)] W_N^{2r[n+(N/2)]}. \tag{9.26}$$

Finally, because of the periodicity of W_N^{2rn},

$$W_N^{2r[n+(N/2)]} = W_N^{2rn} W_N^{rN} = W_N^{2rn} \tag{9.27}$$

and since $W_N^2 = W_{N/2}$, Eq. (9.26) can be expressed as

$$X[2r] = \sum_{n=0}^{(N/2)-1} (x[n] + x[n + (N/2)]) W_{N/2}^{rn}, \qquad r = 0, 1, \dots, (N/2) - 1. \tag{9.28}$$

Equation (9.28) is the $(N/2)$-point DFT of the $(N/2)$-point sequence obtained by adding the first half and the last half of the input sequence. Adding the two halves of the input sequence represents time aliasing, consistent with the fact that in computing only the even-numbered frequency samples, we are undersampling the Fourier transform of $x[n]$.

We can now consider obtaining the odd-numbered frequency points, given by

$$X[2r + 1] = \sum_{n=0}^{N-1} x[n] W_N^{n(2r+1)}, \qquad r = 0, 1, \dots, (N/2) - 1. \tag{9.29}$$

As before, we can rearrange Eq. (9.29) as

$$X[2r + 1] = \sum_{n=0}^{(N/2)-1} x[n] W_N^{n(2r+1)} + \sum_{n=N/2}^{N-1} x[n] W_N^{n(2r+1)}. \tag{9.30}$$

An alternative form for the second summation in Eq. (9.30) is

$$\sum_{n=N/2}^{N-1} x[n] W_N^{n(2r+1)} = \sum_{n=0}^{(N/2)-1} x[n + (N/2)] W_N^{[n+(N/2)](2r+1)}$$

$$= W_N^{(N/2)(2r+1)} \sum_{n=0}^{(N/2)-1} x[n + (N/2)] W_N^{n(2r+1)}$$

$$= - \sum_{n=0}^{(N/2)-1} x[n + (N/2)] W_N^{n(2r+1)}, \tag{9.31}$$

where we have used the fact that $W_N^{(N/2)2r} = 1$ and $W_N^{(N/2)} = -1$. Substituting Eq. (9.31) into Eq. (9.30) and combining the two summations, we obtain

$$X[2r + 1] = \sum_{n=0}^{(N/2)-1} (x[n] - x[n + (N/2)]) W_N^{n(2r+1)} \tag{9.32}$$

or, since $W_N^2 = W_{N/2}$,

$$X[2r + 1] = \sum_{n=0}^{(N/2)-1} \{x[n] - x[n + (N/2)]\} W_N^n W_{N/2}^{nr}, \tag{9.33}$$

$$r = 0, 1, \ldots, (N/2) - 1.$$

Equation (9.33) is the $(N/2)$-point DFT of the sequence obtained by subtracting the second half of the input sequence from the first half and multiplying the resulting sequence by W_N^n. Thus, on the basis of Eqs. (9.28) and (9.33), with $g[n] = x[n] + x[n + N/2]$ and $h[n] = x[n] - x[n + N/2]$, the DFT can be computed by first forming the sequences $g[n]$ and $h[n]$, then computing $h[n]W_N^n$, and finally computing the $(N/2)$-point DFTs of these two sequences to obtain the even-numbered output points and the odd-numbered output points, respectively. The procedure suggested by Eqs. (9.28) and (9.33) is illustrated for the case of an 8-point DFT in Fig. 9.17.

Proceeding in a manner similar to that followed in deriving the decimation-in-time algorithm, we note that since N is a power of 2, $N/2$ is even; consequently, the $(N/2)$-point DFTs can be computed by computing the even-numbered and odd-numbered output points for those DFTs separately. As in the case of the procedure leading to Eqs. (9.28) and (9.33), this is accomplished by combining the first half and the last half of the input points for each of the $(N/2)$-point DFTs and then computing $(N/4)$-point DFTs. The flow chart resulting from taking this step for the 8-point example is shown in Fig. 9.18. For the 8-point example, the computation has now

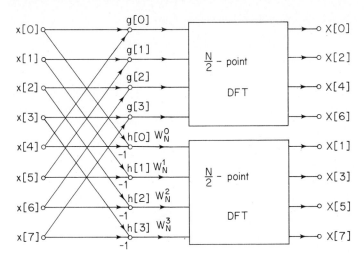

Figure 9.17 Flow graph of decimation-in-frequency decomposition of an N-point DFT computation into two $(N/2)$-point DFT computations ($N = 8$).

been reduced to the computation of 2-point DFTs, which are implemented by adding and subtracting the input points, as discussed previously. Thus, the 2-point DFTs in Fig. 9.18 can be replaced by the computation shown in Fig. 9.19, so the computation of the 8-point DFT can be accomplished by the algorithm depicted in Fig. 9.20.

By counting the arithmetic operations in Fig. 9.20 and generalizing to $N = 2^\nu$, we see that the computation of Fig. 9.20 requires $(N/2)\log_2 N$ complex multiplications and $N \log_2 N$ complex additions. Thus, the total amount of computation is the same for the decimation-in-frequency and the decimation-in-time algorithms.

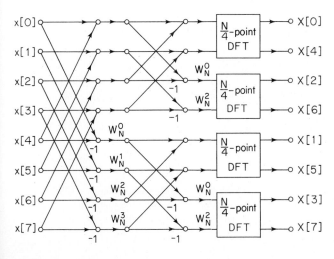

Figure 9.18 Flow graph of decimation-in-frequency decomposition of an 8-point DFT into four 2-point DFT computations.

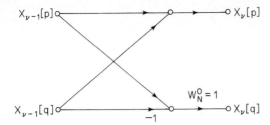

Figure 9.19 Flow graph of a typical 2-point DFT as required in the last stage of decimation-in-frequency decomposition.

9.4.1 In-Place Computation

The flow graph in Fig. 9.20 depicts one FFT algorithm based on decimation-in-frequency. We can observe a number of similarities and also a number of differences in comparing this graph with the flow graphs derived on the basis of decimation-in-time. As with decimation-in-time, of course, the flow graph of Fig. 9.20 corresponds to a computation of the discrete Fourier transform regardless of how the flow graph is drawn, as long as the same nodes are connected to each other with the proper branch transmittances. In other words, the flow graph of Fig. 9.20 is not based on any assumption about the order in which the input sequence values are stored. However, as was done with the decimation-in-time algorithms, we can interpret successive vertical nodes in the flow graph of Fig. 9.20 as corresponding to successive storage registers in a digital memory. In this case, the flow graph in Fig. 9.20 begins with the input sequence in normal order and provides the output DFT in bit-reversed order. The basic computation again has the form of a butterfly computation, although the butterfly is different from that arising in the decimation-in-time algorithms. However, because of the butterfly nature of the computation, the flow graph of Fig. 9.20 can be interpreted as an in-place computation of the discrete Fourier transform.

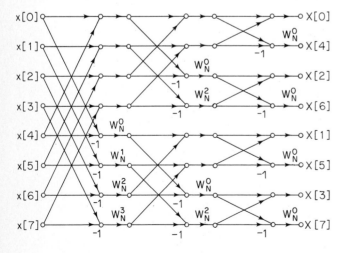

Figure 9.20 Flow graph of complete decimation-in-frequency decomposition of an 8-point DFT computation.

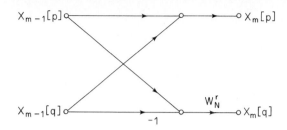

Figure 9.21   Flow graph of a typical butterfly computation required in Fig. 9.20.

9.4.2 Alternative Forms

A variety of alternative forms for the decimation-in-frequency algorithm can be obtained by transposing the decimation-in-time forms developed in Section 9.3.2. If we denote the sequence of complex numbers resulting from the mth stage of the computation as $X_m[\ell]$, where $\ell = 0, 1, \ldots, N - 1$, and $m = 1, 2, \ldots, v$, then the basic butterfly computation shown in Fig. 9.21 has the form

$$X_m[p] = X_{m-1}[p] + X_{m-1}[q], \tag{9.34a}$$

$$X_m[q] = (X_{m-1}[p] - X_{m-1}[q])W_N^r. \tag{9.34b}$$

By comparing Figs. 9.11 and 9.21 or Eqs. (9.21) and (9.34) we see that the butterfly computations are distinctly different for the two classes of FFT algorithms. However, we also note a resemblance between the basic butterfly flow graphs of Figs. 9.11 and 9.21 and between the FFT flow graphs of Figs. 9.10 and 9.20. Specifically, Fig. 9.20 can be obtained from Fig. 9.10 and Fig. 9.21 from Fig. 9.11 by reversing the direction of signal flow and interchanging the input and output. That is, in the terminology of Chapter 6, Fig. 9.20 is the transpose of the flow graph in Fig. 9.10, and Fig. 9.21 is the transpose of Fig. 9.11. The application of the transposition theorem is not straight-forward in this case because the FFT flow graphs have more than one input and output node. Nevertheless, it is true that the input/output characteristics of the flow graphs in Figs. 9.10 and 9.20 are the same. This can be shown by noting that the butterfly equations in Eqs. (9.34) can be solved backward starting with the output array. (Problem 9.6 outlines a proof of this result.) More generally, it is true that for each decimation-in-time FFT algorithm there exists a decimation-in-frequency FFT algorithm that corresponds to interchanging the input and output and reversing the direction of all the arrows in the flow graph.

This result implies that all the flow graphs of Section 9.3 have counterparts in the class of decimation-in-frequency algorithms. This, of course, also corresponds to the fact that, as before, it is possible to rearrange the nodes of a flow graph without altering the final result.

Applying the transposition procedure to Fig. 9.14 leads to Fig. 9.22. In this flow graph, the output is in normal order and the input is in bit-reversed order. Alternatively, the transpose of the flow graph of Fig. 9.15 is the flow graph of Fig. 9.23, where

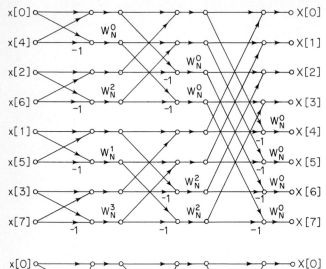

Figure 9.22 Flow graph of a decimation-in-frequency DFT algorithm obtained from Fig. 9.20. Input in bit-reversed order and output in normal order. (Transpose of Fig. 9.14.)

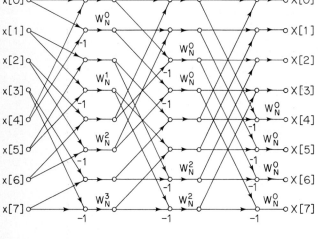

Figure 9.23 Rearrangement of Fig. 9.20 with both input and output in normal order. (Transpose of Fig. 9.15.)

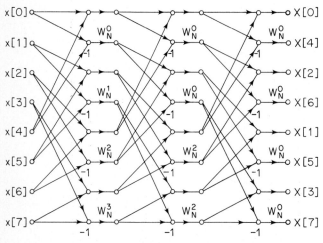

Figure 9.24 Rearrangement of Fig. 9.20 having the same geometry for each stage, thereby permitting sequential data accessing and storage (transpose of Fig. 9.16).

both the input and the output are in normal order. As in the case of Fig. 9.15, this flow graph does not correspond to an in-place computation.

The transpose of Fig. 9.16 is shown in Fig. 9.24. Each stage of Fig. 9.24 has the same geometry, a property which is desirable for computation of large transforms using sequential data storage, as discussed before.

9.5 IMPLEMENTATION OF FFT ALGORITHMS

We have discussed the basic principles of efficient computation of discrete Fourier transforms. In this discussion we have favored the use of signal flow graph representations rather than explicitly writing out in detail the equations that such flow graphs represent. Of necessity we have shown flow graphs for specific values of N. However, by considering a flow graph such as that in Fig. 9.10, for a specific value of N, it is possible to see how to structure a general computational algorithm that would apply to any $N = 2^\nu$.

From a consideration of the flow graphs of previous sections, it should be clear that there are several major considerations in any FFT algorithm. One is the accessing and storing of data in the intermediate arrays. Another is the actual implementation of the butterfly computation once the appropriate data have been obtained. Although it is true that the flow graphs of the previous sections capture the essence of the FFT algorithms that they depict, a multitude of details related to these two basic concerns must be considered in the implementation of a given algorithm. In this section we discuss some of these details. Our emphasis is on algorithms for N a power of 2, but much of the discussion applies to the general case as well. For purposes of illustration, we focus on the decimation-in-time algorithm of Fig. 9.10. Section 9.5.3 contains FORTRAN programs to implement both the decimation-in-time algorithm of Fig. 9.10 and the decimation-in-frequency algorithm of Fig. 9.20. Both programs use the principles discussed in this section.

9.5.1 Indexing

In the algorithm depicted in Fig. 9.10, the input must be in bit-reversed order so that the computation can be performed in place. The resulting DFT is then in normal order. Generally, sequences do not originate in bit-reversed order, so the first step in the implementation of Fig. 9.10 is to sort the input sequence into bit-reversed order. As can be seen from Fig. 9.10 and Eqs. (9.22), bit-reversed sorting can be done in place since samples are only pairwise interchanged; i.e., a sample at a given index is interchanged with the sample in the location specified by the bit-reversed index. This is conveniently done in place by using one index counting in normal order and another index that counts in bit-reversed order and simply interchanging the data in the two positions specified by the normal-order index and the bit-reversed-order index. This is illustrated in the FORTRAN program listed in Fig. 9.25 in Section 9.5.3. Once the input is in bit-reversed order, we can proceed with the first stage of computation. In this case the inputs to the butterflies are adjacent elements of the array $X_0[\cdot]$. In the second stage the inputs to the butterflies are separated by 2. In the mth stage the butterfly inputs are separated by 2^{m-1}. The coefficients are powers of $W_N^{(N/2^m)}$ in the mth stage and are required in normal order if computation of butterflies begins at the top of the flow graph of Fig. 9.10. The above statements define the manner in which data must be accessed in a given stage, which of course depends on the flow graph that is implemented. For example, in the mth stage of Fig. 9.14, the butterfly spacing is 2^{v-m}, and in this case the coefficients are required in bit-reversed order. The input is in normal order; however, the output is in bit-reversed order, so it generally would be necessary to sort the output into normal order by using a normal-order counter and a bit-reversed counter, as discussed previously.

In general, if we consider all the flow graphs in Sections 9.3 and 9.4, we see that each algorithm has its own characteristic indexing problems. The choice of a particular algorithm depends on a number of factors. The algorithms utilizing an in-place computation have the advantage of making efficient use of memory. Two disadvantages, however, are that the kind of memory required is random access rather than sequential memory and that either the input sequence or the output DFT sequence is in bit-reversed order. Furthermore, depending on whether a decimation-in-time or a decimation-in-frequency algorithm is chosen and whether the inputs or the outputs are in bit-reversed order, the coefficients are required to be accessed in either normal order or bit-reversed order. If random access memory is not available but sequential memory is, some fast Fourier transform algorithms utilize sequential memory, as we have shown, but either the inputs or the outputs must be in bit-reversed order. While the flow graph for the algorithm can be arranged so that the inputs, the outputs, and the coefficients are in normal order, the indexing structure required to implement these algorithms is complicated, and twice as much random access memory is required. Consequently, the use of these algorithms does not appear to be advantageous.

The in-place FFT algorithms of Figs. 9.10, 9.14, 9.20, and 9.22 are among the most commonly used. If a sequence is to be transformed only once, then bit-reversed sorting must be implemented on either the input or the output. However, in some situations a sequence is transformed, the result is modified in some way, and then the inverse DFT is computed. For example, in implementing FIR digital filters by block convolution using the discrete Fourier transform, the DFT of a section of the input sequence is multiplied by the DFT of the filter impulse response, and the result is inverse transformed to obtain a segment of the output of the filter. Similarly, in computing an autocorrelation function or cross-correlation function using the discrete Fourier transform, a sequence will be transformed, the DFTs multiplied, and then the resulting product will be inverse transformed. When two transforms are cascaded in this way, it is possible by appropriate choice of the FFT algorithms to avoid the need for bit reversal. For example, in implementing an FIR digital filter using the DFT, we can choose an algorithm for the direct transform that utilizes the data in normal order and provides a DFT in bit-reversed order. Either the flow graph corresponding to Fig. 9.14, based on decimation-in-time, or that of Fig. 9.20, based on decimation-in-frequency, could be used in this way. The difference between these two forms is that the decimation-in-time form requires the coefficients in bit-reversed order, whereas the decimation-in-frequency form requires the coefficients in normal order.

In using either of these algorithms, the transform occurs in bit-reversed order. Consequently, in using the DFT for block convolution, if the DFT of the impulse response of the filter has been stored in bit-reversed order, the DFTs can simply be multiplied point by point in the order stored. For the inverse DFT we can choose a form of the algorithm that requires bit-reversed data at the input and provides normally ordered results. Here, either the flow graph of Fig. 9.10, based on decimation-in-time, or that of Fig. 9.22, based on decimation-in-frequency, can be used. Figure 9.10, however, utilizes coefficients in normal order, whereas Fig. 9.22 requires the coefficients in bit-reversed order. If the decimation-in-time form of the algorithm is chosen for the direct transform, then the decimation-in-frequency form of the algorithm should be chosen for the inverse transform, requiring coefficients in bit-reversed order. Likewise, the decimation-in-frequency algorithm for the direct transform should be paired with the decimation-in-time algorithm for the inverse transform, which would then utilize normally ordered coefficients.

9.5.2 Coefficients

We have observed that the coefficients W_N^r may be required in bit-reversed order or in normal order. In either case we must store a table sufficient to look up all required values or we must compute the values as needed. The first alternative has the advantage of speed but of course requires extra storage. We observe from the flow graphs that we require W_N^r for $r = 0, 1, \ldots, (N/2) - 1$. Thus we require $N/2$ complex storage registers for a complete table of values of W_N^r.† In the case of algorithms in which the coefficients are required in bit-reversed order, we can simply store the table in bit-reversed order.

The computation of the coefficients as they are needed saves storage but is less efficient than storing a look-up table. If the coefficients are to be computed, it is generally most efficient to use a recursion formula. At any given stage, the required coefficients are all powers of a complex number of the form W_N^q, where q depends on the algorithm and the stage. Thus, if the coefficients are required in normal order, we can use the recursion formula

$$W_N^{q\ell} = W_N^q \cdot W_N^{q(\ell - 1)} \tag{9.35}$$

to obtain the ℓth coefficient from the $(\ell - 1)$st coefficient. Clearly, algorithms that require coefficients in bit-reversed order are not well suited to this approach. It should be noted that Eq. (9.35) is essentially the coupled form oscillator of Problem 6.4. When using finite-precision arithmetic, errors can build up in the iteration of this difference equation. Therefore, it is generally necessary to reset the value at prescribed points (e.g., $W_N^{N/4} = -j$) so that errors do not become unacceptable.

---

† This number can be reduced using symmetry at the cost of greater complexity in accessing desired values.

9.5.3 FORTRAN Programs for Implementing the FFT

The FORTRAN program in Fig. 9.25 illustrates an implementation of the decimation-in-time algorithm of Fig. 9.10. This program closely follows the form of a
program originally given by Cooley et al. (1967) as a simple description of the power
of 2 decimation-in-time FFT algorithm. While it is not a highly efficient or general
implementation, its straightforward structure illustrates many of the principles
previously discussed. This program can be used together with the graphical representation of Fig. 9.10 to understand the basic principles of FFT algorithms. Although the
program is not optimized for efficiency, it nevertheless requires only $(N/2)\log_2 N$
complex multiplications. The FORTRAN language explicitly displays much of the
algebraic structure of the equations of the algorithm. (Note that the FORTRAN
indices range from 1 to N rather than from 0 to $(N - 1)$ as we have assumed in all our

```
C     FORTRAN SUBROUTINE FOR DECIMATION-IN-TIME FFT ALGORITHM
C         X IS AN N=2**NU POINT COMPLEX ARRAY THAT INITIALLY
C         CONTAINS THE INPUT AND FINALLY CONTAINS THE DFT.
C
          SUBROUTINE  DITFFT(X,NU)                                 01
          COMPLEX X(1024),U,W,T                                   02
          N=2**NU                                                 03
          NV2=N/2                                                 04
          NM1=N-1                                                 05
          PI=3.14159265358979                                     06
C*****************************************************************
          J=1                                                     07
          DO 7 I=1,NM1                                            08
             IF(I.GE.J) GO TO 5                                   09
                T=X(J)                                            10
                X(J)=X(I)                                         11
                X(I)=T                                            12
    5         K=NV2                                               13
    6         IF(K.GE.J) GO TO 7                                  14
             J=J-K                                                15
             K=K/2                                                16
             GO TO 6                                              17
    7     J=J+K                                                   18
C*****************************************************************
          DO 20 L=1,NU                                            19
             LE=2**L                                              20
             LE1=LE/2                                             21
             U=(1.0,0.0)                                          22
             W=CMPLX(COS(PI/FLOAT(LE1)),-SIN(PI/FLOAT(LE1)))      23
                DO 20 J=1,LE1                                     24
                   DO 10 I=J,N,LE                                 25
                      IP=I+LE1                                    26
                      T=X(IP)*U                                   27
                      X(IP)=X(I)-T                                28
   10                 X(I)=X(I)+T                                 29
   20     U=U*W                                                   30
C*****************************************************************
          RETURN                                                 31
          END                                                    32
```

Figure 9.25   FORTRAN program for decimation-in-time algorithm of Fig. 9.10.

discussions.) The simple structure is easily translated into other programming languages.†

In the program of Fig. 9.25, statements 03 through 06 are concerned with initialization of the subroutine. Bit-reversed sorting is accomplished by running a bit-reversed counter. The instructions for incrementing the bit-reversed counter and for sorting the input sequence are numbered 07 through 18. The structure of Fig. 9.10 is embodied in the nested loop statements numbered 19 through 30. The coefficients required in the butterfly computations (the appropriate powers of W_N) are generated recursively. The variable U represents the coefficient multipliers. Statement number 30 is used to compute the coefficients recursively according to Eq. (9.35). This program does all $(N/2)\log_2 N$ multiplications depicted in Fig. 9.10. No attempt is made to skip multiplications by W_N^0; i.e., in statement number 27 the multiplication is explicitly carried out whether or not the coefficient is unity. An obvious saving of $N/2$ multiplications can be made by doing the first stage of computation separately without multiplications. However, this requires a separate loop, thereby destroying the compactness of the program.

Figure 9.26 shows a FORTRAN program for the decimation-in-frequency algorithm illustrated by Fig. 9.20. As in Fig. 9.25, this program is presented mainly to illustrate the structure of the decimation-in-frequency algorithm; it is by no means an optimized program, since all $(N/2)\log_2 N$ multiplications are done even though some of the multiplications are by W_N^0. In this case, lines 05 through 16 implement the v stages of computation, and lines 17 through 30 implement the bit-reversed sorting of the output into normal order.

9.6 FFT ALGORITHMS FOR COMPOSITE N

Our previous discussion has illustrated the basic principles of decimation-in-time and decimation-in-frequency for the important special case of N a power of 2, i.e., $N = 2^v$. Although this case leads to very simple programs such as those in Figs. 9.25 and 9.26, it is not always possible to deal with sequences whose length is a power of 2. More generally, efficient computation of the discrete Fourier transform is possible if N can be represented as a product of factors. However, the decimation concepts that suggest the power of 2 algorithms are not easily generalized. Thus we will adopt a more formal approach in this section.

9.6.1 Cooley-Tukey Algorithms

As we have seen, efficient algorithms for computing the DFT are achieved by breaking the computation into smaller DFT computations. This strategy is a standard

---

† A variety of efficient and flexible FFT programs in FORTRAN can be found in DSP Committee (1979). Other sources of well-documented FFT programs in FORTRAN are Burrus and Parks (1985) and Press et al. (1986). FFT programs in C can be found in Press et al. (1988).

```
C     FORTRAN SUBROUTINE FOR DECIMATION-IN-FREQUENCY ALGORITHM
C         X IS AN N=2**NU POINT COMPLEX ARRAY THAT INITIALLY
C         CONTAINS THE INPUT, AND FINALLY CONTAINS THE DFT.
C
      SUBROUTINE  DIFFFT(X,NU)                                       01
      COMPLEX X(1024),U,W,T                                         02
      N=2**NU                                                       03
      PI=3.14159265358979                                           04
C**************************************************************
      DO 20 L=1,NU                                                  05
      LE=2**(NU+1-L)                                                06
      LE1=LE/2                                                      07
      U=(1.,0.)                                                     08
      W=CMPLX(COS(PI/FLOAT(LE1)),-SIN(PI/FLOAT(LE1)))              09
            DO 20 J=1,LE1                                           10
                DO 10 I=J,N,LE                                      11
                    IP=I+LE1                                        12
                    T=X(I)+X(IP)                                    13
                    X(IP)=(X(I)-X(IP))*U                            14
 10                 X(I)=T                                          15
 20                 U=U*W                                           16
C**************************************************************
      NV2=N/2                                                       17
      NM1=N-1                                                       18
      J=1                                                           19
      DO 30 I=1,NM1                                                 20
            IF(I.GE.J) GO TO 25                                     21
                    T=X(J)                                          22
                    X(J)=X(I)                                       23
                    X(I)=T                                          24
 25             K=NV2                                               25
 26             IF(K.GE.J) GO TO 30                                 26
                    J=J-K                                           27
                    K=K/2                                           28
            GO TO 26                                                29
 30   J=J+K                                                         30
C**************************************************************
      RETURN                                                        31
      END                                                           32
```

Figure 9.26 FORTRAN program for decimation-in-frequency algorithm of Fig. 9.20.

approach to deriving efficient algorithms for many types of computations, and it is the basic principle that underlies all FFT algorithms. To formalize this approach, let us assume that the sequence length can be expressed as a product of two factors, i.e.,

$$N = N_1 N_2. \tag{9.36}$$

Recall that in deriving both the decimation-in-time and the decimation-in-frequency algorithms for $N = 2^v$, the first step was to divide either the input or the output sequence into two sequences. In that case we represented N as $N = 2(N/2)$. The dividing of the sequence can be formalized through the concept of *index maps*, which was used by Cooley and Tukey (1965) in the original presentation of their algorithm and was used by Burrus (1988) to unify a wide range of FFT algorithms. Suppose that we represent the indices n and k as

$$n = N_2 n_1 + n_2, \qquad \begin{cases} 0 \le n_1 \le N_1 - 1, \\ 0 \le n_2 \le N_2 - 1, \end{cases} \tag{9.37a}$$

$$k = k_1 + N_1 k_2, \qquad \begin{cases} 0 \le k_1 \le N_1 - 1, \\ 0 \le k_2 \le N_2 - 1. \end{cases} \tag{9.37b}$$

It is easy to verify that as n_1 and n_2 take on all possible values in the range indicated, n goes through all possible values from 0 to $(N - 1)$ with no values repeated. This is also true for the frequency index k. Using these index mappings, we can express the DFT as a function of the two indices k_1 and k_2. Substituting Eqs. (9.37) into

$$X[k] = \sum_{n=0}^{N-1} x[n] W_N^{kn}, \qquad 0 \le k \le N - 1, \tag{9.38}$$

we obtain

$$
\begin{aligned}
X[k] &= X[k_1 + N_1 k_2] \\
&= \sum_{n_2=0}^{N_2-1} \sum_{n_1=0}^{N_1-1} x[N_2 n_1 + n_2] W_N^{(k_1 + N_1 k_2)(N_2 n_1 + n_2)} \\
&= \sum_{n_2=0}^{N_2-1} \sum_{n_1=0}^{N_1-1} x[N_2 n_1 + n_2] W_N^{N_2 k_1 n_1} W_N^{k_1 n_2} W_N^{N_1 k_2 n_2} W_N^{N_1 N_2 k_2 n_1}
\end{aligned}
\tag{9.39}
$$

Since $W_N^{N_2 k_1 n_1} = W_{N_1}^{k_1 n_1}$, $W_N^{N_1 k_2 n_2} = W_{N_2}^{k_2 n_2}$, and $W_N^{N_1 N_2 k_2 n_1} = 1$, Eq. (9.39) can be written as

$$X[k_1 + N_1 k_2] = \sum_{n_2=0}^{N_2-1} \left[\left(\sum_{n_1=0}^{N_1-1} x[N_2 n_1 + n_2] W_{N_1}^{k_1 n_1} \right) W_N^{k_1 n_2} \right] W_{N_2}^{k_2 n_2}, \tag{9.40}$$

where $0 \le k_1 \le N_1 - 1$ and $0 \le k_2 \le N_2 - 1$.

To interpret Eq. (9.40), it is helpful to picture the effect of the input index mapping as mapping the input one-dimensional sequence into a two-dimensional sequence that can be represented as a two-dimensional array with n_1 specifying the columns and n_2 specifying the rows of the array. Then the inner parentheses of Eq. (9.40) is seen to be the set of N_1-point DFTs of the N_2 rows; i.e.,

$$G[n_2, k_1] = \sum_{n_1=0}^{N_1-1} x[N_2 n_1 + n_2] W_{N_1}^{k_1 n_1}, \qquad \begin{cases} 0 \le k_1 \le N_1 - 1, \\ 0 \le n_2 \le N_2 - 1. \end{cases} \tag{9.41}$$

Notice that since the input sequence is not needed again, the N_1-point row DFTs can be stored in the same array locations as the original signal samples; i.e., the row DFTs can be done in place. This set of row DFTs is multiplied by the factors $W_N^{k_1 n_2}$ to obtain

$$\tilde{G}[n_2, k_1] = W_N^{k_1 n_2} G[n_2, k_1], \qquad \begin{cases} 0 \le k_1 \le N_1 - 1, \\ 0 \le n_2 \le N_2 - 1. \end{cases} \tag{9.42}$$

This operation can also be done in place. The factors $W_N^{k_1 n_2}$ in Eqs. (9.40) and (9.42) are called *twiddle factors*. If it were not for these factors, Eq. (9.40) would be a two-dimensional DFT. In Section 9.6.2 we will see that the twiddle factors can be eliminated for certain values of N_1 and N_2.

Finally, the outer sum in Eq. (9.40) is seen to be the set of N_2-point DFTs of the columns of the array; i.e.,

$$X[k_1 + N_1 k_2] = \sum_{n_2=0}^{N_2-1} \tilde{G}[n_2, k_1] W_{N_2}^{k_2 n_2}, \qquad \begin{cases} 0 \le k_1 \le N_1 - 1, \\ 0 \le k_2 \le N_2 - 1. \end{cases} \tag{9.43}$$

The column DFTs, $X[k_1 + N_1 k_2]$, can again be stored in the same locations as the input column data $\tilde{G}[n_2, k_1]$. Thus, the computation can be done entirely in place. While the input is entered into the array according to the index map of Eq. (9.37a), the output DFT values must be extracted from the array according to Eq. (9.37b). The following example illustrates this way of looking at the FFT for the familiar case of $N = 2^\nu$.

Example 9.1

Suppose that N is divisible by 2, so we can choose $N_1 = 2$ and therefore $N_2 = N/2$. From Eq. (9.40) it follows that

$$X[k_1 + 2k_2] = \sum_{n_2=0}^{(N/2)-1} \left[\left(\sum_{n_1=0}^{1} x[(N/2)n_1 + n_2] W_2^{k_1 n_1} \right) W_N^{k_1 n_2} \right] W_{N/2}^{k_2 n_2}. \tag{9.44}$$

for $0 \le k_1 \le 1$ and $0 \le k_2 \le (N/2) - 1$. The two-dimensional array representation of the input is

n_2 \ n_1	0	1
0	$x[0]$	$x[N/2]$
1	$x[1]$	$x[N/2 + 1]$
$\vdots$	$\vdots$	$\vdots$
$N/2 - 1$	$x[N/2 - 1]$	$x[N - 1]$

We see that the first column ($n_1 = 0$) of the array is the first half of the input sequence, and the second column ($n_1 = 1$) is the second half of the input sequence. The input sequence can be sequentially reordered by extracting the columns and abutting them. Now the row DFTs are 2-point DFTs, which we represent as

$$G[n_2, k_1] = x[n_2] + (-1)^{k_1} x[N/2 + n_2], \qquad \begin{cases} 0 \le k_1 \le 1, \\ 0 \le n_2 \le (N/2) - 1. \end{cases} \tag{9.45}$$

These 2-point DFTs require no multiplications. The two-dimensional array of row transforms is

n_2 \ k_1	0	1
0	$G[0, 0]$	$G[0, 1]$
1	$G[1, 0]$	$G[1, 1]$
$\vdots$	$\vdots$	$\vdots$
$(N/2) - 1$	$G[(N/2) - 1, 0]$	$G[(N/2) - 1, 1]$

After multiplying by the twiddle factors the array becomes

n_2 \ k_1	0	1
0	$\tilde{G}[0, 0]$	$\tilde{G}[0, 1]$
1	$\tilde{G}[1, 0]$	$\tilde{G}[1, 1]$
$\vdots$	$\vdots$	$\vdots$
$(N/2) - 1$	$\tilde{G}[(N/2) - 1, 0]$	$\tilde{G}[(N/2) - 1, 1]$

where in general

$$\tilde{G}[n_2, k_1] = W_N^{k_1 n_2} G[n_2, k_1], \qquad \begin{cases} 0 \le k_1 \le 1, \\ 0 \le n_2 \le (N/2) - 1. \end{cases} \tag{9.46}$$

Finally, we compute the $(N/2)$-point transforms of the columns

$$X[k_1 + 2k_2] = \sum_{n_2=0}^{(N/2)-1} \tilde{G}[n_2, k_1] W_{N/2}^{k_2 n_2}, \qquad \begin{cases} 0 \le k_1 \le 1, \\ 0 \le k_2 \le (N/2) - 1, \end{cases} \tag{9.47}$$

so that the output two-dimensional array is

k_2 \ k_1	0	1
0	$X[0]$	$X[1]$
1	$X[2]$	$X[3]$
$\vdots$	$\vdots$	$\vdots$
$(N/2) - 1$	$X[N - 2]$	$X[N - 1]$

Since we have carried out the computation in place, the DFT values are in the order specified by the input index map rather than in the order specified by the output index map. That is, if we extract the DFT values in the same order that the input was entered into the array (columnwise), then we get the even-indexed values first and then the odd-indexed values. The flow graph representation of the operations performed in this example is depicted in Fig. 9.27 for the case $N = 16$. The first stage in Fig. 9.27 is identical in form to the first stage of the decimation-in-frequency flow graph of Fig.

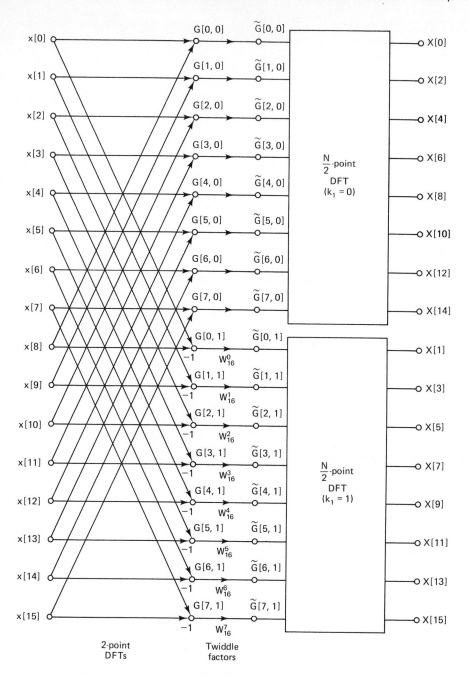

Figure 9.27   Flow graph of decomposition of an *N*-point DFT into two (*N*/2)-point DFTs (*N* = 16).

9.20. As before, if $N = 2^v$ then $N/2$ is also divisible by 2 and we can continue to break the transform down into 2-point DFTs. The complete graph for $N = 8$ is shown in Fig. 9.20. This flow graph would in fact replace the blocks labeled $(N/2)$-point DFT in the 16-point example of Fig. 9.27.

If in Example 9.1 we had chosen $N_1 = N/2$ and $N_2 = 2$, we would have obtained the first stage of the decimation-in-time decomposition of the DFT. Alternatively, we could have defined the index mappings as

$$ n = n_1 + N_1 n_2, \qquad \begin{cases} 0 \le n_1 \le N_1 - 1, \\ 0 \le n_2 \le N_2 - 1, \end{cases} \tag{9.48a} $$

$$ k = N_2 k_1 + k_2, \qquad \begin{cases} 0 \le k_1 \le N_1 - 1, \\ 0 \le k_2 \le N_2 - 1. \end{cases} \tag{9.48b} $$

In this case choosing $N_1 = 2$ and $N_2 = N/2$ leads to the decimation-in-time algorithm while $N_1 = N/2$ and $N_2 = 2$ leads to the decimation-in-frequency algorithm. Algorithms based on index mappings of the form of Eqs. (9.37) or Eqs. (9.48) all involve the use of twiddle factors as an intermediate operation between the smaller DFTs. The original algorithms published by Cooley and Tukey (1965) were of this form; hence, algorithms of this class have come to be called Cooley-Tukey algorithms.

The development leading up to Eq. (9.40) made no assumptions about N_1 and N_2. Either or both could have additional factors. In general, what is required for an efficient algorithm is that N be a highly composite number of the form

$$ N = N_1 N_2 \cdots N_v. \tag{9.49} $$

In this case, we could define more complicated multidimensional index maps for the input and output sequences. In the case $N = 2^v$, repeated decomposition into 2-point transforms leads to a multidimensional index map for the complete decomposition of the DFT of the form

$$ n = 2^{v-1} n_1 + \cdots + 2 n_{v-1} + n_v, \tag{9.50a} $$

$$ k = k_1 + 2 k_2 + \cdots + 2^{v-1} k_v. \tag{9.50b} $$

In this case since the indices n_i and k_i are either 0 or 1, we have seen that if the input is in normal order, the in-place computation puts the output in bit-reversed order and vice versa.

If $N = R^v$, i.e., is the product of identical factors, the corresponding FFT algorithms are called *radix-R algorithms*. If the input is in normal order, the output will occur in a *digit-reversed* order in the radix R number system. When the factors are not all the same, the corresponding FFT algorithms are called *mixed-radix algorithms*.

A formula for the total number of multiplications required for a DFT is useful for comparing algorithms. To obtain such a formula for the general case, we define $\mu(N)$ to be the *number of multiplications required to compute an N-point DFT*. For the 2-factor case, the total number of multiplications is easily determined. Evaluation of the row transforms requires $N_2 \mu(N_1)$ multiplications; the twiddle factors require

$N_1 N_2 = N$ multiplications; and the final set of column transforms requires $N_1 \mu(N_2)$ multiplications. Thus, the total number of multiplications is

$$\mu(N) = N_2\mu(N_1) + N_1\mu(N_2) + N = N\left(\frac{\mu(N_1)}{N_1} + \frac{\mu(N_2)}{N_2} + 1\right). \tag{9.51}$$

If N has v factors as in Eq. (9.49), it can be shown by repeated application of Eq. (9.51) that the total number of multiplications required to compute the DFT by repeatedly breaking the transform down into successively smaller transforms is

$$\mu(N) = N\left(\sum_{i=1}^{v} \frac{\mu(N_i)}{N_i} + (v - 1)\right). \tag{9.52}$$

This equation displays the way that the index mappings allow us to reduce the amount of computation, but it does not tell the whole story. If the smaller transforms are evaluated by the direct method, they would require N_i^2 multiplications for each N_i-point transform. With this assumption, Cooley and Tukey (1965) showed that the least computation is required when $N = 3^v$. However, small transforms of certain sizes do not require N_i^2 multiplications. For example, we have seen that 2-point transforms require *no* multiplications. Equation (9.52) gives the correct result for the case $N = 2^v$ if we remember that 2-point DFTs require no multiplications and that in the last stage of the decimation-in-frequency algorithm, the multiplications are all by W_N^0. Thus Eq. (9.52) gives a total of $N(v - 1)$ multiplications, all of which are due to the twiddle factors. When we recall that half the twiddle factors in any stage are unity, the correct result for either decimation-in-time or decimation-in-frequency is $N(v - 1)/2$, as is evident from Fig. 9.10 or 9.20.†

From the preceding discussion, we see that one way to reduce computation is to obtain efficient algorithms for implementing the smaller N_i-point transforms with fewer than N_i^2 multiplications. This is an important general principle, as we will see in Section 9.7.1. The next example shows that factors of 4 have especially attractive implementations.

Example 9.2

Consider the case when N has a factor of 4, so we can choose $N_1 = 4$ and $N_2 = N/4$. Returning to Eq. (9.40), we see that the first step is to perform $(N/4)$ 4-point transforms. Since $W_4 = -j$, the 4-point transforms have the form

$$\begin{aligned} G[n_2, k_1] = x[n_2] + (-j)^{k_1}x[(N/4) + n_2] \\ + (-1)^{k_1}x[(N/2) + n_2] + (j)^{k_1}x[(3N/4) + n_2] \end{aligned} \tag{9.53}$$

for $k_1 = 0, 1, 2, 3$ and $n_2 = 0, 1, \ldots, (N/4) - 1$. From Eq. (9.53) we see that since multiplication by $\pm j$ can be achieved by interchanging the real and imaginary parts, no multiplications are required for 4-point DFTs.

The 4-point butterfly corresponding to Eq. (9.53) is shown in Fig. 9.28, and Fig. 9.29 shows a complete flow graph for $N = 16$ based on Eq. (9.40). In this graph we have shown only one of the 4-point DFTs in the first stage because otherwise the graph is too

---

† Recall that in our previous discussion we arrived at a total of $Nv/2$, but this included $N/2$ complex multiplications by W_N^0 at the output.

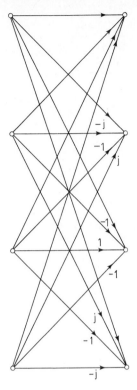

Figure 9.28 Butterfly for 4-point DFT.

complicated. Also, we have not shown the multipliers for each 4-point DFT since they are as shown in Fig. 9.28. We see that a total of only 12 complex multiplications is required for the twiddle factors in a 16-point DFT implemented as two stages involving 4-point DFTs, as in Fig. 9.29. We could also implement a 16-point DFT as four stages involving 2-point DFTs. In this case we would require a total of 24 complex multiplications for the twiddle factors.

Example 9.2 shows that when $N = 4^v$, we can achieve significant savings by using v stages of 4-point butterflies rather than $2v$ stages of 2-point butterflies. These savings occur because we have half the number of stages and therefore half the number of twiddle factor multiplications. Obviously, reducing the number of twiddle factors while not increasing the multiplications in the small DFTs is also an effective way to reduce the total number of multiplications. Since length-8 transforms can be implemented directly with only a few multiplications, even more efficient algorithms can be obtained for $N = 2^v$ using as many factors of 8 as possible and finishing with either a factor of 4 or a factor of 2. A program based on this principle was published by Bergland (1968) and is available in DSP Committee (1979).

Our discussion of the Cooley-Tukey FFT algorithms makes it clear that simply dividing the DFT calculation into many short-length DFTs can result in huge savings of computations. Also, we have seen that certain lengths are advantageous because they require no multiplications and that it is advantageous to reduce the number of twiddle factor multiplications. We now consider how to eliminate the twiddle factors completely.

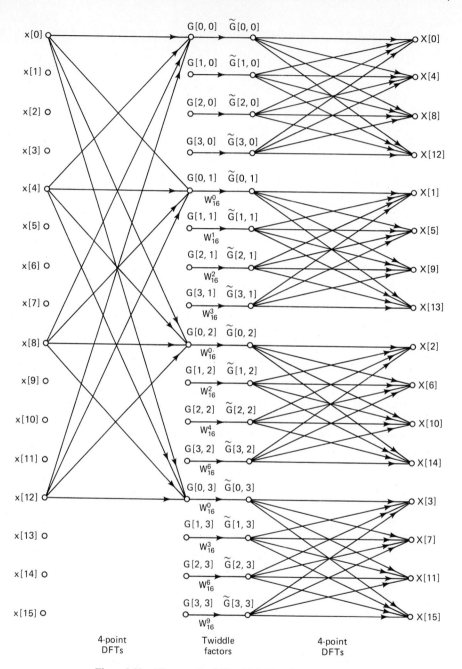

Figure 9.29 Flow graph of $N = 16$ DFT using factors of 4.

9.6.2 Prime Factor Algorithms

To eliminate the twiddle factors it is necessary to define index mappings of the form (Burrus, 1988)

$$n = ((An_1 + Bn_2))_N, \qquad \begin{cases} 0 \le n_1 \le N_1 - 1, \\ 0 \le n_2 \le N_2 - 1, \end{cases} \qquad (9.54a)$$

$$k = ((Ck_1 + Dk_2))_N, \qquad \begin{cases} 0 \le k_1 \le N_1 - 1, \\ 0 \le k_2 \le N_2 - 1, \end{cases} \qquad (9.54b)$$

where, as in Chapter 8, $((\cdot))_N$ denotes evaluation of the index modulo N.

We have seen that the constants $A = N_2$, $B = C = 1$, and $D = N_1$ give the map of Eqs. (9.37), and the constants $A = D = 1$, $B = N_1$, and $C = N_2$ give Eqs (9.48). In both cases, the index maps never exceed N and thus the modulo N reduction is superfluous. These maps produce the Cooley-Tukey decomposition of Eq. (9.40) for any values of N_1 and N_2. However, if N_1 and N_2 are relatively prime (i.e., have no common factors), then a different choice of the constants in Eq. (9.54) completely eliminates the twiddle factors. Specifically, we must choose the constants A, B, C, and D in Eqs. (9.54) so that all values of n and k between 0 and $(N-1)$ are represented once and only once and so that the twiddle factors disappear; i.e.,

$$W_N^{(An_1 + Bn_2)(Ck_1 + Dk_2)} = W_{N_1}^{k_1 n_1} W_{N_2}^{k_2 n_2}.$$

This requires that $((AC))_N = N_2$, $((BD))_N = N_1$, and $((AD))_N = ((BC))_N = 0$. Finding a set of coefficients that satisfies these constraints involves the use of the *Chinese remainder theorem*. A discussion of this and other necessary number theory concepts would take us further into the topic of fast DFT algorithms than is appropriate here. However, one set of coefficients that satisfies the uniqueness condition and eliminates the twiddle factors is (Burrus, 1977, 1988)

$$A = N_2 \qquad\qquad \text{and} \qquad B = N_1, \qquad\qquad (9.55a)$$

$$C = N_2((N_2^{-1}))_{N_1} \qquad \text{and} \qquad D = N_1((N_1^{-1}))_{N_2}, \qquad (9.55b)$$

where $((N_2^{-1}))_{N_1}$ denotes the multiplicative inverse of N_2 reduced modulo N_1. (For example if $N_1 = 4$ and $N_2 = 3$, $((3^{-1}))_4 = 3$ since $((3 \cdot 3))_4 = 1$.) With these choices (which are not unique), the DFT can be expressed as

$$X[((N_2((N_2^{-1}))_{N_1} k_1 + N_1((N_1^{-1}))_{N_2} k_2))_N]$$

$$= \sum_{n_2=0}^{N_2-1} \left[\sum_{n_1=0}^{N_1-1} x[((N_2 n_1 + N_1 n_2))_N] W_{N_1}^{k_1 n_1} \right] W_{N_2}^{k_2 n_2}, \qquad (9.56)$$

where $0 \le k_1 \le N_1 - 1$ and $0 \le k_2 \le N_2 - 1$. Now we have expressed the one-dimensional DFT as a two-dimensional DFT, without the intervening twiddle factors. An algorithm of this form was given by Good (1958). Because of the requirement that N_1 and N_2 be relatively prime, algorithms of the form of Eq. (9.56) are called *prime factor algorithms*. Since the twiddle factor multiplications have been eliminated, the row and column transforms can be done in either order, and the total number of multiplications is given by Eq. (9.52) without the term $(v - 1)$. Thus, given the same efficiency in implementing the N_i-point DFTs, Eq. (9.56) requires $N(v-1)$ fewer complex multiplications than Eq. (9.40).

Example 9.3

Suppose $N = 12$, so we can choose the relatively prime factors $N_1 = 4$ and $N_2 = 3$. (The factors need not be prime numbers.) Using Eqs. (9.54) and (9.55) we obtain the index maps

$$n = ((3n_1 + 4n_2))_{12}, \qquad \begin{cases} 0 \le n_1 \le 3, \\ 0 \le n_2 \le 2, \end{cases} \tag{9.57a}$$

$$k = ((9k_1 + 4k_2))_{12}, \qquad \begin{cases} 0 \le k_1 \le 3, \\ 0 \le k_2 \le 2, \end{cases} \tag{9.57b}$$

which lead to the representation

$$X[((9k_1 + 4k_2))_{12}] = \sum_{n_2=0}^{2} \left[\sum_{n_1=0}^{3} x[((3n_1 + 4n_2))_{12}] W_4^{k_1 n_1} \right] W_3^{k_2 n_2} \tag{9.58}$$

for $0 \le k_1 \le 3$ and $0 \le k_2 \le 2$. The two-dimensional array representation of the input is

n_2 \ n_1	0	1	2	3
0	$x[0]$	$x[3]$	$x[6]$	$x[9]$
1	$x[4]$	$x[7]$	$x[10]$	$x[1]$
2	$x[8]$	$x[11]$	$x[2]$	$x[5]$

The modulo 12 reduction is necessary for these index maps so that the effective values of n and k remain between 0 and 11. The 4-point transforms of the rows lead to the representation

n_2 \ k_1	0	1	2	3
0	$G[0,0]$	$G[0,1]$	$G[0,2]$	$G[0,3]$
1	$G[1,0]$	$G[1,1]$	$G[1,2]$	$G[1,3]$
2	$G[2,0]$	$G[2,1]$	$G[2,2]$	$G[2,3]$

where

$$G[n_2, k_1] = \sum_{n_1=0}^{3} x[((3n_1 + 4n_2))_{12}] W_4^{k_1 n_1}. \tag{9.59}$$

Since there are no twiddle factors, the column transforms

$$X[((9k_1 + 4k_2))_{12}] = \sum_{n_2=0}^{2} G[n_2, k_1] W_3^{k_2 n_2} \tag{9.60}$$

result in the final array

k_2 \ k_1	0	1	2	3
0	$X[0]$	$X[9]$	$X[6]$	$X[3]$
1	$X[4]$	$X[1]$	$X[10]$	$X[7]$
2	$X[8]$	$X[5]$	$X[2]$	$X[11]$

Figure 9.30 shows a complete flow graph for this example. Since the 4-point DFTs shown in Fig. 9.28 require no multiplications and since the 3-point DFTs depicted in Fig. 9.31 require four complex multiplications, this algorithm requires only 12 complex multiplications. A Cooley-Tukey algorithm would have the same structure as in Fig. 9.30, but it would require an additional 9 multiplications for the twiddle factors that

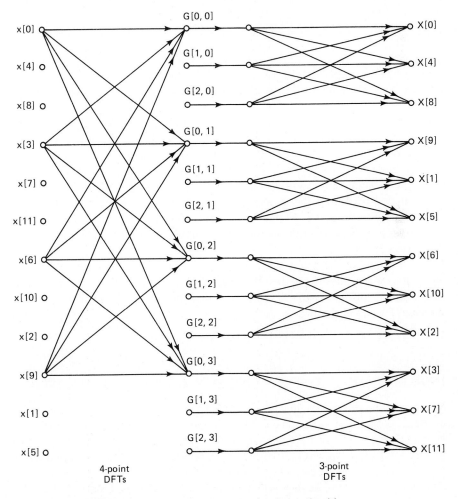

Figure 9.30   Flow graph of $N = 12$ prime factor algorithm.

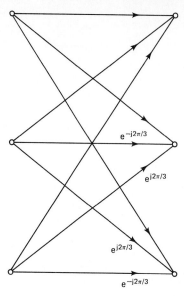

Figure 9.31 Flow graph of 3-point DFT used in Fig. 9.30.

come between the 4-point transforms and the 3-point transforms. Observe that for the prime factor algorithm, neither the input nor the output sequences are in normal order.

The elimination of the twiddle factor multiplications by the prime factor algorithm is at the expense of greater complexity of the indexing and programming of the algorithm. In contrast to radix-2 or radix-4 Cooley-Tukey algorithms, which use a single butterfly (with no multiplications) and a highly nested program structure, a prime factor algorithm requires a different butterfly for each factor and is most easily programmed by taking the factors one at a time. Also, as Example 9.3 shows, the indexing is more complex. FORTRAN subroutines for prime factor algorithms and a discussion of programming details are given by Kolba and Parks (1977) and Burrus and Parks (1985).

The value of the prime factor algorithm is greatly enhanced if highly efficient algorithms are available for the short-length DFTs (butterflies) required for each of the relatively prime factors. The key to this further reduction of computation is discussed briefly in the next section.

9.7 IMPLEMENTATION OF THE DFT USING CONVOLUTION

Because of the dramatic efficiency of the FFT, convolution is often implemented by explicitly computing the inverse DFT of the product of the DFTs of each sequence to be convolved, where an FFT algorithm is used to compute both the forward and the inverse DFTs. In contrast, and even in apparent (but of course not actual) contradiction, it is sometimes preferable to compute the DFT by first reformulating it as a convolution. We have already seen an example of this in the Goertzel algorithm, and a number of other more sophisticated procedures are based on this approach.

9.7.1 Overview of the Winograd Fourier
Transform Algorithm

One procedure proposed and developed by S. Winograd (1978), often referred to as the Winograd Fourier transform algorithm (WFTA), achieves its efficiency by expressing the DFT in terms of polynomial multiplication or, equivalently, convolution. The WFTA uses the same indexing scheme as the prime factor algorithm discussed in Section 9.6.2. As discussed before, this indexing scheme permits the decomposition of the DFT into a multiplicity of short-length DFTs where the lengths are relatively prime. Then the short DFTs are converted into periodic convolutions. A scheme for converting a DFT into a convolution when the number of input samples is prime was proposed by Rader (1968), but its application awaited the development of efficient methods for computing periodic convolutions. Winograd combined all of the above together with highly efficient algorithms for computing cyclic convolutions into a new approach to computing the DFT. The techniques for deriving the efficient algorithms for computing short convolutions are based on relatively advanced number-theoretic concepts such as the Chinese remainder theorem for polynomials, and consequently we do not explore the details here. However, excellent discussions of the details of the WFTA are available in McClellan and Rader (1979), Blahut (1985), and Burrus (1988).

With the WFTA approach, the number of multiplications required for an N-point DFT is proportional to N rather than $N \log N$. Although this approach leads to algorithms that are optimal in terms of minimizing multiplications, the number of additions is significantly increased in comparison with the FFT. Therefore, the WFTA is most advantageous when multiplication is significantly slower than addition, as is often the case with fixed-point digital arithmetic. However, in processors where multiplication and accumulation are tied together, the Cooley-Tukey or prime factor algorithms are generally preferable. Additional difficulties with the WFTA are that indexing is more complicated, in-place computation is not possible, and there are major structural differences in algorithms for different values of N.

Thus, although the WFTA is extremely important as a benchmark for determining how efficient the DFT computation can be (in terms of multiplications), other factors often dominate in determining the speed and efficiency of a hardware or software implementation of the DFT computation.

9.7.2 Chirp Transform Algorithm

Another algorithm based on expressing the DFT as a convolution is referred to as the chirp transform algorithm (CTA). This algorithm is not optimal in minimizing any measure of computational complexity, but it has been useful in a variety of applications, particularly when implemented in technologies that are well suited to doing convolution with a fixed, prespecified impulse response. The CTA is also more flexible than the FFT, since it can be used to compute *any* set of equally spaced samples of the Fourier transform on the unit circle.

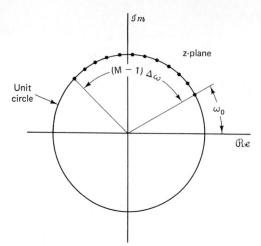

Figure 9.32 Frequency samples for chirp transform algorithm.

To derive the chirp transform algorithm, we let $x[n]$ denote an N-point sequence and $X(e^{j\omega})$ its Fourier transform. We consider the evaluation of M samples of $X(e^{j\omega})$ that are equally spaced in angle on the unit circle, as indicated in Fig. 9.32, i.e., at frequencies

$$\omega_k = \omega_0 + k\Delta\omega, \qquad k = 0, 1, \ldots, M - 1, \tag{9.61}$$

where the starting frequency ω_0 and the frequency increment $\Delta\omega$ can be chosen arbitrarily. (For the specific case of the DFT, $\omega_0 = 0$, $M = N$, and $\Delta\omega = 2\pi/N$.) The Fourier transform corresponding to this more general set of frequency samples is given by

$$X(e^{j\omega_k}) = \sum_{n=0}^{N-1} x[n]e^{-j\omega_k n}, \qquad k = 0, 1, \ldots, M - 1, \tag{9.62}$$

or, with W defined as

$$W = e^{-j\Delta\omega} \tag{9.63}$$

and using Eq. (9.61),

$$X(e^{j\omega_k}) = \sum_{n=0}^{N-1} x[n]e^{-j\omega_0 n}W^{nk}. \tag{9.64}$$

To express $X(e^{j\omega_k})$ as a convolution, we use the identity

$$nk = \tfrac{1}{2}[n^2 + k^2 - (k - n)^2] \tag{9.65}$$

to express Eq. (9.64) as

$$X(e^{j\omega_k}) = \sum_{n=0}^{N-1} x[n]e^{-j\omega_0 n}W^{n^2/2}W^{k^2/2}W^{-(k-n)^2/2}. \tag{9.66}$$

Letting

$$g[n] = x[n]e^{-j\omega_0 n}W^{n^2/2}, \tag{9.67}$$

we can then write

$$X(e^{j\omega_k}) = W^{k^2/2}\left(\sum_{n=0}^{N-1} g[n]W^{-(k-n)^2/2}\right), \qquad k = 0, 1, \ldots, M-1. \qquad (9.68)$$

In preparation for interpreting Eq. (9.68) as the output of a linear time-invariant system, we obtain more familiar notation by replacing k by n and n by k in Eq. (9.68):

$$X(e^{j\omega_n}) = W^{n^2/2}\left(\sum_{k=0}^{N-1} g[k]W^{-(n-k)^2/2}\right), \qquad n = 0, 1, \ldots, M-1. \qquad (9.69)$$

In the form of Eq. (9.69), $X(e^{j\omega_n})$ corresponds to the convolution of the sequence $g[n]$ with the sequence $W^{-n^2/2}$ followed by multiplication by the sequence $W^{n^2/2}$. The output sequence, indexed on the independent variable n, is the sequence of frequency samples $X(e^{j\omega_n})$. With this interpretation, the computation of Eq. (9.69) is as depicted in Fig. 9.33. The sequence $W^{-n^2/2}$ can be thought of as a complex exponential sequence with linearly increasing frequency. In radar systems, such signals are called chirp signals; hence the name *chirp transform*. A system similar to Fig. 9.33 is commonly used in radar and sonar signal processing for pulse compression (Skolnik, 1980).

For the evaluation of the Fourier transform samples specified in Eq. (9.69), we need only compute the output of the system in Fig. 9.33 over a finite interval. In Fig. 9.34 we depict the sequences $g[n]$, $W^{-n^2/2}$, and $g[n] * W^{-n^2/2}$. Since $g[n]$ is of finite duration, only a finite portion of the sequence $W^{-n^2/2}$ is used in obtaining $g[n] * W^{-n^2/2}$ over the interval $n = 0, 1, \ldots, M-1$, specifically that portion from $n = -(N-1)$ to $n = M-1$. Let us define $h[n]$ as

$$h[n] = \begin{cases} W^{-n^2/2}, & -(N-1) \le n \le M-1, \\ 0, & \text{otherwise,} \end{cases} \qquad (9.70)$$

as depicted in Fig. 9.35. It is easily verified by considering the graphical representation of the process of convolution that

$$g[n] * W^{-n^2/2} = g[n] * h[n], \qquad n = 0, 1, \ldots, M-1. \qquad (9.71)$$

Consequently, the infinite-duration impulse response $W^{-n^2/2}$ in the system of Fig. 9.33 can be replaced by the finite-duration impulse response of Fig. 9.35. The system is now as indicated in Fig. 9.36, where $h[n]$ is specified by Eq. (9.70) and the frequency samples $X(e^{j\omega_n})$ are given by

$$X(e^{j\omega_n}) = y[n], \qquad n = 0, 1, \ldots, M-1. \qquad (9.72)$$

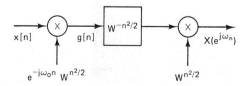

Figure 9.33   Block diagram of chirp transform algorithm.

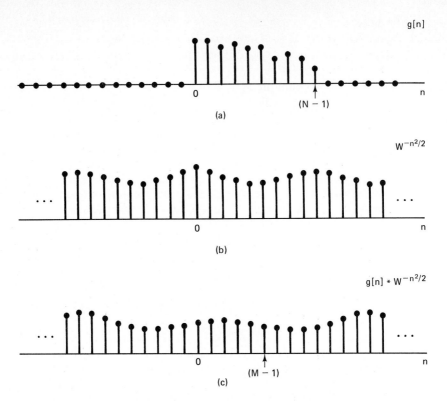

Figure 9.34 An indication of the sequences used in the chirp transform algorithm. Note that the actual sequences involved are complex-valued. (a) $g[n] = x[n]e^{-j\omega_0 n}W^{n^2/2}$. (b) $W^{-n^2/2}$. (c) $g[n] * W^{-n^2/2}$.

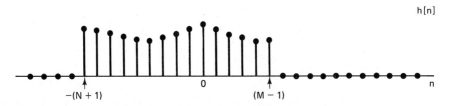

Figure 9.35 An indication of the region of support for the FIR chirp filter. Note that the actual values of $h[n]$ as given by Eq. (9.70) are complex.

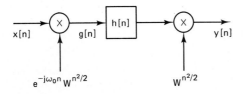

Figure 9.36 Block diagram of chirp transform system for finite-length impulse response.

Evaluation of frequency samples using the procedure indicated in Fig. 9.36 has a number of potential advantages. In general we do not require $N = M$ as in the FFT algorithms, and neither N nor M need be composite numbers. In fact, they may be prime numbers if desired. Furthermore, the parameter ω_0 is arbitrary. This increased flexibility over the FFT does not preclude efficient computation, since the convolution in Fig. 9.36 can be implemented efficiently using an FFT algorithm with the technique of Section 8.8 to compute the convolution. As discussed in Section 8.8, the FFT size must be greater than $(M + N - 1)$, in order that the circular convolution will be equal to $g[n] * h[n]$ for $0 \leq n \leq M - 1$. The FFT size is otherwise arbitrary and can, for example, be chosen to be a power of 2. It is interesting to note that the FFT algorithms used to compute the convolution implied by the chirp transform algorithm could be of the Winograd type. These algorithms themselves use convolution to implement the DFT computation.

In the system of Fig. 9.36, $h[n]$ is noncausal, and for certain real-time implementations it must be modified to obtain a causal system. Since $h[n]$ is of finite duration, this modification is easily accomplished by delaying $h[n]$ by $(N - 1)$ to obtain a causal impulse response:

$$h_1[n] = \begin{cases} W^{-(n-N+1)^2/2}, & n = 0, 1, \ldots, M + N - 2, \\ 0, & \text{otherwise.} \end{cases} \tag{9.73}$$

Since both the chirp demodulation factor at the output and the output signal are also delayed by $(N - 1)$ samples, the Fourier transform values are

$$X(e^{j\omega_n}) = y_1[n + N - 1], \qquad n = 0, 1, \ldots, M - 1. \tag{9.74}$$

Modifying the system of Fig. 9.36 to obtain a causal system results in the causal system in Fig. 9.37. An advantage of the system in Fig. 9.37 stems from the fact that it involves the convolution of the input signal (modulated with a chirp) with a fixed, causal impulse response. Certain technologies such as charge-coupled devices (CCD) and surface acoustic wave devices (SAW) are particularly useful for implementing convolution with a fixed, prespecified impulse response. Such devices can be used for implementation of FIR filters, with the filter impulse response being specified at the time of fabrication by a geometric pattern of electrodes. A similar approach was followed by Hewes et al. (1979) in implementing the chirp transform algorithm with CCD devices.

Further simplification of the chirp transform algorithm results when the frequency samples to be computed correspond to the DFT. In this case, $\omega_0 = 0$ and $W = e^{-j2\pi/N}$, so $\omega_n = 2\pi n/N$. In this case, it is convenient to modify the system of Fig. 9.37. Specifically, let $\omega_0 = 0$ and $W = e^{-j2\pi/N} = W_N$, and consider applying an

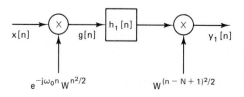

Figure 9.37   Block diagram of chirp transform system for causal finite-length impulse response.

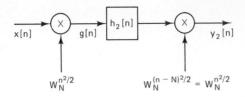

Figure 9.38 Block diagram of chirp transform system for obtaining DFT samples.

additional unit of delay to the impulse response in Fig. 9.37. With N even, $W_N^N = e^{j2\pi} = 1$, so

$$W_N^{-(n-N)^2/2} = W_N^{-n^2/2}. \tag{9.75}$$

Therefore, the system now is as shown in Fig. 9.38, where

$$h_2[n] = \begin{cases} W_N^{-n^2/2}, & n = 1, 2, \dots, M + N - 1, \\ 0, & \text{otherwise.} \end{cases} \tag{9.76}$$

In this case, the chirp signal modulating $x[n]$ and the chirp signal modulating the output of the FIR filter are identical and

$$X(e^{j2\pi n/N}) = y_2[n + N], \qquad n = 0, 1, \dots, M - 1. \tag{9.77}$$

An algorithm similar to the chirp transform algorithm was first proposed by Bluestein (1970), who showed that a recursive realization of Fig. 9.37 can be obtained for the case $\Delta\omega = 2\pi/N$ and N a perfect square (see Problem 9.20). Rabiner et al. (1969) generalized this algorithm to obtain samples of the z-transform equally spaced in angle on a spiral contour in the z-plane. This more general form of the chirp transform was called the chirp z-transform algorithm (CZT). The algorithm that we have called the chirp transform algorithm is a special case of the chirp z-transform algorithm.

9.8 EFFECTS OF FINITE REGISTER LENGTH IN DISCRETE FOURIER TRANSFORM COMPUTATIONS

Since the discrete Fourier transform is widely used for digital filtering and spectrum analysis, it is important to understand the effects of finite register length in DFT calculations. As in the case of digital filters, however, a precise analysis of the effects is difficult, and often a simplified analysis is sufficient for the purpose of choosing the required register length for computing the discrete Fourier transform. The analysis that we will present is similar in style to that carried out in Sections 6.8 and 6.9. Specifically, we will analyze arithmetic roundoff by means of a linear noise model obtained by inserting an additive noise source at each point in the computation algorithm where roundoff occurs. Furthermore, we will make a number of assumptions to simplify the analysis. The results that we obtain lead to several simplified but useful estimates of the effect of arithmetic roundoff. Although the analysis is for rounding, it is generally easy to modify the results for truncation.

We have seen a variety of different algorithmic structures for DFT computation. They are all similar in that they break the computation down into a succession of short-length DFTs. As a result, the effects of roundoff noise are very similar among the different algorithm classes. Therefore, even though we consider only the radix-2 decimation-in-time algorithm, our results can be expected to be representative of other forms as well. We begin our discussion by considering roundoff errors in the direct evaluation of the DFT relation. This will establish the style of analysis and provide a point of comparison for the noise performance of the FFT.

9.8.1 Analysis of Quantization in Direct Computation of the DFT

The discrete Fourier transform is defined by the equation

$$X[k] = \sum_{n=0}^{N-1} x[n]W_N^{kn}, \qquad k = 0, 1, \ldots, N-1, \tag{9.78}$$

where $W_N = e^{-j(2\pi/N)}$. Although Eq. (9.78) is generally evaluated by an FFT algorithm, there are instances where straightforward accumulation of the products in Eq. (9.78) is the most reasonable approach (e.g., when only a few values of k are of interest). The analysis of quantization effects in this computational procedure is quite simple and serves as an introduction to quantization effects in DFT calculations.

We note that for a given value of k, Eq. (9.78) is completely analogous to the convolution sum expression

$$y[n] = \sum_{m=0}^{N-1} h[m]x[n-m] \tag{9.79}$$

for fixed n. In this case the complex quantities W_N^{kn} play the role of the impulse response, $X[k]$ plays the role of the output, and $x[n]$ is the input. Thus, an analysis similar to that of Section 6.9.4 can be used for the direct computation of the DFT, with the additional consideration that the errors in this case are complex sequences. Since the Goertzel algorithm computes the DFT as the output of a second-order discrete-time system, the noise analysis of the Goertzel algorithm is carried out in a similar manner.

For fixed-point arithmetic, one approach to the direct calculation of $X[k]$ is depicted by the flow graph of Fig. 9.39. In the figure, $X'[k]$ represents the results of finite-precision computation of the DFT, and $F[k]$ represents the error in computation of the kth value. The complex quantities $\varepsilon[n, k]$ represent the errors due to rounding of the products $x[n]W_N^{kn}$. It can be seen that these complex errors add directly to the output, so

$$F[k] = \sum_{n=0}^{N-1} \varepsilon[n, k]. \tag{9.80}$$

The product $x[n]W_N^{kn}$ is

$$x[n]W_N^{kn} = \mathscr{R}e\{x[n]\}\cos\left(\frac{2\pi}{N}kn\right) + \mathscr{I}m\{x[n]\}\sin\left(\frac{2\pi}{N}kn\right)$$

$$+ j\left[\mathscr{I}m\{x[n]\}\cos\left(\frac{2\pi}{N}kn\right) - \mathscr{R}e\{x[n]\}\sin\left(\frac{2\pi}{N}kn\right)\right]. \tag{9.81}$$

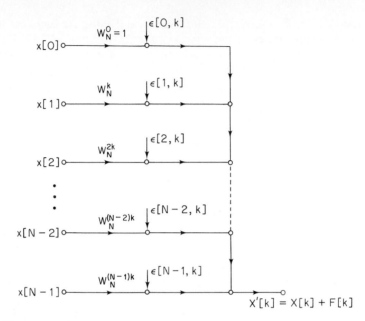

Figure 9.39　Linear noise model for fixed-point roundoff noise in computation of the DFT.

With finite-precision fixed-point arithmetic, the rounded complex product can be represented as

$$Q[x[n]W_N^{kn}] = \mathcal{R}e\{x[n]\}\cos\left(\frac{2\pi}{N}kn\right) + \varepsilon_1[n,k] + \mathcal{I}m\{x[n]\}\sin\left(\frac{2\pi}{N}kn\right)$$

$$+ \varepsilon_2[n,k] + j\left(\mathcal{I}m\{x[n]\}\cos\left(\frac{2\pi}{N}kn\right) + \varepsilon_3[n,k]\right) \qquad (9.82)$$

$$- j\left(\mathcal{R}e\{x[n]\}\sin\left(\frac{2\pi}{N}kn\right) + \varepsilon_4[n,k]\right).$$

That is, each real multiplication contributes a roundoff error.† To facilitate computing the variance of the error in $X'[k]$, we assume that the errors due to each real multiplication have the following properties:

1. The errors are uniformly distributed random variables over the range $-(1/2)\cdot 2^{-B}$ to $(1/2)\cdot 2^{-B}$, where, as defined in Section 6.7, numbers are represented as $(B + 1)$-bit signed fractions. Therefore, each error source has variance $2^{-2B}/12$.

2. The errors are uncorrelated with one another.

3. All the errors are uncorrelated with the input and consequently also with the output.

　† Note that we have assumed that the coefficients W_N^{kn} are represented exactly. The effect of quantizing these coefficients is discussed in Section 9.8.4.

The mean of the error due to rounding a complex multiplication is zero. Since the squared magnitude of the complex error $\varepsilon[n, k]$ is

$$|\varepsilon[n, k]|^2 = (\varepsilon_1[n, k] + \varepsilon_2[n, k])^2 + (\varepsilon_3[n, k] + \varepsilon_4[n, k])^2, \qquad (9.83)$$

the average value of $|\varepsilon[n, k]|^2$ is

$$\mathscr{E}\{|\varepsilon[n, k]|^2\} = 4 \cdot \frac{2^{-2B}}{12} = \frac{1}{3} \cdot 2^{-2B} \qquad (9.84)$$

The average magnitude squared of the output error is

$$\mathscr{E}\{|F[k]|^2\} = \sum_{n=0}^{N-1} \mathscr{E}\{|\varepsilon[n, k]|^2\} = \frac{N}{3} 2^{-2B}. \qquad (9.85)$$

As in the case of the direct form realization of an FIR filter, the output noise is proportional to N.† If a double-length accumulator is available, the roundoff noise in the real and imaginary parts can be reduced to a single noise source by computing the real part first and quantizing the result and then computing the imaginary part and quantizing it. In this case, the output noise is given by Eq. (9.85) with $N = 1$.

As with the fixed-point direct form realization of an FIR filter, the direct DFT calculation is subject to a dynamic range limitation. From Eq. (9.78) we see that

$$|X[k]| \leq \sum_{n=0}^{N-1} |x[n]| < N, \qquad k = 0, 1, \ldots, N-1 \qquad (9.86)$$

if $|x[n]| < 1$. For no overflow to occur, we require that $|X[k]| < 1$. This is assured if

$$\sum_{n=0}^{N-1} |x[n]| < 1. \qquad (9.87)$$

Thus, in the worst case, we may need to divide the input by N to prevent overflow. For example, the sequence $x[n] = 1$, $0 \leq n \leq N - 1$, has the discrete Fourier transform

$$X[k] = \begin{cases} N, & k = 0, \\ 0, & \text{otherwise.} \end{cases} \qquad (9.88)$$

On the other hand, $|X[k]|$ may be less than 1, even though Eq. (9.87) is not satisfied. Consider, for example, the sequence $x[n] = A\delta[n]$, which has DFT $X[k] = A$ for $0 \leq k \leq N - 1$. If $1/N < A < 1$, Eq. (9.87) is not satisfied, but $|X[k]| = A < 1$.

There are a number of solutions to the dynamic range problem. We can divide the input by N, thus increasing the noise-to-signal ratio at the output. We can use a block floating-point scheme where division by 2 is performed whenever an overflow occurs. We can use floating-point arithmetic, in which case overflow is virtually eliminated. We will examine all these solutions for a class of FFT algorithms. In addition, we will comment on the error introduced by quantizing the coefficient values.

---

† Note that Eq. (9.85) is a conservative estimate since some of the multiplications (e.g., by W_N^0) can be done without error.

9.8.2 Analysis of Quantization Effects in Fixed-Point FFT Algorithms

The detailed effects of quantization depend on the specific FFT algorithm used. The most commonly used algorithms are the radix-2 forms, for which the size of the transform is an integer power of 2. For the most part, the following discussion is phrased in terms of the decimation-in-time form of the radix-2 algorithm. The results, however, are applicable with only minor modification to the decimation-in-frequency form. Furthermore, most of the ideas employed in the error analysis of the radix-2 algorithms can be utilized in the analysis of other algorithms.

The flow graph depicting a decimation-in-time algorithm for $N = 8$ was shown in Fig. 9.10 and is reproduced in Fig. 9.40. Some key aspects of this diagram are common to all standard radix-2 algorithms. The DFT is computed in $v = \log_2 N$ stages. At each stage a new array of N numbers is formed from the previous array by linear combinations of the elements taken two at a time. The vth array contains the desired DFT. For radix-2 decimation-in-time algorithms, the basic 2-point DFT computation (with twiddle factor) is of the form

$$X_m[p] = X_{m-1}[p] + W_N^r X_{m-1}[q], \tag{9.89a}$$

$$X_m[q] = X_{m-1}[p] - W_N^r X_{m-1}[q]. \tag{9.89b}$$

Here the subscripts m and $(m-1)$ refer to the mth array and the $(m-1)$st array, respectively, and p and q denote the location of the numbers in each array. (Note that $m = 0$ refers to the input array and $m = v$ refers to the output array.) A flow graph representing the butterfly computation is shown in Fig. 9.41.

The form of the butterfly computation is somewhat different for a radix-2 decimation-in-frequency algorithm, where the basic computation is

$$X_m[p] = X_{m-1}[p] + X_{m-1}[q], \tag{9.90a}$$

$$X_m[q] = (X_{m-1}[p] - X_{m-1}[q])W_N^r. \tag{9.90b}$$

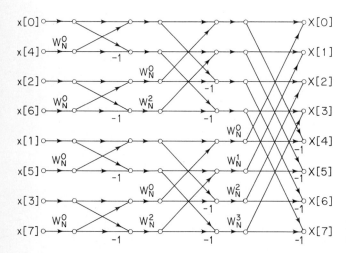

Figure 9.40 Flow graph for decimation-in-time FFT algorithm.

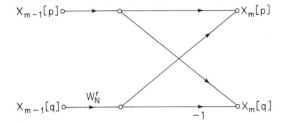

Figure 9.41 Butterfly computation for decimation-in-time.

At each stage, $N/2$ separate butterfly computations are carried out to produce the next array. The integer r varies with p, q, and m in a manner that depends on the specific form of FFT algorithm that is used. Fortunately, our analysis is not tied to the specific way in which r varies. Also, the specific relationship among p, q, and m, which determines how we index through the mth array, is not important for the analysis. The details of the analysis for decimation-in-time and decimation-in-frequency differ somewhat due to the different butterfly forms, but the basic results do not change significantly. In our analysis we assume a butterfly of the form of Eq. (9.89), corresponding to decimation-in-time.

   We model the roundoff noise by associating an additive noise generator with each fixed-point multiplication. With this model the butterfly of Fig. 9.41 is replaced by that of Fig. 9.42 for analyzing the roundoff noise effects. The notation $\varepsilon[m, q]$ explicitly indicates that this quantity represents a complex error introduced in computing the mth array from the $(m - 1)$st array; specifically, it indicates that the error resulted from quantization of multiplication of the qth element of the $(m - 1)$st array by a complex coefficient.

   Since we assume that in general the input to the FFT is a complex sequence, each of the multiplications is complex and thus, in fact, consists of four real multiplications in exactly the same manner discussed in the previous section. We again assume the statistical model of Section 9.8.1 for the noise sources. Since each of the four noise sequences is uncorrelated zero-mean white noise and all have the same variance, we have, as in Eq. (9.84),

$$\mathscr{E}\{|\varepsilon[m, q]|^2\} = \tfrac{1}{3} \cdot 2^{-2B} = \sigma_B^2. \tag{9.91}$$

To determine the mean-square value of the output noise at any output node, we must account for the contribution from each of the noise sources that propagate to that node. We can make the following observations from the flow graph of Fig. 9.40:

1. The transmission function from any node in the flow graph to any other node to which it is connected is multiplication by a complex constant of unity magnitude (because each branch transmittance is either unity or an integer power of W_N).
2. Each output node connects to 7 butterflies in the flow graph. (In general, each output node would connect to $(N - 1)$ butterflies.) For example, Fig. 9.43(a) shows the flow graph with all the butterflies removed that do not connect to $X[0]$, and Fig. 9.43(b) shows the flow graph with all the butterflies removed that do not connect to $X[2]$.

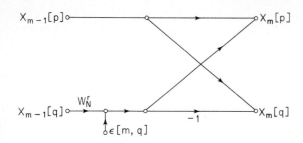

Figure 9.42 Linear noise model for fixed-point roundoff noise in a decimation-in-time butterfly computation.

These observations can be generalized to the case of N an arbitrary power of 2.

As a consequence of the first observation, the mean-square value of the magnitude of the component of the output noise due to each elemental noise source is the same and equal to σ_B^2. The total output noise at each output node is equal to the sum of the noise propagated to that node. Since we assume that all the noise sources are uncorrelated, the mean-square value of the magnitude of the output noise is equal to σ_B^2 times the number of noise sources that propagate to that node. At most one complex noise source is introduced at each butterfly; consequently, from observation 2, at most $(N - 1)$ noise sources propagate to each output node. In fact, not all the butterflies generate roundoff noise since some (for example, all those in the first and second stages) involve only multiplication by unity. However, if we assume that roundoff occurs for each butterfly, we can consider the result as an upper bound on the output noise. With this assumption, then, the mean square value of the output noise in the kth DFT value, $F[k]$, is given by

$$\mathcal{E}\{|F[k]|^2\} = (N - 1)\sigma_B^2, \tag{9.92}$$

which, for large N, we approximate as

$$\mathcal{E}\{|F[k]|^2\} \cong N\sigma_B^2. \tag{9.93}$$

According to this result, the mean-square value of the output noise is proportional to N, the number of points transformed. The effect of doubling N, or adding another stage in the FFT, is to double the mean-square value of the output noise. In Problem 9.24 we consider the modification of this result when we do not insert noise sources for those butterflies that involve only multiplication by unity or j. Note that for FFT algorithms a double-length accumulator does not help us reduce roundoff noise since the outputs of the butterfly computation must be stored in $(B + 1)$-bit registers at the output of each stage.

In implementing an FFT algorithm with fixed-point arithmetic we must ensure against overflow. From Eq. (9.89) it follows that

$$\max(|X_{m-1}[p]|, |X_{m-1}[q]|) \leq \max(|X_m[p]|, |X_m[q]|) \tag{9.94}$$

and also that

$$\max(|X_m[p]|, |X_m[q]|) \leq 2\max(|X_{m-1}[p]|, |X_{m-1}[q]|). \tag{9.95}$$

(See Problem 9.23.) Equation (9.94) implies that the maximum magnitude is non-decreasing from stage to stage. If the magnitude of the output of the FFT is less than

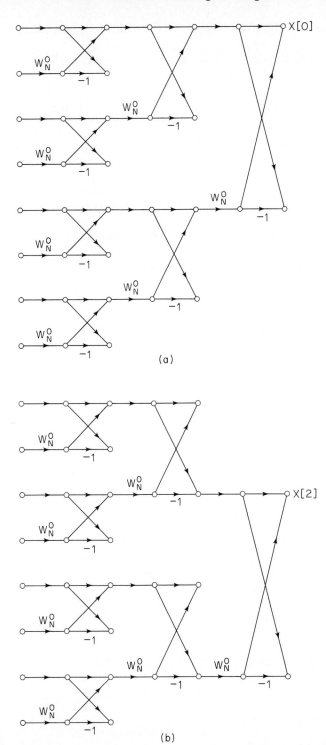

(a)

(b)

Figure 9.43 (a) Butterflies that affect $X[0]$; (b) butterflies that affect $X[2]$.

unity, then the magnitude of the points in each array must be less than unity, i.e., there will be no overflow in any of the arrays.†

To express this constraint as a bound on the input sequence, we recall from Section 9.8.1 that the condition

$$|x[n]| < \frac{1}{N}, \qquad 0 \le n \le N - 1, \tag{9.96}$$

is both necessary and sufficient to guarantee that

$$|X[k]| < 1, \qquad 0 \le k \le N - 1. \tag{9.97}$$

Thus, Eq. (9.96) is sufficient to guarantee no overflow for all stages of the algorithm.

To obtain an explicit expression for the noise-to-signal ratio at the output of the FFT algorithm, consider an input in which successive sequence values are uncorrelated, i.e., a white-noise input signal. Also assume that the real and imaginary parts of the input sequence are uncorrelated and that each has an amplitude density that is uniform between $-1/(\sqrt{2}N)$ and $+1/(\sqrt{2}N)$. (Note that this signal satisfies Eq. 9.96.) Then the average squared magnitude of the complex input sequence is

$$\mathcal{E}\{|x[n]|^2\} = \frac{1}{3N^2} = \sigma_x^2. \tag{9.98}$$

The DFT of the input sequence is

$$X[k] = \sum_{n=0}^{N-1} x[n]W^{kn}, \tag{9.99}$$

from which it can be shown that, under the above assumptions on the input,

$$\mathcal{E}\{|X[k]|^2\} = \sum_{n=0}^{N-1} \mathcal{E}\{|x[n]|^2\}|W^{kn}|^2$$

$$= N\sigma_x^2 = \frac{1}{3N}. \tag{9.100}$$

Combining Eqs. (9.93) and (9.100), we obtain

$$\frac{\mathcal{E}\{|F[k]|^2\}}{\mathcal{E}\{|X[k]|^2\}} = 3N^2\sigma_B^2 = N^2 2^{-2B}. \tag{9.101}$$

According to Eq. (9.101), the noise-to-signal ratio increases as N^2, or 1 bit per stage. That is, if N is doubled, corresponding to adding one additional stage to the FFT, then to maintain the same noise-to-signal ratio, 1 bit must be added to the register length. The assumption of a white-noise input signal is, in fact, not critical here. For a variety of other inputs, the noise-to-signal ratio is still proportional to N^2, with only the constant of proportionality changing.

---

† Actually one should discuss overflow in terms of the real and imaginary parts of the data rather than the magnitude. However, $|x| < 1$ implies that $|\mathcal{R}e\{x\}| < 1$ and $|\mathcal{I}m\{x\}| < 1$, and only a slight increase in allowable signal level is achieved by scaling on the basis of real and imaginary parts.

Equation (9.95) suggests an alternative scaling procedure. Since the maximum magnitude increases by no more than a factor of 2 from stage to stage, we can prevent overflow by requiring that $|x[n]| < 1$ and incorporating an attenuation of $\frac{1}{2}$ at the input to each stage. In this case the output will consist not of the DFT as defined by Eq. (9.78) but of $1/N$ times this DFT. Although the mean-square output signal will be $1/N$ times what it would be if no scaling were introduced, the input amplitude can be N times larger without causing overflow. For the white-noise input signal, this means that we can assume that the real and imaginary parts are uniformly distributed from $-1/\sqrt{2}$ to $1/\sqrt{2}$ so that $|x[n]| < 1$. Thus with the v divisions by 2, the maximum expected value of the magnitude squared of the DFT that can be attained (for the white input signal) is the same as given in Eq. (9.100). However, the output noise level will be much less than in Eq. (9.93) since the noise introduced at early stages of the FFT will be attenuated by the scaling that takes place in the later arrays. Specifically, with scaling by 1/2 introduced at the input to each butterfly, we modify the butterfly of Fig. 9.42 to that of Fig. 9.44, where, in particular, two noise sources are now associated with each butterfly. As before, we assume that the real and imaginary parts of these noise sources are uncorrelated and are also uncorrelated with the other noise sources and that the real and imaginary parts are uniformly distributed between $\pm (1/2) \cdot 2^{-B}$. Thus, as before

$$\mathcal{E}\{|\varepsilon[m, q]|^2\} = \sigma_B^2 = \tfrac{1}{3} \cdot 2^{-2B} = \mathcal{E}\{|\varepsilon[m, p]|^2\}. \qquad (9.102)$$

Because the noise sources are all uncorrelated, the mean-squared magnitude of the noise at each output node is again the sum of the contributions of each noise source in the flow graph. However, unlike the previous case, the attenuation that each noise source experiences through the flow graph depends on the array at which it originates. A noise source originating at the mth array will propagate to the output with multiplication by a complex constant with magnitude $(1/2)^{v-m-1}$. By examination of Fig. 9.40 we see that for the case $N = 8$, each output node connects to

1 butterfly originating at the $(v - 1)$st array,
2 butterflies originating at the $(v - 2)$nd array,
4 butterflies originating at the $(v - 3)$rd array, etc.

For the general case with $N = 2^v$, each output node connects to 2^{v-m-1} butterflies and therefore to 2^{v-m} noise sources that originate at the mth array. Thus,

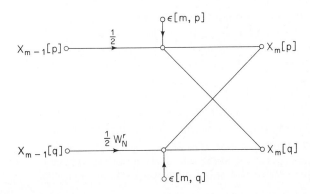

Figure 9.44   Butterfly showing scaling multipliers and associated fixed-point roundoff noise.

at each output node, the mean-square magnitude of the noise is

$$\mathcal{E}\{|F[k]|^2\} = \sigma_B^2 \sum_{m=0}^{v-1} 2^{v-m} \cdot (0.5)^{2v-2m-2}$$

$$= \sigma_B^2 \sum_{m=0}^{v-1} (0.5)^{v-m-2}$$

$$= \sigma_B^2 \cdot 2 \sum_{k=0}^{v-1} 0.5^k$$

$$= 2\sigma_B^2 \frac{1-0.5^v}{1-0.5} = 4\sigma_B^2(1-0.5^v). \qquad (9.103)$$

For large N, we assume that 0.5^v (i.e., $1/N$) is negligible compared with unity, so

$$\mathcal{E}\{|F[k]|^2\} \cong 4\sigma_B^2 = \tfrac{4}{3} \cdot 2^{-2B}, \qquad (9.104)$$

which is much less than the noise variance resulting when all the scaling is carried out on the input data.

Now we can combine Eq. (9.104) with Eq. (9.100) to obtain the output noise-to-signal ratio for the case of step-by-step scaling and white input. We obtain

$$\frac{\mathcal{E}\{|F[k]|^2\}}{\mathcal{E}\{|X[k]|^2\}} = 12N\sigma_B^2 = 4N \cdot 2^{-2B}, \qquad (9.105)$$

a result proportional to N rather than to N^2. An interpretation of Eq. (9.105) is that the output noise-to-signal ratio increases as N, corresponding to half a bit per stage, a result first obtained by Welch (1969). It is important to note again that the assumption of white-noise signal is not essential in the analysis. The basic result of an increase of half a bit per stage holds for a broad class of signals, with only the constant multiplier in Eq. (9.105) being signal dependent.

We should also note that the dominant factor that causes the increase of the noise-to-signal ratio with N is the decrease in signal level (required by the overflow constraint) as we pass from stage to stage. According to Eq. (9.104), very little noise (only a bit or two) is present in the final array. Most of the noise has been shifted out of the binary word by the scalings.

We have assumed straight fixed-point computation in the preceding discussion; i.e., only preset attenuations were allowed, and we were not permitted to rescale on the basis of an overflow test. Clearly, if the hardware or programming facility is such that straight fixed-point computation must be used, we should, if possible, incorporate attenuators of 1/2 at each array rather than use a large attenuation of the input array.

A third approach to avoiding overflow is the use of *block floating point*. In this procedure the original array is normalized to the far left of the computer word, with the restriction that $|x[n]| < 1$; the computation proceeds in a fixed-point manner, except that after every addition there is an overflow test. If overflow is detected, the entire array is divided by 2 and the computation continues. The number of necessary divisions by 2 are counted to determine a scale factor for the entire final array. The output noise-to-signal ratio depends strongly on how many overflows occur and at what stages of the computation they occur. The positions and timing of overflows are determined by the signal being transformed; thus, to analyze the noise-to-signal ratio in a block floating-point implementation of the FFT, we would need to know the input signal.

As we have seen, it is possible to find an input that requires no scaling, and it is also possible to find an input that requires division by N to prevent overflow. The case of a white-noise input signal might be expected to provide an example somewhere between these two extremes; i.e., scaling at all stages might not generally be necessary. This can be illustrated by some experimental results.

Figure 9.45 illustrates the dependence of the output noise-to-signal ratio on N. This figure shows experimentally measured values of output noise-to-signal ratio for block floating-point transforms of white-noise input signals using rounding arithmetic (Oppenheim and Weinstein, 1972). For comparison, the theoretical curve representing the fixed-point noise-to-signal ratio corresponding to Eq. (9.85) is also shown. We see that for this kind of input, block floating point provides some advantages over fixed point, especially for the larger transforms. For $N = 2048$, the noise-to-signal ratio for block floating point is about 1/8 that of fixed point, representing a 3-bit improvement.

Also shown in Fig. 9.45 are the results of an experimental investigation into how the results for block floating point change when truncation rather than rounding is used. Noise-to-signal ratios are generally slightly higher than for rounding. The rate of increase of noise-to-signal ratio with N seems to be about the same as for rounding.

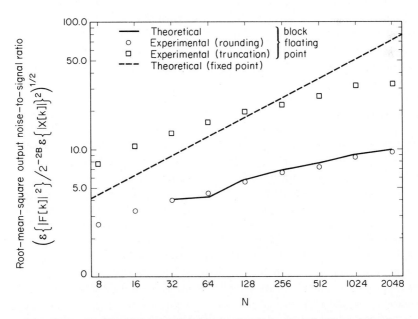

Figure 9.45   Experimental output noise-to-signal ratio for block floating-point realization of decimation-in-time FFT algorithm.

9.8.3 Summary of Quantization Effects in Floating-Point FFT Algorithms

The preceding analysis shows that scaling to avoid overflow is the dominant factor in determining the noise-to-signal ratio of fixed-point implementations of FFT algorithms. Therefore, floating-point arithmetic should improve performance. The effect of floating-point roundoff on the FFT was analyzed both theoretically and experimentally by Gentleman and Sande (1966), Weinstein (1969), and Kaneko and Liu (1970). These investigations show that since scaling is no longer necessary, the decrease of noise-to-signal ratio with increasing N is much less dramatic than for fixed-point arithmetic.

For example, Weinstein (1969a) showed theoretically that the noise-to-signal ratio is proportional to v for $N = 2^v$, rather than proportional to N as in the fixed-point case. Therefore, quadrupling v (raising N to the fourth power) only increases the noise-to-signal ratio by 1 bit. To experimentally verify this result, it was necessary to randomly round up or down when the value of the mantissa was exactly $(1/2) \cdot 2^{-B}$. For normal rounding, the noise-to-signal ratio varied approximately in proportion to v^2.

9.8.4 Effects of Coefficient Quantization in the FFT

As with the implementation of digital filters, the implementation of a fast Fourier transform algorithm requires the use of quantized coefficients. Although the nature of coefficient quantization is inherently nonstatistical, useful results may be obtained by means of a rough statistical analysis (see Oppenheim and Weinstein, 1972). For this analysis, jitter is added to each coefficient; i.e., each coefficient is replaced by its true value plus a white-noise sequence. Although the detailed effect of coefficient error due to quantization is different from that due to jitter, it is reasonable to expect that in a gross sense the magnitudes of the errors are comparable. The result obtained is that the ratio of mean-square magnitude output error to mean-square output signal is $(v/6) \cdot 2^{-2B}$.

Although this does not predict with great accuracy the error in an FFT algorithm due to coefficient quantization, it is helpful as a rough estimate of the error. The key result, which was tested experimentally, is that the error-to-signal ratio increases very mildly with N, being proportional to $v = \log_2 N$, so that doubling N produces only a slight increase in the error-to-signal ratio.

The experimental results are displayed in Fig. 9.46; the quantity plotted is 2^{2B} times the ratio of the mean-square magnitude output error to mean-square output signal. The theoretical curve is shown, and the circles represent measured output error-to-signal ratio for the fixed-point case. The experimental results generally lie below the theoretical curve. No experimental result differs by as much as a factor of 2 from the theoretical result, and since a factor of 2 in error-to signal ratio corresponds to only half a bit difference in the output error, it seems that the analysis provides a reasonably accurate estimate of the effect of coefficient errors. The experimental results increase approximately linearly with v but with smaller slope than given by the analysis.

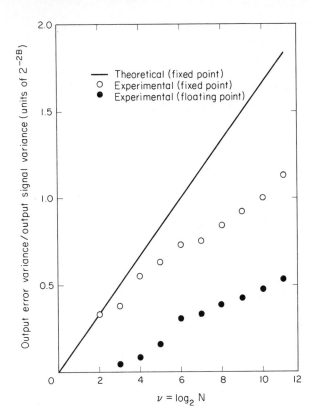

Figure 9.46 Errors due to coefficient quantization in FFT computations.

In these experiments fixed-point arithmetic was assumed. However, since a block floating-point FFT will generally use fixed-point coefficients, the results are also valid for the block floating-point case. With some slight modifications, it is possible to obtain similar results for the floating-point case. Except for a constant factor, the floating- and fixed-point results are the same. Experimental results for the floating-point case, represented by the solid dots in Fig. 9.46, are slightly lower than the results for the fixed-point case.

9.9 SUMMARY

In this chapter we have considered techniques for computation of the discrete Fourier transform, and we have seen how the periodicity and symmetry of the complex factor $e^{-j(2\pi/N)kn}$ can be exploited to increase the efficiency of DFT computations.

We considered the Goertzel algorithm and the direct evaluation of the DFT expression because of the importance of these techniques when not all N of the DFT values are required. However, our major emphasis was on fast Fourier transform (FFT) algorithms. We described the decimation-in-time and decimation-in-frequency classes of FFT algorithms in some detail. Much of the detailed discussion concerned algorithms that require N to be a power of 2, since these algorithms are easy to understand and simple to program. Two FORTRAN programs were given as illustrations of the basic algorithms and of programming techniques.

The concept of index mappings was introduced as a basis for understanding more general FFT algorithms where N is a composite number. We discussed special advantages of certain factors such as 4 and 8, and we discussed in general terms the prime factor algorithms that can be developed when the factors of N are relatively prime numbers.

The use of convolution as the basis for computing the DFT was briefly discussed. We presented a brief overview of the Winograd Fourier transform algorithm, and in somewhat more detail we discussed an algorithm called the chirp transform algorithm.

The final section of this chapter was devoted to a discussion of the effects of quantization in DFT computations. We used linear noise models to show that the noise-to-signal ratio of a DFT computation varies differently with the length of the sequence depending on how scaling is done. We also commented briefly on the use of floating-point representations.

In this chapter our objective was to present the basic principles of efficient computation of the DFT. Our approach was to illustrate the basic principles with examples of FFT algorithms, and therefore much of the discussion is directly concerned with the details of implementing an FFT algorithm. Based on the discussion of this chapter, it should be possible to understand other FFT programs or write new programs for specific computing environments.

PROBLEMS

9.1. Suppose that a computer program is available for computing the DFT,

$$X[k] = \sum_{n=0}^{N-1} x[n]e^{-j(2\pi/N)kn}, \qquad k = 0, 1, \ldots, N-1,$$

i.e., the input to the program is the sequence $x[n]$ and the output is the DFT $X[k]$. Show how the input and/or output sequences may be rearranged such that the program can also be used to compute the inverse DFT,

$$x[n] = \frac{1}{N} \sum_{n=0}^{N-1} X[k]e^{j(2\pi/N)kn}, \qquad n = 0, 1, \ldots, N-1,$$

i.e., the input to the program should be $X[k]$ or a sequence simply related to $X[k]$, and the output should be either $x[n]$ or a sequence simply related to $x[n]$. There are several possible approaches.

9.2. In Section 9.2 we used the fact that $W_N^{-kN} = 1$ to derive a recurrence algorithm for computing a specific DFT value $X[k]$ for a finite-length sequence $x[n]$, $n = 0, 1, \ldots,$ $N-1$.

(a) Using the fact that $W_N^{kN} = W_N^{Nn} = 1$, show that $X[N-k]$ can be obtained as the output after N iterations of the difference equation depicted in Fig. P9.2-1. That is, show that

$$X[N-k] = y_k[N].$$

(b) Show that $X[N-k]$ is also equal to the output after N iterations of the difference equation depicted in Fig. P9.2-2. Note that the system of Fig. P9.2-2 has the same poles as the system in Fig. 9.2, but the coefficient required to implement the complex zero in Fig. P9.2-2 is the complex conjugate of the corresponding coefficient in Fig. 9.2; i.e., $W_N^{-k} = (W_N^k)^*$.

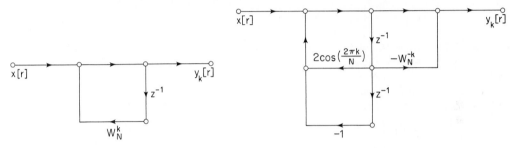

Figure P9.2-1   Figure P9.2-2

9.3. Figure P9.3 shows the flow graph representation of a decimation-in-time FFT algorithm for $N = 8$. The heavy line shows a path from sample $x[7]$ to DFT sample $X[2]$.

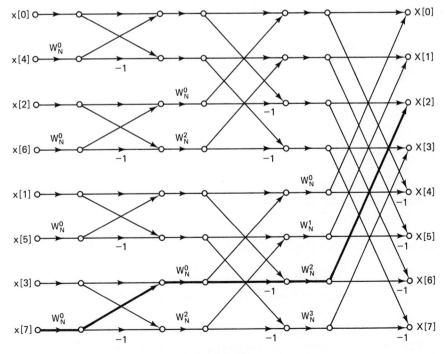

Figure P9.3

(a) What is the "gain" along the path that is emphasized in Fig. P9.3?

(b) How many other paths in the flow graph begin at $x[7]$ and end at $X[2]$? Is this true in general; i.e., how many paths are there between each input sample and each output sample?

(c) Now consider the DFT sample $X[2]$. By tracing paths in the flow graph of Fig. P9.3, show that each input sample contributes the proper amount to the output DFT sample; i.e., verify that

$$X[2] = \sum_{n=0}^{N-1} x[n]e^{-j(2\pi/N)2n}.$$

9.4. Figure P9.4 shows the flow graph for an 8-point decimation-in-time FFT algorithm. Let $x[n]$ be the sequence whose DFT is $X[k]$. In the flow graph of Fig. P9.4, $A[\cdot]$, $B[\cdot]$, $C[\cdot]$, and $D[\cdot]$ represent separate arrays that are indexed consecutively in the same order as the indicated nodes in the flow graph.

(a) Specify how the elements of the sequence $x[n]$ should be placed in the array $A[r]$, $r = 0, 1, \ldots, 7$. Also specify how the elements of the DFT sequence should be extracted from the array $D[r]$, $r = 0, 1, \ldots, 7$.

(b) Without determining the values in the intermediate arrays, $B[\cdot]$ and $C[\cdot]$, determine and sketch the array sequence $D[r]$, $r = 0, 1, \ldots, 7$, if the input sequence is $x[n] = (-W_N)^n$, $n = 0, 1, \ldots, 7$.

(c) Determine and sketch the sequence $C[r]$, $r = 0, 1, \ldots, 7$, if the output Fourier transform is $X[k] = 1$, $k = 0, 1, \ldots, 7$.

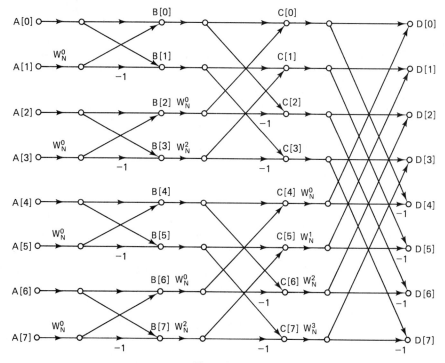

Figure P9.4

9.5. Construct a flow graph for a 16-point radix-2 decimation-in-time FFT algorithm. Label all multipliers in terms of powers of W_{16} and also label any branch transmittances that are equal to -1. Label the input and output nodes with the appropriate values of the input and DFT sequences, respectively. Determine the number of real multiplications and the number of real additions required to implement the flow graph.

9.6. In Section 9.4.2, it was asserted that the transpose of the flow graph of an FFT algorithm is also the flow graph of an FFT algorithm. The purpose of this problem is to develop this result for radix-2 FFT algorithms.

  (a) The basic butterfly for the decimation-in-frequency radix-2 FFT algorithm is depicted in Fig. P9.6-1. This flow graph represents the equations

$$X_m[p] = X_{m-1}[p] + X_{m-1}[q],$$

$$X_m[q] = (X_{m-1}[p] - X_{m-1}[q])W_N^r.$$

  Starting with these equations, show that $X_{m-1}[p]$ and $X_{m-1}[q]$ can be computed from $X_m[p]$ and $X_m[q]$ using the butterfly shown in Fig. P9.6-2.

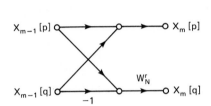 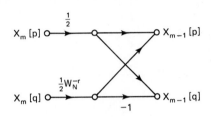

Figure P9.6-1   **Figure P9.6-2**

  (b) In the decimation-in-frequency algorithm of Fig. 9.20, $X_v[r], r = 0, 1, \ldots, N - 1$ is the DFT $X[k]$ arranged in bit-reversed order and $X_0[r] = x[r], r = 0, 1, \ldots, N - 1$; i.e., the zeroth array is the input sequence arranged in normal order. If each butterfly in Fig. 9.20 is replaced by the appropriate butterfly of the form of Fig. P9.6-2, the result would be a flow graph for computing the sequence $x[n]$ (in normal order) from DFT $X[k]$ (in bit-reversed order). Draw the resulting flow graph for $N = 8$.

  (c) The flow graph obtained in part (b) represents an *inverse* DFT algorithm, i.e., an algorithm for computing

$$x[n] = \frac{1}{N} \sum_{n=0}^{N-1} X[k]W_N^{-kn}, \qquad n = 0, 1, \ldots, N - 1.$$

  Modify the flow graph obtained in part (b) so that it computes the DFT

$$X[k] = \sum_{n=0}^{N-1} x[n]W_N^{kn}, \qquad k = 0, 1, \ldots, N - 1,$$

  rather than the inverse DFT.

  (d) Observe that the result in part (c) is the transpose of the decimation-in-frequency algorithm of Fig. 9.20 and that it is identical to the decimation-in-time algorithm depicted in Fig. 9.10. Does it follow that to each decimation-in-time algorithm (e.g., Figs. 9.14–9.16) there corresponds a decimation-in-frequency algorithm that is the transpose of the decimation-in-time algorithm and vice versa? Explain.

9.7. In many applications (such as evaluating frequency responses or interpolation) it is of interest to compute the DFT of a short sequence that is "zero-padded." In such cases, a specialized "pruned" FFT algorithm can be used to increase the efficiency of computation

(Markel, 1971). In this problem we will consider pruning of the radix-2 decimation-in-frequency algorithm when the length of the input sequence is $M \leq 2^{\mu}$ and the length of the DFT is $N = 2^{\nu}$, where $\mu < \nu$.

(a) Draw the complete flow graph of a decimation-in-frequency radix-2 FFT algorithm for $N = 16$. Label all branches appropriately.

(b) Assume that the input sequence is of length $M = 2$; i.e., $x[n] \neq 0$ only for $n = 0$ and $n = 1$. Draw a new flow graph for $N = 16$ that shows how the nonzero input samples propagate to the output DFT; i.e., eliminate or prune all branches in the flow graph of part (a) that represent operations on zero inputs.

(c) In part (b) all of the butterflies in the first three stages of computation should have been effectively replaced by a half-butterfly of the form shown in Fig. P9.7, and in the last stage, all the butterflies should have been of the regular form. For the general case where the length of the input sequence is $M \leq 2^{\mu}$ and the length of the DFT is $N = 2^{\nu}$, where $\mu < \nu$, determine the number of stages in which the pruned butterflies can be used. Also, determine the number of complex multiplications required to compute the N-point DFT of an M-point sequence using the pruned FFT algorithm. Express your answers in terms of ν and μ.

(d) *Optional.* Modify the program in Fig. 9.26 to include pruning.

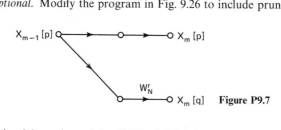

Figure P9.7

9.8. In Section 9.3, we showed that if N is divisible by 2, an N-point DFT may be expressed as

$$X[k] = G[k] + W_N^k H[k], \qquad 0 \leq k \leq N - 1, \tag{P9.8-1}$$

where $G[k]$ is the $N/2$-point DFT of the sequence of even-indexed samples,

$$g[n] = x[2n], \qquad 0 \leq n \leq (N/2) - 1,$$

and $H[k]$ is the $N/2$-point DFT of the odd-indexed samples,

$$h[n] = x[2n + 1], \qquad 0 \leq n \leq (N/2) - 1.$$

When $N = 2^{\nu}$, repeated application of this decomposition leads to the decimation-in-time FFT algorithm depicted for $N = 8$ in Fig. 9.10. As we have seen, such algorithms require complex multiplications by the twiddle factors W_N^k. Rader and Brenner (1976) derived a new algorithm in which the multipliers are purely imaginary, thus requiring only two real multiplications and no real additions. In this algorithm, Eq. (P9.8-1) is replaced by the equations

$$X[0] = G[0] + F[0], \tag{P9.8-2a}$$

$$X[N/2] = G[0] - F[0], \tag{P9.8-2b}$$

$$X[k] = G[k] - \frac{1}{2} j \frac{F[k]}{\sin(2\pi k/N)}, \qquad k \neq 0, N/2. \tag{P9.8-2c}$$

In this case, $F[k]$ is the $N/2$-point DFT of the sequence

$$f[n] = x[2n + 1] - x[2n - 1] + Q,$$

where

$$Q = \frac{2}{N} \sum_{n=0}^{(N/2)-1} x[2n + 1]$$

is a quantity that need be computed only once.

(a) Show that $F[0] = H[0]$ and therefore that Eqs. (P9.8-2a) and (P9.8-2b) give the same result as Eq. (P9.8-1) for $k = 0, N/2$.

(b) Show that

$$F[k] = H[k]W_N^k(W_N^{-k} - W_N^k)$$

for $k = 1, 2, \ldots, (N/2) - 1$. Use this result to obtain Eq. (P9.8-2c). Why must we compute $X[0]$ and $X[N/2]$ using separate equations?

(c) When $N = 2^v$, we can apply Eqs. (P9.8-2) repeatedly to obtain a complete decimation-in-time FFT algorithm. Determine formulas for the number of real multiplications and for the number of real additions as a function of N. In counting operations due to Eq. (P9.8-2c), take advantage of any symmetries and periodicities, but do not exclude "trivial" multiplications by $\pm j/2$.

(d) Rader and Brenner (1976) state that FFT algorithms based on Eqs. (P9.8-2) have "poor noise properties." Explain why this might be true.

9.9. A modified FFT algorithm called the *split-radix FFT* or SRFFT was proposed by Duhamel and Hollman (1984) and Duhamel (1986). The flow graph for the split-radix algorithm is similar to the radix-2 flow graph, but it requires fewer real multiplications. In this problem we illustrate the principles of the SRFFT for computing the DFT $X[k]$ of a sequence $x[n]$ of length N.

(a) Show that the even-indexed terms of $X[k]$ can be expressed as the $N/2$-point DFT

$$X[2k] = \sum_{n=0}^{(N/2)-1} (x[n] + x[n + N/2])W_N^{2kn}$$

for $k = 0, 1, \ldots, (N/2) - 1$.

(b) Show that the odd-indexed terms of the DFT $X[k]$ can be expressed as the $N/4$-point DFTs

$$X[4k + 1]$$

$$= \sum_{n=0}^{(N/4)-1} \{(x[n] - x[n + N/2]) - j(x[n + N/4] - x[n + 3N/4])\}W_N^n W_N^{4kn}$$

for $k = 0, 1, \ldots, (N/4) - 1$, and

$$X[4k + 3]$$

$$= \sum_{n=0}^{(N/4)-1} \{(x[n] - x[n + N/2]) + j(x[n + N/4] - x[n + 3N/4])\}W_N^{3n} W_N^{4kn}$$

for $k = 0, 1, \ldots, (N/4) - 1$.

(c) The flow graph in Fig. P9.9 represents this decomposition of the DFT for a 16-point transform. Redraw this flow graph labeling each branch with the appropriate multiplier coefficient.

(d) Determine the number of real multiplications required to implement the 16-point transform when the SRFFT principle is applied to compute the other DFTs in Fig. P9.9. Compare this number to the number of real multiplications required to

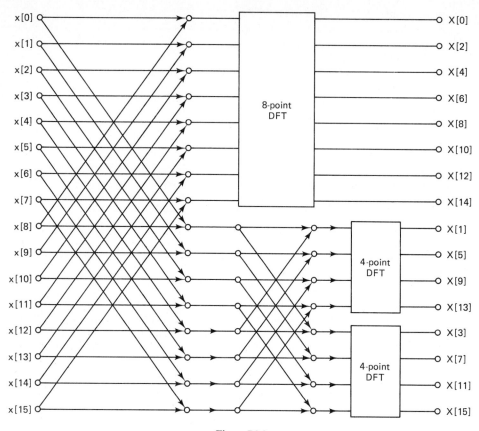

Figure P9.9

implement a 16-point radix-2 decimation-in-frequency algorithm. In both cases assume that multiplications by W_N^0 are not done.

9.10. In implementing an FFT algorithm, it is sometimes useful to generate the powers of W_N recursively, using either a canonic form or coupled form oscillator. In this problem we consider a radix-2 decimation-in-time algorithm for $N = 2^\nu$. Figure 9.10 depicts this type of algorithm for $N = 8$, and Fig. 9.25 is a FORTRAN implementation of the algorithm for any N.

To generate the coefficients efficiently, the frequency of the oscillator would change from stage to stage. Assume that the arrays are numbered 0 through $\nu = \log_2 N$, so the array holding the initial input sequence is the zeroth array and the DFT is in the νth array. In computing the butterflies in a given stage, all butterflies requiring the same coefficients are evaluated before obtaining new coefficients. In indexing through the array, we assume that the data in the array are stored in consecutive complex registers numbered 0 through $(N - 1)$. All the following questions are concerned with the computation of the mth array from the $(m - 1)$st array, where $1 \le m \le \nu$. Answers should be expressed in terms of m.

(a) How many butterflies must be computed in the mth stage? How many different coefficients are required in the mth stage?

(b) What is the frequency of the oscillator used to generate the coefficients; i.e., how many times can it be iterated before repeating?

(c) Assuming that a coupled form oscillator is used to generate the powers of W_N, what are the coefficients in the difference equation for the oscillator?

(d) What is the difference between the array locations of the two complex input points to a butterfly in the mth stage?

(e) What is the difference between the addresses of the first array locations of butterflies that utilize the same coefficients?

9.11. The FORTRAN program in Fig. 9.25 implements a decimation-in-time FFT algorithm of the form depicted in Fig. 9.10.

(a) There are v stages of computation. Which program variable indexes the stages of computation?

(b) Which program variable corresponds to the spacing between the two inputs to a butterfly?

(c) Give the line numbers of the instructions that implement the basic butterfly computation.

(d) Give the line numbers of the instructions that implement the bit-reversed sorting of the input.

(e) What simple modifications can be made so that the subroutine computes the inverse DFT?

9.12. The FORTRAN program in Fig. 9.26 implements a decimation-in-frequency FFT algorithm of the form depicted in Fig. 9.20.

(a) There are v stages of computation. Which program variable indexes the stages of computation?

(b) Which program variable corresponds to the spacing between the two inputs to a butterfly?

(c) Give the line numbers of the instructions that implement the basic butterfly computation.

(d) Give the line numbers of the instructions that implement the bit-reversed unsorting of the output.

(e) What simple modifications can be made so that the subroutine computes the inverse DFT?

9.13. Computing the DFT generally requires complex multiplications. Consider the product $X + jY = (A + jB)(C + jD) = (AC - BD) + j(BC + AD)$. In this form, a complex multiplication requires 4 real multiplications and 2 real additions. Verify that a complex multiplication can be performed with 3 real multiplications and 5 additions using the algorithm

$$X = (A - B)D + (C - D)A,$$

$$Y = (A - B)D + (C + D)B.$$

9.14. In computing the DFT it is necessary to multiply a complex number by another complex number whose magnitude is unity, i.e., $(X + jY)e^{j\theta}$. Clearly, such a complex multiplication changes only the angle of the complex number, leaving the magnitude unchanged. For this reason, multiplications by a complex number $e^{j\theta}$ are sometimes called *rotations*. In DFT or FFT algorithms many different angles θ may be needed. However, it may be undesirable to store a table of all required values of $\sin\theta$ and $\cos\theta$, and computing these functions by a power series requires many multiplications and additions. With the CORDIC algorithm given by Volder (1959), the product $(X + jY)e^{j\theta}$ can be evaluated efficiently by a combination of additions, binary shifts, and table look-up from a small table.

(a) Define $\theta_i = \arctan(2^{-i})$. Show that any angle $0 < \theta < \pi/2$ can be represented as

$$\theta = \sum_{i=0}^{M-1} \alpha_i \theta_i + \epsilon = \hat{\theta} + \epsilon,$$

where $\alpha_i = \pm 1$ and the error ϵ is bounded by

$$|\epsilon| \le \arctan(2^{-M}).$$

(b) The angles θ_i may be computed in advance and stored in a small table of length M. State an algorithm for obtaining the sequence $\{\alpha_i\}$ for $i = 0, 1, \ldots, M-1$ such that $\alpha_i = \pm 1$. Use your algorithm to determine the sequence $\{\alpha_i\}$ for representing the angle $\theta = 100\pi/512$ when $M = 11$.

(c) Using the result of part (a), show that the recursion

$$X_0 = X,$$

$$Y_0 = Y,$$

$$X_i = X_{i-1} - \alpha_{i-1} Y_{i-1} 2^{-i+1}, \qquad i = 1, 2, \ldots, M,$$

$$Y_i = Y_{i-1} + \alpha_{i-1} X_{i-1} 2^{-i+1}, \qquad i = 1, 2, \ldots, M,$$

will produce the complex number

$$(X_M + jY_M) = (X + jY)G_M e^{j\hat{\theta}},$$

where $\hat{\theta} = \sum_{i=0}^{M-1} \alpha_i \theta_i$ and G_M is real, positive, and does not depend on θ. That is, the original complex number is rotated in the complex plane by an angle $\hat{\theta}$ and magnified by the constant G_M.

(d) Determine the magnification constant G_M as a function of M.

9.15. We have seen that an FFT algorithm can be viewed as an interconnection of computational elements called butterflies. For example, the butterfly for a radix-2 decimation-in-frequency FFT algorithm is shown in Fig. P9.15-1. The butterfly takes two complex numbers as input and produces two complex numbers as output. Its implementation requires a complex multiplication by W_N^r, where r is an integer that depends on the location of the butterfly in the flow graph of the algorithm. Since the complex multiplier is of the form $W_N^r = e^{j\theta}$, the CORDIC rotator algorithm discussed in Problem 9.14 can be used to implement the complex multiplication efficiently. Unfortunately, while the CORDIC rotator algorithm accomplishes the desired change of angle, it also introduces a fixed magnification that is independent of the angle θ. Thus, if the CORDIC rotator algorithm were used to implement the multiplications by W_N^r, the butterfly of Fig. P9.15-1 would be replaced by the butterfly of Fig. P9.15-2, where G represents the fixed magnification factor of the CORDIC rotator. (We assume no error in approximating the angle of rotation.) If each butterfly in the flow graph of the decimation-in-frequency FFT algorithm is replaced by the butterfly of Fig. P9.15-2, we obtain a modified FFT

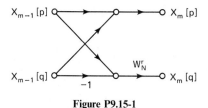

Figure P9.15-1   Figure P9.15-2

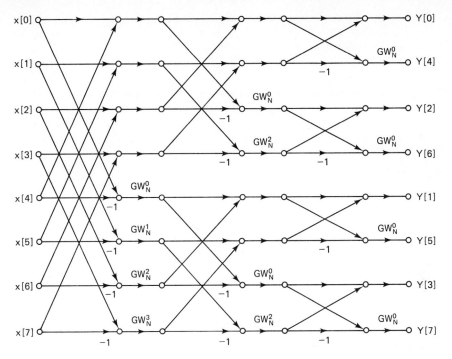

Figure P9.15-3

algorithm for which the flow graph would be as shown in Fig. P9.15-3 for $N = 8$. The output of this modified algorithm would not be the desired DFT.

(a) Show that the output for the modified FFT algorithm is $Y[k] = W[k]X[k]$, where $X[k]$ is the correct DFT of the input sequence $x[n]$ and $W[k]$ is a function of G, N, and k.

(b) The sequence $W[k]$ can be described by a particularly simple rule. Find this rule and indicate its dependence on G, N, and k.

(c) Suppose that we wish to preprocess the input sequence $x[n]$ to compensate for the effect of the modified FFT algorithm. Determine a procedure for obtaining a sequence $\hat{x}[n]$ from $x[n]$ such that if $\hat{x}[n]$ is the input to the modified FFT algorithm, then the output will be $X[k]$, the correct DFT of the original sequence $x[n]$.

9.16. In Example 9.3, we considered a 12-point prime factor FFT algorithm with N represented as the product $N = N_1 N_2$, where $N_1 = 4$ and $N_2 = 3$.

(a) Using Eqs. (9.54) and (9.55), determine the index maps for n and k with $N_1 = 3$ and $N_2 = 4$.

(b) Determine the prime factor decomposition of $X[k]$ using the index maps obtained in part (a).

(c) Draw the flow graph of the 12-point prime factor decomposition determined in part (b).

9.17. The decimation-in-time FFT algorithm was developed in Section 9.3 for radix 2, i.e., $N = 2^\nu$. A similar approach leads to a radix-3 algorithm when $N = 3^\nu$.

(a) Draw a flow graph for a 9-point decimation-in-time FFT algorithm using a 3×3 decomposition of the DFT.

(b) For $N = 3^{\nu}$, how many complex multiplications by powers of W_N are needed to compute the DFT of an N-point complex sequence using a radix-3 decimation-in-time FFT algorithm?

(c) For $N = 3^{\nu}$, is it possible to use in-place computation for the radix-3 decimation-in-time algorithm?

9.18. This problem deals with the efficient computation of samples of the z-transform of a finite-length sequence. Using the chirp transform algorithm, develop a procedure for computing values of $X(z)$ at 25 points spaced uniformly on an arc of a circle of radius 0.5 beginning at an angle of $-\pi/6$ and ending at an angle of $2\pi/3$. The length of the sequence is 100 samples.

9.19. The N-point DFT of the N-point sequence $x[n] = e^{-j(\pi/N)n^2}$, for N even, is

$$X[k] = \sqrt{N}e^{-j\pi/4}e^{j(\pi/N)k^2}.$$

Determine the $2N$-point DFT of the $2N$-point sequence $y[n] = e^{-j(\pi/N)n^2}$, assuming N is even.

9.20. Bluestein (1970) showed that if $N = M^2$, then the chirp transform algorithm has a recursive implementation.

(a) Show that the DFT can be expressed as the convolution

$$X[k] = h^*[k] \sum_{n=0}^{N-1} (x[n]h^*[n])h[k-n],$$

where * denotes complex conjugation and

$$h[n] = e^{j(\pi/N)n^2}, \qquad -\infty < n < \infty.$$

(b) Show that the desired values of $X[k]$ (i.e., for $k = 0, 1, \ldots, N-1$) can also be obtained by evaluating the convolution of part (a) for $k = N, N+1, \ldots, 2N-1$.

(c) Use the result of part (b) to show that $X[k]$ is also equal to the output of the system shown in Fig. P9.20 for $k = N, N+1, \ldots, 2N-1$, where $\hat{h}[k]$ is the finite-duration sequence

$$\hat{h}[k] = \begin{cases} e^{j(\pi/N)k^2}, & 0 \leq k \leq 2N-1, \\ 0, & \text{otherwise.} \end{cases}$$

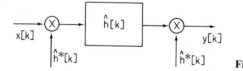

Figure P9.20

(d) Using the fact that $N = M^2$, show that the system function corresponding to the impulse response $\hat{h}[k]$ is

$$\hat{H}(z) = \sum_{k=0}^{2N-1} e^{j(\pi/N)k^2}z^{-k}$$

$$= \sum_{r=0}^{M-1} e^{j(\pi/N)r^2}z^{-r}\frac{1-z^{-2M^2}}{1+e^{j(2\pi/M)r}z^{-M}}.$$

[*Hint*: Express k as $k = r + \ell M$.]

(e) The expression for $\hat{H}(z)$ obtained in part (d) suggests a recursive realization of the FIR system. Draw the flow graph of such an implementation.

(f) Use the result of part (e) to determine the total numbers of complex multiplications and additions required to compute all the N desired values of $X[k]$. Compare those numbers with the numbers required for direct computation of $X[k]$.

9.21. In the Goertzel algorithm for computation of the discrete Fourier transform, $X[k]$ is computed as

$$X[k] = y_k[N],$$

where $y_k[n]$ is the output of the network shown in Fig. P9.21. Consider the implementation of the Goertzel algorithm using fixed-point arithmetic with rounding. Assume that the register length is B bits plus sign and assume that the products are rounded before additions. Also assume that roundoff noise sources are independent.

(a) Assuming that $x[n]$ is real, draw a flow graph of the linear noise model for the finite-precision computation of the real and imaginary parts of $X[k]$. Assume that multiplication by ± 1 produces no roundoff noise.

(b) Compute the variance of the roundoff noise in both the real part and the imaginary part of $X[k]$.

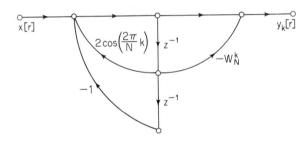

Figure P9.21

9.22. Consider direct computation of the DFT using fixed-point arithmetic with rounding. Assume that the register length is B bits plus sign (i.e., a total of $B + 1$ bits) and that the roundoff noise introduced by any real multiplication is independent of that produced by any other real multiplication. Assuming that $x[n]$ is real, determine the variance of the roundoff noise in both the real part and the imaginary part of each DFT value $X[k]$.

9.23. When implementing a decimation-in-time FFT algorithm, the basic butterfly computation is

$$X_m[p] = X_{m-1}[p] + W_N^r X_{m-1}[q],$$

$$X_m[q] = X_{m-1}[p] - W_N^r X_{m-1}[q].$$

In using fixed-point arithmetic to implement the computations, it is commonly assumed that all numbers are scaled to be less than unity. Therefore, to avoid overflow it is necessary to ensure that the real numbers that result from the butterfly computations do not exceed unity.

(a) Show that if we require

$$|X_{m-1}[p]| < \tfrac{1}{2} \quad \text{and} \quad |X_{m-1}[q]| < \tfrac{1}{2},$$

then overflow cannot occur in the butterfly computation; i.e.,

$$|\mathcal{R}e\{X_m[p]\}| < 1, \qquad |\mathcal{I}m\{X_m[p]\}| < 1,$$

and

$$|\mathcal{R}e\{X_m[q]\}| < 1, \qquad |\mathcal{I}m\{X_m[q]\}| < 1.$$

(b) In practice, it is easier and most convenient to require

$$|\mathcal{R}e\{X_{m-1}[p]\}| < \tfrac{1}{2}, \qquad |\mathcal{I}m\{X_{m-1}[p]\}| < \tfrac{1}{2},$$

and

$$|\mathcal{R}e\{X_{m-1}[q]\}| < \tfrac{1}{2}, \qquad |\mathcal{I}m\{X_{m-1}[q]\}| < \tfrac{1}{2}.$$

Are these conditions sufficient to guarantee that overflow cannot occur in the decimation-in-time butterfly computation? Explain.

9.24. In deriving formulas for the noise-to-signal ratio for the fixed-point radix-2 decimation-in-time FFT algorithm, we assumed that each output node was connected to $(N-1)$ butterfly computations, each of which contributed an amount $\sigma_B^2 = \frac{1}{3}\cdot 2^{-2B}$ to the output noise variance. However, when $W_N^r = \pm 1$ or $\pm j$, the multiplications can in fact be done without error. Thus if the results derived in Section 9.8.2 are modified to account for this fact, we obtain a less pessimistic estimate of quantization noise effects.

(a) For the decimation-in-time algorithm discussed in Section 9.8.2, determine for each stage the number of butterflies that involve multiplication by either ± 1 or $\pm j$.

(b) Use the result of part (a) to modify the estimate of the output noise variance, Eq. (9.92), for the case when all scaling is done at the input. Also obtain a modified expression corresponding to Eq. (9.101) for the noise-to-signal ratio at the output.

(c) Repeat parts (a) and (b) for the case when the output of each stage is attenuated by a factor of $\frac{1}{2}$; i.e., derive modified expressions corresponding to Eq. (9.104) for the output noise variance and Eq. (9.105) for the output noise-to-signal ratio, assuming that multiplications by ± 1 and $\pm j$ do not introduce error.

9.25. In Section 9.8.2 we considered a noise analysis of the decimation-in-time FFT algorithm of Fig. 9.10. Carry out a similar analysis for the decimation-in-frequency algorithm of Fig. 9.20, obtaining equations for the output noise variance and noise-to-signal ratio for scaling at the input and also for scaling by $\frac{1}{2}$ at each stage of computation.

9.26. Consider the computation of the DFT using an FFT algorithm with quantized coefficients. Let $X[k]$ be the desired DFT of a sequence $x[n]$ and let $\hat{X}[k]$ be the output of a decimation-in-time FFT algorithm where the coefficients W_N^r are quantized. Assume that although the coefficients are quantized, multiplications and additions in the FFT algorithm are done without error.

(a) Show that the output $\hat{X}[k]$ can be expressed as

$$\hat{X}[k] = \sum_{n=0}^{N-1} x[n]Q_{nk} = X[k] + F[k],$$

where

$$Q_{nk} = \prod_{i=1}^{v} (W_N^{r_i} + \delta_i) \qquad \text{and} \qquad \prod_{i=1}^{v} W_N^{r_i} = W_N^{kn}.$$

(b) Assume that the coefficients W_N^r are rounded to B bits plus a sign bit (a total of $B+1$ bits) and that the real and imaginary parts of the coefficient errors are uncorrelated and uniformly distributed. Show that the variance of the quantities δ_i is

$$\sigma_\delta^2 = \frac{2^{-2B}}{6}.$$

(c) The error $F[k]$ can be expressed as

$$F[k] = \hat{X}[k] - X[k] = \sum_{n=0}^{N-1} x[n](Q_{nk} - W_N^{kn}), \qquad k = 0, 1, \ldots, N-1.$$

Show that the factor $(Q_{nk} - W_N^{kn})$ can be expressed as

$$(Q_{nk} - W_N^{kn}) = \sum_{i=1}^{\nu} \prod_{\substack{j=1 \\ j \neq i}}^{\nu} W_N^{r_i} + \text{higher-order terms.}$$

(d) Neglecting the higher-order terms and assuming that the quantities δ_i are mutually uncorrelated, show that the variance of $F[k]$ is

$$\sigma_F^2 = \left(\frac{2\nu}{6}\right) 2^{-2B} \sum_{n=0}^{N-1} |x[n]|^2.$$

(e) Use Parseval's theorem for the DFT to show that

$$\frac{\sigma_F^2}{(1/N)\sum_{n=0}^{N-1} |X[k]|^2} = \left(\frac{2\nu}{6}\right) 2^{-2B}.$$

9.27. We are given a finite-length sequence $x[n]$ of length 627 (i.e., $x[n] = 0$ for $n < 0$ and $n > 626$), and we have available an FFT program that will compute the DFT of a sequence of any length $N = 2^\nu$.

For the given sequence we want to compute samples of the discrete-time Fourier transform at frequencies

$$\omega_k = \frac{2\pi}{627} + \frac{2\pi k}{256}, \qquad k = 0, 1, \ldots, 255.$$

Specify how to obtain a new sequence $y[n]$ from $x[n]$ such that the desired frequency samples can be obtained by applying the available FFT program to $y[n]$ with ν *as small as possible.*

9.28. A finite-length signal of length $L = 500$ ($x[n] = 0$ for $n < 0$ and $n > L - 1$) is obtained by sampling a continuous-time signal with sampling rate 10,000 samples per second. We wish to compute samples of the z-transform of $x[n]$ at the N equally spaced points $z_k = (0.8)e^{j2\pi k/N}$, for $0 \leq k \leq N - 1$, with an effective frequency spacing of 50 Hz or less.

(a) Determine the minimum value for N if $N = 2^\nu$.

(b) Determine a sequence $y[n]$ of length N, where N is as determined in part (a), such that its DFT $Y[k]$ is equal to the desired samples of the z-transform of $x[n]$.

9.29. Suppose that a finite-length sequence $x[n]$ has N-point DFT $X[k]$ and suppose that the sequence satisfies the symmetry condition

$$x[n] = -x[((n + N/2))_N], \qquad 0 \leq n \leq N - 1,$$

where N is even and $x[n]$ is complex.

(a) Show that $X[k] = 0$ for $k = 0, 2, \ldots, N - 2$.

(b) Show how to compute the odd-indexed DFT values $X[k]$, $k = 1, 3, \ldots, N - 1$ using only one $N/2$-point DFT plus a small amount of extra computation.

9.30. Consider an N-point sequence $x[n]$ with DFT $X[k]$, $k = 0, 1, \ldots, N - 1$. The following algorithm computes the even-indexed DFT values $X[k]$, $k = 0, 2, \ldots, N - 2$ for N even using only a single $N/2$-point DFT.

    1. Form the sequence $y[n]$ by time aliasing, i.e.,

$$y[n] = \begin{cases} x[n] + x[n + N/2], & 0 \le n \le N/2 - 1, \\ 0, & \text{otherwise.} \end{cases}$$

    2. Compute $Y[r]$, $r = 0, 1, \ldots, (N/2) - 1$, the $N/2$-point DFT of $y[n]$.

    3. Then the even-indexed values of $X[k]$ are $X[k] = Y[k/2]$, for $k = 0, 2, \ldots, N - 2$.

   (a) Show that the above algorithm produces the desired results.

   (b) Now suppose that we form a finite-length sequence $y[n]$ from a sequence $x[n]$ by

$$y[n] = \begin{cases} \sum_{r=-\infty}^{\infty} x[n + rM], & 0 \le n \le M - 1, \\ 0, & \text{otherwise.} \end{cases}$$

Determine the relationship between the M-point DFT $Y[k]$ and $X(e^{j\omega})$, the Fourier transform of $x[n]$. Show that the result of part (a) is a special case of the result of part (b).

   (c) Develop an algorithm similar to the one in part (a) to compute the odd-indexed DFT values $X[k]$, $k = 1, 3, \ldots, N - 1$ for N even using only a single $N/2$-point DFT.

9.31. Suppose that an FFT program is available that computes the DFT of a complex sequence. If we wish to compute the DFT of a real sequence, we may simply specify the imaginary part to be zero and use the program directly. However, the symmetry of the DFT of a real sequence can be used to reduce the amount of computation.

   (a) Let $x[n]$ be a real-valued sequence of length N, and let $X[k]$ be its DFT with real and imaginary parts denoted $X_R[k]$ and $X_I[k]$, respectively; i.e.,

$$X[k] = X_R[k] + jX_I[k].$$

Show that if $x[n]$ is real, then $X_R[k] = X_R[N - k]$ and $X_I[k] = -X_I[N - k]$ for $k = 1, \ldots, N - 1$.

   (b) Now consider two real-valued sequences $x_1[n]$ and $x_2[n]$ with DFTs $X_1[k]$ and $X_2[k]$, respectively. Let $g[n]$ be the complex sequence $g[n] = x_1[n] + jx_2[n]$, with corresponding DFT $G[k] = G_R[k] + jG_I[k]$. Also let $G_{OR}[k]$, $G_{ER}[k]$, $G_{OI}[k]$ and $G_{EI}[k]$ denote, respectively, the odd part of the real part, the even part of the real part, the odd part of the imaginary part, and the even part of the imaginary part of $G[k]$ as defined through Eqs. (8.96). Specifically, for $1 \le k \le N - 1$,

$$G_{OR}[k] = \tfrac{1}{2}\{G_R[k] - G_R[N - k]\},$$
$$G_{ER}[k] = \tfrac{1}{2}\{G_R[k] + G_R[N - k]\},$$
$$G_{OI}[k] = \tfrac{1}{2}\{G_I[k] - G_I[N - k]\},$$
$$G_{EI}[k] = \tfrac{1}{2}\{G_I[k] + G_I[N - k]\},$$

and $G_{OR}[0] = G_{OI}[0] = 0$, $G_{ER}[0] = G_R[0]$, $G_{EI}[0] = G_I[0]$. Determine expressions for $X_1[k]$ and $X_2[k]$ in terms of $G_{OR}[k]$, $G_{ER}[k]$, $G_{OI}[k]$, and $G_{EI}[k]$.

   (c) Assume that $N = 2^\nu$ and that a radix-2 FFT program is available to compute the DFT. Determine the number of real multiplications and the number of real additions required to compute both $X_1[k]$ and $X_2[k]$ by (i) using the program twice (with the imaginary part of the input set to zero) to compute the two complex N-point DFTs $X_1[k]$ and $X_2[k]$ separately, and (ii) using the scheme suggested in part (b), which requires only one N-point DFT to be computed.

(d) Assume that we have only one real N-point sequence $x[n]$, where N is a power of 2. Let $x_1[n]$ and $x_2[n]$ be the two real $N/2$-point sequences $x_1[n] = x[2n]$ and $x_2[n] = x[2n + 1]$, where $n = 0, 1, \ldots, (N/2) - 1$. Determine $X[k]$ in terms of the $(N/2)$-point DFTs $X_1[k]$ and $X_2[k]$.

(e) Using the results of parts (b), (c), and (d), describe a procedure for computing the DFT of the real N-point sequence $x[n]$ using only one $N/2$-point FFT computation. Determine the numbers of real multiplications and real additions required by this procedure, and compare these numbers with the numbers required if the $X[k]$ is computed using one N-point FFT computation with the imaginary part set to zero.

9.32. In this problem we consider a procedure for computing the DFT of four real symmetric or antisymmetric N-point sequences using only one N-point DFT computation. Since we are considering only finite-length sequences, by symmetric and antisymmetric we explicitly mean periodic symmetric and periodic antisymmetric as defined in Section 8.7.4. Let $x_1[n]$, $x_2[n]$, $x_3[n]$, and $x_4[n]$ denote the four real sequences of length N and let $X_1[k]$, $X_2[k]$, $X_3[k]$, and $X_4[k]$ denote the corresponding DFTs. We assume first that $x_1[n]$ and $x_2[n]$ are symmetric and $x_3[n]$ and $x_4[n]$ are antisymmetric; i.e.,

$$x_1[n] = x_1[N - n], \qquad x_2[n] = x_2[N - n],$$

$$x_3[n] = -x_3[N - n], \qquad x_4[n] = -x_4[N - n],$$

for $n = 1, 2, \ldots, N - 1$ and $x_3[0] = x_4[0] = 0$.

(a) Define $y_1[n] = x_1[n] + x_3[n]$ and let $Y_1[k]$ denote the DFT of $y_1[n]$. Determine how $X_1[k]$ and $X_2[k]$ can be recovered from $Y_1[k]$.

(b) $y_1[n]$ as defined in part (a) is a real sequence with symmetric part $x_1[n]$ and antisymmetric part $x_3[n]$. Similarly, we define the real sequence $y_2[n] = x_2[n] + x_4[n]$ and we let $y_3[n]$ be the complex sequence

$$y_3[n] = y_1[n] + jy_2[n].$$

First determine how $Y_1[k]$ and $Y_2[k]$ can be determined from $Y_3[k]$ and then, using the results of part (a), show how to obtain $X_1[k]$, $X_2[k]$, $X_3[k]$, and $X_4[k]$ from $Y_3[k]$.

The result of part (b) shows that we can compute the DFT of four real sequences simultaneously with only one N-point DFT computation if two sequences are symmetric and the other two are antisymmetric. Now consider the case when all four are symmetric; i.e.,

$$x_i[n] = x_i[N - n], \qquad i = 1, 2, 3, 4,$$

for $n = 0, 1, \ldots, N - 1$.

(c) Consider a real symmetric sequence $x_3[n]$. Show that the sequence

$$u_3[n] = x_3[((n + 1))_N] - x_3[((n - 1))_N]$$

is an antisymmetric sequence; i.e., $u_3[n] = -u_3[N - n]$ for $n = 1, 2, \ldots, N - 1$ and $u_3[0] = 0$.

(d) Let $U_3[k]$ denote the N-point DFT of $u_3[n]$. Determine an expression for $U_3[k]$ in terms of $X_3[k]$.

(e) By using the procedure of part (c), we can form the real sequence $y_1[n] = x_1[n] + u_3[n]$, where $x_1[n]$ is the symmetric part and $u_3[n]$ is the antisymmetric part of $y_1[n]$. Determine how $X_1[k]$ and $X_3[k]$ can be recovered from $Y_1[k]$.

(f) Now let $y_3[n] = y_1[n] + jy_2[n]$, where

$$y_1[n] = x_1[n] + u_3[n], \qquad y_2[n] = x_2[n] + u_4[n],$$

with

$$u_3[n] = x_3[((n+1))_N] - x_3[((n-1))_N],$$

$$u_4[n] = x_4[((n+1))_N] - x_4[((n-1))_N],$$

for $n = 0, 1, \ldots, N-1$. Determine how to obtain $X_1[k]$, $X_2[k]$, $X_3[k]$, and $X_4[k]$ from $Y_3[k]$. (Note that $X_3[0]$ and $X_4[0]$ cannot be recovered from $Y_3[k]$, and if N is even, $X_3[N/2]$ and $X_4[N/2]$ also cannot be recovered from $Y_3[k]$.)

9.33. Let $x[n]$ and $h[n]$ be two real finite-length sequences such that

$$x[n] = 0 \qquad \text{for } n \text{ outside the interval } 0 \le n \le L-1,$$

$$h[n] = 0 \qquad \text{for } n \text{ outside the interval } 0 \le n \le P-1.$$

We wish to compute the sequence $y[n] = x[n] * h[n]$, where $*$ denotes ordinary convolution.

(a) What is the length of the sequence $y[n]$?

(b) For direct evaluation of the convolution sum, how many real multiplications are required to compute all of the nonzero samples of $y[n]$? The following identity may be useful:

$$\sum_{k=1}^{N} k = \frac{N(N+1)}{2}.$$

(c) State a procedure for using the DFT to compute all of the nonzero samples of $y[n]$. Determine the minimum size of the DFTs and IDFTs in terms of L and P.

(d) Assume that $L = P = N/2$, where $N = 2^v$ is the DFT size. Determine a formula for the number of real multiplications required to compute all the nonzero values of $y[n]$ using the method of part (c) if the DFTs are computed using a radix-2 FFT algorithm. Use this formula to determine the minimum value of N for which the FFT method requires fewer real multiplications than the direct evaluation of the convolution sum.

9.34. In Section 8.8.3 we showed that linear time-invariant filtering can be implemented by sectioning the input signal into finite-length segments and using the DFT to implement circular convolutions on these segments. The two methods discussed were called the overlap-add and the overlap-save methods. If the DFT's are computed using an FFT algorithm, these sectioning methods can require fewer complex multiplications per output sample than the direct evaluation of the convolution sum.

(a) Assume that the complex input sequence $x[n]$ is of infinite duration and that the complex impulse response $h[n]$ is of length P samples, so that $h[n] \ne 0$ only for $0 \le n \le P-1$. Also assume that the output is computed using the overlap-save method, with the DFTs of length $L = 2^v$ and computed using a radix-2 FFT algorithm. Determine an expression for the number of complex multiplications required per output sample as a function of v and P.

(b) Suppose that the length of the impulse response is $P = 500$. By evaluating the formula obtained in part (a), plot the number of multiplications per output sample as a function of v for the values of $v \leq 20$ such that the overlap-save method applies. For what value of v is the number of multiplications minimum? Compare the number of complex multiplications per output sample for the overlap-save method using the FFT with the number of complex multiplications per output sample required for direct evaluation of the convolution sum.

(c) Show that for large FFT lengths, the number of complex multiplications per output sample is approximately v. Thus, beyond a certain FFT length, the overlap-save method is less efficient than the direct method. If $P = 500$, for what value of v will the direct method be more efficient?

(d) Assume that the FFT length is twice the length of the impulse response (i.e., $L = 2P$) and assume that $L = 2^v$. Using the formula obtained in part (a), determine the smallest value of P such that the overlap-save method using the FFT requires fewer complex multiplications than the direct convolution method.

9.35. The input and output of a linear time-invariant system satisfy a difference equation of the form

$$y[n] = \sum_{k=1}^{N} a_k y[n-k] + \sum_{k=0}^{M} b_k x[n-k].$$

Assume that an FFT program is available for computing the DFT of any finite-length sequence of length $N = 2^v$. Describe a procedure that utilizes the available FFT program to compute

$$H(e^{j(2\pi/512)k}) \qquad \text{for } k = 0, 1, \ldots, 512,$$

where $H(z)$ is the system function of the system.

9.36. Suppose that we wish to multiply two very large numbers (possibly thousands of bits long) on a 16-bit computer. In this problem we will investigate a technique for doing this using FFTs.

(a) Let $p(x)$ and $q(x)$ be the two polynomials

$$p(x) = \sum_{i=0}^{L-1} a_i x^i, \qquad q(x) = \sum_{i=0}^{M-1} b_i x^i.$$

Show that the coefficients of the polynomial $r(x) = p(x)q(x)$ can be computed using circular convolution.

(b) Show how to compute the coefficients of $r(x)$ using a radix-2 FFT program. For what orders of magnitude of $(L + M)$ is this procedure more efficient than direct computation? Assume that $L + M = 2^v$ for some integer v.

(c) Now suppose that we wish to compute the product of two very long positive binary integers u and v. Show that their product can be computed using polynomial multiplication, and describe an algorithm for computing their product using an FFT algorithm. If u is an 8000-bit number and v is a 1000-bit number, approximately how many real multiplications and additions are required to compute the product $u \cdot v$ using this method?

(d) Give a qualitative discussion of the effect of finite-precision arithmetic in implementing the algorithm of part (c).

9.37. A sequence $x[n]$ of length N has a discrete Hartley transform (DHT), defined as

$$X_H[k] = \sum_{n=0}^{N-1} x[n] H_N[nk], \qquad k = 0, 1, \ldots, N-1, \qquad \text{(P9.37-1)}$$

where

$$H_N[a] = C_N[a] + S_N[a],$$

with

$$C_N[a] = \cos(2\pi a/N), \qquad S_N[a] = \sin(2\pi a/N).$$

Originally proposed by R. V. L. Hartley in 1942 for the continuous-time case, the Hartley transform has properties that make it useful and attractive in the discrete-time case as well (Bracewell, 1983, 1984). Specifically, from Eq. (P9.37-1) it is apparent that the DHT of a real sequence is also a real sequence. The DHT also has a convolution property, and fast algorithms exist for its computation.

In complete analogy with the DFT, the DHT has an implicit periodicity that must be acknowledged in its use. That is, if we consider $x[n]$ to be a finite-length sequence such that $x[n] = 0$ for $n < 0$ and $n > N - 1$ then we can form a periodic sequence

$$\tilde{x}[n] = \sum_{r=-\infty}^{\infty} x[n + rN]$$

such that $x[n]$ is simply one period of $\tilde{x}[n]$. The periodic sequence $\tilde{x}[n]$ can be represented by a discrete Hartley series (DHS), which in turn can be interpreted as the DHT by focusing attention on only one period of the periodic sequence.

(a) The DHS analysis equation is defined by

$$\tilde{X}_H[k] = \sum_{n=0}^{N-1} \tilde{x}[n]H_N[nk]. \tag{P9.37-2}$$

Show that the DHS coefficients form a sequence that is also periodic with period N; i.e.,

$$\tilde{X}_H[k] = \tilde{X}_H[k + N] \qquad \text{for all } k.$$

(b) It can also be shown that the sequences $H_N[nk]$ are orthogonal; i.e.,

$$\sum_{n=0}^{N-1} H_N[nk]H_N[mk] = \begin{cases} N, & ((n))_N = ((m))_N, \\ 0, & \text{otherwise.} \end{cases}$$

Using this property and the DHS analysis formula of Eq. (P9.37-2), show that the DHS synthesis formula is

$$\tilde{x}[n] = \frac{1}{N}\sum_{n=0}^{N-1} \tilde{X}_H[k]H_N[nk]. \tag{P9.37-3}$$

Note that the DHT is simply one period of the DHS coefficients, and likewise the DHT synthesis (inverse) equation is identical to the DHS synthesis equation (P9.37-3), except that we simply extract one period of $\tilde{x}[n]$; i.e., the DHT synthesis expression is

$$x[n] = \frac{1}{N}\sum_{n=0}^{N-1} X_H[k]H_N[nk], \qquad n = 0, 1, \ldots, N - 1. \tag{P9.37-4}$$

With Eqs. (P9.37-1) and (P9.37-4) as definitions of the analysis and synthesis relations, respectively, for the DHT, we may now proceed to derive the useful properties of this representation of a finite-length discrete-time signal.

(c) Verify that $H_N[a] = H_N[a + N]$, and verify the following useful property of $H_N[a]$:

$$H_N[a + b] = H_N[a]C_N[b] + H_N[-a]S_N[b]$$
$$= H_N[b]C_N[a] + H_N[-b]S_N[a].$$

(d) Consider a circularly shifted sequence $x_1[n]$ defined as

$$x_1[n] = \begin{cases} \tilde{x}[n - n_0] = x[((n - n_0))_N], & n = 0, 1, \ldots, N - 1, \\ 0, & \text{otherwise.} \end{cases} \quad \text{(P9.37-5)}$$

In other words, $x_1[n]$ is the sequence that is obtained by extracting one period from the shifted periodic sequence $\tilde{x}[n - n_0]$. Using the identity verified in part (c), show that the DHS coefficients for the shifted periodic sequence are

$$\tilde{x}[n - n_0] \xleftrightarrow{\mathscr{DHS}} \tilde{X}_H[k]C_N[n_0 k] + \tilde{X}_H[-k]S_N[n_0 k]. \quad \text{(P9.37-6)}$$

From this we conclude that the DHT of the finite-length circularly shifted sequence $x[((n - n_0))_N]$ is

$$x[((n - n_0))_N] \xleftrightarrow{\mathscr{DHT}} X_H[k]C_N[n_0 k] + X_H[-k]S_N[n_0 k]. \quad \text{(P9.37-7)}$$

(e) Suppose that $x_3[n]$ is the N-point circular convolution of two N-point sequences $x_1[n]$ and $x_2[n]$; i.e.,

$$x_3[n] = x_1[n] \text{Ⓝ} x_2[n]$$

$$= \sum_{m=0}^{N-1} x_1[m]x_2[((n - m))_N], \quad n = 0, 1, \ldots, N - 1. \quad \text{(P9.37-8)}$$

By applying the DHT to both sides of Eq. (P9.37-8) and using Eq. (P9.37-7), show that

$$X_{H3}[k] = \tfrac{1}{2}X_{H1}[k](X_{H2}[k] + X_{H2}[((-k))_N])$$

$$+ \tfrac{1}{2}X_{H1}[((-k))_N](X_{H2}[k] - X_{H2}[((-k))_N]) \quad \text{(P.37-9)}$$

for $k = 0, 1, \ldots, N - 1$. This is the desired convolution property.

Note that a linear convolution can be computed using the DHT in the same way that the DFT can be used to compute a linear convolution. While computing $X_{H3}[k]$ from $X_{H1}[k]$ and $X_{H2}[k]$ requires the same amount of computation as computing $X_3[k]$ from $X_1[k]$ and $X_2[k]$, the computation of the DHT requires only half the number of real multiplications required to compute the DFT.

(f) Suppose that we wish to compute the DHT of an N-point sequence $x[n]$ and we have available the means to compute the N-point DFT. Describe a technique for obtaining $X_H[k]$ from $X[k]$ for $k = 0, 1, \ldots, N - 1$.

(g) Suppose that we wish to compute the DFT of an N-point sequence $x[n]$ and we have available the means to compute the N-point DHT. Describe a technique for obtaining $X[k]$ from $X_H[k]$ for $k = 0, 1, \ldots, N - 1$.

(h) By decomposing $x[n]$ into its even numbered points and odd numbered points and by using the identity derived in part (c), derive a fast DHT algorithm based on the decimation-in-time principle.

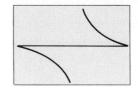

Discrete Hilbert Transforms

10

10.0 INTRODUCTION

In general, the specification of the Fourier transform of a sequence requires complete knowledge of both the real and imaginary parts or the magnitude and phase for all frequencies in the range $-\pi < \omega \leq \pi$. However, we have also seen that under certain conditions there are constraints on the Fourier transform. For example, in Section 2.8 we saw that if $x[n]$ is real, then its Fourier transform is conjugate symmetric, i.e., $X(e^{j\omega}) = X^*(e^{-j\omega})$. From this it follows that for real sequences, specification of $X(e^{j\omega})$ for $0 \leq \omega \leq \pi$ also specifies it for $-\pi \leq \omega \leq 0$. Similarly, we saw in Section 5.4 that under the constraint of minimum phase, the Fourier transform magnitude and phase are not independent, i.e., specification of magnitude determines the phase and specification of phase determines the magnitude to within a scale factor. In Section 8.5 we saw that for sequences of finite length N, specification of $X(e^{j\omega})$ at N equally spaced frequencies determines $X(e^{j\omega})$ at all frequencies.

In this chapter, we will see that the constraint of *causality* of a sequence implies unique relationships between the real and imaginary parts of the Fourier transform. Relationships of this type between the real and imaginary parts of complex functions arise in many fields besides signal processing, and they are commonly known as *Hilbert transform relationships*. In addition to developing these relationships for the Fourier transform of causal sequences, we will develop related results for the DFT and for sequences with one-sided Fourier transforms. Also, in Section 10.3 we will indicate how the relationship between magnitude and phase for minimum-phase sequences can be interpreted in terms of the Hilbert transform.

Although we will take an intuitive approach in this chapter, it is important to be aware that the Hilbert transform relationships follow formally from the properties of analytic functions (see Problem 10.1). Specifically, the complex functions that arise in the mathematical representation of discrete-time signals and systems are generally very well behaved functions. With few exceptions, the z-transforms that have concerned us have had well-defined regions in which the power series is absolutely convergent. Since a power series represents an analytic function within its region of convergence, it follows that z-transforms are analytic functions inside their regions of convergence. By the definition of an analytic function, this means that the z-transform has a well-defined derivative at every point inside the region of convergence. Furthermore, analyticity implies that the z-transform and all its derivatives are continuous functions within the region of convergence.

The properties of analytic functions imply some rather powerful constraints on the behavior of the z-transform within its region of convergence. Since the Fourier transform is the z-transform evaluated on the unit circle, these constraints also restrict the behavior of the Fourier transform. One such constraint is that the real and imaginary parts satisfy the Cauchy-Riemann conditions, which relate the partial derivatives of the real and imaginary parts of an analytic function (see, for example, Churchill and Brown, 1984). Another constraint is the Cauchy integral theorem, through which the value of a complex function is specified everywhere inside a region of analyticity in terms of the values of the function on the boundary of the region. Based on these relations for analytic functions, it is possible, under certain conditions, to derive explicit integral relationships between the real and imaginary parts of a z-transform on a closed contour within the region of convergence. In the mathematics literature these relations are often referred to as *Poisson's formulas*. In the context of system theory, they are known as the *Hilbert transform relations*.

Rather than following the mathematical approach just discussed, we will develop the Hilbert transform relations by exploiting the fact that, on the unit circle, the real and imaginary parts of the z-transform of a causal sequence are the transforms of the even and odd components of the sequence (properties 5 and 6, Table 2.1, page 53). As we will show, a causal sequence is completely specified by its even part, implying that the z-transform of the original sequence is completely specified by its real part on the unit circle. In addition to applying this argument to specifying the z-transform of a causal sequence in terms of its real part on the unit circle, it can also be applied, under certain conditions, to specify the z-transform of a sequence in terms of its magnitude on the unit circle.

The notion of an analytic signal is an important concept in continuous-time signal processing. An analytic signal is a complex time function (which is analytic) having a Fourier transform that vanishes for negative frequencies. A complex sequence cannot be considered in a formal sense to be analytic since it is a function of an integer variable. However, in a style similar to that described in the previous paragraph, it is possible to relate the real and imaginary parts of a complex sequence whose spectrum is zero on the unit circle for $-\pi < \omega < 0$. A similar approach can also be taken in relating the real and imaginary parts of the discrete Fourier transform for a periodic or, equivalently, a finite-length sequence. In this case the "causality" condition is that the periodic sequence be zero in the second half of each period.

Thus in this chapter a notion of causality will be applied to relate the even and odd components of a function or, equivalently, the real and imaginary parts of its transform. We will apply this approach in four situations. First, we relate the real and imaginary parts of the Fourier transform $X(e^{j\omega})$ of a sequence $x[n]$ that is zero for $n < 0$. In the second situation, we obtain a relationship between the real and imaginary parts of the DFT for periodic sequences or, equivalently, for a finite-length sequence considered to be of length N but with the last $(N/2) - 1$ points zero. In the third case we relate the real and imaginary parts of the *logarithm* of the Fourier transform under the condition that the inverse transform of the logarithm of the transform is zero for $n < 0$. Relating the real and imaginary parts of the logarithm of the Fourier transform corresponds to relating the log magnitude and phase of $X(e^{j\omega})$. Finally, we relate the real and imaginary parts of a complex sequence whose Fourier transform, considered as a periodic function of ω, is zero in the second half of each period.

10.1 REAL AND IMAGINARY PART SUFFICIENCY OF THE FOURIER TRANSFORM FOR CAUSAL SEQUENCES

Any sequence can be expressed as the sum of an even sequence and an odd sequence. Specifically, with $x_e[n]$ and $x_o[n]$ denoting the even and odd parts of $x[n]$,† then

$$x[n] = x_e[n] + x_o[n], \tag{10.1}$$

where

$$x_e[n] = \frac{x[n] + x[-n]}{2} \tag{10.2}$$

and

$$x_o[n] = \frac{x[n] - x[-n]}{2}. \tag{10.3}$$

Equations (10.1)–(10.3) apply to an arbitrary sequence whether or not it is causal and whether or not it is real. However, if $x[n]$ is causal, i.e., $x[n] = 0$, $n < 0$, then it is possible to recover $x[n]$ from $x_e[n]$ or to recover $x[n]$ for $n \neq 0$ from $x_o[n]$. Consider, for example, the causal sequence $x[n]$ and its even and odd components as shown in Fig. 10.1. Because $x[n]$ is causal, $x[n] = 0$ for $n < 0$ and $x[-n] = 0$ for $n > 0$. Therefore the nonzero portions of $x[n]$ and $x[-n]$ do not overlap except at $n = 0$. For this reason it follows from Eqs. (10.2) and (10.3) that

$$x[n] = 2x_e[n]u[n] - x_e[0]\delta[n] \tag{10.4}$$

and

$$x[n] = 2x_o[n]u[n] + x[0]\delta[n]. \tag{10.5}$$

---

† If $x[n]$ is real, then $x_e[n]$ and $x_o[n]$ in Eqs. (10.2) and (10.3) are the even and odd parts, respectively, of $x[n]$ as considered in Chapter 2. If $x[n]$ is complex, for the purposes of this discussion we still define $x_e[n]$ and $x_o[n]$ as in Eqs. (10.2) and (10.3), which do not correspond to the conjugate-symmetric and conjugate-antisymmetric parts of a complex sequence as considered in Chapter 2.

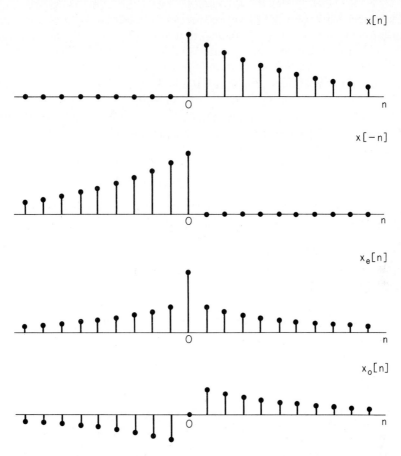

Figure 10.1   Even and odd parts of a real causal sequence.

The validity of these relationships is easily seen in Fig. 10.1. Note that $x[n]$ is completely determined by $x_e[n]$. On the other hand, $x_o[0] = 0$, so we can recover $x[n]$ from $x_o[n]$ only for $n \neq 0$.

Now if $x[n]$ is also stable, i.e., absolutely summable, then its Fourier transform exists. We denote the Fourier transform of $x[n]$ as

$$X(e^{j\omega}) = X_R(e^{j\omega}) + jX_I(e^{j\omega}), \tag{10.6}$$

where $X_R(e^{j\omega})$ is the real part and $X_I(e^{j\omega})$ is the imaginary part of $X(e^{j\omega})$. If $x[n]$ is a *real* sequence, we recall that $X_R(e^{j\omega})$ is the Fourier transform of $x_e[n]$ and that $jX_I(e^{j\omega})$ is the Fourier transform of $x_o[n]$. Therefore for a *causal, stable, real* sequence, $X_R(e^{j\omega})$ completely determines $X(e^{j\omega})$, since if we are given $X_R(e^{j\omega})$ we can find $X(e^{j\omega})$ by the following process:

1. Find $x_e[n]$ as the inverse Fourier transform of $X_R(e^{j\omega})$.
2. Find $x[n]$ using Eq. (10.4).
3. Find $X(e^{j\omega})$ as the Fourier transform of $x[n]$.

This constructive procedure can be interpreted analytically to obtain a general relationship that expresses $X_I(e^{j\omega})$ directly in terms of $X_R(e^{j\omega})$. From Eq. (10.4), the complex convolution theorem, and the fact that $x_e[0] = x[0]$, it follows that

$$X(e^{j\omega}) = \frac{1}{\pi}\int_{-\pi}^{\pi} X_R(e^{j\theta})U(e^{j(\omega-\theta)})\,d\theta - x[0], \tag{10.7}$$

where $U(e^{j\omega})$ is the Fourier transform of the unit step sequence. As stated in Section 2.7, although the unit step is neither absolutely summable nor square summable, it can be represented by the Fourier transform

$$U(e^{j\omega}) = \sum_{k=-\infty}^{\infty} \pi\delta(\omega - 2\pi k) + \frac{1}{1 - e^{-j\omega}} \tag{10.8}$$

or, since the term $1/(1 - e^{-j\omega})$ can be rewritten as

$$\frac{1}{1 - e^{-j\omega}} = \frac{1}{2} - \frac{j}{2}\cot\left(\frac{\omega}{2}\right), \tag{10.9}$$

Eq. (10.8) becomes

$$U(e^{j\omega}) = \sum_{k=-\infty}^{\infty} \pi\delta(\omega - 2\pi k) + \frac{1}{2} - \frac{j}{2}\cot\left(\frac{\omega}{2}\right). \tag{10.10}$$

Using Eq. (10.10), we can express Eq. (10.7) as

$$X(e^{j\omega}) = X_R(e^{j\omega}) + jX_I(e^{j\omega})$$

$$= X_R(e^{j\omega}) + \frac{1}{2\pi}\int_{-\pi}^{\pi} X_R(e^{j\theta})\,d\theta - \frac{j}{2\pi}\int_{-\pi}^{\pi} X_R(e^{j\theta})\cot\left(\frac{\omega-\theta}{2}\right)\,d\theta - x[0]. \tag{10.11}$$

Equating real and imaginary parts in Eq. (10.11) and noting that

$$x[0] = \frac{1}{2\pi}\int_{-\pi}^{\pi} X_R(e^{j\theta})\,d\theta, \tag{10.12}$$

we obtain the relationship

$$X_I(e^{j\omega}) = -\frac{1}{2\pi}\int_{-\pi}^{\pi} X_R(e^{j\theta})\cot\left(\frac{\omega-\theta}{2}\right)\,d\theta. \tag{10.13}$$

A similar procedure can be followed to obtain $x[n]$ and $X(e^{j\omega})$ from $X_I(e^{j\omega})$ and $x[0]$ using Eq. (10.5). This process results in the following equation for $X_R(e^{j\omega})$ in terms of $X_I(e^{j\omega})$:

$$X_R(e^{j\omega}) = x[0] + \frac{1}{2\pi}\int_{-\pi}^{\pi} X_I(e^{j\theta})\cot\left(\frac{\omega-\theta}{2}\right)\,d\theta. \tag{10.14}$$

Equations (10.13) and (10.14), which are called *discrete Hilbert transform relationships*, hold for the real and imaginary parts of the Fourier transform of a causal, stable, real sequence. They are improper integrals since the integrand is singular at $\omega - \theta = 0$. Such integrals must be evaluated carefully to obtain a consistent finite result. This can be done formally by interpreting the integrals as *Cauchy principal values*. That is, Eq. (10.13) becomes

$$X_I(e^{j\omega}) = -\frac{1}{2\pi}\mathscr{P}\int_{-\pi}^{\pi} X_R(e^{j\theta})\cot\left(\frac{\omega - \theta}{2}\right)d\theta \tag{10.15}$$

and Eq. (10.14) becomes

$$X_R(e^{j\omega}) = x[0] + \frac{1}{2\pi}\mathscr{P}\int_{-\pi}^{\pi} X_I(e^{j\theta})\cot\left(\frac{\omega - \theta}{2}\right)d\theta, \tag{10.16}$$

where $\mathscr{P}$ denotes Cauchy principal value of the integral that follows. The meaning of Cauchy principal value in Eq. (10.15), for example, is

$$X_I(e^{j\omega}) = -\frac{1}{2\pi}\lim_{\varepsilon\to 0}\left[\int_{\omega+\varepsilon}^{\pi} X_R(e^{j\theta})\cot\left(\frac{\omega - \theta}{2}\right)d\theta\right.$$
$$\left. + \int_{-\pi}^{\omega-\varepsilon} X_R(e^{j\theta})\cot\left(\frac{\omega - \theta}{2}\right)d\theta\right]. \tag{10.17}$$

Equation (10.17) shows that $X_I(e^{j\omega})$ is obtained by the periodic convolution of $-\cot(\omega/2)$ with $X_R(e^{j\omega})$, with special care being taken in the vicinity of the singularity at $\theta = \omega$. In a similar manner, Eq. (10.16) involves the periodic convolution of $\cot(\omega/2)$ with $X_I(e^{j\omega})$.

The two functions involved in the convolution integral of Eq. (10.15) (or, equivalently, Eq. 10.17) are depicted in Fig. 10.2. The limit in Eq. (10.17) exists because the function $\cot[(\omega - \theta)/2]$ is antisymmetric at the singular point $\theta = \omega$ and the limit is taken symmetrically about the singularity.

The previous discussion can be generalized to show that the real part of the Fourier transform of a causal, stable, real sequence completely determines the z-transform everywhere in its region of convergence. This can be seen by noting that $X(e^{j\omega})$ can be found from $X_R(e^{j\omega})$ by the procedures just applied; then by analytic

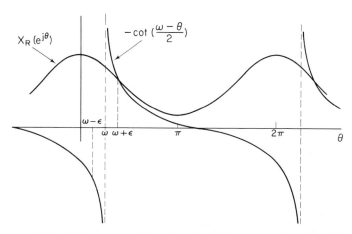

Figure 10.2   Interpretation of the Hilbert transform as a periodic convolution.

continuation (informally, this means letting $e^{j\omega} = z$), $X(z)$ can be obtained from $X(e^{j\omega})$. We can also obtain a direct relationship for $X(z)$ in terms of $X_R(e^{j\omega})$ by applying Eq. (10.4) and the complex convolution theorem of Section 4.6. This mo.` general relation is

$$X(z) = \frac{1}{2\pi j} \oint_C X_R(v) \left(\frac{z+v}{z-v}\right) \frac{dv}{v}, \qquad (10.18)$$

where v, the integration variable, is $v = e^{j\theta}$; i.e., the closed contour C is the unit circle of the v-plane since $X_R(e^{j\theta})$ is all that is assumed to be known. This expression for $X(z)$ holds at least for $|z| \geq 1$ since $x[n]$ is assumed to be a causal, stable sequence. Likewise $X(z)$ can be expressed in terms of $X_I(e^{j\omega})$ as

$$X(z) = \frac{1}{2\pi j} \oint_C X_I(v) \left(\frac{z+v}{z-v}\right) \frac{dv}{v} + x[0], \qquad (10.19)$$

where C is again the unit circle.

Equation (10.18) is particularly useful when $X_R(e^{j\omega})$ can be expressed as a rational function of $e^{j\omega}$, since the contour integral can then be evaluated using Cauchy's residue theorem. When the residues can be found, this is generally much easier than evaluating Eq. (10.15).

Example 10.1

Let $X_R(e^{j\omega})$ be given by

$$X_R(e^{j\omega}) = \frac{1 - \alpha \cos \omega}{1 - 2\alpha \cos \omega + \alpha^2}, \qquad |\alpha| < 1, \qquad (10.20)$$

and α is real. To find $X(z)$ using Eq. (10.18), we first write $X_R(e^{j\omega})$ as a rational function of $e^{j\omega}$:

$$X_R(e^{j\omega}) = \frac{1 - \alpha(e^{j\omega} + e^{-j\omega})/2}{(1 - \alpha e^{-j\omega})(1 - \alpha e^{j\omega})}. \qquad (10.21)$$

Then setting $e^{j\omega} = v$ and substituting into Eq. (10.18), we obtain

$$X(z) = \frac{1}{2\pi j} \oint_C \frac{(1 - \alpha(v + v^{-1})/2)(z+v)dv}{(1 - \alpha v^{-1})(1 - \alpha v)(z-v)v}, \qquad (10.22)$$

where C is the unit circle. Writing this equation so as to display the poles of the integrand, we obtain

$$X(z) = \frac{1}{2\pi j} \oint_C \frac{[v - \alpha(v^2 + 1)/2](z+v)dv}{(v - \alpha)(1 - \alpha v)(z-v)v}. \qquad (10.23)$$

The poles of the integrand are depicted in Fig. 10.3, where we note that only the poles at $v = 0$ and $v = \alpha$ are inside the unit circle. Therefore, using the residue theorem we obtain

$$X(z) = \frac{-(\alpha/2)z}{-\alpha z} + \frac{[\alpha - \alpha(\alpha^2 + 1)/2](z + \alpha)}{(1 - \alpha^2)(z - \alpha)\alpha}$$

$$= \frac{1}{2} + \frac{1}{2}\frac{z + \alpha}{z - \alpha} = \frac{z}{z - \alpha} = \frac{1}{1 - \alpha z^{-1}}. \qquad (10.24)$$

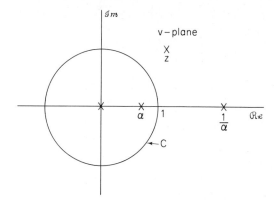

Figure 10.3   Pole locations in Example 10.1, in which the z-transform is computed from the real part using contour integration.

Equation (10.24) was derived assuming $|z| \geq 1$; however, we note that the region of analyticity of the integrand in Eq. (10.23) is $|\alpha| < |v| < \min(|z|, |\alpha^{-1}|)$ and the value of the integral in Eq. (10.23) is the same for all closed contours that enclose the poles at $v = 0$ and $v = \alpha$ but *not* the poles at $v = z$ and $v = \alpha^{-1}$. Thus we have obtained the z-transform everywhere in the region $|z| > |\alpha|$ directly from its real part on the unit circle. The inverse z-transform $x[n]$ is

$$x[n] = \alpha^n u[n]. \tag{10.25}$$

Example 10.2

Let us again consider $X_R(e^{j\omega})$ as given by Eq. (10.20) but this time we first determine $x_e[n]$ and then $x[n]$ using Eq. (10.4). We first express $X_R(e^{j\omega})$ explicitly in terms of $e^{j\omega}$:

$$X_R(e^{j\omega}) = \frac{1 - (\alpha/2)(e^{j\omega} + e^{-j\omega})}{1 - \alpha(e^{j\omega} + e^{-j\omega}) + \alpha^2}, \qquad |\alpha| < 1. \tag{10.26}$$

Next, using analytic continuation, i.e., replacing $e^{j\omega}$ by z, we have

$$X_R(z) = \frac{1 - (\alpha/2)(z + z^{-1})}{1 - \alpha(z + z^{-1}) + \alpha^2} \tag{10.27}$$

$$= \frac{z - (\alpha/2)(z^2 + 1)}{(z - \alpha)(1 - \alpha z)}. \tag{10.28}$$

Since we began with the Fourier transform $X_R(e^{j\omega})$ and obtained $X_R(z)$ by extending this into the z-plane, the ROC of $X_R(z)$ must of course include the unit circle and is then bounded on the inside by the pole at $z = \alpha$ and on the outside by the pole at $z = 1/\alpha$.

From Eq. (10.28), we now want to obtain $x_e[n]$, the inverse z-transform of $X_R(z)$. Using Eq. (4.67), we have

$$x_e[n] = \frac{1}{2\pi j} \oint_C X_R(z) z^{n-1} \, dz, \tag{10.29}$$

where, for example, we can choose C to be the unit circle. For $n \geq 1$ the integrand in Eq. (10.29) has one pole inside the contour of integration, and (after a small amount of algebra) the corresponding value of the integral is

$$x_e[n] = \tfrac{1}{2}\alpha^n, \qquad n \geq 1. \tag{10.30}$$

For $n = 0$ the integrand has two poles, and again evaluating residues we obtain

$$x_e[n] = 1, \qquad n = 0. \tag{10.31}$$

Since $x_e[n] = x_e[-n]$, Eqs. (10.30) and (10.31) completely determine $x_e[n]$. From Eq. (10.4) it then follows that

$$x[n] = \alpha^n u[n] \tag{10.32}$$

and

$$X(z) = \frac{1}{1 - \alpha z^{-1}}, \qquad |z| > |\alpha|. \tag{10.33}$$

10.2 SUFFICIENCY THEOREMS FOR FINITE-LENGTH SEQUENCES

In Section 10.1 we showed that causality or one-sidedness of a real sequence implies some strong constraints on the Fourier transform of the sequence. The results of the previous section apply, of course, to finite-length causal sequences, but since the finite-length property is more restrictive, it is perhaps reasonable to expect the Fourier transform of a finite-length sequence to be more constrained. We will see that this is indeed the case.

One way to take advantage of the finite-length property is to recall that finite-length sequences can be represented by the discrete Fourier transform. Since the DFT involves sums rather than integrals, the problems of improper integrals disappear.

Since the DFT is in reality a representation of a periodic sequence, any results we obtain must be based on corresponding results for periodic sequences. Indeed, it is important to keep the inherent periodicity of the DFT firmly in mind in deriving the desired Hilbert transform relation for finite-length sequences. Therefore we will consider the periodic case first and then discuss the application to the finite-length case.

Consider a periodic sequence $\tilde{x}[n]$ with period N that is related to a finite-length sequence $x[n]$ of length N by

$$\tilde{x}[n] = x[((n))_N]. \tag{10.34}$$

As in Section 10.1, $\tilde{x}[n]$ can be represented over one period as the sum of an even and odd sequence

$$\tilde{x}[n] = \tilde{x}_e[n] + \tilde{x}_o[n], \qquad n = 0, 1, \ldots, (N - 1), \tag{10.35}$$

where

$$\tilde{x}_e[n] = \frac{\tilde{x}[n] + \tilde{x}[-n]}{2}, \qquad n = 0, 1, \ldots, (N - 1), \tag{10.36a}$$

and

$$\tilde{x}_o[n] = \frac{\tilde{x}[n] - \tilde{x}[-n]}{2}, \qquad n = 0, 1, \ldots, (N - 1). \tag{10.36b}$$

A periodic sequence cannot, of course, be causal in the sense used in Section 10.1. We can, however, define a "periodically causal" periodic sequence to be one for which $\tilde{x}[n] = 0$ for $N/2 < n < N$. That is, $\tilde{x}[n]$ is identically zero over the last half of the period. We assume henceforth that N is even; the case of N odd is considered in Problem 10.8. Note that because of the periodicity of $\tilde{x}[n]$, it is also true that $\tilde{x}[n] = 0$ for $-N/2 < n < 0$. For finite-length sequences this restriction means that although the sequence is considered to be of length N, the last $(N/2) - 1$ points are in fact zero. In Fig. 10.4 we show an example of a periodically causal sequence and its even and odd parts with $N = 8$. Because $\tilde{x}[n]$ is zero in the second half of each period, $\tilde{x}[-n]$ is zero in the first half of each period, and, consequently, except for $n = 0$ and $n = N/2$, there is no overlap between the nonzero portions of $\tilde{x}[n]$ and $\tilde{x}[-n]$. Therefore, for "causal" periodic sequences,

$$\tilde{x}[n] = \begin{cases} 2\tilde{x}_e[n], & n = 1, 2, \ldots, (N/2) - 1, \\ \tilde{x}_e[n], & n = 0, N/2, \\ 0, & n = (N/2) + 1, \ldots, N - 1, \end{cases} \qquad (10.37)$$

and

$$\tilde{x}[n] = \begin{cases} 2\tilde{x}_o[n], & n = 1, 2, \ldots, (N/2) - 1, \\ 0, & n = (N/2) + 1, \ldots, N - 1. \end{cases} \qquad (10.38)$$

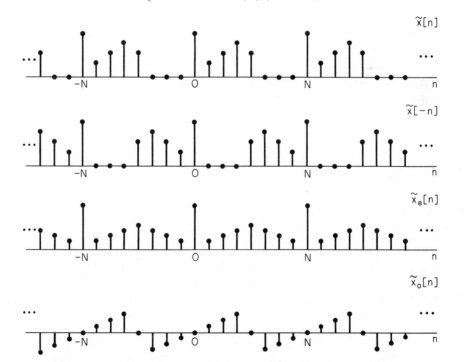

Figure 10.4   Even and odd parts of a "periodically causal," real, periodic sequence of period $N = 8$.

If we define $\tilde{u}_N[n]$ as the periodic sequence

$$\tilde{u}_N[n] = \begin{cases} 1 & n = 0, N/2, \\ 2, & n = 1, 2, \ldots, (N/2) - 1, \\ 0, & n = (N/2) + 1, \ldots, N - 1, \end{cases} \tag{10.39}$$

then it follows that for N even we can express $\tilde{x}[n]$ as

$$\tilde{x}[n] = \tilde{x}_e[n]\tilde{u}_N[n] \tag{10.40}$$

and

$$\tilde{x}[n] = \tilde{x}_o[n]\tilde{u}_N[n] + x[0]\tilde{\delta}[n] + x[N/2]\tilde{\delta}[n - (N/2)], \tag{10.41}$$

where $\tilde{\delta}[n]$ is a periodic unit impulse sequence with period N. Thus, the sequence $\tilde{x}[n]$ can be completely recovered from $\tilde{x}_e[n]$. On the other hand, $\tilde{x}_o[n]$ will always be zero at $n = 0$ and $n = N/2$, and consequently $\tilde{x}[n]$ can be recovered from $\tilde{x}_o[n]$ only for $n \neq 0$ and $n \neq N/2$.

If $\tilde{x}[n]$ is a real periodic sequence of period N with discrete Fourier series $\tilde{X}[k]$, then $\tilde{X}_R[k]$, the real part of $\tilde{X}[k]$, is the DFS of $\tilde{x}_e[n]$ and $j\tilde{X}_I[k]$ is the DFS of $\tilde{x}_o[n]$. Thus Eqs. (10.40) and (10.41) imply that for a periodic sequence of period N, which is causal in the sense defined above, $\tilde{X}[k]$ can be recovered from its real part or (almost) from its imaginary part. Equivalently, $\tilde{X}_I[k]$ can be obtained from $\tilde{X}_R[k]$ and $\tilde{X}_R[k]$ can (almost) be obtained from $\tilde{X}_I[k]$.

Specifically, suppose that we are given $\tilde{X}_R[k]$. Then we can obtain $\tilde{X}_I[k]$ by the following procedure:

1. Compute $\tilde{x}_e[n]$ using the DFS synthesis equation

$$\tilde{x}_e[n] = \frac{1}{N}\sum_{k=0}^{N-1} \tilde{X}_R[k]e^{j(2\pi/N)kn}. \tag{10.42}$$

2. Compute $\tilde{x}[n]$ using Eq. (10.40).
3. Compute $\tilde{X}[k]$ using the DFS analysis equation

$$\tilde{X}[k] = \sum_{k=0}^{N-1} \tilde{x}[n]e^{-j(2\pi/N)kn} = \tilde{X}_R[k] + j\tilde{X}_I[k]. \tag{10.43}$$

In contrast to the general causal case discussed in Section 10.1, the procedure just outlined can be implemented on a computer, since Eqs. (10.42) and (10.43) can be evaluated accurately and efficiently using an FFT algorithm.

To obtain an explicit relation between $\tilde{X}_R[k]$ and $\tilde{X}_I[k]$ we can carry out the procedure analytically. From Eq. (10.40) and Eq. (8.39) it follows that

$$\tilde{X}[k] = \tilde{X}_R[k] + j\tilde{X}_I[k]$$
$$= \frac{1}{N}\sum_{m=0}^{N-1} \tilde{X}_R[m]\tilde{U}_N[k - m], \tag{10.44}$$

i.e., $\tilde{X}[k]$ is the periodic convolution of $\tilde{X}_R[k]$, the DFS of $\tilde{x}_e[n]$, with $\tilde{U}_N[k]$ the DFS of $\tilde{u}_N[n]$. The DFS of $\tilde{u}_N[n]$ can be shown to be (see Problem 10.7)

$$\tilde{U}_N[k] = \begin{cases} N, & k = 0, \\ -j2\cot(\pi k/N), & k \text{ odd}, \\ 0, & k \text{ even}. \end{cases} \tag{10.45}$$

If we define

$$\tilde{V}_N[k] = \begin{cases} -j2\cot(\pi k/N), & k \text{ odd}, \\ 0, & k \text{ even}, \end{cases} \tag{10.46}$$

then Eq. (10.44) can be expressed as

$$\tilde{X}[k] = \tilde{X}_R[k] + \frac{1}{N}\sum_{m=0}^{N-1}\tilde{X}_R[m]\tilde{V}_N[k-m]. \tag{10.47}$$

Therefore

$$j\tilde{X}_I[k] = \frac{1}{N}\sum_{m=0}^{N-1}\tilde{X}_R[m]\tilde{V}_N[k-m], \tag{10.48}$$

which is the desired relation between the real and imaginary parts of the DFS of a periodically causal sequence. Similarly, beginning with Eq. (10.41) we can show that

$$\tilde{X}_R[k] = \frac{1}{N}\sum_{m=0}^{N-1}j\tilde{X}_I[m]\tilde{V}_N[k-m] + \tilde{x}[0] + (-1)^k\tilde{x}[N/2]. \tag{10.49}$$

Equations (10.48) and (10.49) relate the real and imaginary parts of the DFS representation of the periodic sequence $\tilde{x}[n]$. If $\tilde{x}[n]$ is thought of as the periodic repetition of a finite-length sequence $x[n]$ as in Eq. (10.34), then

$$x[n] = \begin{cases} \tilde{x}[n], & 0 \le n \le N-1, \\ 0, & \text{otherwise}. \end{cases} \tag{10.50}$$

If $x[n]$ has the "periodic causality" property with respect to a period N (i.e., $x[n] = 0$ for $n < 0$ and for $n > N/2$), then all of the preceding discussion applies to the discrete Fourier transform of $x[n]$. In other words, we can remove the tildes from Eqs. (10.48) and (10.49), thereby obtaining the DFT relations

$$jX_I[k] = \begin{cases} \dfrac{1}{N}\sum_{m=0}^{N-1}X_R[m]V_N[k-m], & 0 \le k \le N-1, \\ 0, & \text{otherwise}, \end{cases} \tag{10.51}$$

and

$$X_R[k] = \begin{cases} \dfrac{1}{N}\sum_{m=0}^{N-1}jX_I[m]V_N[k-m] + x[0] + (-1)^k x[N/2], & 0 \le k \le N-1, \\ 0, & \text{otherwise} \end{cases} \tag{10.52}$$

Note that the sequence $V_N[k - m]$ given by Eq. (10.46) is periodic with period N so we do not need to worry about computing $((k - m))_N$ in Eqs. (10.51) and (10.52). Equations (10.51) and (10.52) are the desired relations between the real and imaginary parts of the N-point DFT of a real sequence whose actual length is less than or equal to $(N/2) + 1$ (for N even). These equations are circular convolutions, and, for example, Eq. (10.51) can be evaluated efficiently by the following procedure:

1. Compute the inverse DFT of $X_R[k]$ to obtain the sequence

$$x_{ep}[n] = \frac{x[n] + x[((-n))_N]}{2}, \qquad 0 \le n \le N - 1. \tag{10.53}$$

2. Compute the periodic odd part of $x[n]$ by

$$x_{op}[n] = \begin{cases} x_{ep}[n], & 0 < n < N/2, \\ -x_{ep}[n], & N/2 < n \le N - 1, \\ 0, & \text{otherwise.} \end{cases} \tag{10.54}$$

3. Compute the DFT of $x_{op}[n]$ to obtain $jX_I[k]$.

Note that if, instead of computing the odd part of $x[n]$ in step 2, we compute

$$x[n] = \begin{cases} x_{ep}[0], & n = 0, \\ 2x_{ep}[n], & 0 < n < N/2, \\ x_{ep}[N/2], & n = N/2, \\ 0, & \text{otherwise,} \end{cases} \tag{10.55}$$

then the DFT of the resulting sequence would be $X[k]$, the complete DFT of $x[n]$.

10.3 RELATIONSHIPS BETWEEN MAGNITUDE AND PHASE

So far, we have focused on the relationships between the real and imaginary parts of the Fourier transform of a sequence. Often we are interested in relationships between the magnitude and phase of the Fourier transform. In this section we consider the conditions under which these functions might be uniquely related. While it might on the surface appear that a relationship between real and imaginary parts implies a relationship between magnitude and phase, that is not the case. This is clearly demonstrated by Example 5.10 in Section 5.4. The two system functions $H_1(z)$ and $H_2(z)$ in that example were assumed to correspond to causal, stable systems. Therefore the real and imaginary parts of $H_1(e^{j\omega})$ are related through the Hilbert transform relations of Eqs. (10.15) and (10.16), as are the real and imaginary parts of $H_2(e^{j\omega})$. However, $\angle H_1(e^{j\omega})$ could not be obtained from $|H_1(e^{j\omega})|$ since $H_1(e^{j\omega})$ and $H_2(e^{j\omega})$ have the same magnitude but different phase.

The Hilbert transform relationship between real and imaginary parts of the Fourier transform of a sequence $x[n]$ was based on causality of $x[n]$. We can obtain a Hilbert transform relationship between magnitude and phase by imposing causality on a sequence $\hat{x}[n]$ derived from $x[n]$ for which the Fourier transform $\hat{X}(e^{j\omega})$ is the logarithm of the Fourier transform of $x[n]$. Specifically, we define $\hat{x}[n]$ so that

$$x[n] \overset{\mathscr{F}}{\longleftrightarrow} X(e^{j\omega}) = |X(e^{j\omega})|e^{j\arg[X(e^{j\omega})]}, \qquad (10.56a)$$

$$\hat{x}[n] \overset{\mathscr{F}}{\longleftrightarrow} \hat{X}(e^{j\omega}), \qquad (10.56b)$$

where

$$\hat{X}(e^{j\omega}) = \log[X(e^{j\omega})] = \log|X(e^{j\omega})| + j \arg[X(e^{j\omega})]. \qquad (10.57)$$

The sequence $\hat{x}[n]$ is commonly referred to as the *complex cepstrum* of $x[n]$, for reasons to be developed in detail in Chapter 12. Also, Eq. (10.57) requires some careful interpretation since we are considering the logarithm of a complex-valued function. These issues will be discussed in detail in Chapter 12.

If we now require that $\hat{x}[n]$ be causal, then the real and imaginary parts of $\hat{X}(e^{j\omega})$, corresponding to $\log|X(e^{j\omega})|$ and $\arg[X(e^{j\omega})]$, will be related through Eqs. (10.15) and (10.16), i.e.,

$$\arg[X(e^{j\omega})] = -\frac{1}{2\pi} \mathscr{P} \int_{-\pi}^{\pi} \log|X(e^{j\theta})|\cot\left(\frac{\omega - \theta}{2}\right) d\theta \qquad (10.58)$$

and

$$\log|X(e^{j\omega})| = \hat{x}[0] + \frac{1}{2\pi} \mathscr{P} \int_{-\pi}^{\pi} \arg[X(e^{j\theta})]\cot\left(\frac{\omega - \theta}{2}\right) d\theta, \qquad (10.59a)$$

where in Eq. (10.59a), $\hat{x}[0]$ is equal to

$$\hat{x}[0] = \frac{1}{2\pi} \int_{-\pi}^{\pi} \log|X(e^{j\omega})| \, d\omega. \qquad (10.59b)$$

Although it is not at all obvious at this point, we will see in Chapter 12 that the minimum-phase condition defined in Section 5.6, namely that $X(z)$ have all its poles and zeros inside the unit circle, guarantees causality of the complex cepstrum. Thus the minimum-phase condition in Section 5.6 and the condition of causality of the complex cepstrum turn out to be the same constraint developed from different perspectives. Note that when $\hat{x}[n]$ is causal, $\arg[X(e^{j\omega})]$ is completely determined through Eq. (10.58) by $\log|X(e^{j\omega})|$; however, the complete determination of $\log|X(e^{j\omega})|$ by Eq. (10.59) requires both the phase $\arg[X(e^{j\omega})]$ and the quantity $\hat{x}[0]$. If $\hat{x}[0]$ is not known, then $\log|X(e^{j\omega})|$ is determined only to within an additive constant, or, equivalently, $|X(e^{j\omega})|$ is determined only to within a multiplicative (gain) constant.

Minimum phase or causality of the complex cepstrum are not the only constraints that provide a unique relationship between the magnitude and phase of the Fourier transform. As one example of another type of constraint, it has been shown (Hayes et al., 1980) that if a sequence is of finite length and if its z-transform has no zeros in conjugate reciprocal pairs, then to within a scale factor the sequence (and consequently also the Fourier transform magnitude) is uniquely determined by the phase of the Fourier transform.

10.4 HILBERT TRANSFORM RELATIONS FOR COMPLEX SEQUENCES

Thus far we have considered Hilbert transform relations for the Fourier transform of causal sequences and the discrete Fourier transform of periodic sequences that are "periodically causal" in the sense that they are zero in the second half of each period. In this section we consider *complex sequences* for which the real and imaginary components can be related through a discrete convolution similar to the Hilbert transform relations derived in the previous sections. These Hilbert transform relations are particularly useful in representing bandpass signals as complex signals in a manner completely analogous to the analytic signals of continuous-time signal theory.

As in the previous discussions, it is possible to base the derivation of the Hilbert transform relations on a notion of causality or one-sidedness. Since we are interested in relating the real and imaginary parts of a complex sequence, one-sidedness will be applied to the Fourier transform of the sequence. We cannot, of course, require that the Fourier transform be zero for $\omega < 0$ since it must be periodic. Instead, we consider sequences for which the Fourier transform is zero in the second half of each period; i.e., the z-transform is zero on the bottom half $(-\pi \le \omega < 0)$ of the unit circle. Thus, with $x[n]$ denoting the sequence and $X(e^{j\omega})$ its Fourier transform, we require that

$$X(e^{j\omega}) = 0, \qquad -\pi \le \omega < 0. \tag{10.60}$$

(We could just as well assume that $X(e^{j\omega})$ is zero for $0 < \omega \le \pi$.) The sequence $x[n]$ corresponding to $X(e^{j\omega})$ must be complex since if $x[n]$ were real, $X(e^{j\omega})$ would be conjugate symmetric, i.e., $X(e^{j\omega}) = X^*(e^{-j\omega})$. Therefore, we express $x[n]$ as

$$x[n] = x_r[n] + jx_i[n], \tag{10.61}$$

where $x_r[n]$ and $x_i[n]$ are real sequences. In continuous-time signal theory, the comparable signal is an analytic function and thus is called an *analytic signal*. Although analyticity has no formal meaning for sequences, we will nevertheless apply the same terminology to complex sequences whose Fourier transforms are one-sided.

If $X_r(e^{j\omega})$ and $X_i(e^{j\omega})$ denote the Fourier transforms of the real sequences $x_r[n]$ and $x_i[n]$, then

$$X(e^{j\omega}) = X_r(e^{j\omega}) + jX_i(e^{j\omega}), \tag{10.62a}$$

and it follows that

$$X_r(e^{j\omega}) = \frac{1}{2}[X(e^{j\omega}) + X^*(e^{-j\omega})] \tag{10.62b}$$

and

$$jX_i(e^{j\omega}) = \frac{1}{2}[X(e^{j\omega}) - X^*(e^{-j\omega})]. \tag{10.62c}$$

Note that Eq. (10.62c) gives an expression for $jX_i(e^{j\omega})$, which is the Fourier transform of the imaginary signal $jx_i[n]$. Also, note that $X_r(e^{j\omega})$ and $X_i(e^{j\omega})$ are both complex-valued functions in general, and the complex transforms $X_r(e^{j\omega})$ and $jX_i(e^{j\omega})$ play a role similar to that played in the previous sections by the even and odd parts, respectively, of causal sequences. However, $X_r(e^{j\omega})$ is conjugate symmetric, i.e., $X_r(e^{j\omega}) = X_r^*(e^{-j\omega})$. Similarly, $jX_i(e^{j\omega})$ is conjugate antisymmetric, i.e., $jX_i(e^{j\omega}) = -jX_i^*(e^{-j\omega})$.

Figure 10.5 depicts an example of a complex one-sided Fourier transform of a complex sequence $x[n] = x_r[n] + jx_i[n]$ and the corresponding two-sided transforms of the real sequences $x_r[n]$ and $x_i[n]$. This figure shows pictorially the cancellation implied by Eqs. (10.62).

If $X(e^{j\omega})$ is zero for $-\pi \le \omega < 0$, then there is no overlap between the nonzero portions of $X(e^{j\omega})$ and $X^*(e^{-j\omega})$. Thus $X(e^{j\omega})$ can be recovered from either $X_r(e^{j\omega})$ or $X_i(e^{j\omega})$. Since $X(e^{j\omega})$ is assumed to be zero at $\omega = \pm \pi$, $X(e^{j\omega})$ is totally recoverable from $jX_i(e^{j\omega})$. This is in contrast with the situation in Section 10.2, in which the causal sequence could be recovered from its odd part except at the endpoints.

In particular,

$$X(e^{j\omega}) = \begin{cases} 2X_r(e^{j\omega}), & 0 \le \omega < \pi, \\ 0, & -\pi \le \omega < 0, \end{cases} \tag{10.63}$$

and

$$X(e^{j\omega}) = \begin{cases} 2jX_i(e^{j\omega}), & 0 \le \omega < \pi, \\ 0, & -\pi \le \omega < 0. \end{cases} \tag{10.64}$$

Alternatively, we can relate $X_r(e^{j\omega})$ and $X_i(e^{j\omega})$ directly:

$$X_i(e^{j\omega}) = \begin{cases} -jX_r(e^{j\omega}), & 0 < \omega < \pi, \\ jX_r(e^{j\omega}), & -\pi \le \omega < 0, \end{cases} \tag{10.65}$$

or

$$X_i(e^{j\omega}) = H(e^{j\omega})X_r(e^{j\omega}), \tag{10.66a}$$

where

$$H(e^{j\omega}) = \begin{cases} -j, & 0 < \omega < \pi, \\ j, & -\pi < \omega < 0. \end{cases} \tag{10.66b}$$

Equations (10.66) are illustrated by comparing Figs. 10.5(c) and 10.5(d). Now $X_i(e^{j\omega})$ is the Fourier transform of $x_i[n]$, the imaginary part of $x[n]$, and $X_r(e^{j\omega})$ is the Fourier transform of $x_r[n]$, the real part of $x[n]$. Thus according to Eqs. (10.66), $x_i[n]$ can be obtained by processing $x_r[n]$ with a linear time-invariant discrete-time system with frequency response $H(e^{j\omega})$ as given by Eq. (10.66b). This frequency response has

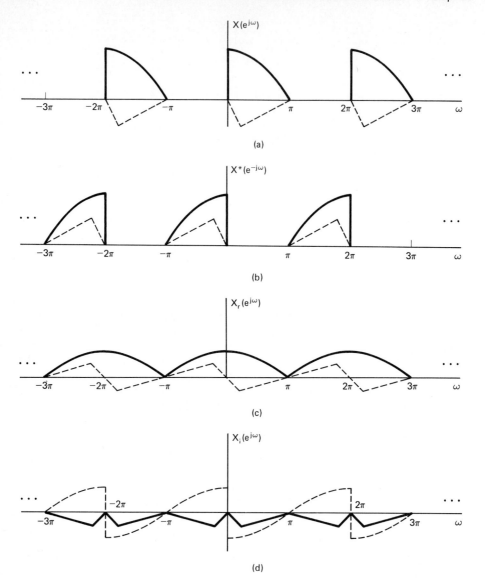

Figure 10.5 Illustration of decomposition of a one-sided Fourier transform. (Solid curves are real parts and dashed curves are imaginary parts.)

unity magnitude, a phase angle of $-\pi/2$ for $0 < \omega < \pi$, and a phase angle of $+\pi/2$ for $-\pi < \omega < 0$. Such a system is called an ideal 90-degree phase shifter. Alternatively, when it is clear that we are considering an operation on a sequence, the 90-degree phase shifter is also called a *Hilbert transformer*. From Eqs. (10.66) it follows that

$$X_r(e^{j\omega}) = \frac{1}{H(e^{j\omega})} X_i(e^{j\omega}) = -H(e^{j\omega})X_i(e^{j\omega}). \qquad (10.67)$$

Thus $-x_r[n]$ can also be obtained from $x_i[n]$ with a 90-degree phase shifter.

The impulse response $h[n]$ of a 90-degree phase shifter, corresponding to the frequency response $H(e^{j\omega})$ given in Eq. (10.66b), is

$$h[n] = \frac{1}{2\pi} \int_{-\pi}^{0} je^{j\omega n} d\omega - \frac{1}{2\pi} \int_{0}^{\pi} je^{j\omega n} d\omega$$

or

$$h[n] = \begin{cases} \dfrac{2}{\pi} \dfrac{\sin^2(\pi n/2)}{n}, & n \neq 0, \\[2ex] 0, & n = 0. \end{cases} \tag{10.68}$$

The impulse response is plotted in Fig. 10.6. Using Eqs. (10.66) and (10.67), we obtain the expressions

$$x_i[n] = \sum_{m=-\infty}^{\infty} h[n-m]x_r[m] \tag{10.69a}$$

and

$$x_r[n] = -\sum_{m=-\infty}^{\infty} h[n-m]x_i[m]. \tag{10.69b}$$

Equations (10.69) are the desired Hilbert transform relations between the real and imaginary parts of a discrete-time analytic signal.

Figure 10.7 shows how a discrete-time Hilbert transformer system can be used to form a complex analytic signal, which is simply a pair of real signals.

10.4.1 Design of Hilbert Transformers

The impulse response of the Hilbert transformer, as given in Eq. (10.68), is not absolutely summable. Consequently,

$$H(e^{j\omega}) = \sum_{n=-\infty}^{\infty} h[n]e^{-j\omega n} \tag{10.70}$$

converges to Eq. (10.66) only in the mean-square sense. Thus the ideal Hilbert transformer or 90-degree phase shifter takes its place alongside the ideal lowpass filter

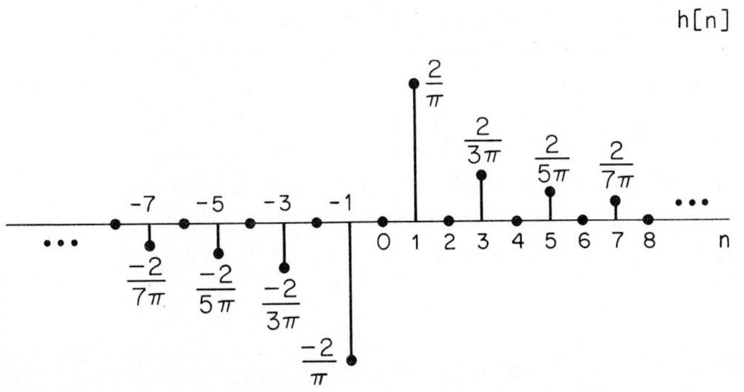

Figure 10.6 Impulse response of an ideal Hilbert transformer or 90-degree phase shifter.

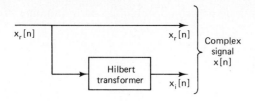

Figure 10.7 Block diagram representation of the creation of a complex sequence whose Fourier transform is one-sided.

and ideal bandlimited differentiator as a valuable theoretical concept that corresponds to a noncausal system and for which the system function exists only in a restricted sense.

Approximations to the ideal Hilbert transformer can, of course, be obtained. FIR approximations with constant group delay can be designed using either the window method or the equiripple approximation method. In such approximations the 90-degree phase shift is realized exactly, with an additional linear phase component required for a causal FIR system. The properties of such approximations are illustrated by examples of Hilbert transformers designed with Kaiser windows.

Example 10.3

The Kaiser window approximation for an FIR discrete Hilbert transformer of order M (length $M + 1$) would be of the form

$$
h[n] = \begin{cases} \dfrac{I_o\{\beta(1 - [(n - n_d)/n_d]^2)^{1/2}\}}{I_o(\beta)} \left\{ \dfrac{2}{\pi} \dfrac{\sin^2[\pi(n - n_d)/2]}{n - n_d} \right\}, & 0 \le n \le M, \\ 0, & \text{otherwise,} \end{cases}
$$

(10.71)

where $n_d = M/2$. If M is even, the system is a type III FIR generalized linear phase system, as discussed in Section 5.7.3.

Figure 10.8(a) shows the impulse response and Fig. 10.8(b) shows the magnitude of the frequency response for $M = 18$ and $\beta = 2.629$. Because $h[n]$ satisfies the symmetry condition $h[n] = -h[M - n]$ for $0 \le n \le M$, the phase is exactly 90° plus a linear component corresponding to a delay of $n_d = 18/2 = 9$ samples, i.e.,

$$
\sphericalangle H(e^{j\omega}) = \frac{-\pi}{2} - 9\omega, \qquad 0 < \omega < \pi.
$$

(10.72)

From Fig. 10.8(b) we see that, as required for a type III system, the frequency response is zero at $z = 1$ and $z = -1$ ($\omega = 0$ and $\omega = \pi$). Thus the magnitude response cannot approximate unity very well except in some middle band $\omega_L < |\omega| < \omega_H$.

If M is an odd integer, we obtain a type IV system as shown in Fig. 10.9, where $M = 17$ and $\beta = 2.44$. For type IV systems the frequency response need be zero only at $z = 1$ ($\omega = 0$). Therefore a better approximation to constant-magnitude response is obtained for frequencies around $\omega = \pi$. The phase response is exactly 90° at all frequencies plus a linear phase component corresponding to $n_d = 17/2 = 8.5$ samples delay, i.e.,

$$
\sphericalangle H(e^{j\omega}) = \frac{-\pi}{2} - 8.5\omega.
$$

(10.73)

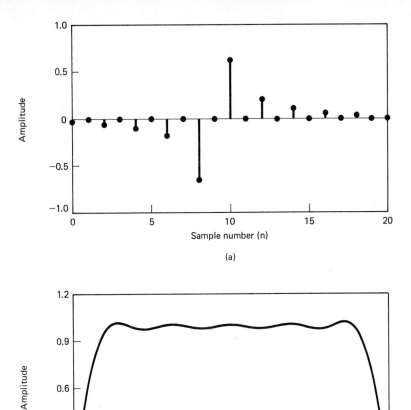

Figure 10.8   (a) Impulse response and (b) magnitude response of an FIR Hilbert transformer designed using the Kaiser window. ($M = 17$ and $\beta = 2.629$.)

From a comparison of Figs. 10.8(a) and 10.9(a) we see that type III FIR Hilbert transformers have significant computational advantage over type IV systems when it is not necessary to approximate constant magnitude at $\omega = \pi$. This is because for type III systems, the even-indexed samples of the impulse response are all exactly zero. Thus, taking advantage of the antisymmetry in both cases, the system with $M = 17$ would require eight multiplications to compute each output sample while the system with $M = 18$ would require only five multiplications per output sample.

Type III and IV FIR linear phase Hilbert transformer approximations with equiripple magnitude approximation and exactly 90° phase can be designed using the

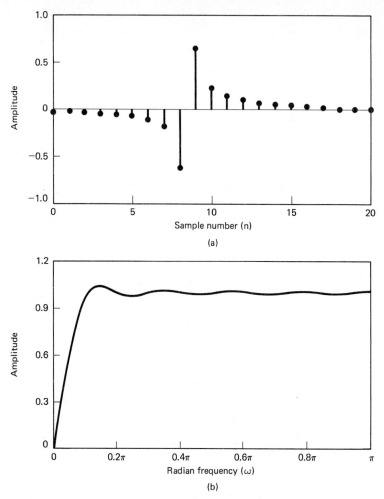

Figure 10.9 (a) Impulse response and (b) magnitude response of an FIR Hilbert transformer designed using the Kaiser window. ($M = 17$ and $\beta = 2.44$.)

Parks-McClellan algorithm as described in Section 7.7, with the expected improvements in magnitude approximation error over window-designed filters of the same length (Rabiner and Schafer, 1974).

The exactness of the phase of type III and IV FIR systems is a compelling motivation for their use in approximating Hilbert transformers. IIR systems must have some phase response error as well as magnitude response error in approximating a Hilbert transformer. The most successful approach for design of IIR Hilbert transformers is to design a "phase splitter," which consists of two allpass systems whose phase responses differ by approximately 90° over some portion of the band $0 < |\omega| < \pi$. Such systems can be designed by using the bilinear transformation to transform a continuous-time phase-splitting system to a discrete-time system. For an example of such a system, see Gold et al. (1970).

Figure 10.10 depicts a 90-degree phase-splitting system. If $x_r[n]$ denotes a real input signal and $x_i[n]$ its Hilbert transform, then the complex sequence $x[n] = x_r[n] + jx_i[n]$ has a Fourier transform that is identically zero for $-\pi \leq \omega < 0$, i.e., $X(z)$ is zero on the bottom half of the unit circle of the z-plane. In the system of Fig. 10.7, a Hilbert transformer was used to form the signal $x_i[n]$ from $x_r[n]$. In Fig. 10.10, we process $x_r[n]$ through two systems, $H_1(e^{j\omega})$ and $H_2(e^{j\omega})$. Now if $H_1(e^{j\omega})$ and $H_2(e^{j\omega})$ in Fig. 10.10 are allpass systems whose phase responses differ by 90°, then the complex signal $y[n] = y_r[n] + jy_i[n]$ has a Fourier transform that also vanishes for $-\pi \leq \omega < 0$. Furthermore, $|Y(e^{j\omega})| = |X(e^{j\omega})|$ since the phase-splitting systems are allpass. The phases of $Y(e^{j\omega})$ and $X(e^{j\omega})$ will differ by the common phase component of $H_1(e^{j\omega})$ and $H_2(e^{j\omega})$.

10.4.2 Representation of Bandpass Signals

Many of the applications of analytic signals concern narrowband communication signals. In such applications it is sometimes convenient to represent a bandpass signal in terms of a lowpass signal. To see how this may be done, consider the complex lowpass signal

$$x[n] = x_r[n] + jx_i[n],$$

where $x_i[n]$ is the Hilbert transform of $x_r[n]$ and

$$X(e^{j\omega}) = 0, \qquad -\pi \leq \omega < 0.$$

The Fourier transforms $X_r(e^{j\omega})$ and $jX_i(e^{j\omega})$ are depicted in Figs. 10.11(a) and 10.11(b), respectively, and the resulting transform $X(e^{j\omega}) = X_r(e^{j\omega}) + jX_i(e^{j\omega})$ is shown in Fig. 10.11(c). (Solid curves are real parts and dashed curves are imaginary parts.) Now consider the sequence

$$s[n] = x[n]e^{j\omega_c n} = s_r[n] + js_i[n], \tag{10.74}$$

where $s_r[n]$ and $s_i[n]$ are real sequences. The corresponding Fourier transform is

$$S(e^{j\omega}) = X(e^{j(\omega - \omega_c)}), \tag{10.75}$$

which is depicted in Fig. 10.11(d). Applying Eqs. (10.62) to $S(e^{j\omega})$ leads to the equations

$$S_r(e^{j\omega}) = \tfrac{1}{2}[S(e^{j\omega}) + S^*(e^{-j\omega})], \tag{10.76a}$$

$$jS_i(e^{j\omega}) = \tfrac{1}{2}[S(e^{j\omega}) - S^*(e^{-j\omega})]. \tag{10.76b}$$

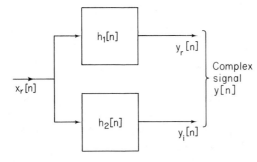

Figure 10.10 Block diagram representation of the allpass phase splitter method for creation of a complex sequence whose Fourier transform is one-sided.

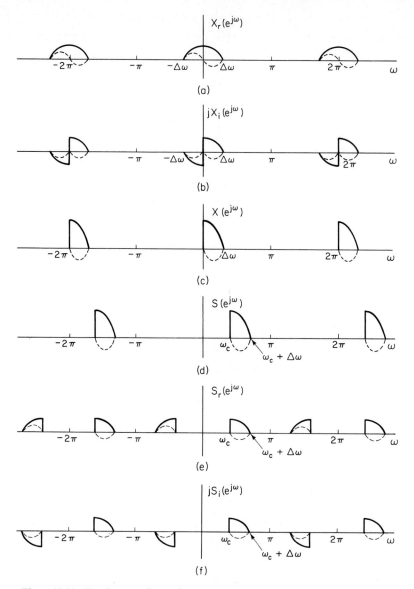

Figure 10.11 Fourier transforms for representation of bandpass signals. (Solid curves are real parts and dashed curves are imaginary parts.) (Note that in parts b and f the functions $jX_i(e^{j\omega})$ and $jS_i(e^{j\omega})$ are plotted, where $X_i(e^{j\omega})$ and $S_i(e^{j\omega})$ are the Fourier transforms of the Hilbert transforms of $x_i[n]$ and $s_i[n]$, respectively.)

For the example of Fig. 10.11, $S_r(e^{j\omega})$ and $jS_i(e^{j\omega})$ are shown in Figs. 10.11(e) and 10.11(f), respectively. It is straightforward to show that if $X_r(e^{j\omega}) = 0$ for $\Delta\omega < |\omega| \leq \pi$, and if $\omega_c + \Delta\omega < \pi$, then $S(e^{j\omega})$ will be a one-sided bandpass signal such that $S(e^{j\omega}) = 0$ *except* in the interval $\omega_c < \omega < \omega_c + \Delta\omega$. As the example of Fig. 10.11 illustrates, and as can be shown using Eqs. (10.61) and (10.62), $S_i(e^{j\omega}) = H(e^{j\omega})S_r(e^{j\omega})$, or, in other words, $s_i[n]$ is the Hilbert transform of $s_r[n]$.

An alternative representation of a complex signal is in terms of magnitude and phase; i.e., $x[n]$ can be expressed as

$$x[n] = A[n]e^{j\phi[n]}, \tag{10.77a}$$

where

$$A[n] = (x_r^2[n] + x_i^2[n])^{1/2} \tag{10.77b}$$

and

$$\phi[n] = \arctan\left(\frac{x_i[n]}{x_r[n]}\right). \tag{10.77c}$$

Therefore, from Eqs. (10.74) and (10.77) we can express $s[n]$ as

$$s[n] = (x_r[n] + jx_i[n])e^{j\omega_c n} \tag{10.78a}$$

$$= A[n]e^{j(\omega_c n + \phi[n])}, \tag{10.78b}$$

from which we obtain the expressions

$$s_r[n] = x_r[n]\cos \omega_c n - x_i[n]\sin \omega_c n \tag{10.79a}$$

or

$$s_r[n] = A[n]\cos(\omega_c n + \phi[n]) \tag{10.79b}$$

and

$$s_i[n] = x_r[n]\sin \omega_c n + x_i[n]\cos \omega_c n \tag{10.80a}$$

or

$$s_i[n] = A[n]\sin(\omega_c n + \phi[n]). \tag{10.80b}$$

Equations (10.79a) and (10.80a) are depicted in Figs. 10.12(a) and 10.12(b), respectively. These diagrams illustrate how a complex bandpass (single-sideband) signal can be formed from a real lowpass signal.

Taken together, Eqs. (10.79) and (10.80) are the desired time-domain representations of a general complex bandpass signal $s[n]$ in terms of the real and imaginary parts of a complex lowpass signal $x[n]$. Generally, this complex representation is a convenient mechanism for representing a real bandpass signal. For example, Eq. (10.79a) provides a time-domain representation for the real bandpass signal in terms of an "in-phase" component $x_r[n]$ and a "quadrature" (90-degree phase-shifted) component $x_i[n]$. Indeed, as illustrated in Fig. 10.11(e), Eq. (10.79a) permits the representation of real bandpass signals (or filter impulse responses) whose Fourier transforms are not conjugate symmetric about the center of the passband (as would be the case for signals of the form $x_r[n]\cos \omega_c n$).

It is clear from the form of Eqs. (10.79) and (10.80) and from Fig. 10.12 that a general bandpass signal has the form of a sinusoid that is both amplitude and phase modulated. The sequence $A[n]$ is called the envelope and $\phi[n]$ the phase. This narrowband signal representation can be used to represent a variety of amplitude and

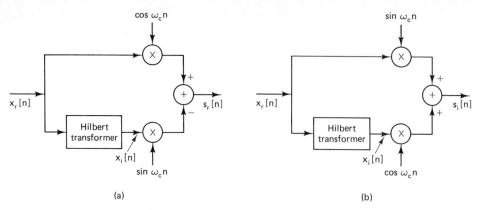

Figure 10.12   Block diagram representation of Eqs. (10.79a) and (10.80a) for obtaining a single-sideband signal.

phase modulation systems. The example of Fig. 10.11 is an illustration of single-sideband modulation. If we consider the real signal $s_r[n]$ as resulting from single-sideband modulation with the lowpass real signal $x_r[n]$ as the input, then Fig. 10.12(a) represents a scheme for implementing the single-sideband modulation system. Single-sideband modulation systems are useful in frequency-division multiplexing since they can represent a real bandpass signal with minimum bandwidth.

10.4.3 Bandpass Sampling

Another important use of analytic signals is in the sampling of bandpass signals. In Chapter 3 we saw that in general if a continuous-time signal has a bandlimited Fourier transform such that $S_c(j\Omega) = 0$ for $|\Omega| \geq \Omega_N$, then the signal is exactly represented by its samples if the sampling rate satisfies the inequality $2\pi/T \geq 2\Omega_N$. The key to the proof of this result is to avoid overlapping the replicas of $S_c(j\Omega)$ that form the discrete-time Fourier transform of the sequence of samples. A bandpass continuous-time signal has a Fourier transform such that $S_c(j\Omega) = 0$ for $0 \leq |\Omega| \leq \Omega_c$ and for $|\Omega| \geq \Omega_c + \Delta\Omega$. Thus its bandwidth, or region of support, is really only $2\Delta\Omega$ rather than $2(\Omega_c + \Delta\Omega)$, and with a proper sampling strategy the region $-\Omega_c \leq \Omega \leq \Omega_c$ can be filled with images of the nonzero part of $S_c(j\Omega)$ without overlapping. This is greatly facilitated by using a complex representation of the bandpass signal.

As an illustration, consider the system of Fig. 10.13 and the example signal of Fig. 10.14(a). The highest frequency of the input signal is $\Omega_c + \Delta\Omega$, as shown in Fig. 10.14(a). If this signal is sampled at exactly the Nyquist rate, $2\pi/T = 2(\Omega_c + \Delta\Omega)$, then the resulting sequence of samples, $s_r[n] = s_c(nT)$, has the Fourier transform $S_r(e^{j\omega})$ plotted in Fig. 10.14(b). Using a discrete-time Hilbert transformer, we can form the complex sequence $s[n] = s_r[n] + js_i[n]$ whose Fourier transform is $S(e^{j\omega})$ in Fig. 10.14(c). The width of the nonzero region of $S(e^{j\omega})$ is $\Delta\omega = (\Delta\Omega)T$. Defining M as the largest integer less than or equal to $2\pi/\Delta\omega$, M copies of $S(e^{j\omega})$ would fit into the interval $-\pi < \omega < \pi$. (In the example of Fig. 10.14c, $2\pi/\Delta\omega = 5$.) Thus the sampling rate of $s[n]$ can be reduced by decimation as shown in Fig. 10.13, thereby obtaining

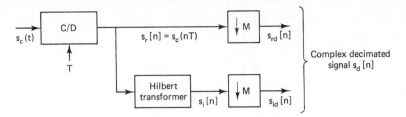

Figure 10.13 System for reduced-rate sampling of a real bandpass signal by decimation of the equivalent complex bandpass signal.

the reduced-rate complex sequence $s_d[n] = s_{rd}[n] + js_{id}[n] = s[Mn]$ whose Fourier transform is

$$S_d(e^{j\omega}) = \frac{1}{M} \sum_{k=0}^{M-1} S(e^{j[(\omega - 2\pi k)/M]}). \tag{10.81}$$

Figure 10.14(d) shows $S_d(e^{j\omega})$ with $M = 5$ in Eq. (10.81). $S(e^{j\omega})$ and two of the frequency-scaled and translated copies of $S(e^{j\omega})$ are indicated explicitly in Fig. 10.14(d). It is clear that aliasing has been avoided and that all the information necessary to reconstruct the original sampled real bandpass signal now resides in the discrete-time frequency interval $-\pi < \omega \leq \pi$. A complex filter applied to $s_d[n]$ can transform this information in useful ways such as further bandlimiting, amplitude or phase compensation, etc., or the complex signal can be coded for transmission or digital storage. This processing takes place at the low sampling rate, and this is, of course, the motivation for reducing the sampling rate.

The original real bandpass signal $s_r[n]$ can be reconstructed ideally by the following procedure:

1. Expand the complex sequence by a factor M; i.e., obtain

$$s_e[n] = \begin{cases} s_{rd}[n/M] + js_{id}[n/M], & n = 0,\ \pm M,\ \pm 2M, \ldots, \\ 0, & \text{otherwise} \end{cases} \tag{10.82}$$

2. Filter the signal $s_e[n]$ using an ideal *complex* bandpass filter with impulse response $h_i[n]$ and frequency response

$$H_i(e^{j\omega}) = \begin{cases} 0, & -\pi < \omega < \omega_c, \\ M, & \omega_c < \omega < \omega_c + \Delta\omega, \\ 0, & \omega_c + \Delta\omega < \omega < \pi. \end{cases} \tag{10.83}$$

(In our example, $\omega_c + \Delta\omega = \pi$.)

3. Obtain $s_r[n] = \mathcal{R}e\{s_e[n] * h_i[n]\}$.

A useful exercise is to plot the Fourier transform $S_u(e^{j\omega})$ for the example of Fig. 10.14 and verify that the filter of Eq. (10.83) does indeed recover $s[n]$.

Another useful exercise is to consider a complex continuous-time signal with a one-sided Fourier transform equal to $S_c(j\Omega)$ for $\Omega \geq 0$. It can be shown that such a signal can be sampled with sampling rate $2\pi/T = \Delta\Omega$, thereby obtaining directly the complex sequence $s_d[n]$.

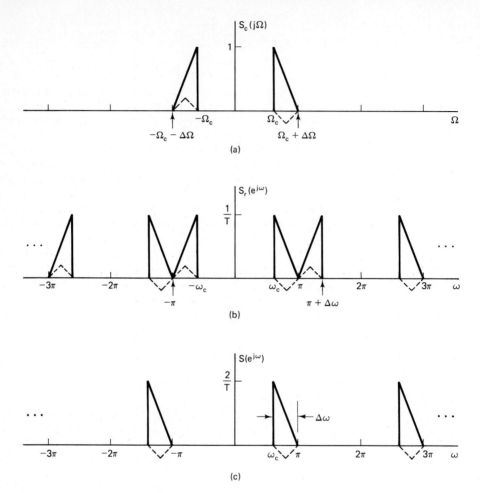

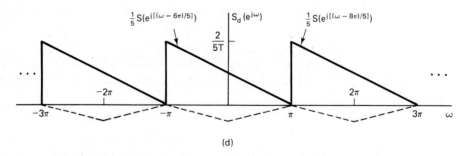

Figure 10.14 Example of reduced-rate sampling of a bandpass signal using the system of Fig. 10.13. (a) Fourier transform of continuous-time bandpass signal. (b) Fourier transform of sampled signal. (c) Fourier transform of complex bandpass discrete-time signal derived from the signal of part (a). (d) Fourier transform of decimated complex bandpass signal of part (c). (Solid curves are real parts and dashed curves are imaginary parts.)

10.5 SUMMARY

In this chapter we have discussed a variety of relations between the real and imaginary parts of Fourier transforms and the real and imaginary parts of complex sequences. These relationships are collectively referred to as *Hilbert transform relationships*.

Our approach to deriving all the Hilbert transform relations was to apply a basic causality principle that allows a sequence or function to be recovered from its even part. We showed that for a causal sequence, the real and imaginary parts of the Fourier transform are related through a convolution-type integral. Also, for the special case when the complex cepstrum of a sequence is causal or, equivalently, both the poles and zeros of its z-transform lie inside the unit circle (the minimum-phase condition), we stated that the logarithm of the magnitude and the phase of the Fourier transform are a Hilbert transform pair of each other. This result is explored in more detail in Chapter 12.

Hilbert transform relations were derived for periodic sequences that satisfy a modified causality constraint and for complex sequences whose Fourier transforms vanish on the bottom half of the unit circle. Applications of complex analytic signals to the representation and efficient sampling of bandpass signal were also discussed.

PROBLEMS

10.1. Let $x[n]$ be a real causal sequence for which $|x[n]| < \infty$. The z-transform of $x[n]$ is

$$X(z) = \sum_{n=0}^{\infty} x[n]z^{-n},$$

which is a Taylor's series in the variable z^{-1} and therefore converges to an analytic function everywhere outside some circular disc centered at $z = 0$. (The region of convergence includes the point $z = \infty$ and, in fact, $X(\infty) = x[0]$.) The statement that $X(z)$ is analytic (in its region of convergence) implies strong constraints on the function $X(z)$ (see Churchill and Brown, 1984). Specifically, its real and imaginary parts each satisfy Laplace's equation, and the real and imaginary parts are related by the Cauchy-Riemann equations. We will use these properties to determine $X(z)$ from its real part when $x[n]$ is a real, finite-valued, causal sequence.

Let the z-transform of such a sequence be

$$X(z) = X_R(z) + jX_I(z),$$

where $X_R(z)$ and $X_I(z)$ are real-valued functions of z. Suppose that $X_R(z)$ is

$$X_R(\rho e^{j\omega}) = \frac{\rho + \alpha \cos \omega}{\rho}, \quad \alpha \text{ real},$$

for $z = \rho e^{j\omega}$. Then find $X(z)$ (as an explicit function of z) assuming that $X(z)$ is analytic everywhere except at $z = 0$. Do this using both of the following methods.

(a) *Method 1, Frequency Domain:* Use the fact that the real and imaginary parts of $X(z)$ must satisfy the Cauchy-Riemann equations everywhere that $X(z)$ is analytic. The Cauchy-Riemann equations are the following:

1. In Cartesian coordinates:

$$\frac{\partial U}{\partial x} = \frac{\partial V}{\partial y}, \qquad \frac{\partial V}{\partial x} = -\frac{\partial U}{\partial y},$$

where $z = x + jy$ and $X(x + jy) = U(x, y) + jV(x, y)$.

2. In polar coordinates:

$$\frac{\partial U}{\partial \rho} = \frac{1}{\rho}\frac{\partial V}{\partial \omega}, \qquad \frac{\partial V}{\partial \rho} = -\frac{1}{\rho}\frac{\partial U}{\partial \omega},$$

where $z = \rho e^{j\omega}$ and $X(\rho e^{j\omega}) = U(\rho, \omega) + jV(\rho, \omega)$.

Since we know that $U = X_R$, we can integrate these equations to find $V = X_I$ and hence X. (Be careful to treat the constant of integration properly.)

(b) *Method 2, Time Domain:* The sequence $x[n]$ can be represented as $x[n] = x_e[n] + x_o[n]$, where $x_e[n]$ is real and even with Fourier transform $X_R(e^{j\omega})$, and the sequence $x_o[n]$ is real and odd with Fourier transform $jX_I(e^{j\omega})$. Find $x_e[n]$ and, using causality, find $x_o[n]$ and hence $x[n]$ and $X(z)$.

10.2. Consider a sequence $x[n]$ with discrete-time Fourier transform $X(e^{j\omega})$. The sequence $x[n]$ is real-valued and causal and

$$\mathcal{R}e\{X(e^{j\omega})\} = 2 - 2a \cos \omega.$$

Determine $\mathcal{I}m\{X(e^{j\omega})\}$.

10.3. This problem is concerned with time- and frequency-domain constraints.

(a) Consider a sequence $x[n]$ and its discrete-time Fourier transform $X(e^{j\omega})$. The following is known:

$$x[n] \text{ is real and causal,}$$

$$\mathcal{R}e\{X(e^{j\omega})\} = \tfrac{5}{4} - \cos \omega.$$

Determine a sequence $x[n]$ consistent with the given information.

(b) Consider a sequence $x[n]$ and its discrete-time Fourier transform $X(e^{j\omega})$. The following is known:

$$x[n] \text{ is real,}$$

$$x[0] = 0,$$

$$x[1] > 0,$$

$$|X(e^{j\omega})|^2 = \tfrac{5}{4} - \cos \omega.$$

Determine two distinct sequences $x_1[n]$ and $x_2[n]$ consistent with the given information.

10.4. Derive Eq. (10.18) by starting with Eq. (10.4) and applying the complex convolution theorem of Section 4.6.

10.5. Derive an integral expression for $H(z)$ *inside* the unit circle in terms of $\mathcal{R}e\{H(e^{j\omega})\}$, when $h[n]$ is a real, stable sequence such that $h[n] = 0$ for $n > 0$.

10.6. In Section 10.1, we observed that the z-transform is completely determined outside the unit circle by the value of its imaginary part on the unit circle and the value of $x[0]$.

(a) Beginning with Eq. (10.5), derive Eq. (10.19) by applying the complex convolution theorem of Section 4.6.

(b) Use Eq. (10.19) and contour integration techniques to find $X(z)$ when $x[0] = 1$ and

$$X_I(e^{j\omega}) = \frac{-\alpha \sin \omega}{1 + \alpha^2 - 2\alpha \cos \omega}.$$

10.7. Show that the sequence of discrete Fourier series coefficients for the sequence

$$\tilde{u}_N[n] = \begin{cases} 1, & n = 0, \ N/2, \\ 2, & n = 1, 2, \ldots, N/2 - 1, \\ 0, & n = N/2 + 1, \ldots, N - 1, \end{cases}$$

is

$$\tilde{U}_N[k] = \begin{cases} N, & k = 0, \\ -j2 \cot(\pi k/N), & k \text{ odd}, \\ 0, & k \text{ even}. \end{cases}$$

[*Hint*: Find the z-transform of the sequence

$$u_N[n] = 2u[n] - 2u[n - N/2] - \delta[n] + \delta[n - N/2]$$

and sample it to obtain $\tilde{U}[k]$.]

10.8. Consider a real-valued finite-duration sequence $x[n]$ of length N. Specifically, $x[n] = 0$ for $n < 0$ and $n > N - 1$, where N is an *odd* integer. Let $X[k]$ denote the M-point DFT of $x[n]$ with $M \geq N$; i.e.,

$$X[k] = \sum_{n=0}^{N-1} x[n]e^{-j(2\pi/M)kn}, \qquad k = 0, 1, \ldots, M - 1.$$

The real part of $X[k]$ is denoted $X_R[k]$.

(a) Determine, in terms of N, the smallest value of M that will permit $X[k]$ to be uniquely determined from $X_R[k]$.

(b) With M satisfying the condition determined in part (a), $X[k]$ can be expressed as the circular convolution of $X_R[k]$ with a sequence $U_M[k]$. Determine $U_M[k]$.

10.9. Consider a complex sequence $x[n] = x_r[n] + jx_i[n]$, where $x_r[n]$ and $x_i[n]$ are the real part and imaginary part, respectively. The z-transform, $X(z)$, of the sequence $x[n]$ is zero on the bottom half of the unit circle; i.e., $X(e^{j\omega}) = 0$ for $\pi \leq \omega \leq 2\pi$. The real part of $x[n]$ is

$$x_r[n] = \begin{cases} 1/2, & n = 0, \\ -1/4, & n = \pm 2, \\ 0, & \text{otherwise}. \end{cases}$$

Determine the real and imaginary parts of $X(e^{j\omega})$.

10.10. Consider a complex sequence $h[n] = h_r[n] + jh_i[n]$, where $h_r[n]$ and $h_i[n]$ are both real sequences, and let $H(e^{j\omega}) = H_R(e^{j\omega}) + jH_I(e^{j\omega})$ denote the Fourier transform of $h[n]$, where $H_R(e^{j\omega})$ and $H_I(e^{j\omega})$ are the real and imaginary parts, respectively, of $H(e^{j\omega})$.

Let $H_{ER}(e^{j\omega})$ and $H_{OR}(e^{j\omega})$ denote the even and odd parts, respectively, of $H_R(e^{j\omega})$, and let $H_{EI}(e^{j\omega})$, and $H_{OI}(e^{j\omega})$ denote the even and odd parts, respectively, of $H_I(e^{j\omega})$. Furthermore, let $H_A(e^{j\omega})$ and $H_B(e^{j\omega})$ denote the real and imaginary parts of the Fourier transform of $h_r[n]$, and let $H_C(e^{j\omega})$ and $H_D(e^{j\omega})$ denote the real and imaginary parts of the Fourier transform of $h_i[n]$. Express $H_A(e^{j\omega})$, $H_B(e^{j\omega})$, $H_C(e^{j\omega})$, and $H_D(e^{j\omega})$, in terms of $H_{ER}(e^{j\omega})$, $H_{OR}(e^{j\omega})$, $H_{EI}(e^{j\omega})$, and $H_{OI}(e^{j\omega})$.

10.11. Let $\mathcal{H}\{\cdot\}$ denote the (ideal) operation of Hilbert transformation:

$$\mathcal{H}\{x[n]\} = \sum_{k=-\infty}^{\infty} x[k]h[n-k],$$

where $h[n]$ is

$$h[n] = \begin{cases} \dfrac{2\sin^2(\pi n/2)}{\pi n}, & n \neq 0, \\ 0, & n = 0. \end{cases}$$

Prove the following properties of the ideal Hilbert transform operator.
(a) $\mathcal{H}\{\mathcal{H}\{x[n]\}\} = -x[n]$

(b) $\displaystyle\sum_{n=-\infty}^{\infty} x[n]\mathcal{H}\{x[n]\} = 0$ [*Hint:* Use Parseval's theorem.]

(c) $\mathcal{H}\{x[n]*y[n]\} = \mathcal{H}\{x[n]\}*y[n] = x[n]*\mathcal{H}\{y[n]\}$, where $x[n]$ and $y[n]$ are any sequences

10.12. Find the Hilbert transforms $x_i[n] = \mathcal{H}\{x_r[n]\}$ of the following sequences:
(a) $x_r[n] = \cos \omega_0 n$
(b) $x_r[n] = \sin \omega_0 n$
(c) $x_r[n] = \dfrac{\sin(\omega_c n)}{\pi n}$

10.13. An ideal Hilbert transformer with impulse response

$$h[n] = \begin{cases} \dfrac{2\sin^2(\pi n/2)}{\pi n}, & n \neq 0, \\ 0, & n = 0, \end{cases}$$

has input $x_r[n]$ and output $x_i[n] = x_r[n] * h[n]$, where $x_r[n]$ is a discrete-time random signal.
(a) Find an expression for the autocorrelation sequence $\phi_{x_i x_i}[m]$ in terms of $h[n]$ and $\phi_{x_r x_r}[m]$.
(b) Find an expression for the cross-correlation sequence $\phi_{x_r x_i}[m]$. Show that in this case $\phi_{x_r x_i}[m]$ is an odd function of m.
(c) Find an expression for the autocorrelation function of the complex analytic signal $x[n] = x_r[n] + jx_i[n]$.
(d) Determine the power spectrum $P_{xx}(\omega)$ for the complex signal in part (c).

10.14. The ideal Hilbert transformer (90-degree phase shifter) has frequency response (over one period)

$$H(e^{j\omega}) = \begin{cases} -j, & \omega > 0, \\ j, & \omega < 0. \end{cases}$$

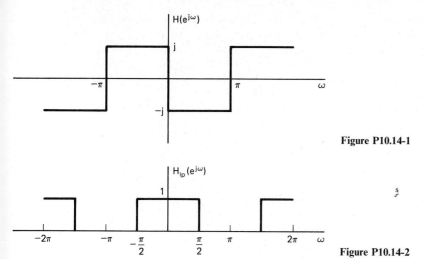

Figure P10.14-1

Figure P10.14-2

Figure P10.14-1 shows $H(e^{j\omega})$, and Fig. P10.14-2 shows the frequency response of an ideal lowpass filter $H_{lp}(e^{j\omega})$ with cutoff frequency $\omega_c = \pi/2$. These frequency responses are clearly similar, each having discontinuities separated by π.

(a) Obtain a relationship that expresses $H(e^{j\omega})$ in terms of $H_{lp}(e^{j\omega})$. Solve this equation for $H_{lp}(e^{j\omega})$ in terms of $H(e^{j\omega})$.

(b) Use the relationships in part (a) to obtain expressions for $h[n]$ in terms of $h_{lp}[n]$ and for $h_{lp}[n]$ in terms of $h[n]$.

The relationships obtained in parts (a) and (b) were based on definitions of ideal systems with zero phase. However, similar relationships hold for nonideal systems with generalized linear phase.

(c) Use the results of part (b) to obtain a relationship between the impulse response of a causal FIR approximation to the Hilbert transformer and the impulse response of a causal FIR approximation to the lowpass filter, both of which are designed by (1) incorporating appropriate linear phase, (2) determining the corresponding ideal impulse response, and (3) multiplying by the same window of length $(M + 1)$ samples, i.e., by the window method discussed in Chapter 7. (If necessary, consider the cases of M even and M odd separately.)

(d) For the Hilbert transformer approximations of Example 10.3, sketch the magnitude of the frequency responses of the corresponding lowpass filters.

10.15. In Section 10.4.3, we discussed an efficient scheme for sampling a bandpass continuous-time signal with Fourier transform such that

$$S_c(j\Omega) = 0 \qquad \text{for} \quad |\Omega| \leq \Omega_c \quad \text{and} \quad |\Omega| \geq \Omega_c + \Delta\Omega.$$

In that discussion, it was assumed that the signal was initially sampled with sampling frequency $2\pi/T = 2(\Omega_c + \Delta\Omega)$, i.e., at the lowest possible frequency that avoids aliasing. The bandpass sampling scheme is depicted in Fig. 10.13. After forming a complex bandpass discrete-time signal $s[n]$ with one-sided Fourier transform $S(e^{j\omega})$, the complex signal is decimated by a factor M, which is assumed to be the largest integer less than or equal to $2\pi/(\Delta\Omega T)$.

(a) By carrying through an example such as the one depicted in Fig. 10.14, show that if the quantity $2\pi/(\Delta\Omega T)$ is not an integer for the initial sampling rate chosen, then the resulting decimated signal $s_d[n]$ will have regions of nonzero length where its Fourier transform $S_d(e^{j\omega})$ is identically zero.

(b) How should the initial sampling frequency $2\pi/T$ be chosen so that a decimation factor M can be found such that the decimated sequence $s_d[n]$ in the system of Fig. 10.13 will have a Fourier transform $S_d(e^{j\omega})$ that is not aliased yet has no regions where it is zero over an interval of nonzero length?

10.16. In Section 10.4.3, we discussed an efficient scheme for sampling a bandpass continuous-time signal with Fourier transform such that

$$S_c(j\Omega) = 0 \quad \text{for} \quad |\Omega| \le \Omega_c \quad \text{and} \quad |\Omega| \ge \Omega_c + \Delta\Omega.$$

The bandpass sampling scheme is depicted in Fig. 10.13. At the end of Section 10.4.3, a scheme for reconstructing the original sampled signal $s_r[n]$ was given. The original continuous-time signal $s_c(t)$ in Fig. 10.13 can, of course, be reconstructed from $s_r[n]$ by ideal bandlimited interpolation (ideal D/C conversion). Figure P10.16-1 shows a block diagram of the system for reconstructing a real continuous-time bandpass signal from a decimated complex signal. The complex bandpass filter $H_i(e^{j\omega})$ in Fig. P10.16-1 has frequency response given by Eq. (10.83).

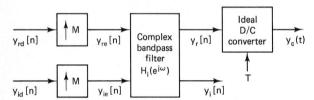

Figure P10.16-1

(a) Using the example depicted in Fig. 10.14, show that the system of Fig. P10.16-1 will reconstruct the original real bandpass signal (i.e., $y_c(t) = s_c(t)$) if the inputs to the reconstruction system are $y_{rd}[n] = s_{rd}[n]$ and $y_{id}[n] = s_{id}[n]$.

(b) Determine the impulse response $h_i[n] = h_{ri}[n] + jh_{ii}[n]$ of the complex bandpass filter in Fig. P10.16-1.

(c) Draw a more detailed block diagram of the system of Fig. P10.16-1 in which only real operations are shown. Eliminate any parts of the diagram that are not necessary to compute the final output.

(d) Now consider placing a complex linear time-invariant system between the system of Fig. 10.13 and the system of Fig. P10.16-1. This is depicted in Fig. P10.16-2, where the frequency response of the system is denoted $H(e^{j\omega})$. Determine how $H(e^{j\omega})$ should be chosen if it is desired that

$$Y_c(j\Omega) = H_{eff}(j\Omega)S_c(j\Omega),$$

where

$$H_{eff}(j\Omega) = \begin{cases} 1, & \Omega_c < |\Omega| < \Omega_c + \Delta\Omega/2, \\ 0, & \text{otherwise.} \end{cases}$$

$s_{rd}[n]$ → Complex LTI system $H(e^{j\omega})$ → $y_{rd}[n]$

$s_{id}[n]$ → → $y_{id}[n]$ **Figure P10.16-2**

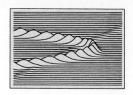

Fourier Analysis of Signals Using the Discrete Fourier Transform

11

11.0 INTRODUCTION

In Chapter 8, we developed the discrete Fourier transform (DFT) as a Fourier representation of finite-length signals. Because the DFT can be explicitly computed, it plays a central role in a wide variety of signal processing applications, including filtering and spectral analysis. In this chapter, we take an introductory look at Fourier analysis of signals using the DFT. In Chapter 12 we will discuss the use of the DFT in the implementation of a class of nonlinear filtering techniques referred to as homomorphic signal processing.

In applications and algorithms based on explicit evaluation of the Fourier transform, it is ideally the discrete-time Fourier transform that is desired, while it is the DFT that can actually be computed. For finite-length signals, the DFT provides frequency-domain samples of the discrete-time Fourier transform, and the implications of this sampling must be clearly understood and accounted for. For example, as considered in Section 8.8, in linear filtering or convolution implemented by multiplying DFTs rather than Fourier transforms, a circular convolution is implemented and special care must be taken to ensure that the results will be equivalent to a linear convolution. In addition, in many filtering and spectral analysis applications, the signals do not inherently have finite length. As we will discuss, this inconsistency between the finite-length requirement of the DFT and the reality of indefinitely long signals can be accommodated exactly or approximately through the concepts of *windowing*, *block processing*, and the *time-dependent Fourier transform*.

11.1 FOURIER ANALYSIS OF SIGNALS USING THE DFT

One of the major applications of the DFT is in analyzing the frequency content of continuous-time signals. For example, as we describe in Section 11.5.1, in speech analysis and processing, frequency analysis of speech signals is particularly useful in identifying and modeling the resonances of the vocal cavity. Another example, introduced in Section 11.5.2, is Doppler radar systems, where the velocity of a target is represented by the frequency shift between the transmitted and received signal.

The basic steps in applying the DFT to continuous-time signals are indicated in Fig. 11.1. The anti-aliasing filter is incorporated to eliminate or minimize the effect of aliasing when the continuous-time signal is converted to a sequence. The need for multiplication of $x[n]$ by $w[n]$, i.e., windowing, is a consequence of the finite-length requirement of the DFT. In many cases of practical interest, $s_c(t)$ and, consequently, $x[n]$ are very long or even indefinitely long (such as with speech or music). Therefore, a finite-duration window $w[n]$ is applied to $x[n]$ prior to computation of the DFT. Figure 11.2 illustrates the Fourier transforms of the signals in Fig. 11.1. Specifically, Fig. 11.2(a) illustrates a continuous-time spectrum that tapers off at high frequencies but is not bandlimited. It also indicates the presence of some narrowband signal energy as represented by the narrow, sharp peaks. The frequency response of an anti-aliasing filter is illustrated in Fig. 11.2(b). As indicated in Fig. 11.2(c), the resulting continuous-time Fourier transform $X_c(j\Omega)$ contains little useful information about $S_c(j\Omega)$ for frequencies above the cutoff frequency of the filter. Since $H_{aa}(j\Omega)$ cannot be ideal, the Fourier components of the input in the passband and the transition band also will be modified by the frequency response of the filter.

The conversion of $x_c(t)$ to the sequence of samples $x[n]$ is represented in the frequency domain by periodic replication and frequency normalization, i.e.,

$$X(e^{j\omega}) = \frac{1}{T} \sum_{r=-\infty}^{\infty} X_c\left(j\frac{\omega}{T} + j\frac{2\pi r}{T}\right), \tag{11.1}$$

as illustrated in Fig. 11.2(d). Since in a practical implementation the anti-aliasing filter cannot have infinite attenuation in the stopband, some nonzero overlap of the terms in Eq. (11.1), i.e., aliasing, can be expected; however, this source of error can be made negligibly small either with a high-quality continuous-time filter or through the use of initial oversampling followed by more effective lowpass filtering and decimation, as discussed in Section 3.7.5. If $x[n]$ is a digital signal so that A/D conversion is incorporated in the second system in Fig. 11.1, then quantization error is also introduced. As we have seen in Chapter 3, this error can be modeled as a noise

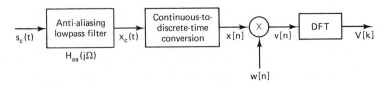

Figure 11.1 Processing steps in the discrete-time Fourier analysis of a continuous-time signal.

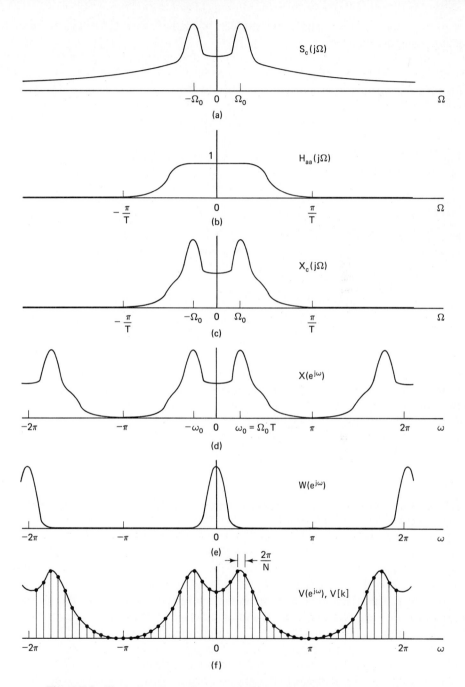

Figure 11.2   Illustration of the Fourier transforms in the system of Fig. 11.1. (a) Fourier transform of continuous-time input signal. (b) Frequency response of anti-aliasing filter. (c) Fourier transform of output of anti-aliasing filter. (d) Fourier transform of sampled signal. (e) Fourier transform of window sequence. (f) Fourier transform of windowed signal segment and frequency samples obtained using DFT samples.

sequence added to $x[n]$. This noise also can be made negligible through the use of fine-grained quantization.

As indicated, the sequence $x[n]$ is typically multiplied by a finite-duration window $w[n]$ since the input to the DFT must be of finite duration. The effect in the frequency domain is a periodic convolution, i.e.,

$$V(e^{j\omega}) = \frac{1}{2\pi} \int_{-\pi}^{\pi} X(e^{j\theta}) W(e^{j(\omega-\theta)}) \, d\theta. \tag{11.2}$$

Figure 11.2(e) illustrates the Fourier transform of a typical window sequence. Note that the mainlobe is assumed to be concentrated around $\omega = 0$. If $w[n]$ is constant over the range of n for which it is nonzero, it is referred to as a *rectangular window*. However, as we will see, there are good reasons to taper the window at its edges. The properties of windows such as the Bartlett, Hamming, Hanning, Blackman, and Kaiser windows are discussed in Chapter 7 and in Section 11.2. At this point it is sufficient to observe that convolution of $W(e^{j\omega})$ with $X(e^{j\omega})$ will tend to smooth sharp peaks and discontinuities in $X(e^{j\omega})$. This is depicted by the continuous curve plotted in Fig. 11.2(f).

The final operation in Fig. 11.1 is the DFT. The DFT of the windowed sequence is

$$V[k] = \sum_{n=0}^{N-1} v[n] e^{-j(2\pi/N)kn}, \qquad k = 0, 1, \ldots, N-1, \tag{11.3}$$

where we assume that the window length L is less than or equal to the DFT length N. $V[k]$, the DFT of the finite-length sequence $v[n]$, corresponds to equally spaced samples of the Fourier transform of $v[n]$, i.e.,

$$V[k] = V(e^{j\omega}) \Big|_{\omega = 2\pi k/N}. \tag{11.4}$$

Figure 11.2(f) also shows $V[k]$ as the samples of $V(e^{j\omega})$. Since the spacing between DFT frequencies is $2\pi/N$ and the relationship between the normalized discrete-time frequency variable and the continuous-time frequency variable is $\omega = \Omega T$, the DFT frequencies correspond to the continuous-time frequencies Ω_k given by

$$\Omega_k = \frac{2\pi k}{NT}. \tag{11.5}$$

Many commercial real-time spectrum analyzers are based on the principles embodied in Figs. 11.1 and 11.2. It should be clear from the preceding discussion, however, that many factors affect the interpretation of the continuous-time Fourier transform of the input in terms of the DFT of a windowed segment of the sampled signal. To accommodate these factors, care must be taken in filtering and sampling the input signal. Furthermore, to interpret the results correctly, the effects of the time-domain windowing and of the frequency-domain sampling inherent in the DFT must be clearly understood. For the remainder of the discussion, we will assume that the issues of anti-aliasing filtering and continuous-to-discrete-time conversion have been satisfactorily handled and are negligible. In the next section we concentrate specifically on the effects of windowing and of the frequency-domain sampling imposed by the DFT. We choose sinusoidal signals as the specific class of examples to discuss, but most of the issues raised apply generally.

11.2 DFT ANALYSIS OF SINUSOIDAL SIGNALS

The discrete-time Fourier transform of a sinusoidal signal $A\cos(\omega_0 n + \phi)$ is a pair of impulses at $+\omega_0$ and $-\omega_0$ (repeating periodically with period 2π). In analyzing sinusoidal signals using the DFT, windowing and spectral sampling have an important effect. As we will see in Section 11.2.1, the windowing smears or broadens the impulses in the theoretical Fourier representation and thus the exact frequency is less sharply defined. Windowing also reduces the ability to resolve sinusoidal signals that are closely spaced in frequency. The spectral sampling inherent in the DFT has the effect of potentially giving a misleading or inaccurate picture of the true spectrum of the sinusoidal signal. This effect is discussed in Section 11.2.2.

11.2.1 The Effect of Windowing

Let us consider a continuous-time signal consisting of the sum of two sinusoidal components, i.e.

$$s_c(t) = A_0\cos(\Omega_0 t + \theta_0) + A_1\cos(\Omega_1 t + \theta_1), \qquad -\infty < t < \infty. \qquad (11.6)$$

Assuming ideal sampling with no aliasing and no quantization error, we obtain the discrete-time signal

$$x[n] = A_0\cos(\omega_0 n + \theta_0) + A_1\cos(\omega_1 n + \theta_1), \qquad -\infty < n < \infty, \qquad (11.7)$$

where $\omega_0 = \Omega_0 T$ and $\omega_1 = \Omega_1 T$. The windowed sequence $v[n]$ in Fig. 11.1 is then

$$v[n] = A_0 w[n]\cos(\omega_0 n + \theta_0) + A_1 w[n]\cos(\omega_1 n + \theta_1). \qquad (11.8)$$

To obtain the Fourier transform of $v[n]$, we can expand Eq. (11.8) in terms of complex exponentials and utilize the frequency-shifting property of Eq. (2.142) in Section 2.9.2. Specifically, we rewrite $v[n]$ as

$$
\begin{aligned}
v[n] = {} & \frac{A_0}{2} w[n]e^{j\theta_0}e^{j\omega_0 n} + \frac{A_0}{2} w[n]e^{-j\theta_0}e^{-j\omega_0 n} \\
& + \frac{A_1}{2} w[n]e^{j\theta_1}e^{j\omega_1 n} + \frac{A_1}{2} w[n]e^{-j\theta_1}e^{-j\omega_1 n},
\end{aligned}
\qquad (11.9)
$$

from which, with Eq. (2.142), it follows that the Fourier transform of the windowed sequence is

$$
\begin{aligned}
V(e^{j\omega}) = {} & \frac{A_0}{2} e^{j\theta_0} W(e^{j(\omega-\omega_0)}) + \frac{A_0}{2} e^{-j\theta_0} W(e^{j(\omega+\omega_0)}) \\
& + \frac{A_1}{2} e^{j\theta_1} W(e^{j(\omega-\omega_1)}) + \frac{A_1}{2} e^{-j\theta_1} W(e^{j(\omega+\omega_1)}).
\end{aligned}
\qquad (11.10)
$$

According to Eq. (11.10), the Fourier transform of the windowed signal consists of the Fourier transform of the window replicated at the frequencies $\pm\omega_0$ and $\pm\omega_1$ and scaled by the complex amplitudes of the individual complex exponentials that make up the signal.

Example 11.1

Let us consider the system of Fig. 11.1 and in particular $W(e^{j\omega})$ and $V(e^{j\omega})$ for the specific case of a sampling rate $1/T = 10\,\text{kHz}$ and a rectangular window $w[n]$ of length 64 with $A_0 = 1$ and $A_1 = 0.75$. For convenience, we choose $\theta_0 = \theta_1 = 0$. To illustrate the essential features, we specifically display only the magnitudes of the Fourier transforms.

In Fig. 11.3(a) we show $|W(e^{j\omega})|$ and in Figs. 11.3(b), (c), (d), and (e) we show $|V(e^{j\omega})|$ for several choices of Ω_0 and Ω_1 in Eq. (11.6) or, equivalently, ω_0 and ω_1 in Eq. (11.7). In Fig. 11.3(b), $\Omega_0 = (2\pi/6) \times 10^4$ and $\Omega_1 = (2\pi/3) \times 10^4$ or, equivalently,

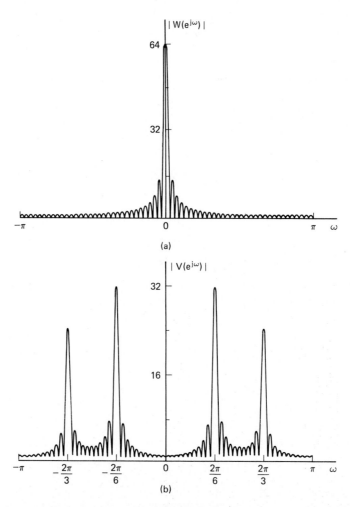

(a)

(b)

Figure 11.3 Illustration of Fourier analysis of windowed cosines with a rectangular window. (a) Fourier transform of window. (b)–(e) Fourier transform of windowed cosines as the frequency spacing becomes progressively smaller. (b) $\Omega_0 = (2\pi/6) \times 10^4$, $\Omega_1 = (2\pi/3) \times 10^4$.

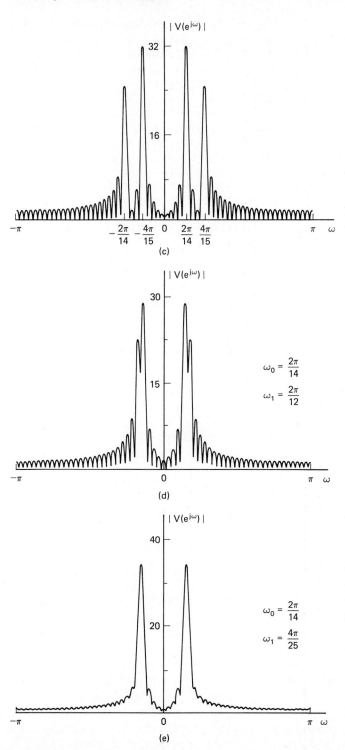

Figure 11.3 *(continued)* (c) $\Omega_0 = (2\pi/14) \times 10^4$, $\Omega_1 = (4\pi/15) \times 10^4$. (d) $\Omega_0 = (2\pi/14) \times 10^4$, $\Omega_1 = (2\pi/12) \times 10^4$. (e) $\Omega_0 = (2\pi/14) \times 10^4$, $\Omega_1 = (4\pi/25) \times 10^4$.

$\omega_0 = 2\pi/6$ and $\omega_1 = 2\pi/3$. In Fig. 11.3(c)–(e), the frequencies become progressively closer. For the parameters in Fig. 11.3(b), the frequency and amplitude of the individual components are evident. Specifically, Eq. (11.10) suggests that with no overlap between the replicas of $W(e^{j\omega})$ at ω_0 and ω_1 there will be a peak of height $32A_0$ at ω_0 and $32A_1$ at ω_1 since $W(e^{j\omega})$ has a peak height of 64. In Fig. 11.3(b), the two peaks are at approximately $\omega_0 = 2\pi/6$ and $\omega_1 = 2\pi/3$ with peak amplitudes in the correct ratio. In Fig. 11.3(c), there is more overlap between the window replicas at ω_0 and ω_1 and while two distinct peaks are present, the amplitude of the spectrum at $\omega = \omega_0$ is affected by the amplitude of the sinusoidal signal at frequency ω_1 and vice versa. This interaction is called *leakage*: the component at one frequency leaks into the vicinity of another component due to the spectral smearing introduced by the window. Figure 11.3(d) shows the case where the leakage is even greater. Notice how sidelobes adding out of phase can *reduce* the heights of the peaks. In Fig. 11.3(e), the overlap between the spectral windows at ω_0 and ω_1 is so significant that the two peaks visible in (b)–(d) have merged into one. In other words, with this window, the two frequencies corresponding to Fig. 11.3(e) will not be *resolved* in the spectrum.

Reduced resolution and leakage are the two primary effects on the spectrum as a result of applying a window to the signal. The resolution is primarily influenced by the width of the mainlobe of $W(e^{j\omega})$, while the degree of leakage depends on the relative amplitude of the mainlobe and the sidelobes of $W(e^{j\omega})$. In Chapter 7, in a filter design context, we showed that the width of the mainlobe and the relative sidelobe amplitude depend primarily on the window length L and the shape (amount of tapering) of the window. The rectangular window, which has Fourier transform

$$W_r(e^{j\omega}) = \sum_{N=0}^{L-1} e^{-j\omega n} = e^{-j\omega(L-1)/2} \frac{\sin(\omega L/2)}{\sin(\omega/2)}, \qquad (11.11)$$

has the narrowest mainlobe for a given length, but it has the largest sidelobes of all the commonly used windows. As defined in Chapter 7, the Kaiser window is

$$w_K[n] = \begin{cases} \dfrac{I_0[\beta(1 - [(n-\alpha)/\alpha]^2)^{1/2}]}{I_0(\beta)}, & 0 \le n \le L-1, \\[2mm] 0, & \text{otherwise,} \end{cases} \qquad (11.12)$$

where $\alpha = (L-1)/2$ and $I_0(\cdot)$ is the zeroth-order modified Bessel function of the first kind. (Note that the notation of Eq. 11.12 differs slightly from that of Eq. 7.89 in that L denotes the length of the window in Eq. 11.12 while the length of the filter design window in Eq. 7.89 is denoted $M + 1$.) We have already seen in the context of the filter design problem that this window has two parameters, β and L, which can be used to trade between mainlobe width and relative sidelobe amplitude. (Recall that the Kaiser window reduces to the rectangular window when $\beta = 0$.) The mainlobe width Δ_{ml} is defined as the symmetric distance between the central zero crossings. The relative sidelobe level A_{sl} is defined as the ratio in dB of the amplitude of the mainlobe to the amplitude of the largest sidelobe. Figure 11.4, which is a duplicate of Fig. 7.32, shows Fourier transforms of Kaiser windows for different lengths and different values of β. In designing a Kaiser window for spectrum analysis, we want to specify a desired value of A_{sl} and determine the required value of β. Figure 11.4(c) shows that the

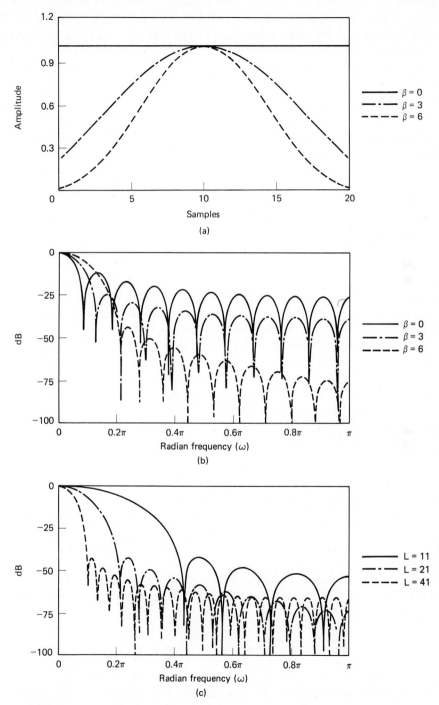

Figure 11.4 (a) Kaiser windows for $\beta = 0$, 3, and 6 and $L = 21$. (b) Fourier transforms corresponding to windows in (a). (c) Fourier transforms of Kaiser windows with $\beta = 6$ and $L = 11$, 21, and 41.

relative sidelobe amplitude is essentially independent of the window length and thus depends only on β. This was confirmed by Kaiser and Schafer (1980), who obtained the following least-squares approximation to β as a function of A_{sl}:

$$\beta = \begin{cases} 0, & A_{sl} < 13.26, \\ 0.76609(A_{sl} - 13.26)^{0.4} + 0.09834(A_{sl} - 13.26), & 13.26 < A_{sl} < 60, \quad (11.13) \\ 0.12438(A_{sl} + 6.3), & 60 < A_{sl} < 120. \end{cases}$$

Using values of β from Eq. (11.13) gives windows with actual values of A_{sl} that differ by less than 0.36% from the desired value across the range $13.26 < A_{sl} < 120$. (Note that the value 13.26 is the relative sidelobe amplitude of the rectangular window, to which the Kaiser window reduces for $\beta = 0$.)

Figure 11.4(c) also shows that the mainlobe width is inversely proportional to the length of the window. The tradeoff between mainlobe width, relative sidelobe amplitude, and window length is displayed by the approximate relationship

$$L \simeq \frac{24\pi(A_{sl} + 12)}{155\Delta_{ml}} + 1, \qquad (11.14)$$

which was also given by Kaiser and Schafer (1980).

Equations (11.12), (11.13), and (11.14) are the necessary equations for determining a Kaiser window with desired values of mainlobe width and relative sidelobe amplitude. To design a window for prescribed values of A_{sl} and Δ_{ml} requires simply the computation of β from Eq. (11.13); the computation of L from Eq. (11.14); and the computation of the window using Eq. (11.12) and the subroutine for $I_0(\cdot)$ given in Chapter 7. Many of the remaining examples of this chapter use the Kaiser window. Other spectral analysis windows are considered by Harris (1978).

11.2.2 The Effect of Spectral Sampling

As mentioned previously, the DFT of the windowed sequence $v[n]$ provides samples of $V(e^{j\omega})$ at the N equally spaced discrete-time frequencies $\omega_k = 2\pi k/N$, $k = 0, 1, \ldots,$ $N - 1$, or equivalently at the continuous-time frequencies $\Omega_k = (2\pi k)/(NT)$, $k = 0, 1,$ $\ldots, N - 1$. Spectral sampling as imposed by the DFT can sometimes produce misleading results. This effect is best illustrated by example.

Example 11.2

Let us consider the same parameters as in Fig. 11.3(c) in Example 11.1, i.e., $A_0 = 1$, $A_1 = 0.75$, $\omega_0 = 2\pi/7.5$, and $\omega_1 = 2\pi/14$ in Eq. (11.8). $w[n]$ is a rectangular window of length 64. Then

$$v[n] = \begin{cases} \cos\left(\frac{2\pi}{14}n\right) + 0.75\cos\left(\frac{4\pi}{15}n\right), & 0 \le n \le 63, \\ 0, & \text{otherwise.} \end{cases} \qquad (11.15)$$

Figure (11.5a) shows the windowed sequence $v[n]$. Figures 11.5(b), (c), (d), and (e) show the corresponding real part, imaginary part, magnitude, and phase, respectively, of the DFT of length $N = 64$. Figure 11.5(f) shows $|V(e^{j\omega})|$, corresponding to Fig. 11.3(c) but plotted from $\omega = 0$ to $\omega = 2\pi$ for comparison to the DFT in Fig. 11.5(d).

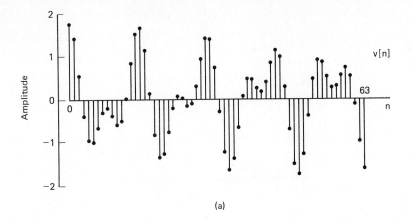

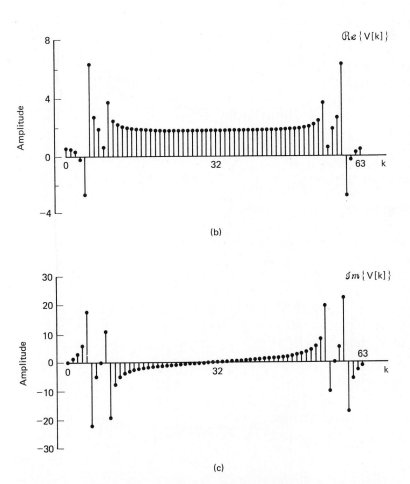

Figure 11.5   Cosine sequence and discrete Fourier transform with a rectangular window. (a) Windowed signal. (b) Real part of DFT. (c) Imaginary part of DFT.

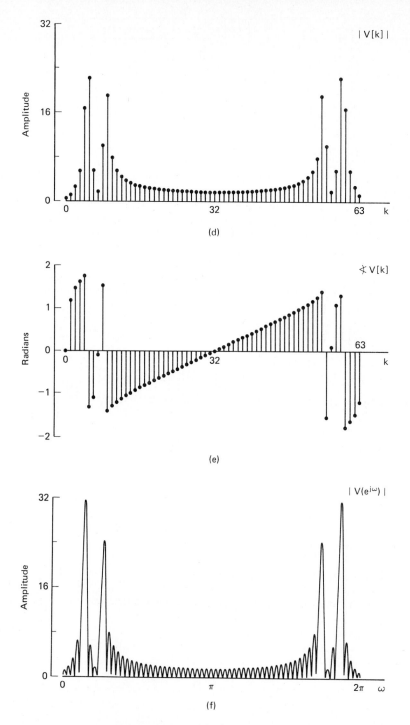

Figure 11.5 (*continued*) (d) Magnitude of DFT. (e) Phase of DFT. (f) Magnitude of discrete-time Fourier transform.

In Figs. 11.5(b)–(e), the horizontal (frequency) axis is labeled in terms of the DFT index or frequency sample number k. The value $k = 32$ corresponds to $\omega = \pi$ or, equivalently, $\Omega = \pi/T$. As is the usual convention in displaying the DFT of a time sequence, we display the DFT values in the range from $k = 0$ to $k = N - 1$, corresponding to displaying samples of the discrete-time Fourier transform in the frequency range 0 to 2π. Because of the inherent periodicity of the discrete-time Fourier transform, the first half of this range corresponds to the positive continuous-time frequencies, i.e., Ω between zero and π/T, and the second half of this range to the negative frequencies, i.e., Ω between $-\pi/T$ and zero. Note the even periodic symmetry of the real part and the magnitude and the odd periodic symmetry of the imaginary part and the phase.

The magnitude of the DFT in Fig. 11.5(d) corresponds to samples of the magnitude of the spectrum displayed in Fig. 11.5(f) and shows the expected concentration around $\omega = 2\pi/7.5$ and $\omega = 2\pi/14$, the frequencies of the two sinusoidal components of the input. Specifically, the frequency $\omega_1 = 4\pi/15 = 2\pi(8.533\ldots)/64$ lies between the DFT samples corresponding to $k = 8$ and $k = 9$. Likewise, the frequency $\omega_0 = 2\pi/14 = 2\pi(4.5714\ldots)/64$ lies between the DFT samples corresponding to $k = 4$ and $k = 5$. Note that the frequency locations of the peaks in Fig. 11.5(f) are between spectral samples obtained from the DFT. In general, the locations of peaks in the DFT values do not necessarily coincide with the exact frequency locations of the peaks in the Fourier transform since the true spectral peaks can lie between spectral samples. Correspondingly, as evidenced in comparing Figs. 11.5(f) and 11.5(d), the relative amplitudes of peaks in the DFT will not necessarily reflect the relative amplitudes of the true spectral peaks.

Example 11.3

Let us now choose the parameters corresponding to Fig. 11.3(d) so that

$$
v[n] = \begin{cases} \cos\left(\dfrac{2\pi}{16}n\right) + 0.75\cos\left(\dfrac{2\pi}{8}n\right), & 0 \le n \le 63, \\ 0, & \text{otherwise,} \end{cases} \tag{11.16}
$$

as shown in Fig. 11.6(a). Again, a rectangular window is used with $N = L = 64$. This is very similar to the previous example except that in this case, the frequencies of the cosines coincide exactly with two of the DFT frequencies. Specifically, the frequency $\omega_1 = 2\pi/8 = 2\pi 8/64$ corresponds exactly to the DFT sample $k = 8$, and the frequency $\omega_0 = 2\pi/16 = 2\pi 4/64$ to the DFT sample $k = 4$.

The magnitude of the 64-point DFT of $v[n]$ for this example is shown in Fig. 11.6(b) and corresponds to samples of $|V(e^{j\omega})|$ (which is plotted in Fig. 11.6c) at a frequency spacing of $2\pi/64$. Although the signal parameters for this example and Example 11.2 are very similar, the appearance of the DFT is strikingly different. In particular, for this example the DFT has two strong spectral lines at the frequencies of the two sinusoidal components in the signal and no frequency content at the other DFT values. In fact, this clean appearance of the DFT in Fig. 11.6(b) is largely an illusion resulting from the sampling of the spectrum. Comparing Figs. 11.6(b) and (c), we can see that the reason for the clean appearance of Fig. 11.6(b) is that for this choice of parameters, the Fourier transform is exactly zero at the frequencies that are sampled by the DFT except those corresponding to $k = 4, 8, 56$, and 60. Although the signal of Fig. 11.6(a) has significant content at almost all frequencies, as evidenced by Fig. 11.6(c), we do not see this in the DFT because of the sampling of the spectrum.

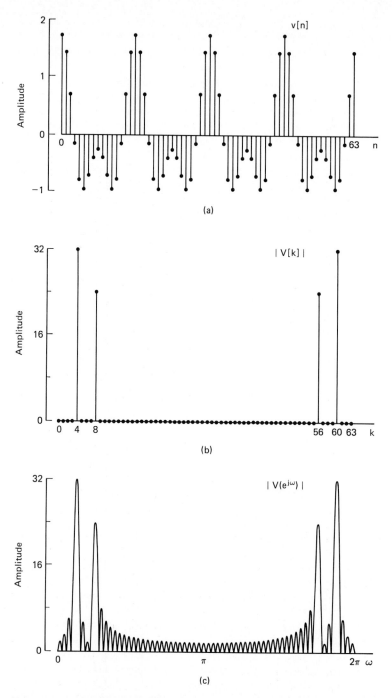

Figure 11.6 Discrete Fourier analysis of the sum of two sinusoids for a case in which the Fourier transform is zero at all DFT frequencies except those corresponding to the frequencies of the two sinusoidal components. (a) Windowed signal. (b) Magnitude of DFT. (c) Magnitude of discrete-time Fourier transform ($|V(e^{j\omega})|$).

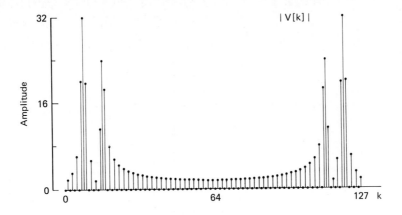

Figure 11.7 DFT of the same signal as Fig. 11.6 but with twice the number of frequency samples.

To illustrate this point further, let us extend $v[n]$ in Eq. (11.16) by zero padding to obtain a 128-point sequence. The corresponding 128-point DFT is shown in Fig. 11.7. With this finer sampling of the spectrum, the presence of significant content at other frequencies becomes apparent.

In Figs. 11.5, 11.6, and 11.7, the windows were rectangular. In the next set of examples we illustrate the effect of different choices for the window.

Example 11.4

Let us return to the frequency, amplitude, and phase parameters of Example 11.1, but now with a Kaiser window applied so that

$$v[n] = w_K[n] \cos\left(\frac{2\pi}{14} n\right) + 0.75 w_K[n] \cos\left(\frac{4\pi}{15} n\right), \tag{11.17}$$

where $w_K[n]$ is the Kaiser window as given by Eq. (11.12). We will select the Kaiser window parameter β to be equal to 5.48, which according to Eq. (11.13) results in a window for which the relative sidelobe amplitude is $A_{sl} = -40\,\text{dB}$. Figure 11.8(a) shows the windowed sequence $v[n]$ for a window length of $L = 64$, and Fig. 11.8(b) shows the magnitude of the corresponding DFT. From Eq. (11.15) we see that the difference between the two frequencies is $\omega_1 - \omega_0 = 2\pi/7.5 - 2\pi/14 = 0.389$. From Eq. (11.14) it follows that the width of the mainlobe of the Fourier transform of the Kaiser window with $L = 64$ and $\beta = 5.48$ is $\Delta_{ml} = 0.401$. Thus, the mainlobes of the two replicas of $W_K(e^{j\omega})$ centered at ω_0 and ω_1 will just slightly overlap in the frequency interval between the two frequencies. This is evident in Fig. 11.8(b), where we see that the two frequency components are clearly resolved.

Figure 11.8(c) shows the same signal multiplied by a Kaiser window with $L = 32$ and $\beta = 5.48$. Since the window is half as long, we expect the width of the mainlobe of the Fourier transform of the window to double, and Fig. 11.8(d) confirms this. Specifically, Eqs. (11.13) and (11.14) confirm that for $L = 32$ and $\beta = 5.48$, the mainlobe width is $\Delta_{ml} = 0.815$. Now the mainlobes of the two copies of the Fourier transform of the window overlap throughout the region between the two cosine frequencies and we do not see two distinct peaks.

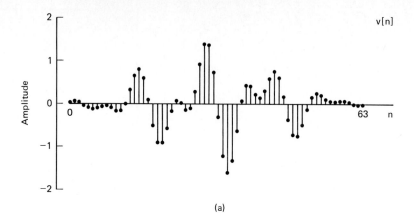

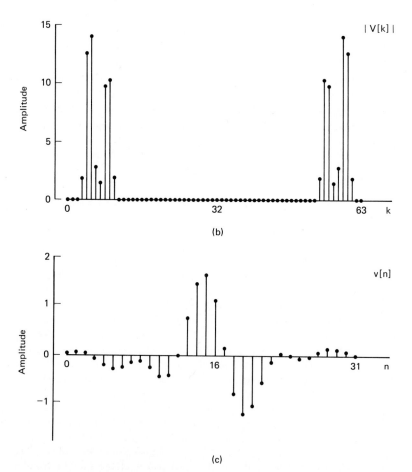

Figure 11.8   Discrete Fourier analysis with Kaiser window. (a) Windowed sequence for $L = 64$. (b) Magnitude of DFT for $L = 64$. (c) Windowed sequence for $L = 32$.

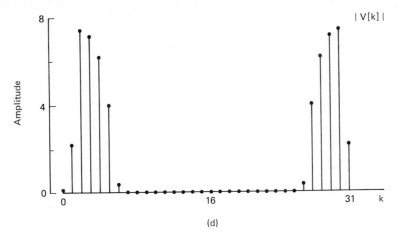

Figure 11.8   *(continued)* (d) Magnitude of DFT for $L = 32$.

In all the previous examples except in Fig. 11.7, the DFT length N was equal to the window length L. In Fig. 11.7, zero padding was applied to the windowed sequence before computing the DFT to obtain the Fourier transform on a more finely divided set of frequencies. However, we must realize that this zero padding will not improve the resolution; the resolution depends on the length and shape of the window. This is illustrated by the next example.

Example 11.5

In this example we repeat Example 11.4 using the Kaiser window with $L = 32$ and $\beta = 5.48$ but with the DFT length varying. Figure 11.9(a) shows the DFT magnitude for $N = L = 32$ as in Fig. 11.8(d), and Figs. 11.9(b), (c), (d) show the DFT magnitude again with $L = 32$ but with DFT lengths $N = 64$, $N = 128$, and $N = 1024$, respectively. As with Example 11.3, this zero padding of the 32-point sequence results in finer spectral sampling of the Fourier transform. The underlying envelope of each DFT magnitude in Fig. 11.9 is the same. Consequently, increasing the DFT size by zero padding does not change the resolution of the two sinusoidal frequency components, but it does change the spacing of the frequency samples.

For a complete representation of a sequence of length L, the L-point DFT is sufficient since the original sequence can be completely recovered from it. However, as we saw in the preceding examples, simple examination of the L-point DFT can result in misleading interpretations. For this reason, it is common to apply zero padding so that the spectrum is sufficiently oversampled and important features are therefore readily apparent. With a high degree of time-domain zero padding or frequency-domain oversampling, simple interpolation (e.g., linear interpolation) between the DFT values provides a reasonably accurate picture of the Fourier spectrum, which can then be used, for example, to estimate the locations and amplitudes of spectral peaks.

Figure 11.10 shows the 1024-point DFT of the signal of Example 11.4 for the 64-point Kaiser window. Comparing this figure to Fig. 11.8(b) suggests that the 1024-point DFT would be much more effective for precisely locating the peak of the windowed Fourier transform. Also, the amplitudes of the two peaks in Fig. 11.10 are very close to being in the correct ratio of 0.75 to 1.

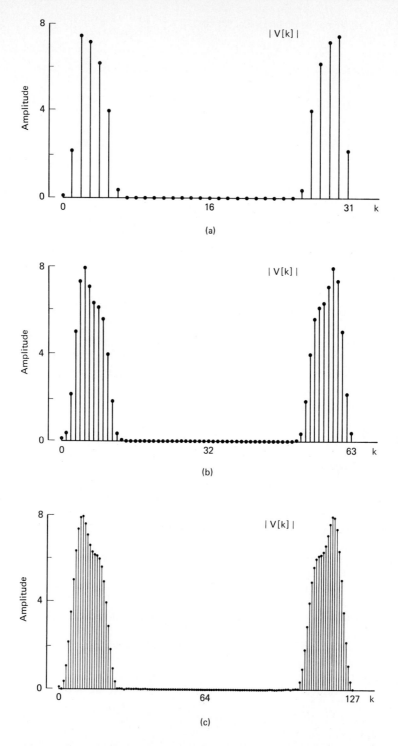

Figure 11.9 Illustration of effect of DFT length for Kaiser window of length $L = 32$. (a) Magnitude of DFT for $N = 32$. (b) Magnitude of DFT for $N = 64$. (c) Magnitude of DFT for $N = 128$.

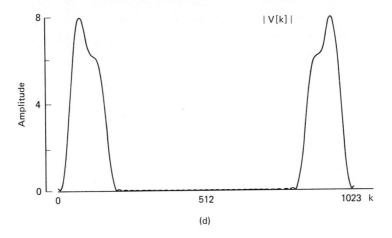

(d)

Figure 11.9 (*continued*) (d) Magnitude of DFT for $N = 1024$. (DFT values are linearly interpolated to obtain a smooth curve.)

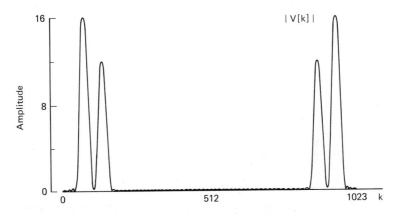

Figure 11.10 Illustration of the computation of the DFT for $N \gg L$ with linear interpolation to create a smooth curve ($N = 1024$, $L = 64$).

11.3 THE TIME-DEPENDENT FOURIER TRANSFORM

The previous section illustrated the use of the DFT for obtaining a frequency-domain representation of a signal composed of sinusoidal components. In that discussion, we assumed that the frequencies of the cosines did not change with time so that no matter how long the window, the signal properties would be the same from the beginning to the end of the window. Often, in practical applications of sinusoidal signal models, the signal properties (amplitudes, frequencies, and phases) will change with time. For example, nonstationary signal models of this type are required to describe radar, sonar, speech, and data communication signals. A single DFT estimate is not sufficient to describe such signals, and as a result we are led to the concept of the *time-dependent Fourier transform*, also referred to as the short-time Fourier transform.†

The time-dependent Fourier transform of a signal $x[n]$ is defined as

$$X[n, \lambda] = \sum_{m=-\infty}^{\infty} x[n + m]w[m]e^{-j\lambda m}, \qquad (11.18)$$

where $w[n]$ is a window sequence. In the time-dependent Fourier representation, the one-dimensional sequence $x[n]$, a function of a single discrete variable, is converted into a two-dimensional function of the time variable n, which is discrete, and the frequency variable λ, which is continuous.‡ Note that the time-dependent Fourier transform is periodic in λ with period 2π, and therefore we need consider only values of λ for $0 \leq \lambda < 2\pi$ or any other interval of length 2π.

Equation (11.18) can be interpreted as the Fourier transform of $x[n + m]$ as viewed through the window $w[m]$. The window has a stationary origin and as n changes, the signal slides past the window so that at each value of n, a different portion of the signal is viewed. This is depicted in Fig. 11.11 for the signal

$$x[n] = \cos(\omega_0 n^2), \qquad \omega_0 = 2\pi \times 14 \times 10^{-6}, \qquad (11.19)$$

corresponding to a linear frequency modulation (i.e., the "instantaneous frequency" is $\omega_0 n$). As we saw in Chapter 9 in the context of the chirp transform algorithm, a signal of this type is often referred to as a linear chirp. Typically, $w[m]$ in Eq. (11.18) has finite length around $m = 0$ so that $X[n, \lambda]$ displays the frequency characteristics of the signal around time n. As an example, in Fig. 11.12 we show a display of the magnitude of the time-dependent Fourier transform of the signal of Eq. (11.19) and Fig. 11.11 with $w[m]$ a Hamming window of length 400. In this display, referred to as a *spectrogram*, the vertical dimension is frequency (λ) and the horizontal dimension is time (n). The magnitude of the time-dependent Fourier transform is represented by the darkness of the markings. In Fig. 11.12, the linear progression of the frequency with time is clear.

Since $X[n, \lambda]$ is the discrete-time Fourier transform of $x[n + m]w[m]$, the time-dependent Fourier transform is invertible if the window has at least one nonzero sample. Specifically, from the Fourier transform synthesis equation (2.112),

$$x[n + m]w[m] = \frac{1}{2\pi} \int_0^{2\pi} X[n, \lambda]e^{j\lambda m} \, d\lambda, \qquad -\infty < m < \infty, \qquad (11.20)$$

from which it follows that

$$x[n] = \frac{1}{2\pi w[0]} \int_0^{2\pi} X[n, \lambda] \, d\lambda \qquad (11.21)$$

---

† Further discussion of the time-dependent Fourier transform can be found in a variety of references, including Allen and Rabiner (1977), Rabiner and Schafer (1978), Crochiere and Rabiner (1983), and Nawab and Quatieri (1988).

‡ We denote the frequency variable of the time-dependent Fourier transform by λ to maintain a distinction from the frequency variable of the conventional discrete-time Fourier transform, which will be denoted ω. We use the mixed bracket-parenthesis notation $X[n, \lambda]$ as a reminder that n is a discrete variable and λ a continuous variable.

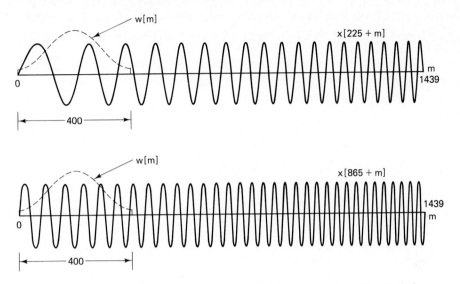

Figure 11.11   Two segments of the linear chirp signal $x[n] = \cos(2\pi \times 14 \times 10^{-6})n^2$ with the window superimposed. $X[n, \lambda)$ at $n = 225$ is the discrete-time Fourier transform of the top trace multiplied by the window. $X[865, \lambda)$ is the discrete-time Fourier transform of the bottom trace multiplied by the window.

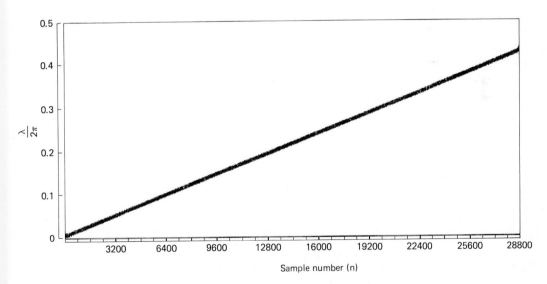

Figure 11.12   The magnitude of the time-dependent Fourier transform of $x[n] = \cos(2\pi \times 14 \times 10^{-6})n^2$ using a Hamming window of length 400.

if $w[0] \neq 0$.† Clearly not just the single sample $x[n]$ but all of the samples that are multiplied by nonzero samples of the window can be recovered in a similar manner using Eq. (11.20).

A rearrangement of the sum in Eq. (11.18) leads to another useful interpretation of the time-dependent Fourier transform. If we make the substitution $m' = n + m$ in Eq. (11.18), then $X[n, \lambda)$ can be written as

$$X[n, \lambda) = \sum_{m'=-\infty}^{\infty} x[m']w[-(n-m')]e^{j\lambda(n-m')}. \tag{11.22}$$

Equation (11.22) can be interpreted as the convolution

$$X[n, \lambda) = x[n] * h_\lambda[n], \tag{11.23a}$$

where

$$h_\lambda[n] = w[-n]e^{j\lambda n}. \tag{11.23b}$$

From Eq. (11.23a) we see that the time-dependent Fourier transform as a function of n with λ fixed can be interpreted as the output of a linear time-invariant filter with impulse response $h_\lambda[n]$, or, equivalently, with frequency response

$$H_\lambda(e^{j\omega}) = W(e^{j(\lambda-\omega)}). \tag{11.24}$$

In general a window that is nonzero for positive time will be called a *noncausal window* since the computation of $X[n, \lambda)$ using Eq. (11.18) requires samples that *follow* sample n in the sequence. Equivalently, in the linear filtering interpretation, the impulse response $h_\lambda[n] = w[-n]e^{j\lambda n}$ is noncausal.

In the definition of Eq. (11.18), the time origin of the window is held fixed and the signal is shifted past the interval of support of the window. This effectively redefines the time origin for Fourier analysis to be at sample n of the signal. Another possibility is to shift the window as n changes, keeping the time origin for Fourier analysis fixed at the original time origin of the signal. This leads to a definition for the time-dependent Fourier transform of the form

$$\check{X}[n, \lambda) = \sum_{m=-\infty}^{\infty} x[m]w[m-n]e^{-j\lambda m}. \tag{11.25}$$

The relationship between the definitions of Eqs. (11.18) and (11.25) is easily shown to be

$$\check{X}[n, \lambda) = e^{-j\lambda n} X[n, \lambda) \tag{11.26}$$

The definition of Eq. (11.18) is particularly convenient when we consider using the DFT to obtain samples in λ of the time-dependent Fourier transform since if $w[m]$ is of finite length in the range $0 \leq m \leq (L-1)$, then so is $x[n+m]w[m]$. On the other hand, the definition of Eq. (11.25) has some advantages for the interpretation of Fourier analysis in terms of filter banks. Since our primary interest is in applications of the DFT, we will base our discussion on Eq. (11.18).

---

† Since $X[n, \lambda)$ is periodic in λ with period 2π, the integration in Eqs. (11.20) and (11.21) can be over any interval of length 2π.

11.3.1 The Effect of the Window

The primary purpose of the window in the time-dependent Fourier transform is to limit the extent of the sequence to be transformed so that the spectral characteristics are reasonably stationary over the duration of the window. The more rapidly the signal characteristics change, the shorter the window should be. As we saw in Section 11.2, as the window becomes shorter, frequency resolution decreases. The same effect is true, of course, for $X[n, \lambda)$. On the other hand, as the window length decreases, the ability to resolve changes with time increases. Consequently, the choice of window length becomes a tradeoff between frequency resolution and time resolution.

The effect of the window on the properties of the time-dependent Fourier transform can be seen by assuming that the signal $x[n]$ has a conventional discrete-time Fourier transform $X(e^{j\omega})$. First let us assume that the window is unity for all m; i.e., assume that there is no window at all. Then from Eq. (11.18),

$$X[n, \lambda) = X(e^{j\lambda})e^{j\lambda n}. \tag{11.27}$$

Of course, a typical window for spectrum analysis tapers to zero so as to select only a portion of the signal for analysis. As discussed in Section 11.2, the length and shape of the window are chosen so that the Fourier transform of the window is narrow in frequency compared with changes in the Fourier transform of the signal. The Fourier transform of a typical window is depicted in Fig. 11.13(a).

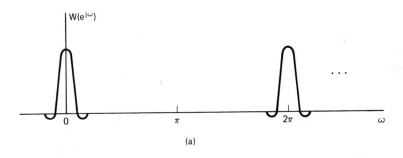

(a)

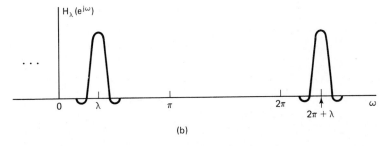

(b)

Figure 11.13   (a) Fourier transform of window in time-dependent Fourier analysis. (b) Equivalent bandpass filter for time-dependent Fourier analysis.

If we consider the time-dependent Fourier transform for fixed n, then it follows from the properties of Fourier transforms that

$$X[n, \lambda] = \frac{1}{2\pi} \int_0^{2\pi} e^{j\theta n} X(e^{j\theta}) W(e^{j(\lambda - \theta)}) \, d\theta, \qquad (11.28)$$

i.e., the Fourier transform of the shifted signal is convolved with the Fourier transform of the window. This is similar to Eq. (11.2) except that in Eq. (11.2) we assumed that the signal was not successively shifted relative to the window. Here we compute a Fourier transform for each value of n. In Section 11.2 we saw that the ability to resolve two narrowband signal components depends on the width of the mainlobe of the Fourier transform of the window, while the degree of leakage of one component into the vicinity of the other depends on the relative sidelobe amplitude.

In the linear filtering interpretation of Eqs. (11.23) and (11.24), $W(e^{j\omega})$ typically has the lowpass characteristics depicted in Fig. 11.13(a), and consequently $H_\lambda(e^{j\omega})$ is a bandpass filter whose passband is centered at $\omega = \lambda$, as depicted in Fig. 11.13(b). Clearly, the width of the passband of this filter is approximately equal to the width of the mainlobe of the Fourier transform of the window. The degree of rejection of adjacent frequency components depends on the relative sidelobe amplitude.

The preceding discussion suggests that if we are using the time-dependent Fourier transform to obtain a·time-dependent estimate of the frequency spectrum of a signal, it is desirable to taper the window to lower the sidelobes and to use as long a window as feasible to improve the frequency resolution. We will consider some examples in Section 11.5. However, before doing so, we first discuss the use of the DFT in explicitly evaluating the time-dependent Fourier transform.

11.3.2 Sampling in Time and Frequency

Explicit computation of $X[n, \lambda]$ can be done only at a finite set of values of λ, corresponding to sampling the time-dependent Fourier transform in the frequency variable domain. Just as finite-length signals can be exactly represented through samples of the discrete-time Fourier transform, signals of indeterminate length can be represented through samples of the time-dependent Fourier transform if the window in Eq. (11.18) has finite length. As an example, suppose that the window has length L with samples beginning at $m = 0$, i.e.,

$$w[m] = 0 \qquad \text{outside the interval } 0 \le m \le L - 1. \qquad (11.29)$$

If we sample $X[n, \lambda]$ at N equally spaced frequencies $\lambda_k = 2\pi k/N$, with $N \ge L$, then we can still recover the original sequence from the sampled time-dependent Fourier transform. Specifically, if we define $X[n, k]$ to be

$$X[n, k] = X[n, 2\pi k/N] = \sum_{m=0}^{L-1} x[n + m]w[m]e^{-j(2\pi/N)km}, \qquad 0 \le k \le N - 1, \quad (11.30)$$

then $X[n, k]$ is the DFT of the windowed sequence $x[n + m]w[m]$. Using the inverse DFT,

$$x[n + m]w[m] = \frac{1}{N} \sum_{k=0}^{N-1} X[n, k]e^{j(2\pi/N)km}, \qquad 0 \le m \le L - 1. \qquad (11.31)$$

Since we assume that the window $w[m] \ne 0$ for $0 \le m \le L - 1$, the sequence values can be recovered in the interval from n through $(n + L - 1)$ using the equation

$$x[n + m] = \frac{1}{Nw[m]} \sum_{k=0}^{N-1} X[n, k]e^{j(2\pi/N)km}, \qquad 0 \le m \le L - 1, \qquad (11.32)$$

where it is assumed that $w[m] \ne 0$ for $0 \le m \le L - 1$. The important point is that the window has finite length and that we take at least as many samples in the λ dimension as there are nonzero samples in the window; i.e., $N \ge L$. While Eq. (11.29) corresponds to a noncausal window, we could have used a causal window with $w[m] \ne 0$ for $-(L - 1) \le m \le 0$ or a symmetric window such that $w[m] = w[-m]$ for $|m| \le (L - 1)/2$, with L an odd integer. The use of a noncausal window in Eq. (11.30) is simply more convenient for our analysis since it leads very naturally to the interpretation of the sampled time-dependent Fourier transform as the DFT of the windowed sequence beginning with sample n.

Since Eq. (11.30) corresponds to sampling Eq. (11.18) in λ, it also corresponds to sampling Eqs. (11.22) and (11.23) in λ. Specifically, Eq. (11.30) can be rewritten as

$$X[n, k] = x[n] * h_k[n], \qquad 0 \le k \le N - 1, \qquad (11.33a)$$

where

$$h_k[n] = w[-n]e^{j(2\pi/N)kn}. \qquad (11.33b)$$

Equations (11.33) can be viewed as a bank of N filters, as depicted in Fig. 11.14, with the kth filter having frequency response

$$H_k(e^{j\omega}) = W(e^{j[(2\pi k/N) - \omega]}). \qquad (11.34)$$

Our discussion suggests that $x[n]$ for $-\infty < n < \infty$ can be reconstructed if $X[n, \lambda)$ or $X[n, k]$ is sampled in the time dimension as well. Specifically, using Eq. (11.32) we can reconstruct the signal in the interval $n_0 \le n \le n_0 + L - 1$ from $X[n_0, k]$, and we can reconstruct the signal in the interval $n_0 + L \le n \le n_0 + 2L - 1$

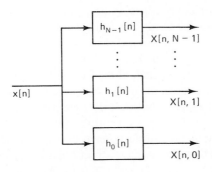

Figure 11.14 Filter bank representation of the time-dependent Fourier transform.

from $X[n_0 + L, k]$, etc. Thus, $x[n]$ can be reconstructed exactly from the time-dependent Fourier transform sampled in both the frequency and the time dimension. In general for the region of support of the window as specified in Eq. (11.29), we define this sampled time-dependent Fourier transform as

$$X[rR, k] = X[rR, 2\pi k/N) = \sum_{m=0}^{L-1} x[rR + m]w[m]e^{-j(2\pi/N)km}, \qquad (11.35)$$

where r and k are integers such that $-\infty < r < \infty$ and $0 \le k \le N - 1$. To further simplify our notation, we define

$$X_r[k] = X[rR, k] = X[rR, \lambda_k), \qquad -\infty < r < \infty, \quad 0 \le k \le N - 1, \quad (11.36)$$

where $\lambda_k = 2\pi k/N$. This notation denotes explicitly that the sampled time-dependent Fourier transform is simply a sequence of N-point DFTs of the windowed signal segments

$$x_r[m] = x[rR + m]w[m], \qquad -\infty < r < \infty, \quad 0 \le m \le L - 1. \quad (11.37)$$

Figure 11.15 shows lines in the (n, λ)-plane corresponding to $X[n, \lambda)$ and the grid of sampling points in the (n, λ)-plane for the case $N = 10$ and $R = 3$. As we have shown, it is possible to uniquely reconstruct the original signal from such a two-dimensional discrete representation.

Equation (11.35) involves the following integer parameters: window length L; the number of samples in the frequency dimension or DFT length N; and the sampling interval in the time dimension R. However, not all choices of these parameters will permit exact reconstruction of the signal. The choice $L \le N$ guarantees that we can reconstruct the windowed segments $x_r[m]$ from the block transforms $X_r[k]$. If $R < L$, the segments overlap, but if $R > L$, some of the samples of the signal are not used and therefore cannot be reconstructed from $x_r[k]$. Thus, in general, the three sampling parameters should satisfy the relation $N \ge L \ge R$. Notice that each block of R samples of the signal is represented by N complex numbers in the sampled time-dependent Fourier representation; or, if the signal is real, only N real numbers are required due to the symmetry of the DFT. As shown above, the signal can be reconstructed exactly from the sampled time-dependent Fourier transform for the special case $N = L = R$. In this case, N samples of a real signal are represented by N real numbers, and this is the minimum that we could expect to achieve for an arbitrarily chosen signal.

Another way to see that the time-dependent Fourier transform can be sampled in the time dimension is to recall that for fixed λ, or equivalently for fixed k, the time-dependent Fourier transform is a one-dimensional sequence that is the output of a bandpass filter with frequency response as in Eq. (11.24). Thus, we expect that the sampling rate of the sequences representing each of the frequencies could be reduced by a factor $2\pi/\Delta_{ml}$, where Δ_{ml} is the width of the mainlobe of the Fourier transform of the window. This argument would suggest that we might encounter a problem with aliasing in the time dimension since the Fourier transform of any finite-length window will *not* be an ideal filter response. The previous argument based on block processing ideas, however, shows conclusively that we *can* reconstruct the original signal exactly

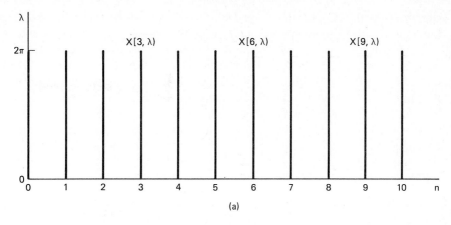

(a)

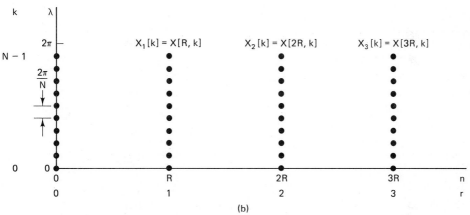

(b)

Figure 11.15   Grid of sampling points in the (n, λ)-plane for the sampled time-dependent Fourier transform with $N = 10$ and $R = 3$.

from the time- and frequency-sampled time-dependent Fourier transform. A more detailed analysis of the linear filtering point of view shows that the aliasing distortion can be canceled in the reconstruction process. This point is considered in Problem 11.19 and is discussed in detail in Rabiner and Schafer (1978) and Crochiere and Rabiner (1983).

11.4 BLOCK CONVOLUTION USING THE TIME-DEPENDENT FOURIER TRANSFORM

One of the uses of the time-dependent Fourier transform is as a basis for processing a discrete-time signal by performing the modifications in the frequency domain. This is done by computing a time-dependent Fourier transform representation, modifying that representation, and then reconstructing a discrete-time signal. This approach is widely used in digital speech coding where instead of quantizing the samples of the speech signal, the sampled time-dependent Fourier transform is quantized and coded for either transmission or storage. Discussion of applications of this type would take us too far afield; however, such block processing techniques for discrete-time signals were also introduced in Chapter 8 when we discussed the use of the DFT for implementing the convolution of a finite-length impulse response with input signal of indefinite length. This method of implementation of linear time-invariant systems has a useful interpretation in terms of the definitions and concepts of Section 11.3.

Assume that $x[n] = 0$ for $n < 0$ and suppose that we compute the time-dependent Fourier transform for $R = L$ and a rectangular window. In other words, the sampled time-dependent Fourier transform $X_r[k]$ consists of a set of N-point DFTs of segments of the input sequence

$$x_r[m] = x[rL + m], \qquad 0 \le m \le L - 1. \tag{11.38}$$

Since each sample of the signal $x[n]$ is included and the blocks do not overlap, it follows that

$$x[n] = \sum_{r=0}^{\infty} x_r[n - rL]. \tag{11.39}$$

Now suppose that we define a new time-dependent Fourier transform

$$Y_r[k] = H[k]X_r[k], \qquad 0 \le k \le N - 1, \tag{11.40}$$

where $H[k]$ is the N-point DFT of a finite-length unit sample sequence $h[n]$ such that $h[n] = 0$ for $n < 0$ and for $n > P - 1$. If we compute the inverse DFT of $Y_r[k]$, we obtain

$$y_r[m] = \frac{1}{N}\sum_{k=0}^{N-1} Y_r[k]e^{j(2\pi/N)km} = \sum_{\ell=0}^{N-1} x_r[\ell]h[((m - \ell))_N] \tag{11.41}$$

That is, $y_r[m]$ is the N-point circular convolution of $h[m]$ and $x_r[m]$. Since $h[m]$ has length P samples and $x_r[m]$ has length L samples, it follows from the discussion of Section 8.8 that if $N \ge L + P - 1$, then $y_r[m]$ will be identical to the linear convolution of $h[m]$ with $x_r[m]$ in the interval $0 \le m \le L + P - 2$, and it will be zero otherwise. Thus, it follows that if we construct an output signal $y[n]$ according to

$$y[n] = \sum_{r=0}^{\infty} y_r[n - rL], \tag{11.42}$$

then $y[n]$ is the output of a linear time-invariant system with impulse response $h[n]$. The procedure just described corresponds exactly to the *overlap-add* method of block convolution. The overlap-save method discussed in Section 8.8 can also be applied within the framework of the time-dependent Fourier transform.

In general, the frequency-domain representation provides great flexibility that is only hinted at by this example. In Chapter 12, we will see another example of discrete-time signal processing utilizing the time-dependent Fourier transform.

11.5 FOURIER ANALYSIS OF NONSTATIONARY SIGNALS

In Section 11.4 we considered a simple example of how the time-dependent Fourier transform can be used to implement linear filtering. In such applications, we are not so much interested in spectral resolution as in whether it is possible to reconstruct a modified signal from the modified time-dependent Fourier transform. On the other hand, the concept of the time-dependent Fourier transform is perhaps most widely used as a framework for a variety of techniques for obtaining spectral estimates for nonstationary discrete-time signals, and in these applications, spectral resolution, time variation, and other issues are the most important.

A nonstationary signal is one for which the signal properties vary with time, for example, a sum of sinusoidal components with time-varying amplitudes, frequencies, or phases. As we will illustrate in Section 11.5.1 for speech signals and in Section 11.5.2 for Doppler radar signals, the time dependent Fourier transform often provides a useful description of how the signal properties change with time.

When we apply time-dependent Fourier analysis to a sampled signal, the entire discussion of Section 11.1 holds for each DFT that is computed. In other words, for each segment $x_r[n]$ of the signal, the sampled time-dependent Fourier transform $X_r[k]$ would be related to the Fourier transform of the original continuous-time signal by the processes described in Section 11.1. Furthermore, if we were to apply the time-dependent Fourier transform to sinusoidal signals with constant (i.e., non-time-varying) parameters, the discussion of Section 11.2 should also apply to each of the DFTs that we compute. When the signal frequencies do not change with time, it is tempting to assume that the time-dependent Fourier transform would vary only in the frequency dimension in the manner described in Section 11.2, but this would be true only in very special cases. For example, the time-dependent Fourier transform will be constant in the time dimension if the signal is periodic with period N_p and $L = \ell_0 N_p$ and $R = r_0 N_p$, where ℓ_0 and r_0 are integers; i.e., the window includes exactly ℓ_0 periods and the window is moved by exactly r_0 periods between computations of the DFT. In general, even if the signal is exactly periodic, the varying phase relationships that would result as different segments of the waveform are shifted into the analysis window would cause the time-dependent Fourier transform to vary in the time dimension. However, for stationary signals if we use a window that tapers to zero at its ends, the magnitude $|X_r[k]|$ will vary only slightly from segment to segment, with most of the variation of the complex time-dependent Fourier transform occurring in the phase.

11.5.1 Time-Dependent Fourier Analysis of Speech Signals

Speech is produced by excitation of an acoustic tube, the *vocal tract*, which is terminated on one end by the lips and on the other end by the glottis. There are three basic classes of speech sounds:

- *Voiced sounds* are produced by exciting the vocal tract with quasi-periodic pulses of airflow caused by the opening and closing of the glottis.

- *Fricative sounds* are produced by forming a constriction somewhere in the vocal tract and forcing air through the constriction so that turbulence is created, thereby producing a noiselike excitation.
- *Plosive sounds* are produced by completely closing off the vocal tract, building up pressure behind the closure, and then abruptly releasing it.

Detailed discussions of models for the speech signal and applications of the time-dependent Fourier transform may be found in texts such as Flanagan (1972), Rabiner and Schafer (1978), O'Shaughnessy (1987), and Parsons (1986).

Section 12.9.1 will describe a more detailed analytical model of the speech signal. Here we note that with a constant vocal tract shape, speech can be modeled as the response of a linear time-invariant system (the vocal tract) to a quasi-periodic pulse train for voiced sounds or wideband noise for unvoiced sounds. The vocal tract is an acoustic transmission system characterized by its natural frequencies, called *formants*, which correspond to resonances in its frequency response. In normal speech, the vocal tract changes shape relatively slowly with time as the tongue and lips perform the gestures of speech, and thus it can be modeled as a slowly time-varying filter that imposes its frequency-response properties on the spectrum of the excitation. A typical speech waveform is shown in Fig. 11.16.

From this brief description of the process of speech production, we see that speech is clearly a nonstationary signal. However, as illustrated in Fig. 11.16, the characteristics of the signal can be assumed to remain essentially constant over time intervals on the order of 30 or 40 ms. The frequency content of the speech signal may range up to 15 kHz or higher, but speech is highly intelligible even when bandlimited to frequencies below about 3 kHz. Commercial telephone systems, for example, typically limit the highest transmitted frequency to the range of 3–4 kHz. The standard sampling rate for digital telephone communication systems is 8000 samples/s. At this sampling rate, a 40-ms time interval is spanned by 320 samples.

Figure 11.16 illustrates that the waveform consists of a sequence of quasi-periodic *voiced* segments interspersed with noise-like *unvoiced* segments. This figure suggests that if the window length L is not too long, the properties of the signal will not change appreciably from the beginning of the segment to the end. Thus, the DFT of a windowed speech segment should display the frequency-domain properties of the signal. For example, if the window length is long enough so that the harmonics are resolved, the DFT of a windowed segment of voiced speech should show a series of peaks at integer multiples of the fundamental frequency of the signal in that interval. This would normally require that the window span several periods of the waveform. If the window is too short, then the harmonics will not be resolved, but the general spectral shape will still be evident. This is typical of the tradeoff between frequency resolution and time resolution that is required in the analysis of nonstationary signals. If the window is too long, the signal properties may change too much across the window; if the window is too short, resolution of narrowband components will be sacrificed. This tradeoff is illustrated in the following discussion.

Figure 11.17(a) shows a spectrogram display of the time-dependent Fourier transform of the sentence in Fig. 11.16. The time waveform is also shown on the same

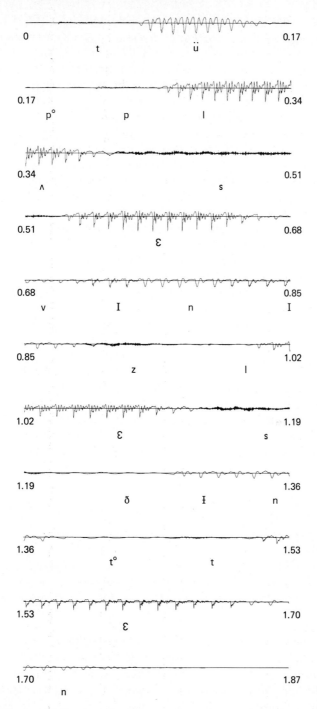

Figure 11.16 Waveform of the speech utterance "Two plus seven is less than ten." Each line is 0.17 s in duration. The time-aligned phonemic transcript is indicated below the waveform.

time scale, below the spectrogram. More specifically, Fig. 11.17(a) is a *wideband spectrogram*. A wideband spectrogram representation results from a window that is relatively short in time and is characterized by poor resolution in the frequency dimension and good resolution in the time dimension. The frequency axis is labeled in terms of continuous-time frequency. Since the sampling rate of the signal was 16,000 samples/s, it follows that the frequency $\lambda = \pi$ corresponds to 8 kHz. The specific window used in Fig. 11.17(a) was a Hamming window of duration 6.7 ms, corresponding to $L = 108$. The value of R was $R = 16$, representing 1-ms time increments. The broad dark bars that move horizontally across the spectrogram correspond to the resonance frequencies of the vocal tract, which, as we see, change with time. The vertically striated appearance of the spectrogram is due to the quasi-periodic nature of voiced portions of the waveform, as is evident by comparing the variations in the waveform display and the spectrogram. Since the length of the analysis window is on the order of the length of a period of the waveform, as the window slides along in time it alternately covers high-energy segments of the waveform and then lower-energy segments between, thereby producing the vertical striations in the plot during voiced intervals.

In a *narrowband* time-dependent Fourier analysis, a longer window is used to provide higher frequency resolution, with a corresponding decrease in time resolution. Such a narrowband analysis of speech is illustrated by the display in Fig. 11.17(b). In this case the window was a Hamming window of duration 45 ms. This corresponds to $L = 720$. The value of R was $R = 16$.

This discussion only hints at the many reasons that the time-dependent Fourier transform is so important in speech analysis and processing. Indeed, this concept is used directly and indirectly as the basis for acoustic-phonetic analysis and for many fundamental speech processing applications such as digital coding, noise and reverberation removal, speech recognition, speaker verification, and speaker identification. For present purposes, our discussion simply serves as an introductory illustration.

11.5.2 Time-Dependent Fourier Analysis of Radar Signals

Another application area in which the time-dependent Fourier transform plays an important role is radar signal analysis. The following are elements of a typical radar system:

- An antenna used for both transmitting and receiving.
- A transmitter that generates an appropriate signal at microwave frequencies. In our discussion, we will assume this signal to be sinusoidal pulses. While this is often the case, other signals may be used, depending on the specific radar objectives and design.
- A receiver that amplifies and detects echoes of the transmitted pulses that have been reflected from objects illuminated by the antenna.

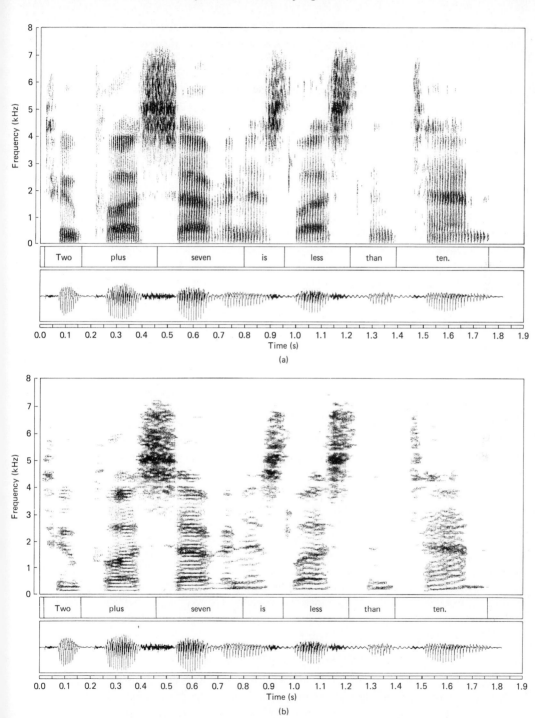

Figure 11.17 (a) Wideband spectrogram of waveform of Fig. 11.16. (b) Narrowband spectrogram.

In such a radar system, the transmitted sinusoidal signal propagates with the speed of light, reflects off the object, and returns with the speed of light to the antenna, thereby experiencing a time delay of the round-trip travel time from the antenna to the object. If we assume that the transmitted signal is a sinusoidal pulse of the form $\cos(\Omega_0 t)$ and the distance from the antenna to the object is $\rho(t)$, then the received signal is a pulse of the form

$$s(t) = \cos[\Omega_0(t - 2\rho(t)/c)], \tag{11.43}$$

where c is the velocity of light. If the object is not moving relative to the antenna, then $\rho(t) = \rho_0$, where ρ_0 is the *range*. Since the time delay between the transmitted and received pulses is $2\rho_0/c$, a measurement of the time delay may be used to estimate the range. If, however, $\rho(t)$ is not constant, the received signal is an angle-modulated sinusoid and the phase difference contains information about both the range and the relative motion of the object with respect to the antenna. Specifically, let us represent the time-varying range in a Taylor's series expansion as

$$\rho(t) = \rho_0 + \dot{\rho}_0 t + \frac{1}{2!}\ddot{\rho}_0 t^2 + \cdots, \tag{11.44}$$

where ρ_0 is the nominal range, $\dot{\rho}_0$ is the velocity, $\ddot{\rho}_0$ is the acceleration, and so on. Assuming that the object moves with constant velocity (i.e., $\ddot{\rho}_0 = 0$) and substituting Eq. (11.44) into Eq. (11.43), we obtain

$$s(t) = \cos[(\Omega_0 - 2\Omega_0\dot{\rho}_0/c)t - 2\Omega_0\rho_0/c]. \tag{11.45}$$

In this case, the frequency of the received signal differs from the frequency of the transmitted signal by the *Doppler frequency*, defined as

$$\Omega_d = -2\Omega_0\dot{\rho}_0/c. \tag{11.46}$$

Thus, the time delay can still be used to estimate the range and we can determine the speed of the object relative to the antenna if we can determine the Doppler frequency.

In a practical setting, the received signal is generally very weak, and thus a noise term should be added to Eq. (11.45). We will neglect the effects of noise in the simple analysis of this section. Also, in most radar systems, the signal of Eq. (11.45) would be frequency shifted to a lower nominal frequency in the detection process. However, the Doppler shift will still satisfy Eq. (11.46) even if $s(t)$ is demodulated to a lower center frequency.

To apply time-dependent Fourier analysis to such signals, we first bandlimit the signal to a frequency band that includes the expected Doppler frequency shifts and then sample the resulting signal with an appropriate sampling period T, thereby obtaining a discrete-time signal of the form

$$x[n] = \cos[(\omega_0 - 2\omega_0\dot{\rho}_0/c)n - 2\omega_0\rho_0/c], \tag{11.47}$$

where $\omega_0 = \Omega_0 T$. In many cases, the object motion would be more complicated than we have assumed, requiring the incorporation of higher-order terms from Eq. (11.44) and thereby producing a more complicated angle modulation in the received signal. Another way to represent this more complicated variation of the frequency of the echoes is to use the time-dependent Fourier transform with a window that is short

Short-time Fourier analysis

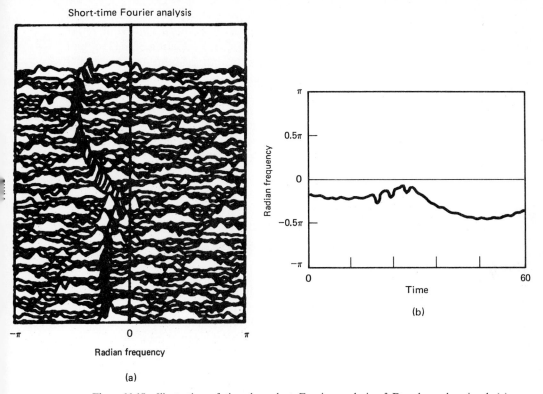

(a)

(b)

Figure 11.18 Illustration of time-dependent Fourier analysis of Doppler radar signal. (a) Sequence of Fourier transforms of Doppler radar signal. (b) Doppler frequency estimated by picking the largest peak in the time-dependent Fourier transform.

enough so that the assumption of constant Doppler-shifted frequency is valid across the entire window interval but not so short as to sacrifice adequate resolution when two or more moving objects create Doppler-shifted return signals that are superimposed at the receiver.

An example of time-dependent Fourier analysis of Doppler radar signals is shown in Fig. 11.18. (See Schaefer et al., 1979.) The radar data had been preprocessed to remove low-velocity Doppler shift, leaving the variations displayed in Fig. 11.18. The window for the time-dependent Fourier transform was a Kaiser window with $N = L = 64$ and $\beta = 4$. In this figure, $|X_r[k]|$ is plotted with time as the vertical dimension (increasing upward) and frequency as the horizontal dimension.† In this case the successive DFTs are plotted close together. A hidden-line elimination algorithm is used to create a two-dimensional view of the time-dependent Fourier transform. To the left of the center line is a strong peak that moves in a smooth path through the time-frequency plane. This corresponds to a moving object whose

---

† The plot shows the negative frequencies on the left of the line through the center of the plot and positive frequencies on the right. This can be achieved by computing the DFT of $(-1)^n x_r[n]$ and noting that this effectively shifts the origin of the DFT index to $k = N/2$. Alternatively, the DFT of $x_r[n]$ can be computed and then re-indexed.

velocity is varying in a regular manner. The other broad peaks in the time-dependent Fourier transform are due to noise and spurious returns called *clutter* in radar terminology. An example of motion that might create such a variation of the Doppler frequency is a rocket moving at constant velocity but rotating about its longitudinal axis. A peak moving through the time-dependent Fourier transform might correspond to reflections from a fin on the rocket that is alternately moving toward and then away from the antenna because of the spinning of the rocket. Figure 11.18(b) shows an estimate of the Doppler frequency as a function of time. This estimate was obtained by simply locating the highest peak in each DFT. The cover illustration is an example with simulated signals. Notice the effect of aliasing as the Doppler frequency exceeds half the sampling rate.

11.6 FOURIER ANALYSIS OF STATIONARY RANDOM SIGNALS: THE PERIODOGRAM

In the previous sections, we have discussed and illustrated Fourier analysis for sinusoidal signals with stationary (non-time-varying) parameters and for nonstationary signals such as speech and radar. In cases where the signal can be modeled by a sum of sinusoids or a linear system excited by a periodic pulse train, the Fourier transforms of finite-length segments of the signal have a convenient and natural interpretation in terms of Fourier transforms, windowing, and linear system theory. However, more noise-like signals such as the example of unvoiced speech in Section 11.5.1 are best modeled as random signals.

As we discussed in Section 2.10 and as shown in Appendix A, random processes are often used as signal models when the process that generates the signal is too complex for a reasonable deterministic model. Typically, when the input to a linear time-invariant system is modeled as a stationary random process, many of the essential characteristics of the input and output are adequately represented by average properties such as the mean value (dc level), variance (average power), autocorrelation function, or power density spectrum. Consequently, it is of particular interest to estimate these for a given signal. As discussed in Appendix A, a typical estimate of the mean value of a stationary random process from a finite-length segment of data is the *sample mean*, defined as

$$\hat{m}_x = \frac{1}{L} \sum_{n=0}^{L-1} x[n]. \tag{11.48}$$

Similarly, a typical estimate of the variance is the *sample variance*, defined as

$$\hat{\sigma}_x^2 = \frac{1}{L} \sum_{n=0}^{L-1} (x[n] - \hat{m}_x)^2. \tag{11.49}$$

Both of these are *unbiased* estimators; i.e., the expected value of $\hat{m}_x$ is the true mean m_x and the expected value of $\hat{\sigma}_x^2$ is the true variance σ_x^2. Furthermore, they are both *consistent* estimators, i.e., the variance of these estimates both approach zero as M approaches ∞. Estimates that are unbiased and consistent are generally desirable since they become increasingly better as the number of data samples increases.

In the remainder of this chapter we study the estimation of the power spectrum† of a random signal using the DFT. As we will see, there are two basic approaches to estimating the power spectrum. One approach, which we develop in this section, is referred to as *periodogram analysis* and is based on direct Fourier transformation of finite-length segments of the signal. The second approach, developed in Section 11.7, is to first estimate the autocovariance sequence and then compute the Fourier transform of this estimate. In either case, we are typically interested in obtaining unbiased consistent estimators. Unfortunately, the analysis of such estimators is very difficult, and generally only approximate analyses can be accomplished. Even approximate analyses are beyond the scope of this text, and we refer to the results of such analyses only in a qualitative way. Detailed discussions are given in Blackman and Tukey (1958), Hannan (1960), Jenkins and Watts (1968), Koopmanns (1974), Kay and Marple (1981), Marple (1987), and Kay (1988).

11.6.1 The Periodogram

Let us consider the problem of estimating the power density spectrum $P_{ss}(\Omega)$ of a continuous-time signal $s_c(t)$. An intuitive approach to the estimation of the power spectrum is suggested by Fig. 11.1 and the associated discussion in Section 11.1, where we now assume that the input signal $s_c(t)$ is a stationary random signal. The anti-aliasing lowpass filter creates a new stationary random signal whose power spectrum is bandlimited so that the signal can be sampled without aliasing. Then $x[n]$ is a stationary discrete-time random signal whose power density spectrum $P_{xx}(\omega)$ is proportional to $P_{ss}(\Omega)$ over the bandwidth of the anti-aliasing filter, i.e.,

$$P_{xx}(\omega) = \frac{1}{T} P_{ss}\left(\frac{\omega}{T}\right), \qquad |\omega| < \pi, \tag{11.50}$$

where we have assumed that the cutoff frequency of the anti-aliasing filter is π/T and that T is the sampling period. (See Problem 11.8 for a further consideration of sampling of random signals.) Consequently, a reasonable estimate of $P_{xx}(\omega)$ will provide a reasonable estimate of $P_{ss}(\Omega)$. The window $w[n]$ in Fig. 11.1 selects a finite-length segment (L samples) of $x[n]$, which we denote $v[n]$, the Fourier transform of which is

$$V(e^{j\omega}) = \sum_{n=0}^{L-1} w[n]x[n]e^{-j\omega n}. \tag{11.51}$$

Consider as an estimate of the power spectrum the quantity

$$I(\omega) = \frac{1}{LU} |V(e^{j\omega})|^2, \tag{11.52}$$

---

† The term *power spectrum* is commonly used interchangeably with the more precise term *power density spectrum*.

where the constant U anticipates a need for normalization to remove bias in the spectral estimate. When the window $w[n]$ is the rectangular window sequence, this estimator for the power spectrum is called the *periodogram*. If the window is not rectangular, it is called the *modified periodogram*. Clearly, the periodogram has some of the basic properties of the power spectrum. It is nonnegative, and for real signals it is a real and even function of frequency. Furthermore, it can be shown (Problem 11.11) that

$$I(\omega) = \frac{1}{LU} \sum_{m=-(L-1)}^{L-1} c_{vv}[m]e^{-j\omega m}, \tag{11.53}$$

where

$$c_{vv}[m] = \sum_{n=0}^{L-1} x[n]w[n]x[n+m]w[n+m]. \tag{11.54}$$

We note that the sequence $c_{vv}[m]$ is the aperiodic correlation sequence for the finite-length sequence $v[n] = w[n]x[n]$. Consequently, the periodogram is in fact the Fourier transform of the aperiodic correlation of the windowed data sequence.

Explicit computation of the periodogram can be carried out only at discrete frequencies. From Eqs. (11.51) and (11.52) we see that if the Fourier transform of $w[n]x[n]$ is replaced by its DFT, we will obtain samples at the DFT frequencies $\omega_k = 2\pi k/N$, $k = 0, 1, \ldots, N-1$. Specifically, samples of the periodogram are given by

$$I(\omega_k) = \frac{1}{LU}|V[k]|^2, \tag{11.55}$$

where $V[k]$ is the N-point DFT of $w[n]x[n]$. If we want to choose N to be greater than the window length L, appropriate zero padding would be applied to the sequence $w[n]x[n]$.

If a random signal has a nonzero mean, its power spectrum has an impulse at zero frequency. If the mean is relatively large, this component will dominate the spectrum estimate, causing low-amplitude, low-frequency components to be obscured by leakage. Therefore, in practice the mean is often estimated using Eq. (11.48), and the resulting estimate is subtracted from the random signal before computing the power spectrum estimate. Although the sample mean is only an approximate estimate of the zero frequency component, subtracting it from the signal often leads to better estimates at neighboring frequencies.

11.6.2 Properties of the Periodogram

The nature of the periodogram estimate of the power spectrum can be determined by recognizing that for each value of ω, $I(\omega)$ is a random variable. By computing the mean and variance of $I(\omega)$, we can determine whether the estimate is biased and whether it is consistent.

From Eq. (11.53), the expected value of $I(\omega)$ is

$$\mathcal{E}\{I(\omega)\} = \frac{1}{LU} \sum_{m=-(L-1)}^{L-1} \mathcal{E}\{c_{vv}[m]\}e^{-j\omega m}. \tag{11.56}$$

The expected value of $c_{vv}[m]$ can be expressed as

$$\mathcal{E}\{c_{vv}[m]\} = \sum_{n=0}^{L-1} \mathcal{E}\{x[n]w[n]x[n+m]w[n+m]\}$$

$$= \sum_{n=0}^{L-1} w[n]w[n+m]\mathcal{E}\{x[n]x[n+m]\}. \tag{11.57}$$

Since we are assuming that $x[n]$ is stationary,

$$\mathcal{E}\{x[n]x[n+m]\} = \phi_{xx}[m], \tag{11.58}$$

and Eq. (11.57) can then be rewritten as

$$\mathcal{E}\{c_{vv}[m]\} = c_{ww}[m]\phi_{xx}[m], \tag{11.59}$$

where $c_{ww}[m]$ is the aperiodic autocorrelation of the window, i.e.,

$$c_{ww}[m] = \sum_{n=0}^{L-1} w[n]w[n+m]. \tag{11.60}$$

From Eq. (11.56), Eq. (11.59), and the modulation/windowing property of Fourier transforms (Section 2.9.7), it follows that

$$\mathcal{E}\{I(\omega)\} = \frac{1}{2\pi LU} \int_{-\pi}^{\pi} P_{xx}(\theta)C_{ww}(e^{j(\omega-\theta)})\,d\theta, \tag{11.61}$$

where $C_{ww}(e^{j\omega})$ is the Fourier transform of the aperiodic autocorrelation of the window, i.e.,

$$C_{ww}(e^{j\omega}) = |W(e^{j\omega})|^2. \tag{11.62}$$

According to Eq. (11.61), the (modified) periodogram is a biased estimate of the power spectrum since $\mathcal{E}\{I(\omega)\}$ is not equal to $P_{xx}(\omega)$. Indeed, we see that the bias arises as a result of convolution of the true power spectrum with the Fourier transform of the aperiodic autocorrelation of the data window. If we increase the window length, we expect that $W(e^{j\omega})$ should become more concentrated around $\omega = 0$, and thus $C_{ww}(e^{j\omega})$ should look increasingly like a periodic impulse train. If the scale factor $1/(LU)$ is correctly chosen, then $\mathcal{E}\{I(\omega)\}$ should approach $P_{xx}(\omega)$ as $W(e^{j\omega})$ approaches a periodic impulse train. The scale can be adjusted by choosing the normalizing constant U so that

$$\frac{1}{2\pi LU} \int_{-\pi}^{\pi} |W(e^{j\omega})|^2\,d\omega = \frac{1}{LU} \sum_{n=0}^{L-1} (w[n])^2 = 1, \tag{11.63}$$

or

$$U = \frac{1}{L} \sum_{n=0}^{L-1} (w[n])^2. \tag{11.64}$$

For the rectangular window, we would then choose $U = 1$, while other data windows would require a value of $0 < U < 1$ if $w[n]$ is normalized to a maximum value of 1. Alternatively, the normalization can be absorbed into the amplitude of $w[n]$. In summary, then, if properly normalized, the (modified) periodogram is asymptotically unbiased; i.e., the bias approaches zero as the window length increases.

To examine whether the periodogram is a consistent estimate or becomes a consistent estimate as the window length increases, it is necessary to consider the behavior of the variance of the periodogram. An expression for the variance of the periodogram is very difficult to obtain even in the simplest cases. However, it has been shown (see Jenkins and Watts, 1968) that over a wide range of conditions, as the window length increases,

$$\text{var}[I(\omega)] \simeq P_{xx}^2(\omega). \tag{11.65}$$

That is, the variance of the periodogram estimate is approximately the same size as the power spectrum that we are estimating. Therefore, since the variance does not asymptotically approach zero with increasing window length, the periodogram is not a consistent estimate.

The properties of the periodogram estimate of the power spectrum just discussed are illustrated in Fig. 11.19, which shows periodogram estimates of white noise using rectangular windows of lengths $L = 16$, 64, 256, and 1024. The sequence $x[n]$ was obtained from a pseudorandom number generator whose output was scaled so that $|x[n]| \leq \sqrt{3}$. A good random number generator produces a uniform distribution of amplitudes, and the sample-to-sample correlation is small. Thus, the power spectrum of the output of the random number generator could be modeled in this case by $P_{xx}(\omega) = \sigma_x^2 = 1$ for all ω. For each of the four rectangular windows, the periodogram was computed with normalizing constant $U = 1$ and at frequencies $\omega_k = 2\pi k/N$ for $N = 1024$ using the DFT. That is,

$$I[k] = I(\omega_k) = \frac{1}{L}|V[k]|^2 = \frac{1}{L}\left|\sum_{n=0}^{L-1} w[n]x[n]e^{-j(2\pi/N)kn}\right|^2. \tag{11.66}$$

In Fig. 11.19, the DFT values are connected by straight lines for purposes of display. We note that the spectral estimate fluctuates more rapidly as the window length L increases. This behavior can be understood by recalling that although we view the periodogram method as a direct computation of the spectral estimate, we have seen that the underlying correlation estimate of Eq. (11.54) is in effect Fourier transformed to obtain the periodogram. Figure 11.20 illustrates a windowed sequence, $x[n]w[n]$, and a shifted version, $x[n + m]w[n + m]$, as required in Eq. (11.54). From this figure we see that $L - m - 1$ signal values are involved in computing a particular correlation lag value $c_{vv}[m]$. Thus, when m is close to L, only a few values of $x[n]$ are involved in the computation and we expect that the estimate of the correlation sequence would be considerably more inaccurate for these values of m and consequently will also show considerable variation between adjacent values of m. On the other hand, when m is small, many more samples are involved and the variability of $c_{vv}[m]$ with m should not be as great. The variability at large values of m manifests itself in the Fourier transform as fluctuations at all frequencies and thus, for

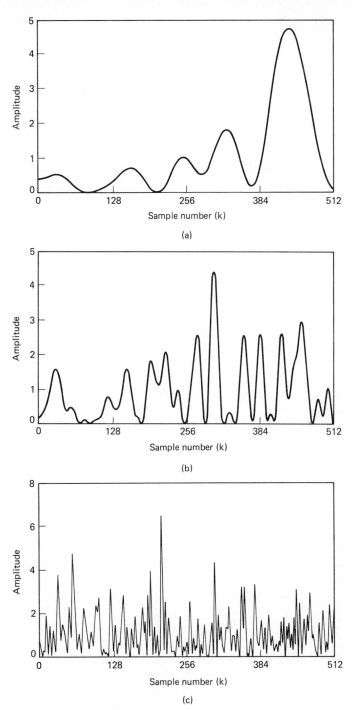

Figure 11.19 Periodograms of pseudorandom white-noise sequence. (a) Window length $L = 16$ and DFT length $N = 1024$. (b) $L = 64$ and $N = 1024$. (c) $L = 256$ and $N = 1024$.

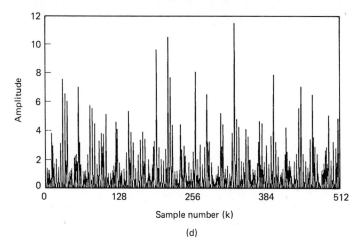

(d)

Figure 11.19 *(continued)* (d) $L = 1024$ and $N = 1024$.

large L, the periodogram estimate tends to vary rapidly with frequency. Indeed, it can be shown (see Jenkins and Watts, 1968) that if $N = L$, the periodogram estimates at the DFT frequencies $2\pi k/N$ become uncorrelated. Since as N increases, the DFT frequencies get closer together, this behavior is inconsistent with our goal of obtaining a good estimate of the power spectrum. We would prefer to obtain a smooth spectrum estimate without random variations resulting from the estimation process. This can be accomplished by averaging multiple independent periodogram estimates to reduce the fluctuations.

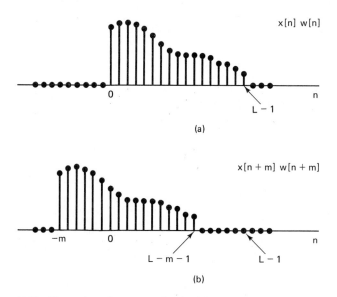

Figure 11.20 Illustration of sequences involved in Eq. (11.54). (a) A finite-length sequence. (b) Shifted sequence for $m > 0$.

11.6.3 Periodogram Averaging

The averaging of periodograms in spectrum estimation was first studied extensively by Bartlett (1953); later, after fast algorithms for computing the DFT were developed, Welch (1970) combined these computational algorithms with the use of a data window $w[n]$ to develop the method of averaging modified periodograms. In periodogram averaging, a data sequence $x[n]$, $0 \leq n \leq Q - 1$, is divided into segments of length L samples, with a window of length L applied to each; i.e., we form the segments

$$x_r[n] = x[rR + n]w[n], \qquad 0 \leq n \leq L - 1. \tag{11.67}$$

If $R < L$ the segments overlap, and for $R = L$ the segments are contiguous. Note that Q denotes the length of the available data. The total number of segments depends on the values of and relationship among R, L, and Q. Specifically, there will be K full-length segments where K is the largest integer for which $(K - 1)R + (L - 1) \leq Q - 1$. The periodogram of the rth segment is

$$I_r(\omega) = \frac{1}{LU}|X_r(e^{j\omega})|^2. \tag{11.68}$$

Each $I_r(\omega)$ has the properties of a periodogram as described previously. Periodogram averaging consists of averaging together the K periodogram estimates $I_r(\omega)$; i.e., we form the time average periodogram defined as

$$\bar{I}(\omega) = \frac{1}{K} \sum_{r=0}^{K-1} I_r(\omega). \tag{11.69}$$

To examine the bias and variance of $\bar{I}(\omega)$, let us take $L = R$ so that the segments do not overlap, and assume that $\phi_{xx}[m]$ is small for $m > L$; i.e., signal samples more than L apart are approximately uncorrelated. Then it is reasonable to assume that the periodograms $I_r(\omega)$ will be identically distributed independent random variables. Under this assumption, the expected value of $\bar{I}(\omega)$ is

$$\mathcal{E}\{\bar{I}(\omega)\} = \frac{1}{K} \sum_{r=0}^{K-1} \mathcal{E}\{I_r(\omega)\}, \tag{11.70}$$

or, since we assume that the periodograms are independent and identically distributed,

$$\mathcal{E}\{\bar{I}(\omega)\} = \mathcal{E}\{I_r(\omega)\} \qquad \text{for any } r. \tag{11.71}$$

From Eq. (11.61) it follows that

$$\mathcal{E}\{\bar{I}(\omega)\} = \mathcal{E}\{I_r(\omega)\} = \frac{1}{2\pi LU} \int_{-\pi}^{\pi} P_{xx}(\theta)C_{ww}(e^{j(\omega - \theta)}) \, d\theta, \tag{11.72}$$

where L is the window length. When the window $w[n]$ is the rectangular window, the method of averaging periodograms is called *Bartlett's procedure*, and in this case it can be shown that

$$c_{ww}[m] = \begin{cases} L - |m|, & |m| \leq (L - 1), \\ 0 & \text{otherwise,} \end{cases} \tag{11.73}$$

and, therefore,

$$C_{ww}(e^{j\omega}) = \left(\frac{\sin(\omega L/2)}{\sin(\omega/2)}\right)^2. \tag{11.74}$$

That is, the expected value of the average periodogram spectrum estimate is the convolution of the true power spectrum with the Fourier transform of the triangular sequence $c_{ww}[n]$ that results as the autocorrelation of the rectangular window. Thus, the average periodogram is also a biased estimate of the power spectrum.

To examine the variance, we use the fact that in general, the variance of the sum of K independent identically distributed random variables is $1/K$ times the variance of each individual random variable (see Papoulis, 1984). Therefore, the variance of the average periodogram is

$$\text{var}[\bar{I}(\omega)] = \frac{1}{K}\text{var}[I_r(\omega)], \tag{11.75}$$

or with Eq. (11.65) it follows that

$$\text{var}[\bar{I}(\omega)] \simeq \frac{1}{K}P_{xx}^2(\omega). \tag{11.76}$$

Consequently, the variance of $\bar{I}(\omega)$ is inversely proportional to the number of periodograms averaged, and as K increases, the variance approaches zero.

From Eq. (11.74) we see that as L, the length of the segment $x_r[n]$, increases, the mainlobe of $C_{ww}(e^{j\omega})$ decreases in width and consequently, from Eq. (11.72), $\mathcal{E}\{\bar{I}(\omega)\}$ more closely approximates $P_{xx}(\omega)$. However, for fixed total data length Q, the total number of segments (assuming $L = R$) is Q/L; therefore, as L increases, K decreases. Correspondingly, from Eq. (11.76), the variance of $\bar{I}(\omega)$ will increase. Thus, as is typical in statistical estimation problems, for fixed data length there is a tradeoff between bias and variance. However, as the data length Q increases, both L and K can be allowed to increase, so that as Q approaches ∞, the bias and variance of $\bar{I}(\omega)$ can approach zero. Consequently, periodogram averaging provides an asymptotically unbiased, consistent estimate of $P_{xx}(\omega)$.

The preceding discussion assumed that nonoverlapping rectangular windows were used in computing the time-dependent periodograms. If a different window shape is used, Welch (1970) showed that the variance of the average periodogram still behaves as in Eq. (11.76). Welch also considered the case of overlapping windows and showed that if the overlap is one-half the window length, the variance is further reduced by almost a factor of 2 due to the doubling of the number of sections. Greater overlap does not continue to reduce the variance because the segments become less and less independent as the overlap increases.

11.6.4 Computation of Average Periodograms Using the DFT

As with the periodogram, the average periodogram can be explicitly evaluated only at a discrete set of frequencies. Because of the availability of the fast Fourier transform algorithms for computing the DFT, a particularly convenient and widely used choice are the frequencies $\omega_k = 2\pi k/N$ for an appropriate choice of N. From Eq. (11.69) we see that if the DFT of $x_r[n]$ is substituted for the Fourier transform of $x_r[n]$ in Eq. (11.68), we obtain samples of $\bar{I}(\omega)$ at the DFT frequencies $\omega_k = 2\pi k/N$, $k = 0, 1, \ldots, N-1$. Specifically, with $X_r[k]$ denoting the DFT of $x_r[n]$,

$$I_r[k] = I_r(\omega_k) = \frac{1}{LU}|X_r[k]|^2, \tag{11.77a}$$

$$\bar{I}[k] = \bar{I}(\omega_k) = \frac{1}{K}\sum_{r=0}^{K-1} I_r[k]. \tag{11.77b}$$

We will denote $I_r(2\pi k/N)$ as the sequence $I_r[k]$ and $\bar{I}(2\pi k/N)$ as the sequence $\bar{I}[k]$. According to Eqs. (11.77), the average periodogram estimate of the power spectrum is computed at N equally spaced frequencies by averaging the DFTs of the windowed data segments with the normalizing factor LU. This method of power spectrum estimation provides a very convenient framework within which to trade between resolution and variance of the spectral estimate. It is particularly simple and efficient to implement using the fast Fourier transform algorithms discussed in Chapter 9. An important advantage of this method over the methods to be discussed in Section 11.7 is that the spectrum estimate is always nonnegative.

11.6.5 An Example of Periodogram Analysis

Power spectrum analysis is a valuable tool for signal modeling, and it also can be used for signal detection, particularly for finding hidden periodicities in sampled signals. As an example of this type of application of the average periodogram method, consider the sequence

$$x[n] = A\cos(\omega_0 n + \theta) + e[n], \tag{11.78}$$

where θ is a random variable uniformly distributed between 0 and 2π and $e[n]$ is a zero-mean white-noise sequence that has a power spectrum approximately constant, i.e., $P_{ee}(\omega) = \sigma_e^2$ for all ω. In signal models of this form, the cosine is generally the desired component and $e[n]$ is an undesired noise component. Often in practical signal detection problems we are interested in the case for which the power in the cosine signal is small compared with the noise power. It can be shown (see Problem 11.9) that over one period in frequency, the power spectrum for this signal is

$$P_{xx}(\omega) = \frac{A^2\pi}{2}[\delta(\omega - \omega_0) + \delta(\omega + \omega_0)] + \sigma_e^2 \qquad \text{for } |\omega| \le \pi. \tag{11.79}$$

From Eqs. (11.72) and (11.79) it follows that the expected value of the average periodogram is

$$\mathcal{E}\{\bar{I}(\omega)\} = \frac{A^2}{4LU}[C_{ww}(e^{j(\omega-\omega_0)}) + C_{ww}(e^{j(\omega+\omega_0)})] + \sigma_e^2. \tag{11.80}$$

Figures 11.21 and 11.22 show the use of the averaging method for a signal of the form of Eq. (11.78), with $A = 0.5$, $\omega_0 = 2\pi/21$, and $\sigma_e^2 = 1$. The mean of the noise component is zero. Figure 11.21 shows 101 samples of the sequence $x[n]$. Since the noise component $e[n]$ has maximum amplitude $\sqrt{3}$, the cosine component in the sequence $x[n]$ is not visually apparent. Figure 11.22 shows average periodogram

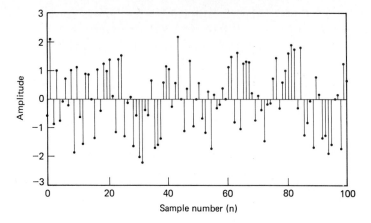

Figure 11.21   Cosine sequence with white noise as in Eq. (11.78).

estimates of the power spectrum for rectangular windows with amplitude 1, so that $U = 1$, and of lengths $L = 1024, 256, 64$, and 16, with the total record length $Q = 1024$ in all cases. Except for Fig. 11.22(a), the windows overlap by one-half the window length. Figure 11.22(a) is the periodogram of the entire record, and Figs. 11.22(b), (c), and (d) show the average periodogram for $K = 7, 31$, and 127 segments, respectively. In all cases the average periodogram was evaluated using 1024-point DFTs at frequencies $\omega_k = 2\pi k/1024$. Therefore, the frequency $\omega_0 = 2\pi/21$ lies between DFT frequencies $\omega_{48} = 2\pi 48/1024$ and $\dot{\omega}_{49} = 2\pi 49/1024$.

In using such estimates of the power spectrum to detect the presence and/or the frequency of the cosine component, we might search for the highest peaks in the spectral estimate and compare their size to the surrounding spectral values. Using Eq. (11.74) and (11.80), the expected value of the average periodogram at frequency ω_0 is

$$\mathscr{E}\{\bar{I}(\omega_0)\} = \frac{A^2 L}{4} + \sigma_e^2. \tag{11.81}$$

Thus, if the peak due to the cosine component is to stand out against the variability of the average periodogram, then in this special case we must choose L so that $A^2 L/4 \gg \sigma_e^2$. This is illustrated by Fig. 11.22(a), where L is as large as it can be for the record length Q. We see that $L = 1024$ gives a very narrow mainlobe of the Fourier transform of the autocorrelation of the rectangular window, so it would be possible to resolve very closely spaced sinusoidal signals. Note that for the parameters of this example, the peak amplitude in the average periodogram at frequency $2\pi/21$ is close but not equal to the expected value of 65. We also observe additional peaks in the average periodogram with amplitudes greater than 10. Clearly, if the cosine amplitude A had been smaller by only a factor of 2, it is likely that its peak would have been confused with the inherent variability of the periodogram.

To reduce the variability while keeping the record length constant, we can use shorter windows and average over more sections. This is illustrated by parts (b), (c), and (d) of Fig. 11.22. Note that as more sections are used, the variance of the spectral

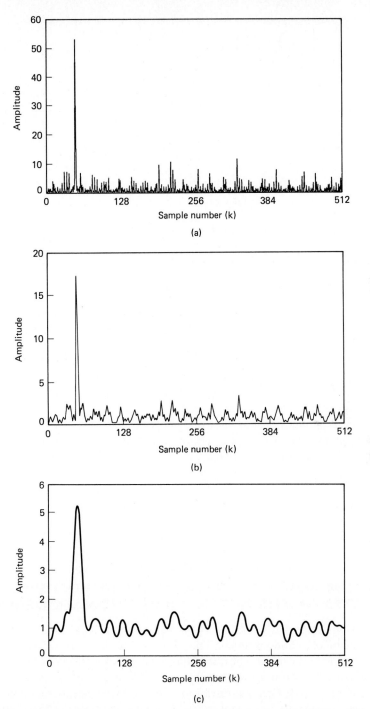

Figure 11.22   Example of average periodogram for signal of length $Q = 1024$. (a) Periodogram for window length $L = Q = 1024$ (only one segment). (b) $K = 7$ and $L = 256$ (overlap by $L/2$). (c) $K = 31$ and $L = 64$.

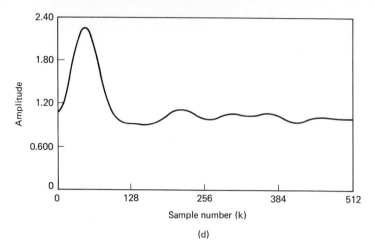

Figure 11.22   (*continued*) (d) $K = 127$ and $L = 16$.

estimate decreases, but in accordance with Eq. (11.81) so does the amplitude of the peak due to the cosine. Recall that the idealized power spectrum of the model for the pseudorandom noise generator is a constant ($\sigma_e^2 = 1$) for all frequencies. In Fig. 11.22(d), the window is very short and thus the fluctuations of the spectrum estimate are reduced, but the spectral peak due to the cosine is very broad; if the length is any shorter, spectral leakage from the negative frequency component would cause there to be no distinct peak in the low-frequency region.

This example confirms that the average periodogram provides a straightforward method of trading off between spectral resolution and reduction of the variance of the spectral estimate. Although the theme of the example was the detection of a sinusoid in noise, the average periodogram could also be used in signal modeling. The spectral estimates of Fig. 11.22 clearly suggest a signal model of the form of Eq. (11.78), and most of the parameters of the model could be estimated from the average periodogram power spectrum estimate.

11.7 SPECTRUM ANALYSIS OF RANDOM SIGNALS USING ESTIMATES OF THE AUTOCORRELATION SEQUENCE

In the previous section we considered the periodogram as a direct estimate of the power spectrum of a random signal. The periodogram or the average periodogram is a direct estimate in the sense that it is obtained directly by Fourier transformation of the samples of the random signal. Another approach, based on the fact that the power density spectrum is the Fourier transform of the autocorrelation function, is to first estimate the autocorrelation function $\phi_{xx}[m]$ and then compute the Fourier transform of this estimate. In this section we explore some of the important facets of this approach and show how the DFT can be used to implement it.

Let us assume as before that we are given a finite record of a random signal $x[n]$. This sequence is denoted

$$v[n] = \begin{cases} x[n] & \text{for } 0 \le n \le Q - 1, \\ 0 & \text{otherwise.} \end{cases} \tag{11.82}$$

Consider an estimate of the autocorrelation sequence as

$$\hat{\phi}_{xx}[m] = \frac{1}{Q} c_{vv}[m], \tag{11.83}$$

where, since $c_{vv}[-m] = c_{vv}[m]$,

$$c_{vv}[m] = \sum_{m=0}^{Q-1} v[n]v[n + m] = \begin{cases} \displaystyle\sum_{n=0}^{Q-|m|-1} x[n]x[n + |m|], & |m| \le Q - 1, \\ 0 & \text{otherwise,} \end{cases} \tag{11.84}$$

corresponding to the aperiodic correlation of a rectangularly windowed segment of $x[n]$.

To determine the properties of this estimate of the autocorrelation sequence, we consider the mean and variance of the random variable $\hat{\phi}_{xx}[m]$. From Eqs. (11.83) and (11.84) it follows that

$$\mathcal{E}\{\hat{\phi}_{xx}[m]\} = \frac{1}{Q} \sum_{n=0}^{Q-|m|-1} \mathcal{E}\{x[n]x[n + |m|]\} = \frac{1}{Q} \sum_{n=0}^{Q-|m|-1} \phi_{xx}[m], \tag{11.85}$$

and since $\phi_{xx}[m]$ does not depend on n for a stationary random process,

$$\mathcal{E}\{\hat{\phi}_{xx}[m]\} = \begin{cases} \left(\dfrac{Q - |m|}{Q}\right)\phi_{xx}[m], & |m| \le Q - 1, \\ 0 & \text{otherwise.} \end{cases} \tag{11.86}$$

From Eq. (11.86) we see that $\hat{\phi}_{xx}[m]$ is a biased estimate of $\phi_{xx}[m]$, since $\mathcal{E}\{\hat{\phi}_{xx}[m]\}$ is not equal to $\phi_{xx}[m]$, but the bias is small if $|m| \ll Q$. We see also that an unbiased estimator of the autocorrelation sequence for $|m| \le Q - 1$ is

$$\breve{\phi}_{xx}[m] = \left(\frac{1}{Q - |m|}\right) c_{vv}[m], \tag{11.87}$$

i.e., the estimator is unbiased if we divide by the number of nonzero terms in the sum of lagged products rather than by the total number of samples in the data record.

The variance of the autocorrelation function estimates is difficult to compute even with simplifying assumptions. However, approximate formulas for the variance of both $\hat{\phi}_{xx}[m]$ and $\breve{\phi}_{xx}[m]$ can be found in Jenkins and Watts (1968). For our purposes here, it is sufficient to observe from Eq. (11.84) that as $|m|$ approaches Q, fewer and fewer samples of $x[n]$ are involved in the computation of the autocorrelation estimate, and therefore the variance of the autocorrelation estimate can be expected to increase with increasing $|m|$. In the case of the periodogram, this increased variance affects the spectrum estimate at all frequencies because all the autocorrelation lag values are implicitly involved in the computation of the periodogram. However, by explicitly computing the autocorrelation estimate, we are free to choose which correlation lag values to include in estimating the power spectrum. Thus, we define the power spectrum estimate

$$S(\omega) = \sum_{m=-(M-1)}^{M-1} \hat{\phi}_{xx}[m]w_c[m]e^{-j\omega m}, \qquad (11.88)$$

where $w_c[m]$ is a symmetric window of length $(2M-1)$ applied to the estimated autocorrelation function. We require that the product of the autocorrelation sequence and the window be an even sequence when $x[n]$ is real so that the power spectrum estimate will be a real even function of ω. Therefore, the correlation window must be an even sequence. By limiting the length of the correlation window so that $M \ll Q$, we include only autocorrelation estimates for which the variance is low.

The mechanism by which windowing the autocorrelation sequence reduces the variance of the power spectrum estimate is best understood in the frequency domain. From Eqs. (11.53), (11.54), and (11.84) it follows that with $w[n] = 1$ for $0 \leq n \leq (Q-1)$, i.e., a rectangular window, the periodogram is the Fourier transform of the autocorrelation estimate $\hat{\phi}_{xx}[m]$, i.e.,

$$\hat{\phi}_{xx}[m] = \frac{1}{Q}c_{vv}[m] \overset{\mathscr{F}}{\leftrightarrow} \frac{1}{Q}|V(e^{j\omega})|^2 = I(\omega). \qquad (11.89)$$

Therefore, from Eq. (11.88), the spectrum estimate $S(\omega)$ is the convolution

$$S(\omega) = \frac{1}{2\pi}\int_{-\pi}^{\pi} I(\theta)W_c(e^{j(\omega-\theta)})d\theta. \qquad (11.90)$$

From Eq. (11.90) we see that the effect of applying the window $w_c[m]$ to the autocorrelation estimate is to convolve the periodogram with the Fourier transform of the autocorrelation window. This will have the effect of smoothing the rapid fluctuations of the periodogram spectrum estimate. The shorter the correlation window, the smoother the spectrum estimate and vice versa.

The power spectrum $P_{xx}(\omega)$ is a nonnegative function of frequency, and the periodogram and the average periodogram automatically have this property by definition. In contrast, from Eq. (11.90) it is evident that nonnegativity is not guaranteed for $S(\omega)$ unless we impose the further condition that

$$W_c(e^{j\omega}) \geq 0 \qquad \text{for } -\pi < \omega \leq \pi. \qquad (11.91)$$

This condition is satisfied by the Fourier transform of the triangular (Bartlett) window, but it is not satisfied by the rectangular, Hanning, Hamming, or Kaiser windows. Therefore, although these latter windows have smaller sidelobes than the triangular window, spectral leakage may cause negative spectral estimates in low-level regions of the spectrum.

The expected value of the smoothed periodogram is

$$\mathcal{E}\{S(\omega)\} = \sum_{m=-(M-1)}^{M-1} \mathcal{E}\{\hat{\phi}_{xx}[m]\} w_c[m] e^{-j\omega m}$$

$$= \sum_{m=-(M-1)}^{M-1} \phi_{xx}[m]\left(\frac{Q-|m|}{Q}\right) w_c[m] e^{-j\omega m}. \tag{11.92}$$

If $Q \gg M$, the term $(Q - |m|)/Q$ in Eq. (11.92) can be neglected,† so we obtain

$$\mathcal{E}\{S(\omega)\} \cong \sum_{m=-(M-1)}^{M-1} \phi_{xx}[m] w_c[m] e^{-j\omega m} = \frac{1}{2\pi}\int_{-\pi}^{\pi} P_{xx}(\theta) W_c[e^{j(\omega-\theta)}] d\theta. \tag{11.93}$$

Thus, the windowed autocorrelation estimate leads to a biased estimate of the power spectrum. Just as with the average periodogram, it is possible to trade spectral resolution for reduced variance of the spectrum estimate. If the length of the data record is fixed, we can have lower variance if we are willing to accept poorer resolution of closely spaced narrowband spectral components, or we can have better resolution if we accept higher variance. If we are free to observe the signal for a longer time (i.e., increase the length Q of the data record), then both the resolution and the variance can be improved. The spectrum estimate $S(\omega)$ is asymptotically unbiased if the correlation window is normalized so that

$$\frac{1}{2\pi}\int_{-\pi}^{\pi} W_c(e^{j\omega})\, d\omega = 1 = w_c[0]. \tag{11.94}$$

With this normalization, as we increase Q together with the length of the correlation window, the Fourier transform of the correlation window approaches a periodic impulse train and the convolution of Eq. (11.93) duplicates $P_{xx}(\omega)$.

The variance of $S(\omega)$ has been shown (see Jenkins and Watts, 1968) to be of the form

$$\text{var}[S(\omega)] \simeq \left(\frac{1}{Q}\sum_{m=-(M-1)}^{M-1} w_c^2[m]\right) P_{xx}^2(\omega). \tag{11.95}$$

Comparing Eq. (11.95) with the corresponding result in Eq. (11.65) for the periodogram leads to the conclusion that to reduce the variance of the spectrum estimate we should choose M and the window shape, possibly subject to the condition of Eq. (11.91), so that the factor

$$\left(\frac{1}{Q}\sum_{m=-(M-1)}^{M-1} w_c^2[m]\right) \tag{11.96}$$

is as small as possible. Problem 11.17 deals with the computation of this variance reduction factor for several commonly used windows.

Estimation of the power spectrum based on the Fourier transform of an estimate of the autocorrelation function is a clear alternative to the method of averaging periodograms. It is not necessarily better in any general sense; it simply has different

---

† More precisely, we could define an effective correlation window $w_e[m] = w_c[m](Q - |m|)/Q$.

features and its implementation would be different. In some situations it may be desirable to compute estimates of both the autocorrelation sequence and the power spectrum, in which case it would be natural to use the method of this section. Problem 11.13 explores the issue of determining an autocorrelation estimate from the average periodogram.

11.7.1 Computing Correlation and Power Spectrum Estimates Using the DFT

The autocorrelation estimate

$$\hat{\phi}_{xx}[m] = \frac{1}{Q} \sum_{n=0}^{Q-|m|-1} x[n]x[n + |m|] \tag{11.97}$$

is required for $|m| \leq M - 1$ in the method of power spectrum estimation that we are considering. Since $\hat{\phi}_{xx}[-m] = \hat{\phi}_{xx}[m]$, it is necessary to compute Eq. (11.97) only for nonnegative values of m, i.e., for $0 \leq m \leq M - 1$. The DFT and its associated fast computational algorithms can be used to advantage in the computation of $\hat{\phi}_{xx}[m]$ if we observe that $\hat{\phi}_{xx}[m]$ is the aperiodic discrete convolution of the finite-length sequence $x[n]$ with $x[-n]$. If we compute $X[k]$, the N-point DFT of $x[n]$, and multiply by $X^*[k]$, we obtain $|X[k]|^2$, which corresponds to the circular convolution of the finite-length sequence $x[n]$ with $x[((-n))_N]$, i.e., a *circular autocorrelation*. As our discussion in Section 8.8 suggests, and as developed in Problem 11.14, it should be possible to augment the sequence $x[n]$ with zero-valued samples and force the circular autocorrelation to be equal to the desired aperiodic autocorrelation over the interval $0 \leq m \leq M - 1$.

To see how to choose N for the DFT, consider Fig. 11.23. Figure 11.23(a) shows the two sequences $x[n]$ and $x[n + m]$ for a particular positive value of m. Figure 11.23(b) shows the sequences $x[n]$ and $x[((n + m))_N]$ that are involved in the circular autocorrelation corresponding to $|X[k]|^2$. Clearly, the circular autocorrelation will be equal to $Q\hat{\phi}_{xx}[m]$ for $0 \leq m \leq M - 1$ if $x[((n + m))_N]$ does not wrap around and overlap $x[n]$ when $0 \leq m \leq M - 1$. From Fig. 11.23(b) it follows that this will be the case whenever $N - (M - 1) \geq Q$ or $N \geq Q + M - 1$.

In summary, we can compute $\hat{\phi}_{xx}[m]$ for $0 \leq m \leq M - 1$ by the following procedure:

1. Form an N-point sequence by augmenting $x[n]$ with $(M - 1)$ zero samples.
2. Compute the N-point DFT,

$$X[k] = \sum_{n=0}^{N-1} x[n]e^{-j(2\pi/N)kn} \qquad \text{for } k = 0, 1, \ldots, N - 1.$$

3. Compute

$$|X[k]|^2 = X[k]X^*[k] \qquad \text{for } k = 0, 1, \ldots, N - 1.$$

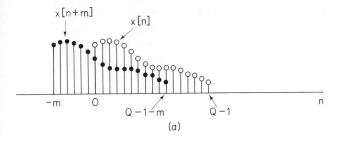

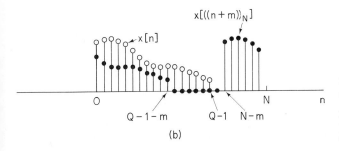

Figure 11.23 Computation of the circular autocorrelation. (a) $x[n]$ and $x[n + m]$ for a finite-length sequence of length Q. (b) $x[n]$ and $x[((n + m))_N]$ as in circular correlation.

4. Compute the inverse DFT of $|X[k]|^2$ to obtain

$$\tilde{c}_{vv}[m] = \frac{1}{N} \sum_{k=0}^{N-1} |X[k]|^2 e^{j(2\pi/N)km} \qquad \text{for } m = 0, 1, \ldots, N - 1$$

5. Divide the resulting sequence by Q to obtain the autocorrelation estimate

$$\hat{\phi}_{xx}[m] = \frac{1}{Q} \tilde{c}_{vv}[m] \qquad \text{for } m = 0, 1, \ldots, M - 1.$$

This is the desired set of autocorrelation values, which can be extended symmetrically for negative values of m.

If M is small, it may be more efficient to simply evaluate Eq. (11.97) directly. In this case, the amount of computation is proportional to $Q \cdot M$. In contrast, if the DFTs in this procedure are computed using one of the FFT algorithms discussed in Chapter 9 with $N \geq Q + M - 1$, the amount of computation will be approximately proportional to $N \log_2 N$ for N a power of 2. Consequently, for sufficiently large values of M, use of the FFT is more efficient than direct evaluation of Eq. (11.97). The exact break-even value of M will depend on the particular implementation of the DFT computations; however, as shown by Stockham (1966), this value would probably be less than $M = 100$.

We should remember that in order to reduce the variance of the estimate of the autocorrelation sequence or the power spectrum estimated from it, we must use large values of the record length Q. In such cases it may be inconvenient or impossible to efficiently compute the $(N = Q + M - 1)$-point DFTs called for by the procedure outlined above. However, since M is generally much less than Q, in such cases it is possible to section the sequence $x[n]$ in a manner similar to the procedures that were discussed in Section 8.8 for convolution of a finite-length impulse response with an indefinitely long input sequence. Rader (1970) presented a particularly efficient and flexible procedure that uses many of the properties of the DFT of real sequences to reduce the amount of computation required. The development of this technique is the basis for Problem 11.15.

Once the autocorrelation estimate has been computed, samples of the power spectrum estimate $S(\omega)$ can be computed at frequencies $\omega_k = 2\pi k/N$ by forming the finite-length sequence

$$
s[m] = \begin{cases} \hat{\phi}_{xx}[m]w_c[m], & 0 \le m \le M - 1, \\ 0, & M \le m \le N - M, \\ \hat{\phi}_{xx}[N - m]w_c[N - m], & N - M + 1 \le m \le N - 1, \end{cases} \tag{11.98}
$$

where $w_c[m]$ is the symmetric correlation window. Then the DFT of $s[m]$ is

$$
S[k] = S(\omega)|_{\omega = 2\pi k/N}, \qquad k = 0, 1, \ldots, N - 1, \tag{11.99}
$$

where $S(\omega)$ is the Fourier transform of the windowed autocorrelation sequence as defined by Eq. (11.88). Note that N can be chosen as large as is convenient and practical, thereby providing samples of $S(\omega)$ at closely spaced frequencies. However, the frequency resolution is always determined by the length and shape of the window $w_c[m]$.

11.7.2 An Example of Power Spectrum Estimation Based on Estimation of the Autocorrelation Sequence

In Chapter 3 we assumed that the error introduced by quantization is a white-noise random process. We can use the techniques of this section to test the validity of this assumption by estimating the autocorrelation sequence and power spectrum of quantization noise.

As an example, consider the experiment depicted in Fig. 11.24. A lowpass-filtered speech signal $x_c(t)$ was sampled at a 10-KHz rate, yielding a sequence of samples $x[n]$. (Although the samples were quantized to 12 bits by the A/D converter, for purposes of this experiment we assume that the samples are unquantized.) The samples were rescaled so that $|x[n]| \le 16,000$. These samples were first quantized with an 8-bit linear quantizer, and the corresponding error sequence $e[n] = Q[x[n]] - x[n]$ was computed as shown in Fig. 11.24. Figure 11.25(a) shows 400 consecutive samples of the speech signal, and Fig. 11.25(b) shows the corresponding error sequence. (The samples are connected with straight lines for convenience in plotting.) Visual inspection and comparison of these two plots tends to strengthen our

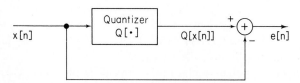

Figure 11.24   Procedure for obtaining quantization noise sequence.

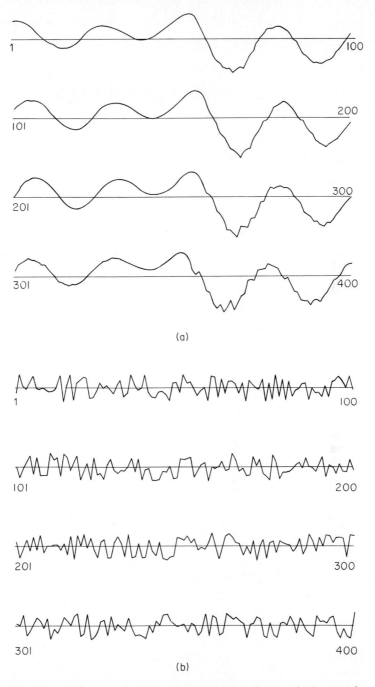

(a)

(b)

Figure 11.25   (a) Speech waveform and (b) the corresponding quantization error for 8-bit quantization (magnified 66 times with respect to part a). Each line corresponds to 100 consecutive samples connected by straight lines for convenience in plotting.

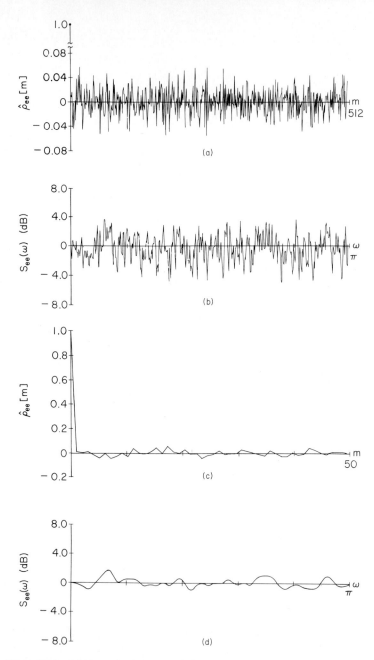

Figure 11.26   (a) Normalized autocorrelation estimate for 8-bit quantization noise; record length $Q = 2000$. (b) Power spectrum estimate using Bartlett window with $M = 512$. (c) Normalized autocorrelation estimate, $0 \leq m \leq 50$. (d) Power spectrum estimate using Bartlett window with $M = 50$.

belief in the previously assumed model; however, the flatness of the quantization noise spectrum can be verified only by estimating the power spectrum.

Figure 11.26 shows estimates of the normalized autocorrelation and power spectrum for a record length of $Q = 2000$ samples. The mean of the noise sequence was estimated and subtracted before the autocorrelation sequence estimate was calculated for $M = 512$. The resulting autocorrelation estimate was divided by $\hat{\phi}_{ee}[0]$ to obtain the normalized estimate $\hat{\rho}_{ee}[m] = \hat{\phi}_{ee}[m]/\hat{\phi}_{ee}[0]$, which is plotted in Figs. 11.26(a) and (c). Note that the normalized autocorrelation is 1.0 at $m = 0$ and much smaller elsewhere. Indeed, $-0.0548 \le \hat{\rho}_{xx}[m] \le 0.0579$ for $1 \le m \le 512$. This seems to support our assumption that the error sequence is uncorrelated from sample to sample.

The power spectrum was estimated by windowing the normalized autocorrelation with a Bartlett window, as discussed in Section 11.7.1, and with $M = 512$. The result, shown in Fig. 11.26(b) plotted in dB, shows rather erratic fluctuations about 0 dB (the value of the normalized power spectrum of white noise). A smoother estimate is shown in Fig. 11.26(d). In this case a Bartlett window with $M = 50$ was used. The resulting smoothing, corresponding to a loss of resolution, is very evident in comparing Fig. 11.26(b) with Fig. 11.26(d). We see from Fig. 11.26(d) that the spectrum estimate is between -1.097 dB and $+1.631$ dB for all frequencies. Thus, we are again encouraged to believe that the white-noise model is appropriate for this case of quantization.

Although we have computed quantitative estimates of the autocorrelation and the power spectrum, our interpretation of these measurements has been only qualitative. It is reasonable now to ask how small the autocorrelation would be if $e[n]$ were really a white-noise process? To give quantitative answers to such questions, confidence intervals for our estimates could be computed and statistical decision theory applied. (See Jenkins and Watts, 1968, for some tests for white noise.) In many cases, however, this additional statistical treatment is not necessary. In a practical setting, we are often comfortable and content simply with the observation that the normalized autocorrelation is very small everywhere except at $m = 0$.

One of the most important insights of this chapter is that the estimate of the autocorrelation and power spectrum of a stationary random process should improve if the record length is increased. This is illustrated by Fig. 11.27, which corresponds to Fig. 11.26 except that Q is increased to 14,000 samples. Recall that the variance of the autocorrelation estimate is proportional to $1/Q$. Thus, increasing Q from 2000 to 14,000 should bring about a sevenfold reduction in the variance of the estimate. A comparison of Figs. 11.26(a) and 11.27(a) seems to verify this result. For $Q = 2000$, the estimate falls between the limits -0.0548 and $+0.0579$, while for $Q = 14,000$ the limits are -0.0254 and $+0.02311$. Intuitively, this is consistent with the sevenfold variance reduction that we expected. We note from Eq. (11.96) that a similar reduction in variance of the spectrum estimate is also expected. This is again evident in comparing Figs. 11.26(b) and (d) with Figs. 11.27(b) and (d), respectively.

In Chapter 3 we argued that the white-noise model was reasonable as long as the quantization step size was small. When the number of bits is small, this condition does not hold. To see the effect on the quantization noise spectrum, the previous experiment was repeated using only 8 quantization levels, or 3 bits. Figure 11.28

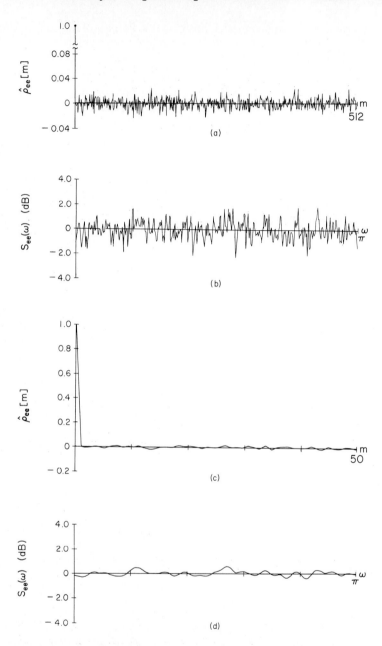

Figure 11.27   (a) Normalized autocorrelation estimate for 8-bit quantization noise; record length $Q = 14000$. (b) Power spectrum estimate using Bartlett window with $M = 512$. (c) Normalized autocorrelation estimate, $0 \leq m \leq 50$. (d) Power spectrum estimate using Bartlett window with $M = 50$.

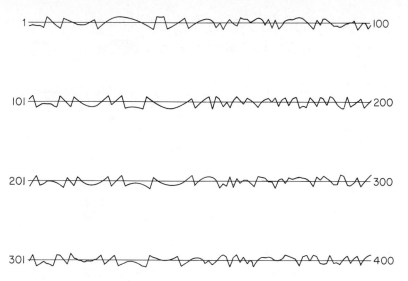

Figure 11.28   Quantization error waveform for 3-bit quantization. (Same scale as original signal, which is shown in Fig. 11.25a.)

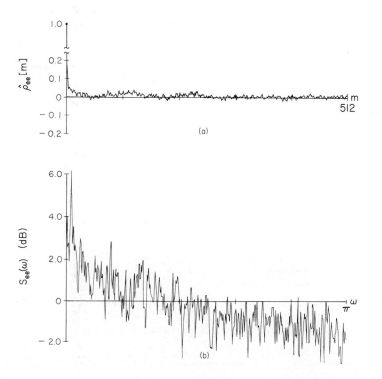

Figure 11.29   (a) Normalized autocorrelation estimate for 3-bit quantization noise; record length $Q = 14000$. (b) Power spectrum estimate using Bartlett window with $M = 512$.

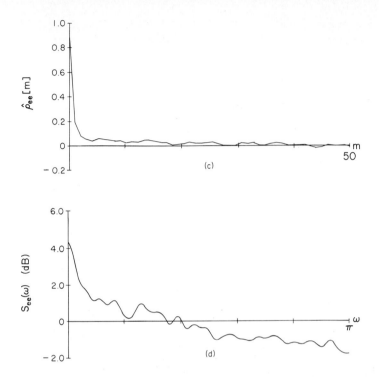

Figure 11.29 (*continued*) (c) Normalized autocorrelation estimate, $0 \le m < 50$. (d) Power spectrum estimate using Bartlett window with $M = 50$.

shows the quantization error for 3-bit quantization of the speech waveform segment shown in Fig. 11.25(a). Note that portions of the error waveform tend to look very much like the original speech waveform. We would expect this to be reflected in the estimate of the power spectrum.

Figure 11.29 shows the autocorrelation and power spectrum estimates of the error sequence for 3-bit quantization for a record length of 14,000 samples. In this case the autocorrelation shown in Figs. 11.29(a) and 11.29(c) is much less like the ideal autocorrelation for white noise.

Figures 11.29(b) and (d) show the power spectrum estimates for Bartlett windows with $M = 512$ and $M = 50$, respectively. In this case, the spectrum is clearly not flat. (In fact, it tends to have the general shape of the speech spectrum.) Thus the white-noise model for quantization noise can be viewed only as a rather crude approximation in this case.

This example illustrates how autocorrelation and power spectrum estimates are often used to bolster our confidence in theoretical models. Specifically, we have demonstrated the validity of some of our basic assumptions in Chapter 3, and we have given an indication of how these assumptions break down for very crude quantization. This is only a rather simple but useful example of how the techniques of this chapter are often applied in practice.

11.8 SUMMARY

One of the important application areas in signal processing is in the spectral analysis of signals. Because of the computational efficiency of the FFT, many of the techniques for spectral analysis of continuous-time or discrete-time signals utilize the DFT either directly or indirectly. In this chapter, we explored and illustrated some of these techniques.

Many of the issues associated with spectral analysis are best understood in the context of the analysis of sinusoidal signals. Since the use of the DFT requires finite-length signals, windowing must be applied in advance of the analysis. For sinusoidal signals, the width of the spectral peak observed in the DFT is dependent on the window length, with an increasing window length resulting in the sharpening of the peak. Consequently, the ability to resolve closely spaced sinusoids in the spectral estimate decreases as the window becomes shorter. A second, independent effect inherent in spectral analysis using the DFT is the associated spectral sampling. Specifically, since the spectrum can be computed only at a set of sample frequencies, the observed spectrum can be misleading if we are not careful in the interpretation. For example, important features in the spectrum may not be directly evident in the sampled spectrum. To avoid this, the spectral sample spacing can be reduced by increasing the DFT size in one of two ways: by zero padding the windowed sequence, thereby keeping the spectral resolution fixed, or by increasing the length of the window applied to the data and the DFT size, which will correspondingly sharpen resolution.

While increased window length and resolution are typically beneficial in spectral analysis of stationary data, for time-varying data it is generally preferable to keep the window length sufficiently short so that over the window duration, the signal characteristics are approximately stationary. This leads to the concept of the time-dependent Fourier transform, which, in effect, is a sequence of Fourier transforms obtained as the signal sequence slides past a finite-duration window. A common and useful interpretation of the time-dependent Fourier transform is as a bank of filters, with the frequency response of each filter corresponding to the transform of the window, frequency shifted to one of the DFT frequencies. The time-dependent Fourier transform has important applications both as an intermediate step in filtering signals and for analyzing and interpreting time-varying signals such as speech and radar signals. Spectral analysis of nonstationary signals typically involves a tradeoff between time and frequency resolution. Specifically, our ability to track spectral characteristics in time increases as the length of the analysis window decreases. However, a shorter analysis window results in decreased frequency resolution.

The DFT also plays an important role in the analysis of stationary random signals. An intuitive approach to estimating the power spectrum of random signals is to compute the squared magnitude of the DFT of a segment of the signal. The resulting estimate, called the periodogram, is asymptotically unbiased. The variance of the periodogram estimate does not decrease to zero as the length of the segment increases, and consequently it is not a good estimate. However, by dividing the available signal sequence into shorter segments and averaging the associated periodograms, we can obtain a well-behaved estimate. An alternative approach is to first estimate the autocorrelation function. This can be done either directly or with the DFT. If a window is then applied to the autocorrelation estimates followed by the DFT, the result, referred to as the smoothed periodogram, is a good spectral estimate.

PROBLEMS

11.1. A real continuous-time signal $x_c(t)$ is bandlimited to frequencies below 5 kHz; i.e., $X_c(j\Omega) = 0$ for $|\Omega| > 2\pi(5000)$. The signal $x_c(t)$ is sampled with a sampling rate of 10,000 samples per second (10 kHz) to produce a sequence $x[n] = x_c(nT)$ with $T = 10^{-4}$. The N-point DFT $X[k]$ of $N = 1000$ samples of $x[n]$ is computed.
(a) To what continuous-time frequency does index $k = 150$ in $X[k]$ correspond?
(b) To what continuous-time frequency does index $k = 800$ in $X[k]$ correspond?

11.2. A continuous-time signal $x_c(t)$ is bandlimited to 5 kHz; i.e., $X_c(j\Omega) = 0$ for $|\Omega| \geq 2\pi(5000)$. $x_c(t)$ is sampled with period T, producing the sequence $x[n] = x_c(nT)$. To examine the spectral properties of the signal, we compute the N-point DFT of a segment of N samples of $x[n]$ using a computer program that requires $N = 2^v$, where v is an integer.

Determine the *minimum* value for N and the range of sampling rates

$$F_{min} < \frac{1}{T} < F_{max}$$

such that aliasing is avoided and the effective spacing between DFT values is *less* than 5 Hz; i.e., the equivalent continuous-time frequencies at which the Fourier transform is evaluated are separated by less than 5 Hz.

11.3. Consider a real time-limited continuous-time signal $x_c(t)$ whose duration is 100 msec. Assume that this signal has a bandlimited Fourier transform such that $X_c(j\Omega) = 0$ for $|\Omega| \geq 2\pi(10000)$ rad/s; i.e., we assume aliasing is negligible. We want to compute samples of $X_c(j\Omega)$ with 5-Hz spacing over the interval $0 \leq \Omega < 2\pi(10000)$. This can be done with a 4000-point DFT. Specifically, we want to obtain a 4000-point sequence $x[n]$ for which the 4000-point DFT is related to $X_c(j\Omega)$ by

$$X[k] = \alpha X_c(j2\pi \cdot 5 \cdot k), \qquad k = 0, 1, \ldots, 2000,$$

where α is a known scale factor. Three methods are proposed to obtain a 4000-point sequence whose DFT gives the desired samples of $X_c(j\Omega)$.

METHOD 1: $x_c(t)$ is sampled with a sampling period $T = 25\,\mu s$; i.e., we compute $X_1[k]$, the DFT of the sequence

$$x_1[n] = \begin{cases} x_c(nT), & n = 0, 1, \ldots, 3999, \\ 0, & \text{otherwise.} \end{cases}$$

Since $x_c(t)$ is timelimited to 100 ms, $x_1[n]$ is a finite-length sequence of length 4000 (100 ms/25 μs).

METHOD 2: $x_c(t)$ is sampled with a sampling period of $T = 50\,\mu s$. Since $x_c(t)$ is timelimited to 100 ms, the resulting sequence will have only 2000 (100 ms/50 μs)

nonzero samples; i.e.,

$$x_2[n] = \begin{cases} x_c(nT), & n = 0, 1, \ldots, 1999, \\ 0, & \text{otherwise.} \end{cases}$$

In other words, the sequence is "padded" with zero samples to create a 4000-point sequence for which the 4000-point DFT $X_2[k]$ is computed.

METHOD 3: $x_c(t)$ is sampled with a sampling period of $T = 50\ \mu s$ as in method 2. The resulting 2000-point sequence is used to form the sequence $x_3[n]$ as follows:

$$x_3[n] = \begin{cases} x_c(nT), & 0 \leq n \leq 1999, \\ x_c((n - 2000)T), & 2000 \leq n \leq 3999, \\ 0, & \text{otherwise.} \end{cases}$$

The 4000-point DFT $X_3[k]$ of this sequence is computed.

For each of the three methods, determine how each 4000-point DFT is related to $X_c(j\Omega)$. Indicate this relationship in a sketch for a "typical" Fourier transform $X_c(j\Omega)$. State explicitly which method(s) provide the desired samples of $X_c(j\Omega)$.

11.4. A continuous-time finite-duration signal $x_c(t)$ is sampled at a rate of 20,000 samples/s, yielding a 1000-point finite-length sequence $x[n]$ that is nonzero in the interval $0 \leq n \leq 999$. Assume for this problem that the continuous-time signal is also band-limited such that $X_c(j\Omega) = 0$ for $|\Omega| \geq 2\pi(10000)$; i.e., assume that negligible aliasing distortion occurs in sampling. Assume also that a device or program is available for computing 1000-point DFTs and inverse DFTs.

(a) If $X[k]$ denotes the 1000-point DFT of the sequence $x[n]$, how is $X[k]$ related to $X_c(j\Omega)$? What is the effective frequency spacing between DFT samples?

The following procedure is proposed for obtaining an expanded view of the Fourier transform $X_c(j\Omega)$ in the interval $|\Omega| \leq 2\pi(5000)$ starting with the 1000-point DFT $X[k]$.

STEP 1: Form the new 1000-point DFT

$$W[k] = \begin{cases} X[k], & 0 \leq k \leq 250, \\ 0, & 251 \leq k \leq 749, \\ X[k], & 750 \leq k \leq 999. \end{cases}$$

STEP 2: Compute the inverse 1000-point DFT of $W[k]$, obtaining $w[n]$ for $n = 0, 1, \ldots, 999$.

STEP 3: Decimate the sequence $w[n]$ by a factor of 2 and augment the result with 500 consecutive zero samples, obtaining the sequence

$$y[n] = \begin{cases} w[2n], & 0 \leq n \leq 499, \\ 0, & 500 \leq n \leq 999. \end{cases}$$

STEP 4: Compute the 1000-point DFT of $y[n]$, obtaining $Y[k]$.

(b) The designer of this procedure asserts that

$$Y[k] = \alpha X_c(j2\pi \cdot 5 \cdot k), \qquad k = 0, 1, \ldots, 500,$$

where α is a constant of proportionality. Is this assertion correct? If not, explain why not.

11.5. Show that the time-dependent Fourier transform, as defined by Eq. (11.18), has the following properties:

(a) *Linearity Property:*

$$\text{If} \quad x[n] = ax_1[n] + bx_2[n], \quad \text{then} \quad X[n, \lambda) = aX_1[n, \lambda) + bX_2[n, \lambda).$$

(b) *Shifting Property:* If $y[n] = x[n - n_0]$, then $Y[n, \lambda) = X[n - n_0, \lambda)$.

(c) *Modulation Property:* If $y[n] = e^{j\omega_0 n}x[n]$, then $Y[n, \lambda) = X[n, \lambda - \omega_0)$.

(d) *Conjugate Symmetry Property:* If $x[n]$ is real, then $X[n, \lambda) = X^*[n, -\lambda)$.

11.6. Suppose that $y[n]$ is the output of a linear time-invariant FIR system with input $x[n]$; i.e.,

$$y[n] = \sum_{k=0}^{M} h[k]x[n - k].$$

(a) Obtain a relationship between the time-dependent Fourier transform $Y[n, \lambda)$ of the output of the linear system and the time-dependent Fourier transform $X[n, \lambda)$ of the input.

(b) Under what conditions would

$$Y[n, \lambda) \simeq H(e^{j\lambda})X\lfloor n, \lambda),$$

where $H(e^{j\omega})$ is the frequency response of the linear system?

11.7. A speech signal is sampled with a sampling rate of 16,000 samples/s (16 kHz). A window of 20-ms duration is used in time-dependent Fourier analysis of the signal as described in Section 11.3, with the window being advanced by 40 samples between computations of the DFT. Assume that the length of each DFT is $N = 2^v$.

(a) How many samples are there in each segment of speech selected by the window?

(b) What is the "frame rate" of the time-dependent Fourier analysis; i.e., how many DFT computations are done per second?

(c) What is the minimum size N of the DFT such that the original input signal can be reconstructed from the time-dependent Fourier transform?

(d) What is the spacing (in Hz) between the DFT samples?

11.8. Suppose that $x_c(t)$ is a continuous-time stationary random signal with autocorrelation function

$$\phi_c(\tau) = \mathcal{E}\{x_c(t)x_c(t + \tau)\}$$

and power density spectrum

$$P_c(\Omega) = \int_{-\infty}^{\infty} \phi_c(\tau)e^{-j\Omega\tau} \, d\tau.$$

Consider a discrete-time stationary random signal $x[n]$ that is obtained by sampling $x_c(t)$ with sampling period T; i.e., $x[n] = x_c(nT)$.

(a) Show that $\phi[m]$, the autocorrelation sequence for $x[n]$, is

$$\phi[m] = \phi_c(mT).$$

(b) What is the relationship between the power density spectrum $P_c(\Omega)$ for the continuous-time random signal and the power density spectrum $P(\omega)$ for the discrete-time random signal?

(c) What condition is necessary such that

$$P(\omega) = \frac{1}{T} P_c\left(\frac{\omega}{T}\right), \qquad |\omega| < \pi?$$

11.9. In Section 11.6.5, we considered the estimation of the power spectrum of a sinusoid plus white noise. In this problem we will determine the true power spectrum of such a signal. Suppose that

$$x[n] = A \cos(\omega_0 n + \theta) + e[n],$$

where θ is a random variable that is uniformly distributed on 0 to 2π, and $e[n]$ is a sequence of zero-mean random variables that are uncorrelated with each other and also uncorrelated with θ. In other words, the cosine component has a randomly selected phase, and $e[n]$ represents white noise.

(a) Show that for the above assumptions, the autocorrelation function for $x[n]$ is

$$\phi_{xx}[m] = \mathcal{E}\{x[n]x[m+n]\} = \frac{A^2}{2} \cos(\omega_0 m) + \sigma_e^2 \delta[m],$$

where $\sigma_e^2 = \mathcal{E}\{(e[n])^2\}$.

(b) From the result of part (a), show that over one period in frequency, the power spectrum of $x[n]$ is

$$P_{xx}(\omega) = \frac{A^2 \pi}{2} [\delta(\omega - \omega_0) + \delta(\omega + \omega_0)] + \sigma_e^2, \qquad |\omega| \le \pi.$$

11.10. Consider a discrete-time signal $x[n]$ of length N samples that was obtained by sampling a stationary, white, zero-mean continuous-time signal. It follows that

$$\mathcal{E}\{x[n]x[m]\} = \sigma_x^2 \delta[n-m],$$

$$\mathcal{E}\{x[n]\} = 0.$$

Suppose that we compute the DFT of the finite-length sequence $x[n]$, thereby obtaining $X[k]$ for $k = 0, 1, \ldots, N-1$.

(a) Determine the variance of $|X[k]|^2$.

(b) Determine the cross-correlation between values of the DFT; i.e., determine $\mathcal{E}\{X[k]X^*[r]\}$ as a function of k and r.

11.11. The periodogram $I(\omega)$ of a discrete-time random signal $x[n]$ was defined in Eq. (11.52) as

$$I(\omega) = \frac{1}{LU} |V(e^{j\omega})|^2,$$

where $V(e^{j\omega})$ is the discrete-time Fourier transform of the finite-length sequence $v[n] = w[n]x[n]$, with $w[n]$ a finite-length window sequence of length L, and U is a normalizing constant.

Show that the periodogram is also equal to $1/LU$ times the Fourier transform of the aperiodic autocorrelation sequence of $v[n]$; i.e.,

$$I(\omega) = \frac{1}{LU} \sum_{m=-(L-1)}^{L-1} c_{vv}[m] e^{-j\omega m},$$

where $c_{vv}[m]$ is

$$c_{vv}[m] = \sum_{n=0}^{L-1} v[n]v[n+m].$$

11.12. A bandlimited continuous-time signal has a bandlimited power spectrum that is zero for $|\Omega| \ge 2\pi(10^4)$ rad/s. The signal is sampled at a rate of 20,000 samples/s over a time interval of 10 s. The power spectrum of the signal is estimated by the method of averaging periodograms as described in Section 11.6.3.

(a) What is the length Q (number of samples) of the data record?

(b) If a radix-2 FFT program is used to compute the periodograms, what is the minimum length N if we wish to obtain estimates of the power spectrum at equally spaced frequencies no more than 10 Hz apart?

(c) If the segment length L is equal to the FFT length N in part (b), how many segments K are available if the segments do not overlap?

(d) Suppose that we wish to reduce the variance of the spectrum estimates by a factor of 10 while maintaining the frequency spacing of part (b). Give two methods of doing this. Do these two methods give the same results? If not, explain how they differ.

11.13. Suppose that an estimate of the power spectrum of a signal is obtained by the method of averaging periodograms, as discussed in Section 11.6.3. That is, the spectrum estimate is

$$\bar{I}(\omega) = \frac{1}{K} \sum_{r=0}^{K-1} I_r(\omega),$$

where the K periodograms $I_r(\omega)$ are computed from L-point segments of the signal using Eqs. (11.67) and (11.68). We define an estimate of the autocorrelation function as the inverse Fourier transform of $\bar{I}(\omega)$; i.e.,

$$\bar{\phi}[m] = \frac{1}{2\pi} \int_{-\pi}^{\pi} \bar{I}(\omega) e^{j\omega m} \, d\omega.$$

(a) Show that

$$\mathcal{E}\{\bar{\phi}[m]\} = \frac{1}{LU} c_{ww}[m] \phi_{xx}[m],$$

where L is the length of the segments, U is a normalizing factor given by Eq. (11.64), and $c_{ww}[m]$, given by Eq. (11.62), is the aperiodic autocorrelation function of the window that is applied to the signal segments.

(b) In application of periodogram averaging, we normally use an FFT algorithm to compute $\bar{I}(\omega)$ at N equally spaced frequencies; i.e.,

$$\bar{I}[k] = \bar{I}(2\pi k/N), \qquad k = 0, 1, \dots, N-1,$$

where $N \geq L$. Suppose that we compute an estimate of the autocorrelation function by computing the inverse DFT of $\bar{I}[k]$ as in

$$\bar{\phi}_p[m] = \frac{1}{N} \sum_{k=0}^{N-1} \bar{I}[k] e^{j(2\pi/N)km}, \qquad m = 0, 1, \dots, N-1.$$

Obtain an expression for $\mathcal{E}\{\bar{\phi}_p[m]\}$.

(c) How should N be chosen so that

$$\mathcal{E}\{\bar{\phi}_p[m]\} = \mathcal{E}\{\bar{\phi}[m]\}, \qquad m = 0, 1, \dots, L-1?$$

11.14. Consider a finite-length sequence $x[n]$ such that $x[n] = 0$ for $n < 0$ and $n \geq L$. Let $X[k]$ be the N-point DFT of the sequence $x[n]$, where $N > L$. Define $c_{xx}[m]$ to be the aperiodic autocorrelation function of $x[n]$; i.e.,

$$c_{xx}[m] = \sum_{n=-\infty}^{\infty} x[n]x[n+m].$$

Define $\tilde{c}_{xx}[m]$ as

$$\tilde{c}_{xx}[m] = \frac{1}{N} \sum_{m=0}^{N-1} |X[k]|^2 e^{j(2\pi/N)km}, \qquad m = 0, 1, \ldots, N-1.$$

(a) Determine the minimum value of N that can be used for the DFT if we require that

$$c_{xx}[m] = \tilde{c}_{xx}[m], \qquad 0 \le m \le L-1.$$

(b) Determine the minimum value of N that can be used for the DFT if we require that

$$c_{xx}[m] = \tilde{c}_{xx}[m], \qquad 0 \le m \le M-1,$$

where $M < L$.

11.15. Consider the computation of the autocorrelation estimate

$$\hat{\phi}_{xx}[m] = \frac{1}{Q} \sum_{n=0}^{Q-|m|-1} x[n]x[n+|m|], \tag{P11.15-1}$$

where $x[n]$ is a real sequence. Since $\hat{\phi}_{xx}[-m] = \hat{\phi}_{xx}[m]$, it is necessary only to evaluate Eq. (P11.15-1) for $0 \le m \le M-1$ to obtain $\hat{\phi}_{xx}[m]$ for $-(M-1) \le m \le M-1$, as required to estimate the power density spectrum using Eq. (11.88).

(a) When $Q \gg M$, it may not be feasible to compute $\hat{\phi}_{xx}[m]$ using a single FFT computation. In such cases, it is convenient to express $\hat{\phi}_{xx}[m]$ as a sum of correlation estimates based on shorter sequences. Show that if $Q = KM$,

$$\hat{\phi}_{xx}[m] = \frac{1}{Q} \sum_{i=0}^{K-1} c_i[m], \tag{P11.15-2}$$

where

$$c_i[m] = \sum_{n=0}^{M-1} x[n+iM]x[n+iM+m], \tag{P11.15-3}$$

for $0 \le m \le M-1$.

(b) Show that the correlations $c_i[m]$ can be obtained by computing the N-point *circular* correlations

$$\tilde{c}_i[m] = \sum_{n=0}^{N-1} x_i[n]y_i[((n+m))_N], \tag{P11.15-4}$$

where the sequences $x_i[n]$ and $y_i[n]$ are

$$x_i[n] = \begin{cases} x[n+iM], & 0 \le n \le M-1, \\ 0, & M \le n \le N-1, \end{cases} \tag{P11.15-5}$$

and

$$y_i[n] = x[n+iM], \qquad 0 \le n \le N-1. \tag{P11.15-6}$$

What is the *minimum* value of N (in terms of M) such that $c_i[m] = \tilde{c}_i[m]$ for $0 \le m \le M-1$?

(c) State a procedure for computing $\hat{\phi}_{xx}[m]$ for $0 \le m \le M-1$ that involves the computation of $2K$ N-point DFTs of real sequences and *one* N-point inverse DFT. How many complex multiplications are required to compute $\hat{\phi}_{xx}[m]$ for $0 \le m \le M-1$ if a radix-2 FFT is used?

(d) What modifications to the procedure developed in part (c) would be necessary to compute the cross-correlation estimate

$$\hat{\phi}_{xy}[m] = \frac{1}{Q} \sum_{n=0}^{Q-|m|-1} x[n]y[n+m], \qquad -(M-1) \le m \le M-1, \quad \text{(P11.15-7)}$$

where $x[n]$ and $y[n]$ are real sequences known for $0 \le n \le Q-1$?

(e) Rader (1970) showed that for computing the autocorrelation estimate $\hat{\phi}_{xx}[m]$ for $0 \le m \le M-1$, significant savings of computation can be achieved if $N = 2M$. Show that the N-point DFT of a segment $y_i[n]$ as defined in Eq. (P11.15-6) can be expressed as

$$Y_i[k] = X_i[k] + (-1)^k X_{i+1}[k], \qquad k = 0, 1, \ldots, N-1. \quad \text{(P11.15-8)}$$

State a procedure for computing $\hat{\phi}_{xx}[m]$ for $0 \le m \le M-1$ that involves the computation of K N-point DFTs and one N-point inverse DFT. Determine the total number of complex multiplications in this case if a radix-2 FFT is used.

11.16. The symmetric Bartlett window arises in many aspects of power spectrum estimation. It is defined as

$$w_B[m] = \begin{cases} 1 - |m|/M, & |m| \le M-1, \\ 0, & \text{otherwise.} \end{cases} \quad \text{(P11.16-1)}$$

The Bartlett window is particularly attractive for obtaining estimates of the power spectrum by windowing an estimated autocorrelation function as discussed in Section 11.7. This is because its Fourier transform is nonnegative, which guarantees that the smoothed spectrum estimate will be nonnegative at all frequencies.

(a) Show that the Bartlett window as defined in Eq. (P11.16-1) is the aperiodic autocorrelation of the sequence $(u[n] - u[n-M])$.

(b) From the result of part (a) show that the Fourier transform of the Bartlett window is

$$W_B(e^{j\omega}) = \frac{1}{M} \left[\frac{\sin(\omega M/2)}{\sin(\omega/2)} \right]^2, \quad \text{(P11.16-2)}$$

which is clearly nonnegative.

(c) Describe a procedure for generating other finite-length window sequences that have nonnegative Fourier transforms.

11.17. In Section 11.7 we showed that a smoothed estimate of the power spectrum can be obtained by windowing an estimate of the autocorrelation sequence. It was stated (see Eq. 11.95) that the variance of the smoothed spectrum estimate is

$$\text{var}[S(\omega)] \simeq P_{xx}^2(\omega),$$

where F, the *variance ratio* or *variance reduction factor*, is

$$F = \frac{1}{Q} \sum_{m=-(M-1)}^{M-1} (w_c[m])^2 = \frac{1}{2\pi Q} \int_{-\pi}^{\pi} |W_c(e^{j\omega})|^2 \, d\omega.$$

As discussed in Section 11.7, Q is the length of the sequence $x[n]$ and $(2M-1)$ is the length of the symmetric window $w_c[m]$ that is applied to the autocorrelation estimate. Thus, if Q is fixed, the variance of the smoothed spectrum estimate can be reduced by adjusting the shape and duration of the window applied to the correlation function.

In this problem we will show that F decreases as the window length decreases, but we also know from the previous discussion of windows in Chapter 7 that the width of the mainlobe of $W_c(e^{j\omega})$ increases with decreasing window length, so that the ability to resolve two adjacent frequency components is reduced as the window width decreases. Thus, there is a tradeoff between variance reduction and resolution. In this problem we will study this tradeoff for the following commonly used windows.

Rectangular

$$w_R[m] = \begin{cases} 1, & |m| \le M - 1, \\ 0, & \text{otherwise} \end{cases}$$

Bartlett (triangular)

$$w_B[m] = \begin{cases} 1 - |m|/M, & |m| \le M - 1, \\ 0, & \text{otherwise} \end{cases}$$

Hanning/Hamming

$$w_H[m] = \begin{cases} \alpha + \beta \cos[\pi m/(M - 1)], & |m| \le M - 1, \\ 0, & \text{otherwise} \end{cases}$$

($\alpha = \beta = 0.5$ for the Hanning window and $\alpha = 0.54$ and $\beta = 0.46$ for the Hamming window.)

(a) Find the Fourier transform of each of the above windows; i.e., compute $W_R(e^{j\omega})$, $W_B(e^{j\omega})$, and $W_H(e^{j\omega})$. Sketch each of these Fourier transforms as functions of ω.

(b) For each of these windows, show that the entries in the following table are approximately true when $M \gg 1$.

Window Name	Approximate Mainlobe Width	Approximate Variance Ratio (F)
Rectangular	$2\pi/M$	$2M/Q$
Bartlett	$4\pi/M$	$2M/(3Q)$
Raised cosine	$3\pi/M$	$2M(\alpha^2 + \beta^2/2)/Q$

11.18. In Section 11.3 we defined the time-dependent Fourier transform of the signal $x[m]$ so that for fixed n it is equivalent to the regular discrete-time Fourier transform of the sequence $x[n + m]w[m]$, where $w[m]$ is a window sequence. It is also useful to define a time-dependent autocorrelation function for the sequence $x[n]$ such that for fixed n its regular Fourier transform is the magnitude squared of the time-dependent Fourier transform. Specifically, the time-dependent autocorrelation function is defined as

$$c[n, m] = \frac{1}{2\pi} \int_{-\pi}^{\pi} |X[n, \lambda)|^2 e^{j\lambda m} d\lambda, \tag{P11.18-1}$$

where $X[n, \lambda)$ is defined by Eq. (11.18).

(a) Show that

$$c[n, m] = \sum_{r = -\infty}^{\infty} x[n + r]w[r]x[m + n + r]w[m + r], \tag{P11.18-2}$$

i.e., for fixed n, $c[n, m]$ is the aperiodic autocorrelation of the sequence $x[n + r]w[r]$, $-\infty < r < \infty$.

(b) Show that the time-dependent autocorrelation function is an even function of m for n fixed, and use this fact to obtain the equivalent expression

$$c[n, m] = \sum_{r=-\infty}^{\infty} x[r]x[r - m]h_m[n - r], \qquad \text{(P11.18-3)}$$

where

$$h_m[r] = w[-r]w[-(m + r)]. \qquad \text{(P11.18-4)}$$

(c) What condition must the window $w[r]$ satisfy so that Eq. (P11.18-4) can be used to compute $c[n, m]$ for fixed m and $-\infty < n < \infty$ by causal operations?

(d) Suppose that $w[-r]$ is

$$w[-r] = \begin{cases} a^r, & r \ge 0, \\ 0, & r < 0. \end{cases} \qquad \text{(P11.18-5)}$$

Find the impulse response $h_m[r]$ for computing the mth autocorrelation lag value, and find the corresponding system function $H_m(z)$. From the system function, draw the block diagram of a causal system for computing the mth autocorrelation lag value $c[n, m]$ for $-\infty < n < \infty$ for the window of Eq. (P11.18-5).

(e) Repeat part (d) for

$$w[-r] = \begin{cases} ra^r, & r \ge 0, \\ 0, & r < 0. \end{cases} \qquad \text{(P11.18-6)}$$

11.19. Time-dependent Fourier analysis is sometimes implemented as a bank of filters, and even when FFT methods are used, the filter bank interpretation may provide useful insight. This problem examines the filter bank interpretation. The basis for this interpretation is the fact that when λ is fixed, the time-dependent Fourier transform $X[n, \lambda)$, defined by Eq. (11.18), is simply a sequence that can be viewed as the result of a combination of filtering and modulation operations.

(a) Show that $X[n, \lambda)$ is the output of the system of Fig. P11.19-1 if the impulse response of the linear time-invariant system is $h_0[n] = w[-n]$. Also show that if λ is fixed, the overall system in Fig. P11.19-1 behaves as a linear time-invariant system, and determine the impulse response and frequency response of the equivalent LTI system.

(b) Assuming λ fixed in Fig. P11.19-1, show that for typical window sequences and for fixed λ, the sequence $s[n] = \check{X}[n, \lambda)$ has a discrete-time Fourier transform that is lowpass. Also show that for typical window sequences, the frequency response of the overall system in Fig. P11.19-1 is a bandpass filter centered at $\omega = \lambda$.

(c) Figure P11.19-2 shows a bank of N bandpass filter channels where each channel is implemented as in Fig. P11.19-1. The center frequencies of the channels are $\lambda_k = 2\pi k/N$, and $h_0[n] = w[-n]$ is the impulse response of a lowpass filter. Show that the individual outputs $y_k[n]$ are samples (in the λ-dimension) of the time-dependent Fourier transform. Also show that the overall output is $y[n] = w[0]x[n]$; i.e., show that the system of Fig. P11.19-2 reconstructs the input exactly (within a scale factor) from the sampled time-dependent Fourier transform.

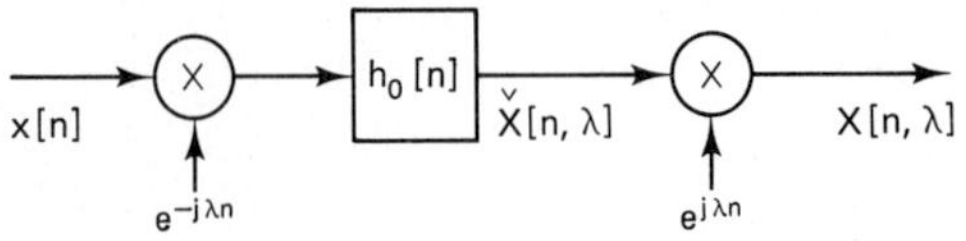

Figure P11.19-1

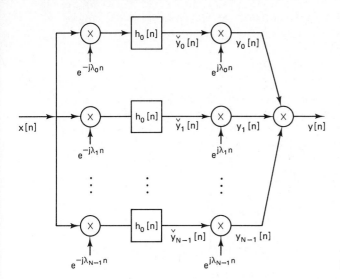

Figure P11.19-2

The system of Fig. P11.19-2 converts the single input sequence $x[n]$ into N sequences, thereby increasing the total number of samples per second by the factor N. As shown in part (b), for typical window sequences, the channel signals $\check{y}_k[n]$ have lowpass Fourier transforms. Thus, it should be possible to reduce the sampling rate of these signals as shown in Fig. P11.19-3. In particular, if the sampling rate is reduced by a factor $R = N$, the total number of samples per second is the same as for $x[n]$. In this case, the filterbank is said to be *critically sampled* (see Crochiere and Rabiner, 1983). Reconstruction of the original signal from the decimated channel signals requires interpolation as shown. Clearly, it is of interest to determine how well the original input $x[n]$ can be reconstructed by the system.

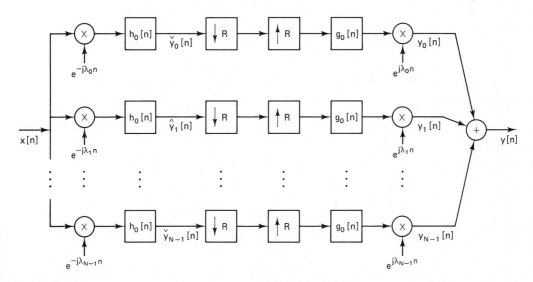

Figure P11.19-3

(d) For the system of Fig. P11.19-3, show that the regular discrete-time Fourier transform of the output is given by the relation

$$Y(e^{j\omega}) = \frac{1}{R}\sum_{\ell=0}^{R-1}\sum_{k=0}^{N-1} G_0(e^{j(\omega-\lambda_k)})H_0(e^{j(\omega-\lambda_k-2\pi\ell/R)})X(e^{j(\omega-2\pi\ell/R)}),$$

where $\lambda_k = 2\pi k/N$. This expression clearly shows the aliasing resulting from the decimation of the channel signals $\check{y}[n]$. From this expression, determine a relation or set of relations that must be satisfied jointly by $H_0(e^{j\omega})$ and $G_0(e^{j\omega})$ such that the aliasing cancels and $y[n] = x[n]$.

(e) Assume that $R = N$ and the frequency response of the lowpass filter is an ideal lowpass filter with frequency response

$$H_0(e^{j\omega}) = \begin{cases} 1, & |\omega| < \pi/N, \\ 0, & \pi/N < |\omega| \le \pi. \end{cases}$$

For this frequency response $H_0(e^{j\omega})$, determine if it is possible to find a frequency response of the interpolation filter $G_0(e^{j\omega})$ such that the condition derived in part (d) is satisfied. If so, determine $G_0(e^{j\omega})$.

(f) *Optional*: Explore the possibility of exact reconstruction when the frequency response of the lowpass filter $H_0(e^{j\omega})$ (Fourier transform of $w[-n]$) is nonideal and nonzero in the interval $|\omega| < 2\pi/N$.

(g) Show that the output of the system of Fig. P11.19-3 is

$$y[n] = N\sum_{r=-\infty}^{\infty} x[n-rN]\sum_{\ell=-\infty}^{\infty} g_0[n-\ell R]h_0[\ell R + rN - n].$$

From this expression, determine a relation or set of relations that must be satisfied jointly by $h_0[n]$ and $g_0[n]$ such that $y[n] = x[n]$.

(h) Assume that $R = N$ and the impulse response of the lowpass filter is

$$h_0[n] = \begin{cases} 1, & -(N-1) \le n \le 0, \\ 0, & \text{otherwise}. \end{cases}$$

For this impulse response $h_0[n]$, determine if it is possible to find an impulse response of the interpolation filter $g_0[n]$ such that the condition derived in part (f) is satisfied. If so, determine $g_0[n]$.

(i) *Optional*: Explore the possibility of exact reconstruction when the impulse response of the lowpass filter $h_0[n] = w[-n]$ is a tapered window with length greater than N.

11.20. Consider a stable linear time-invariant system as depicted in Fig. P11.20, where the input $x[n]$ is assumed to be white noise with zero mean and variance σ_x^2. Assume that the system has system function

$$H(z) = \frac{\displaystyle\sum_{k=0}^{M} b_k z^{-k}}{1 - \displaystyle\sum_{k=1}^{N} a_k z^{-k}},$$

so that the input and output satisfy the following difference equation with constant coefficients:

$$y[n] = \sum_{k=1}^{N} a_k y[n-k] + \sum_{k=0}^{M} b_k x[n-k].$$

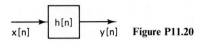

Figure P11.20

If all the a_k's are zero, $y[n]$ is called a *moving average* (MA) linear random process. If all the b_k's are zero, except $b_0 \neq 0$, then $y[n]$ is called an *autoregressive* (AR) linear random process. If both N and M are nonzero, then $y[n]$ is an *autoregressive moving average* (ARMA) linear random process.

(a) Express the autocorrelation of $y[n]$ in terms of the impulse response $h[n]$ of the linear system.

(b) Use the result of part (a) to express the power density spectrum of $y[n]$ in terms of the frequency response of the system.

(c) Show that the autocorrelation sequence $\phi_{yy}[m]$ of an MA process is nonzero only in the interval $|m| \leq M$.

(d) Find a general expression for the autocorrelation sequence for an AR process.

(e) Show that if $b_0 = 1$, the autocorrelation function of an AR process satisfies the difference equation

$$\phi_{yy}[0] = \sum_{k=1}^{N} a_k \phi_{yy}[k] + \sigma_x^2,$$

$$\phi_{yy}[m] = \sum_{k=1}^{N} a_k \phi_{yy}[m-k], \qquad m \geq 1.$$

(f) Use the result of part (e) and the symmetry of $\phi_{yy}[m]$ to show that

$$\sum_{k=1}^{N} a_k \phi_{yy}[|m-k|] = \phi_{yy}[m], \qquad m = 1, 2, \ldots, N.$$

It can be shown that given $\phi_{yy}[m]$ for $m = 0, 1, \ldots, N$, we can always solve uniquely for the values of the a_k's and σ_x^2 for the random process model. These values may be used in the result in part (b) to obtain an expression for the power density spectrum of $y[n]$. This approach is the basis for a number of parametric spectrum estimation techniques. (For more discussion of these methods see Gardner, 1988; Kay, 1988; Marple, 1987.)

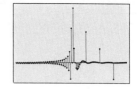

Cepstrum Analysis and Homomorphic Deconvolution

12

12.0 INTRODUCTION

In the earlier chapters, we discussed and illustrated a variety of linear signal processing methods. In this chapter, we introduce a class of nonlinear techniques referred to as *cepstrum analysis* and *homomorphic deconvolution*. These methods have proved to be extremely effective and useful in certain applications. In addition, they further illustrate the considerable flexibility and sophistication offered by discrete-time signal processing technologies.

In 1963, Bogert, Healy, and Tukey published a paper with the unusual title "The Quefrency Alanysis of Time Series for Echoes: Cepstrum, Pseudoautocovariance, Cross-Cepstrum, and Saphe Cracking." They observed that the logarithm of the power spectrum of a signal containing an echo has an additive periodic component due to the echo, and thus the Fourier transform of the logarithm of the power spectrum should exhibit a peak at the echo delay. They called this function the *cepstrum*, interchanging letters in the word *spectrum* because "in general, we find ourselves operating on the frequency side in ways customary on the time side and vice versa." Bogert et al. went on to define an extensive vocabulary to describe this new signal processing technique; however, only the term *cepstrum* has been widely used.

At about the same time, Oppenheim (1967, 1969a) proposed a new class of systems called *homomorphic systems*. Although nonlinear in the classical sense, these systems satisfy a generalization of the principle of superposition; i.e., input signals and their corresponding responses are superimposed (combined) by an operation having the same algebraic properties as addition. The concept of homomorphic filtering is very general, but it has been studied most extensively for the combining operations of multiplication and convolution because many signal models involve these operations.

The transformation of a signal into its cepstrum is a homomorphic transformation, and the concept of the cepstrum is a fundamental part of the theory of homomorphic systems for processing signals that have been combined by convolution.

Since introduction of the cepstrum, the concepts of the cepstrum and homomorphic systems have proved useful in signal analysis and have been applied with success in processing speech signals (Oppenheim, 1969b; Schafer and Rabiner, 1970), seismic signals (Ulrych, 1971; Tribolet, 1979), biomedical signals (Senmoto and Childers, 1972), old acoustic recordings (Stockham et al., 1975), and sonar signals (Reut et al., 1985). This chapter provides a detailed treatment of the properties and computational issues associated with the cepstrum and with deconvolution based on homomorphic filtering. A number of these concepts are illustrated in Section 12.9 in the context of speech processing.

12.1 DEFINITION OF THE COMPLEX CEPSTRUM

Consider a stable sequence $x[n]$ whose z-transform expressed in polar form is

$$X(z) = |X(z)|e^{j \angle X(z)}, \tag{12.1}$$

where $|X(z)|$ and $\angle X(z)$ are the magnitude and angle, respectively, of the complex number $X(z)$. Since $x[n]$ is stable, the region of convergence for $X(z)$ includes the unit circle, and the Fourier transform of $x[n]$ exists and is equal to $X(e^{j\omega})$. The *complex cepstrum* corresponding to $x[n]$ is defined to be the stable sequence $\hat{x}[n]$† whose z-transform is

$$\hat{X}(z) = \log[X(z)]. \tag{12.2}$$

Although in defining the complex cepstrum any base can be used for the logarithm, the natural logarithm (i.e., base e) is typically used and will be assumed throughout the remainder of the discussion. The logarithm of a complex quantity $X(z)$ expressed as in Eq. (12.1) is defined as

$$\log[X(z)] = \log[|X(z)|e^{j \angle X(z)}] = \log|X(z)| + j\angle X(z). \tag{12.3}$$

Since in the polar representation of a complex number the angle is unique only to within integer multiples of 2π, the imaginary part of Eq. (12.3) is not yet well defined. We will address that issue shortly; for now we assume that an appropriate definition has been chosen.

The complex cepstrum exists if $\log[X(z)]$ has a convergent power series representation of the form

$$\hat{X}(z) = \log[X(z)] = \sum_{n=-\infty}^{\infty} \hat{x}[n]z^{-n}, \qquad |z| = 1, \tag{12.4}$$

---

† In a somewhat more general definition of the complex cepstrum, $x[n]$ and $\hat{x}[n]$ need not be restricted to be stable. However, with this restriction the important concepts can be illustrated with simpler notation than in the general case.

i.e., $\hat{X}(z) = \log[X(z)]$ must have all the properties of the z-transform of a stable sequence. Specifically, the region of convergence for the power series representation of $\log[X(z)]$ must be of the form

$$r_R < |z| < r_L, \tag{12.5}$$

where $0 < r_R < 1$ and $r_L > 1$. If this is the case, $\hat{x}[n]$, the sequence of coefficients of the power series, corresponds to the *complex cepstrum* of $x[n]$ and is also given by the inverse z-transform integral

$$\hat{x}[n] = \frac{1}{2\pi j} \oint_C \log[X(z)] z^{n-1} \, dz, \tag{12.6}$$

where the contour of integration is within the region of convergence. Since we require $\hat{x}[n]$ to be stable, the region of convergence includes the unit circle, and the complex cepstrum can be represented using the inverse Fourier transform as

$$
\begin{aligned}
\hat{x}[n] &= \frac{1}{2\pi} \int_{-\pi}^{\pi} \log[X(e^{j\omega})] e^{j\omega n} \, d\omega \\
&= \frac{1}{2\pi} \int_{-\pi}^{\pi} [\log|X(e^{j\omega})| + j \sphericalangle X(e^{j\omega})] e^{j\omega n} \, d\omega.
\end{aligned}
\tag{12.7}
$$

The term *complex cepstrum* distinguishes our more general definition from the original definition of the cepstrum by Bogert et al. (1963). The use of the word *complex* in this context implies that the complex logarithm is used in the definition. It does not imply that the complex cepstrum is necessarily a sequence of complex numbers. Indeed, as we will see shortly, the definition we choose for the complex logarithm ensures that the complex cepstrum of a real sequence will also be a real sequence.

In contrast to the complex cepstrum, the cepstrum $c_x[n]$ of a signal (sometimes referred to as the *real cepstrum*) is defined as the inverse Fourier transform of the logarithm of the magnitude of the Fourier transform; i.e.,

$$c_x[n] = \frac{1}{2\pi} \int_{-\pi}^{\pi} \log|X(e^{j\omega})| e^{j\omega n} \, d\omega. \tag{12.8}$$

Since the Fourier transform magnitude is real and nonnegative, no special considerations are involved in defining the logarithm in Eq. (12.8). By comparing Eq. (12.8) and Eq. (12.7), we see that $c_x[n]$ is the inverse transform of the real part of $\hat{X}(e^{j\omega})$. Consequently $c_x[n]$ is equal to the conjugate-symmetric part of $\hat{x}[n]$; i.e.,

$$c_x[n] = \frac{\hat{x}[n] + \hat{x}^*[-n]}{2}. \tag{12.9}$$

The cepstrum is useful in many applications, and since it does not depend on the phase of $X(e^{j\omega})$ it is much easier to compute than the complex cepstrum. However, since it is based on only the Fourier transform magnitude it is not invertible, i.e., $x[n]$ cannot be recovered from $c_x[n]$. The complex cepstrum is somewhat more difficult to compute, but it is invertible. Since the complex cepstrum is a more general concept than the cepstrum, and since using Eq. (12.9) the properties of the cepstrum can be derived from the properties of the complex cepstrum, we will emphasize the complex cepstrum in this chapter.

The additional difficulties encountered in defining and computing the complex cepstrum are worthwhile for a variety of reasons. First, we see from Eq. (12.3) that the complex logarithm has the effect of creating a new Fourier transform whose real and imaginary parts are $\log|X(e^{j\omega})|$ and $\sphericalangle X(e^{j\omega})$, respectively. Thus we can obtain Hilbert transform relations between these two quantities when the complex cepstrum is causal. We discuss this point further in Section 12.6.1 and see in particular how it relates to minimum-phase sequences. A second, more general motivation, developed in Section 12.2, stems from the role that the complex cepstrum plays in defining a class of systems for generalized linear filtering of signals that are combined by convolution.

12.2 HOMOMORPHIC DECONVOLUTION

12.2.1 Generalized Superposition

The principle of superposition as stated for linear systems requires that, if T is the system transformation, then for any two inputs $x_1[n]$ and $x_2[n]$ and any scalar c,

$$T\{x_1[n] + x_2[n]\} = T\{x_1[n]\} + T\{x_2[n]\} \tag{12.10a}$$

and

$$T\{cx_1[n]\} = cT\{x_1[n]\}. \tag{12.10b}$$

To generalize this principle, we denote by $\square$ a rule for combining inputs with each other (e.g., addition, multiplication, convolution, etc.) and by $:$ a rule for combining inputs with scalars. Similarly, $\circ$ will denote a rule for combining system outputs and ι a rule for combining outputs with scalars. Then, with H denoting the system transformation, Eqs. (12.10) can be generalized by requiring that

$$H\{x_1[n] \,\square\, x_2[n]\} = H\{x_1[n]\} \circ H\{x_2[n]\} \tag{12.11a}$$

and

$$H\{c : x_1[n]\} = c \,\iota\, H\{x_1[n]\}. \tag{12.11b}$$

Such systems are depicted as in Fig. 12.1 and are said to obey a generalized principle of superposition with an input operation $\square$ and an output operation $\circ$. Systems satisfying this generalized principle of superposition are referred to as *homomorphic* systems since they can be represented by algebraically linear (homomorphic) mappings between input and output signal spaces. Clearly, linear systems are a special case for which $\square$ and $\circ$ are addition and $:$ and ι are multiplication.

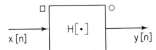

Figure 12.1 Representation of a system satisfying generalized superposition with input operation $\square$, output operation $\circ$, and system transformation H.

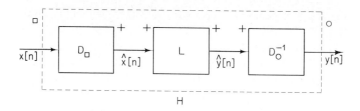

Figure 12.2   Canonic representation of homomorphic systems.

It can be shown (Oppenheim, 1967, 1969a) that any homomorphic system can be decomposed as a cascade of three homomorphic systems, as indicated in Fig. 12.2, with the first system depending only on the input operation □, the last system depending only on the output operation ○ and the middle system corresponding to a linear system. The system $D_\square[\cdot]$ is referred to as the characteristic system associated with the input operation □, and $D_\circ^{-1}[\cdot]$ is the inverse of $D_\circ[\cdot]$, the characteristic system associated with the output operation ○. The overall decomposition in Fig. 12.2 is referred to as the canonic representation for homomorphic systems, and we will find it very useful in applying homomorphic systems to deconvolution.

12.2.2 The Characteristic System for Convolution

Figure 12.3 shows the canonic representation of a class of systems that obey a generalized principle of superposition for convolution. Specifically, if the input is

$$x[n] = x_1[n] * x_2[n] = \sum_{k=-\infty}^{\infty} x_1[k]x_2[n-k], \qquad (12.12)$$

then the corresponding output is

$$y[n] = y_1[n] * y_2[n], \qquad (12.13)$$

where $y_1[n]$ and $y_2[n]$ are the responses of the system to $x_1[n]$ and $x_2[n]$, respectively. The use of a system of the form of Fig. 12.3 to remove or alter one of the components of a convolution is called *homomorphic deconvolution*.

The first system in the canonic system of Fig. 12.3 is the characteristic system for convolution. It is a homomorphic system with convolution as the input operation and addition as the output operation, so that with

$$x[n] = x_1[n] * x_2[n],$$

then

$$\hat{x}[n] = \hat{x}_1[n] + \hat{x}_2[n].$$

In particular, $D_*[\cdot]$ can be chosen such that $\hat{x}[n]$ is the complex cepstrum of $x[n]$, i.e.,

$$\hat{X}(z) = \log[X(z)]. \qquad (12.14)$$

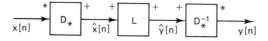

Figure 12.3   Canonic form for homomorphic systems with convolution as the input and the output operations.

As will be discussed in Section 12.3, if the complex logarithm is appropriately defined, when $X(z) = X_1(z)X_2(z)$, then

$$\hat{X}(z) = \log[X(z)] = \log[X_1(z)] + \log[X_2(z)]$$
$$= \hat{X}_1(z) + \hat{X}_2(z) \tag{12.15}$$

and

$$\hat{x}[n] = \hat{x}_1[n] + \hat{x}_2[n]. \tag{12.16}$$

Thus, the system $D_*[\cdot]$ satisfies the generalized principle of superposition, with convolution as the operation of superposition at the input and addition as the operation of superposition at the output.

In Section 12.1, the operations that defined the complex cepstrum were the same as those shown in block diagram form in Figure 12.4(a). The cascade of z-transform, complex logarithm, and inverse z-transform can be thought of as a representation of the characteristic system $D_*[\cdot]$. Since we are assuming throughout that all sequences and their complex cepstra are stable, the associated z-transforms always include the unit circle in their regions of convergence; consequently the z-transforms in Fig. 12.4 can also be specialized to Fourier transforms as in Eq. (12.7). Each of the three basic component transformations is also homomorphic, and the corresponding input and output operations are indicated in Fig. 12.4(a). The z-transform maps convolution to multiplication; the complex logarithm converts multiplication to addition; and the inverse transform is a linear transformation in the ordinary sense.

The third system in the cascade of Fig. 12.3 is the inverse of the characteristic system for convolution. Consequently, its input must be the complex cepstrum of its output, i.e.,

$$y[n] = D_*^{-1}[\hat{y}[n]]. \tag{12.17}$$

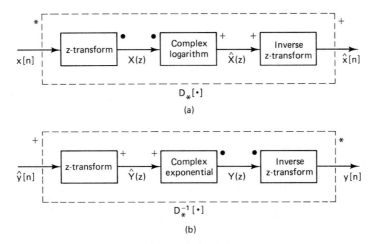

Figure 12.4   Representations of (a) characteristic system for convolution and (b) its inverse.

This means that if $\hat{Y}(z)$ is the z-transform of the input to $D_*^{-1}[\cdot]$, then the z-transform of the output is

$$Y(z) = \exp[\hat{Y}(z)]. \tag{12.18}$$

If the input to $D_*^{-1}[\cdot]$ is the sum of two complex cepstra, $\hat{y}[n] = \hat{y}_1[n] + \hat{y}_2[n]$, then the output is $y[n] = y_1[n] * y_2[n]$. This follows from Eq. (12.18) since

$$\exp[\hat{Y}_1(z) + \hat{Y}_2(z)] = \exp[\hat{Y}_1(z)] \cdot \exp[\hat{Y}_2(z)]$$
$$= Y_1(z) Y_2(z). \tag{12.19}$$

The basic operations that define the inverse characteristic system for convolution are depicted in Fig. 12.4(b). In this case, the linearity of the z-transform takes a sum of complex cepstrums into a sum of transforms; the complex exponential maps a sum into a product; and the inverse transform maps a product into a convolution.

12.2.3 Deconvolution Based on Generalized Superposition

Figure 12.3 depicts the canonic representation of homomorphic systems with convolution as the input and output operations. As discussed, the systems $D_*[\cdot]$ and $D_*^{-1}[\cdot]$ depend only on the specific class of homomorphic systems, i.e., on the fact that the input and output operations are convolution. The system L is a linear system in the usual sense.

To suggest how this class of homomorphic systems might be useful for deconvolution, consider a signal that is the convolution of two components, i.e.,

$$x[n] = x_1[n] * x_2[n]. \tag{12.20}$$

For example, $x_2[n]$ might be the impulse response of a stable linear time-invariant system, the effects of which we are interested in removing; i.e., we would like to recover $x_1[n]$ from $x[n]$. The complex cepstrum of $x[n]$ is the sum of the complex cepstra of $x_1[n]$ and $x_2[n]$, i.e.,

$$\hat{x}[n] = \hat{x}_1[n] + \hat{x}_2[n]. \tag{12.21}$$

Now if we can find a choice for the linear system L in Fig. 12.3 such that its output is

$$\hat{y}[n] = \hat{x}_1[n] \tag{12.22}$$

when its input is $\hat{x}[n] = \hat{x}_1[n] + \hat{x}_2[n]$, then the output of the overall system in Fig. 12.3 will be

$$y[n] = x_1[n]. \tag{12.23}$$

In other words, if the linear system removes $\hat{x}_2[n]$ completely from the additive combination of Eq. (12.21), then $x_2[n]$ is removed from the convolutional combination of Eq. (12.20).

Although in principle any linear system can be used for L in Fig. 12.3, the class of linear *frequency-invariant* systems has proved to be most useful for deconvolution in practical applications. For this class of systems, the output sequence $\hat{y}[n]$ is related to the input sequence $\hat{x}[n]$ by a relation of the form

$$\hat{y}[n] = \ell[n]\hat{x}[n]. \tag{12.24}$$

Equivalently, the Fourier transform of the output is related to the Fourier transform of the input by the periodic convolution

$$\hat{Y}(e^{j\omega}) = \frac{1}{2\pi} \int_{-\pi}^{\pi} \hat{X}(e^{j\theta}) L(e^{j(\omega - \theta)}) \, d\theta, \tag{12.25}$$

where $L(e^{j\omega})$ is the Fourier transform of $\ell[n]$. This class of systems is particularly useful when $\hat{x}_1[n]$ and $\hat{x}_2[n]$ have nonoverlapping regions of support, either exactly or approximately. For example, suppose that $\hat{x}_1[n] = 0$ for $n \geq n_0$ and $\hat{x}_2[n] = 0$ for $n < n_0$. Then a frequency-invariant linear filter such that

$$\ell[n] = \begin{cases} 1, & n < n_0, \\ 0, & n \geq n_0, \end{cases} \tag{12.26}$$

would remove $\hat{x}_2[n]$ from $\hat{x}[n]$, leaving $\hat{y}[n] = \hat{x}_1[n]$.

At this point we are reminded of the quotation from Bogert et al. (1963): "In general, we find ourselves operating on the frequency side in ways customary on the time side and vice versa." Indeed, frequency-invariant linear systems operate on the logarithm of the Fourier transform in exactly the way that time-invariant systems operate on the time function. For this reason Bogert et al. referred to frequency-invariant linear systems as "lifters," and they associated the name "quefrency" with the time index independent variable of the cepstrum.

In the remainder of this chapter, we will see many examples of how the concepts of homomorphic deconvolution can be applied. In some cases, it is possible to separate exactly the components of the complex cepstrum. In other cases, only approximate separation is possible. To carry these ideas further, however, we must first carefully examine the properties of the complex cepstrum. We must also consider the computation of the complex cepstrum, i.e., the implementation of the system $D_*[\cdot]$.

12.3 PROPERTIES OF THE COMPLEX LOGARITHM

Since the complex logarithm plays a key role in the definition of the complex cepstrum, it is important to understand its definition and properties. Recall that a sequence has a complex cepstrum if the logarithm of its z-transform has a power series expansion as in Eq. (12.4), where we have specified the region of convergence to include the unit circle. This means that the Fourier transform

$$\hat{X}(e^{j\omega}) = \log|X(e^{j\omega})| + j \measuredangle X(e^{j\omega}) \tag{12.27}$$

must be a continuous periodic function of ω, and consequently both $\log|X(e^{j\omega})|$ and $\measuredangle X(e^{j\omega})$ must be continuous functions of ω. Provided that $X(z)$ does not have zeros on the unit circle, the continuity of $\log|X(e^{j\omega})|$ is guaranteed since $X(e^{j\omega})$ is assumed to be analytic on the unit circle. However, as discussed in Section 5.3, $\measuredangle X(e^{j\omega})$ is ambiguous since at each ω any integer multiple of 2π can be added, and continuity of $\measuredangle X(e^{j\omega})$ is dependent on how it is specified.

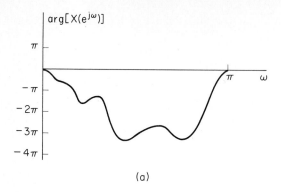

(a)

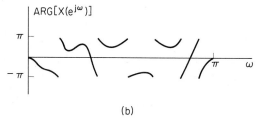

(b)

Figure 12.5 (a) Typical phase curve for a z-transform evaluated on the unit circle. (b) Principal value of the phase in part (a).

The problem of uniqueness and analyticity of $\measuredangle X(e^{j\omega})$ (and consequently also the complex logarithm) is illustrated in Fig. 12.5. In Fig. 12.5(a) we show a continuous-phase curve $\arg[X(e^{j\omega})]$. Figure 12.5(b) shows $\mathrm{ARG}[X(e^{j\omega})]$, corresponding to the principal value of the phase curve in Fig. 12.5(a) as would typically be obtained from an arctangent subroutine. By definition,

$$-\pi < \mathrm{ARG}[X(e^{j\omega})] \leq \pi \tag{12.28}$$

and

$$\arg[X(e^{j\omega})] = \mathrm{ARG}[X(e^{j\omega})] + 2\pi r(\omega), \tag{12.29}$$

where $r(\omega)$ takes on the appropriate integer values to "unwrap" the principal value of the phase resulting in a continuous curve.

Since $\mathrm{ARG}[X(e^{j\omega})]$ can be discontinuous, it generally will not satisfy the continuity requirement. Consequently, it is necessary to explicitly specify $\measuredangle X(e^{j\omega})$ in Eq. (12.27) as the continuous phase curve $\arg[X(e^{j\omega})]$. It is important to note that if $X(z) = X_1(z)X_2(z)$, then

$$\arg[X(e^{j\omega})] = \arg[X_1(e^{j\omega})] + \arg[X_2(e^{j\omega})]. \tag{12.30}$$

A similar additive property will not hold for $\mathrm{ARG}[X(e^{j\omega})]$, i.e.,

$$\mathrm{ARG}[X(e^{j\omega})] \neq \mathrm{ARG}[X_1(e^{j\omega})] + \mathrm{ARG}[X_2(e^{j\omega})]. \tag{12.31}$$

Therefore, in order that $\hat{X}(e^{j\omega})$ be analytic and have the property that if $X(e^{j\omega}) = X_1(e^{j\omega})X_2(e^{j\omega})$, then

$$\hat{X}(e^{j\omega}) = \hat{X}_1(e^{j\omega}) + \hat{X}_2(e^{j\omega}), \tag{12.32}$$

we must define $\hat{X}(e^{j\omega})$ as

$$\hat{X}(e^{j\omega}) = \log|X(e^{j\omega})| + j\arg[X(e^{j\omega})]. \tag{12.33}$$

With $x[n]$ real, $\arg[X(e^{j\omega})]$ can always be specified so that it is an odd periodic function of ω. With $\arg[X(e^{j\omega})]$ an odd function of ω and $\log|X(e^{j\omega})|$ an even function of ω, the complex cepstrum $\hat{x}[n]$ is guaranteed to be real.

The approach outlined above to the problems presented by the complex logarithm can be developed more formally through the concept of the Riemann surface (Churchill and Brown, 1984). An alternative approach to defining the complex logarithm of a Fourier transform or z-transform is to assume that the continuous complex logarithm is obtained by integration of its derivative. If we assume a single-valued differentiable complex logarithm, then

$$\frac{d\hat{X}(z)}{dz} = \frac{d}{dz}\{\log[X(z)]\} = \frac{1}{X(z)}\frac{dX(z)}{dz}. \tag{12.34}$$

If $x[n]$ is a stable sequence (so that the region of convergence of $X(z)$ includes the unit circle), then both $X(e^{j\omega})$ and its derivative $X'(e^{j\omega})$ exist and are continuous functions of ω. Therefore, the unit circle evaluation of the logarithmic derivative on the unit circle is

$$\hat{X}'(e^{j\omega}) = \frac{X'(e^{j\omega})}{X(e^{j\omega})} = \hat{X}'_R(e^{j\omega}) + j\hat{X}'_I(e^{j\omega}), \tag{12.35}$$

where the prime denotes differentiation with respect to ω. From Eq. (12.33),

$$\hat{X}'_I(e^{j\omega}) = \frac{d}{d\omega}\{\arg[X(e^{j\omega})]\} \tag{12.36}$$

and since

$$X(e^{j\omega}) = X_R(e^{j\omega}) + jX_I(e^{j\omega}), \tag{12.37}$$

we obtain

$$\hat{X}'_I(e^{j\omega}) = \frac{d}{d\omega}\{\arg[X(e^{j\omega})]\} = \frac{X_R(e^{j\omega})X'_I(e^{j\omega}) - X_I(e^{j\omega})X'_R(e^{j\omega})}{X_R^2(e^{j\omega}) + X_I^2(e^{j\omega})}. \tag{12.38}$$

Therefore $\arg[X(e^{j\omega})]$ can be obtained by integration:

$$\arg[X(e^{j\omega})] = \int_0^\omega \frac{X_R(e^{j\theta})X'_I(e^{j\theta}) - X_I(e^{j\theta})X'_R(e^{j\theta})}{X_R^2(e^{j\theta}) + X_I^2(e^{j\theta})}\,d\theta. \tag{12.39}$$

Note that the condition

$$\arg[X(e^{j\omega})]_{\omega=0} = 0 \tag{12.40}$$

is implicit in Eq. (12.39), thereby ensuring that $\arg[X(e^{j\omega})]$ is an odd function of ω.

We have carefully considered the complex logarithm for two reasons. First, it is important to fully understand the complex logarithm so that our formal manipulations can be made with confidence in their validity. Second, the problems of ambiguity in the definition of the complex logarithm manifest themselves as important computational issues. These will be considered in detail in Section 12.7.

12.4 ALTERNATIVE EXPRESSIONS FOR THE COMPLEX CEPSTRUM

So far we have defined the complex cepstrum as the sequence of coefficients in the power series representation of $\hat{X}(z) = \log[X(z)]$, and we have also given integral formulas in Eqs. (12.6) and (12.7) for specifying $\hat{x}[n]$ from $\hat{X}(z)$ and $\hat{X}(e^{j\omega})$, respectively, where we now interpret $\angle X(e^{j\omega})$ as the continuous (unwrapped) phase function $\arg[X(e^{j\omega})]$. The logarithmic derivative can be used to derive other relations for the complex cepstrum that do not explicitly involve the complex logarithm. Assuming that $\log[X(z)]$ is analytic, then

$$\hat{X}'(z) = \frac{X'(z)}{X(z)}, \tag{12.41}$$

where the prime now denotes differentiation with respect to z. From property 4 in Table 4.2 (page 180), $z\hat{X}'(z)$ is the z-transform of $-n\hat{x}[n]$, i.e.,

$$-n\hat{x}[n] = \frac{1}{2\pi j} \oint_C z\hat{X}'(z) z^{n-1} \, dz. \tag{12.42}$$

Consequently, from Eq. (12.41),

$$-n\hat{x}[n] = \frac{1}{2\pi j} \oint_C \frac{zX'(z)}{X(z)} z^{n-1} \, dz, \tag{12.43}$$

where C denotes a closed contour (such as the unit circle) in the region of convergence of $\hat{X}(z)$. Solving for $\hat{x}[n]$, we obtain

$$\hat{x}[n] = \frac{-1}{2\pi j n} \oint_C \frac{zX'(z)}{X(z)} z^{n-1} \, dz, \qquad n \neq 0. \tag{12.44}$$

The value of $\hat{x}[0]$ can be obtained by noting that by definition

$$\hat{x}[0] = \frac{1}{2\pi} \int_{-\pi}^{\pi} \hat{X}(e^{j\omega}) \, d\omega. \tag{12.45}$$

Since the imaginary part of $\hat{X}(e^{j\omega})$ is an odd function of ω, Eq. (12.45) becomes

$$\hat{x}[0] = \frac{1}{2\pi} \int_{-\pi}^{\pi} \log|X(e^{j\omega})| \, d\omega. \tag{12.46}$$

Beginning with Eq. (12.41) we can also derive a difference equation that is satisfied by $x[n]$ and $\hat{x}[n]$. Rearranging Eq. (12.41) and multiplying by z, we obtain

$$zX'(z) = z\hat{X}'(z) \cdot X(z). \tag{12.47}$$

The inverse z-transform of this equation is

$$nx[n] = \sum_{k=-\infty}^{\infty} k\hat{x}[k]x[n-k]. \tag{12.48}$$

Dividing both sides by n, we obtain

$$x[n] = \sum_{k=-\infty}^{\infty} \left(\frac{k}{n}\right)\hat{x}[k]x[n-k], \qquad n \neq 0. \tag{12.49}$$

Thus, a signal and its complex cepstrum satisfy a nonlinear difference equation. (Equivalently, the input and output of the characteristic system $D_*[\cdot]$ satisfy this nonlinear difference equation.) Under certain conditions, this implicit relation between $\hat{x}[n]$ and $x[n]$ can be rearranged into a recursion formula that can be used in computation. Formulas of this type are discussed in Section 12.6.

12.5 THE COMPLEX CEPSTRUM OF EXPONENTIAL SEQUENCES

If a sequence $x[n]$ consists of a sum of complex exponential sequences, its z-transform $X(z)$ is a rational function of z. Such sequences are both useful and amenable to analysis. In this section we consider the complex cepstrum for stable sequences $x[n]$ whose z-transforms are of the form

$$X(z) = \frac{Az^r \prod_{k=1}^{M_i}(1 - a_k z^{-1}) \prod_{k=1}^{M_o}(1 - b_k z)}{\prod_{k=1}^{N_i}(1 - c_k z^{-1}) \prod_{k=1}^{N_o}(1 - d_k z)}, \tag{12.50}$$

where $|a_k|$, $|b_k|$, $|c_k|$, and $|d_k|$ are all less than unity, so that factors of the form $(1 - a_k z^{-1})$ and $(1 - c_k z^{-1})$ correspond to the M_i zeros and the N_i poles inside the unit circle, and the factors $(1 - b_k z)$ and $(1 - d_k z)$ correspond to the M_o zeros and the N_o poles outside the unit circle. Such z-transforms are characteristic of sequences composed of a sum of stable exponential sequences. In the special case where there are no poles (i.e., the denominator of Eq. 12.50 is unity), then the corresponding sequence $x[n]$ is a sequence of finite length ($M_o + M_i + 1$).

If $\log[X(z)]$ is computed so that Eqs. (12.15) and (12.16) hold, the product of terms in Eq. (12.50) is transformed to the sum of logarithmic terms:

$$\hat{X}(z) = \log(A) + \log(z^r) + \sum_{k=1}^{M_i} \log(1 - a_k z^{-1}) + \sum_{k=1}^{M_o} \log(1 - b_k z)$$

$$- \sum_{k=1}^{N_i} \log(1 - c_k z^{-1}) - \sum_{k=1}^{N_o} \log(1 - d_k z). \tag{12.51}$$

The properties of $\hat{x}[n]$ depend on the composite properties of the inverse transforms of each term.

For real sequences, A is real, and if A is positive, the first term $\log(A)$ simply contributes only to $\hat{x}[0]$. Specifically (see Problem 12.15),

$$\hat{x}[0] = \log|A|. \tag{12.52}$$

If A is negative, it is more difficult to determine the contribution to the complex cepstrum due to the term $\log[A]$. The term z^r corresponds only to a delay or advance of the sequence $x[n]$. If $r = 0$, this term vanishes from Eq. (12.51). However, if $r \neq 0$, then the unwrapped phase function $\arg[X(e^{j\omega})]$ will include a linear term with slope r. Consequently, with $\arg[X(e^{j\omega})]$ defined to be odd and periodic in ω and continuous for $|\omega| < \pi$, this linear phase term will force a discontinuity in $\arg[X(e^{j\omega})]$ at $\omega = \pm\pi$, and $\hat{X}(z)$ will no longer be analytic on the unit circle. Although the cases of A negative and/or $r \neq 0$ can be formally accommodated, doing so seems to offer no real advantage because if two transforms of the form of Eq. (12.50) are multiplied together, we would not expect to determine how much of either A or r was contributed by each component. This is analogous to the situation in ordinary linear filtering where two signals, each with dc levels, have been added. Therefore, this question is avoided in practice by determining the algebraic sign of A and the value of r and then altering the input so that its z-transform is of the form

$$X(z) = \frac{|A| \prod\limits_{k=1}^{M_i}(1 - a_k z^{-1}) \prod\limits_{k=1}^{M_o}(1 - b_k z)}{\prod\limits_{k=1}^{N_i}(1 - c_k z^{-1}) \prod\limits_{k=1}^{N_o}(1 - d_k z)}. \tag{12.53}$$

Likewise, Eq. (12.51) becomes

$$\begin{aligned}
\hat{X}(z) = \log|A| &+ \sum_{k=1}^{M_i} \log(1 - a_k z^{-1}) + \sum_{k=1}^{M_o} \log(1 - b_k z) \\
&- \sum_{k=1}^{N_i} \log(1 - c_k z^{-1}) - \sum_{k=1}^{N_o} \log(1 - d_k z).
\end{aligned} \tag{12.54}$$

With the exception of the term $\log|A|$, which we have already considered, all the terms in Eq. (12.54) are of the form $\log(1 - \alpha z^{-1})$ and $\log(1 - \beta z)$. Bearing in mind that these factors represent z-transforms with regions of convergence that include the unit circle, we can make the power series expansions

$$\log(1 - \alpha z^{-1}) = -\sum_{n=1}^{\infty} \frac{\alpha^n}{n} z^{-n}, \qquad |z| > |\alpha|, \tag{12.55}$$

$$\log(1 - \beta z) = -\sum_{n=1}^{\infty} \frac{\beta^n}{n} z^{n}, \qquad |z| < |\beta^{-1}|. \tag{12.56}$$

Using these expressions, we see that for signals with rational z-transforms as in Eq. (12.53), $\hat{x}[n]$ has the general form

$$\hat{x}[n] = \begin{cases}
\log|A|, & n = 0, & (12.57\text{a}) \\[2mm]
-\sum\limits_{k=1}^{M_i} \dfrac{a_k^n}{n} + \sum\limits_{k=1}^{N_i} \dfrac{c_k^n}{n}, & n > 0, & (12.57\text{b}) \\[2mm]
\sum\limits_{k=1}^{M_o} \dfrac{b_k^{-n}}{n} - \sum\limits_{k=1}^{N_o} \dfrac{d_k^{-n}}{n}, & n < 0. & (12.57\text{c})
\end{cases}$$

Note that for the special case of a finite-length sequence, the second term would be missing in each of Eqs. (12.57b) and (12.57c). Equations (12.57) suggest the following general properties of the complex cepstrum:

PROPERTY 1:   The complex cepstrum decays at least as fast as $1/|n|$: Specifically,

$$|\hat{x}[n]| < C \left| \frac{\alpha^n}{n} \right|, \qquad -\infty < n < \infty,$$

where C is a constant and α equals the maximum of $|a_k|, |b_k|, |c_k|,$ and $|d_k|$. This property is also suggested by Eq. (12.44).

PROPERTY 2: $\hat{x}[n]$ will have infinite duration, even if $x[n]$ has finite duration.

PROPERTY 3:   If $x[n]$ is real, $\hat{x}[n]$ is also real.

Properties 1 and 2 follow directly from Eqs. (12.57). We have suggested property 3 earlier on the basis that for $x[n]$ real, $\log|X(e^{j\omega})|$ is even and $\arg[X(e^{j\omega})]$ is odd so that the inverse transform of

$$\hat{X}(e^{j\omega}) = \log|X(e^{j\omega})| + j \arg[X(e^{j\omega})]$$

is real. To see property 3 in the context of this section, we note that if $x[n]$ is real, then the poles and zeros of $X(z)$ are in complex conjugate pairs. Therefore, for every complex term of the form α^n/n in Eqs. (12.57) there will be a complex conjugate term $(\alpha^*)^n/n$ so that their sum will be real.

12.6 MINIMUM-PHASE AND MAXIMUM-PHASE SEQUENCES

As discussed in Chapter 5, a minimum-phase sequence is a real, causal, and stable sequence whose poles and zeros are inside the unit circle. Note that $\log[X(z)]$ has singularities at both the poles and the zeros of $X(z)$. Since we require that the region of convergence of $\log[X(z)]$ include the unit circle so that $\hat{x}[n]$ is stable, and since causal sequences have a region of convergence of the form $r_R < |z|$, it follows that there can be no singularities of $\log[X(z)]$ on or outside the unit circle if $\hat{x}[n] = 0$ for $n < 0$. Conversely, if all the singularities of $\hat{X}(z) = \log[X(z)]$ are inside the unit circle, then it follows that $\hat{x}[n] = 0$ for $n < 0$. Since the singularities of $\hat{X}(z)$ are the poles and the zeros of $X(z)$, the complex cepstrum of $x[n]$ will be causal ($\hat{x}[n] = 0$ for $n < 0$) if and only if the poles and zeros of $X(z)$ are inside the unit circle. In other words, $x[n]$ is a minimum-phase sequence if and only if its complex cepstrum is causal.

This is easily seen for the case of exponential or finite-length sequences by considering Eqs. (12.57). Clearly, all terms in Eq. (12.57c) will be zero if all the b_k's and d_k's are zero, i.e., if there are no poles or zeros outside or on the unit circle. Thus, another property of the complex cepstrum is

PROPERTY 4:   The complex cepstrum $\hat{x}[n] = 0$ for $n < 0$ if and only if $x[n]$ is minimum phase, i.e., $X(z)$ has all its poles and zeros inside the unit circle.

Therefore, causality of the complex cepstrum is equivalent to the minimum phase lag, minimum group delay, and minimum energy delay properties that also characterize minimum-phase sequences.

Maximum-phase sequences are stable sequences whose poles and zeros are all *outside* the unit circle. Thus, maximum-phase sequences are left-sided, and, by analogous arguments, it follows that the complex cepstrum of a maximum-phase sequence is also left-sided. Thus, still another property of the complex cepstrum is

> **PROPERTY 5:**   The complex cepstrum $\hat{x}[n] = 0$ for $n > 0$ if and only if $x[n]$ is maximum phase, i.e., $X(z)$ has all its poles and zeros outside the unit circle.

This property of the complex cepstrum is easily verified for exponential or finite-length sequences by noting that if all the c_k's and a_k's are zero (i.e., no poles or zeros *inside* the unit circle), then Eq. (12.57b) shows that $\hat{x}[n] = 0$ for $n > 0$.

12.6.1 Hilbert Transform Relations

In Chapter 10 we showed that causality (or, alternatively, anticausality) of a sequence places powerful constraints on the Fourier transform. As discussed in Section 10.3, when applied to the cepstrum and its Fourier transform, causality implies that the log magnitude and phase of the Fourier transform of a minimum-phase signal satisfy a Hilbert transform relationship. Specifically, recall that the Fourier transform of any real, causal sequence is almost completely determined by either the real part or the imaginary part of the Fourier transform. For the complex cepstrum of a real sequence, $\log|X(e^{j\omega})|$ is the real part of $\hat{X}(e^{j\omega})$ and $\arg[X(e^{j\omega})]$ is the imaginary part. Thus, if $\hat{x}[n]$ is causal, i.e., $x[n]$ is minimum phase, we can obtain Hilbert transform relations between $\log|X(e^{j\omega})|$ and $\arg[X(e^{j\omega})]$. Indeed, Eqs. (10.15) and (10.16) were applied directly to $\hat{X}(e^{j\omega})$ to obtain Eqs. (10.58) and (10.59). However, it is instructive to review the steps in the derivation following the approach in Section 10.1 and using the terminology of this chapter.

First note that the cepstrum $c_x[n]$ is the inverse Fourier transform of $\log|X(e^{j\omega})|$, i.e.,

$$c_x[n] = \frac{1}{2\pi} \int_{-\pi}^{\pi} \log|X(e^{j\omega})| e^{j\omega n} \, d\omega. \qquad (12.58)$$

Also, from Eq. (12.9), for real sequences the cepstrum as defined in Eq. (12.58) is the even part of the complex cepstrum, i.e.,

$$c_x[n] = \frac{\hat{x}[n] + \hat{x}[-n]}{2}. \qquad (12.59)$$

Thus, if $\hat{x}[n]$ is causal, then $\hat{x}[n]$ can be recovered from $c_x[n]$ by frequency-invariant linear filtering; i.e.,

$$\hat{x}[n] = c_x[n]\ell_{min}[n], \qquad (12.60a)$$

where

$$\ell_{min}[n] = 2u[n] - \delta[n]. \tag{12.60b}$$

The desired Hilbert transform relations result from the frequency-domain representation of Eqs. (12.60). Specifically, carrying out an analysis identical to that of Section 10.1 leads to the equations

$$\arg[X(e^{j\omega})] = -\frac{1}{2\pi}\mathscr{P}\int_{-\pi}^{\pi}\log|X(e^{j\theta})|\cot\left(\frac{\omega-\theta}{2}\right)d\theta, \tag{12.61a}$$

$$\log|X(e^{j\omega})| = \hat{x}[0] + \frac{1}{2\pi}\mathscr{P}\int_{-\pi}^{\pi}\arg[X(e^{j\theta})]\cot\left(\frac{\omega-\theta}{2}\right)d\theta, \tag{12.61b}$$

which of course are identical to Eqs. (10.58) and (10.59a). Note that without knowledge of $\hat{x}[0]$, $|X(e^{j\omega})|$ is determined only to within a constant multiplier by $\arg[X(e^{j\omega})]$.

Equations (12.60) indicate how the complex cepstrum can be computed from the cepstrum and consequently also from the log magnitude alone if $x[n]$ is known to be minimum phase. Figure 12.6 shows a block diagram of a system for computing the complex cepstrum under these conditions.

The derivation of the Hilbert transform relations for minimum-phase sequences can be repeated with only minor modifications to obtain a pair of Hilbert transform relations for the log magnitude and phase of the Fourier transform of a maximum-phase sequence. In this case, the complex cepstrum is related to the cepstrum through

$$\hat{x}[n] = c_x[n]\ell_{max}[n], \tag{12.62a}$$

where

$$\ell_{max}[n] = 2u[-n] - \delta[n]. \tag{12.62b}$$

The derivation of the corresponding Hilbert transform relations proceeds in much the same way. The details are outlined in Problem 12.18.

12.6.2 Minimum-Phase/Allpass Decomposition by Homomorphic Filtering

If $x[n]$ is mixed phase, then the system of Fig. 12.6 with input $x[n]$ and $\ell_{min}[n]$ given by Eq. (12.60b) produces the complex cepstrum of the minimum-phase sequence that has the same Fourier transform magnitude as $x[n]$. If $\ell_{max}[n]$ as in Eq. (12.62b) is

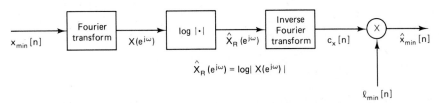

Figure 12.6   Realization of the system $D_*[\cdot]$ for minimum-phase inputs.

used, the output will be the complex cepstrum of the maximum-phase sequence having the same Fourier transform magnitude as $x[n]$.

As discussed in Section 5.6, any nonminimum-phase causal sequence can be represented as

$$x[n] = x_{min}[n] * x_{ap}[n], \tag{12.63}$$

where $x_{min}[n]$ is a minimum-phase sequence such that

$$|X(e^{j\omega})| = |X_{min}(e^{j\omega})| \tag{12.64}$$

and $x_{ap}[n]$ is an allpass sequence such that

$$|X_{ap}(e^{j\omega})| = 1.$$

We can obtain the complex cepstrum $\hat{x}_{min}[n]$ through the operations of Fig. 12.6. The complex cepstrum $\hat{x}_{ap}[n]$ can be obtained from $\hat{x}[n]$ by subtracting $\hat{x}_{min}[n]$ from $\hat{x}[n]$, i.e.,

$$\hat{x}_{ap}[n] = \hat{x}[n] - \hat{x}_{min}[n].$$

To obtain $x_{min}[n]$ and $x_{ap}[n]$, we apply the transformation D_*^{-1} to $\hat{x}_{min}[n]$ and $\hat{x}_{ap}[n]$.

While the approach outlined above to obtain $x_{min}[n]$ and $x_{ap}[n]$ is theoretically correct, explicit evaluation of the complex cepstrum $\hat{x}[n]$ is required in its implementation. If we are interested only in obtaining $x_{min}[n]$ and $x_{ap}[n]$, evaluation of the complex cepstrum and the associated need for phase unwrapping can be avoided. The basic strategy is incorporated in the block diagram of Fig. 12.7. This system relies on the fact that

$$X_{ap}(e^{j\omega}) = \frac{X(e^{j\omega})}{X_{min}(e^{j\omega})}. \tag{12.65a}$$

The magnitude of $X_{ap}(e^{j\omega})$ is therefore

$$|X_{ap}(e^{j\omega})| = \frac{|X(e^{j\omega})|}{|X_{min}(e^{j\omega})|} = 1 \tag{12.65b}$$

and

$$\measuredangle X_{ap}(e^{j\omega}) = \measuredangle X(e^{j\omega}) - \measuredangle X_{min}(e^{j\omega}). \tag{12.65c}$$

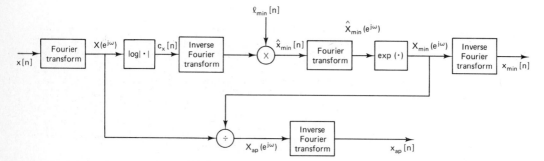

Figure 12.7   The use of homomorphic filtering to separate a sequence into minimum-phase and allpass components.

Since $x_{ap}[n]$ is obtained as the inverse Fourier transform of $e^{j \sphericalangle X_{ap}(e^{j\omega})}$ ($|X_{ap}(e^{j\omega})| = 1$), each of the phase functions in Eq. (12.65c) need only be known or specified to within integer multiples of 2π. Although as a natural consequence of the procedure outlined in Fig. 12.7, $\sphericalangle X_{min}(e^{j\omega})$ will be an unwrapped phase function, $\sphericalangle X(e^{j\omega})$ can be computed modulo 2π.

12.6.3 Recursive Realizations of D_*

For minimum-phase sequences, the difference equation (12.49) can be rearranged to provide a recursion formula for $\hat{x}[n]$. Since for minimum-phase sequences both $\hat{x}[n] = 0$ and $x[n] = 0$ for $n < 0$, Eq. (12.49) becomes

$$x[n] = \sum_{k=0}^{n} \left(\frac{k}{n}\right)\hat{x}[k]x[n-k], \qquad n > 0,$$

$$= \hat{x}[n]x[0] + \sum_{k=0}^{n-1} \left(\frac{k}{n}\right)\hat{x}[k]x[n-k]. \qquad (12.66)$$

Solving for $\hat{x}[n]$ yields the recursion formula

$$\hat{x}[n] = \begin{cases} 0, & n < 0, \\ \dfrac{x[n]}{x[0]} - \displaystyle\sum_{k=0}^{n-1} \left(\frac{k}{n}\right)\hat{x}[k]\dfrac{x[n-k]}{x[0]}, & n > 0. \end{cases} \qquad (12.67)$$

The value of $\hat{x}[0]$ can be shown to be (see Problem 12.15)

$$\hat{x}[0] = \log(A) = \log(x[0]). \qquad (12.68)$$

Therefore, Eqs. (12.67) and (12.68) constitute a representation of the system $D_*[\cdot]$ for minimum-phase signals. It also follows from Eq. (12.67) that the system $D_*[\cdot]$ behaves as a causal system for minimum-phase inputs; i.e., the output at time n_0 is dependent only on the input for $n \le n_0$, where n_0 is arbitrary (see Problem 12.21). Similarly, Eqs. (12.66) and (12.68) represent the inverse characteristic system $D_*^{-1}[\cdot]$.

For maximum-phase signals, $\hat{x}[n] = 0$ and $x[n] = 0$ for $n > 0$. Thus in this case Eq. (12.49) becomes

$$x[n] = \sum_{k=n}^{0} \left(\frac{k}{n}\right)\hat{x}[k]x[n-k], \qquad n < 0,$$

$$= \hat{x}[n]x[0] + \sum_{k=n+1}^{0} \left(\frac{k}{n}\right)\hat{x}[k]x[n-k]. \qquad (12.69)$$

Solving for $\hat{x}[n]$, we have

$$\hat{x}[n] = \begin{cases} \dfrac{x[n]}{x[0]} - \displaystyle\sum_{k=n+1}^{0} \left(\frac{k}{n}\right)\hat{x}[k]\dfrac{x[n-k]}{x[0]}, & n < 0, \\ \log(x[0]), & n = 0, \\ 0, & n > 0. \end{cases} \qquad (12.70)$$

Equations (12.70) and (12.69) serve as a representation of the characteristic system and its inverse for maximum-phase inputs.

12.6.4 Minimum-Phase/Maximum-Phase Decomposition by Homomorphic Filtering

In general, if a z-transform of a signal has no poles or zeros on the unit circle, then it can be expressed as

$$X(z) = X_{mn}(z) \cdot X_{mx}(z), \tag{12.71}$$

where $X_{mn}(z)$ is minimum phase and $X_{mx}(z)$ is maximum phase. Equivalently, $x[n]$ can be expressed as

$$x[n] = x_{mn}[n] * x_{mx}[n] \tag{12.72}$$

and the corresponding complex cepstrum is

$$\hat{x}[n] = \hat{x}_{mn}[n] + \hat{x}_{mx}[n]. \tag{12.73}$$

It should be noted that the representation of $x[n]$ according to Eq. (12.72) is distinctly different from that of Eq. (12.63). Specifically, the two minimum-phase sequences are different, i.e., $x_{min}[n] \neq x_{mn}[n]$. From the above properties it follows that $x_{mn}[n]$ and $x_{mx}[n]$ can be extracted (deconvolved) from $x[n]$ by homomorphic filtering. Specifically, we specify $\hat{x}_{mn}[n]$ as

$$\hat{x}_{mn}[n] = \ell_{mn}[n]\hat{x}[n] \tag{12.74a}$$

where

$$\ell_{mn}[n] = u[n]. \tag{12.74b}$$

Similarly, we specify $\hat{x}_{mx}[n]$ as

$$\hat{x}_{mx}[n] = \ell_{mx}[n]\hat{x}[n] \tag{12.75a}$$

where

$$\ell_{mx}[n] = u[-n - 1]. \tag{12.75b}$$

Obviously, $x_{mn}[n]$ and $x_{mx}[n]$ can be obtained from $\hat{x}_{mn}[n]$ and $\hat{x}_{mx}[n]$, respectively, as the output of the inverse characteristic system $D_*^{-1}[\cdot]$. The operations required for the decomposition of Eq. (12.72) are depicted in Fig. 12.8. This method of factoring a

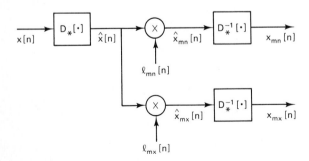

Figure 12.8 The use of homomorphic filtering to separate a sequence into minimum-phase and maximum-phase components.

sequence into its minimum- and maximum-phase parts has been used by Smith and Barnwell (1986) in the design of filter banks. Note that we have arbitrarily assigned all of $\hat{x}[0]$ to $\hat{x}_{mn}[0]$ and we have set $\hat{x}_{mx}[0] = 0$. Obviously, other combinations are possible since all that is required is that $\hat{x}_{mn}[0] + \hat{x}_{mx}[0] = \hat{x}[0]$.

The recursion formulas of Section 12.6.3 can be combined with the representation of Eq. (12.71) to yield an interesting result for finite-length sequences. Specifically, in spite of the infinite extent of the complex cepstrum of a finite-length sequence, we can show that for an input sequence of length N, we need only N samples of $\hat{x}[n]$ to determine $x[n]$. To see this, consider the z-transform of Eq. (12.71), where

$$X_{mn}(z) = A \prod_{k=1}^{M_i} (1 - a_k z^{-1}), \tag{12.76a}$$

$$X_{mx}(z) = \prod_{k=1}^{M_o} (1 - b_k z), \tag{12.76b}$$

with $|a_k| < 1$ and $|b_k| < 1$. Correspondingly, $x[n]$ has the form of Eq. (12.72), where $x_{mn}[n] = 0$ outside the interval $0 \le n \le M_i$ and $x_{mx}[n] = 0$ outside the interval $-M_o \le n \le 0$. Thus the sequence $x[n]$ is nonzero in the interval $-M_o \le n \le M_i$. Using the previous recursion formulas, we can write

$$x_{mn}[n] = \begin{cases} 0, & n < 0, \\ e^{\hat{x}[0]}, & n = 0, \\ \hat{x}[n]\hat{x}_{mn}[0] + \sum_{k=0}^{n-1} \left(\frac{k}{n}\right)\hat{x}[k]x_{mn}[n-k], & n > 0, \end{cases} \tag{12.77}$$

and

$$x_{mx}[n] = \begin{cases} \hat{x}[n] + \sum_{k=n+1}^{0} \left(\frac{k}{n}\right)\hat{x}[k]x_{mx}[n-k], & n < 0, \\ 1, & n = 0, \\ 0, & n > 0. \end{cases} \tag{12.78}$$

Clearly we require $M_i + 1$ values of $\hat{x}[n]$ to compute $x_{mn}[n]$ and M_o values of $\hat{x}[n]$ to compute $x_{mx}[n]$. Thus only $M_i + M_o + 1$ values of the infinite sequence $\hat{x}[n]$ are required to completely recover the components of the finite-length sequence $x[n]$.

12.7 REALIZATIONS OF THE CHARACTERISTIC SYSTEM $D_*[\cdot]$

So far our discussion of the complex cepstrum has been concerned with definitions, properties, and other theoretical issues. The application of the complex cepstrum requires that we develop accurate and efficient computational methods. Implicit in all of the previous discussions was the assumption of uniqueness and continuity of the complex logarithm of the Fourier transform of the input signal. If the mathematical representations obtained above are to serve as the basis for computation of the complex cepstrum, or equivalently as the basis for realizations of the system $D_*[\cdot]$, then we must deal with the problems of computing the Fourier transform and the complex logarithm.

12.7.1 Realization Using the Complex Logarithm

The system $D_*[\cdot]$ is represented in terms of the Fourier transform by the equations

$$X(e^{j\omega}) = \sum_{n=-\infty}^{\infty} x[n]e^{-j\omega n}, \tag{12.79a}$$

$$\hat{X}(e^{j\omega}) = \log[X(e^{j\omega})], \tag{12.79b}$$

$$\hat{x}[n] = \frac{1}{2\pi} \int_{-\pi}^{\pi} \hat{X}(e^{j\omega})e^{j\omega n} \, d\omega. \tag{12.79c}$$

These equations correspond to the cascade of three systems as depicted in Fig. 12.4(a) with $z = e^{j\omega}$. As noted there, each of the three systems is a homomorphic system with the appropriate input and output operations.

Since digital computers perform finite computations, we are limited to finite-length input sequences, and we can compute the Fourier transform at only a finite number of points. That is, instead of using the Fourier transform, we must use the discrete Fourier transform. Thus, instead of Eqs. (12.79) we have the computational realization

$$X[k] = X(e^{j\omega})\Big|_{\omega=(2\pi/N)k} = \sum_{n=0}^{N-1} x[n]e^{-j(2\pi/N)kn}, \tag{12.80a}$$

$$\hat{X}[k] = \log[X(e^{j\omega})]\Big|_{\omega=(2\pi/N)k} = \log(X[k]), \tag{12.80b}$$

$$\hat{x}_p[n] = \frac{1}{N} \sum_{k=0}^{N-1} \hat{X}[k]e^{j(2\pi/N)kn}. \tag{12.80c}$$

These operations are depicted in Fig. 12.9(a), and the corresponding operations for realizing the inverse system are depicted in Fig. 12.9(b).

In writing Eq. (12.80b) it is assumed that sampling of the Fourier transform commutes with computation of the continuous complex logarithm; i.e., we assume that $X[k]$ can be computed using an FFT algorithm, and then the samples of $\log[X(e^{j\omega})]$ can be computed from $X[k]$. With this assumption it follows from the sampling theorem for the z-transform derived in Chapter 8 that $\hat{x}_p[n]$ is related to the desired $\hat{x}[n]$ by

$$\hat{x}_p[n] = \sum_{r=-\infty}^{\infty} \hat{x}[n + rN]. \tag{12.81}$$

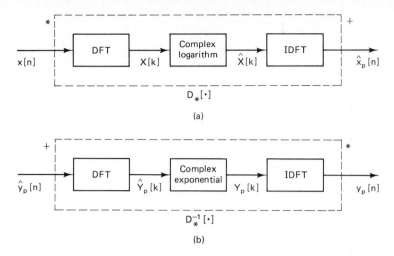

Figure 12.9   Approximate realization using the DFT of (a) the characteristic system for convolution and (b) its inverse.

Since the complex cepstrum in general has infinite duration, $\hat{x}_p[n]$ will be a time-aliased version of $\hat{x}[n]$; however, we noted in property 1 in Section 12.5 that $\hat{x}[n]$ decays faster than an exponential sequence, so it is to be expected that the approximation would become increasingly better as N increases. By appending zeros to an input sequence it is generally possible to increase the sampling rate of the complex logarithm of the Fourier transform so that severe time aliasing does not occur in the computation of the complex cepstrum.

In writing Eqs. (12.80) and Eq. (12.81), we have assumed that $\hat{X}[k]$ represents a sampled version of the continuous complex logarithm. Thus we must consider means for computing samples of $\arg[X(e^{j\omega})]$ from the DFT, $X[k]$. A simple algorithm is based on the computation of the phase modulo 2π, i.e.,

$$-\pi < \mathrm{ARG}(X[k]) \le \pi, \tag{12.82}$$

which is achieved using standard inverse tangent routines available on most computers. This sampled principal value of the phase is then "unwrapped" to obtain samples of the continuous-phase curve and, if present, a linear phase term is removed prior to computing $\hat{x}_p[n]$. For example, consider a finite-length causal input sequence whose Fourier transform is

$$
\begin{aligned}
X(e^{j\omega}) &= \sum_{n=0}^{M} x[n]e^{-j\omega n} \\
&= Ae^{-j\omega M_o} \prod_{k=1}^{M_i}(1 - a_k e^{-j\omega}) \prod_{k=1}^{M_o}(1 - b_k e^{j\omega}),
\end{aligned}
\tag{12.83}
$$

where $|a_k|$ and $|b_k|$ are less than unity, $M = M_o + M_i$, A is positive, and $M_o = 3$. A continuous-phase curve for a sequence of this form is shown in Fig. 12.10(a). The dots indicate samples at $\omega = (2\pi/N)k$. (N is assumed to be even.) Figure 12.10(b) shows

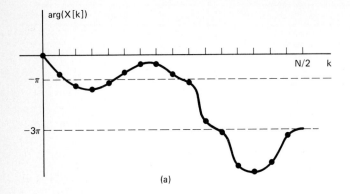

(a)

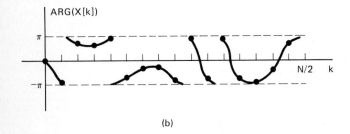

(b)

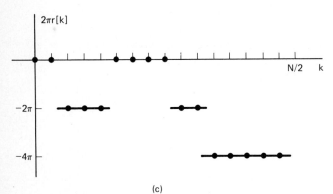

(c)

Figure 12.10   (a) Samples of $\arg[X(e^{j\omega})]$.
(b) Principal value of part (a).
(c) Correction sequence for obtaining arg
from ARG.

the principal value and its samples as computed from the DFT of the input
sequence. One approach to unwrapping the principal-value phase is based on the
relation

$$\arg(X[k]) = \text{ARG}(X[k]) + 2\pi r[k], \tag{12.84}$$

where $r[k]$ denotes an integer that determines the appropriate multiple of 2π to add to
the principal value at frequency $\omega_k = 2\pi k/N$. Figure 12.10(c) shows $2\pi r[k]$ required
to obtain Fig. 12.10(a) from 12.10(b). This example suggests the following algorithm
for computing $r[k]$ from $\text{ARG}(X[k])$ starting with $r[0] = 0$:

    1. If $\text{ARG}(X[k]) - \text{ARG}(X[k-1]) > 2\pi - \varepsilon_1$, then $r[k] = r[k-1] - 1$.
    2. If $\text{ARG}(X[k]) - \text{ARG}(X[k-1]) < -(2\pi - \varepsilon_1)$, then $r[k] = r[k-1] + 1$.

3. Otherwise $r[k] = r[k-1]$.
4. Repeat steps 1–3 for $1 \le k \le N/2$.

After $r[k]$ is determined, Eq. (12.84) can be used to compute $\arg(X[k])$ for $0 \le k \le N/2$. The values of $\arg(X[k])$ for $N/2 < k \le N-1$ can be obtained by using symmetry.

The above algorithm works well if the samples of $\text{ARG}(X[k])$ are close enough together to detect the discontinuities. The parameter ε_1 is a tolerance recognizing that the magnitude of the difference between adjacent samples of the principal-value phase will always be less than 2π. If ε_1 is too large, a discontinuity will be indicated where there is none. If ε_1 is too small, the algorithm will miss a discontinuity falling between two adjacent samples of a rapidly varying unwrapped phase function $\arg[X(e^{j\omega})]$. Obviously, increasing the sampling rate of the DFT by increasing N improves the chances of correctly detecting discontinuities and thus correctly computing $\arg(X[k])$. If $\arg[X(e^{j\omega})]$ varies rapidly, then we expect $\hat{x}[n]$ to decay less rapidly than if $\arg[X(e^{j\omega})]$ varied more slowly. Therefore, aliasing of $\hat{x}[n]$ is more of a problem for rapidly varying phase. Increasing the value of N reduces the aliasing of the complex cepstrum and also improves the chances of being able to unwrap the phase of $X[k]$.

In many practical cases, the simple algorithm we just developed fails because it is impossible or impractical to use a large enough value for N. Often the aliasing for a given N is acceptable, but principal-value discontinuities cannot be reliably detected. Tribolet (1977) proposed a modification of the algorithm that uses both the principal value of the phase and the phase derivative to compute the unwrapped phase. As above, Eq. (12.84) gives the set of permissible values at frequency $\omega_k = (2\pi/N)k$, and we seek to determine $r[k]$. It is assumed that we know the phase derivative,

$$\arg'(X[k]) = \left. \frac{d}{d\omega} \arg[X(e^{j\omega})] \right|_{\omega = 2\pi k/N}$$

at all values of k. (A procedure for computing these samples of the phase derivative will be developed in Section 12.7.2.) Now to compute $\arg(X[k])$ we further assume that $\arg(X[k-1])$ is known. Then $\widetilde{\arg}(X[k])$, the estimate of $\arg(X[k])$, is defined as

$$\widetilde{\arg}(X[k]) = \arg(X[k-1]) + \frac{\Delta\omega}{2}\{\arg'(X[k]) + \arg'(X[k-1])\}. \quad (12.85)$$

Equation (12.85) is obtained by applying trapezoidal numerical integration to the samples of the phase derivative. This estimate is said to be *consistent* if for some ε_2 an integer $r[k]$ exists such that

$$|\widetilde{\arg}(X[k]) - \text{ARG}(X[k]) - 2\pi r[k]| < \varepsilon_2 < \pi. \quad (12.86)$$

Obviously, the estimate improves with decreasing numerical integration step size $\Delta\omega$. Initially, $\Delta\omega = 2\pi/N$ as provided by the DFT. If Eq. (12.86) cannot be satisfied by an integer $r[k]$, then $\Delta\omega$ is halved and a new estimate of $\arg(X[k])$ is computed with the new step size. Then Eq. (12.86) is evaluated with the new estimate. Increasingly accurate estimates of $\arg(X[k])$ are computed by numerical integration until Eq. (12.86) can be satisfied by an integer $r[k]$. That resulting $r[k]$ is used in Eq. (12.84) to finally compute $\arg(X[k])$. This unwrapped phase is then used to compute $\arg(X[k+1])$, and so on. Tribolet's algorithm is available in FORTRAN code (DSP Committee, 1979).

12.7.2 Realization Using the Logarithmic Derivative

As an alternative to the explicit computation of the complex logarithm, a mathematical representation based on the logarithmic derivative can be derived. For real sequences, the derivative of $\hat{X}(e^{j\omega})$ can be represented in the equivalent forms

$$\hat{X}'(e^{j\omega}) = \frac{d\hat{X}(e^{j\omega})}{d\omega} = \frac{d}{d\omega} \log|X(e^{j\omega})| + j\frac{d}{d\omega}\arg[X(e^{j\omega})] \qquad (12.87a)$$

and

$$\hat{X}'(e^{j\omega}) = \frac{X'(e^{j\omega})}{X(e^{j\omega})}, \qquad (12.87b)$$

where the prime represents differentiation with respect to ω. Since the Fourier transform of $x[n]$ is

$$X(e^{j\omega}) = \sum_{n=-\infty}^{\infty} x[n]e^{-j\omega n}, \qquad (12.88)$$

its derivative with respect to ω is

$$X'(e^{j\omega}) = \sum_{n=-\infty}^{\infty} (-jnx[n])e^{-j\omega n}, \qquad (12.89)$$

i.e., $X'(e^{j\omega})$ is the Fourier transform of $-jnx[n]$. Likewise, $\hat{X}'(e^{j\omega})$ is the Fourier transform of $-jn\hat{x}[n]$. Thus $\hat{x}[n]$ can be determined for $n \neq 0$ from

$$\hat{x}[n] = \frac{-1}{2\pi nj} \int_{-\pi}^{\pi} \frac{X'(e^{j\omega})}{X(e^{j\omega})} e^{j\omega n}\, d\omega, \qquad n \neq 0. \qquad (12.90)$$

The value of $\hat{x}[0]$ can be determined from the log magnitude as

$$\hat{x}[0] = \frac{1}{2\pi} \int_{-\pi}^{\pi} \log|X(e^{j\omega})|\, d\omega. \qquad (12.91)$$

Equations (12.88)–(12.91) represent the complex cepstrum in terms of the Fourier transforms of $x[n]$ and $nx[n]$ and thus do not explicitly involve the unwrapped phase. For finite-length sequences, these transforms can be computed using the DFT, thereby leading to the corresponding equations

$$X[k] = \sum_{n=0}^{N-1} x[n]e^{-j(2\pi/N)kn} = X(e^{j\omega})\Big|_{\omega=(2\pi/N)k}, \tag{12.92a}$$

$$X'[k] = -j\sum_{n=0}^{N-1} nx[n]e^{-j(2\pi/N)kn} = X'(e^{j\omega})\Big|_{\omega=(2\pi/N)k}, \tag{12.92b}$$

$$\hat{x}_{dp}[n] = -\frac{1}{jnN}\sum_{k=0}^{N-1} \frac{X'[k]}{X[k]} e^{j(2\pi/N)kn}, \qquad 1 \le n \le N-1, \tag{12.92c}$$

$$\hat{x}_{dp}[0] = \frac{1}{N}\sum_{k=0}^{N-1} \log|X[k]|, \tag{12.92d}$$

where the subscript d refers to the use of the logarithmic derivative and the subscript p reminds us of the inherent periodicity of the DFT calculations. With the use of Eqs. (12.92) we avoid the problems of computing the complex logarithm, at the cost, however, of more severe aliasing since now

$$\hat{x}_{dp}[n] = \frac{1}{n}\sum_{r=-\infty}^{\infty} (n+rN)\hat{x}[n+rN]. \tag{12.93}$$

Thus, assuming that the sampled phase curve is accurately computed, we would expect that for a given value of N, $\hat{x}_p[n]$ in Eq. (12.80c) would be a better approximation to $\hat{x}[n]$ than would $\hat{x}_{dp}[n]$ in Eq. (12.92c).

For finite-length sequences with Fourier transform as in Eq. (12.83) it can be shown that

$$M_o = \frac{-1}{2\pi j}\int_{-\pi}^{\pi} \frac{X'(e^{j\omega})}{X(e^{j\omega})}\, d\omega. \tag{12.94}$$

Approximating this expression using the inverse DFT, we obtain

$$M_{op} = \frac{-1}{jN}\sum_{k=0}^{N-1} \frac{X'[k]}{X[k]}. \tag{12.95}$$

The quantity M_{op} will, in general, not be an integer; however, for large N, we would expect M_{op} to approach M_o, the number of zeros of $X(z)$ outside the unit circle.

Another comment about the computation of the continuous complex logarithm concerns the sign of A and the linear phase component due to the factor $e^{-j\omega M_o}$. The sign of A can be easily determined since it is identical to the sign of $X[k]$ at $k=0$. The value of M_o can be determined from the result of adding the correction term $r[k]$ to $\text{ARG}[X(k)]$ since it follows from Eq. (12.83) that for $x[n]$ real,

$$\arg[X(e^{j\pi})] = -M_o\pi \tag{12.96}$$

and, if the sampling rate is high enough,

$$r[N/2 - 1] \simeq \frac{-M_o}{2}. \tag{12.97}$$

This linear phase component can be subtracted from the unwrapped phase, and if $X[0]$ is negative, the sign of the DFT samples is changed prior to computing the phase, thereby making A positive.

Finally, recall that Tribolet's algorithm for phase unwrapping requires the phase derivative. From Eq. (12.87) it follows that for finite-length sequences, the phase derivative can be computed exactly at frequencies $\omega_k = 2\pi k/N$ using the relation

$$\arg'(X[k]) = \mathscr{Im}\{\hat{X}'[k]\}$$

$$= \mathscr{Im}\left\{\frac{X'[k]}{X[k]}\right\}. \tag{12.98}$$

Specifically, $\arg'(X[k])$ is the imaginary part of the ratio of the DFT of $-jnx[n]$ to the DFT of $x[n]$. These DFTs can, of course, be computed at N points using an FFT algorithm. Such a computation is the first stage of Tribolet's algorithm. Subsequent phase derivative computations involved in the adaptive integration part of the algorithm are done using a direct DFT calculation at the required frequencies between the basic DFT frequencies $\omega_k = 2\pi k/N$. The Goertzel algorithm, as described in Section 9.2, can be used for efficient computation of the small number of Fourier transform values required in the adaptive integration part of the algorithm.

12.7.3 Minimum-Phase Realizations for Minimum-Phase Sequences

In the special case of minimum-phase inputs, the mathematical representation is simplified as indicated in Fig. 12.6. A computational realization based on using the DFT in place of the Fourier transform in Fig. 12.6 is given by the equations

$$X[k] = \sum_{n=0}^{N-1} x[n]e^{-j(2\pi/N)kn}, \tag{12.99a}$$

$$c_{xp}[n] = \frac{1}{N}\sum_{k=0}^{N-1} \log|X[k]|e^{j(2\pi/N)kn}. \tag{12.99b}$$

In this case it is the cepstrum that is aliased; i.e.,

$$c_{xp}[n] = \sum_{r=-\infty}^{\infty} c_x[n+rN]. \tag{12.100}$$

To compute the complex cepstrum from $c_{xp}[n]$ based on Fig. 12.6, we write

$$\hat{x}_{cp}(n) = \begin{cases} c_{xp}[n], & n = 0, \ N/2, \\ 2c_{xp}[n], & 1 \le n < N/2, \\ 0, & N/2 < n \le N - 1. \end{cases} \tag{12.101}$$

Clearly $\hat{x}_{cp}[n] \ne \hat{x}_p[n]$ since it is the even part of $\hat{x}[n]$ that is aliased rather than $\hat{x}[n]$ itself. Nevertheless, for large N, $\hat{x}_{cp}[n]$ can be expected to be a reasonable approximation to $\hat{x}[n]$. Similarly, if $x[n]$ is maximum phase, an approximation to the complex cepstrum would be obtained from

$$\hat{x}_{cp}[n] = \begin{cases} c_{xp}[n], & n = 0, \quad N/2, \\ 0, & 1 \le n < N/2, \\ 2c_{xp}[n], & N/2 < n \le N - 1. \end{cases} \tag{12.102}$$

In the case of minimum-phase or maximum-phase sequences, we also have the recursion formulas of Eq. (12.66)–(12.70) as possible realizations of the characteristic system and its inverse. These equations can be quite useful when the input sequence is very short or when only a few samples of the complex cepstrum are desired. With these formulas there is, of course, no aliasing error.

The systems of Figs. 12.7 and 12.8 for minimum-phase/allpass decomposition and for minimum-phase/maximum-phase decomposition, respectively, can be approximated by substituting the approximate realizations of Fig. 12.9 for the ideal systems $D_*[\cdot]$ and $D_*^{-1}[\cdot]$.

12.7.4 Polynomial Rooting

Another approach to computing the complex cepstrum without phase unwrapping is suggested by Section 12.5. If the sequence $x[n]$ has finite length, then only the numerator remains in Eq. (12.50), so its z-transform is a polynomial in z^{-1}. This representation can be found using a numerical polynomial rooting algorithm on the polynomial whose coefficients are the sequence $x[n]$ to find the zeros a_k and $1/b_k$ that lie inside and outside the unit circle, respectively. Then the complex cepstrum can be computed from Eqs. (12.57) as

$$\hat{x}[n] = \begin{cases} \log|A|, & n = 0, \\ -\displaystyle\sum_{k=1}^{M_i} \frac{a_k^n}{n}, & n > 0, \\ \displaystyle\sum_{k=1}^{M_o} \frac{b_k^{-n}}{n}, & n < 0. \end{cases} \tag{12.103}$$

This method of computation is particularly useful when $N = M_o + M_i + 1$ is small. Steiglitz and Dickenson (1977) studied this method and reported successful rooting of polynomials with degree as high as $N = 256$. The advantages of this method are that there is no aliasing and no uncertainty about whether a phase discontinuity was either undetected or falsely detected. One application of this method is in testing of cepstrum computations that involve phase unwrapping.

12.7.5 The Use of Exponential Weighting

Exponential weighting of a sequence can be used to avoid or mitigate some of the problems encountered in computing the complex cepstrum. Exponential weighting of a sequence $x[n]$ is defined by

$$w[n] = \alpha^n x[n]. \tag{12.104}$$

The corresponding z-transform is

$$W(z) = X(\alpha^{-1}z). \tag{12.105}$$

If the region of convergence of $X(z)$ is $r_R < |z| < r_L$, then the region of convergence of $W(z)$ is $|\alpha|r_R < |z| < |\alpha|r_L$, and the poles and zeros of $X(z)$ are shifted radially by the factor $|\alpha|$; i.e., if z_0 is a pole or zero of $X(z)$, then $z_0\alpha$ is the corresponding pole or zero of $W(z)$.

An important property of exponential weighting is that it commutes with convolution. That is, if $x[n] = x_1[n] * x_2[n]$ and $w[n] = \alpha^n x[n]$, then

$$W(z) = X(\alpha^{-1}z) = X_1(\alpha^{-1}z)X_2(\alpha^{-1}z), \tag{12.106}$$

so that

$$w[n] = (\alpha^n x_1[n]) * (\alpha^n x_2[n]) \\ = w_1[n] * w_2[n]. \tag{12.107}$$

Thus in computing the complex cepstrum,

$$\hat{W}(z) = \log[W(z)] \\ = \log[W_1(z)] + \log[W_2(z)]. \tag{12.108}$$

Exponential weighting can be exploited with cepstrum computation in a variety of ways. For example, poles or zeros of $X(z)$ on the unit circle require special care in computing the complex cepstrum. It can be shown (Carslaw, 1952) that a factor $\log(1 - e^{j\theta}e^{-j\omega})$ has a Fourier series

$$\log(1 - e^{j\theta}e^{-j\omega}) = -\sum_{n=1}^{\infty} \frac{e^{j\theta n}}{n} e^{-j\omega n} \tag{12.109}$$

and thus the contribution of such a term to the complex cepstrum is $(e^{j\theta n}/n)u[n-1]$. However, the log magnitude is infinite and the phase is discontinuous with a jump of π radians at $\omega = \theta$. This presents obvious computational difficulties that we would prefer to avoid. By exponential weighting with $0 < \alpha < 1$, all poles and zeros are moved radially inward. Therefore a pole or zero on the unit circle will move inside the unit circle.

As another example, consider a signal $x[n]$ that is nonminimum phase. The exponentially weighted signal, $w[n] = \alpha^n x[n]$, can be converted into a minimum-phase sequence if α is chosen so that $|z_{max}\alpha| < 1$, where z_{max} is the location of the zero with the greatest magnitude.

Still another interesting use of exponential weighting is in computing the complex cepstrum without phase unwrapping. Assume that $X(z)$ has no poles and zeros on the unit circle. Then it is possible to find an exponential weighting factor α such that none of the poles or zeros of $X(z)$ are shifted across the unit circle in forming $W(z) = X(\alpha^{-1}z)$. From Eq. (12.105), if no poles or zeros move across the unit circle, then since the region of convergence of $W(z)$ includes the unit circle,

$$\hat{w}[n] = \alpha^n \hat{x}[n]. \tag{12.110}$$

Now suppose that instead of the complex cepstrum, we compute the cepstrum of $x[n]$ and $w[n]$. $c_x[n]$ is

$$c_x[n] = \frac{\hat{x}[n] + \hat{x}[-n]}{2} \tag{12.111a}$$

and from Eq. (12.110),

$$c_w[n] = \frac{\hat{w}[n] + \hat{w}[-n]}{2} = \frac{\alpha^n \hat{x}[n] + \alpha^{-n} \hat{x}[-n]}{2}. \tag{12.111b}$$

From Eqs. (12.111) it follows that

$$\hat{x}[n] = \frac{2(c_x[n] - \alpha^n c_w[n])}{1 - \alpha^{2n}}, \qquad n \neq 0. \tag{12.112}$$

Since $c_x[n]$ and $c_w[n]$ can be computed from $\log|X(e^{j\omega})|$ and $\log|W(e^{j\omega})|$, respectively, we have the basis for computing the complex cepstrum without computing the phase of $X(e^{j\omega})$. The further the poles and zeros of $X(z)$ are from the unit circle, the more flexibility there is in choosing α. An obvious difficulty with this method is the need to ensure that no roots cross the unit circle due to the exponential weighting. This may require values of α that are close to 1, in which case the denominator of Eq. (12.112) may be quite small for small values of n.

12.8 EXAMPLES OF HOMOMORPHIC FILTERING

In the previous sections we have discussed the theoretical and computational aspects of cepstrum analysis and of homomorphic filtering of signals that are combined by convolution. In this section we present an extended example of our previous discussions. Our approach will be to consider a relatively simple signal model that involves only rational z-transforms so that exact analytical results can be obtained and compared to results computed using the DFT. In Section 12.9, we illustrate cepstral analysis and homomorphic filtering in the context of a more realistic practical application, speech processing.

12.8.1 Signal Model

Throughout this section we will consider a discrete-time signal $x[n]$ that is the convolution of two components, i.e.,

$$x[n] = v[n] * p[n] \tag{12.113a}$$

or, in the z-transform domain,

$$X(z) = V(z)P(z). \tag{12.113b}$$

We choose $p[n]$ to be of the form

$$p[n] = \delta[n] + \beta\delta[n - N_0] + \beta^2\delta[n - 2N_0], \tag{12.114}$$

and its z-transform is then

$$P(z) = 1 + \beta z^{-N_0} + \beta^2 z^{-2N_0} = \frac{1 - \beta^3 z^{-3N_0}}{1 - \beta z^{-N_0}}. \tag{12.115}$$

For example, $p[n]$ might correspond to the impulse response of a multipath channel or other system that generates multiple echoes at a spacing of N_0 and $2N_0$. The component $v[n]$ will be taken to be the response of a second-order system such that

$$V(z) = \frac{b_0 + b_1 z^{-1}}{(1 - re^{j\theta}z^{-1})(1 - re^{-j\theta}z^{-1})}. \tag{12.116}$$

In the time domain $v[n]$ can be expressed as

$$v[n] = b_0 w[n] + b_1 w[n-1], \tag{12.117}$$

where

$$w[n] = \frac{r^n}{4\sin^2\theta}\{\cos(\theta n) - \cos[\theta(n+2)]\}u[n], \qquad \theta \neq 0, \pi. \tag{12.118}$$

Figure 12.11 shows the pole-zero plot of the z-transform $X(z) = V(z)P(z)$ for the specific set of parameters $b_0 = 0.98$, $b_1 = 1$, $\beta = r = 0.9$, $\theta = \pi/6$, and $N_0 = 15$. Figure 12.12 shows the signals $v[n]$, $p[n]$, and $x[n]$ for these parameters. As seen in Fig. 12.12, the convolution of the pulse-like signal $v[n]$ with the impulse train $p[n]$ results in a series of superimposed delayed copies (echoes) of $v[n]$.

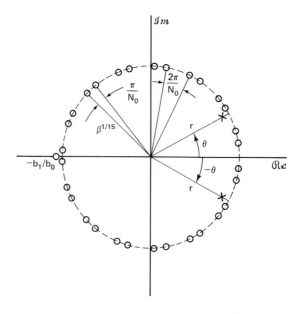

Figure 12.11   Pole-zero plot of the z-transform $X(z) = V(z)P(z)$ for the example signal of Fig. 12.12.

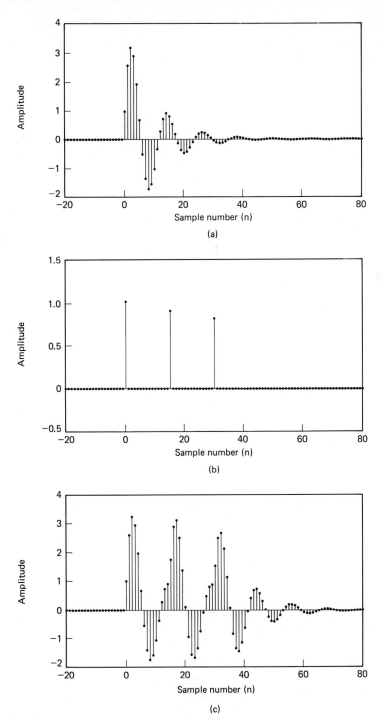

Figure 12.12   The sequences $v[n]$, $p[n]$, and $x[n]$ corresponding to the pole-zero plot of Fig. 12.11.

This signal model is a simplified version of models that are used in the analysis and processing of signals in a variety of contexts, including communications systems, speech processing, sonar, and seismic data analysis. In a communications context, $v[n]$ in Eq. (12.113) might represent a signal transmitted over a multipath channel, $x[n]$ the received signal, and $p[n]$ the channel impulse response. In speech processing, $v[n]$ would represent the combined effects of the glottal pulse shape and the resonance effects of the human vocal tract, while $p[n]$ would represent the periodicity of the vocal excitation during voiced speech such as a vowel sound (Flanagan, 1972; Rabiner and Schafer, 1978). Equation (12.116) incorporates only one resonance, while in the general speech model the denominator would generally include at least ten complex poles. With seismic data analysis, $v[n]$ would represent the waveform of an acoustic pulse propagating in the earth due to a dynamite explosion or similar disturbance. The impulsive component $p[n]$ would represent reflections at boundaries between layers having different propagation characteristics. In the practical use of such a model, there would be more impulses in $p[n]$ than we assumed in Eq. (12.114) and they would be unequally spaced. Also, the component $V(z)$ would generally involve many more zeros, and often no poles are included in the model (Ulrych, 1971; Tribolet, 1979; Robinson and Treitel, 1980).

Although the model defined above is much simpler than is commonly used in practical problems, we will see that this simple model displays all the important properties of the cepstrum of a signal with a rational z-transform.

12.8.2 Analytical Computation of the Complex Cepstrum

To analytically determine $\hat{x}[n]$, the complex cepstrum of $x[n]$, we use the relations

$$\hat{x}[n] = \hat{v}[n] + \hat{p}[n], \tag{12.119a}$$

$$\hat{X}(z) = \hat{V}(z) + \hat{P}(z), \tag{12.119b}$$

where

$$\hat{X}(z) = \log[X(z)], \tag{12.120a}$$

$$\hat{V}(z) = \log[V(z)], \tag{12.120b}$$

and

$$\hat{P}(z) = \log[P(z)]. \tag{12.120c}$$

To determine $\hat{v}[n]$, we can directly apply the results in Section 12.5. Specifically, to express $V(z)$ in the form of Eq. (12.50), we first note that for the specific signal $x[n]$ in Fig. 12.12, the poles of $V(z)$ are inside the unit circle and the zero is outside ($r = 0.9$ and $b_0/b_1 = 0.98$), so that in accordance with Eq. (12.50) we rewrite $V(z)$ as

$$V(z) = \frac{b_1 z^{-1}[1 + (b_0/b_1)z]}{(1 - re^{j\theta}z^{-1})(1 - re^{-j\theta}z^{-1})}. \tag{12.121}$$

As discussed in Section 12.5, the factor z^{-1} contributes a linear component to the unwrapped phase that will force a discontinuity at $\omega = \pm\pi$ in the Fourier transform of $\hat{v}[n]$, so $\hat{V}(z)$ will not be analytic on the unit circle. To avoid this problem, we can alter $v[n]$ (and therefore also $x[n]$) with a one-sample time shift so that we evaluate instead the complex cepstrum of $v[n+1]$ and consequently also $x[n+1]$. If $x[n]$ or $v[n]$ is to be resynthesized after some processing of the complex cepstrum, we can remember this time shift and compensate for it at the final output.

With $v[n]$ replaced by $v[n+1]$, and correspondingly $V(z)$ replaced by $zV(z)$, we now consider $V(z)$ to have the form

$$V(z) = \frac{b_1[1 + (b_0/b_1)z]}{(1 - re^{j\theta}z^{-1})(1 - re^{-j\theta}z^{-1})}.$$

(12.122)

From Eqs. (12.57) we can write $\hat{v}[n]$ directly as

$$\hat{v}[0] = \log b_1,$$

(12.123a)

$$\hat{v}[n] = \frac{1}{n}[(re^{j\theta})^n + (re^{-j\theta})^n], \qquad n > 0,$$

(12.123b)

$$\hat{v}[n] = \frac{1}{n}\left(\frac{-b_0}{b_1}\right)^{-n}, \qquad n < 0.$$

(12.123c)

To determine $\hat{p}[n]$, we can evaluate the inverse z-transform of $\hat{P}(z)$, which, from Eq. (12.115), is

$$\hat{P}(z) = \log(1 - \beta^3 z^{-3N_0}) - \log(1 - \beta z^{-N_0}),$$

(12.124)

where for our example $\beta = 0.9$ and consequently $|\beta| < 1$. One approach to determining the inverse z-transform of Eq. (12.124) is to use the power series expansion of $\hat{P}(z)$. Specifically, with $|\beta| < 1$,

$$\hat{P}(z) = -\sum_{k=1}^{\infty} \frac{\beta^{3k}}{k} z^{-3N_0k} + \sum_{k=1}^{\infty} \frac{\beta^k}{k} z^{-N_0k},$$

(12.125)

from which it follows that $\hat{p}[n]$ is

$$\hat{p}[n] = -\sum_{k=1}^{\infty} \frac{\beta^{3k}}{k} \delta[n - 3N_0k] + \sum_{k=1}^{\infty} \frac{\beta^k}{k} \delta[n - N_0k].$$

(12.126)

An alternative approach to obtaining $\hat{p}[n]$ is to use the property developed in Problem 12.29.

From Eq. (12.119a), the complex cepstrum of $x[n]$ is

$$\hat{x}[n] = \hat{v}[n] + \hat{p}[n],$$

(12.127)

where $\hat{v}[n]$ and $\hat{p}[n]$ are given by Eqs. (12.123) and (12.126), respectively. The sequences $\hat{v}[n]$, $\hat{p}[n]$, and $\hat{x}[n]$ are shown in Fig. 12.13.

The cepstrum of $x[n]$, $c_x[n]$, is the even part of $\hat{x}[n]$, i.e.,

$$c_x[n] = \tfrac{1}{2}(\hat{x}[n] + \hat{x}[-n])$$

(12.128)

and furthermore

$$c_x[n] = c_v[n] + c_p[n].$$

(12.129)

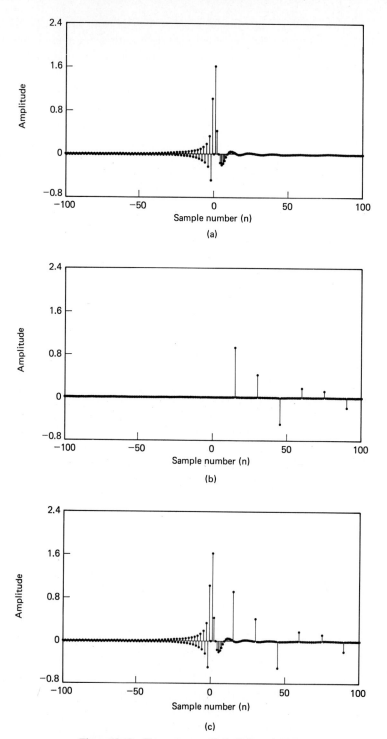

Figure 12.13 The sequences $\hat{v}[n]$, $\hat{p}[n]$, and $\hat{x}[n]$.

From Eqs. (12.123),

$$c_v[n] = \log(b_1)\delta[n] + \sum_{k=1}^{\infty} \frac{(-1)^k (b_0/b_1)^{-k}}{2k} (\delta[n-k] + \delta[n+k])$$

$$+ \sum_{k=1}^{\infty} \frac{r^k \cos(\theta k)}{k} (\delta[n-k] + \delta[n+k]). \tag{12.130a}$$

and from Eq. (12.126),

$$c_p[n] = -\frac{1}{2} \sum_{k=1}^{\infty} \frac{\beta^{3k}}{k} \{\delta[n - 3N_0 k] + \delta[n + 3N_0 k]\}$$

$$+ \frac{1}{2} \sum_{k=1}^{\infty} \frac{\beta^{k}}{k} \{\delta[n - N_0 k] + \delta[n + N_0 k]\}. \tag{12.130b}$$

The sequences $c_v[n]$, $c_p[n]$, and $c_x[n]$ for this example are shown in Fig. 12.14.

12.8.3 Computation of the Cepstrum Using the DFT

In Figs. 12.13 and 12.14 we showed the complex cepstra and the cepstra corresponding to evaluating the analytical expressions obtained in Section 12.8.2. In most applications we do not have analytical expressions for the signal and consequently we cannot analytically determine $\hat{x}[n]$ or $c_x[n]$. In this section, we illustrate the computation of the complex cepstrum and the cepstrum of $x[n]$ using the DFT.

To compute the complex cepstrum or the cepstrum using the DFT as in Fig. 12.9(a), it is necessary that the input be of finite extent. Thus, for the signal model discussed in Section 12.8.1, $x[n]$ must be truncated. In the examples discussed in this section, the signal $x[n]$ in Fig. 12.12(c) was truncated to $N = 1024$ samples and 1024-point DFTs were used in the system of Fig. 12.9(a) to compute the complex cepstrum and the cepstrum of the signal. Figure 12.15 shows the Fourier transforms that are involved in the computation of the complex cepstrum. Figure 12.15(a) shows the logarithm of the magnitude of the DFT of 1024 samples of $x[n]$ in Fig. 12.12, with the DFT samples connected in the plot to suggest the appearance of the Fourier transform of the finite-length input sequence. Figure 12.15(b) shows the principal value of the phase. Note the discontinuities as the phase exceeds $\pm\pi$ and wraps around modulo 2π. Figure 12.15(c) shows the continuous "unwrapped" phase curve obtained as discussed in Section 12.7.1. As noted in Section 12.8.2 and as is evident by carefully comparing Figs. 12.15(b) and 12.15(c), a linear phase component has been removed so that the unwrapped phase curve is continuous at 0 and π. Thus the unwrapped phase of Fig. 12.15(c) corresponds to $x[n+1]$ rather than $x[n]$.

Figures 12.15(a) and 12.15(c) correspond to the computation of the real and imaginary parts, respectively, of the Fourier transform of the complex cepstrum. Only the frequency range $0 \le \omega \le \pi$ is shown, since the function of Fig. 12.15(a) is even and periodic with period 2π, and the function of Fig. 12.15(c) is odd and periodic with period 2π. In examining the curves in Figs. 12.15(a) and 12.15(c), we note that they have the general appearance of a rapidly varying, periodic (in frequency) component added to a more slowly varying component. The periodically varying component in fact corresponds to $\hat{P}(e^{j\omega})$ and the more slowly varying component to $\hat{V}(e^{j\omega})$.

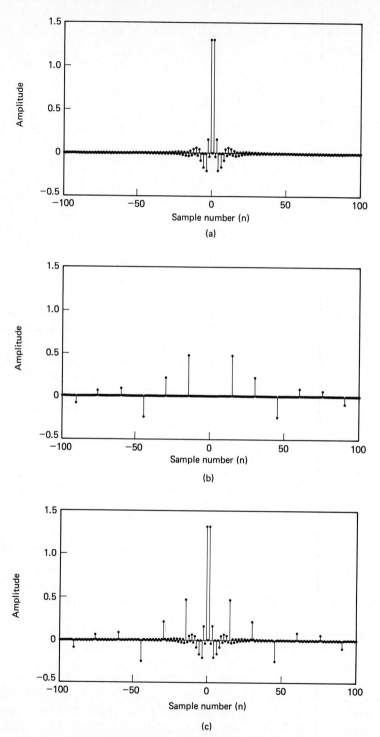

Figure 12.14 The sequences $c_v[n]$, $c_p[n]$, and $c_x[n]$.

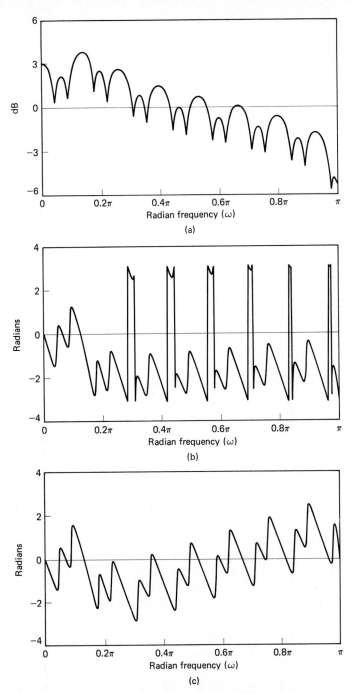

Figure 12.15 Fourier transforms of $x[n]$ in Fig. 12.12. (a) Log magnitude. (b) Principal value of the phase. (c) Continuous "unwrapped" phase after removing a linear phase component from part (b). The DFT samples are connected by straight lines.

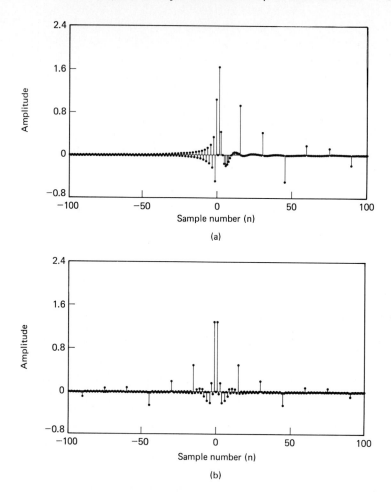

Figure 12.16   (a) Complex cepstrum $\hat{x}_p[n]$ of sequence in Fig. 12.12(c). (b) Cepstrum $c_x[n]$ of sequence in Fig. 12.12(c).

In Fig. 12.16(a) we show the inverse Fourier transform of the complex logarithm of the DFT, i.e., the time-aliased complex cepstrum $\hat{x}_p[n]$. Note the impulses at integer multiples of $N_0 = 15$. These are contributed by $\hat{p}[n]$ and correspond to the rapidly varying periodic component observed in the logarithm of the DFT. We see also that since the input signal is not minimum phase, the complex cepstrum is nonzero for $n < 0$.†

Since a large number of points were used in computing the DFTs, the time-aliased complex cepstrum differs very little from the exact values that would be obtained by evaluating Eqs. (12.123), (12.126), and (12.127) for the specific values of the parameters used to generate the input signal of Fig. 12.12.

---

† In using the DFT to obtain the inverse Fourier transform of Figs. 12.15(a) and 12.15(c), the values associated with $n < 0$ would normally appear in the interval $N/2 < n \leq N - 1$. Traditionally, time sequences are displayed with $n = 0$ in the center, so we have repositioned $\hat{x}_p[n]$ accordingly.

The time-aliased cepstrum $c_{xp}[n]$ for this example is shown in Fig. 12.16(b). As with the complex cepstrum, impulses at multiples of 15 are evident, corresponding to the periodic component of the logarithm of the magnitude of the Fourier transform.

As mentioned in Section 12.8.1, convolution of a signal $v[n]$ with an impulse train such as $p[n]$ is a model for a signal containing echoes. Since $x[n]$ is a convolution of $v[n]$ and $p[n]$, the echo times are often not easily detected by examining $x[n]$. In the cepstral domain, however, the effect of $p[n]$ is present as an additive impulse train, and consequently the presence and location of the echoes are often more evident. It was this observation that motivated the proposal by Bogert et al. (1963) that the cepstrum be used as a means for detecting echoes. This same idea was later used by Noll (1967) as a basis for detecting vocal pitch in speech signals.

12.8.4 Homomorphic Filtering

For the example in Section 12.8.3, the slowly varying component of the complex logarithm, and equivalently the "low-time" portion of the complex cepstrum, were mainly due to $v[n]$. Correspondingly, the more rapidly varying component of the complex logarithm and the "high-time" portion of the complex cepstrum were due primarily to $p[n]$. This suggests that the two convolved components of $x[n]$ can be separated by applying linear filtering to the logarithm of the Fourier transform, or, as discussed in Section 12.2, the complex cepstrum components can be separated by time gating, i.e., by frequency-invariant linear filtering.

Figure 12.17(a) depicts the operations involved in separation of the components of a convolution by frequency-invariant filtering of the complex logarithm of the Fourier transform of a signal. These are the operations that define the technique of *homomorphic deconvolution.* The frequency-invariant linear filter can be implemented by convolution in the frequency domain or, as indicated in Fig. 12.17(b), by multiplication in the time domain. Figure 12.18(a) shows the time response of a lowpass frequency-invariant linear system as required for recovering an approximation to $v[n]$, and Fig. 12.18(b) shows the time response of a highpass frequency-invariant linear system for recovering an approximation to $p[n]$.

Figure 12.19 shows the result of lowpass frequency-invariant filtering. The more rapidly varying curves in Figs. 12.19(a) and 12.19(b) are the complex logarithm of the Fourier transform of the input signal, i.e., the Fourier transform of the complex

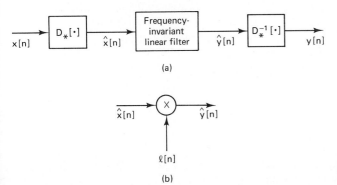

(a)

(b)

Figure 12.17   (a) System for homomorphic deconvolution. (b) Time-domain representation of frequency-invariant filtering.

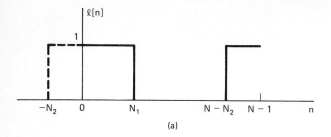

(a)

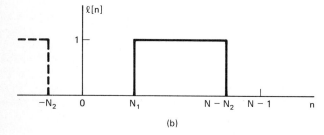

(b)

Figure 12.18 Time response of frequency-invariant linear systems for homomorphic deconvolution. (a) Lowpass system. (b) Highpass system. (Solid line indicates envelope of the sequence $\ell[n]$ as it would be applied in a DFT implementation. The dashed line indicates the periodic extension.)

cepstrum. The slowly varying curves in Figs. 12.19(a) and 12.19(b) are the real and imaginary parts, respectively, of the Fourier transform of $\hat{y}[n]$, when the frequency-invariant linear system $\ell[n]$ is of the form of Fig. 12.18(a) with $N_1 = 14$, $N_2 = 14$, and with the system of Fig. 12.17 implemented using DFTs of length $N = 1024$. Figure 12.19(c) shows the corresponding output $y[n]$. This sequence is the approximation to $v[n]$ obtained by homomorphic deconvolution. To relate this output $y[n]$ to $v[n]$, recall that in computing the unwrapped phase, a linear phase component was removed, corresponding to a one-sample time shift of $v[n]$. Consequently, $y[n]$ in Fig. 12.19(c) corresponds to an approximation to $v[n + 1]$ obtained by homomorphic deconvolution. This type of filtering has been successfully used in speech processing to recover the vocal tract response information (Oppenheim, 1969b; Schafer and Rabiner, 1970) and in seismic signal analysis to recover seismic wavelets (Ulrych, 1971; Tribolet, 1979).

Figure 12.20 shows the result of highpass frequency-invariant filtering. The rapidly varying curves in Figs. 12.20(a) and (b) are the real and imaginary parts, respectively, of the Fourier transform of $\hat{y}[n]$ when the frequency-invariant linear system $\ell[n]$ is of the form of Fig. 12.18(b) with $N_1 = 14$ and $N_2 = 512$ (i.e., the negative-time parts are removed). Again, the system is implemented using a 1024-point DFT. Figure 12.20(c) shows the corresponding output $y[n]$. This sequence is the approximation to $p[n]$ obtained by homomorphic deconvolution. In contrast to the use of the cepstrum to *detect* echoes or periodicity, this approach seeks to obtain the impulse train that specifies the location and size of the repeated copies of $v[n]$.

12.8.5 Minimum-Phase Decomposition

In Section 12.6 we discussed ways that homomorphic deconvolution could be used to decompose a sequence into minimum-phase and allpass components or minimum-phase and maximum-phase components. We will apply these techniques to the signal

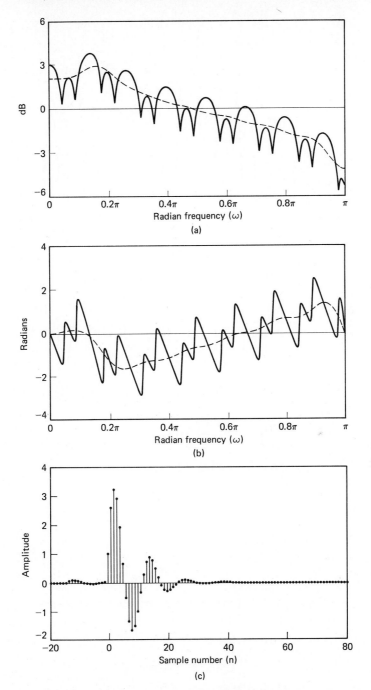

Figure 12.19   Illustration of lowpass frequency-invariant linear filtering in the system of Fig. 12.17. (a) Real part of the Fourier transform of the input (solid line) and output (dashed line) of the lowpass frequency-invariant system with $N_1 = 14$ and $N_2 = 14$ in Fig. 12.18(a). (b) Imaginary part of the Fourier transform of the input (solid line) and output (dashed line) of the lowpass frequency-invariant system with $N_1 = 14$ and $N_2 = 14$. (c) Output sequence $y[n]$ for the input of Fig. 12.12(c).

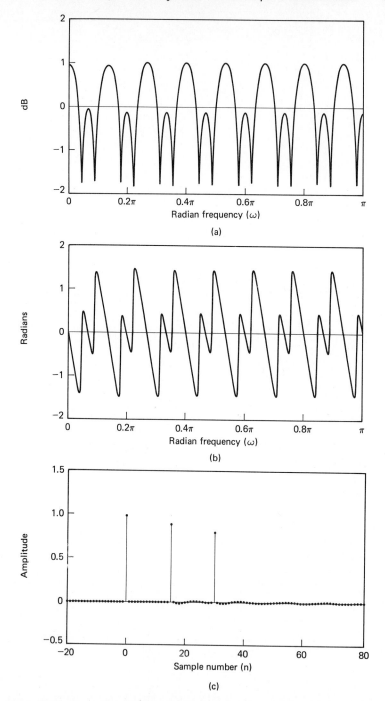

Figure 12.20 Illustration of highpass frequency-invariant linear filtering in the system of Fig. 12.17. (a) Real part of the Fourier transform of the output of the highpass frequency-invariant system with $N_1 = 14$ and $N_2 = 512$ in Fig. 12.18(b). (b) Imaginary part for conditions of part (a). (c) Output sequence $y[n]$ for the input of Fig. 12.12.

model of Section 12.8.1. Specifically, for the parameters of the example, the z-transform of the input is

$$X(z) = V(z)P(z) = \frac{(0.98 + z^{-1})(1 + 0.9z^{-15} + 0.81z^{-30})}{(1 - 0.9e^{j\pi/6}z^{-1})(1 - 0.9e^{-j\pi/6}z^{-1})}. \tag{12.131}$$

First, we can write $X(z)$ as the product of a minimum-phase z-transform and an allpass z-transform; i.e.,

$$X(z) = X_{min}(z)X_{ap}(z), \tag{12.132}$$

where

$$X_{min}(z) = \frac{(1 + 0.98z^{-1})(1 + 0.9z^{-15} + 0.81z^{-30})}{(1 - 0.9e^{j\pi/6}z^{-1})(1 - 0.9e^{-j\pi/6}z^{-1})} \tag{12.133}$$

and

$$X_{ap}(z) = \frac{0.98 + z^{-1}}{1 + 0.98z^{-1}}. \tag{12.134}$$

The sequences $x_{min}[n]$ and $x_{ap}[n]$ can be found using the partial fraction expansion methods of Chapter 4, and the corresponding complex cepstra $\hat{x}_{min}[n]$ and $\hat{x}_{ap}[n]$ can be found using the power series technique of Section 12.5 (see Problem 12.27). Alternatively, $\hat{x}_{min}[n]$ and $\hat{x}_{ap}[n]$ can be obtained exactly from $\hat{x}[n]$ by the operations discussed in Section 12.6.2 and as depicted in Fig. 12.7. If the characteristic systems in Fig. 12.7 are implemented using the DFT, then the separation is only approximate, but the approximation error can be small if the DFT length is large. Figure 12.21(a) shows the complex cepstrum for $x[n]$ as computed using a 1024-point DFT, again with a one-sample time delay removed from $v[n]$ so that the phase is continuous at π. Figure 12.21(b) shows the complex cepstrum of the minimum-phase component $\hat{x}_{min}[n]$, and Fig. 12.21(c) shows the complex cepstrum of the allpass component $\hat{x}_{ap}[n]$ as obtained by the operations of Fig. 12.7 with $D_*[\cdot]$ implemented as in Fig. 12.9(a).

Using the DFT as in Fig. 12.9(b) to implement the system $D_*^{-1}[\cdot]$ gives the approximations to the minimum-phase and allpass components shown in Figs. 12.22(a) and 12.22(b), respectively. Since all the zeros of $P(z)$ are inside the unit circle, all of $P(z)$ is included in the minimum-phase z-transform or, equivalently, $\hat{p}[n]$ is entirely included in $\hat{x}_{min}[n]$. Thus, the minimum-phase component consists of delayed and scaled replicas of the minimum-phase component of $v[n]$. Therefore, the minimum-phase component of Fig. 12.22(a) appears very similar to the input shown in Fig. 12.12. From Eq. (12.134), the allpass component can be shown to be

$$x_{ap}[n] = 0.98\delta[n] + 0.0396(-0.98)^{n-1}u[n-1]. \tag{12.135}$$

The result of Fig. 12.22(b) is very close to this ideal result for small values of n where the sequence values are of significant amplitude. This example illustrates a technique of decomposition that has been applied by Bauman et al. (1985) in the analysis and

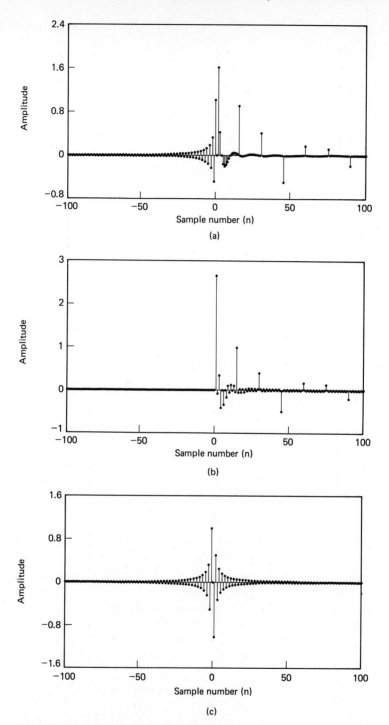

Figure 12.21   (a) Complex cepstrum of $x[n] = x_{min}[n] * x_{ap}[n]$. (b) Complex cepstrum of $x_{min}[n]$. (c) Complex cepstrum of $x_{ap}[n]$.

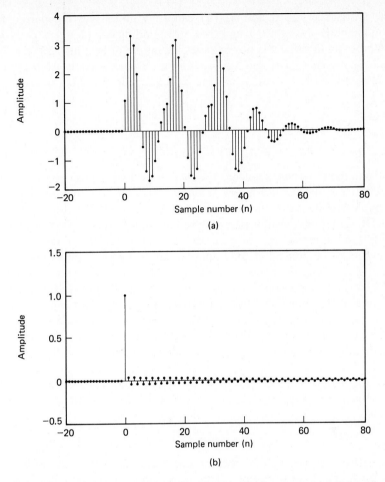

Figure 12.22   (a) Minimum-phase output. (b) Allpass output obtained as depicted in Fig. 12.7.

characterization of the response of electroacoustic transducers. A similar decomposition technique can be used to factor magnitude-squared functions as required in digital filter design (see Problem 12.28).

As an alternative to the minimum-phase/allpass decomposition, we can express $X(z)$ as the product of a minimum-phase z-transform and a maximum-phase z-transform; i.e.,

$$X(z) = X_{mn}(z)X_{mx}(z), \tag{12.136}$$

where

$$X_{mn}(z) = \frac{(1 + 0.9z^{-15} + 0.81z^{-30})}{(1 - 0.9e^{j\pi/6}z^{-1})(1 - 0.9e^{-j\pi/6}z^{-1})} \tag{12.137}$$

and

$$X_{mx}(z) = 0.98 + z^{-1}. \tag{12.138}$$

The sequences $x_{mn}[n]$ and $x_{mx}[n]$ can be found using the partial fraction expansion methods of Chapter 4, and the corresponding complex cepstra $\hat{x}_{mn}[n]$ and $\hat{x}_{mx}[n]$ can be found using the power series technique of Section 12.5 (see Problem 12.27). Alternatively, $\hat{x}_{mn}[n]$ and $\hat{x}_{mx}[n]$ can be obtained exactly from $\hat{x}[n]$ by the operations discussed in Section 12.6.4 and as depicted in Fig. 12.8, where

$$\ell_{mn}[n] = u[n] \tag{12.139}$$

and

$$\ell_{mx}[n] = u[-n-1]. \tag{12.140}$$

That is, the minimum-phase sequence is now defined by the positive time part of the complex cepstrum and the maximum-phase part is defined by the negative time part of the complex cepstrum. If the characteristic systems in Fig. 12.8 are implemented using the DFT, the negative time part of the complex cepstrum is positioned in the last half of the DFT interval. In this case, the separation of the minimum-phase and maximum-phase components is only approximate because of time aliasing, but the time-aliasing error can be made small by choosing a sufficiently large DFT length. Figure 12.21(a) shows the complex cepstrum of $x[n]$ as computed using a 1024-point DFT. Figure 12.23 shows the two output sequences that are obtained from the complex cepstrum of Fig. 12.21(a) using Eqs. (12.137) and (12.138) with the inverse characteristic system being implemented using the DFT as in Fig. 12.9(b). As before, since $\hat{p}[n]$ is entirely included in $\hat{x}_{mn}[n]$, the corresponding output $x_{mn}[n]$ consists of delayed and scaled replicas of a minimum-phase sequence and thus it also looks very much like the input sequence. However, a careful comparison of Figs. 12.22(a) and 12.23(a) shows that $x_{min}[n] \neq x_{mn}[n]$. From Eq. (12.138), the maximum-phase sequence is

$$x_{mx}[n] = 0.98\delta[n] + \delta[n-1]. \tag{12.141}$$

Figure 12.23(b) is very close to this ideal result. (Note the shift due to the linear phase removed in the phase unwrapping.) This technique of minimum-phase/maximum-phase decomposition was used by Smith and Barnwell (1986) in the design and implementation of exact reconstruction filter banks for speech analysis and coding.

12.8.6 Generalizations

The example in Section 12.8.5 considered a simple complex exponential signal that was convolved with an impulse train to produce a series of delayed and scaled replicas of the complex exponential signal. This model illustrates many of the features of the complex cepstrum and of homomorphic filtering.

In particular, in more general models associated with speech, communications, and seismic applications an appropriate signal model consists of the convolution of two components. One component has the characteristics of $v[n]$, specifically a Fourier transform that is slowly varying in frequency. The second has the characteristics of $p[n]$, i.e., an echo pattern or impulse train for which the Fourier transform is more rapidly varying and quasi-periodic in frequency. Thus, the contributions of the two components would be separated in the complex cepstrum or the cepstrum, and

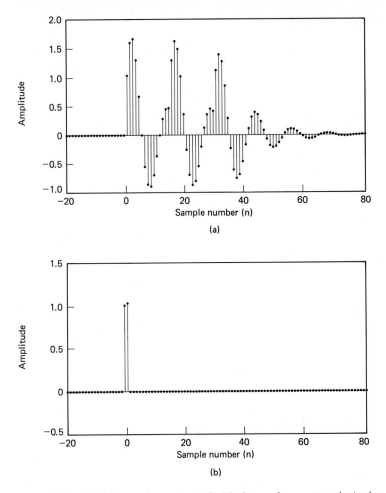

Figure 12.23   (a) Minimum-phase output. (b) Maximum-phase output obtained as depicted in Fig. 12.8.

furthermore the complex cepstrum or the cepstrum would contain impulses at multiples of the echo delays. Thus, homomorphic filtering can be used to separate the convolutional components of the signal, or the cepstrum can be used to detect echo delays. In the next section we will illustrate the use of these general properties of the cepstrum in applications to speech analysis.

12.9 APPLICATIONS TO SPEECH PROCESSING

Cepstrum techniques have been applied quite successfully to speech analysis. The previous theoretical discussion and the extended example of Section 12.8 generalize in a relatively straightforward way to this application, discussed briefly in this section.

12.9.1 The Speech Model

As we briefly described in Section 11.5.1, there are three basic classes of speech sounds corresponding to different forms of excitation of the vocal tract. Specifically:

- *Voiced sounds* are produced by exciting the vocal tract with quasi-periodic pulses of airflow caused by the opening and closing of the glottis.
- *Fricative sounds* are produced by forming a constriction somewhere in the vocal tract and forcing air through the constriction so that turbulence is created, thereby producing a noise-like excitation.
- *Plosive sounds* are produced by completely closing off the vocal tract, building up pressure behind the closure, and then abruptly releasing the pressure.

In each case, the speech signal is produced by exciting the vocal tract system (an acoustic transmission system) with a wideband excitation. The vocal tract changes shape relatively slowly with time, and thus it can be modeled as a slowly time-varying filter that imposes its frequency-response properties on the spectrum of the excitation. The vocal tract is characterized by its natural frequencies (called *formants*), which correspond to resonances in its frequency response.

If we assume that the excitation sources and the vocal tract shape are independent, we arrive at the discrete-time model of Fig. 12.24 as a representation of the sampled speech waveform. In this model, samples of the speech signal are assumed to be the output of a time-varying discrete-time system that models the resonances of the vocal tract system. The mode of excitation of the system switches between periodic impulses and random noise depending on the type of sound being produced.

Since the vocal tract changes shape rather slowly in continuous speech, it is reasonable to assume that the discrete-time system in the model has fixed properties over a time interval on the order of 10 ms. Thus, the discrete-time system may be characterized in each such time interval by an impulse response or a frequency response or a set of coefficients for an IIR system. Specifically, a model for the system function of the vocal tract takes the form

$$
V(z) = \frac{\displaystyle\sum_{k=0}^{K} b_k z^{-k}}{\displaystyle\sum_{k=0}^{P} a_k z^{-k}}
\tag{12.142}
$$

or, equivalently,

$$
V(z) = \frac{A z^{-K_o} \displaystyle\prod_{k=1}^{K_i} (1 - \alpha_k z^{-1}) \prod_{k=1}^{K_o} (1 - \beta_k z)}{\displaystyle\prod_{k=1}^{[P/2]} (1 - r_k e^{j\theta_k} z^{-1})(1 - r_k e^{-j\theta_k} z^{-1})},
\tag{12.143}
$$

where the quantities $r_k e^{j\theta_k}$ are the complex natural frequencies of the vocal tract, which of course are dependent on the vocal tract shape and consequently are time

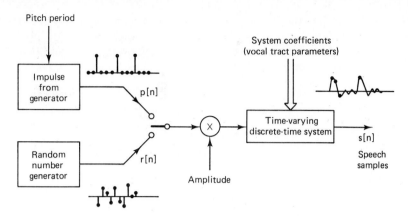

Figure 12.24   Discrete-time model of speech production.

varying. The zeros of $V(z)$ account for the finite-duration glottal pulse waveform and for the zeros of transmission caused by the constrictions of the vocal tract in the creation of nasal voiced sounds and fricatives. Such zeros are often not included because it is very difficult to estimate their locations from only the speech waveform and because it has been shown (Atal and Hanauer, 1971) that the spectral shape of the speech signal can be accurately modeled using no zeros if we include extra poles beyond the number needed just to account for the vocal tract resonances. The zeros are included in our analysis because they are necessary for an accurate representation of the complex cepstrum of speech. Note that we include the possibility of zeros outside the unit circle.

The vocal tract system is excited by an excitation sequence $p[n]$, which is a train of impulses when modeling voiced speech sounds and a pseudorandom noise sequence when modeling unvoiced speech sounds such as fricatives and plosives.

Many of the fundamental problems of speech processing reduce to the estimation of the parameters of the model of Fig. 12.24. These parameters are as follows:

- The coefficients of $V(z)$ in Eq. (12.142) or the pole and zero locations in Eq. (12.143)
- The mode of excitation of the vocal tract system; i.e., a *periodic impulse train* or *random noise*
- The amplitude of the excitation signal
- The period of the speech excitation when it is periodic

Homomorphic deconvolution can be applied to the estimation of the parameters if it is assumed that the model is valid over a short time interval so that a short segment of length L samples of the sampled speech signal can be thought of as the convolution

$$s[n] = v[n] * p[n] \qquad \text{for } 0 \le n \le L - 1, \tag{12.144}$$

where $v[n]$ is the impulse response of the vocal tract and $p[n]$ is either periodic (for voiced speech) or random noise (for unvoiced speech). Obviously, the model of Eq. (12.144) is not valid at the edges of the interval because of pulses that occur before the beginning of the analysis interval and pulses that end after the end of the interval. To mitigate the effect of the "discontinuities" of the model at the ends of the interval, the speech signal $s[n]$ can be multiplied by a window $w[n]$ that tapers smoothly to zero at both ends. Thus, the input to the homomorphic deconvolution system is

$$x[n] = w[n]s[n]. \qquad (12.145)$$

Let us first consider the case of voiced speech. If $w[n]$ varies slowly with respect to the variations of $v[n]$, then the analysis is greatly simplified if we assume that

$$x[n] = v[n] * p_w[n], \qquad (12.146)$$

where

$$p_w[n] = w[n]p[n]. \qquad (12.147)$$

(See Oppenheim and Schafer, 1968.) A more detailed analysis without this assumption leads to essentially the same conclusions (Verhelst and Steenhaut, 1986). For voiced speech, $p[n]$ is a train of impulses of the form

$$p[n] = \sum_{k=0}^{M-1} \delta[n - kN_0] \qquad (12.148)$$

so that

$$p_w[n] = \sum_{k=0}^{M-1} w[kN_0]\delta[n - kN_0], \qquad (12.149)$$

where we have assumed that the pitch period is N_0 and that M periods are spanned by the window.

The complex cepstra of $x[n]$, $v[n]$, and $p_w[n]$ are related by

$$\hat{x}[n] = \hat{v}[n] + \hat{p}_w[n]. \qquad (12.150)$$

To obtain $\hat{p}_w[n]$, we define a sequence

$$w_{N_0}[k] = \begin{cases} w[kN_0], & k = 0, 1, \ldots, M - 1, \\ 0, & \text{otherwise,} \end{cases} \qquad (12.151)$$

whose Fourier transform is

$$P_w(e^{j\omega}) = \sum_{k=0}^{M-1} w[kN_0]e^{-j\omega k} = W_{N_0}(e^{j\omega N_0}). \qquad (12.152)$$

Thus, $P_w(e^{j\omega})$ and $\hat{P}_w(e^{j\omega})$ are both periodic with period $2\pi/N_0$, and the complex cepstrum of $p_w[n]$ is

$$\hat{p}_w[n] = \begin{cases} \hat{w}_{N_0}[n/N_0], & n = 0, \pm N_0, \pm 2N_0, \ldots, \\ 0, & \text{otherwise.} \end{cases} \qquad (12.153)$$

The periodicity of the complex logarithm resulting from the periodicity of the voiced speech signal is manifest in the complex cepstrum as impulses spaced at integer multiples of N_0 samples (the pitch period). If the sequence $w_{N_0}[n]$ is minimum phase, then $\hat{p}_w[n]$ will be zero for $n < 0$. Otherwise, $\hat{p}_w[n]$ will have impulses spaced at intervals of N_0 samples for both positive and negative values of n. In either case, the contribution of $\hat{p}_w[n]$ to $\hat{x}[n]$ will be found in the interval $|n| \geq N_0$.

From the power series expansion of the complex logarithm of $V(z)$ it can be shown that the contribution to the complex cepstrum due to $v[n]$ is

$$
\hat{v}[n] = \begin{cases}
\displaystyle\sum_{k=1}^{K_0} \frac{\beta_k^{-n}}{n}, & n < 0, \\[3mm]
\log|A|, & n = 0, \\[3mm]
\displaystyle -\sum_{k=1}^{K_i} \frac{\alpha_k^n}{n} + \sum_{k=1}^{[P/2]} \frac{2r_k^n}{n}\cos(\theta_k n), & n > 0.
\end{cases}
\tag{12.154}
$$

As with the simpler example in Section 12.8.1, the term z^{-K_0} in Eq. (12.143) represents a linear phase factor that would be removed in obtaining the unwrapped phase and the complex cepstrum. Consequently, $\hat{v}[n]$ in Eq. (12.154) more accurately is the complex cepstrum of $v[n + K_0]$.

From Eq. (12.154) we see that the contributions of the vocal tract response to the complex cepstrum occupy the full range $-\infty < n < \infty$, but they are concentrated around the origin. We note also that since the vocal tract resonances are represented by poles inside the unit circle, their contribution to the complex cepstrum is zero for $n < 0$.

12.9.2 An Example of Homomorphic Deconvolution of Speech

For speech sampled at 10,000 samples/s, the pitch period N_0 will range from about 25 samples for a high-pitched voice up to about 150 samples for a very low-pitched voice. Since the vocal tract component of the complex cepstrum $\hat{v}[n]$ decays rapidly, the peaks of $\hat{p}_w[n]$ stand out from $\hat{v}[n]$. In other words, in the complex logarithm, the vocal tract components are slowly varying and the excitation components are rapidly varying. This is illustrated in Fig. 12.25. Figure 12.25(a) shows a segment of speech weighted by a Hamming window, and Fig. 12.25(b) shows the complex logarithm (log magnitude and unwrapped phase) of the DFT of the signal in Fig. 12.25(a).† Note the rapidly varying, almost periodic component due to $p_w[n]$ and the slowly varying component due to $v[n]$. These properties are manifest in the complex cepstrum of Fig. 12.25(c) in the form of impulses at multiples of approximately 8 ms (the period of the input speech segment) due to $\hat{p}_w[n]$ and in the samples in the region $|nT| < 5$ ms, which we attribute to $\hat{v}[n]$.

As in the previous section, frequency-invariant filtering can be used to separate the components of the convolutional model of the speech signal. Lowpass filtering of the complex logarithm can be used to recover an approximation to $v[n]$, and highpass filtering can be used to obtain $p_w[n]$. Figure 12.26 shows an example (Oppenheim et al., 1968). Figure 12.26(a) shows 256 samples of a vowel sound. This segment was multiplied by a Hamming window and the complex cepstrum was computed using the DFT as depicted in Fig. 12.9(a). The complex cepstrum is shown in Fig. 12.26(b). Figure 12.26(c) is an approximation to $p_w[n]$ obtained by applying to the complex

† In all the figures of this section, the samples of all sequences were connected for ease in plotting.

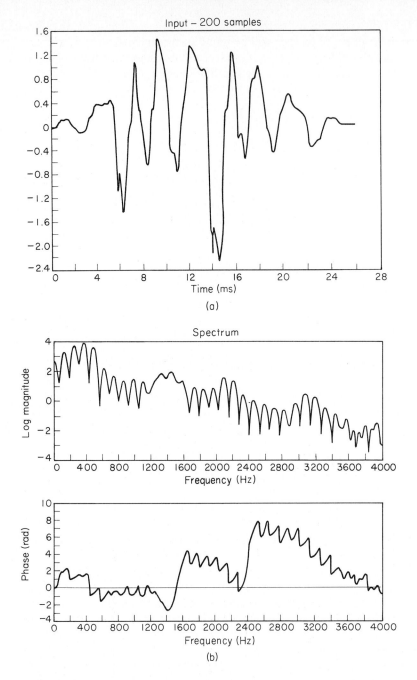

Figure 12.25   (a) Segment of speech weighted by a Hamming window. (b) Complex logarithm of the discrete Fourier transform of part (a).

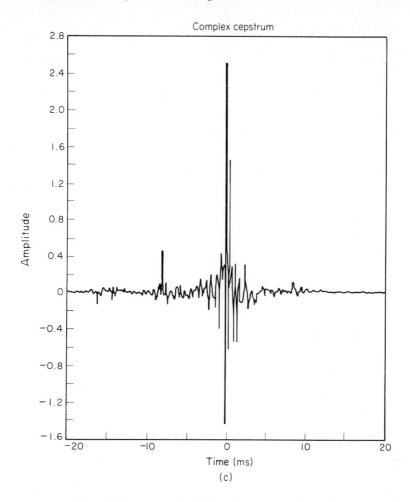

Figure 12.25 *(continued)* (c) Complex cepstrum of part (a).

cepstrum a symmetrical highpass frequency-invariant filter as in Fig. 12.18(b) with $N_1 = 40$ and $N_2 = 40$. On the other hand, Fig. 12.26(d) shows an approximation to $v[n]$ obtained by using a lowpass frequency-invariant filter as in Fig. 12.18(a) with the same cutoff points. In both cases, the inverse characteristic system was implemented using 1024-point DFTs as in Fig. 12.9(b). Finally, to illustrate the validity of the deconvolution, Fig. 12.26(e) shows the result of convolving the sequence of Fig. 12.26(d) with an impulse train of equal-amplitude impulses occurring at the locations of the peaks in Fig. 12.26(c). As we see by comparing Figs. 12.26(a) and 12.26(e), the reconstructed waveform is very close to the original.

12.9.3 Estimating the Parameters of the Speech Model

Although homomorphic deconvolution can be successfully applied in *separating* the components of a speech waveform, in many speech processing applications we are

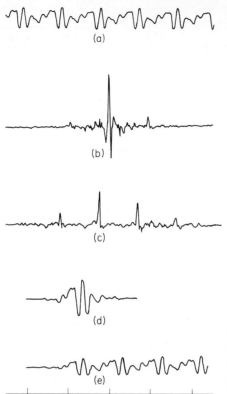

(a)

(b)

(c)

(d)

(e)

-10   -5   0   5   10

Time (ms)

Figure 12.26 (a) A segment of a vowel waveform. (b) Complex cepstrum of part (a). (c) Recovered excitation function $p_w[n]$. (d) Recovered vocal tract impulse response $v[n]$. (e) Synthesized speech using impulse response of part (d) and pitch period as measured from part (b).

interested only in *estimating* the parameters in a parametric representation of the speech signal. Since the properties of the speech signal change rather slowly with time, it is common to estimate the parameters of the model of Fig. 12.24 at intervals of about 10 ms. In this case, the time-dependent Fourier transform discussed in Chapter 11 serves as the basis for time-dependent homomorphic analysis. For example, it may be sufficient to examine segments of speech selected about every 100 samples to determine the mode of excitation of the model (voiced or unvoiced) and, for voiced speech, the pitch period. Or we may wish to track the variation of the vocal tract resonances (formants). For such problems, the difficult phase computation can be avoided by using the cepstrum, which requires only the logarithm of the magnitude of the Fourier transform. Since the cepstrum is the even part of the complex cepstrum, our previous discussion suggests that the low-time portion of $c_x[n]$ should correspond to the slowly varying components of the log magnitude of the Fourier transform of the speech segment, and for voiced speech the cepstrum should contain impulses at multiples of the pitch period. An example is shown in Fig. 12.27.

Figure 12.27(a) shows the operations involved in estimating the speech parameters using the cepstrum. Figure 12.27(b) shows a typical result for voiced speech. The windowed speech signal is labeled A, $\log|X[k]|$ is labeled C, and the

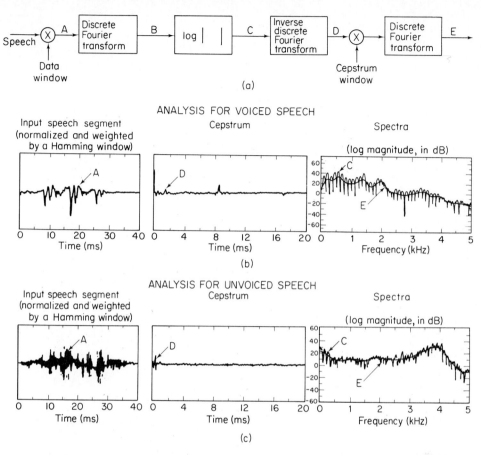

Figure 12.27   (a) System for cepstrum analysis of speech signals. (b) Analysis for voice speech. (c) Analysis for unvoiced speech.

cepstrum $c_x[n]$ is labeled D. The peak in the cepstrum at about 8 ms indicates that this segment of speech is voiced with that period. The smoothed spectrum, or *spectrum envelope*, obtained by frequency-invariant lowpass filtering with cutoff below 8 ms, is labeled E and is superimposed on C. The situation for unvoiced speech, shown in Fig. 12.27(c), is similar, except that the random nature of the excitation component of the input speech segment causes a rapidly varying random component in $\log|X[k]|$ instead of a periodic component. Thus, in the cepstrum the low-time components correspond as before to the vocal tract system function; however, since the rapid variations in $\log|X[k]|$ are not periodic, no strong peak appears in the cepstrum. Therefore, the presence or absence of a peak in the cepstrum in the normal pitch period range serves as a very good voiced/unvoiced detector and pitch period estimator. The result of lowpass frequency-invariant filtering in the unvoiced case is similar to that in the voiced case. A smoothed spectrum envelope estimate is obtained as in E.

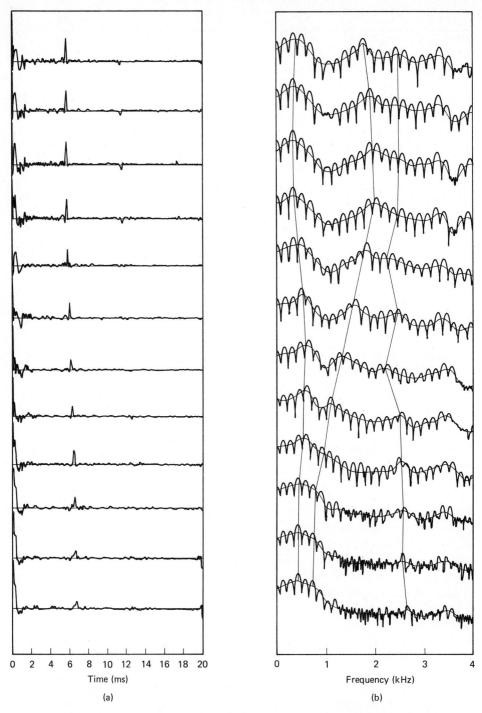

Time (ms)

(a)

Frequency (kHz)

(b)

Figure 12.28 (a) Cepstra and (b) log spectra for sequential segments of voiced speech.

In speech analysis applications, the operations of Fig. 12.27(a) are applied repeatedly to sequential segments of the speech waveform. The length of the segments must be carefully selected. If the segments are too long, the properties of the speech signal will change too much across the segment. If the segments are too short, there will not be enough of the signal to obtain a strong indication of periodicity. Usually the segment length is set at about 3–4 times the average pitch period of the speech signal. Figure 12.28 shows an example of how the cepstrum can be used for pitch detection and for estimation of the vocal tract resonance frequencies. Figure 12.28(a) shows a sequence of cepstra computed for speech waveform segments selected at 20-ms intervals. The existence of a prominent peak throughout the sequence of speech segments indicates that the speech was voiced throughout. The location of the cepstrum peak indicates the value of the pitch period in each corresponding time interval. Figure 12.28(b) shows the log magnitude with the corresponding smoothed spectra superimposed. The lines connect estimates of the vocal tract resonances obtained by a heuristic peak-picking algorithm. (See Schafer and Rabiner, 1970.)

12.9.4 Applications

As indicated previously, cepstrum analysis methods have found widespread application in speech processing problems. One of the most successful applications is in pitch detection (Noll, 1967). They also have been used successfully in speech analysis/synthesis systems for low-bit-rate coding of the speech signal (Oppenheim, 1969b; Schafer and Rabiner, 1970).

Cepstrum representations of speech have also been used with considerable success in pattern recognition problems associated with speech processing such as speaker identification (Atal, 1976), speaker verification (Furui, 1981) and speech recognition (Davis and Mermelstein, 1980). Although the technique of linear predictive analysis is the most widely used method of obtaining a representation of the vocal tract component of the speech model, the linear predictive model representation is often transformed to a cepstrum representation for use in pattern recognition problems (Schroeder, 1981; Juang et al., 1987). This transformation is explored in Problem 12.32.

12.10 SUMMARY

In this chapter we discussed the technique of cepstrum analysis and homomorphic deconvolution. We focused primarily on definitions and properties of the complex cepstrum and on the practical problems in the computation of the complex cepstrum. An idealized example was discussed to illustrate the use of cepstrum analysis and homomorphic deconvolution for separating components of a convolution. The application of cepstrum analysis techniques to speech processing problems was discussed in some detail as an illustration of their use in a real application.

PROBLEMS

12.1. Each of the following operators or systems is homomorphic, with the input operation indicated. Determine the output operation for each.

System Definition $T\{x[n]\}$	Input Operation
(a) $y[n] = 2x[n]$	Addition
(b) $y[n] = 2x[n]$	Multiplication
(c) $X(z) = \sum\limits_{n=-\infty}^{\infty} x[n]z^{-n}$	Addition
(d) $X(z) = \sum\limits_{n=-\infty}^{\infty} x[n]z^{-n}$	Convolution
(e) $X(z) = \sum\limits_{n=-\infty}^{\infty} x[n]z^{-n}$	Multiplication
(f) $y[n] = (x[n])^2$	Multiplication
(g) $y[n] = |x[n]|$	Multiplication
(h) $y[n] = e^{x[n]}$	Addition
(i) $y[n] = e^{x[n]}$	Multiplication

12.2. (a) Consider a discrete-time system that is linear in the conventional sense. If $y[n] = T\{x[n]\}$ is the output when the input is $x[n]$, then the *zero signal* $\mathbf{0}[n]$ is the signal that can be added to $x[n]$ such that $T\{x[n] + \mathbf{0}[n]\} = y[n] + T\{\mathbf{0}[n]\} = y[n]$. What is the zero signal for conventional linear systems?

(b) Consider a discrete-time system $y[n] = T\{x[n]\}$ that is homomorphic, with multiplication as the operation for combining signals at both the input and the output. What is the zero signal for such a system; i.e., what is the signal $\mathbf{0}[n]$ such that $T\{x[n] \cdot \mathbf{0}[n]\} = y[n] \cdot T\{\mathbf{0}[n]\} = y[n]$?

(c) Consider a discrete-time system $y[n] = T\{x[n]\}$ that is homomorphic, with convolution as the operation for combining signals at both the input and the output. What is the zero signal for such a system; i.e., what is the signal $\mathbf{0}[n]$ such that $T\{x[n] * \mathbf{0}[n]\} = y[n] * T\{\mathbf{0}[n]\} = y[n]$?

12.3. Consider the following systems.

$$\text{Fourier transform:} \quad X(e^{j\omega}) = \sum_{n=-\infty}^{\infty} x[n]e^{-j\omega n}$$

$$\text{nth power:} \quad y[n] = (x[n])^n$$

$$\text{Modulator:} \quad y[n] = g[n]x[n]$$

(a) Can any of these systems be homomorphic? For those that are, specify the input and output operations and draw the respective canonic decompositions as in Fig. 12.2.

(b) Is the cascade of homomorphic systems a homomorphic system? State the general conditions under which the cascade of homomorphic systems is a homomorphic system. [*Hint*: Consider cascading the above systems.]

12.4. A discrete-time system is defined by the input/output relation

$$y[n] = (x[n])^\gamma,$$

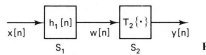

$$x[n] \quad (\cdot)^\gamma \quad y[n] \quad \text{Figure P12.4}$$

where γ is a real number. This system is depicted in Fig. P12.4. Assuming that $(x[n])^\gamma$ can be uniquely defined, can you find operations $\square$, o, $:$, and ι such that

$$(a:x_1[n] \square b:x_2[n])^\gamma = a\,\iota\,(x_1[n])^\gamma \, o\, b\,\iota\,(x_2[n])^\gamma?$$

If not, explain why not.

12.5. Consider the system depicted in Fig. P12.5, where S_1 is an LTI system with impulse response $h_1[n]$ and S_2 is a homomorphic system with convolution as the input and output operations; i.e., the transformation $T_2\{\cdot\}$ satisfies

$$T_2\{w_1[n] * w_2[n]\} = T_2\{w_1[n]\} * T_2\{w_2[n]\}.$$

Suppose that the complex cepstrum of the input $x[n]$ is $\hat{x}[n] = \delta[n] + \delta[n-1]$. Find a closed-form expression for $h_1[n]$ such that the output is $y[n] = \delta[n]$.

$$x[n] \quad h_1[n] \quad w[n] \quad T_2\{\cdot\} \quad y[n]$$
$$S_1 \qquad\qquad S_2 \qquad \text{Figure P12.5}$$

12.6. If we take the point of view that the purpose of a conventional linear system is to separate a signal from an additive combination, what is the purpose of the system in Fig. P12.6? The systems L_1 and L_2 represent conventional linear systems.

$$x[n] \quad \log \quad L_1 \quad \exp \quad L_2 \quad y[n] \quad \text{Figure P12.6}$$

12.7. Let $x_1[n]$ and $x_2[n]$ denote two sequences and $\hat{x}_1[n]$ and $\hat{x}_2[n]$ their corresponding complex cepstra. If $x_1[n] * x_2[n] = \delta[n]$, determine the relationship between $\hat{x}_1[n]$ and $\hat{x}_2[n]$.

12.8. Consider the class of sequences that are real and stable and whose z-transforms are of the form

$$X(z) = |A| \frac{\displaystyle\prod_{k=1}^{M_i} (1 - a_k z^{-1}) \prod_{k=1}^{M_o} (1 - b_k z)}{\displaystyle\prod_{k=1}^{N_i} (1 - c_k z^{-1}) \prod_{k=1}^{N_o} (1 - d_k z)},$$

where $|a_k|, |b_k|, |c_k|, |d_k| < 1$. Let $\hat{x}[n]$ denote the complex cepstrum of $x[n]$.
(a) Suppose that $\hat{x}[n] = 0$ for all n. Determine $x[n]$.
(b) Let $y[n] = x[-n]$. Determine $\hat{y}[n]$ in terms of $\hat{x}[n]$.
(c) If $x[n]$ is causal, is it also mininum phase? Explain.
(d) Suppose that $x[n]$ is a finite-duration sequence such that

$$X(z) = |A| \prod_{k=1}^{M_i} (1 - a_k z^{-1}) \prod_{k=1}^{M_o} (1 - b_k z),$$

with $|a_k| < 1$ and $|b_k| < 1$. The function $X(z)$ has zeros inside and outside the unit circle. Suppose that we wish to determine $y[n]$ such that $|Y(e^{j\omega})| = |X(e^{j\omega})|$ and

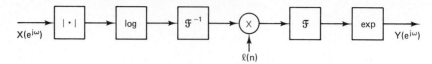

Figure P12.8

$Y(z)$ has no zeros outside the unit circle. One approach that achieves this objective is depicted in Fig. P12.8. Determine the required sequence $\ell[n]$. A possible application of the system in Fig. P12.8 is to stabilize an unstable system by applying the transformation of Fig. P12.8 to the sequence of coefficients of the denominator of the system function.

12.9. In considering the implementation of homomorphic systems for convolution, we restricted our attention to input signals with rational z-transforms of the form of Eq. (12.53). If an input sequence $x[n]$ has a rational z-transform but has either a negative gain constant or an amount of delay not represented by Eq. (12.53), then we can obtain a z-transform of the form of Eq. (12.53) by shifting $x[n]$ appropriately and multiplying by -1. The complex cepstrum may then be computed using Eq. (12.54).

Suppose that $x[n] = \delta[n] - 2\delta[n-1]$, and define $y[n] = \alpha x[n-r]$, where $\alpha = \pm 1$ and r is an integer. Find α and r such that $Y(z)$ is in the form of Eq. (12.53), and then find $\hat{y}[n]$.

12.10. In Section 12.5 we stated that linear phase contributions should be removed from the unwrapped phase curve before computation of the complex cepstrum. This problem is concerned with the effect of not removing the linear phase component due to the factor z^r in Eq. (12.50).

Specifically, assume that the input to the characteristic system for convolution in Fig. 12.4(a) is $x[n] = \delta[n + r]$. Show that formal application of the operations in Fig. 12.4(a) leads to

$$\hat{x}[n] = \begin{cases} r\dfrac{\cos(\pi n)}{n}, & n \neq 0, \\ 0, & n = 0. \end{cases}$$

The advantage of removing the linear phase component of the phase is clear from this result, since for large r such a component would dominate the complex cepstrum.

12.11. Suppose that the z-transform of $s[n]$ is

$$S(z) = \frac{(1 - \frac{1}{2}z^{-1})(1 - \frac{1}{4}z)}{(1 - \frac{1}{3}z^{-1})(1 - \frac{1}{5}z)}.$$

Determine the pole locations of the z-transform of $n\hat{s}[n]$, other than poles at $|z| = 0$ or ∞.

12.12. Suppose that the complex cepstrum of $y[n]$ is $\hat{y}[n] = \hat{s}[n] + 2\delta[n]$. Determine $y[n]$ in terms of $s[n]$.

12.13. Determine the complex cepstrum of $x[n] = 2\delta[n] - 2\delta[n-1] + 0.5\delta[n-2]$. You may shift $x[n]$ or change its sign if necessary.

12.14. Suppose that the z-transform of a stable sequence $x[n]$ is given by

$$X(z) = \frac{1 - \frac{1}{2}z^{-1}}{1 + \frac{1}{2}z},$$

and suppose that a stable sequence $y[n]$ has complex cepstrum $\hat{y}[n] = \hat{x}[-n]$, where $\hat{x}[n]$ is the complex cepstrum of $x[n]$. Determine $y[n]$.

12.15. In Chapter 4 it was shown that if $x[n] = 0$ for $n < 0$, then

$$x[0] = \lim_{z \to \infty} X(z).$$

This result was called the *inital value theorem for right-sided sequences.*

(a) Prove a similar result for *left-sided sequences*, i.e., for sequences such that $x[n] = 0$ for $n > 0$.

(b) Use the initial value theorems to prove that $\hat{x}[0] = \log(x[0])$ if $x[n]$ is a minimum-phase sequence.

(c) Use the initial value theorems to prove that $\hat{x}[0] = \log(x[0])$ if $x[n]$ is a maximum-phase sequence.

(d) Use the initial value theorems to prove that $\hat{x}[0] = \log|A|$ when $X(z)$ is given by Eq. (12.53). Is this result consistent with the results of parts (b) and (c)?

12.16. Consider a sequence $x[n]$ with complex cepstrum $\hat{x}[n]$ such that $\hat{x}[n] = -\hat{x}[-n]$. Determine the quantity

$$E = \sum_{n=-\infty}^{\infty} x^2[n].$$

12.17. Consider a real, stable, even, two-sided sequence $h[n]$. The Fourier transform of $h[n]$ is positive for all ω, i.e.,

$$H(e^{j\omega}) > 0, \qquad -\pi < \omega \le \pi.$$

Assume that the z-transform of $h[n]$ exists. Do not assume that $H(z)$ is rational.

(a) Show that there exists a minimum-phase signal $g[n]$ such that

$$H(z) = G(z)G(z^{-1}),$$

where $G(z)$ is the z-transform of a sequence $g[n]$, which has the property that $g[n] = 0$ for $n < 0$. State explicitly the relationship between $\hat{h}[n]$ and $\hat{g}[n]$, the complex cepstra of $h[n]$ and $g[n]$, respectively.

(b) Given a stable signal $s[n]$, with rational z-transform

$$S(z) = \frac{(1 - 2z^{-1})(1 - \tfrac{1}{2}z^{-1})}{(1 - 4z^{-1})(1 - \tfrac{1}{3}z^{-1})}.$$

Define $h[n] = s[n] * s[-n]$. Find $G(z)$ (as in part a) in terms of $S(z)$.

(c) Consider the system in Fig. P12.17, where $\ell[n]$ is defined as

$$\ell[n] = u[n - 1] + (-1)^n u[n - 1].$$

Determine the *most general* conditions on $x[n]$ such that $y[n] = x[n]$ for all n.

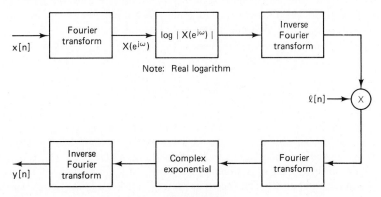

Note: Real logarithm

Figure P12.17

12.18. Consider a maximum-phase signal $x[n]$.

(a) Show that the complex cepstrum $\hat{x}[n]$ of a maximum-phase signal is related to its cepstrum $c_x[n]$ by

$$\hat{x}[n] = c_x[n]\ell_{max}[n],$$

where $\ell_{max}[n] = 2u[-n] - \delta[n]$.

(b) Using the relationships in part (a), show that

$$\arg\{X(e^{j\omega})\} = \frac{1}{2\pi}\mathscr{P}\int_{-\pi}^{\pi} \log|X(e^{j\theta})|\cot\left(\frac{\omega - \theta}{2}\right)d\theta.$$

(c) Also show that

$$\log|X(e^{j\omega})| = \hat{x}[0] - \frac{1}{2\pi}\mathscr{P}\int_{-\pi}^{\pi} \arg\{X(e^{j\theta})\}\cot\left(\frac{\omega - \theta}{2}\right)d\theta.$$

12.19. Consider a sequence $x[n]$ with Fourier transform $X(e^{j\omega})$ and complex cepstrum $\hat{x}[n]$. A new signal $y[n]$ is obtained by homomorphic filtering where

$$\hat{y}[n] = \{\hat{x}[n] - \hat{x}[-n]\}u[n-1].$$

(a) Show that $y[n]$ is a minimum-phase sequence.

(b) What is the phase of $Y(e^{j\omega})$?

(c) Obtain a relationship between $\arg[Y(e^{j\omega})]$ and $\log|Y(e^{j\omega})|$.

(d) If $x[n]$ is minimum phase, how is $y[n]$ related to $x[n]$?

12.20. Equations (12.67) and (12.70) are recursive relationships that can be used to compute the complex cepstrum $\hat{x}[n]$ when the input sequence $x[n]$ is minimum phase and maximum phase, respectively.

(a) Use Eq. (12.67) to compute recursively the complex cepstrum of the sequence $x[n] = a^n u[n]$, where $|a| < 1$.

(b) Use Eq. (12.70) to compute recursively the complex cepstrum of the sequence $x[n] = \delta[n] - a\delta[n+1]$, where $|a| < 1$.

12.21. Equation (12.67) represents a recursive relationship between a sequence $x[n]$ and its complex cepstrum $\hat{x}[n]$. Show from Eq. (12.67) that the characteristic system $D_*[\cdot]$ behaves as a causal system for minimum-phase inputs; i.e., show that for minimum-phase inputs, $\hat{x}[n]$ is dependent only on $x[k]$ for $k \leq n$.

12.22. Describe an exact procedure for computing a causal sequence $x[n]$ for which

$$X(z) = -z^3\frac{(1 - 0.95z^{-1})^{2/5}}{(1 - 0.9z^{-1})^{7/13}}.$$

12.23. Let $\text{ARG}\{X(e^{j\omega})\}$ represent the principal value of the phase of $X(e^{j\omega})$, and let $\arg\{X(e^{j\omega})\}$ represent the continuous phase of $X(e^{j\omega})$. Suppose that $\text{ARG}\{X(e^{j\omega})\}$ has been sampled at frequencies $\omega_k = 2\pi k/N$ to obtain $\text{ARG}\{X[k]\} = \text{ARG}\{X(e^{j(2\pi/N)k})\}$ as shown in Fig. P12.23. Assuming that $|\arg\{X[k]\} - \arg\{X[k-1]\}| < \pi$ for all k, determine and plot the sequence $r[k]$ as in Eq. (12.84) and $\arg\{X[k]\}$ for $0 \leq k \leq 10$.

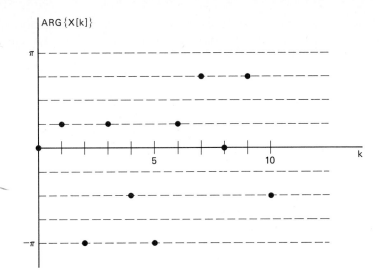

Figure P12.23

12.24. The sequence

$$h[n] = \delta[n] + \alpha\delta[n - n_0]$$

is a simplified model for the impulse response of a system that introduces echo.
(a) Determine the complex cepstrum $\hat{h}[n]$ for this sequence. Sketch the result.
(b) Determine and sketch the cepstrum $c_h[n]$.
(c) Suppose that an approximation to the complex cepstrum is computed using N-point DFTs as in Eqs. (12.80). Obtain a closed-form expression for the approximation $\hat{h}_p[n], 0 \le n \le N - 1$, for the case $n_0 = N/6$. Assume that phase unwrapping can be accurately done. What happens if N is not divisible by n_0?
(d) Repeat part (c) for the cepstrum approximation $c_{xp}[n], 0 \le n \le N - 1$, as computed using Eqs. (12.99).
(e) If the largest impulse in the cepstrum approximation $c_{xp}[n]$ is to be used to detect the value of the echo delay n_0, how large must N be to avoid ambiguity? Assume that accurate phase unwrapping can be achieved with this value of N.

12.25. Let $x[n]$ be a *finite-length* minimum-phase sequence with complex cepstrum $\hat{x}[n]$, and define another sequence

$$y[n] = \alpha^n x[n]$$

having complex cepstrum $\hat{y}[n]$.
(a) If $0 < \alpha < 1$, how is $\hat{y}[n]$ related to $\hat{x}[n]$?
(b) How should α be chosen so that $y[n]$ is no longer minimum phase?
(c) How should α be chosen so that if linear phase terms are removed before computing the complex cepstrum, then $\hat{y}[n] = 0$ for $n > 0$?

12.26. Consider a minimum-phase sequence $x[n]$ with z-transform $X(z)$ and complex cepstrum $\hat{x}[n]$. A new complex cepstrum is defined by the relation

$$\hat{y}[n] = (\alpha^n - 1)\hat{x}[n].$$

Determine the z-transform $Y(z)$. Is the result also minimum phase?

12.27. Section 12.8.5 contains an example of how the complex cepstrum can be used to obtain two different decompositions involving convolution of a minimum-phase sequence with another sequence. In that example,

$$X(z) = \frac{(0.98 + z^{-1})(1 + 0.9z^{-15} + 0.81z^{-30})}{(1 - 0.9e^{j\pi/6}z^{-1})(1 - 0.9e^{-j\pi/6}z^{-1})}.$$

(a) In one decomposition, $X(z) = X_{min}(z)X_{ap}(z)$ where

$$X_{min}(z) = \frac{(1 + 0.98z^{-1})(1 + 0.9z^{-15} + 0.81z^{-30})}{(1 - 0.9e^{j\pi/6}z^{-1})(1 - 0.9e^{-j\pi/6}z^{-1})}$$

and

$$X_{ap}(z) = \frac{(0.98 + z^{-1})}{(1 + 0.98z^{-1})}.$$

Use the power series expansion of the logarithmic terms to find the complex cepstra $\hat{x}_{min}[n]$, $\hat{x}_{ap}[n]$, and $\hat{x}[n]$. Plot these sequences and compare your plots with those in Fig. 12.21.

(b) In the second decomposition, $X(z) = X_{mn}(z)X_{mx}(z)$ where

$$X_{mn}(z) = \frac{(1 + 0.9z^{-15} + 0.81z^{-30})}{(1 - 0.9e^{j\pi/6}z^{-1})(1 - 0.9e^{-j\pi/6}z^{-1})}$$

and

$$X_{mx}(z) = (0.98 + z^{-1}).$$

Use the power series expansion of the logarithmic terms to find the complex cepstra and show that $\hat{x}_{mn}[n] \neq \hat{x}_{min}[n]$ but that $\hat{x}[n] = \hat{x}_{mn}[n] + \hat{x}_{mx}[n]$ is the same as in part (a). Note that

$$(1 + 0.9z^{-15} + 0.81z^{-30}) = \frac{(1 - (0.9)^3 z^{-45})}{(1 - 0.9z^{-15})}.$$

12.28. Let $x[n]$ be a sequence with z-transform $X(z)$ and complex cepstrum $\hat{x}[n]$. The magnitude-squared function for $X(z)$ is

$$V(z) = X(z)X^*(1/z^*).$$

Since $V(e^{j\omega}) = |X(e^{j\omega})|^2 \geq 0$, the complex cepstrum $\hat{v}[n]$ corresponding to $V(z)$ can be computed without phase unwrapping.

(a) Obtain a relationship between the complex cepstrum $\hat{v}[n]$ and the complex cepstrum $\hat{x}[n]$.

(b) Express the complex cepstrum $\hat{v}[n]$ in terms of the cepstrum $c_x[n]$.

(c) Determine the sequence $\ell[n]$ such that

$$\hat{x}_{min}[n] = \ell[n]\hat{v}[n]$$

is the complex cepstrum of a minimum-phase sequence $x_{min}[n]$ for which

$$|X_{min}(e^{j\omega})|^2 = V(e^{j\omega}).$$

(d) Suppose that $X(z)$ is as given by Eq. (12.53). Use the result of part (c) and Eqs. (12.57) to find the complex cepstrum of the minimum-phase sequence, and work backward to find $X_{min}(z)$.

The technique employed in part (d) may be used in general to obtain a minimum-phase factorization of a magnitude-squared function.

12.29. Let $\hat{x}[n]$ be the complex cepstrum of $x[n]$. Define a sequence $x_e[n]$ to be

$$x_e[n] = \begin{cases} x[n/N], & n = 0, \pm N, \pm 2N, \ldots, \\ 0, & \text{otherwise.} \end{cases}$$

Show that the complex cepstrum of $x_u[n]$ is given by

$$\hat{x}_e[n] = \begin{cases} \hat{x}[n/N], & n = 0, \pm N, \pm 2N, \ldots, \\ 0, & \text{otherwise.} \end{cases}$$

12.30. Suppose that $s[n] = h[n] * g[n] * p[n]$, where $h[n]$ is a minimum-phase sequence, $g[n]$ is a maximum-phase sequence, and $p[n]$ is

$$p[n] = \sum_{k=0}^{4} \alpha_k \delta[n - k n_0],$$

where α_k and n_0 are not known. Develop a method to separate $h[n]$ from $s[n]$.

12.31. In speech analysis, synthesis, and coding, the speech signal is commonly modeled over a short time interval as the response of a linear time-invariant system excited by an excitation that switches between a train of equally spaced pulses for voiced sounds and a wideband random noise source for unvoiced sounds. To use homomorphic deconvolution to separate the components of the speech model, the speech signal $s[n] = v[n] * p[n]$ is multiplied by a window sequence $w[n]$ to obtain $x[n] = s[n]w[n]$. To simplify the analysis, $x[n]$ is approximated by

$$x[n] = (v[n] * p[n]) \cdot w[n] \simeq v[n] * (p[n] \cdot w[n]) = v[n] * p_w[n]$$

where $p_w[n] = p[n]w[n]$ as in Eq. (12.147).
(a) Give an example of $p[n]$, $v[n]$, and $w[n]$ for which the above assumption may be a poor approximation.
(b) One approach to estimating the excitation parameters (voiced/unvoiced decision and pulse spacing for voiced speech) is to compute the real cepstrum $c_x[n]$ of the windowed segment of speech $x[n]$ as depicted in Fig. P12.31-1. For the model of Section 12.9.1, express $c_x[n]$ in terms of the complex cepstrum $\hat{x}[n]$. How would you use $c_x[n]$ to estimate the excitation parameters?

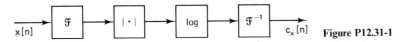

Figure P12.31-1

(c) Suppose that we replace the log operation in Fig. P12.31-1 with the "squaring" operation so that the resulting system is as depicted in Fig. P12.31-2. Can the new "cepstrum" $q_x[n]$ be used to estimate the excitation parameters? Explain.

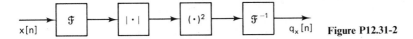

Figure P12.31-2

12.32. Consider a stable linear time-invariant system with impulse response $h[n]$ and all-pole system function

$$H(z) = \frac{G}{1 - \sum_{k=1}^{N} a_k z^{-k}}.$$

Such all-pole systems arise in linear predictive analysis. It is of interest to compute the complex cepstrum directly from the coefficients of $H(z)$.

(a) Determine $\hat{h}[0]$.

(b) Show that

$$\hat{h}[n] = a_n + \sum_{k=1}^{n-1} \left(\frac{k}{n}\right)\hat{h}[k]a_{n-k}, \qquad n \geq 1.$$

With the relations in parts (a) and (b), the complex cepstrum can be computed without phase unwrapping and without solving for the roots of the denominator of $H(z)$.

12.33. A somewhat more realistic model for echo than the system in Problem 12.24 is the system depicted in Fig. P12.33. The impulse response of this system is

$$h[n] = \delta[n] + \alpha g[n - n_0],$$

where $\alpha g[n]$ is the impulse response of the echo path.

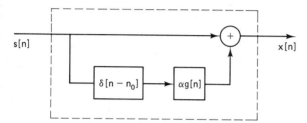

Figure P12.33

(a) Assuming that

$$\max_{-\pi < \omega < \pi} \; |\alpha G(e^{j\omega})| < 1,$$

show that the complex cepstrum $\hat{h}[n]$ has the form

$$\hat{h}[n] = \sum_{k=1}^{\infty} (-1)^{k+1} \frac{\alpha^k}{k} g_k[n - kn_0],$$

and determine an expression for $g_k[n]$ in terms of $g[n]$.

(b) For the conditions of part (a), determine and sketch the complex cepstrum $\hat{h}[n]$ when $g[n] = \delta[n]$.

(c) For the conditions of part (a), determine and sketch the complex cepstrum $\hat{h}[n]$ when $g[n] = a^n u[n]$. What condition must be satisfied by α and a so that the result of part (a) applies?

(d) For the conditions of part (a), determine and sketch the complex cepstrum $\hat{h}[n]$ when $g[n] = a_0\delta[n] + a_1\delta[n - n_1]$. What condition must be satisfied by α, a_0, a_1, and n_1 so that the result of part (a) applies?

Appendix A
Random Signals

In this appendix, we collect and summarize a number of results and establish the notation relating to the representation of stochastic signals. We make no attempt here to provide a detailed discussion of the difficult and subtle mathematical issues of the theory of random processes. Although our approach is not rigorous, we have summarized the important results and the mathematical assumptions implicit in their derivation.

A.1 DISCRETE-TIME RANDOM PROCESSES

The fundamental concept in the mathematical representation of stochastic signals is that of a *random process*. In our discussion of random processes as models for discrete-time signals, we assume that the reader is familiar with fundamental concepts of the theory of probability such as random variables, probability distributions, and averages. Readers who require further background in probability theory are referred to one of the basic texts listed in the Bibliography.

In utilizing the random process model in practical signal processing applications, we consider a particular sequence to be one of an ensemble of sample sequences. Given a discrete-time signal, the structure, i.e., the underlying probability law, of the corresponding random process is generally not known and must somehow be inferred. It may be possible to make reasonable assumptions about the structure of the process, or it may be possible to estimate the properties of a random process representation from a finite segment of a typical sample sequence.

Formally, a random process is an indexed family of random variables $\{\mathbf{x}_n\}$. The family of random variables is characterized by a set of probability distribution functions that in general may be a function of the index n. In using the concept of a random process as a model for discrete-time signals, the index n is associated with the time index. An individual random variable $\mathbf{x}_n$ is described by the probability distribution function

$$P_{\mathbf{x}_n}(x_n, n) = \text{Probability}[\mathbf{x}_n \le x_n], \tag{A.1}$$

where $\mathbf{x}_n$ denotes the random variable and x_n is a particular value of $\mathbf{x}_n$.† If $\mathbf{x}_n$ takes on a continuous range of values, it is equivalently specified by the *probability density function*

$$p_{\mathbf{x}_n}(x_n, n) = \frac{\partial P_{\mathbf{x}_n}(x_n, n)}{\partial x_n} \tag{A.2}$$

or

$$P_{\mathbf{x}_n}(x_n, n) = \int_{-\infty}^{x_n} p_{\mathbf{x}_n}(x, n)\, dx. \tag{A.3}$$

The interdependence of two random variables $\mathbf{x}_n$ and $\mathbf{x}_m$ of a random process is described by the joint probability distribution function

$$P_{\mathbf{x}_n, \mathbf{x}_m}(x_n, n, x_m, m) = \text{Probability } [\mathbf{x}_n \le x_n \text{ and } \mathbf{x}_m \le x_m] \tag{A.4}$$

and by the joint probability density

$$p_{\mathbf{x}_n, \mathbf{x}_m}(x_n, n, x_m, m) = \frac{\partial^2 P_{\mathbf{x}_n, \mathbf{x}_m}(x_n, n, x_m, m)}{\partial x_n \partial x_m}. \tag{A.5}$$

Two random variables are *statistically independent* if knowledge of the value of one does not affect the probability density of the other. If the random variables $\{\mathbf{x}_n\}$ are statistically independent, then

$$P_{\mathbf{x}_n, \mathbf{x}_m}(x_n, x_m, m) = P_{\mathbf{x}_n}(x_n, n) \cdot P_{\mathbf{x}_m}(x_m, m). \tag{A.6}$$

A complete characterization of a random process requires the specification of all possible joint probability distributions. As we have indicated, these probability distribution functions may be a function of the time index n. In the case where all the probability functions are independent of a shift of time origin, the random process is said to be *stationary*. For example, the second-order distribution of a stationary process satisfies

$$P_{\mathbf{x}_{n+k}, \mathbf{x}_{m+k}}(x_{n+k}, n + k, x_{m+k}, m + k) = P_{\mathbf{x}_n \mathbf{x}_m}(x_n, n, x_m, m). \tag{A.7}$$

In many of the applications of discrete-time signal processing, random processes serve as models for signals in the sense that a particular signal can be considered a sample sequence of a random process. Although the details of such signals are unpredictable — making a deterministic approach to signal representation inappropriate — certain average properties of the ensemble can be determined given the probability law of the process. The average properties often serve as a useful, although incomplete, characterization of such signals.

---

† In this appendix, boldface type is used to denote the random variables and regular type denotes dummy variables of probability functions.

A.2 AVERAGES

It is often useful to characterize a random variable by averages such as mean and variance. Since a random process is an indexed set of random variables, we may likewise characterize the process by statistical averages of the random variables comprising the random process. Such averages are called *ensemble averages*. We begin the discussion of averages with some definitions.

A.2.1 Definitions

The average or mean of a random process is defined as

$$m_{\mathbf{x}_n} = \mathscr{E}\{\mathbf{x}_n\} = \int_{-\infty}^{\infty} x p_{\mathbf{x}_n}(x, n)\, dx, \tag{A.8}$$

where $\mathscr{E}$ denotes mathematical expectation. In general the mean (expected value) may depend on n. In addition, if $g(\cdot)$ is a single-valued function, then $g(\mathbf{x}_n)$ is a random variable, and the set of random variables $\{g(\mathbf{x}_n)\}$ defines a new random process. To compute averages of the new random process, we can derive probability distributions of the new random variables, or it can be shown that

$$\mathscr{E}\{g(\mathbf{x}_n)\} = \int_{-\infty}^{\infty} g(x) p_{\mathbf{x}_n}(x, n)\, dx. \tag{A.9}$$

If the random variables are discrete, the integrals become summations over all possible values of the random variable, i.e., $\mathscr{E}\{g(x)\}$ has the form

$$\mathscr{E}\{g(\mathbf{x}_n)\} = \sum_{x} g(x) \hat{p}_{\mathbf{x}_n}(x, n). \tag{A.10}$$

In cases where we are interested in the relationship between two (or more) random processes, we must be concerned with two sets of random variables $\{\mathbf{x}_n\}$ and $\{\mathbf{y}_m\}$. For example, the expected value of a function of two random variables is defined as

$$\mathscr{E}\{g(\mathbf{x}_n, \mathbf{y}_m)\} = \int_{-\infty}^{\infty} \int_{-\infty}^{\infty} g(x, y) p_{\mathbf{x}_n, \mathbf{y}_m}(x, n, y, m)\, dx\, dy, \tag{A.11}$$

where $p_{\mathbf{x}_n, \mathbf{y}_m}(x_m, n, y_m, m)$ is the joint probability density of the random variables $\mathbf{x}_n$ and $\mathbf{y}_m$.

A number of simple properties of averages are particularly useful. It is easily shown that:

1. $\mathscr{E}\{\mathbf{x}_n + \mathbf{y}_m\} = \mathscr{E}\{\mathbf{x}_n\} + \mathscr{E}\{\mathbf{y}_m\}$, i.e., the average of a sum is the sum of the averages;
2. $\mathscr{E}\{a\mathbf{x}_n\} = a\mathscr{E}\{\mathbf{x}_n\}$, i.e., the average of a constant times $\mathbf{x}_n$ is equal to the constant times the average of $\mathbf{x}_n$.

In general, the average of a product of two random variables is not equal to the product of the averages. If this is the case, however, the two random variables are said to be *linearly independent* or *uncorrelated*. That is, $\mathbf{x}_n$ and $\mathbf{y}_m$ are linearly independent if

$$\mathscr{E}\{\mathbf{x}_n\mathbf{y}_m\} = \mathscr{E}\{\mathbf{x}_m\} \cdot \mathscr{E}\{\mathbf{y}_m\}. \tag{A.12}$$

It is easy to see from Eqs. (A.11) and (A.12) that a sufficient condition for linear independence is

$$p_{\mathbf{x}_n, \mathbf{y}_m}(x_n, n, y_m, m) = p_{\mathbf{x}_n}(x_n, n) \cdot p_{\mathbf{y}_m}(y_m, m). \tag{A.13}$$

However, it can be shown that Eq. (A.13) is a stronger statement of independence than Eq. (A.12). As previously stated, random variables satisfying Eq. (A.13) are said to be *statistically independent*. If Eq. (A.13) holds for all values of n and m, the random processes $\{\mathbf{x}_n\}$ and $\{\mathbf{y}_m\}$ are said to be statistically independent. Statistically independent random processes are also linearly independent; however, linear independence does not imply statistical independence.

It can be seen from Eqs. (A.9)–(A.11) that averages generally are functions of time. In the case of stationary processes, this is not true. For stationary processes, the mean is the same for all the random variables that constitute the process; i.e., the mean of a stationary process is a constant, which we denote simply m_x.

In addition to the mean of a random process, as defined in Eq. (A.8), a number of averages are particularly important within the context of signal processing. These are defined below. For notational convenience we assume that the probability distributions are continuous. Corresponding definitions for discrete random processes can be obtained by applying Eq. (A.10).

The *mean-square value* of $\mathbf{x}_n$ is the average of $|\mathbf{x}_n|^2$, i.e.,

$$\mathscr{E}\{|\mathbf{x}_n|^2\} = \text{mean square} = \int_{-\infty}^{\infty} |x|^2 p_{\mathbf{x}_n}(x, n)\, dx. \tag{A.14}$$

The mean-square value is sometimes referred to as the *average power*.

The *variance* of $\mathbf{x}_n$ is the mean-square value of $[\mathbf{x}_n - m_{x_n}]$; i.e.,

$$\text{var}[\mathbf{x}_n] = \mathscr{E}\{|(\mathbf{x}_n - m_{x_n})|^2\} = \sigma_{\mathbf{x}_n}^2. \tag{A.15}$$

Since the average of a sum is the sum of the averages, it can easily be shown that Eq. (A.15) can be written as

$$\text{var}[\mathbf{x}_n] = \mathscr{E}\{|\mathbf{x}_n|^2\} - |m_{x_n}|^2. \tag{A.16}$$

In general, the mean-square value and the variance are functions of time; however, they are constant for stationary processes.

The mean, mean square, and variance are simple averages that provide only a small amount of information about the process. A more useful average is the *autocorrelation sequence,* which is defined as

$$\phi_{xx}[n, m] = \mathcal{E}\{\mathbf{x}_n \mathbf{x}_m^*\}$$

$$= \int_{-\infty}^{\infty} \int_{-\infty}^{\infty} x_n x_m^* p_{\mathbf{x}_n, \mathbf{x}_m}(x_n, n, x_m, m) \, dx_n \, dx_m, \tag{A.17}$$

where * denotes complex conjugation. The autocovariance sequence of a random process is defined as

$$\gamma_{xx}[n, m] = \mathcal{E}\{(\mathbf{x}_n - m_{x_n})(\mathbf{x}_m - m_{x_m})^*\}, \tag{A.18}$$

which can be written as

$$\gamma_{xx}[n, m] = \phi_{xx}[n, m] - m_{x_n} m_{x_m}^*. \tag{A.19}$$

Note that in general, both the autocorrelation and autocovariance are two-dimensional sequences, i.e., functions of two variables.

The autocorrelation sequence is a measure of the dependence between values of the random processes at different times. In this sense it partially describes the time variation of a random signal. A measure of the dependence between two different random signals is obtained from the cross-correlation sequence. If $\{\mathbf{x}_n\}$ and $\{\mathbf{y}_m\}$ are two random processes, their cross-correlation is

$$\phi_{xy}[n, m] = \mathcal{E}\{\mathbf{x}_n \mathbf{y}_m^*\}$$

$$= \int_{-\infty}^{\infty} \int_{-\infty}^{\infty} xy^* p_{\mathbf{x}_n, \mathbf{y}_m}(x, n, y, m) \, dx \, dy, \tag{A.20}$$

where $p_{\mathbf{x}_n, \mathbf{y}_m}(x, n, y, m)$ is the joint probability density of $\mathbf{x}_n$ and $\mathbf{y}_m$. The cross-covariance function is defined as

$$\gamma_{xy}[n, m] = \mathcal{E}\{(\mathbf{x}_n - m_{x_n})(\mathbf{y}_m - m_{y_m})^*\}$$
$$= \phi_{xy}[n, m] - m_{x_n} m_{y_m}^*. \tag{A.21}$$

As we have pointed out, the statistical properties of a random process generally vary with time. However, a stationary random process is characterized by an equilibrium condition in which the statistical properties are invariant to a shift of time origin. This means that the first-order probability distribution is independent of time. Similarly, all the joint probability functions are also invariant to a shift of time origin; i.e., the second-order joint probability distributions depend only on the time difference $(m - n)$. First-order averages such as mean and variance are independent of time; second-order averages, such as the autocorrelation $\phi_{xx}[n, m]$, are dependent on the time difference $(m - n)$. Thus for a stationary process we can write

$$m_x = \mathcal{E}\{\mathbf{x}_n\}, \tag{A.22}$$

$$\sigma_x^2 = \mathcal{E}\{|(\mathbf{x}_n - m_x)|^2\} \tag{A.23}$$

independent of n, and now if we denote the time difference by m,

$$\phi_{xx}[n + m, n] = \phi_{xx}[m] = \mathcal{E}\{\mathbf{x}_{n+m} \mathbf{x}_n^*\}. \tag{A.24}$$

That is, the autocorrelation sequence of a stationary random process is a one-dimensional sequence, a function of the time difference m.

In many instances we encounter random processes that are not stationary in the *strict sense*, i.e., their probability distributions are not time-invariant, but Eqs. (A.22)–(A.24) still hold. Such random processes are said to be *wide sense stationary*.

A.2.2 Time Averages

In a signal processing context, the notion of an ensemble of signals is a convenient mathematical concept that allows us to use the theory of probability in their representation. However, in a practical sense, we would prefer to deal with a single sequence rather than an infinite ensemble of sequences. For example, we might wish to infer the probability law or certain averages of the random process representation from measurements on a single member of the ensemble. When the probability distributions are independent of time, we might feel intuitively that the amplitude distribution of a long segment of a single-sample sequence should be approximately equal to the probability density. Similarly, the arithmetic average of a large number of samples of a single sequence should be very close to the mean of the process. To formalize these intuitive notions, we define the time average of a random process as

$$\langle \mathbf{x}_n \rangle = \lim_{L \to \infty} \frac{1}{2L + 1} \sum_{n=-L}^{L} \mathbf{x}_n. \tag{A.25}$$

Similarly, the time autocorrelation sequence is defined as

$$\langle \mathbf{x}_{n+m}\mathbf{x}_n^* \rangle = \lim_{L \to \infty} \frac{1}{2L + 1} \sum_{n=-L}^{L} \mathbf{x}_{n+m}\mathbf{x}_n^*. \tag{A.26}$$

It can be shown that the preceding limits exist if $\{\mathbf{x}_n\}$ is a stationary process with finite mean. As defined in Eqs. (A.25) and (A.26), these time averages are functions of an infinite set of random variables and thus are properly viewed as random variables themselves. However, under the condition known as *ergodicity*, the time averages in Eqs. (A.25) and (A.26) are equal to constants in the sense that the time averages of almost all possible sample sequences are equal to the same constant. Furthermore, they are equal to the corresponding ensemble average.† That is, for any single sample sequence $\{x[n]\}$ for $-\infty < n < \infty$,

$$\langle x[n] \rangle = \lim_{L \to \infty} \frac{1}{2L + 1} \sum_{n=-L}^{L} x[n] = \mathscr{E}\{\mathbf{x}_n\} = m_x \tag{A.27}$$

and

$$\langle x[n+m]x^*[n] \rangle = \lim_{L \to \infty} \frac{1}{2L + 1} \sum_{n=-L}^{L} x[n+m]x^*[n] = \mathscr{E}\{\mathbf{x}_{n+m}\mathbf{x}_n^*\} = \phi_{xx}[m]. \tag{A.28}$$

---

† A more precise statement is that the random variables $\langle \mathbf{x}_n \rangle$ and $\langle \mathbf{x}_{n+m}\mathbf{x}_n^* \rangle$ have means equal to m_x and $\phi_{xx}[m]$, respectively, and their variances are zero.

The time average operator $\langle\cdot\rangle$ has the same properties as the ensemble average operator $\mathscr{E}\{\cdot\}$. Thus we generally do not distinguish between the random variable $\mathbf{x}_n$ and its value in a sample sequence, $x[n]$. For example, the expression $\mathscr{E}\{x[n]\}$ should be interpreted as $\mathscr{E}\{\mathbf{x}_n\} = \langle x[n]\rangle$. In general, a random process for which time averages equal ensemble averages is called an *ergodic process*.

In practice, it is common to assume that a given sequence is a sample sequence of an ergodic random process so that averages can be computed from a single sequence. Of course, we generally cannot compute with the limits in Eqs. (A.27) and (A.28), but the quantities

$$\hat{m}_x = \frac{1}{L}\sum_{n=0}^{L-1} x[n], \tag{A.29}$$

$$\hat{\sigma}_x^2 = \frac{1}{L}\sum_{n=0}^{L-1} |x[n] - \hat{m}_x|^2, \tag{A.30}$$

and

$$\langle x[n+m]x^*[n]\rangle_L = \frac{1}{L}\sum_{n=0}^{L-1} x[n+m]x^*[n] \tag{A.31}$$

or similar quantities are often computed as *estimates* of the mean, variance, and autocorrelation. $\hat{m}_x$ and $\hat{\sigma}_x^2$ are referred to as the sample mean and sample variance, respectively. The estimation of averages of a random process from a finite segment of data is a problem of statistics, which we touch on briefly in Chapter 11.

A.3 PROPERTIES OF CORRELATION AND COVARIANCE SEQUENCES

Several useful properties of correlation and covariance functions follow in a simple way from the definitions. These properties are given in this section.

Consider two real stationary random processes $\{\mathbf{x}_n\}$ and $\{\mathbf{y}_n\}$ with autocorrelation, autocovariance, cross-correlation, and cross-covariance being given by, respectively,

$$\phi_{xx}[m] = \mathscr{E}\{\mathbf{x}_{n+m}\mathbf{x}_n^*\}, \tag{A.32}$$

$$\gamma_{xx}[m] = \mathscr{E}\{(\mathbf{x}_{n+m} - m_x)(\mathbf{x}_n - m_x)^*\}, \tag{A.33}$$

$$\phi_{xy}[m] = \mathscr{E}\{\mathbf{x}_{n+m}\mathbf{y}_n^*\}, \tag{A.34}$$

$$\gamma_{xy}[m] = \mathscr{E}\{(\mathbf{x}_{n+m} - m_x)(\mathbf{y}_n - m_y)^*\}, \tag{A.35}$$

where m_x and m_y are the means of the two processes. The following properties are easily derived by simple manipulations of the definitions.

Property 1

$$\gamma_{xx}[m] = \phi_{xx}[m] - |m_x|^2 \qquad\qquad (A.36a)$$

$$\gamma_{xy}[m] = \phi_{xy}[m] - m_x m_y^* \qquad\qquad (A.36b)$$

These results follow directly from Eqs. (A.19) and (A.21), and they indicate that the correlation and covariance sequences are identical for zero-mean processes.

Property 2

$$\phi_{xx}[0] = E[|\mathbf{x}_n|^2] = \text{Mean-square value} \qquad\qquad (A.37a)$$

$$\gamma_{xx}[0] = \sigma_x^2 = \text{Variance} \qquad\qquad (A.37b)$$

Property 3

$$\phi_{xx}[-m] = \phi_{xx}^*[m] \qquad\qquad (A.38a)$$

$$\gamma_{xx}[-m] = \gamma_{xx}^*[m] \qquad\qquad (A.38b)$$

$$\phi_{xy}[-m] = \phi_{yx}^*[m] \qquad\qquad (A.38c)$$

$$\gamma_{xy}[-m] = \gamma_{yx}^*[m] \qquad\qquad (A.38d)$$

Property 4

$$|\phi_{xy}[m]|^2 \le \phi_{xx}[0]\phi_{yy}[0] \qquad\qquad (A.39a)$$

$$|\gamma_{xy}[m]|^2 \le \gamma_{xx}[0]\gamma_{yy}[0] \qquad\qquad (A.39b)$$

In particular,

$$|\phi_{xx}[m]| \le \phi_{xx}[0] \qquad\qquad (A.40a)$$

$$|\gamma_{xx}[m]| \le \gamma_{xx}[0] \qquad\qquad (A.40b)$$

Property 5. If $\mathbf{y}_n = \mathbf{x}_{n-n_0}$, then

$$\phi_{yy}[m] = \phi_{xx}[m], \qquad\qquad (A.41a)$$

$$\gamma_{yy}[m] = \gamma_{xx}[m]. \qquad\qquad (A.41b)$$

Property 6. For many random processes, the random variables become uncorrelated as they become more separated in time. If this is true,

$$\lim_{m \to \infty} \gamma_{xx}[m] = 0, \qquad\qquad (A.42a)$$

$$\lim_{m \to \infty} \phi_{xx}[m] = |m_x|^2, \qquad\qquad (A.42b)$$

$$\lim_{m \to \infty} \gamma_{xy}[m] = 0, \qquad\qquad (A.42c)$$

$$\lim_{m \to \infty} \phi_{xy}[m] = m_x m_y^*. \qquad\qquad (A.42d)$$

The essence of these results is that the correlation and covariance are finite-energy sequences that tend to die out for large values of m. Thus it is often possible to represent these sequences in terms of their Fourier transforms or z-transforms.

A.4 TRANSFORM REPRESENTATIONS OF RANDOM SIGNALS

Although the Fourier transform and z-transform of a random signal do not exist, the autocovariance and autocorrelation sequences of such a signal are aperiodic sequences for which these transforms often do exist. The spectral representation of these averages plays an important role in describing the input/output relations for a linear time-invariant system when the input is a stochastic signal. Therefore, it is of interest to consider the properties of correlation and covariance sequences and their corresponding Fourier and z-transforms.

Let us define $\Phi_{xx}(e^{j\omega})$, $\Gamma_{xx}(e^{j\omega})$, $\Phi_{xy}(e^{j\omega})$, and $\Gamma_{xy}(e^{j\omega})$ as the Fourier transforms of $\phi_{xx}[m]$, $\gamma_{xx}[m]$, $\phi_{xy}[m]$ and $\gamma_{xy}[m]$, respectively. From Eqs. (A.36) it follows that

$$\Phi_{xx}(e^{j\omega}) = \Gamma_{xx}(e^{j\omega}) + 2\pi|m_x|^2\delta(\omega), \qquad |\omega| < \pi, \tag{A.43a}$$

and

$$\Phi_{xy}(e^{j\omega}) = \Gamma_{xy}(e^{j\omega}) + 2\pi m_x m_y^*\delta(\omega), \qquad |\omega| < \pi. \tag{A.43b}$$

In the case of zero-mean processes ($m_x = 0$ and $m_y = 0$), $\Phi_{xx}(e^{j\omega}) = \Gamma_{xx}(e^{j\omega})$ and $\Phi_{xy}(e^{j\omega}) = \Gamma_{xy}(e^{j\omega})$.

From the inverse Fourier transform equation, it follows that

$$\gamma_{xx}[m] = \frac{1}{2\pi}\int_{-\pi}^{\pi}\Gamma_{xx}(e^{j\omega})e^{j\omega m}\,d\omega, \tag{A.44a}$$

$$\phi_{xx}[m] = \frac{1}{2\pi}\int_{-\pi}^{\pi}\Phi_{xx}(e^{j\omega})e^{j\omega m}\,d\omega, \tag{A.44b}$$

and, consequently,

$$\mathcal{E}\{|x[n]|^2\} = \phi_{xx}[0] = \frac{1}{2\pi}\int_{-\pi}^{\pi}\Phi_{xx}(e^{j\omega})\,d\omega, \tag{A.45a}$$

$$\sigma_x^2 = \gamma_{xx}[0] = \frac{1}{2\pi}\int_{-\pi}^{\pi}\Gamma_{xx}(e^{j\omega})\,d\omega. \tag{A.45b}$$

It is convenient to define the quantity $P_{xx}(\omega)$ as

$$P_{xx}(\omega) = \Phi_{xx}(e^{j\omega}), \tag{A.46}$$

in which case Eqs. (A.45) are expressed as

$$\mathcal{E}\{|x[n]|^2\} = \frac{1}{2\pi}\int_{-\pi}^{\pi}P_{xx}(\omega)\,d\omega, \tag{A.47a}$$

$$\sigma_x^2 = \frac{1}{2\pi}\int_{-\pi}^{\pi}P_{xx}(\omega)\,d\omega - |m_x|^2. \tag{A.47b}$$

Thus the area under $P_{xx}(\omega)$ for $-\pi \leq \omega \leq \pi$ is proportional to the average power in the signal. In fact, as we discuss in Section 2.10, the integral of $P_{xx}(\omega)$ over a band of frequencies is proportional to the power in the signal in that band. For this reason the function $P_{xx}(\omega)$ is called the *power density spectrum*, or simply the *power spectrum*. When $P_{xx}(\omega)$ is a constant independent of ω, it is referred to as a white process or white noise. When $P_{xx}(\omega)$ is constant over a band and zero otherwise, we refer to it as bandlimited white noise.

From Eq. (A.38b), it follows that $P_{xx}(\omega)$ is always real-valued, and for real random processes, $\phi_{xx}[m] = \phi_{xx}[-m]$, so $P_{xx}(\omega)$ is both real and even; i.e.,

$$P_{xx}(\omega) = P_{xx}(-\omega). \tag{A.48}$$

An additional important property is that the power density spectrum is nonnegative. This point is discussed in Section 2.10.

The *cross power density spectrum* is defined as

$$P_{xy}(\omega) = \Phi_{xy}(e^{j\omega}). \tag{A.49}$$

This function is generally complex, and from Eq. (A.38c) it follows that

$$P_{xy}(\omega) = P_{yx}^*(\omega). \tag{A.50}$$

If the z-transforms of $\phi_{xx}[m]$ and $\phi_{xy}[m]$ exist, a condition that requires, among other things, that the processes be zero mean, then Eqs. (A.47), (A.48), and (A.50) may be generalized to z-transform properties. Specifically, Eq. (A.47) corresponds to the following:

Property 1

$$\sigma_x^2 = \frac{1}{2\pi j} \oint_C \Phi_{xx}(z) z^{-1}\, dz, \tag{A.51}$$

where C is a counterclockwise closed contour in the region of convergence of $\Phi_{xx}(z)$. Equations (A.48) and (A.50) correspond to the following:

Property 2

$$\Phi_{xx}(z) = \Phi_{xx}^*(1/z^*) \tag{A.52a}$$

$$\Phi_{xy}(z) = \Phi_{yx}^*(1/z^*) \tag{A.52b}$$

Since $\phi_{xx}[m]$ is two-sided and symmetric, it follows that the region of convergence of $\Phi_{xx}(z)$ must be of the form

$$R_a < |z| < \frac{1}{R_a}.$$

In the important case when $\Phi_{xx}(z)$ is a rational function of z, Eqs. (A.52) imply that the poles and zeros of $\Phi_{xx}(z)$ must occur in complex-conjugate reciprocal pairs. The major advantage of the z-transform representation is that integrals such as Eq. (A.51) may be evaluated using the residue theorem.

Appendix B
Continuous-Time Filters

The techniques discussed in Sections 7.1 and 7.2 for designing IIR digital filters rely on the availability of appropriate continuous-time filter designs. In this appendix, we briefly summarize the characteristics of several classes of lowpass filter approximations that we refer to in Chapter 7. A more detailed discussion of these filter classes is available (see Guillemin, 1957; Weinberg, 1975; and Parks and Burrus, 1987). Extensive design tables and formulas are available in Zverev (1967), and a variety of computer programs are available for computer-aided design of IIR digital filters based on analog filter approximations. (See, for example, Gray and Markel, 1976; DSP Committee, 1979; and Mersereau et al., 1984.)

B.1 BUTTERWORTH LOWPASS FILTERS

Butterworth lowpass filters are defined by the property that the magnitude response is maximally flat in the passband. For an Nth-order lowpass filter, this means that the first $(2N - 1)$ derivatives of the magnitude-squared function are zero at $\Omega = 0$. Another property is that the magnitude response is monotonic in the passband and the stopband. The magnitude-squared function for a continuous-time Butterworth lowpass filter is of the form

$$|H_c(j\Omega)|^2 = \frac{1}{1 + (j\Omega/j\Omega_c)^{2N}}. \tag{B.1}$$

This function is sketched in Fig. B.1.

As the parameter N in Eq. (B.1) increases, the filter characteristics become sharper; that is, they remain close to unity over more of the passband and become close to zero more rapidly in the stopband, although the magnitude-squared function at the cutoff frequency Ω_c will always be equal to one-half because of the nature of Eq. (B.1). The dependence of the Butterworth filter characteristic on the parameter N is indicated in Fig. B.2.

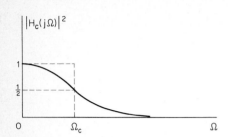

Figure B.1 Magnitude squared for continuous-time Butterworth filter.

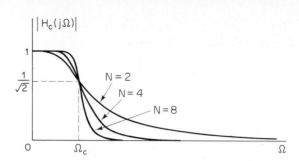

Figure B.2 Dependence of Butterworth magnitude characteristic on the order N.

From the magnitude-squared function in Eq. (B.1), we observe that $H_c(s)H_c(-s)$ must be of the form

$$H_c(s)H_c(-s) = \frac{1}{1 + (s/j\Omega_c)^{2N}}. \tag{B.2}$$

The roots of the denominator polynomial (the poles of the magnitude-squared function) are therefore located at

$$s_k = (-1)^{1/2N}(j\Omega_c) = \Omega_c e^{(j\pi/2N)(2k+N-1)}, \qquad k = 0, 1, \ldots, 2N - 1. \tag{B.3}$$

Thus there are $2N$ poles equally spaced in angle on a circle of radius Ω_c in the s-plane. The poles are symmetrically located with respect to the imaginary axis. A pole never falls on the imaginary axis, and one occurs on the real axis for N odd but not for N even. The angular spacing between the poles on the circle is π/N radians. For example, for $N = 3$, the poles are spaced by $\pi/3$ radians, or 60 degrees, as indicated in Fig. B.3. To determine the system function of the analog filter to associate with the Butterworth magnitude-squared function, we must perform the factorization $H_c(s)H_c(-s)$. The poles of the magnitude-squared function always occur in pairs, i.e., if there is a pole at $s = s_k$, then a pole also occurs at $s = -s_k$. Consequently, to construct $H_c(s)$ from the magnitude-squared function, we would choose one pole from each such pair. To obtain a stable and causal filter, we should choose the poles on the left-half-plane part of the s-plane.

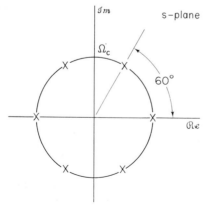

Figure B.3 s-plane pole locations for a third-order Butterworth filter.

B.2 CHEBYSHEV FILTERS

In a Butterworth filter the magnitude response is monotonic in both the passband and the stopband. Consequently, if the filter specifications are in terms of maximum passband and stopband approximation error, the specifications are exceeded toward the low-frequency end of the passband and above the stopband cutoff frequency. A more efficient approach, which usually leads to a lower-order filter, is to distribute the accuracy of the approximation uniformly over the passband or the stopband or both. This is accomplished by choosing an approximation that has an equiripple behavior rather than a monotonic behavior. The class of Chebyshev filters has the property that the magnitude of the frequency response is either equiripple in the passband and monotonic in the stopband (referred to as a type I Chebyshev filter) or monotonic in the passband and equiripple in the stopband (a type II Chebyshev filter). A type I Chebyshev filter is shown in Fig. B.4. The magnitude-squared function for a type I filter is

$$|H_c(j\Omega)|^2 = \frac{1}{1 + \varepsilon^2 V_N^2(\Omega/\Omega_c)}, \tag{B.4}$$

where $V_N(x)$ is the Nth-order Chebyshev polynomial defined as

$$V_N(x) = \cos(N \cos^{-1} x). \tag{B.5}$$

For example, for $N = 0$, $V_0(x) = 1$; for $N = 1$, $V_1(x) = \cos(\cos^{-1} x) = x$; for $N = 2$, $V_2(x) = \cos(2 \cos^{-1} x) = 2x^2 - 1$; and so on.

From Eq. (B.5), which defines the Chebyshev polynomials, it is straightforward to obtain a recurrence formula from which $V_{N+1}(x)$ can be obtained from $V_N(x)$ and $V_{N-1}(x)$. By applying trigonometric identities to Eq. (B.5), it follows that

$$V_{N+1}(x) = 2xV_N(x) - V_{N-1}(x). \tag{B.6}$$

From Eq. (B.5) we note that $V_N^2(x)$ varies between zero and unity for $0 < x < 1$. For $x > 1$, $\cos^{-1} x$ is imaginary, so $V_N(x)$ behaves as a hyperbolic cosine and consequently increases monotonically. Referring to Eq. (B.4), $|H_a(j\Omega)|^2$ ripples between 1 and

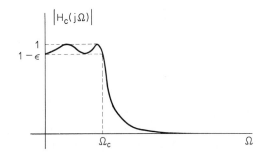

Figure B.4   Type I Chebyshev lowpass filter approximation.

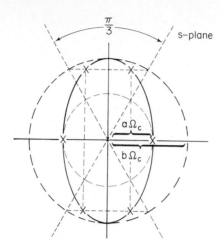

Figure B.5 Location of poles for a third-order type I lowpass Chebyshev filter.

$1/(1 + \varepsilon^2)$ for $0 \leq \Omega/\Omega_c \leq 1$ and decreases monotonically for $\Omega/\Omega_c > 1$. Three parameters are required to specify the filter: ε, Ω_c, and N. In a typical design, ε is specified by the allowable passband ripple and Ω_c is specified by the desired passband cutoff frequency. The order N is then chosen so that the stopband specifications are met.

The poles of the Chebyshev filter lie on an ellipse in the s-plane. Referring to Fig. B.5, the ellipse is defined by two circles whose diameters are equal to the minor and major axes of the ellipse. The length of the minor axis is $2a\Omega_c$, where

$$a = \tfrac{1}{2}(\alpha^{1/N} - \alpha^{-1/N}) \tag{B.7}$$

with

$$\alpha = \varepsilon^{-1} + \sqrt{1 + \varepsilon^{-2}}. \tag{B.8}$$

The length of the major axis is $2b\Omega_c$, where

$$b = \tfrac{1}{2}(\alpha^{1/N} + \alpha^{-1/N}). \tag{B.9}$$

To locate the poles of the Chebyshev filter on the ellipse, we first identify the points on the major and minor circles equally spaced in angle with a spacing of π/N in such a way that the points are symmetrically located with respect to the imaginary axis and such that a point never falls on the imaginary axis and a point occurs on the real axis for N odd but not for N even. This division of the major and minor circles corresponds exactly to the manner in which the circle is divided in locating the poles of a Butterworth filter as in Eq. (B.3). The poles of a Chebyshev filter fall on the ellipse with the ordinate specified by the points identified on the major circle and the abscissa specified by the points identified on the minor circle. In Fig. B.5 the poles are shown for $N = 3$.

A type II Chebyshev lowpass filter can be related to a type I filter through a transformation. Specifically, if in Eq. (B.4) we replace the term $\varepsilon^2 V_N^2(\Omega/\Omega_c)$ by its

reciprocal and also replace the argument of $V_N^2(\cdot)$ by its reciprocal, we ϵ

$$|H_a(j\Omega)|^2 = \frac{1}{1 + [\varepsilon^2 V_N^2(\Omega_c/\Omega)]^{-1}}.$$

This is the analytic form for the type II Chebyshev lowpass filter. One ϵ designing a type II Chebyshev filter is to first design a type I filter and th ϵ above transformation.

B.3 ELLIPTIC FILTERS

If we distribute the error uniformly across the entire passband or acros ϵ stopband as in the Chebyshev case, we are able to meet the design specifica ϵ lower-order filter than if we permit a monotonically increasing error in th ϵ as in the Butterworth case. We note that in the type I Chebyshev and ϵ approximations, the stopband error decreases monotonically with freque ϵ the possibility of further improvements if we distribute the stopband erro ϵ across the stopband. This suggests a lowpass filter approximation as in F ϵ deed, it can be shown (Papoulis, 1957) that this type of approximation (i.e. in the passband and the stopband) is the best that can be achieved for a ϵ order N, in the sense that for given values of Ω_p, δ_1, and δ_2, the band $(\Omega_s - \Omega_p)$ is as small as possible.

This class of approximations, referred to as elliptic filters, has the fo ϵ

$$|H_a(j\Omega)|^2 = \frac{1}{1 + \varepsilon^2 U_N^2(\Omega)},$$

where $U_N(\Omega)$ is a Jacobian elliptic function. To obtain equiripple error i ϵ passband and the stopband, elliptic filters must have both poles and zeros. seen from Fig. B.6, such a filter will have zeros on the $j\Omega$-axis of the discussion of elliptic filter design, even on a superficial level, is beyond the sc ϵ appendix. The reader is referred to the texts by Guillemin (1957), Storer (1 ϵ and Rader (1969), and Parks and Burrus (1987) for more detailed discuss ϵ

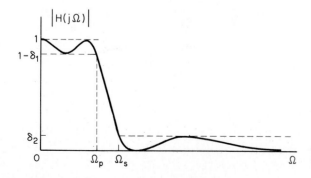

Figure B.6 Equiripple appro both passband and stopband.

Bibliography

This bibliography contains all the books and articles that are referenced in the text as well as some additional books of general interest in the study of signal processing. The information is presented in two lists. The first list, which is ordered alphabetically by authors' names, gives complete authors, title, and publication information. All references in the text can be found in this list. The second list contains all the entries in the first list sorted into categories—from general topics such as mathematical and system theory background to more specific topics that coincide with chapters or major areas covered in this book. The symbol ◆ in the topical list indicates general reference texts.

REFERENCES CITED IN THE TEXT

Ahlfors, L. V., *Complex Analysis*, 2nd ed., McGraw-Hill Book Company, New York, 1966.

Allen, J. B., and Rabiner, L. R., "A Unified Approach to Short-Time Fourier Analysis and Synthesis," *Proc. IEEE*, Vol. 65, pp. 1558–1564, Nov. 1977.

Andrews, H. C., and Hunt, B. R., *Digital Image Restoration*, Prentice-Hall, Englewood Cliffs, NJ, 1977.

Antoniou, A., *Digital Filters: Analysis and Design*, McGraw-Hill Book Company, New York, 1979.

Arsac, J., *Fourier Transforms and the Theory of Distributions*, translated by Allen Nussbaum and Gretchen C. Heim, Prentice-Hall, Englewood Cliffs, NJ, 1966.

Atal, B. S., "Automatic Recognition of Speakers from Their Voices," *Proc. IEEE*, Vol. 64, No. 4, pp. 460–475, April 1976.

Atal, B. S., and Hanauer, S. L., "Speech Analysis and Synthesis by Linear Prediction of the Speech Wave," *J. Acoustical Society of America*, Vol. 50, pp. 637–655, Aug. 1971.

Barnes, C. W., and Fam, A. T., "Minimum Norm Recursive Digital Filters That Are Free of Overflow Limit Cycles," *IEEE Trans. on Circuits and Systems*, Vol. CAS-24, pp. 569–574, Oct. 1977.

Bartlett, M. S., *An Introduction to Stochastic Processes with Special Reference to Methods and Applications*, Cambridge University Press, Cambridge, 1953.

Bauman, P. D., Lipshitz, S. P., and Vanderkooy, J., "Cepstral Analysis of Electro-acoustic Transducers," *Proc. 1985 International Conference on Acoustics, Speech, and Signal Processing*, Tampa, FL, pp. 1832–1835, Mar. 1985.

Bennett, W. R., "Spectra of Quantized Signals," *Bell System Technical J.*, Vol. 27, pp. 446–472, 1948.

Bergland, G. D., "A Fast Fourier Transform Algorithm Using Base 8 Iterations," *Mathematics of Computation*, Vol. 22, pp. 275–279, April 1968.

Berkhout, A. J., "On the Minimum Phase Criterion of Sampled Signals," *IEEE Trans. Geoscience Electronics*, Vol. GE-11, No. 4, pp. 186–198, Oct. 1973.

Birkhoff, G., and Rota, G., *Ordinary Differential Equations*, 3rd ed., John Wiley and Sons, New York, 1978.

Black, H. S., *Modulation Theory*, Van Nostrand, New York, 1953.

Blackman, R. B., *Linear Data-Smoothing and Prediction in Theory and Practice*, Addison-Wesley Publishing Company, Reading, MA, 1965.

Blackman, R. B., and Tukey, J. W., *The Measurement of Power Spectra*, Dover Publications, New York, 1958.

Blahut, R. E., *Fast Algorithms for Digital Signal Processing*, Addison-Wesley Publishing Company, Reading, MA, 1985.

Bluestein, L. I., "A Linear Filtering Approach to the Computation of Discrete Fourier Transform," *IEEE Trans. Audio Electroacoustics*, Vol. AU-18, pp. 451–455, Dec. 1970.

Bogert, B. P., Healy, M. J. R., and Tukey, J. W., "The Quefrency Alanysis of Time Series for Echoes: Cepstrum, Pseudo-autocovariance, Cross-Cepstrum, and Saphe Cracking," *Proc. Symposium Time Series Analysis*, M. Rosenblatt, Ed., John Wiley and Sons, New York, pp. 209–243, 1963.

Box, G. E. P., and Jenkins, G. M., *Time Series Analysis Forecasting and Control*, Holden-Day, San Francisco, 1970.

Bracewell, R. N., "The Discrete Hartley Transform," *J. Optical Society of America*, Vol. 73, pp. 1832–1835, Dec. 1983.

Bracewell, R. N., "The Fast Hartley Transform," *Proc. IEEE*, Vol. 72, No. 8, pp. 1010–1018, Aug. 1984.

Bracewell, R. N., *The Fourier Transform and Its Applications*, 2nd ed. revised, McGraw-Hill Book Company, New York, 1986.

Brigham, E. O., *The Fast Fourier Transform and Its Applications*, Prentice Hall, Englewood Cliffs, NJ, 1988.

Burrus, C. S., "Index Mappings for Multidimensional Formulation of the DFT and Convolution," *IEEE Trans. on Acoustics, Speech, and Signal Processing*, Vol. ASSP-25, pp. 239–242, June 1977.

Burrus, C. S., "Efficient Fourier Transform and Convolution Algorithms," in *Advanced Topics in Signal Processing*, J. S. Lim and A. V. Oppenheim, Eds., Prentice Hall, Englewood Cliffs, NJ, 1988.

Burrus, C. S., and Parks, T. W., *DFT/FFT and Convolution Algorithms Theory and Implementation*, John Wiley and Sons, New York, 1985.

Carlson, B. A., *Communications Systems*, 2nd ed., McGraw-Hill Book Company, New York, 1975.

Carslaw, H. S., *Introduction to the Theory of Fourier's Series and Integrals*, 3rd ed., Dover Publications, New York, 1952.

Castleman, K. R., *Digital Image Processing*, Prentice-Hall, Englewood Cliffs, NJ, 1979.

Cavanagh, J. F., *Digital Computer Arithmetic: Design and Implementation*, McGraw-Hill Book Company, New York, 1984.

Chan, D. S. K., and Rabiner, L., "Theory of Roundoff Noise in Cascade Realizations of Finite Impulse Response Digital Filters," *Bell System Technical J.*, Vol. 52, No. 3, pp. 329–345, Mar. 1973a.

Chan, D. S. K., and Rabiner, L., "An Algorithm for Minimizing Roundoff Noise in Cascade Realizations of Finite Impulse Response Digital Filters," *Bell System Technical J.*, Vol. 52, No. 3, pp. 347–385, Mar. 1973b.

Chan, D. S. K., and Rabiner, L., "Analysis of Quantization Errors in the Direct Form for Finite Impulse Response Digital Filters," *IEEE Trans. Audio Electroacoustics*, Vol. AU-21, pp. 354–366, Aug. 1973c.

Chen, C., *Introduction to Linear System Theory*, Holt, Rinehart and Winston, New York, 1970.

Chen, X., and Parks, T. W., "Design of FIR Filters in the Complex Domain," *IEEE Trans. on Acoustics, Speech, and Signal Processing*, Vol. ASSP-35, pp. 144–153, 1987.

Cheney, E. W., *Introduction to Approximation Theory*, McGraw-Hill Book Company, New York, 1966.

Chow, Y., and Cassignol, E., *Linear Signal Flow Graphs and Applications*, John Wiley and Sons, New York, 1962.

Christian, E., and Eisenmann, E., *Filter Design Tables and Graphs*, John Wiley and Sons, 1966.

Churchill, R. V., and Brown, J. W., *Introduction to Complex Variables and Applications*, 4th ed., McGraw-Hill Book Company, New York, 1984.

Čížek, V. "Discrete Hilbert Transform," *IEEE Trans. Audio Electroacoustics*, Vol. AU-18, No. 4, pp. 340–343, Dec. 1970.

Claasen, T. A. C. M., Mecklenbrauker, W. F. G., and Peek, J. B. H., "Second-Order Digital Filter with Only One Magnitude-Truncation Quantizer and Having Practically No Limit Cycles," *Electronics Letters*, Vol. 9, No. 2, pp. 531–532, Nov. 1973.

Clements, M. A., and Pease, J., "On Causal Linear Phase IIR Digital Filters," *IEEE Trans. on Acoustics, Speech, and Signal Processing*, Vol. ASSP-3, Mar. 1989.

Constantinides, A. G., "Spectral Transformations for Digital Filters," *Proc. IEE*, Vol. 117, No. 8, pp. 1585–1590, Aug. 1970.

Cooley, J. W., Lewis, P. A. W., and Welch, P. D., "Historical Notes on the Fast Fourier Transform," *IEEE Trans. Audio Electroacoustics*, Vol. AU-15, pp. 76–79, June 1967.

Cooley, J. W., and Tukey, J. W., "An Algorithm for the Machine Computation of Complex Fourier Series," *Mathematics of Computation*, Vol. 19, pp. 297–301, Apr. 1965.

Cooper, G. R., and McGillem, C. D., *Methods of Signal and System Analysis*, Holt, Rinehart and Winston, New York, 1967.

Crochiere, R. E., "Digital Ladder Structures and Coefficient Sensitivity," *IEEE Trans. Audio Electroacoustics*, Vol. AU-20, pp. 240–246, Oct. 1972.

Crochiere, R. E., and Oppenheim, A. V., "Analysis of Linear Digital Networks," *Proc. IEEE*, Vol. 63, pp. 581–595, April 1975.

Crochiere, R. E., and Rabiner, L. R., *Multirate Digital Signal Processing*, Prentice-Hall, Englewood Cliffs, NJ, 1983.

Cruz, J. B., and Van Valkenburg, M. E., *Signals in Linear Circuits*, Houghton Mifflin Company, Boston, 1974.

Daniels, R. W., *Approximation Methods for Electronic Filter Design*, McGraw-Hill Book Company, New York, 1974.

Danielson, G. C., and Lanczos, C., "Some Improvements in Practical Fourier Analysis and Their Application to X-Ray Scattering from Liquids," *J. Franklin Inst.*, Vol. 233, pp. 365–380, 435–452, April and May 1942.

Davenport, W. B., *Probability and Random Processes: An Introduction for Applied Scientists and Engineers*, McGraw-Hill Book Company, New York, 1970.

Davis, S. B., and Mermelstein, P., "Comparison of Parametric Representations for Monosyllabic Word Recognition in Continuously Spoken Sentence," *IEEE Trans. on Acoustics, Speech, and Signal Processing*, Vol. ASSP-28, No. 4, pp. 357–366, Aug. 1980.

Deczky, A. G., "Synthesis of Recursive Digital Filters Using the Minimum P Error Criterion," *IEEE Trans. Audio Electroacoustics*, Vol. AU-20, pp. 257–263, Oct. 1972.

Drake, A. W., *Fundamentals of Applied Probability Theory*, McGraw-Hill Book Company, New York, 1967.

DSP Committee, IEEE ASSP, Eds., *Selected Papers in Digital Signal Processing II*, IEEE Press, New York, 1976.

DSP Committee, IEEE ASSP, Eds., *Programs for Digital Signal Processing*, IEEE Press, New York, 1979.

Dudgeon, D. E., and Mersereau, R. M., *Two-Dimensional Digital Signal Processing*, Prentice-Hall, Englewood Cliffs, NJ, 1983.

Duhamel, P., "Implementation of 'Split-Radix' FFT Algorithms for Complex, Real, and Real-Symmetric Data," *IEEE Trans. on Acoustics, Speech, and Signal Processing*, Vol. ASSP-34, pp. 285–295, April 1986.

Duhamel, P., and Hollmann, H., "Split Radix FFT Algorithm," *Electronics Letters*, Vol. 20, pp. 14–16, Jan. 5, 1984.

Dym, H., and McKean, H. P., *Fourier Series and Integrals*, Academic Press, New York, 1972.

Ebert, P. M., Mazo, J. E., and Taylor, M. C., "Overflow Oscillations in Digital Filters," *Bell System Technical J.*, Vol. 48, pp. 2999–3020, 1969.

Feller, W., *An Introduction to Probability Theory and Its Applications*, 3rd ed., Vol. 1, 1968, Vol. 2, 1966, John Wiley and Sons, New York.

Fettweis, A., "Wave Digital Filters: Theory and Practice," *Proc. IEEE*, Vol. 74, No. 2, pp. 270–327, Feb. 1986.

Flanagan, J. L., *Speech Analysis, Synthesis and Perception*, 2nd ed., Springer-Verlag, New York, 1972.

Flanagan, J. L., Coker, C. H., Rabiner, L. R., Schafer, R. W., Umeda, N., "Synthetic Voices for Computers," *IEEE Spectrum*, Vol. 7, pp. 22–45, 1970.

Fletcher, R., and Powell, M. J. D., "A Rapidly Convergent Descent Method for Minimization," *Computer J.*, Vol. 6, No. 2, pp. 163–168, 1963.

Freeman, H., *Discrete-Time Systems*, John Wiley and Sons, New York, 1965.

Furui, S., "Cepstral Analysis Technique for Automatic Speaker Verification," *IEEE Trans. on Acoustics, Speech, and Signal Processing*, Vol. ASSP-29, No. 2, pp. 254–272, April 1981.

Gabel, R. A., and Roberts, R. A., *Signals and Linear Systems*, John Wiley and Sons, New York, 1973.

Gardner, W. A., *Statistical Spectral Analysis: A Nonprobabilistic Theory*, Prentice Hall, Englewood Cliffs, NJ, 1988.

Gel'fand, I. M., et al., *Generalized Functions*, 5 vols., translated by E. Saletan et al., Academic Press, New York, 1964–1968.

Gentleman, W. M., and Sande, G., "Fast Fourier Transforms for Fun and Profit," in *Proc. 1966 Fall Joint Computer Conf.*, AFIPS Conf. Proc., Vol. 29, pp. 563–578, Spartan Books, Washington, DC, 1966.

Goertzel, G., "An Algorithm for the Evaluation of Finite Trigonometric Series," *American Math. Monthly*, Vol. 65, pp. 34–35, Jan. 1958.

Gold, B., and Rader, C. M., *Digital Processing of Signals*, McGraw-Hill Book Company, New York, 1969.

Gold, B., Oppenheim, A. V., and Rader, C. M., "Theory and Implementation of the Discrete Hilbert Transform," *Proc. Symp. Computer Processing in Communications*, Vol. 19, Polytechnic Press, New York, 1970.

Golden, R. M., and Kaiser, J. F., "Design of Wideband Sampled-Data Filters," *Bell System Technical J.*, Vol. 43, No. 4, Pt. 2, pp. 1533–1545, July 1964.

Good, I. J., "The Interaction Algorithm and Practical Fourier Analysis," *J. Royal Stat. Soc.*, Vol. B-20, pp. 361–372, 1958.

Gray, A. H., and Markel, J. D., "Digital Lattice and Ladder Filter Synthesis," *IEEE Trans. Acoustics, Speech, and Signal Processing*, Vol. ASSP-21, pp. 491–500, Dec. 1973.

Gray, A. H., and Markel, J. D., "A Normalized Digital Filter Structure," *IEEE Trans. on Acoustics, Speech, and Signal Processing*, Vol. ASSP-23, pp. 258–277, June 1975.

Gray, A. H., and Markel, J. D., "A Computer Program for Designing Digital Elliptic Filters," *IEEE Trans. on Acoustics, Speech, and Signal Processing*, Vol. ASSP-24, pp. 529–538, Dec. 1976.

Grenander, U., and Rosenblatt, M., *Statistical Analysis of Stationary Time Series*, John Wiley and Sons, New York, 1957.

Griffin, D., and Lim, J. S., "Signal Estimation from Modified Short-Time Fourier Transform," *IEEE Trans. on Acoustics, Speech, and Signal Processing*, Vol. ASSP-32, pp. 236–243, April 1984.

Grossman, S. I., *Calculus Part 2*, Academic Press, New York, 1982.

Guillemin, E. A., *Synthesis of Passive Networks*, John Wiley and Sons, New York, 1957.

Guillemin, E. A., *Theory of Linear Physical Systems*, John Wiley and Sons, New York, 1963.

Hamming, R. W., *Digital Filters*, 3rd ed., Prentice Hall, Englewood Cliffs, NJ, 1988.

Hannan, E. J., *Time Series Analysis*, Methuen and Company, London, 1960.

Harris, F. J., "On the Use of Windows for Harmonic Analysis with the Discrete Fourier Transform," *Proc. IEEE*, Vol. 66, pp. 51–83, Jan. 1978.

Hayes, M. H., Lim, J. S., and Oppenheim, A. V., "Signal Reconstruction from Phase and Magnitude," *IEEE Trans. on Acoustics, Speech, and Signal Processing*, Vol. ASSP-28, No. 6, pp. 672–680, Dec. 1980.

Haykin, S. S., *Communication Systems*, John Wiley and Sons, New York, 1978.

Haykin, S. S., *Adaptive Filter Theory*, Prentice Hall, Englewood Cliffs, NJ, 1986.

Heideman, M. T., Johnson, D. H., and Burrus, C. S., "Gauss and the History of the Fast Fourier Transform," *IEEE ASSP Magazine*, Vol. 1, No. 4, pp. 14–21, Oct, 1984.

Helms, H. D., "Fast Fourier Transform Method of Computing Difference Equations and Simulating Filters," *IEEE Trans. Audio Electroacoustics*, Vol. 15, No. 2, pp. 85–90, 1967.

Herrmann, O., "On the Design of Nonrecursive Digital Filters with Linear Phase," *Elec. Lett.*, Vol. 6, No. 11, pp. 328–329, 1970.

Herrmann, O., Rabiner, L. R., and Chan, D. S. K., "Practical Design Rules for Optimum Finite Impulse Response Lowpass Digital Filters," *Bell System Technical J.*, Vol. 52, No. 6, pp. 769–799, July–Aug. 1973.

Herrmann, O., and Schüssler, W., "Design of Nonrecursive Digital Filters with Minimum Phase," *Elec. Lett.*, Vol. 6, No. 11, pp. 329–330, 1970a.

Herrmann, O., and Schüssler, W., "On the Accuracy Problem in the Design of Nonrecursive Digital Filters," *Arch. Electronik Uber tragungstechnik*, Vol. 24, pp. 525–526, 1970b.

Hewes, C. R., Broderson, R. W., and Buss, D. D., "Applications of CCD and Switched Capacitor Filter Technology," *Proc. IEEE*, Vol. 67, No. 10, pp. 1403–1415, Oct. 1979.

Hildebrand, F. B., *Finite Difference Equations and Simulation*, Prentice-Hall, Englewood Cliffs, NJ, 1968.

Hildebrand, F. B., *Introduction to Numerical Analysis*, 2nd ed., McGraw-Hill Book Company, New York, 1973.

Hildebrand, F. B., *Advanced Calculus with Applications*, 2nd ed., Prentice-Hall, Englewood Cliffs, NJ, 1976.

Hnatek, E. R., *A User's Handbook of D/A and A/D Conversions*, John Wiley and Sons, New York, 1976.

Hofstetter, E., Oppenheim, A. V., and Siegel, J., "On Optimum Nonrecursive Digital Filters," *Proc. 9th Allerton Conf. Circuit System Theory*, Oct. 1971.

Huelsman, L. P., and Allen, P. E., *Introduction to the Theory and Design of Active Filters*, McGraw-Hill Book Company, New York, 1980.

Hwang, S. Y., "On Optimization of Cascade Fixed Point Digital Filters," *IEEE Trans. on Circuits and Systems*, Vol. CAS-21, No. 1, pp. 163–166, Jan. 1974.

Jackson, L., "An Analysis of Limit Cycles Due to Multiplicative Rounding in Recursive Digital Filters," *Proc. 7th Allerton Conf. Circuit System Theory*, pp. 69–78, 1969.

Jackson, L. B., "On the Interaction of Roundoff Noise and Dynamic Range in Digital Filters," *Bell System Technical J.*, Vol. 49, pp. 159–184, Feb. 1970a.

Jackson, L. B., "Roundoff-Noise Analysis for Fixed-Point Digital Filters Realized in Cascade or Parallel Form," *IEEE Trans. Audio Electroacoustics*, Vol. AU-18, pp. 107–122, June 1970b.

Jackson, L. B., *Digital Filters and Signal Processing*, Kluwer Academic Publishers, Hingham, MA, 1986.

Jayant, N. S., and Noll, P., *Digital Coding of Waveforms*, Prentice Hall, Englewood Cliffs, NJ, 1984.

Jenkins, G. M., and Watts, D. G., *Spectral Analysis and Its Applications*, Holden-Day, San Francisco, 1968.

Johnson, D. E., *Introduction to Filter Theory*, Prentice-Hall, Englewood Cliffs, NJ, 1976.

Johnson, D. E., Johnson, J. R., and Moore, H. P. *A Handbook of Active Filters*, Prentice-Hall, Englewood Cliffs, NJ, 1980.

Jolley, L. B. W., *Summation of Series*, Dover, New York, 1961.

Juang, B.-H., Rabiner, L. R., and Wilpon, J. G., "On the Use of Bandpass Liftering in Speech Recognition," *IEEE Trans. on Acoustics, Speech, and Signal Processing*, Vol. ASSP-35, No. 7, pp. 947–954, July 1987.

Jury, E. I., *Sampled-Data Control Systems*, John Wiley and Sons, New York, 1958.

Jury, E. I., *Theory and Application of the z-Transform Method*, John Wiley and Sons, New York, 1964.

Kailath, T., *Linear Systems*, Prentice-Hall, Englewood Cliffs, NJ, 1980.

Kaiser, J. F., "Digital Filters," Chap. 7 in *System Analysis by Digital Computer*, F. F. Kuo and J. F. Kaiser, Eds., John Wiley and Sons, New York, 1966.

Kaiser, J. F., "Nonrecursive Digital Filter Design Using the I_0-Sinh Window Function," *Proc. 1974 IEEE International Symp. on Circuits and Systems*, San Francisco, pp. 20–23, April 1974.

Kaiser, J. F., and Hamming, R. W., "Sharpening the Response of a Symmetric Nonrecursive Filter by Multiple Use of the Same Filter," *IEEE Trans. on Acoustics, Speech, and Signal Processing*, Vol. ASSP-25, No. 5, pp. 415–422, Oct. 1977.

Kaiser, J. F., and Schafer, R. W., "On the Use of the I_0-Sinh Window for Spectrum Analysis," *IEEE Trans. on Acoustics, Speech, and Signal Processing*, Vol. ASSP-28, No. 1, pp. 105–107, Feb. 1980.

Kan, E. P. F., and Aggarwal, J. K., "Error Analysis of Digital Filters Employing Floating Point Arithmetic," *IEEE Trans. Circuit Theory*, Vol. CT-18, pp. 678–686, Nov. 1971.

Kaneko, T., and Liu, B., "Accumulation of Roundoff Error in Fast Fourier Transforms," *J. Assoc. Comput. Mach.*, Vol. 17, pp. 637–654, Oct. 1970.

Kay, S. M., *Modern Spectral Estimation Theory and Application*, Prentice Hall, Englewood Cliffs, NJ, 1988.

Kay, S. M., and Marple, S. L., "Spectrum Analysis: A Modern Perspective," *Proc. IEEE*, Vol. 69, pp. 1380–1419, Nov. 1981.

Kelly, J. L., Jr., and Lochbaum, C., "Speech Synthesis," *Proc. Stockholm Speech Communications Seminar*, R.I.T., Stockholm, Sweden, Sept. 1962.

Knowles, J. B., and Olcayto, E. M., "Coefficient Accuracy and Digital Filter Response," *IEEE Trans. Circuit Theory*, Vol. CT-15, pp. 31–41, Mar. 1968.

Knuth, D. E., *The Art of Computer Programming, Seminumerical Algorithms*, 2nd ed., Addison-Wesley Publishing Co., Reading, MA, 1981.

Kolba, D. P., and Parks, T. W., "A Prime Factor FFT Algorithm Using High Speed Convolution," *IEEE Trans. on Acoustics, Speech, and Signal Processing*, Vol. ASSP-25, pp. 281–294, Aug. 1977.

Koopmanns, L. H., *Spectral Analysis of Time Series*, Academic Press, New York, 1974.

Lam, H. Y.-F., *Analog and Digital Filters: Design and Realization*, Prentice-Hall, Englewood Cliffs, NJ, 1979.

Lathi, B. P., *Signals, Systems and Communication*, John Wiley and Sons, New York, 1965.

Lighthill, M. J., *Introduction to Fourier Analysis and Generalized Functions*, Cambridge University Press, Cambridge, 1958.

Lim, J. S., *Two-Dimensional Digital Signal Processing*, Prentice Hall, Englewood Cliffs, NJ, 1989.

Lim, J. S., and Oppenheim, A. V., Eds., *Advanced Topics in Signal Processing*, Prentice Hall, Englewood Cliffs, NJ, 1988.

Liu, B., "Effect of Finite Word Length on the Accuracy of Digital Filters: A Review," *IEEE Trans. Circuit Theory*, Vol. CT-18, pp. 670–677, Nov. 1971.

Liu, B., and Kaneko, T., "Error Analysis of Digital Filters Realized in Floating-Point Arithmetic," *Proc. IEEE*, Vol. 57, pp. 1735–1747, Oct. 1969.

Liu, C. L., and Liu, J. W. S., *Linear System Analysis*, McGraw-Hill Book Company, New York, 1975.

Liu, B., and Peled, A., "Heuristic Optimization of the Cascade Realization of Fixed Point Digital Filters," *IEEE Trans. on Acoustics, Speech, and Signal Processing*, Vol. ASSP-23, pp. 464–473, 1975.

Macovski, A., *Medical Image Processing*, Prentice-Hall, Englewood Cliffs, NJ, 1983.

Makhoul, J., "Linear Prediction: A Tutorial Review," *Proc. IEEE*, Vol. 62, pp. 561–580, April 1975.

Markel, J. D., "FFT Pruning," *IEEE Trans. Audio and Electroacoustics*, Vol. AU-19, pp. 305–311, Dec. 1971.

Markel, J. D., and Gray, A. H., Jr., *Linear Prediction of Speech*, Springer-Verlag, New York, 1976.

Marple, S. L., *Digital Spectral Analysis with Applications*, Prentice Hall, Englewood Cliffs, NJ, 1987.

Martinez, H. G., and Parks, T. W., "Design of Recursive Digital Filters with Optimum Magnitude and Attenuation Poles on the Unit Circle," *IEEE Trans. on Acoustics, Speech, and Signal Processing*, Vol. ASSP-26, pp. 150–156, 1978.

Mason, S., and Zimmermann, H. J., *Electronic Circuits, Signals and Systems*, John Wiley and Sons, New York, 1960.

McClellan, J. H., and Parks, T. W., "A Unified Approach to the Design of Optimum FIR Linear Phase Digital Filters," *IEEE Trans. Circuit Theory*, Vol. CT-20, pp. 697–701, November 1973.

McClellan, J. H., Parks, T. W., and Rabiner, L. R., "A Computer Program for Designing Optimum FIR Linear Phase Digital Filters," *IEEE Trans. Audio Electroacoustics*, Vol. AU-21, pp. 506–526, Dec. 1973.

McClellan, J. H., and Rader, C. M., *Number Theory in Digital Signal Processing*, Prentice-Hall, Englewood Cliffs, NJ, 1979.

McGillem, C. D., and Cooper, G. R., *Continuous and Discrete Signal and System Analysis*, Holt, Rinehart and Winston, New York, 1974.

Mersereau, R. M., Schafer, R. W., Barnwell, T. P., and Smith, D. L., "A Digital Filter Design Package for PCs and TMS320s," *Proc. MIDCON*, Dallas, TX, 1984.

Mills, W. L., Mullis, C. T., and Roberts, R. A., "Digital Filter Realizations Without Overflow Oscillations," *IEEE Trans. on Acoustics, Speech, and Signal Processing*, Vol. ASSP-26, pp. 334–338, Aug. 1978.

Morse, P. M., and Feshbach, H., *Methods of Theoretical Physics*, McGraw-Hill Book Company, New York, 1953.

Mullis, C. T., and Roberts, R. A., "Synthesis of Minimum Roundoff Noise Fixed Point Digital Filters," *IEEE Trans. Circuits and Systems*, Vol. CAS-23, pp. 551–562, Sept. 1976.

Nawab, S. N., and Quatieri, T. F., "Short-Time Fourier Transform," in *Advanced Topics in Signal Processing*, J. S. Lim and A. V. Oppenheim, Eds., Prentice Hall, Englewood Cliffs, NJ, 1988.

Noll, A. M., "Cepstrum Pitch Determination," *J. Acoustical Society of America*, Vol. 41, pp. 293–309, Feb. 1967.

Nyquist, H., "Certain Topics in Telegraph Transmission Theory," *AIEE Trans.*, pp. 617–644, 1928.

Oetken, G., Parks, T. W., and Schüssler, H. W., "New Results in the Design of Digital Interpolators," *IEEE Trans. on Acoustics, Speech, and Signal Processing*, Vol. ASSP-23, pp. 301–309, June 1975.

Oppenheim, A. V., "Generalized Superposition," *Information Control*, Vol. 11, Nos. 5-6, pp. 528–536, Nov.-Dec. 1967.

Oppenheim, A. V., "Generalized Linear Filtering," Chap. 8 in *Digital Processing of Signals*, B. Gold and C. M. Rader, Eds., McGraw-Hill Book Company, New York, 1969a.

Oppenheim, A. V., "A Speech Analysis-Synthesis System Based on Homomorphic Filtering," *J. Acoustical Society of America*, Vol. 45, pp. 458–465, Feb. 1969b.

Oppenheim, A. V., "Realization of Digital Filters Using Block-Floating-Point Arithmetic," *IEEE Trans. Audio Electroacoustics*, Vol. AU-18, pp. 130–136, Jan. 1970.

Oppenheim, A. V., Ed., *Applications of Digital Signal Processing*, Prentice-Hall, Englewood Cliffs, NJ, 1978.

Oppenheim, A. V., and Johnson, D. H., "Discrete Representation of Signals," *Proc. IEEE*, Vol. 60, No. 6, pp. 681–691, June 1972.

Oppenheim, A. V., and Schafer, R. W., "Homomorphic Analysis of Speech," *IEEE Trans. Audio Electroacoustics*, Vol. AU-16, No. 2, pp. 221–226, June 1968.

Oppenheim, A. V., and Schafer, R. W., *Digital Signal Processing*, Prentice-Hall, Englewood Cliffs, NJ, 1975.

Oppenheim, A. V., Schafer, R. W., and Stockam T. G., Jr., "Nonlinear Filtering of Multiplied and Convolved Signals," *Proc. IEEE*, Vol. 56, No. 8, pp. 1264–1291, Aug. 1968.

Oppenheim, A. V., and Weinstein, C. J., "Effects of Finite Register Length in Digital Filtering and the Fast Fourier Transform," *Proc. IEEE*, pp. 957–976, Aug. 1972.

Oppenheim, A. V., and Willsky, A. S., with I. T. Young, *Signals and Systems*, Prentice-Hall, Englewood Cliffs, NJ, 1983.

O'Shaughnessy, D., *Speech Communication, Human and Machine*, Addison-Wesley Publishing Company, Reading, MA, 1987.

Papoulis, A., "On the Approximation Problem in Filter Design," *IRE Conv. Record*, Pt. 2, pp. 175–185, 1957.

Papoulis, A., *The Fourier Integral and Its Applications*, McGraw-Hill Book Company New York, 1962.

Papoulis, A., *Systems and Transforms with Applications in Optics*, McGraw-Hill Book Company, New York, 1968.

Papoulis, A., *Signal Analysis*, McGraw-Hill Book Company, New York, 1977.

Papoulis, A., *Circuits and Systems*, Holt, Rinehart and Winston, New York, 1980.

Papoulis, A., *Probability, Random Variables and Stochastic Processes*, 2nd ed., McGraw-Hill Book Company, New York, 1984.

Parks, T. W., and Burrus, C. S., *Digital Filter Design*, John Wiley, New York, 1987.

Parks, T. W., and McClellan, J. H., "Chebyshev Approximation for Nonrecursive Digital Filters with Linear Phase," *IEEE Trans. Circuit Theory*, Vol. CT-19, pp. 189–194, Mar. 1972a.

Parks, T. W., and McClellan, J. H., "A Program for the Design of Linear Phase Finite Impulse Response Filters," *IEEE Trans. Audio Electroacoustics*, Vol. AU-20, No. 3, pp. 195–199, Aug. 1972b.

Parsons, T. *Voice and Speech Processing*, McGraw-Hill Book Company, New York, 1986.

Parzen, E., *Modern Probability Theory and Its Applications*, John Wiley and Sons, New York, 1960.

Parzen, E., *Stochastic Processes*, Holden-Day, San Francisco, 1962.

Peled, A., and Liu, B., *Digital Signal Processing: Theory, Design and Implementation*, John Wiley and Sons, New York, 1976.

Phillips, C. L., and Nagle, H. T., Jr., *Digital Control System Analysis and Design*, Prentice Hall, Englewood Cliffs, NJ, 1984.

Pratt, W., *Digital Image Processing*, John Wiley and Sons, New York, 1978.

Press, W. H., Flannery, B. P., Teukolsky, S. A., and Vetterling, W. T., *Numerical Recipes*, Cambridge University Press, Cambridge, 1986.

Press, W. H., Flannery, B. P., Teukolsky, S. A., and Vetterling, W. T., *Numerical Recipes in C*, Cambridge University Press, Cambridge, 1988.

Proakis, J. G., and Manolakis, D. G., *Introduction to Digital Signal Processing*, Macmillan Publishing Company, New York, 1988.

Rabiner, L. R., "The Design of Finite Impulse Response Digital Filters Using Linear Programming Techniques," *Bell System Technical J.*, pp. 1177–1198, July–Aug. 1972a.

Rabiner, L. R., "Linear Program Design of Finite Impulse Response (FIR) Digital Filters," *IEEE Trans. on Audio and Electroacoustics*, Vol. AU-20, No. 4, pp. 280–288, Oct. 1972b.

Rabiner, L. R., and Gold, B., *Theory and Application of Digital Signal Processing*, Prentice-Hall, Englewood Cliffs, NJ, 1975.

Rabiner, L. R., Graham, N.Y., and Helms, H. D., "Linear Programming Design of IIR Digital Filters with Arbitrary Magnitude," *IEEE Trans. on Acoustics, Speech, and Signal Processing*, Vol. ASSP-22, pp. 117–123, April 1974.

Rabiner, L. R., Kaiser, J. F., Herrmann, O., and Dolan, M. T., "Some Comparisons Between FIR and IIR Digital Filters," *Bell System Technical J.*, Vol. 53, No. 2, pp. 305–331, Feb. 1974.

Rabiner, L. R., McClellan, J. H., and Parks, T. W., "FIR Digital Filter Design Techniques Using Weighted Chebyshev Approximation," *Proc. IEEE*, Vol. 63, pp. 595–610, April 1975.

Rabiner, L. R., and Rader, C. M., Eds., *Digital Signal Processing*, IEEE Press, New York, 1972.

Rabiner, L. R., and Schafer, R. W., "On the Behavior of Minimax FIR Digital Hilbert Transformers," *Bell System Technical J.*, Vol. 53, No. 2, pp. 361–388, Feb. 1974.

Rabiner, L. R., and Schafer, R. W., *Digital Processing of Speech Signals*, Prentice-Hall, Englewood Cliffs, NJ, 1978.

Rabiner, L. R., Schafer, R. W., and Rader, C. M., "The Chirp z-Transform Algorithm," *IEEE Trans. Audio Electroacoustics*, Vol. AU-17, pp. 86–92, June 1969.

Rader, C. M., "Discrete Fourier Transforms When the Number of Data Samples Is Prime," *Proc. IEEE*, Vol. 56, pp. 1107–1108, June 1968.

Rader, C. M., "An Improved Algorithm for High-Speed Autocorrelation with Applications to Spectral Estimation," *IEEE Trans. Audio Electroacoustics*, Vol. AU-18, pp. 439–441, Dec. 1970.

Rader, C. M., and Brenner, N. M., "A New Principle for Fast Fourier Transformation," *IEEE Trans. on Acoustics, Speech, and Signal Processing*, Vol. ASSP-25, pp. 264–265, June 1976.

Rader, C. M., and Gold, B., "Digital Filter Design Techniques in the Frequency Domain," *Proc. IEEE*, Vol. 55, pp. 149–171, Feb. 1967.

Ragazzini, J. R., and Franklin, G. F., *Sampled Data Control Systems*, McGraw-Hill Book Company, New York, 1958.

Rainville, E. D., *The Laplace Transform: An Introduction*, Macmillan Publishing Company, New York, 1963.

Rao, S. K., and Kailath, T., "Orthogonal Digital Filters for VLSI Implementation," *IEEE Trans. Circuits and Systems*, Vol. CAS-31, No. 11, pp. 933–945, Nov. 1984.

Reut, Z., Pace, N. G., and Heaton, M. J. P., "Computer Classification of Sea Beds by Sonar," *Nature*, Vol. 314, pp. 426–428, Apr. 4, 1985.

Richards, M. A., "Applications of Deczky's Program for Recursive Filter Design to the Design of Recursive Decimators," *IEEE Trans. on Acoustics, Speech, and Signal Processing*, Vol. ASSP-30, pp. 811–814, Oct. 1982.

Roberts, R. A., and Mullis, C. T., *Digital Signal Processing*, Addison-Wesley Publishing Company, Reading, MA, 1987.

Robinson, E. A., and Durrani, T. S., *Geophysical Signal Processing*, Prentice Hall, Englewood Cliffs, NJ, 1985.

Robinson, E. A., and Treitel, S., *Geophysical Signal Analysis*, Prentice-Hall, Englewood Cliffs, NJ, 1980.

Runge, C., "Über die Zerlegung Empirisch Gegebener Periodischer Functionen in Sinuswellen," *Z. Math. Physik*, Vol. 48, pp. 443–456, 1903; Vol. 53, pp. 117–123, 1905.

Sandberg, I. W., "Floating-Point-Roundoff Accumulation in Digital Filter Realization," *Bell System Technical J.*, Vol. 46, pp. 1775–1791, Oct. 1967.

Schaefer, R. T., Schafer, R. W., and Mersereau, R. M., "Digital Signal Processing for Doppler Radar Signals," *Proc. 1979 IEEE Int. Conf. on Acoustics, Speech, and Signal Processing*, pp. 170–173, 1979.

Schafer, R. W., and Rabiner, L. R., "System for Automatic Formant Analysis of Voiced Speech," *J. Acoustical Society of America*, Vol. 47, No. 2, Pt. 2, pp. 634–648, Feb. 1970.

Schafer, R. W., and Rabiner, L. R., "A Digital Signal Processing Approach to Interpolation," *Proc. IEEE*, Vol. 61, pp. 692–702, June 1973.

Schmid, H., *Electronic Analog/Digital Conversions*, John Wiley and Sons, New York, 1976.

Schroeder, M. R., "Direct (Nonrecursive) Relations Between Cepstrum and Predictor Coefficients," *IEEE Trans. on Acoustics, Speech, and Signal Processing*, Vol. ASSP-29, No. 2, pp. 297–301, April 1981.

Schüssler, H. W., and Steffen, P., "Some Advanced Topics in Filter Design," in *Advanced Topics in Signal Processing*, J. S. Lim and A. V. Oppenheim, Eds., Prentice Hall, Englewood Cliffs, NJ, 1988.

Schwartz, M., *Information Transmission, Modulation and Noise: A Unified Approach to Communications Systems*, 3rd ed., McGraw-Hill Book Company, New York, 1980.

Senmoto, S., and Childers, D. G., "Adaptive Decomposition of a Composite Signal of Identical Unknown Wavelets in Noise," *IEEE Trans. on Systems, Man, and Cybernetics*, Vol. SMC-2, No. 1, p. 59, Jan. 1972.

Shannon, C. E., "Communication in the Presence of Noise," *Proc. IRE*, pp. 10–12, Jan. 1949.

Siebert, W. M., *Circuits, Signals, and Systems*, McGraw-Hill Book Company, New York, 1986.

Singleton, R. C., "A Method for Computing the Fast Fourier Transform with Auxiliary Memory and Limited High-Speed Storage," *IEEE Trans. Audio Electroacoustics*, Vol. AU-15, pp. 91–97, June 1967.

Singleton, R. C., "An Algorithm for Computing the Mixed Radix Fast Fourier Transform," *IEEE Trans. Audio Electroacoustics*, Vol. AU-17, pp. 93–103, June 1969.

Skolnik, M. I., *Introduction to Radar Systems*, 2nd ed., McGraw-Hill Book Company, New York, 1980.

Slepian, D., Landau, H. T., and Pollack, H. O., "Prolate Spheroidal Wave Functions, Fourier Analysis, and Uncertainty Principle (I and II)," *Bell System Technical J.*, Vol. 40, No. 1, pp. 43–80, Jan. 1961.

Smith, M. J. T., and Barnwell, T. P. III, "Exact Reconstruction Techniques for Tree-Structured Subband Coders," *IEEE Trans. on Acoustics, Speech, and Signal Processing*, Vol. ASSP-34, pp. 434–441, June 1986.

Stearns, S. D., and David, R. A., *Signal Processing Algorithms*, Prentice Hall, Englewood Cliffs, NJ, 1988.

Steiglitz, K., "The Equivalence of Analog and Digital Signal Processing," *Information and Control*, Vol. 8, No. 5, pp. 455–467, Oct. 1965.

Steiglitz, K., "Computer-Aided Design of Recursive Digital Filters," *IEEE Trans. Audio Electroacoustics*, Vol. AU-18, June 1970.

Steiglitz, K., and Dickenson, B., "Computation of the Complex Cepstrum by Factorization of the z-Transform," *Proc. 1977 IEEE Int. Conf. on Acoustics, Speech, and Signal Processing*, Hartford, CT, May 1977.

Stockham, T. G., Jr., "High Speed Convolution and Correlation," *1966 Spring Joint Computer Conference, AFIPS Proc.*, Vol. 28, pp. 229–233, 1966.

Stockham, T. G., Jr., Cannon, T. M., and Ingebretsen, R. B., "Blind Deconvolution Through Digital Signal Processing" *Proc. IEEE*, Vol. 63, pp. 678–692, April 1975.

Storer, J. E., *Passive Network Synthesis*, McGraw-Hill Book Company, New York, 1957.

Strum, R. D., and Kirk, D. E., *First Principles of Discrete Systems and Digital Signal Processing*, Addison-Wesley Publishing Company, Reading, MA, 1988.

Thomas, G. B., Jr., and Finney, R. L., *Calculus and Analytic Geometry*, 6th ed., Addison-Wesley Publishing Company, Reading, MA, 1984.

Thomas, J. B., *An Introduction to Statistical Communication Theory*, John Wiley and Sons, New York, 1969.

Tretter, S. A., *Introduction to Discrete-Time Signal Processing*, John Wiley and Sons, New York, 1976.

Tribolet, J. M., "A New Phase Unwrapping Algorithm," *IEEE Trans. on Acoustics, Speech, and Signal Processing*, Vol. ASSP-25, No. 2, pp. 170–177, April 1977.

Tribolet, J. M., *Seismic Applications of Homomorphic Signal Processing*, Prentice-Hall, Englewood Cliffs, NJ, 1979.

Tukey, J. W., *Exploratory Data Analysis*, Addison-Wesley Publishing Company, Reading, MA, 1977.

Ulrych, T. J., "Application of Homomorphic Deconvolution to Seismology," *Geophys.*, Vol. 36, No. 4, pp. 650–660, Aug. 1971.

Vaidyanathan, P. P., "A Unified Approach to Orthogonal Digital Filters and Wave Digital Filters, Based on LBR Two-Pair Extraction," *IEEE Trans. Circuits and Systems*, Vol. CAS-32, pp. 673–686, July 1985.

Verhelst, W., and Steenhaut, O., "A New Model for the Short-Time Complex Cepstrum of Voiced Speech," *IEEE Trans. on Acoustics, Speech, and Signal Processing*, Vol. ASSP-34, No. 1, pp. 43–51, Feb. 1986.

Vernet, J. L., "Real Signals Fast Fourier Transform: Storage Capacity and Step Number Reduction by Means of an Odd Discrete Fourier Transform," *Proc. IEEE*, pp. 1531–1532, Oct. 1971.

Volder, J. E., "The CORDIC Trigonometric Computing Technique," *IRE Trans. Electronic Computers*, Vol. EC-8, pp. 330–334, Sept. 1959.

Weinberg, L., *Network Analysis and Synthesis*, R. E. Kreiger, Huntington, NY, 1975.

Weinstein, C. J., "Roundoff Noise in Floating Point Fast Fourier Transform Computation," *IEEE Trans. Audio Electroacoustics*, Vol. AU-17, pp. 209–215, Sept. 1969.

Weinstein, C. J., and Oppenheim, A. V., "A Comparison of Roundoff Noise in Floating Point and Fixed Point Digital Filter Realizations," *Proc. IEEE*, Vol. 57, pp. 1181–1183, June 1969.

Welch, P. D., "A Fixed-Point Fast Fourier Transform Error Analysis," *IEEE Trans. Audio Electroacoustics*, Vol. AU-17, pp. 153–157, June 1969.

Welch, P. D., "The Use of the Fast Fourier Transform for the Estimation of Power Spectra," *IEEE Trans. Audio Electroacoustics*, Vol. AU-15, pp. 70–73, June 1970.

Widrow, B., "A Study of Rough Amplitude Quantization by Means of Nyquist Sampling Theory," *IRE Trans. Circuit Theory*, Vol. CT-3, pp. 266–276, Dec. 1956.

Widrow, B., "Statistical Analysis of Amplitude-Quantized Sampled-Data Systems," *AIEE Trans. (Applications and Industry)*, Vol. 81, pp. 555–568, Jan. 1961.

Winograd, S., "On Computing the Discrete Fourier Transform," *Mathematics of Computation*, Vol. 32, No. 141, pp. 175–199, Jan. 1978.

Wozencraft, J. M., and Jacobs, I. M., *Principles of Communication Engineering*, John Wiley and Sons, New York, 1965.

Zemanian, A. H., *Distribution Theory and Transform Analysis: An Introduction to Generalized Functions, with Applications*, McGraw-Hill Book Company, New York, 1965.

Zverev, A. I., *Handbook of Filter Synthesis*, John Wiley and Sons, New York, 1967.

REFERENCES BY TOPIC

Applications of Discrete-Time Signal Processing

- Andrews and Hunt (1977)
  Atal (1976)
  Atal and Hanauer (1971)
- Castleman (1979)
- Flanagan (1972)
  Flanagan, Coker, Rabiner, Schafer, and Umeda (1970)
- Jayant and Noll (1984)
- Macovski (1983)
  Makhoul (1975)

- Markel and Gray (1976)
- Oppenheim (1978)
- O'Shaughnessy (1987)
- Parsons (1986)
- Pratt (1978)
- Rabiner and Schafer (1978)
- Robinson and Durrani (1985)
- Robinson and Treitel (1980)
- Skolnik (1980)

Background and Basic Mathematics

- Ahlfors (1966)
- Birkhoff and Rota (1978)
- Cheney (1966)
- Churchill and Brown (1984)
- Gel'fand et al. (1964–1968)
- Grossman (1982)

- Hildebrand (1968, 1973, 1976)
- Jolley (1961)
- Morse and Feshbach (1953)
- Thomas and Finney (1984)
- Zemanian (1965)

Difference Equations and Filter Structures

- Chow and Cassignol (1962)
  Clements and Pease (1988)
  Crochiere (1972)
  Crochiere and Oppenheim (1975)
  Fettweis (1986)

  Gray and Markel (1973, 1975)
  Kaiser and Hamming (1977)
  Kelly and Lochbaum (1962)
  Rao and Kailath (1984)
  Vaidyanathan (1985)

Digital Signal Processing

- Blackman (1965)
- DSP Committee (1976, 1979)
- Dudgeon and Mersereau (1983)
- Freeman (1965)
- Gold and Rader (1969)
- Hamming (1988)
- Haykin (1986)
- Jackson (1986)
- Lim (1989)
- Lim and Oppenheim (1988)

- Oppenheim and Schafer (1975)
- Peled and Liu (1976)
- Proakis and Manolakis (1988)
- Rabiner and Gold (1975)
- Rabiner and Rader (1972)
- Roberts and Mullis (1987)
- Stearns and David (1988)
- Strum and Kirk (1988)
- Tretter (1976)
- Tukey (1977)

Discrete Fourier Transforms

  Bergland (1968)
- Blahut (1985)
  Bluestein (1970)
  Bracewell (1983, 1984)

- Brigham (1988)
  Burrus (1977, 1988)
- Burrus and Parks (1985)
  Cooley and Tukey (1965)

Cooley, Lewis, and Welch (1967)
Danielson and Lanczos (1942)
Duhamel (1986)
Duhamel and Hollmann (1984)
Gentleman and Sande (1966)
Goertzel (1958)
Good (1958)
Heideman, Johnson, and Burrus (1984)
Helms (1967)
Hewes, Broderson, and Buss (1979)
Kolba and Parks (1977)
Markel (1971)

♦ McClellan and Rader (1979)
♦ Press, Flannery, Teukolsky, and
      Vetterling (1986, 1988)
Rabiner, Schafer, and Rader (1969)
Rader (1968)
Rader and Brenner (1976)
Runge (1903)
Singleton (1967, 1969)
Stockham (1966)
Vernet (1971)
Volder (1959)
Winograd (1978)

Filter Design (Continuous Time)

♦ Christian and Eisenmann (1966)
♦ Daniels (1974)
♦ Guillemin (1957)
♦ Huelsman and Allen (1980)
♦ Johnson (1976)
♦ Johnson, Johnson, and Moore (1980)

♦ Lam (1979)
♦ Papoulis (1957)
♦ Storer (1957)
♦ Weinberg (1975)
♦ Zverev (1967)

Filter Design (Discrete Time)

♦ Antoniou (1979)
Chen and Parks (1987)
Constantinides (1970)
Deczky (1972)
Fletcher and Powell (1963)
Golden and Kaiser (1964)
Gray and Markel (1976)
Herrmann (1970)
Herrmann, Rabiner, and Chan (1973)
Herrmann and Schüssler (1970a, 1970b)
Hofstetter, Oppenheim, and Siegel (1971)
Kaiser (1966, 1974)
Martinez and Parks (1978)
McClellan and Parks (1973)
McClellan, Parks, and Rabiner (1973)

Mersereau, Schafer, Barnwell,
   and Smith (1984)
♦ Parks and Burrus (1987)
Parks and McClellan (1972a, 1972b)
Rabiner (1972a, 1972b)
Rabiner, Graham, and Helms (1974)
Rabiner, Kaiser, Herrmann, and Dolan
   (1974)
Rabiner, McClellan, and Parks (1975)
Rader and Gold (1967)
Richards (1982)
Schüssler and Steffen (1988)
Slepian, Landau, and Pollack (1961)
Steiglitz (1970)

Fourier Series and Fourier, Laplace, and z-Transforms

♦ Arsac (1966)
♦ Bracewell (1986)
♦ Carslaw (1952)
♦ Dym and McKean (1972)

♦ Jury (1964)
♦ Lighthill (1958)
♦ Papoulis (1962)
♦ Rainville (1963)

Hilbert Transforms

Berkhout (1973)
Cïzek (1970)
Gold, Oppenheim, and Rader (1970)

Hayes, Lim, and Oppenheim (1980)
Rabiner and Schafer (1974)

Homomorphic Systems

Bauman, Lipshitz, and Vanderkooy
  (1985)
Bogert, Healy, and Tukey (1963)
Davis and Mermelstein (1980)
Furui (1981)
Juang, Rabiner, and Wilpon (1987)
Noll (1967)
Oppenheim (1967, 1969a, 1969b)
Oppenheim and Schafer (1968)
Oppenheim, Schafer, and
  Stockham (1968)
Reut, Pace, and Heaton (1985)

Schafer and Rabiner (1970)
Schroeder (1981)
Senmoto and Childers (1972)
Smith and Barnwell (1986)
Steiglitz and Dickenson (1977)
Stockham, Cannon, and
  Ingebretsen (1975)
Tribolet (1977)
♦ Tribolet (1979)
Ulrych (1971)
Verhelst and Steenhaut (1986)

Quantization Effects

Barnes and Fam (1977)
Bennett (1948)
♦ Cavanagh (1984)
Chan and Rabiner (1973a, 1973b, 1973c)
Claasen, Mecklenbrauker, and
  Peek (1973)
Ebert, Mazo, and Taylor (1969)
Hwang (1974)
Jackson (1969, 1970a, 1970b)
Kan and Aggarwal (1971)
Kaneko and Liu (1970)
♦ Knuth (1981)
Knowles and Olcayto (1968)

Liu (1971)
Liu and Kaneko (1969)
Liu and Peled (1975)
Mills, Mullis, and Roberts (1978)
Mullis and Roberts (1976)
Oppenheim (1970)
Oppenheim and Weinstein (1972)
Sandberg (1967)
Weinstein (1969)
Weinstein and Oppenheim (1969)
Welch (1969)
Widrow (1956, 1961)

Random Signals and Probability

♦ Davenport (1970)
♦ Drake (1967)
♦ Feller (1968)

♦ Papoulis (1984)
♦ Parzen (1960)
♦ Parzen (1962)

Sampling and Multirate Signal Processing

♦ Crochiere and Rabiner (1983)
♦ Hnatek (1976)
  Nyquist (1928)
  Oetken, Parks, and Schüssler (1975)
  Oppenheim and Johnson (1972)

Schafer and Rabiner (1973)
♦ Schmid (1970)
  Shannon (1949)
  Steiglitz (1965)

Signals and Systems

♦ Black (1953)
♦ Carlson (1975)
♦ Chen (1970)
♦ Cooper and McGillem (1967)
♦ Cruz and Van Valkenburg (1974)

♦ Gabel and Roberts (1973)
♦ Guillemin (1963)
♦ Haykin (1978)
♦ Jury (1958)
♦ Kailath (1980)

- Lathi (1965)
- Liu and Liu (1975)
- Mason and Zimmermann (1960)
- McGillem and Cooper (1974)
- Oppenheim and Willsky (1983)
- Papoulis (1968)
- Papoulis (1977)

- Papoulis (1980)
- Phillips and Nagle (1984)
- Ragazzini and Franklin (1958)
- Schwartz (1980)
- Siebert (1986)
- Thomas (1969)
- Wozencraft and Jacobs (1965)

Spectral Analysis

  Allen and Rabiner (1977)
- Bartlett (1953)
- Blackman and Tukey (1958)
- Box and Jenkins (1970)
- Grenander and Rosenblatt (1957)
  Griffin and Lim (1984)
- Hannan (1960)
  Harris (1978)
- Jenkins and Watts (1968)

  Kaiser and Schafer (1980)
- Kay (1988)
  Kay and Marple (1981)
- Koopmanns (1974)
- Marple (1987)
  Nawab and Quatieri (1988)
  Rader (1970)
  Schaefer, Schafer, and Mersereau (1979)
  Welch (1970)

Index